Formulas
for Mechanical and Structural
SHOCK and IMPACT

Gregory Szuladziński

CRC Press
Taylor & Francis Group
Boca Raton London New York

CRC Press is an imprint of the
Taylor & Francis Group, an **informa** business

CRC Press
Taylor & Francis Group
6000 Broken Sound Parkway NW, Suite 300
Boca Raton, FL 33487-2742

First issued in paperback 2019

© 2010 by Taylor and Francis Group, LLC
CRC Press is an imprint of Taylor & Francis Group, an Informa business

No claim to original U.S. Government works

ISBN-13: 978-1-4200-6556-5 (hbk)
ISBN-13: 978-0-367-38507-1 (pbk)

Library of Congress Cataloging-in-Publication Data

Szuladzinski, Gregory, 1940-
    Formulas for mechanical and structural shock and impact / author, Gregory Szuladzinski.
        p. cm.
    "A CRC title."
    Includes bibliographical references and index.
    ISBN 978-1-4200-6556-5 (alk. paper)
    1. Impact--Mathematical models. 2. Shock (Mechanics)--Mathematical models. 3. Engineering mathematics--Formulae. I. Title.

TA354.S98 2010
624.1'76--dc22
                                                                    2009013703

Visit the Taylor & Francis Web site at
http://www.taylorandfrancis.com

and the CRC Press Web site at
http://www.crcpress.com

Contents

Preface

This book is essentially a collection of formulas describing dynamic responses to shock loads. The presentation is inspired by Roark's classic *Formulas for Stress and Strain*, which presents equations and explanatory sketches in a compact manner. The theoretical basis is presented in a concise, although somewhat superficial manner, as appropriate for a reference work rather than a text intended to teach the subject. Still, this book is a reasonably self-contained tool. Although it is written for engineers in general, experienced dynamicists may also find some new ideas and approaches.

The objective of this book is to provide a meaningful reference in today's computer-oriented environment. Nowadays, due to the development and availability of general-purpose structural programs, any large analytical task can be performed by computers. This limits the importance of manual calculations, in general, and of certain analytical methods, in particular, as compared, for example, with the 1960s. Some of the methods in which manual work is and will remain useful in the foreseeable future are

1. Obtaining preliminary figures on the anticipated dynamic response of a system that is in an early stage of design and for which a full-scale computation is not practical.
2. The preparatory phase of large-scale calculations where a dynamic model is generated. At this stage, prudent analysts conduct a number of checks to ensure that they neither miss anything important nor incorporate too many unnecessary details, which is wasteful.
3. An indirect verification of computer-generated results by using a simplified model or a calculation method. (This may not only explain some unbelievable results, but may also help to guard against hidden errors, for which there seem to be unlimited opportunities.)
4. Work in preparation for physical testing.

Thus, it is quite clear that the only sensible options for today's engineers embarking on a hand calculation are the methods and approaches that are concise and relatively easy. Anything more mundane might as well be done by a major finite-element (FE) code. This was the major criterion in the selection of problems and methodology. Simplicity is therefore the main purpose and the keynote of this book. Still, not all situations can be explained in an elementary manner. In such cases, it is preferable to provide a set of two or three simple equations, each of them corresponding to some definite concept, instead of building a large, complex, single equation. In many instances, the effort to simplify things goes into offering simple approximate equations instead of more complicated, but also more accurate, solutions. (One should note that terms like "manual" or "hand" calculations in a modern setting mean not only the use of a calculator, as they did in the 1970s, but also the use of such simple programs as a spreadsheet or Mathcad.)

There is another advantage in being able to do simplified, although not very accurate, estimates. This is often evident with regard to extreme loads or structures. For example, one can ask: "Is the peak strain to be expected closer to 5% or to 50%?" with one of the answers implying safety and the other indicating danger. To be able to answer such a question early on has a strong impact on the engineering approach to be adopted.

This book is written for engineers and for all others who want to have an insight into how objects and structures respond to sudden, strong impulses that are usually of short duration. Since there is no emphasis on any particular type of structure and the scope is quite broad, those involved in the mechanical aspects of aeronautical, automotive, nuclear, and civil engineering, as well as those in general machine design, may find it equally beneficial.

Some of the more recent results presented here may be hypothetical or of unproven reliability, and must therefore be used with caution. The decision to include such materials was based on two reasons. The first was the lack of better and practical solutions. The second was an old saying that if everyone waited until things could be done perfectly, nothing would ever be done.

In order to benefit most from this book, the reader should be familiar with the theory of the strength (mechanics) of materials and engineering mathematics. Some familiarity with the concepts of dynamics can be helpful, but this is not an absolute necessity as the introductory part of every chapter should explain the setting.

The author is indebted to several people who helped to bring this book to its present shape. Verl A. Stanford carefully checked most of the material and made many valuable suggestions. Miecia (May) Paszkiewicz carried out the important task of cross-referencing numerous items in the text. Professor Ali Saleh read Chapters 10 and 11 and made useful comments. John Quinn read several chapters and helped to remove some flaws.

All comments and criticisms regarding the contents of this book can be sent as e-mails to gsgsgsgs87@hotmail.com. All such comments will be most gratefully received and answered, if possible.

Dr. Gregory Szuladziński
Sydney

Introduction

Every typical chapter of this book is divided into three parts: a theoretical outline, a tabulation, and examples. The tabulation is an attempt to present the subject matter more clearly by breaking it down into simple blocks. The examples are an essential part of this book, because it is well known that most people learn best when numerical examples are provided.

Because of a difficult subject matter as well as due to style of presentation, there are many cross-references in the text linking the three major parts listed above. The symbols used are listed at the outset.

Author

Dr. Gregory Szuladziński received his master's degree in mechanical engineering from Warsaw University of Technology in 1965 and his doctoral degree in structural mechanics from the University of Southern California in 1973.

From 1966 to 1980, he worked in the United States in the fields of aerospace, nuclear engineering, and shipbuilding. He has done extensive work in computer simulations of seismic events and accidental dynamic conditions as related to the safety of nuclear plants and military hardware.

From 1981 until the present time, he has been working in Australia in the fields of aerospace, railway, power, offshore, automotive, and process industries, as well as in rock mechanics, underground blasting, infrastructure protection, and military applications. He has a number of publications to his credit in the area of nonlinear mechanics. His first book on the subject, *Dynamics of Structures and Machinery: Problems and Solutions*, was published by John Wiley Interscience in 1982.

Dr. Szuladziński has been involved with the finite-element method of simulation of structural problems since 1966. In 1978–1979, he worked as the principal analyst for Control Data in Los Angeles in support of finite-element analysis (FEA) codes.

Since the early 1990s he has been working on computer simulations of such violent phenomena as rock breaking with the use of explosives, fragmentation of metallic objects, shock damage to buildings, structural collapse, fluid–structure interaction, blast protection, and aircraft impact protection. He has conducted a number of state-of-the-art studies showing explicit fragmentations of structures and other objects.

Dr. Szuladziński is a fellow of the Institute of Engineers Australia, a member of its Structural and Mechanical College, a member of the American Society of Mechanical Engineers and of the American Society of Civil Engineers.

Symbols and Abbreviations (General)

SYMBOLS

a	acceleration (m/s^2), a shorter edge length of a rectangular plate (m), inside radius (m)
α	angular displacement (rad)
A	area (m^2), amplitude
A_s	shear area (m^2)
b	coefficient for bending deflection of beams and plates, outside radius (m)
C	damping coefficient (N-s/m), torsional constant of cross section (m^4)
C_c	critical damping coefficient (N-s/m)
c	In bending, distance from neutral axis to extreme fiber (m)
c_0	speed of propagation, axial waves (m/s)
c_1	speed of propagation, twisting waves (m/s)
c_2	speed of propagation, lateral wave in a string (m/s)
c_3	speed of propagation, shear waves in beams (m/s)
c_p	speed of propagation, pressure waves in a continuum (m/s)
c_{pl}	speed of propagation, wave in plastic range (m/s)
c_R	speed of propagation, Rayleigh waves (m/s)
c_s	speed of propagation, shear waves in a continuum (m/s)
d	diameter (m)
D	energy dissipated in damping (J), plate bending stiffness parameter (N-m), detonation velocity (m/s)
$\bar{\varepsilon}$	logarithmic strain (m/m)
e	eccentricity, static shaft unbalance (m)
$\check{e}$	angular acceleration (rad/s^2)
E	Young's modulus of elasticity (Pa)
E_c	modulus of elasticity of concrete
E_{ef}	effective modulus of in presence of lateral constraint (Pa)
E_k	kinetic energy (N-m)
E_p	plastic modulus (Pa)
f	natural frequency (Hz = cycle/s)
$f(t)$	function of time
F_y, F_u	yield strength, ultimate strength (Pa)
g	a gram of mass (kg/1000), acceleration of gravity (9.81 m/s)
G	modulus of elasticity in shear (Pa)
h	height or thickness (m)
$H(t)$	unit step function
i	radius of inertia (m)
$I = A\rho c$	impedance of a bar (kg/s)
I	second area moment (area moment of inertia (m^4))
I_0	cross-sectional polar moment of inertia (m^4)
J	mass moment of inertia about axis of revolution (kg-m^2)

k	stiffness of translatory spring (N/m)
k^*, k_{ef}	effective stiffness (N/m)
$\bar{k}$	stiffness modified by presence of axial force (N/m)
k_f	stiffness of elastic foundation (N/m^2)
K	stiffness of a rotary spring (N-m/rad)
K_σ	geometric stress concentration factor
L	length (m)
l	length (m)
$\mathcal{L}$	work performed (N-m)
$\mathcal{L}_e$	work performed in elastic straining (N-m)
$\mathcal{L}_p$	work performed in plastic straining (N-m)
m	distributed mass (kg/m or kg/m^2)
M	lumped mass (kg)
M^*	reduced mass (kg)
$\mathcal{M}$	bending moment (N-m)
$\mathcal{M}_y$	bending moment at onset of yield (N-m)
$\mathcal{M}_0$	moment capacity for perfectly plastic material (N-m)
N	one Newton of force
n	stress–strain curve parameter, equivalent polytropic exponent
P	axial force (N)
P_{cr}	buckling force (N)
P_e	Euler force, $\pi^2 EI/L^2$ (N)
p	pressure (Pa)
q	distributed load (N/m), specific energy content (J/kg)
q_0	distributed edge load (N/m)
Q	beam or plate shear force (N or N/m), energy content (J)
r	current radius (m)
R	force of resistance (N), (mean) radius (m)
S	engineering stress (MPa), true stress (MPa), impulse (N-s)
s	coefficient for shear deflection of beams and plates, impulse per unit length of beam (N-s/m)
S_n, S_t	normal and tangential impulse (N-s)
t	time (s)
t_m	loading phase duration (s)
t_0	duration (s)
T	twisting moment (N-m)
T_y	twisting moment at yield (N-m)
T_u	ultimate twisting moment (N-m)
u	displacement (m)
u_{st}	static displacement (m)
u_c	increase of cavity radius caused by deformation of intact medium (m)
u_d	dynamic displacement (m)
U	strong shock wave velocity (m/s)
v	velocity (m/s)
V	velocity following rebound (m/s), volume (m^3)
w	distributed load (N/m or N/m^2)
W	force (N)

x, y, z	coordinates (m)
X	function of spatial variable x
Y	yield parameter (MPa)
α	angular displacement (rad), mass-proportional damping coefficient
β	stiffness-proportional damping coefficient
γ	shear angle
Γ	energy expression coefficient
ω	circular frequency (rad/s)
$\bar{\omega}$	frequency modified by presence of axial force
ω_d	damped circular frequency (rad/s)
δ	displacement (m)
ε	(engineering) strain (m/m), angular acceleration (rad/s^2)
ε_y	strain at onset of yielding (m/m)
ε_u	ultimate strain, at peak stress (m/m)
ε_r	maximum strain, at rupture (m/m)
$\dot{\varepsilon}$	strain rate (1/s)
ζ	damping ratio
κ	coefficient of restitution
λ	angular velocity (rad/s), elongation of a structural member (m)
Λ	half-wave length (m)
ν	Poisson's ratio
ξ	distance to moving joint (m)
Π	strain energy (N-m)
σ	stress, direct (MPa)
σ_0	yield stress of EPP material (MPa)
σ_I	incoming wave stress (MPa)
σ_R	reflected wave stress (MPa)
σ_T	transmitted wave stress (MPa)
ρ	specific mass or mass density (kg/m^3), radius of curvature (m)
τ	shear stress (MPa), natural period of vibrations (s)
τ_y	shear stress at yield (MPa)
τ_0	shear stress at yield for EPP material (MPa)
τ_u	ultimate shear stress (MPa)
$\bar{\tau}$	natural period modified by presence of axial force (s)
χ	damping coefficient
ψ	coefficient of friction, shear deflection coefficient
Ω	angular velocity (rad/s)
η	distance to stationary joint (m)
Ψ	energy coefficient, material toughness (N-m)
Θ	work expression coefficient

ABBREVIATIONS

2D	two-dimensional
3D	three-dimensional
BL	bilinear material model
CC	clamped ends

CF	one end clamped, the other free (cantilever)
CG	one end clamped, the other guided, center of gravity
CL	centerline
DF	dynamic (magnification) factor
DOF	degree of freedom
EPP	elastic, perfectly plastic material model
FE	finite elements
FEA	finite element analysis
ln	logarithm to the base e
MDOF	multiple degrees of freedom
NL	nonlinear
PL	power law material model
PPV	peak particle velocity
RC	reinforced concrete
RO	Ramberg–Osgood material model
RSH	rigid-strain hardening material model
SC	one end supported, one clamped
SDOF	single degree of freedom
SF	safety factor
SS	simply-supported ends
VOD	velocity of detonation

1 Concepts and Definitions

THEORETICAL OUTLINE

Static or quasistatic loading means that the load is applied slowly. On the other hand, dynamic or shock loading implies fast or abrupt application. Those adjectives are relevant with respect to the natural period of a structural element under consideration. When the load increases to its maximum value over five or six natural periods, it is a quasistatic load. When it does so over a fraction of the period, it is a shock loading. (Broader definitions exist, where anything with up to two periods duration is a shock loading.)

When speaking of a shock load, engineers usually have in mind a load that is of a large magnitude, but with a very short duration. But this is not the only circumstance where the term applies. The load can last for a very long time, but if it is suddenly applied or suddenly removed, the term "shock" is appropriate as well. The broadest definition seems to be that a shock is any abrupt change of a force, a position, a velocity, or an acceleration affecting the body under consideration, Silva [78].

This book is not a study of periodic motion, but the subject of the natural period (or natural frequency of vibration) is given its due attention. This is because that quantity is an important parameter when assessing a shock response of a structural component.

STIFFNESS AND NATURAL FREQUENCY

Figure 1.1 presents two simple systems capable of periodic motion: (a) translational oscillator and (b) rotational oscillator. In (a) mass M is attached to the ground by a spring of stiffness k (N/m). In (b) a disk with mass moment of inertia J is connected to the ground by a shaft of an angular stiffness K (N-m/rad). The unit for M is kg and for J it is kg-m^2. The terms *stiffness*, *rigidity*, and *spring constant* can be used interchangeably when describing the resistance capability of an elastic member. The reciprocal of stiffness is called *flexibility* although the term *compliance* is also used by some authors. For a translational motion, stiffness is defined as a force that causes a unit (1 m) translation. Flexibility is a displacement caused by a unit (1 N) force. In angular motion, the definitions are analogous with "moment" replacing "force" and "rotation" instead of "translation."

According to the elementary vibration theory, the motion of an oscillator in Figure 1.1a, when disturbed from the initial position, may be described by the following equation:

$$M\ddot{u} + ku = 0 \tag{1.1}$$

where $\ddot{u}$ stands for acceleration. The displacement relative to the initial position is given as a solution to the above:

$$u = A\cos(\omega t - \beta) \tag{1.2}$$

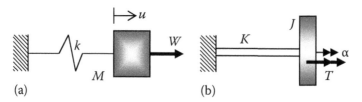

FIGURE 1.1   (a) Translational oscillator and (b) rotational oscillator.

where A and β are constants that depend on the initial conditions and

$$\omega = \sqrt{\frac{k}{M}} \qquad (1.3)$$

depends only on system properties. According to this equation, mass M returns to the same point after every interval of time τ called the *period* of motion:

$$\tau = \frac{2\pi}{\omega} = 2\pi\sqrt{\frac{M}{k}} \; (\text{s}) \qquad (1.4)$$

The number of periods, or cycles of vibrations per second, is called the *frequency* or *natural frequency* and is designated by f:

$$f = \frac{1}{\tau} = \frac{\omega}{2\pi} \; (\text{s}^{-1}) \qquad (1.5)$$

The basic unit of this frequency is one cycle per second, or hertz (Hz). The units of ω are radians per second; ω is also referred to as the natural frequency of the system, but often the term *circular frequency is* used to distinguish it from f. In translational vibration, ω has little physical significance. It is used because it is convenient in the mathematical description of motion. The nature of the disturbance dictates the magnitude of the constants A and β in Equation 1.2.

The equation of motion of the rotational system in Figure 1.1b is

$$J\ddot{\alpha} + K\alpha = 0 \qquad (1.6)$$

in which α is the angle of rotation measured from initial position. This angle of rotation is shown as a straight-line vector throughout this book. (It is distinguished from translation by a double arrow. A curved arrow is used to represent rotation only if the plane of rotation is parallel to the plane of a drawing.) The solution is analogous to Equation 1.2:

$$\alpha = A\cos(\omega t - \beta) \qquad (1.7)$$

The circular frequency ω is not any more meaningful than it is for translation, but is often used for convenience:

$$\omega = 2\pi f = \sqrt{\frac{K}{J}} \qquad (1.8)$$

THE ANALOGY BETWEEN TRANSLATIONAL AND ROTATIONAL MOTION

This analogy is complete in every respect, as shown in Table 1.1.

UNITS

The discussions relating to the convenience of various systems of units are probably as old as the units themselves. One way to look at this is to say: *the most convenient system is such that the quantities, frequently seen in daily life, are expressed by numbers between 1 and 10*. This makes the English or imperial system a leader in the field of length measurement. However, that system had been replaced (with the notable exception of the United States) by the metric or SI system. Although m (meter), kg (kilogram), and s (second) are the basic units, the use of subunits is also allowed, which gives the user a fair amount of freedom to choose the most convenient set.

Any unit system can be adopted for calculations, provided it is consistent. That consistency can partially be checked by invoking Newton's law: $W = Ma$, which must be satisfied. The question of convenience can be answered only for specific circumstances. It is certainly desirable to see frequently used quantities as numbers that are neither too large nor too small. In the analysis of violent events, system (2) seems to make it possible more often than others. For this reason this "milli" system will be used here for most illustrative examples.

Prior to general acceptance of the SI system in many countries, the unit of kG or kgf (kilogram-force) equal to 9.81 N was used. In older texts a force unit of the dyne = $N/10^5$ may be encountered. Also, a unit of pressure, bar = 0.1 MPa, a favorite of weathermen and physicists, may occasionally be found. The gravitational constant may slightly vary

TABLE 1.1
Analogy between Translational (1) and Rotational (2) Vibration

(1) Mass	M	kg
(2) Mass moment of inertia	J	kg-m^2
(1) Force	W	N
(2) Torque	T	N-m
(1) Displacement	u	m
(2) Rotation	α	rad
(1) Velocity	$v = \dot{u}$	m/s
(2) Angular velocity	$\lambda = \dot{\alpha}$	rad/s
(1) Acceleration	$a = \dot{v}$	m/s^2
(2) Angular acceleration	$\breve{e} = \dot{\lambda}$	rad/s^2
(1) Stiffness	k	N/m
(2) Rotational stiffness	K	N-m/rad
(1) Momentum	Mv	N-s
(2) Angular momentum	$J\lambda$	kg-m^2-rad/s
(1) Kinetic energy	$Mv^2/2$	N-m
(2) Kinetic energy	$J\lambda^2/2$	N-m
(1) Strain energy	$ku^2/2$	N-m
(2) Strain energy	$K\alpha^2/2$	N-m

TABLE 1.2
Metric Unit Systems

	(1)	(2) ("Milli")	(3)
Length	Meter	Millimeter	Millimeter
Mass	Kilogram	Gram (g)	Tonne (1000 kg)
Time	Second (s)	Millisecond (ms)	Second (s)
Force	Newton (N)	Newton	Newton
Pressure or stress	Pascal (Pa)	MPa (10^6 Pa)	MPa
Energy	Joule (J)	N-mm	N-mm
Acceleration of gravity	9.81 m/s^2	0.00981 mm/ms^2	9810 mm/s^2
Density of steel	7850 kg-m^3	0.00785 g/mm^3	7.85×10^{-9} tonne/mm^3
Young's modulus of steel	200×10^9 Pa	$200,000$ MPa	$200,000$ MPa
Yield stress, mild steel	250×10^6 Pa	250 MPa	250 MPa

between geographical locations, but $g = 9.81$ m/s^2 seems to be an accepted level in most publications.

Although the examples in this book are typically written in metric units, a need to recalculate to the English system may often arise.* The format of Table 1.3 will, hopefully, assist that recalculation. If the acceleration of gravity is taken as 9.81 m/s^2, it corresponds to 386.22 in./s^2, while a more accurate, standard 9.80665 m/s^2 yields close to 386 in./s^2.

The unit of 1 pound is often used as a unit of force (lbf) and unit of mass (lb). Unfortunately, the latter is inconsistent in terms of Newton's law, as quoted above. To keep consistency and avoid using artificial units, the mass unit in the English system is a quotient of a unit of force and a unit of acceleration, $M = W/a$. To illustrate one of the ways of changing units, an operation, which is a frequent source of errors, consider a pressure quoted as 1000 lbf/ft^2 and its equivalent:

$$1,000 \text{ lbf/ft}^2 = 1,000 \frac{4.45 \text{ N}}{(0.3048 \text{ m})^2} = 47,899 \text{ N/m}^2 = 47,899 \text{ Pa} = 0.0479 \text{ MPa}$$

TABLE 1.3
English to Metric Units (1 g = 1 Gram)

The Unit of	Equals to	Or to
1 inch (in.)	25.4 mm	0.0254 m
1 foot (ft)	304.8 mm	0.3048 m
1 mile	1609.3 m	
1 n. mile (knot)	1852 m	
1 av. ounce (oz)	28.35 g	
1 pound (lb)	453.6 g	0.4536 kg
1 ton (short)	907.2 kg	
1 lbf (pound-force)	4.45 N	
1 lbf/in.^{2} (psi)	1/145.04 MPa	6.8948 KPa

* The United States is still using English units in engineering practice.

A compact way of stating unit dependence was selected for this book. When, for example, length L is intended to be expressed in meters, we simply say $L \sim$ m.

INERTIA FORCE AND DYNAMIC EQUILIBRIUM

Newton's (second) law of motion can be expressed as $W = Ma$, which means that the applied force W is proportional to the acceleration a, which the mass M is experiencing. In relation to Figure 1.1a, for example, W would be the driving force less the spring resistance. In a more general setting, W is the resultant of all external forces applied to M. The above relationship can also be written as $W - Ma = 0$. In this form Newton's law is usually called *d'Alembert's principle* as it regards force $-Ma$ as the inertia force. The principle says that an accelerating mass can be viewed as being in a special state of static equilibrium, where the total of the external forces is balanced by the inertia force.

Returning to Figure 1.1a, let W act for a very short time and let us observe the movement of M to follow. Due to spring resistance, M will briefly stop, in time, and reverse its motion. At that stopping point or *stagnation point*, there are two forces in equilibrium: the spring force Ku and the inertia force Ma. If the driving force W still acts when a stagnation point is reached, then the expression of dynamic equilibrium is

$$W - Ku = Ma \qquad (1.9)$$

At that point M is in an instantaneous equilibrium and no kinetic energy is present. This observation is often used when applying energy methods. One can then equate the strain energy of the spring with the energy supplied from outside, usually as the work of the driving force W or the initial kinetic energy arising from the initial velocity of M. This procedure is also extended to large and complex systems, although determination of a stagnation point may then become less than straightforward.

EQUIVALENT STRESS AND A SAFETY FACTOR

When there are several stress components present, i.e., tension in one direction, compression at right angles to it and a shear stress, there is a problem of calculating a single quantity, which would be a combined measure of all the stress components and which could be compared to the (uniaxial) allowable material stress. This is discussed in detail in Chapter 6. With regard to metals, two theories clearly dominate. One is associated with the name of Tresca and it relates to the maximum difference between the principal stress components. The other one is the energy-of-distortion theory, which is often, but inaccurately attributed to Mises [62], while Huber [37] published it 9 years earlier. For this reason the energy-of-distortion theory is referred to as Huber–Mises theory in this book.

A safety factor (SF) of a structural element is defined as a ratio of strength of that element to the applied load. Instead of using the word *strength*, one often reads terms like the *ultimate load* or the *breaking load*, all having the same meaning. The SF of the whole structure is the smallest factor of any of its components. The need for a SF comes from incomplete knowledge of the designer; the applied loads are known with a limited accuracy only and so are the material properties. The response of a structure to a particular loading is usually a matter to be established by analysis.

The material supplier provides minimum-guaranteed properties of a structural material. If a sample of that material is tested at random, it will most likely be stronger than what is guaranteed. Mild steel, for example, can have its average yield strength 30% higher and the ultimate strength 20% larger than the "catalog" values. While the statistical averages can often be established, most engineers are content with the minimum-guaranteed values, as this builds an additional SF in their predictions. After all, the objective of the designer is to create *safe* structures. In aerospace design, for example, the SF can be as small as 1.5. However, using "catalog" values of material constants increases that factor, although in an invisible way.

It is probably quite clear to the reader that loads and material properties should be treated in a statistical manner, because of variations involved. This had been tried, but did not achieve popularity so far, because even the simplest problems have a rather involved formulation then. Extending such analyses to nonlinear dynamics may become truly mind-boggling. However, if the statistical input is known, some advanced finite-element analysis (FEA) programs offer help in that respect.

The aim of this book is to present methods for determining dynamic response to, typically, transient loading and comparing the result with a dynamic strength of a member in question. If that strength is, in fact, a statistical variable, then using average material property values offers *the most likely outcome* in a statistical sense. This outcome is, much of the time, the objective in this book. The application of SFs, as appropriate for a situation at hand, is left to the designer/analyst.

2 Natural Frequency

THEORETICAL OUTLINE

The concepts of stiffness and natural frequency, first introduced in Chapter 1, are being continued here. For a simple oscillator, the equation of motion is the first-order differential equation, and the determination of natural frequency is trivial. When a body has a continuous mass distribution, the description of motion is accomplished by partial differential equations. In this chapter, essentially the first natural frequency is of interest, which makes the task relatively simple. Much of the material relates to frequencies of very basic elements, but a few methods combining those are also provided, so that more complex arrangements can be approximated.

DETERMINATION OF NATURAL FREQUENCY BY THE DIRECT METHOD AND BY THE ENERGY METHOD

The direct method is based on a simple definition. The stiffness k is defined as a force that is needed to move a point by one unit of length in the direction of motion. This parameter is used, together with a value for mass, in Equation 1.3. Sometimes the weight, or gravity force, $G = Mg$, with $g = 9.81$ m/s^2, is used instead of mass, as it may be more convenient. The symbol W is employed as a designation of force in this book, and its similarity with weight is quite incidental. (If gravity is a part of the picture, a statement is made to that effect and the g symbol placed in tabulations.) For a rotational system, Equation 1.8 is used, with the angular quantities replacing the translational values.

The principle of energy conservation may also be used to find a natural frequency. It says that if there are no damping forces and no external forces involved (beginning at some time point), we have

$$\Pi + E_k = C \tag{2.1}$$

in which

C is a constant

Π is the energy of deformation or strain energy

For a spring in Figure 1.1a, the energy of deformation is $\Pi = ku^2/2$, where k is the spring rigidity and u is the deflection from the unstrained position. The kinetic energy is $E_k = mv^2/2$. The incremental form of Equation 2.1 is

$$\Delta\Pi + \Delta E_k = 0 \tag{2.2}$$

which means that there is no net increase in the total energy. If gravity forces are involved, we have, in place of the above

$$\Delta\Pi + \Delta E_k = U_g \tag{2.3}$$

That is, if we consider the system moving from position 1 to position 2, the sum of increases of strain energy and of kinetic energy is equal to the work performed by forces of gravity U_g:

$$U_g = Mg(z_1 - z_2)$$ (2.4)

in which Mg is the weight of a body under consideration. The terms z_1 and z_2 denote the locations of the center of gravity (CG) of mass M. The value of U_g is positive if $z_1 > z_2$ (i.e., the CG moves from a higher to a lower position). Regardless of the gravity force involvement, a vibratory motion takes place between two stationary (zero velocity) positions, for which the kinetic energy is zero. Equation 2.1 has two positive functions on the left side, and if one becomes zero, the other must attain its maximum. Thus, for any stationary point, we have $\Pi_m = C$. Midway between these extreme points, there is a location at which no strain exists (neutral position) in the linear-elastic elements, and then $E_{km} = C$. Comparing the last two equalities gives

$$\Pi_m = E_{km}$$ (2.5)

which is a restatement of Equation 2.1. Once a particular form of Equation 2.5 is known, it is sufficient to use the relation, $v_m^2 = \omega^2 u_m^2$, between the maximum velocity and maximum displacement to determine ω. (Refer to Chapter 1.)

BODIES WITH CONTINUOUS MASS DISTRIBUTION

Equations of Motion: Bar and Shaft

A segment of an axial bar, shown in Figure 2.1, is in the state of dynamic equilibrium. The external load of intensity, q (N/m), is balanced by the increment of the internal force, P, and by the inertia force of magnitude, $m\ddot{u}\,dx$, where $m = A\rho$ is the mass per unit length (kg/m). The rate of change of the stretching force P is $\partial P/\partial x$, therefore its value changes by $(\partial P/\partial x)dx$ over the segment dx. Projecting all forces on the x-axis, we obtain

$$\left(P + \frac{\partial P}{\partial x}dx\right) - P + q\,dx - m\ddot{u}\,dx = 0$$ (2.6)

The elongation of the segment is, by an elementary relation

$$\frac{\partial u}{\partial x} = \frac{P}{EA}$$ (2.7a)

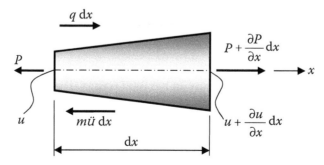

FIGURE 2.1   Dynamic equilibrium of a bar element.

or

$$P = EAu'$$ (2.7b)

If a bar is of variable cross section, A is a function of x. The derivative of P is in this case

$$\frac{\partial P}{\partial x} = \frac{\partial}{\partial x}\left(\frac{\partial u}{\partial x}EA\right) \equiv (EAu')'$$ (2.8)

The equation of motion now becomes

$$m\ddot{u} - (EAu')' = q$$ (2.9)

in which the prime denotes differentiation with respect to x. For a constant axial stiffness, $EA = $ constant, the equation simplifies to

$$\ddot{u} - c_0^2 u'' = \frac{q}{m}$$ (2.10)

where both u and q are functions of time and the spatial variable x, and c_0 can be shown to be the speed of propagation of a longitudinal elastic disturbance along the bar:

$$c_0 = \left(\frac{EA}{m}\right)^{1/2} = \left(\frac{E}{\rho}\right)^{1/2}$$ (2.11)

with the last equality being true only if the structural mass and the total mass are equal.

Figure 2.2 shows a segment of a shaft in dynamic equilibrium; T is the resultant twisting moment at a certain location, while the external distributed load has the intensity, m_t (N-m/m). The moment of inertia about the shaft axis is $I_0 \rho \, dx$, in which I_0 is the polar moment of inertia of the cross section. The independent variable is the angle of rotation, α. There is a complete analogy between a bar and a shaft. Projecting the vectors pertaining to the angular quantities on the x-axis, we obtain

$$\frac{\partial T}{\partial x} + m_t - I_0 \rho \ddot{\alpha} = 0$$ (2.12)

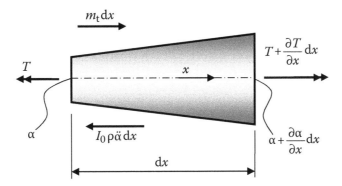

FIGURE 2.2   Dynamic equilibrium of a shaft element.

If the elementary equation for the torsional deformation is applied to a shaft segment of length dx, one gets

$$\frac{\partial \alpha}{\partial x} = \frac{T}{GC} \tag{2.13a}$$

or

$$T = GC\alpha' \tag{2.13b}$$

in which
  α is the angle of twist
  C is the torsional constant of the cross section

Limiting ourselves to shafts with a constant section, we obtain

$$\frac{\partial T}{\partial x} = GC \frac{\partial^2 \alpha}{\partial x^2} \tag{2.14}$$

The equation of motion now becomes

$$\ddot{\alpha} - c_1^2 \alpha'' = \frac{m_t}{\rho I_0} \quad \text{with} \quad c_1 = \left(\frac{GC}{\rho I_0} \right)^{1/2} \tag{2.15}$$

where c_1 is the speed of propagation of twisting disturbances. For circular cross sections, $C = I_0$.

Although the material to follow refers to a bar, everything stated is also valid for a shaft if the notation is suitably changed. To solve Equation 2.10, we must know the initial conditions in the form

$$u(x,0) = u_0(x) \tag{2.16a}$$

and

$$\dot{u}(x,0) = v_0(x) \tag{2.16b}$$

that is, the initial position and velocity of any point of the bar axis at $t = 0$. Also the *end conditions*, which are displacements or forces (or combinations thereof) at the ends, must be given. Our concern at this point is limited to solving Equation 2.10 in its homogeneous form, that is, when the right side is zero. A general solution is assumed as

$$u(x,t) = X(x) \cdot f(t) \tag{2.17}$$

where X is a function of x only and is referred to as a *mode shape*, while f depends only on time. The detailed expressions for X and f are

$$X(x) = D_1 \cos \frac{\omega x}{c_0} + D_2 \sin \frac{\omega x}{c_0} \tag{2.18a}$$

and

$$f(t) = B_1 \cos \omega t + B_2 \sin \omega t \qquad (2.18b)$$

Substituting the end conditions into this relation gives us a frequency equation for a particular system, with which we can establish the set of natural frequencies, ω_i. There are infinitely many such frequencies and the like number of natural modes, resulting from Equation 2.18a. In the process, we also establish the relationship between the constants D_1 and D_2. For example, suppose that we have found $D_2 = 0.5D_1$. This allows us to express a modal shape, i, as

$$X_i = D_1 \left(\cos \frac{\omega_i x}{c_0} + 0.5 \sin \frac{\omega_i x}{c_0} \right) \qquad (2.19)$$

Finally, one can set $D_1 = 1$, since a scale factor of a mode shape has no meaning.

The equations of motion for bars and shafts are of the second order with respect to the x variable, which is the reason for describing them as the *second-order elements*.

Equations of Motion: Cable and Shear Beam

These two also belong to second-order elements. The segment of a cable (string) in Figure 2.3 is subjected to a laterally distributed load, w (N/m), to a stretching force, P, which is assumed constant, and to the distributed inertia force of magnitude, $m\ddot{u}\,dx$. The lateral displacement, u, is assumed to be very small in comparison with the length of the cable. By projecting all forces on the vertical axis we obtain

$$\ddot{u} - c_2^2 u'' = \frac{w}{m} \quad \text{with} \quad c_2 = \left(\frac{P}{m} \right)^{1/2} \qquad (2.20)$$

which is analogous to Equations 2.10 and 2.15. The constant, c_2, is the speed of propagation of lateral displacements. The solution of Equation 2.20, when the right side equals zero, is analogous to that for bars and shafts. But the cable motion is not limited to lateral displacements only. The element has a longitudinal stiffness as well, so it may be viewed also as a bar, especially when its initial shape is straight. Accordingly, Equation 2.11 again defines the speed of longitudinal waves. In fact, both components of motion are almost invariably

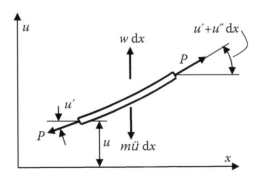

FIGURE 2.3 Dynamic equilibrium of a cable element.

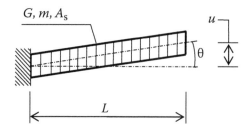

FIGURE 2.4   Shear beam deformation.

coupled in any real cable, although they do not necessarily have the same importance. This matter is treated in greater detail in Chapter 9.

The *shear beam* is the fourth member of the group. The lateral deflection of a two-dimensional (2D) beam has two components: bending and shear. The first is due to curving of the axis, while the second results from distortion of elements normal to the axis. When only the latter is accounted for, we speak of a shear beam, as depicted in Figure 2.4.

This model can be thought of as consisting of thin layers capable of shear distortion, but infinitely rigid in the direction normal to their thickness. The shear angle, θ, and the associated tip deflection are defined by

$$\theta = \frac{Q}{GA_s} \tag{2.21a}$$

and

$$u = \theta L = \frac{QL}{GA_s} \tag{2.21b}$$

where
  G is the shear modulus
  A_s is the shear area

When the lateral force is constant along the length, as in Figure 2.4, Q is the same for every vertical slice of the beam.

Although some elements behave nearly like ideal shear beams, the main usefulness of this concept is in visualizing the shear component of deflection of actual beams. The influence of shear is likely to be substantial when one or more of the following conditions take place:

1. The beam is short, say, a cantilever whose length is only twice its depth.
2. The cross section is hollow or branched (as opposed to a compact section like a solid circle or a rectangle).
3. The beam is sandwich type with a lightweight core.

By definition, the elastic rotation of a cross section cannot take place in a shear beam. From this viewpoint, it makes no difference whether the end section is completely fixed or merely restrained against translation. The fixity in Figure 2.4 is needed for rigid-body equilibrium.

The equation of motion is analogous to that of a bar, a shaft, and a cable. The speed of propagation of a shear disturbance is

$$c_3 = \left(\frac{GA_s}{m} \right)^{1/2} \tag{2.22}$$

Equations of Motion: Flexural Beam

The beam element in Figure 2.5a is in the state of dynamic equilibrium. The bending stiffness, EI (N-m^2), is assumed constant, and so is the mass per unit length, $m = A\rho$. The equation of motion is of the fourth order with respect to x:

$$m\ddot{u} + EIu^{iv} = w \tag{2.23}$$

where both u and w are functions of time, t, and position, x. The other important equations for the beam segment are

$$EIu'' = M \tag{2.24}$$

$$EIu''' = Q \tag{2.25}$$

Equation 2.24 is the relation between the bending moment and the curvature of the beam axis. Equation 2.25 may be obtained from the previous one by differentiation and by noting that $M' = Q$, if the higher order terms are ignored in the angular equilibrium of the beam element. Equation 2.23 may be obtained by differentiation of Equation 2.25 and by writing the equation of vertical equilibrium for Figure 2.5a.

When the equation of motion includes the terms relating to shear deformation and rotary inertia, it is called a *Timoshenko beam* equation. It is generally agreed that the rotary effects are small in most practical applications, while the additional effects of shearing distortions seem to be best included by applying one of the approximate methods.

The boundary conditions involve deflection and slope (Figure 2.5b) as well as shear forces and bending moments. The general form of the solution for free vibration ($w = 0$) is the same as that given by Equation 2.17. Likewise, the solution procedure is the same.

A beam element may be thought of as a cable, additionally endowed with a flexural stiffness. In a conventional design, where small deflections are involved, the longitudinal and

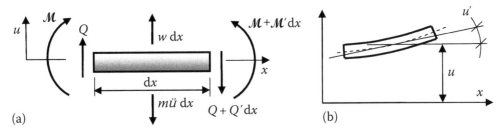

FIGURE 2.5 Flexural beam element: (a) dynamic equilibrium; (b) displacement and slope.

the geometric stiffness are not very important and are ignored. For intermediate deflections, a cable-like resistance may become predominant. (Refer to Chapter 11.)

Beams on Elastic Foundation

An elastic foundation means, in effect, a dense row of springs with stiffness k_f per unit length of beam. The springs are independent from one another (much like a railroad track supported on wooden ties, which lie on the ground). As a result, a beam element in Figure 2.5 experiences a downward reaction, $k_f u\, dx$, and the equation of motion becomes

$$m\ddot{u} + EIu^{iv} = w - k_f u \tag{2.26}$$

where the exponent "iv" means fourth-order integration with respect to u. Addition of the foundation increases the natural frequencies. It can be shown, for example, that for a simply supported beam the circular frequencies can be expressed by

$$\omega_i^2 = \left(\frac{i\pi}{L}\right)^4 \frac{EI}{m} \tag{2.27}$$

for $i = 1, 2, 3\ldots$, while, after addition of the foundation, the frequencies become [111]

$$\omega_{if}^2 = \omega_i^2\left(1 + \frac{k_f}{EI}\left(\frac{L}{i\pi}\right)^4\right) \tag{2.28}$$

while our interest is usually limited to the first mode, $i = 1$. The foundation modifies natural frequencies, but has no influence on the mode shapes. The relation between natural frequencies with and without a foundation, expressed by Equation 2.28, applies to beams with other end conditions as well. A more detailed discussion of beams on elastic foundations is available in Chapter 10.

Continuous Elastic Medium

When a rigid block is attached to the surface of a semi-infinite medium, it can be treated as supported by elastic springs. The constants of these springs are given in Cases 2.48 and 2.49. The basic formula for a natural frequency can then be used with an appropriate mass (be it for translation or rotation) and a prescribed spring constant. This, however, is only the first approximation. The medium also has some inertia, which gives rise to the increase in the effective mass of the block. The reader can find more information in Ref. [17].

TYPES OF STRUCTURAL ARRANGEMENTS

There are a variety of structural arrangements, that is, ways in which the elements can be connected to each other. The two cases that are found most often are referred to as *in series*

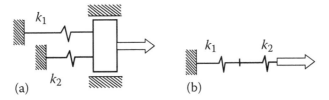

FIGURE 2.6   Structural arrangements: (a) parallel and (b) in series.

and *in parallel*. In the first instance, the load is the same in all members, and therefore, the deflections (due to flexibilities) are additive. The resultant flexibility, $1/k^*$, is thus

$$\frac{1}{k^*} = \frac{1}{k_1} + \frac{1}{k_2} + \cdots + \frac{1}{k_n} \tag{2.29}$$

with the summation extended to all n members of the system. In a parallel connection, the displacement of all members is the same and their loads and rigidities (stiffnesses) are additive. The resultant stiffness of a parallel connection is

$$k^* = k_1 + k_2 + \cdots + k_n \tag{2.30}$$

As the cases presented later show, structural elements in a series connection are often placed along one continuous line, while in a parallel connection they are located side by side, as in Figure 2.6. The appearance, however, may sometimes be misleading, and the definitions given above must always be kept in mind.

A TWO-MASS OSCILLATOR

A two-mass oscillator represents an important system, to which many real configurations are reducible. The two masses in Figure 2.7 are connected with a weightless elastic bar of stiffness k. There is no external force applied, therefore each mass is subjected to only the resistance of the elastic element during vibrations. If, at any instant, the stretching force is P_s, then one has

$$M_1 \ddot{u}_1 = P_s \tag{2.31a}$$

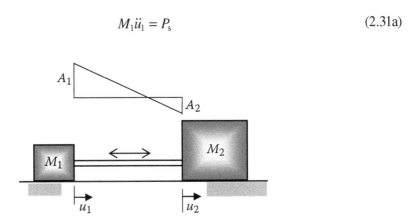

FIGURE 2.7   A vibrating two-mass oscillator. (A_1 and A_2 stand for amplitudes of displacement.)

and

$$M_2 \ddot{u}_2 = -P_s \tag{2.31b}$$

This indicates that the ratio of accelerations is inversely proportional to the ratio of masses, that is

$$\frac{\ddot{u}_1}{\ddot{u}_2} = -\frac{M_2}{M_1} \tag{2.32}$$

For a system vibrating with a frequency ω, one can put

$$u_1 = A_1 \sin \omega t \tag{2.33a}$$

and

$$u_2 = -A_2 \sin \omega t \tag{2.33b}$$

From this and Equation 2.32, after differentiation, the ratio of amplitudes becomes

$$\frac{A_1}{A_2} = \frac{M_2}{M_1} \tag{2.34}$$

as illustrated in Figure 2.7. In this way, the system is equivalent to two oscillators, both having the base at the stagnation point determined by Equation 2.34. If the system is initially at rest, then not only accelerations, but also displacements and velocities are proportional to the same ratio of masses. For a series connection, Equation 2.29 holds, so one can find, for the left spring, for example, k_1 as a function of k and M_1/M_2, which leads to the determination of the natural frequency as

$$\omega^2 = \frac{k_1}{M_1} = \frac{k}{M^*} \quad \text{where} \quad \frac{1}{M^*} = \frac{1}{M_1} + \frac{1}{M_2} \tag{2.35}$$

The resulting effective mass, M^*, is smaller than any of the two component masses. This system has two degrees of freedom (DOFs), therefore two natural frequencies, at least nominally, must be present. One of them is associated with vibrations, as determined above, and the other one relates to rigid-body motion, which means $\omega = 0$.

GEOMETRIC STIFFENING

There is an important effect associated with slender structural elements, namely, elements that have (at least) one dimension considerably smaller than the other two. If the loading of a member of this type is carried out in two steps, the magnitude of deflections in the second step may depend on the magnitude of loads in the first. This effect is called *geometric stiffening*, and is best illustrated by the behavior of a rod that has one end pinned and the

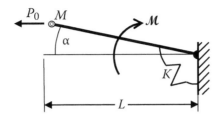

FIGURE 2.8   A rigid, pin-ended bar.

other end free. As long as the element has no axial loading, it can be freely rotated. If, however, a horizontal stretching load is applied first and an attempt is then made to deflect the free end in the direction perpendicular to the initial force, a resistance is encountered. This axial force, which remains here parallel to the original direction, is called the *pre-load* or *prestress*.

Consider a rigid, pin-ended bar shown in Figure 2.8, which illustrates one possible arrangement. When the horizontal force P_0 is zero, the angular stiffness of the system is K.

The presence of P_0 makes it more difficult to rotate the bar because of the additional moment, $P_0 L \alpha$, opposing the rotation. The total apparent stiffness is thus

$$\bar{K} = K + P_0 L = K + K_g \tag{2.36}$$

where K_g is called the geometric stiffness. When the axial force is directed opposite, so that it places the bar in compression, the apparent stiffness $\bar{K}$ is less than the elastic stiffness K. If the magnitude of the compressive force is sufficient, $\bar{K}$ may become zero, which means that no resistance is offered to rotation. We then say that the system becomes unstable and write

$$K - P_{cr} L = 0 \tag{2.37a}$$

or

$$P_{cr} = K/L \tag{2.37b}$$

where P_{cr} is the critical force or the *buckling force*. Using this concept, we can put the expression for the apparent stiffness in a different form:

$$\bar{K} = \left(1 + \frac{P_0}{P_{cr}}\right) K \tag{2.38}$$

in which positive P_0 means tension. When J is the mass moment of inertia of the bar with respect to the pivot point, the natural frequency in the absence of P_0 is calculated from $\omega^2 = K/J$. Taking the axial force into account merely modifies the stiffness; therefore,

$$\bar{\omega}^2 = \left(1 + \frac{P_0}{P_{cr}}\right) \omega^2 \tag{2.39}$$

where $\bar{\omega}$ is the natural frequency calculated in the presence of the axial force, P_0, which is positive when tensile. The equation of motion of the system in Figure 2.8, rotating about the pivot point, is

$$J\ddot{\alpha} = -C'\dot{\alpha} - K\alpha - P_0 L\alpha + \mathcal{M} \qquad (2.40)$$

where the term $C'\dot{\alpha}$, not visualized in Figure 2.8, represents the damping resistance. This is essentially the same as the equation for a translatory motion, except for the added axial force term. When written as

$$J\ddot{\alpha} + C'\dot{\alpha} + \bar{K}\alpha = \mathcal{M}(t) \qquad (2.41)$$

with $\bar{K}$ defined by Equation 2.36, the similarity is even more pronounced. (Note that a damper is not shown in Figure 2.8 and the damping term was added to the equation for the sake of completeness.) The conclusion is that once we replace the elastic stiffness, K, by its apparent value, $\bar{K}$, the axial force will not show up in the equation motion of a single degree of freedom (SDOF) system.

BUCKLING FORCE, CRITICAL FORCE, STIFFNESS, AND FREQUENCY

When a beam is under the influence of an axial compressive force P, one can solve its differential equation of equilibrium and request the value of P for which the lateral displacements become infinitely large. This yields a series of forces

$$P_{cr1}, P_{cr2}, P_{cr3}, \ldots$$

These special values of the axial load are called critical forces. The name stems from the fact that for P approaching any P_{cr}, the calculated deflections tend to become infinite. The smallest of all critical forces (in terms of absolute values), P_{cr1}, is called the *buckling force*, because it is associated with the lateral instability, or buckling, during a static compression test. A beam cannot carry a sustained compressive load larger than the buckling force, and for this reason the remaining critical values are of lesser importance (unless the axial loading lasts only for a very short time).

A beam under lateral loads, in the presence of an axial force, is commonly referred to as a *beam-column*. A dynamic definition of a critical force may be obtained by creating an expression for natural frequency in the presence of a compressive force, P, and then requesting that the natural frequency becomes zero. The natural frequency of a beam-column vibrating in the rth mode may be approximately expressed as

$$\bar{\omega}_r^2 = \left(1 + \frac{P_0}{P_{crr}}\right)\omega_r^2 \qquad (2.42a)$$

or

$$\bar{\omega}_r^2 = \left(1 - \frac{P_0}{P_{crr}}\right)\omega_r^2 \qquad (2.42b)$$

where

   ω_r is the natural frequency in the absence of P_0
   $\bar{\omega}_r$ is the actual frequency of the rth mode
   P_{crr} is the magnitude of the critical force associated with the rth mode

Equation 2.42a is used for the tensile and Equation 2.42b for the compressive P. (Writing two equations instead of one allows us to treat all symbols as positive quantities.) According to Blevins [11], Equation 2.42 represents the exact solution for some classical support cases and offers a close approximation for others. For a given value of P, the first mode frequency is most significantly affected, while the influence on the higher modes is typically much less important. In shock and impact analyses the situation is relatively simple in that usually the first mode is of principal interest (Equation 2.42 reverts then to Equation 2.39).

A convenient way of specifying P or P_{cr} is by expressing it as a multiple of *Euler force*, P_e:

$$P_e = \frac{\pi^2 EI}{L^2} \tag{2.43}$$

where P_e is the buckling force for a simply supported beam made of linearly elastic material.

For most structures, even those that look quite slender, the buckling load calculated on an elastic basis exceeds the true value determined with the use of actual material properties. From our viewpoint, however, this is of little concern, because as long as the structure is deforming within the elastic range, it is the elastic value of P_{cr} that influences stiffness and natural frequency.

In most cases, geometric stiffening only modifies the behavior and response of a structural member. In some instances, however, this phenomenon is necessary for proper functioning of a member. The rod, pinned with one end only, is a simple example. Also, members like cables or membranes, which, ideally, have no bending stiffness, rely on geometric stiffening to resist lateral loads. A three-dimensional (3D) example is provided by a toy rubber balloon, which must be inflated to have some rigidity normal to its surface.

STATIC DEFLECTION METHOD, DUNKERLEY'S METHOD, AND ADDITION OF FLEXIBLILITIES

The static deflection method is developed by extrapolating a certain SDOF system approach to a multiple degree of freedom (MDOF) structure. The natural frequency ω of a mass M restrained by a spring k is found from

$$\omega^2 = \frac{k}{M} = \frac{gk}{Mg} = \frac{g}{u_{st}} \tag{2.44}$$

where the quotient, Mg/k, is the static deflection, u_{st}, of the mass, M, under the force of gravity.

If a static deflection pattern of a particular MDOF system subjected to gravity is known, one can use this formula to calculate an approximate value of the natural frequency. It is assumed, of course, that the vibratory mode is similar to the gravity-deflected shape. The static deflection, u_{st}, should be the absolute value of the maximum deflection.

This approximate method is less accurate than the previous two described at the beginning of the chapter, but often more convenient. Some static analyses usually precede the dynamic calculations in any engineering project, and the maximum static deflection may be readily available. In computer-aided analysis, it should also be the standard practice to perform a static analysis first. By the use of Equation 2.44, one may then get a quick estimate of ω, thereby making it easier to plan the scope of the dynamics work.

For very simple problems, such as those typically used in this chapter, this method usually underestimates the frequency. In large systems, however, this may not necessarily be the case, because it is often difficult to decide which deflection component to use as u_{st}.

The Dunkerley's method, in its very essence, is the result of the following approach. Suppose there are three lumped masses in the structure, 1, 2, and 3, performing vibrations in a fundamental mode. Ignore masses 2 and 3 and assume only mass 1 exists, the rest of the structure being weightless. Find the corresponding natural frequency, ω_1. Repeat the operation for each of the two remaining masses, obtaining ω_2 and ω_3, respectively. The approximation of the fundamental frequency of the whole structure will be

$$\frac{1}{\omega^2} = \frac{1}{\omega_1^2} + \frac{1}{\omega_2^2} + \frac{1}{\omega_3^2} \qquad (2.45)$$

or, in terms of a natural period

$$\tau^2 = \tau_1^2 + \tau_2^2 + \tau_3^2 \qquad (2.46)$$

A complex structure may have a number of flexibilities in series. This means each of those flexibilities adds up to the result in the final deflection. For this arrangement, Equation 2.29 representing a series connection is applicable. When each side of it is multiplied by the system mass M, we get a similar relationship, as represented by Equations 2.45 and 2.46.

This can be interpreted as follows: If a component natural period, τ_i, is calculated by suppressing all other flexibilities except the ith one and if the process is repeated for all component flexibilities, then the system period (squared) is approximated by summing the component periods (squared). This is applicable to a system with all flexibilities being additive. One should also note that while the last two methods give essentially the same equations, the results differ somewhat due to a different approach being used.

This approach can also be used for a parallel connection expressed by Equation 2.30. When both sides are divided by M, the resulting frequency (squared) is the sum of the component frequencies (squared). This applies only when all the component rigidities oppose the contemplated movement.

Distributed Mass System with a Single Lumped Mass

Consider two beams with the same properties except that Beam 1 has only a distributed mass, m, and Beam 2 has only a lumped mass, M_m. The fundamental frequency of Beam 1 is known from the theory, so one can calculate the lumped mass, M_m, so that Beam 2 has the same frequency. This procedure is presented as Example 2.3, where it is shown that if

TABLE 2.1
Modal Mass for Various End Conditions

Configuration	Modal Mass, $M_m/(mL)$
Cantilever, tip impact	0.243
Simply supported, center impact	0.493
Supported-clamped, center impact	0.451
Clamped, center impact	0.384
Clamped-guided, end impact	0.384

a mass of $M_m \approx 0.243\,mL$ is placed at the tip of the massless cantilever, it gives the proper natural frequency. The lumped mass, M_m, calculated in this manner, is called the *modal mass*. Table 2.1 lists the modal masses for the end conditions of interest.

One application of this definition is, when there is a distributed mass and a lumped mass M_1 present and neither of these two mass components prevails. One can then expect that the equivalent mass, $M_e = M_m + M_1$, placed on an otherwise massless beam, gives a good approximation of the natural frequency. This indeed is the case, as Blevins [11] states, where the approximate formulas based on such reasoning are accurate within 1% of the exact solutions. This is how frequencies were obtained for Cases 2.24 through 2.28.

There are reasonable limits as to how this procedure can be used, the most important one being that the additional lumped mass is placed at or near the location of the maximum deflection in the fundamental mode.

SYMMETRY PROPERTIES

A structure is said to be symmetric with respect to a plane, if one-half of it is a mirror image of the other half. When the material properties are also symmetrically distributed, we speak of *elastic symmetry*. An arbitrary loading system applied to a structure can be resolved into a symmetric and an antisymmetric part. To achieve this, each individual force is treated as schematically shown in Figure 2.9. The formulas used for this operation are

$$W'' = \frac{1}{2}(W_1 + W_2) \quad \text{(symm.);} \tag{2.47a}$$

$$W' = \frac{1}{2}(W_1 - W_2) \quad \text{(antisym.)} \tag{2.47b}$$

FIGURE 2.9 (a) General loading; (b) symmetric forces; and (c) antisymmetric forces.

With the loading so resolved we may apply the following theorem:

When a structure, symmetric about a plane, is subjected to a symmetric set of forces, its response is also symmetric. Similarly, an antisymmetric load would produce only an antisymmetric response.

Here, by the term *response*, we mean displacements, internal forces, reactions, and so on. Figure 2.10 shows the symmetric and the antisymmetric parts of the internal forces in cross sections of a 3D beam and a 2D beam. (Note that when applying the above theorem to a 3D problem, one should draw the moments in the form of circular arrows.)

Structural symmetry may be used to reduce the size of a problem, as shown in Figure 2.11. The arbitrary loading acting on a frame is resolved into a symmetric and an antisymmetric component. Each is analyzed separately, with only one-half of the frame modeled

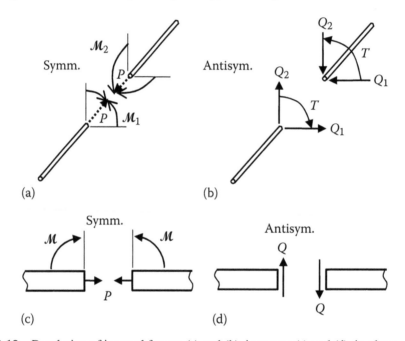

FIGURE 2.10   Resolution of internal forces: (a) and (b), in space; (c) and (d), in plane.

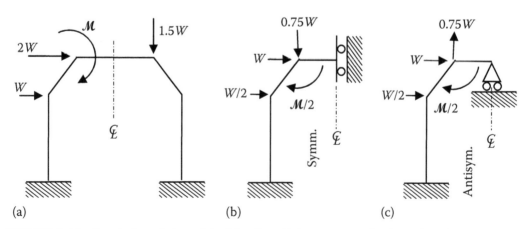

FIGURE 2.11   Resolution of the original loading (a) into symmetric (b) and antisymmetric (c) parts for a plane case.

and the other half being simulated by the end condition at the plane of cut. The static response is obtained by a superposition of both subcases. In spite of the additional operations involved, reducing the problem size by one-half may often be advantageous.

The above statements apply to statics and dynamics as well. Additionally, a symmetric structure will have modes of vibrations, which are either symmetric or antisymmetric. Besides the basic symmetry described here, there are cases of multiple symmetries as well. For example, a cube has three planes of symmetry. If loaded with uniform pressure, this triple symmetry allows a treatment of only one-eighth of that cube.

In real structures, there seldom is a perfect symmetry, whether with respect to elastic properties or the applied load. In such cases, a near-symmetry may be used to conclude about the near-symmetry of the response.

Nonlinear problems may also exhibit symmetry. While the above theorem applies to these as well, the resolution of loading into symmetric and antisymmetric components is not generally involved, because superposition is not allowed in nonlinear cases. Still, such a resolution may, at times, provide a reasonable approximation, or even an accurate solution.

CLOSING REMARKS

Natural frequency is usually of interest when dealing with prolonged vibration. In shock- and impact-type problems, the knowledge of natural frequency, or equivalently, the natural period, is important as far as the timescale of an event is concerned. This statement is true for nonlinear problems as well, although the concept of natural frequency is less clear then.

TABULATION OF CASES

COMMENTS

All structural members in this chapter are either linear-elastic or rigid bodies.

CASE 2.1   TWO-DOF SYSTEM

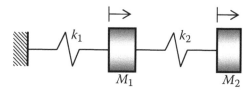

$$\omega_1^2 = \frac{k_1}{M_1}; \quad \omega_2^2 = \frac{k_2}{M_2};$$

Put

$$\omega_m^2 = \frac{k_1}{M_1} + \frac{k_2}{M^*} \quad \text{with} \quad \frac{1}{M^*} = \frac{1}{M_1} + \frac{1}{M_2}$$

Then the upper and lower frequencies are

$$\omega_{1,2} = \left[\frac{1}{2}\left[\omega_m^2 \pm \sqrt{\omega_m^4 - \omega_1^2\omega_2^2}\right]\right]^{1/2}$$

and natural periods are

$$\tau_{1,2} = \frac{2\pi}{\omega_{1,2}}$$

CASE 2.2   TWO MASSES, FREE TO SLIDE, CONNECTED BY A SPRING

$$\omega^2 = \frac{k}{M^*} \quad \text{with} \quad \frac{1}{M^*} = \frac{1}{M_1} + \frac{1}{M_2}; \quad \tau = 2\pi\sqrt{\frac{M^*}{k}}$$

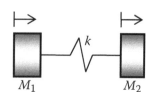

CASE 2.3 PENDULUM WITH A RIGID, MASSLESS ROD, AND GRAVITY PRESENT

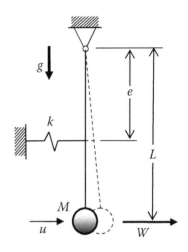

$\bar{k} = W/u$; stiffness definition

$$\bar{k} = \frac{Mg}{L} + \frac{ke^2}{L^2}; \quad \bar{\omega} = \sqrt{\frac{ke^2}{ML^2} + \frac{g}{L}}; \quad \bar{\tau} = \frac{2\pi}{\bar{\omega}}$$

CASE 2.4 RIGID BEAM WITH A PIVOT POINT AND AN ANGULAR SPRING

Lateral vibrations in the presence of a compressive force.
   (The arrow indicates the stiffness direction.)

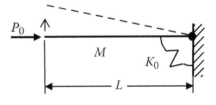

$$\bar{k} = \frac{K_0}{L^2}\left(1 - \frac{P_0}{P_{cr}}\right); \quad \bar{\omega} = \sqrt{\frac{K_0}{J}\left(1 - \frac{P_0}{P_{cr}}\right)}; \quad \bar{\tau} = \frac{2\pi}{\bar{\omega}}$$

$$P_{cr} = K_0/L; \quad J = ML^2/3 \ (P < P_{cr})$$

CASE 2.5 TWO RIGID, PIVOTING BEAMS, WITH ANGULAR SPRINGS

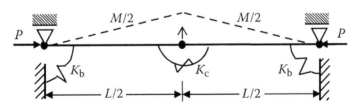

Lateral vibrations in the presence of a compressive force.
   (The arrow indicates the stiffness direction.)

$$\bar{k} = \frac{8}{L^2}(K_b + 2K_c)\left(1 - \frac{P_0}{P_{cr}}\right); \quad \bar{\omega} = \sqrt{\frac{(K_b + 2K_c)}{J}\left(1 - \frac{P_0}{P_{cr}}\right)}; \quad \bar{\tau} = \frac{2\pi}{\bar{\omega}}$$

$$P_{cr} = \frac{2}{L}(K_b + 2K_c); \quad J = ML^2/24$$

CASE 2.6   MASSLESS CABLE WITH A LUMPED MASS, PRELOADED WITH P_0, VIBRATING LATERALLY

$$(a + b = L)$$

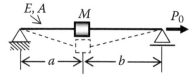

$$k = \frac{(a+b)P_0}{ab}; \quad \omega = \sqrt{\frac{k}{M}}; \quad \tau = 2\pi\sqrt{\frac{M}{k}}$$

Special case: $a = b = L/2$

$$k = \frac{4P_0}{L}; \quad \omega = \sqrt{\frac{4P_0}{ML}}; \quad \tau = \pi\sqrt{\frac{ML}{P_0}}$$

Note: After the preload is applied, the right end is fixed, Ref. [11].

CASE 2.7   CABLE VIBRATING LATERALLY, PRELOADED WITH P_0

$$\omega = \frac{\pi}{L}\sqrt{\frac{P_0}{m}}; \quad \tau = 2L\sqrt{\frac{m}{P_0}} \quad \text{Ref. [11]}$$

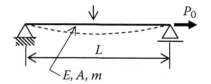

Note: After the preload is applied, the end is fixed and there is no sag.

CASE 2.8   SAGGING CABLE VIBRATING IN A PLANE

Small sag: $(\delta/L)^2 \ll 1$
   δ is the center deflection, when at rest.

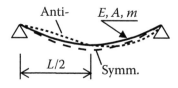

$$P = EA\frac{\Delta L}{L}; \text{ definition of axial stiffness } EA$$

$$\alpha^2 = \left(\frac{8\delta}{L}\right)^2\frac{EA}{P_{av}}\frac{L}{L_e}; \quad \text{sag parameter, } L_e \approx \left[1+8\left(\frac{\delta}{L}\right)^2\right]$$

P_{av} is the average axial load in a cable when at rest.

$$\omega = \frac{2\pi}{L}\sqrt{\frac{P_{av}}{m}}; \quad \tau = L\sqrt{\frac{m}{P_{av}}}; \text{ first antisymmetric mode}$$

$$\omega = \frac{\pi\lambda}{L}\sqrt{\frac{P_{av}}{m}}; \quad \tau = \frac{2L}{\lambda}\sqrt{\frac{m}{P_{av}}}; \text{ first symmetric mode}$$

For $\alpha^2 \gg 100$, $\lambda = 2.861$; for $\alpha^2 \ll 1$, $\lambda = 1.0$,

$$\tan\left(\frac{\pi\lambda}{2}\right) = \frac{\pi\lambda}{2}\left(1 - \frac{4}{\alpha^2}\left(\frac{\pi\lambda}{2}\right)^2\right); \quad \text{general expression for } \lambda, \text{ Ref. } [11].$$

CASE 2.9   MASSLESS CANTILEVER BEAM WITH A LUMPED-TIP MASS (CF)

$$k = \frac{3EI}{L^3}; \quad \omega = \sqrt{\frac{3EI}{ML^3}}; \quad \tau = 2\pi\sqrt{\frac{ML^3}{3EI}}$$

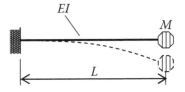

k is the static stiffness when a lateral force is applied to M.

CASE 2.10   MASSLESS BEAM WITH A LUMPED MASS AT THE MIDPOINT (SS)

$$k = \frac{48EI}{L^3}; \quad \omega = 4\sqrt{\frac{3EI}{ML^3}}; \quad \tau = \frac{\pi}{2}\sqrt{\frac{ML^3}{3EI}}$$

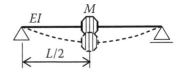

CASE 2.11   MASSLESS BEAM WITH A LUMPED MASS AT THE MIDPOINT (SC)

$$k = \frac{107.3EI}{L^3}; \quad \omega = 5.98\sqrt{\frac{3EI}{ML^3}}; \quad \tau = 1.051\sqrt{\frac{ML^3}{3EI}}$$

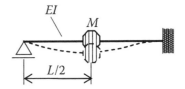

CASE 2.12   MASSLESS BEAM WITH A LUMPED MASS AT THE MIDPOINT (CC)

$$k = \frac{192EI}{L^3}; \quad \omega = 8\sqrt{\frac{3EI}{ML^3}}; \quad \tau = \frac{\pi}{4}\sqrt{\frac{ML^3}{3EI}}$$

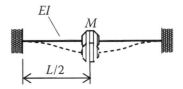

CASE 2.13   MASSLESS BEAM WITH A LUMPED MASS AT THE END (CG)

$$k = \frac{12EI}{L^3}; \quad \omega = 2\sqrt{\frac{3EI}{ML^3}}; \quad \tau = \pi\sqrt{\frac{ML^3}{3EI}}$$

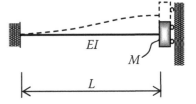

CASE 2.14　MASSLESS BEAM WITH A ROTATORY MASS AT THE MIDPOINT (SS)

$$k = \frac{12EI}{L}; \quad \omega = \sqrt{\frac{12EI}{JL}}; \quad \tau = \pi\sqrt{\frac{JL}{3EI}}$$

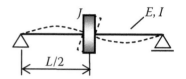

Note: A thin disk with diameter D has $J = MD^2/16$.

CASE 2.15　MASSLESS BEAM WITH A ROTATORY MASS AT THE MIDPOINT (CC)

$$k = \frac{16EI}{L}; \quad \omega = 4\sqrt{\frac{EI}{JL}}; \quad \tau = \frac{\pi}{2}\sqrt{\frac{JL}{EI}}$$

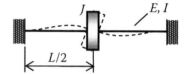

Note: A thin disk with diameter D has $J = MD^2/16$.

CASE 2.16　MASS *M* SUPPORTED BY *n* MASSLESS BEAMS

(Illustration for $n = 2$)

$$k = \frac{12EI}{H^3}n; \quad \omega = \sqrt{\frac{12EI}{MH^3}n}; \quad \tau = \pi\sqrt{\frac{MH^3}{3EIn}}$$

Note: Stiffness k is relative to the horizontal load applied to the cap.

CASE 2.17　BUILDING WITH RIGID FLOORS

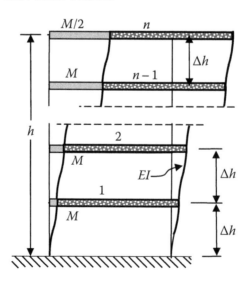

There are p weightless columns on each floor ($p = 2$ is shown).

$$\tau_0 = 4h\sqrt{\frac{M(\Delta h)}{12EIp}};\ \text{fundamental lateral mode period}$$

Note: The building is treated as a shear beam with mass M per floor. The equivalent shear stiffness per floor is $GA_s = 12EIp/(\Delta h)^2$. (Refer to Case 2.21.) If the columns are unequal, but are made of the same material, Ip should be replaced by the sum, ΣI_i, for a floor.

CASE 2.18   BUILDING WITH RIGID FLOORS AND FUNDAMENTAL PERIOD τ_0

(As in Case 2.17), additional mass, M_t, placed on the top level (roof)

$$\tau_t = 2\pi\sqrt{\frac{n(\Delta h)^3 M_t}{12EIp}}\ ;\ \text{fictitious, period-like term to be used below}$$

$$\tau^2 \approx \tau_0^2 + \tau_t^2;\ \text{fundamental period, including the effect of the additional mass}$$

Note: Comments of Case 2.17 apply. If, for example, there is mass M on the roof rather than mass $M/2$, as prescribed in Case 2.17, then use $M_t = M/2$.

CASE 2.19   WEIGHTLESS AXIAL BAR WITH LUMPED MASSES ATTACHED

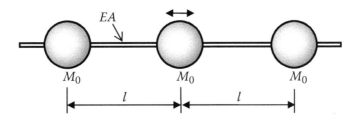

Natural frequencies of two highest modes

  (a) Vibration pattern: adjacent nodes are moving against one another, and midpoints between nodes are stationary. (Not illustrated.)

$$\omega = 2\sqrt{\frac{EA}{M_0 l}};\quad \tau = \pi\sqrt{\frac{M_0 l}{EA}};$$

$M_0 = \dfrac{4}{\pi^2}ml \approx 0.4053\,ml$: this value is used to obtain the natural frequency of a bar with a continuous mass distribution, vibrating in this mode.

  (b) Vibration pattern: one node is oscillating and the adjacent two are stationary, as illustrated in the sketch.

$$\omega = \sqrt{2}\sqrt{\frac{EA}{M_0 l}};\quad \tau = \frac{\pi}{\sqrt{2}}\sqrt{\frac{M_0 l}{EA}};$$

$M_0 = \dfrac{8}{\pi^2} ml \approx 0.8106\ ml$: this value is used to obtain the natural frequency of a bar with a continuous mass distribution, vibrating in this mode.

CASE 2.20   AXIAL BAR VIBRATING LONGITUDINALLY

(a) $\omega = \dfrac{\pi}{L}\sqrt{\dfrac{EA}{m}};\ \tau = 2L\sqrt{\dfrac{m}{EA}}$

(b) $\omega = \dfrac{\pi}{2L}\sqrt{\dfrac{EA}{m}};\ \tau = 4L\sqrt{\dfrac{m}{EA}};$ Ref. [11]

(c) $\omega = \dfrac{\pi}{L}\sqrt{\dfrac{EA}{m}};\ \tau = 2L\sqrt{\dfrac{m}{EA}}$

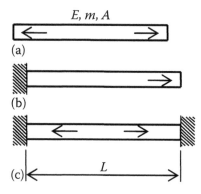

CASE 2.21   SHEAR BEAM VIBRATING LATERALLY

(a) $\omega = \dfrac{\pi}{L}\sqrt{\dfrac{GA_s}{m}};\ \tau = 2L\sqrt{\dfrac{m}{GA_s}}$

(b) $\omega = \dfrac{\pi}{2L}\sqrt{\dfrac{GA_s}{m}};\ \tau = 4L\sqrt{\dfrac{m}{GA_s}};$ Ref. [11]

(c) $\omega = \dfrac{\pi}{L}\sqrt{\dfrac{GA_s}{m}};\ \tau = 2L\sqrt{\dfrac{m}{GA_s}}$

Note: For shear area, A_s, refer to Chapter 10.

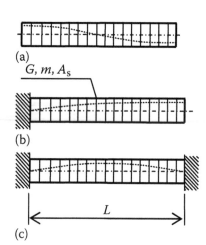

CASE 2.22   SHAFT PERFORMING TWISTING VIBRATIONS

(a) $\omega = \dfrac{\pi}{L}\sqrt{\dfrac{GC}{\rho I_0}};\ \tau = 2L\sqrt{\dfrac{\rho I_0}{GC}}$

(b) $\omega = \dfrac{\pi}{2L}\sqrt{\dfrac{GC}{\rho I_0}};\ \tau = 4L\sqrt{\dfrac{\rho I_0}{GC}}$

(c) $\omega = \dfrac{\pi}{L}\sqrt{\dfrac{GC}{\rho I_0}};\ \tau = 2L\sqrt{\dfrac{\rho I_0}{GC}}$

Note: For circular-section shafts, $C = I_0$, Ref. [11].

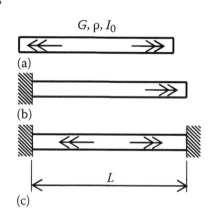

CASE 2.23   FREE-FREE BEAM VIBRATING LATERALLY (FF)

$$\omega = \frac{22.373}{L^2}\sqrt{\frac{EI}{m}}; \quad \tau = \frac{L^2}{3.561}\sqrt{\frac{m}{EI}}; \quad P_{cr} = \frac{\pi^2 EI}{L^2}; \quad \text{Ref. [11]}$$

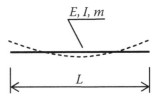

CASE 2.24   CANTILEVER BEAM (CF)

(a) With a distributed mass m only:

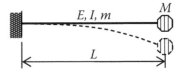

$$\omega = \frac{3.516}{L^2}\sqrt{\frac{EI}{m}}; \quad \tau = 1.787 L^2\sqrt{\frac{m}{EI}}; \quad P_{cr} = \frac{\pi^2 EI}{4L^2}$$

(b) With an additional lumped-tip mass (see Case 2.9):

$$\omega = \sqrt{\frac{3EI}{(M+0.243mL)L^3}}; \quad \tau = 2\pi\sqrt{\frac{(M+0.243mL)L^3}{3EI}}; \quad \text{Ref. [11].}$$

CASE 2.25   SIMPLY SUPPORTED BEAM (SS)

(a) With a distributed mass m only:

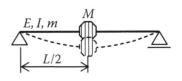

$$\omega = \frac{\pi^2}{L^2}\sqrt{\frac{EI}{m}}; \quad \tau = \frac{2L^2}{\pi}\sqrt{\frac{m}{EI}}; \quad P_{cr} = \frac{\pi^2 EI}{L^2}; \quad \text{Ref. [11]}$$

(b) With an additional M at the midpoint (see Case 2.10):

$$\omega = 4\sqrt{\frac{3EI}{(M+0.493mL)L^3}}; \quad \tau = \frac{\pi}{2}\sqrt{\frac{(M+0.493mL)L^3}{3EI}}$$

CASE 2.26   SUPPORTED-CLAMPED BEAM (SC)

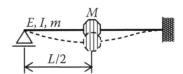

(a) With a distributed mass m only:

$$\omega = \frac{15.418}{L^2}\sqrt{\frac{EI}{m}}; \quad \tau = \frac{L^2}{2.454}\sqrt{\frac{m}{EI}}; \quad P_{cr} = \frac{2.05\pi^2 EI}{L^2}; \quad \text{Ref. [11].}$$

(b) With an additional M at the midpoint (see Case 2.11):

$$\omega = 5.98 \sqrt{\frac{3EI}{(M + 0.451mL)L^3}}; \quad \tau = 1.051 \sqrt{\frac{(M + 0.451mL)L^3}{3EI}}$$

CASE 2.27   CLAMPED BEAM (CC)

(a) With a distributed mass m only:

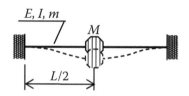

$$\omega = \frac{22.373}{L^2}\sqrt{\frac{EI}{m}}; \quad \tau = \frac{L^2}{3.561}\sqrt{\frac{m}{EI}}; \quad P_{cr} = \frac{4\pi^2 EI}{L^2}; \quad \text{Ref. [11]}$$

(b) With an additional M at the midpoint (see Case 2.12):

$$\omega = 8\sqrt{\frac{3EI}{(M + 0.384mL)L^3}}; \quad \tau = \frac{\pi}{4}\sqrt{\frac{(M + 0.384mL)L^3}{3EI}}$$

CASE 2.28   CLAMPED-GUIDED BEAM (CG)

(a) With a distributed mass m only:

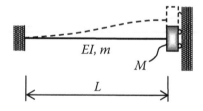

$$\omega = \frac{5.5932}{L^2}\sqrt{\frac{EI}{m}}; \quad \tau = 1.1234L^2\sqrt{\frac{m}{EI}}; \quad P_{cr} = \frac{\pi^2 EI}{L^2}$$

(b) With an additional M at the end (see Case 2.13):

$$\omega = 2\sqrt{\frac{3EI}{(M + 0.384mL)L^3}}; \quad \tau = \pi\sqrt{\frac{(M + 0.384mL)L^3}{3EI}}; \quad \text{Ref. [11]}$$

CASE 2.29   PORTAL FRAME WITH A HEAVY AND STIFF CAPPING BEAM

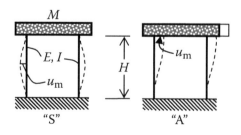

m is the distributed mass of a column.

Symmetric mode "S": $\dfrac{wH}{u_\mathrm{m}} = \dfrac{384EI}{H^3};\quad \omega = \dfrac{22.373}{H^2}\sqrt{\dfrac{EI}{m}};\quad \tau = \dfrac{H^2}{3.561}\sqrt{\dfrac{m}{EI}}$

Antisymmetric mode "A": $\dfrac{wH}{u_\mathrm{m}} = \dfrac{24EI}{H^3};\quad \omega = 2\sqrt{\dfrac{3EI}{(M+0.384mH)H^3}};$

$$\tau = \pi\sqrt{\dfrac{(M+0.384mH)H^3}{3EI}}$$

Note: These two modes of deformation need not be related. If a distributed load w is involved, it may be different in each mode. See Chapter 10 for beam deflections and responses.

CASE 2.30   PORTAL FRAME WITH EACH COLUMN OF THE SAME PROPERTIES AS THE CAPPING BEAM

Each of the three members has a distributed mass m. Also, $B = H$.

Symmetric mode: $\omega = \dfrac{12.65}{H^2}\sqrt{\dfrac{EI}{m}};\quad \tau = 0.4968H^2\sqrt{\dfrac{m}{EI}}$

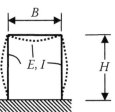

CASE 2.31   CIRCULAR MEMBRANE, PRELOADED WITH q_0 (N/m) PER UNIT LENGTH OF EDGE, VIBRATING LATERALLY

$$\omega = \dfrac{2.404}{R}\sqrt{\dfrac{q_0}{m}};\quad \tau = 2.614R\sqrt{\dfrac{m}{q_0}};$$

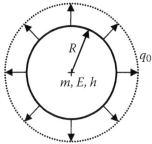

where $m \sim$ kg/m² is the mass per unit surface

Note: After the preload is applied, the edge is fixed, Ref. [92].

CASE 2.32   RECTANGULAR MEMBRANE, PRELOADED WITH q_0 (N/mm) PER UNIT LENGTH OF EDGE, VIBRATING LATERALLY

$$\omega = \pi\left(\dfrac{1}{a^2}+\dfrac{1}{b^2}\right)^{1/2}\sqrt{\dfrac{q_0}{m}};\quad \tau = \dfrac{2\pi}{\omega};$$

where $m \sim$ kg/m²

Note: After the preload is applied, the edge is fixed, Ref. [111].

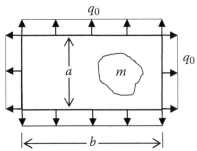

CASE 2.33   CIRCULAR PLATE WITH SEVERAL TYPES OF EDGE SUPPORTS

$$\omega = \frac{5.253}{R^2}\sqrt{\frac{D}{m}}; \quad \tau = 1.1961R^2\sqrt{\frac{m}{D}} \quad \text{(free)}$$

$$\omega = \frac{4.977}{R^2}\sqrt{\frac{D}{m}}; \quad \tau = 1.2624R^2\sqrt{\frac{m}{D}} \quad \text{(supported);} \quad \text{Ref. [11]}$$

$$\omega = \frac{10.22}{R^2}\sqrt{\frac{D}{m}}; \quad \tau = 0.6148R^2\sqrt{\frac{m}{D}} \quad \text{(clamped)}$$

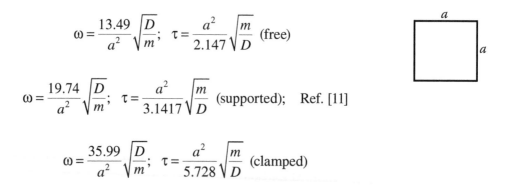

CASE 2.34   SQUARE PLATE WITH SEVERAL TYPES OF EDGE SUPPORTS

$$\omega = \frac{13.49}{a^2}\sqrt{\frac{D}{m}}; \quad \tau = \frac{a^2}{2.147}\sqrt{\frac{m}{D}} \quad \text{(free)}$$

$$\omega = \frac{19.74}{a^2}\sqrt{\frac{D}{m}}; \quad \tau = \frac{a^2}{3.1417}\sqrt{\frac{m}{D}} \quad \text{(supported);} \quad \text{Ref. [11]}$$

$$\omega = \frac{35.99}{a^2}\sqrt{\frac{D}{m}}; \quad \tau = \frac{a^2}{5.728}\sqrt{\frac{m}{D}} \quad \text{(clamped)}$$

CASE 2.35   RECTANGULAR PLATE WITH A SUPPORTED OR CLAMPED EDGE

$$\omega = \frac{12.34}{a^2}\sqrt{\frac{D}{m}}; \quad \tau = 0.5092a^2\sqrt{\frac{m}{D}} \quad \text{(supported)}$$

$$\omega = \frac{24.57}{a^2}\sqrt{\frac{D}{m}}; \quad \tau = 0.2557a^2\sqrt{\frac{m}{D}} \quad \text{(clamped);} \quad \text{Ref. [11]}$$

CASE 2.36   LONG PLATE WITH A SUPPORTED OR CLAMPED EDGE

$$\omega = \frac{\pi^2}{a^2}\sqrt{\frac{D}{m}}; \quad \tau = \frac{2a^2}{\pi}\sqrt{\frac{m}{D}} \quad \text{(supported)}$$

$$\omega = \frac{22.373}{a^2}\sqrt{\frac{D}{m}}; \quad \tau = \frac{a^2}{3.5608}\sqrt{\frac{m}{D}} \quad \text{(clamped);} \quad \text{Ref. [11]}$$

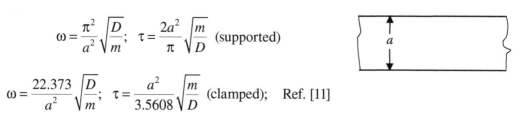

CASE 2.37   RECTANGULAR PLATE WITH POINT-SUPPORTED CORNERS

$$\omega = \frac{\alpha^2}{b^2}\sqrt{\frac{D}{m}}; \quad \tau = 2\pi\frac{b^2}{\alpha^2}\sqrt{\frac{m}{D}}$$

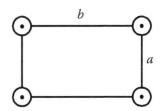

Note: This is for ideal point-supports. If the columns play the role of these supports, the bending stiffness of a column must be small compared to the bending stiffness of a tributary segment of the plate, Ref. [111].

b/a	α²
1.0	7.12
1.5	8.92
2.0	9.29

CASE 2.38   CIRCULAR PLATE WITH A CENTER HOLE, WITH THE INNER EDGE SUPPORTED, AND THE OUTER EDGE FREE (SF)

$$\omega = \frac{4.11}{R^2}\sqrt{\frac{D}{m}}; \quad \tau = 1.529R^2\sqrt{\frac{m}{D}}; \quad \text{Ref. [11]}$$

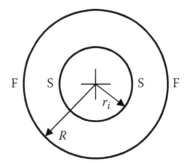

Note: The coefficient is valid only for $(r_i/R) = 0.5$.

CASE 2.39   CIRCULAR PLATE WITH A CENTER HOLE, WITH THE INNER EDGE FREE, AND THE OUTER EDGE SUPPORTED (FS)

$$\omega = \frac{5.07}{R^2}\sqrt{\frac{D}{m}}; \quad \tau = 1.239R^2\sqrt{\frac{m}{D}}; \quad \text{Ref. [11]}$$

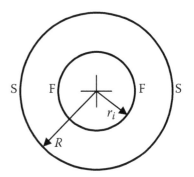

Note: The coefficient is valid only for $(r_i/R) = 0.5$. For other ratios, consult Young [118].

CASE 2.40   CIRCULAR PLATE WITH A CENTER HOLE, WITH THE INNER EDGE CLAMPED, AND THE OUTER EDGE FREE (CF)

$$\omega = \frac{13}{R^2}\sqrt{\frac{D}{m}}; \quad \tau = 0.4833R^2\sqrt{\frac{m}{D}}; \quad \text{Ref. [11]}$$

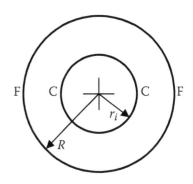

Note: The coefficient is valid only for $(r_i/R) = 0.5$.

CASE 2.41   CIRCULAR PLATE WITH A CENTER HOLE, WITH THE INNER EDGE FREE, AND THE OUTER EDGE CLAMPED (FC)

$$\omega = \frac{17.7}{R^2}\sqrt{\frac{D}{m}}; \quad \tau = 0.355 R^2 \sqrt{\frac{m}{D}}; \quad \text{Ref. [11]}$$

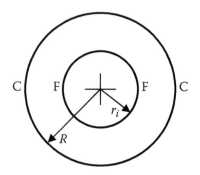

Note: The coefficient is valid only for $(r_i/R) = 0.5$.

CASE 2.42   ARBITRARILY SHAPED PLATE WITH AN ARBITRARY EDGE SUPPORT

u_{st} is the peak static deflection of the plate under gravity g.

$$\omega = 1.277\sqrt{\frac{g}{u_{st}}}$$

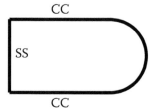

Note: This is an adaptation of Equation 2.15 for plates. According to Ref. [11], the error should be within 3%. The static deflection pattern must be similar to that imposed by gravity.

CASE 2.43   CIRCULAR, MODERATELY THICK RING, VIBRATING RADIALLY

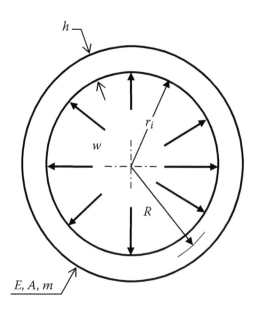

$$k_{st} = \frac{w}{u} = \frac{EA}{r_i R} \quad \text{(static stiffness, for expansion with a radial load, } w\text{)}$$

$$\omega = \frac{1}{R}\sqrt{\frac{EA}{m}} \quad \text{and} \quad \tau = 2\pi R\sqrt{\frac{m}{EA}}$$

A is the cross-sectional area, $m = A\rho$.

CASE 2.44 CIRCULAR THIN RING, WITH THE FIRST BENDING MODE, VIBRATING IN A PLANE

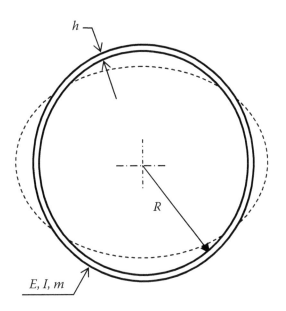

$$\omega = \frac{2.683}{R^2}\sqrt{\frac{EI}{m}} \quad \text{and} \quad \tau = 2.342R^2\sqrt{\frac{m}{EI}}$$

$$m = A\rho, \quad \text{Ref. [111].}$$

Note: I is the section moment of inertia about an axis normal to a paper.

CASE 2.45   THIN CIRCULAR SHELL, VIBRATING IN THREE BASIC MODES (R IS THE MEAN RADIUS)

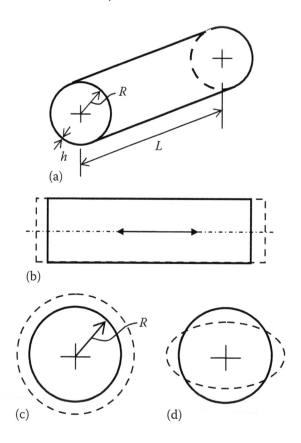

(a)

(b)

(c)   (d)

(b) Longitudinal (axial)

$$\omega = \frac{\pi}{L}\sqrt{\frac{EA}{m}}; \quad \tau = 2L\sqrt{\frac{m}{EA}};$$

$$A = 2\pi Rh; \quad m = A\rho$$

(c) Radial (axial movement constrained)

$$k = \frac{p}{u} = \frac{Eh}{R^2(1-v^2)} \quad \text{(under internal pressure } p\text{)}$$

$$\omega = \frac{1}{R}\sqrt{\frac{E}{\rho(1-v^2)}} \quad \text{and} \quad \tau = 2\pi R\sqrt{\frac{\rho(1-v^2)}{E}}$$

(d) Squashing

$$\omega = \frac{0.7746h}{R^2}\sqrt{\frac{E}{\rho(1-\nu^2)}} \quad \text{and} \quad \tau = \frac{8.112R^2}{h}\sqrt{\frac{\rho(1-\nu^2)}{E}}, \quad \text{Ref. [11].}$$

CASE 2.46   SPHERICAL SHELL, VIBRATING RADIALLY

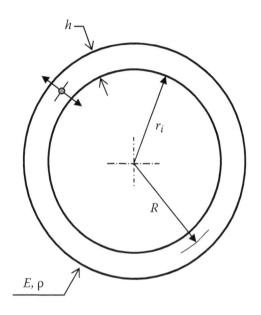

$$k = \frac{p}{u} = \frac{2Eh}{(1-\nu)r_i^2} \quad \text{(under internal pressure } p\text{)}$$

$$\omega = \frac{1}{R}\sqrt{\frac{2E}{\rho(1-\nu)\left(1+\dfrac{h^2}{12R^2}\right)}} \quad \text{and} \quad \tau = \frac{2\pi}{\omega}$$

For thin shells ($h \ll R$):

$$\omega = \frac{1}{R}\sqrt{\frac{2E}{\rho(1-\nu)}} \quad \text{and} \quad \tau = 2\pi R\sqrt{\frac{\rho(1-\nu)}{2E}}, \quad \text{Ref. [11].}$$

CASE 2.47   SOLID, UNCONSTRAINED SPHERE, FREELY VIBRATING IN A PURELY RADIAL MODE

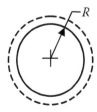

$$\omega = \frac{0.816\pi}{R}\sqrt{\frac{E''}{\rho}}; \quad \tau = 2.451R\sqrt{\frac{\rho}{E''}}$$

$$E'' = \frac{(1-v)E}{(1+v)(1-2v)}, \quad \text{Ref. [44].}$$

Note: The 0.816 coefficient was derived assuming $v = 0.25$.

CASE 2.48 CIRCULAR RIGID BLOCK OF MASS M ATTACHED TO AN ELASTIC HALF-SPACE

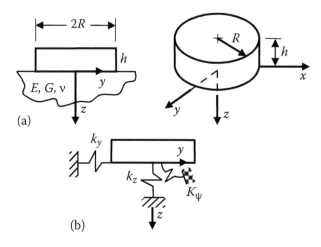

(a)

(b)

E, G, and v; elastic constants, $G = \dfrac{E}{2(1+v)}$

$$k_z = \frac{4GR}{1-v}; \quad k_y = \frac{32(1-v)GR}{7-8v}; \quad K_\psi = \frac{8GR^3}{3(1-v)}; \quad K_\varphi = \frac{16GR^3}{3};$$

with $k_x = k_y$;   Ref. [72].

K_ψ relates to rocking about a horizontal axis, while K_φ is for twisting about the z-axis.

Note: The first sketch shows how the medium is replaced by springs, while the block itself is shown in the second. See comments in text.

CASE 2.49 RECTANGULAR RIGID BLOCK OF MASS M ATTACHED TO AN ELASTIC HALF-SPACE

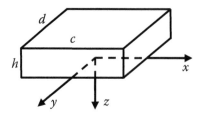

Use formulas for a circular block after the following changes are made:

$$R = \sqrt{\frac{cd}{\pi}} \text{ in } k_z \text{ and } k_y$$

$$R' = \left(\frac{cd}{6\pi} \left(c^2 + d^2 \right) \right)^{1/4} \quad \text{instead of } R \text{ in } K_\varphi$$

$K_\psi = \dfrac{Gcd^2}{1-\nu} \left(0.3923 + \dfrac{d}{10.59c} \right)$ rocking about the x-axis. When rocking takes place about the y-axis, the symbols c and d are interchanged.

Note: See comments in text, Ref. [72].

EXAMPLES

EXAMPLE 2.1   COMBINATIONS OF SERIES AND PARALLEL CONNECTIONS

One of the blocks has a mass M, the other may be considered massless. What is the circular frequency for this mixed arrangement of springs?

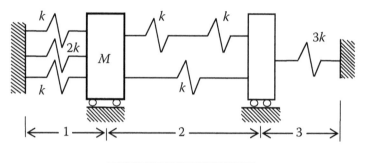

Divide the springs into three groups, as shown.

1. $k^* = k + 2k + k = 4k$; as this group works in parallel.
2. First, for two springs in series:

$$\frac{1}{k_s^*} = \frac{1}{k} + \frac{1}{k} \quad \text{or} \quad k_s^* = k/2;$$

  then $k_2^* = k_s^* + k = 1.5k$ for k working in parallel with the above.
3. $k_3^* = 3k$.

Since the block on the right has no mass, it also has no inertia forces during free vibrations and serves merely as a connector of the springs, k_2^* and k_3^*. The latter assemblies are, in effect, joined in series, and their effective stiffness is k_4^*:

$$\frac{1}{k_4^*} = \frac{1}{k_2^*} + \frac{1}{k_3^*} = \frac{1}{1.5k} + \frac{1}{3k} = \frac{1}{k}$$

This assembly, along with k_1^*, resists the motion of M, therefore it works in parallel with k_1^*. The resultant for the whole system is

$$k^* = k_1^* + k_4^* = 4k + k = 5k; \quad \text{therefore,} \; \omega = \sqrt{\frac{5k}{M}}.$$

EXAMPLE 2.2   PORTAL FRAME VIBRATION

Consider a frame, 4 m wide and 4 m high, with the feet clamped. The material is concrete, $E = 30\,\text{GPa}$, and has a specific gravity of 2.4. The section is square, 400 mm × 400 mm. Calculate the natural frequency of the symmetric mode, f_0 (Hz). Also, find that f_1 would result, if the capping beam were treated as rigid.

Establish properties of the beam:

$$A = 400 \times 400 = 160,000 \text{ mm}^2; \quad I = H^4/12 = 2.133 \times 10^9 \text{ mm}^4,$$

$$m = A\rho = 160,000 \times 0.0024 = 384 \text{ g/mm}.$$

Use Case 2.30 to get the period:

$$\tau = 0.4968\, H^2 \sqrt{\frac{m}{EI}} = 0.4968 \times 4,000^2 \sqrt{\frac{384}{30,000 \times 2.133 \times 10^9}} = 19.47 \text{ ms} \approx 0.0195 \text{ s}.$$

The natural frequency is $f_0 = 1/\tau = 51.35$ Hz. The other frame, with a stiff capping beam is done according to Case 2.29:

$$\tau = \frac{H^2}{3.561} \sqrt{\frac{m}{EI}} = 11.01 \text{ ms} \approx 0.011 \text{ s}, \quad \text{or} \quad f_1 = 90.9 \text{ Hz}.$$

This shows a substantial effect of the capping member on the natural frequency.

EXAMPLE 2.3 MODAL MASS DETERMINATION

Given a cantilever beam with a distributed mass, m, find the equivalent modal mass, M_m, as a fraction of the beam mass, mL. By definition, modal mass is a lumped mass, replacing the prescribed distributed mass, and selected to preserve the natural frequency.

For a bending-only deformation, the exact value of a fundamental frequency is, per Case 2.24,

$$\omega = \frac{3.516}{L^2} \sqrt{\frac{EI}{m}}$$

A massless beam, with M_m at the tip, should have the same frequency. This beam can be viewed as an oscillator with a stiffness, $k = 3EI/L^3$, according to Table 10.2. (The stiffness is with respect to a lateral force at the tip). This frequency can be written as

$$\omega^2 = \frac{k}{M_m} = \frac{3EI}{M_m L^3} \quad \text{while, from the above,} \quad \omega^2 = \frac{3.516^2}{L^4} \frac{EI}{m}.$$

Equating the right sides of both expressions, the modal mass is calculated: $M_m = 0.2427 mL$.

EXAMPLE 2.4   NATURAL FREQUENCY WITH ADDITIVE FLEXIBILITIES

The cantilever shown here forms one continuous element with its base. According to Young [118], the rotational flexibility at the base is as follows: if a force, W, is applied at the tip, then the base will rotate by an angle α given by

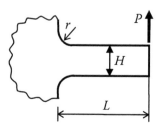

$$\frac{\alpha}{WL} = \frac{16.67}{\pi E H_1^2} + \frac{1-v}{ELH_1}$$

with $H_1 = H + 1.5r$.

Find the approximate natural frequency by a superposition of three flexibility components: bending, base rotation, and shear.

$$L = 200 \text{ mm}, \quad H = 50 \text{ mm}, \quad r = 12 \text{ mm}, \quad \text{Width} = 1 \text{ mm},$$

$$E = 200,000 \text{ MPa}, \quad \rho = 0.00785 \text{ g/mm}^3, \quad v = 0.3, \quad \text{and} \quad G = 76,920 \text{ Mpa}.$$

Section properties: $A = 50 \text{ mm}^2$, $A_s = A/1.2 = 41.67 \text{ mm}^2$, $I = 10,417 \text{ mm}^4$, $m = A\rho = 0.3925 \text{ g/mm}$ (distributed mass).

All three deflection components, namely, bending, base rotation, and shear are additive here. Equation 2.29 takes the following form:

$$\frac{1}{\omega^2} = \frac{1}{\omega_b^2} + \frac{1}{\omega_r^2} + \frac{1}{\omega_s^2}$$

The bending frequency comes from Case 2.24:

$$\omega_b^2 = \frac{3.516^2}{L^4} \frac{EI}{m} = \frac{3.516^2}{200^4} \frac{200,000 \times 10,417}{0.3925} = 41.012$$

The angular flexibility (angle over moment), with $H_1 = 50 + 1.5 \times 12 = 68 \text{ mm}$, is

$$\frac{1}{K} = \frac{16.67}{\pi E H_1^2} + \frac{1-v}{ELH_1} = \frac{16.67}{\pi E \cdot 68^2} + \frac{1-0.3}{E \cdot 200 \times 68} = \frac{1}{166.8 \times 10^6}$$

To find a frequency, associated with the base rotation of an otherwise rigid beam, Case 2.4 may be invoked, with $J = ML^2/3 = mL^3/3 = 0.3925 \times 200^3/3 = 1.047 \times 10^6 \text{ g-mm}^2$:

$$\omega_r^2 = \frac{K}{J} = \frac{166.8 \times 10^6}{1.047 \times 10^6} = 159.4$$

Finally, the shear-related frequency comes from Case 2.21b:

$$\omega_s^2 = \left(\frac{\pi}{2L}\right)^2 \frac{GA_s}{m} = \left(\frac{\pi}{2 \times 200}\right)^2 \frac{76,920 \times 41.67}{0.3925} = 503.7$$

Finally: $\dfrac{1}{\omega^2} = \dfrac{1}{41.012} + \dfrac{1}{159.4} + \dfrac{1}{503.7}$ or $\omega = 5.535\,\text{rad/ms}$, after Equation 2.45.

The above shows that the effect of the shear component is minor. However, ignoring the base flexibility would increase the calculated frequency to 6.405 and would amount to a sizeable omission. Ignoring this effect seems to be a very common flaw in the work of practicing engineers.

EXAMPLE 2.5   NATURAL FREQUENCY WITH ADDITIVE STIFFNESS

Modify Equation 2.30 by dividing each side by a system mass M. The result says that when rigidities work in parallel, the natural frequency of the system (squared) is the sum of (squared) component frequencies. Apply this to determine the fundamental frequency, ω_{1f}, of a beam on an elastic foundation with a stiffness, k_f. The frequency without the foundation is known to be ω_1.

The effects of lateral supports and the foundation rigidity are additive in that both items resist elastic deflection. Alone, the beam vibrates with ω_1. Consider the same beam, but without supports; its motion resisted by the foundation only. The frequency now has a very simple expression:

$$\omega_f^2 = \frac{k_f}{m}$$

which is true for a single segment of unit length as well as for the beam as a whole. The system frequency, per modified Equation 2.30, becomes

$$\omega^2 = \omega_1^2 + \omega_f^2 = \omega_1^2\left(1 + \frac{\omega_f^2}{\omega_1^2}\right)$$

This is essentially the same, as expressed by Equation 2.28, where the distributed mass, m, is canceled out.

EXAMPLE 2.6   SIX-STORY, GENERIC CONCRETE BUILDING

A horizontal section is given in (a) below. The slabs are 300 mm thick, and each slab (including the roof slab) carries a distributed mass of $300\,\text{kg/m}^2$. A partial vertical section is shown in (b).

The ground story is 5.3 m high, and each of the remaining five stories is 3.6 m high. The corner columns are $0.45\,\text{m} \times 0.45\,\text{m}$ and the midside columns have $0.6\,\text{m} \times 0.6\,\text{m}$ sections, as they carry a larger load. Determine the natural period of the horizontal mode of vibration, treating the slabs as undeformable. $E = 31$ GPa, and the specific gravity is 2.4.

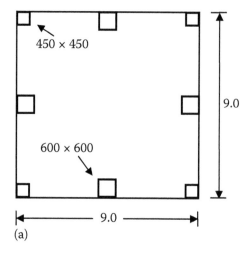

(a)

In spite of the building structure being quite simple, it still needs some adjustments to make it possible to use Cases 2.17 and 2.18 for the purpose of frequency determination. The net height of the building is $h = 23,300 - 300/2 = 23,150$ mm. The average floor height is $\Delta h = h/6 = 23,150/6 = 3,858$ mm. The sum of column sections is $A = 4 \times 0.45^2 + 4 \times 0.6^2 = 2.25$ m^2. Assign the column mass to floors. In terms of volume: $V_c = A(\Delta h) = 2.25 \times 3.858 = 8.681$ m^3 and $V_s = 9 \times 9 \times 0.3 = 24.3$ m^3 is the single slab volume.

Total effective mass of each floor (including added mass):

$$M = (24.3 + 8.681)2,400 + 9 \times 9 \times 300 = 103,454 \text{ kg.}$$

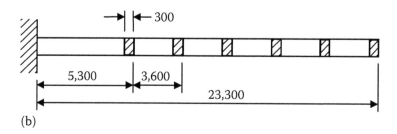

(b)

Case 2.17 was written for p columns, each with a section area moment, I, the sum of moduli being pI. Here we have

$$I_1 = \frac{H_1^4}{12} = \frac{450^4}{12} = 3.417 \times 10^9 \text{ mm}^4; \quad I_2 = \frac{H_2^4}{12} = \frac{600^4}{12} = 10.8 \times 10^9 \text{ mm}^4.$$

The sum is therefore $pI = 4(3.417 + 10.8)10^9 = 56.87 \times 10^9$ mm^4. From Case 2.17:

$$\tau_0 = 4h\sqrt{\frac{M(\Delta h)}{12EpI}} = 4 \times 23,150\sqrt{\frac{103.454 \times 10^6(3858)}{12 \times 31,000 \times 56.87 \times 10^9}} = 402.3 \text{ ms} \approx 0.402 \text{ s.}$$

This, however, would hold for a structure where the mass of the roof would be only half of that of a floor below. In this design, the roof is the same as that of any floor below, therefore Case 2.18 must be employed to calculate the effect of the added top mass, $M_t = 103{,}454/2 = 51{,}727\,\mathrm{kg}$.

$$\tau_t = 2\pi\sqrt{\frac{n(\Delta h)^3 M_t}{12EIp}} = 2\pi\sqrt{\frac{6(3{,}858)^3\,51.727\times10^6}{12\times31{,}000\times56.87\times10^9}} = 182.4\,\mathrm{ms}$$

$$\tau^2 \approx 402.3^2 + 182.4^2 \quad \text{or} \quad \tau = 442\,\mathrm{ms}.$$

The increase in the natural period is noticeable. It becomes quite important in configurations where the roof structure is massive compared to a floor below. One should also note that the calculated period is an underestimate because the floors were moved closer to the ground so that the building model may conform to the formula.

3 Simple Linear Systems

THEORETICAL OUTLINE

The basic mass–spring oscillator, with or without a damper, is the principal subject of investigation in this chapter. Both the translational and the rotational versions are considered. Physical simplicity of such oscillators is often misleading because in practical applications they not only may represent the behavior of single structural elements, but they often may approximate the responses of entire structures, or at least some aspects of such responses.

EQUATIONS OF MOTION

Understanding of the response of a simple oscillator to various types of shock loading is fundamental to comprehending more general problems. First of all, any structure with a dominating, single dynamic degree of freedom (SDOF) is reducible to such an oscillator. Second, complex systems can often be approximated by such an oscillator, which certainly justifies devoting attention to the subject. To make the relationships more general the influence of damping needs to be included.

Any of the two arrangements in Figure 3.1 will be referred to as a simple, damped oscillator. In Figure 3.1a, mass M is restrained by a spring of stiffness k and a viscous damper, which offers a resisting force $-C\dot{u}$, proportional to the magnitude of velocity $\dot{u}$, but opposite to its direction. Displacement u is measured from a position in which the spring is unstretched. Figure 3.1b illustrates the basic oscillator for a rotational motion. A disk with mass moment of inertia J is restrained by a shaft with rotational stiffness K. The blades attached to the shaft are in contact with a viscous medium, which provides a resisting moment $-C\dot{\alpha}$.

Newton's second law applied to the translational oscillator gives

$$M\ddot{u} + C\dot{u} + ku = W(t) \tag{3.1}$$

Not only the displacement u, but also force $W(t)$ is, in general, function of time. When $W(t)=0$, we have the so-called free motion and Equation 3.1 may now be written as

$$\ddot{u} + 2\omega\zeta\dot{u} + \omega^2 u = 0 \tag{3.2}$$

where

$$\omega^2 = \frac{k}{M}; \quad \zeta = \frac{C}{C_c}; \quad C_c = 2\sqrt{kM} = 2M\omega \tag{3.3}$$

In most technical applications one has $C < C_c$ or $\zeta < 1$ and our interest is therefore limited to this range of *moderate damping*.* The *damping ratio* ζ is another practical and convenient

---

* For $C > C_c$ the motion changes its character to nonperiodic. C_c is called the critical damping.

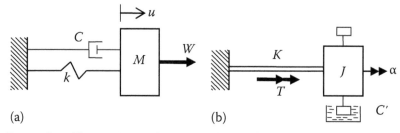

FIGURE 3.1   Damped oscillators: (a) translatory and (b) rotational.

measure of damping. When $\zeta = 0$, we obtain the expression of free, undamped motion, analogous to the previously quoted Equation 1.1, which can also be written as

$$\ddot{u} + \omega^2 u = 0 \tag{3.4}$$

FREE, UNDAMPED MOTION

To solve either Equation 3.2 or 3.4 for $u(t)$, the initial values of displacement $u(0) = u_0$ and/or velocity $\dot{u}(0) = v_0$ must be prescribed. The solution of Equation 3.4 is

$$u = u_0 \cos \omega t + (v_0 / \omega) \sin \omega t \tag{3.5}$$

or, after using some trigonometric identities

$$u = A \cos(\omega t - \alpha) \tag{3.6}$$

in which

$$A = \left[u_0^2 + \left(\frac{v_0}{\omega} \right)^2 \right]^{1/2} \tag{3.7a}$$

and

$$\tan \alpha = \frac{v_0}{\omega u_0} \tag{3.7b}$$

(Notice that there are two values of α differing by 180° and both giving the same $\tan \alpha$ in a 360° angular segment. Only one of those, however, will satisfy the initial conditions.) The motion represented by Equations 3.5 and 3.6 is periodic with the period $\tau = 2\pi/\omega$, which means that every τ seconds the mass begins to repeat its path. Successive differentiation of Equation 3.6 yields velocity $v = \dot{u}$ and acceleration $a = \ddot{u}$:

$$\dot{u} = -A\omega \sin(\omega t - \alpha) \tag{3.8a}$$

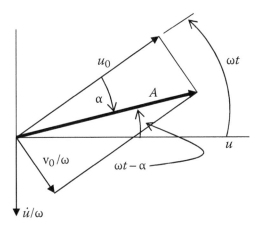

FIGURE 3.2 Representing harmonic motion as rotating vectors.

$$\ddot{u} = -A\omega^2 \cos(\omega t - \alpha) \tag{3.8b}$$

The relationships between the extreme values of velocity v and acceleration a are

$$\left|v_m\right| = \omega \left|u_m\right| \tag{3.9a}$$

and

$$\left|a_m\right| = \omega \left|v_m\right| \tag{3.9b}$$

The graphical representation of this motion is given in Figure 3.2. When displacement is a sine or cosine function of time, as in Equation 3.6, it is called simple harmonic motion. The displacement at any time point is a projection of vector A, rotating with a constant angular speed ω, on the horizontal axis.

The initial displacement and/or velocity are imposed in the following manner. Suppose we move the mass M in Figure 3.1a from the neutral, unstrained position by a distance u_0 and allow it to vibrate freely, setting $t = 0$ at the instant just before the release. This is what it means to impose an initial displacement. The initial velocity v_0 may be induced by impacting the mass and letting it vibrate afterward. The time $t = 0$ is chosen at the end of impact. If no initial displacement is to be introduced, the impact must last for a short interval of time only.

FORCED, UNDAMPED MOTION

When a load W is applied to an undamped translational system, Equation 3.1 may be written as

$$\ddot{u} + \omega^2 u = W/M \tag{3.10}$$

which is Equation 3.2 with the damping term missing. The solution has the form

$$u = B_1 \cos \omega t + B_2 \sin \omega t + u_p(t) \tag{3.11a}$$

which, in essence, differs from Equation 3.5 or 3.6 only by the last term. Velocity and acceleration are obtained by differentiation:

$$\dot{u} = -B_1 \omega \sin \omega t + B_2 \omega \cos \omega t + \dot{u}_p(t) \tag{3.11b}$$

$$\ddot{u} = -B_1 \omega^2 \cos \omega t - B_2 \omega^2 \sin \omega t + \ddot{u}_p(t) \tag{3.11c}$$

When the term *response* is used, it may mean any of the variables like displacement, velocity, acceleration, or a spring force resulting from some external action. When an analytical expression for displacement is available, the spring force is obtained after multiplying that displacement by a spring constant. The two remaining variables result from differentiation of displacement.

The coefficients B_1 and B_2 are the integration constants to be determined from the initial conditions. The first two terms in Equation 3.11a are of the same form as a general solution of Equation 3.4, while the additional function $u_p(t)$ is called the particular solution of Equation 3.10. There are several methods of finding $u_p(t)$. The most common one is to assume that $u_p(t)$ has the same form as W/M, substitute the assumed expression in place of u in Equation 3.10, and use the resulting identity to find the unknown coefficients.

Once the calculation is complete and u is known for any instant of time, one can divide the extreme deflection u_m by the value u_{st} obtained from static application of the same force W. The ratio of the absolute values of the two will be called the dynamic factor (for deflections):

$$DF(u) = \frac{|u_m|}{|u_{st}|} \tag{3.12}$$

If W changes with time, $W = W(t)$, the peak value W_m should be used for calculating u_{st}. For the mass–spring system, which we are considering at this moment, $DF(u)$ is the same for displacement as well as for the force in the spring, $DF(P)$.*

FREE, DAMPED MOTION

The form, which solution of Equation 3.1, with $W(t) = 0$, will take depends on the magnitude of the damping ratio ζ. Two ranges of values are of practical interest:

Small damping, or $\zeta \ll 1.0$. This is the most frequent case with ζ typically not exceeding a few percent.

Moderate damping, or $\zeta < 1.0$. This happens in special situations, including structures under extreme loading.

---

* Many authors use the term "dynamic *magnification* factor." It does not seem purposeful, since a reduction as well as an increase can also take place.

The solutions given here will be valid for both of the above ranges, unless otherwise mentioned. (Large damping, $\zeta > 1.0$, requires different formulas, but has very limited applications.) For Equation 3.1:

$$u = \left(u_0 \cos \omega_\mathrm{d} t + \frac{v_0 + \omega \zeta u_0}{\omega_\mathrm{d}} \sin \omega_\mathrm{d} t \right) e^{-\omega \zeta t} \tag{3.13}$$

where

$$\omega = \sqrt{\frac{k}{M}} \tag{3.14a}$$

$$\omega_\mathrm{d} = \omega \sqrt{1 - \zeta^2} \tag{3.14b}$$

The new constant ω_d is called *damped natural frequency*. The main difference between Equations 3.6 and 3.13 is the presence of the exponential term in the latter. Consequently, the amplitude of damped vibrations diminishes with time until it becomes insignificantly small. Equation 3.13 may also be written in a more compact form:

$$u = A e^{-\omega \zeta t} \cos(\omega_\mathrm{d} t - \alpha_\mathrm{d}) \tag{3.15}$$

in which

$$A^2 = u_0^2 + \frac{(v_0 + \omega \zeta u_0)^2}{\omega_\mathrm{d}^2} \tag{3.16a}$$

$$\tan \alpha_\mathrm{d} = \frac{v_0 + \omega \zeta u_0}{\omega_\mathrm{d} u_0} \tag{3.16b}$$

(As mentioned in the section "Free, undamped motion," some caution is needed in determining which of the two possible angles α_d will fit the initial conditions.)

The cosine function is ± 1 at the extreme points. This means that the deflection described by Equation 3.15 has the absolute value not larger than $A e^{-\omega \zeta t}$ for any given t. The motion is not exactly periodic, because it does not repeat itself after each damped period $\tau_\mathrm{d} = 2\pi/\omega_\mathrm{d}$. Yet, it may be shown that not only zero points, but also the extreme points of $u(t)$ are attained after every τ_d. For this reason the term *pseudoperiodic* may be used to describe this type of oscillation. One may also notice that u from Equation 3.15 can also be represented by projection of a rotating vector, as in Figure 3.2, but the length of this vector is now decreasing with time.

Suppose that for a certain time point t_1 a local maximum displacement u_1 is obtained. After another cycle of motion, at time $t_1 + \tau_\mathrm{d}$ another maximum u_2 is reached. The ratio of these maxima is (by Equation 3.15)

$$\frac{u_1}{u_2} = \frac{\exp(-\omega \zeta t_1)}{\exp(-\omega \zeta (t_1 + \tau_\mathrm{d}))} = \exp(\omega \zeta \tau_\mathrm{d}) \tag{3.17a}$$

This means that the ratio of any two adjacent maximum deflections is always the same. One can easily show that after n cycles we have

$$\frac{u_1}{u_{1+n}} = \exp(n\omega\zeta\tau_d) \equiv e^{n\omega\zeta\tau_d} \qquad (3.17b)$$

The term "ratio of amplitudes" may be used instead of "ratio of maximum displacements" for the sake of brevity, since the amplitude is not a parameter of motion as it is in case of simple harmonic vibrations.

The logarithm of the displacement ratio in Equation 3.17a is called *logarithmic decrement* Δ:

$$\Delta = \log\left(\frac{u_1}{u_2}\right) = \omega\zeta\tau_d \approx 2\pi\zeta \qquad (3.18)$$

(In some texts Δ is used as the basic damping constant.) The near-equality sign used in the above expression is valid for a small damping ratio. In this case τ_d is only very slightly larger than τ. The case with $\zeta < 1.0$ just discussed is the only situation where vibratory motion is possible. For $\zeta \geq 1.0$, the motion loses its oscillatory character.

FORCED, DAMPED MOTION

This type of motion is described by Equation 3.1 with all the terms present, or by an alternative form:

$$\ddot{u} + 2\omega\zeta\dot{u} + \omega^2 u = W/M \qquad (3.19)$$

When damping is less than critical (i.e., $\zeta < 1.0$), the solution of this equation is

$$u = (B_1 \cos\omega_d t + B_2 \sin\omega_d t)e^{-\omega\zeta t} + u_p(t) \qquad (3.20)$$

The first term is the general solution of Equation 3.19. The second term, $u_p(t)$, is the particular solution. The method of finding $u_p(t)$ is the same as that described for free, damped motion.

Because of the presence of an exponential expression, which is responsible for the gradual diminishing of the amplitude, the first term is also called the *transient component*. After some time it becomes insignificantly small, and then we have $u \approx u_p$. This is why u_p is referred to as the *steady-state component*.

Since u_p does not contain the integration constants, it is independent of the initial conditions and therefore it is a function of forcing and system properties only. When the phrase *transient response* is used in connection with dynamic analysis, it usually refers to the response just after the application of load, when the transient component is to be included along with the steady-state component. From a physical point of view, the transient vibratory component is a disturbance that takes place because of the transition from rest to motion and decays as time goes on, leaving only the steady-state component.

(Having read all this one must remember that the terms "transient" and "steady state" are meaningful only for prolonged vibrations, while in shock response analysis the distinction has little practical value.)

Equation 3.20 was discussed only for the sake of completeness. It is rarely used in its full form for practical reasons. Once we have determined the formula for $u_p(t)$, we will proceed to find the integration constants B_1 and B_2. The resulting response expression is quite lengthy. If we want to know what is happening during the early stage of motion, an alternate procedure is recommended:

1. Calculate the steady-state component $u_p(t)$ using Equation 3.19.
2. Instead of using Equation 3.20 to find B_1 and B_2 for the given initial conditions, use Equation 3.11a.

The second step is equivalent to ignoring damping in the transient component and is justified only when ζ is small, otherwise the calculated response is excessive. Of all system properties, damping is usually known with the least accuracy, and by ignoring and obtaining somewhat larger response the analyst errs on the side of caution.

RESPONSE OF OSCILLATOR TO STEP LOADING

The basic undamped oscillator, initially at rest, is subjected to a *step loading* shown in Figure 3.3a, which can be expressed as $W(t) = W_0 H(t)$. The function $H(t)$ is known in mathematics as the *Heaviside step function*. It has a value of unity for $t > 0$ and nil for $t \leq 0$. This means that during the time of interest $W(t)$ is a constant, which makes the response calculations quite simple. In determination of the displacement $u(t)$ Equation 3.10 is used and the particular solution is assumed as constant B:

$$u_p(t) = B \tag{3.21}$$

Then, after substituting this into Equation 3.10 we obtain

$$B = \frac{W_0}{\omega^2 M} = \frac{W_0}{k} = u_{st} \tag{3.22}$$

which is the static displacement. When this is substituted in Equation 3.11b along with the initial condition $\dot{u}(0) = 0$, one gets $B_2 = 0$. Then, from Equation 3.11a: $B_1 = -u_{st}$. The end result is

$$u_1(t) = u_{st}(1 - \cos \omega t) \tag{3.23}$$

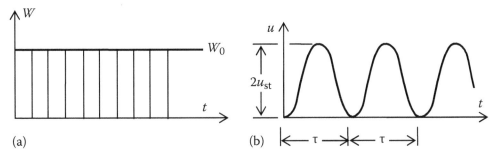

(a)   (b)

FIGURE 3.3   Step load (a) and oscillator response (b).

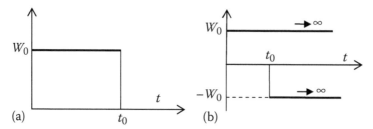

FIGURE 3.4   Truncated step load (a) as a superposition of two steps in (b).

The mass oscillates between $u = 0$ and $u = 2u_{st}$, which means the amplitude is u_{st} and the vibration is taking place about the position of static equilibrium, Figure 3.3b.

The above load was implied to be lasting for indefinitely long time. If it is removed after time t_0, as shown in Figure 3.4a, we deal with a *truncated step load* or a *rectangular pulse*.

Now the extreme displacement will depend on the magnitude of t_0. The pattern may be treated as a superposition of two step loads as shown in Figure 3.4b.

The first step originating at $t=0$ gives a deflection u_1 defined by Equation 3.20. The second step beginning at $t = t_0$ will result in

$$u_2 = -u_{st}[1 - \cos\omega(t - t_0)] \tag{3.24}$$

For $t < t_0$, $u_1(t)$ prescribed by Equation 3.23 is the deflection response. For $t > t_0$ both terms are superposed and the total response is

$$u = u_1 + u_2 = 2u_{st} \sin\frac{\omega t_0}{2} \sin\omega\left(t - \frac{t_0}{2}\right) \tag{3.25}$$

$$\dot{u} = 2\omega u_{st} \sin\frac{\omega t_0}{2} \cos\omega\left(t - \frac{t_0}{2}\right) \tag{3.26}$$

and the extreme values are

$$\left|u\right|_m = 2u_{st} \tag{3.27a}$$

and

$$\left|\dot{u}\right|_m = \omega\left|u\right|_m \tag{3.27b}$$

Only the second trigonometric term in displacement and velocity expressions is time-dependent and oscillates in the range between +1 and −1. The first term is a function of t_0 and the natural period τ. Its absolute value reaches unity for $t_0 = \tau/2, 3\tau/2, 5\tau/2,...$. This means that if the load W_0 is removed after any of those values of t_0, the amplitude of displacement during the free vibrations that follow will be $2u_{st}$. An inspection of Figure 3.3b reveals why this happens. The amplitude has doubled in comparison with what is obtained from the step loading in Figure 3.3a, because there are two shock loads applied here: sudden loading one way followed by sudden unloading when deflection reaches maximum.

If t_0 is different from an odd multiple of $\tau/2$, the amplitude of deflection will be less than $2u_{st}$. In fact, when t_0 is equal to a multiple of τ, no vibration takes place for $t > t_0$ so the mass will be at rest thereafter.

The damped response of an oscillator can be found in Case 3.9.

RESPONSE OF OSCILLATOR TO IMPULSIVE LOAD

When a plot of the applied force vs. time is drawn, the area under the $W-t$ curve in Figure 3.5 represents the impulse S acting on the system. If the load that acts for a very short time, say $t_0 < 0.1\tau$ or less, will be referred to as *impulsive*. The peak force may be very large, the duration quite small, but the impulse will have some prescribed value. Mathematically, a function acting in the interval from t to $t+dt$, with the integral of that function over dt remaining constant (regardless of how small dt becomes), is known as *Dirac delta function* $\delta(t)$. (In fact, an idealized impulse of a specified value S, with infinitely short duration, implies an infinitely large load.) When this impulse is applied to mass M, the mass acquires its initial velocity v_0 instantly. The latter is found by the application of the impulse–momentum principle

$$S = W_0 t_0 = M v_0 \quad \text{or} \quad v_0 = S/M \tag{3.28}$$

assuming of course that the mass is initially at rest. (One may note that $\delta(t)$ is a derivative of $H(t)$ defined before, but this appears to be of little practical significance.)

With the initial velocity given by Equation 3.28 one can apply Equation 3.5 to get the maximum displacement:

$$u_m = \frac{v_0}{\omega} \tag{3.29a}$$

or

$$u_m = \frac{S\omega}{k} \tag{3.29b}$$

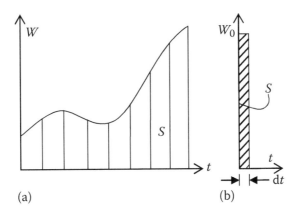

(a)   (b)

FIGURE 3.5   Applied force vs. time (a) and a short impulse (b).

When the rectangular pulse is of somewhat longer duration, $0.1\tau < t_0 < 0.25\tau$, a more accurate formula for amplitude, derived from Equation 3.25, must be used:

$$u_m = 2u_{st} \sin \frac{\omega t_0}{2} \tag{3.30}$$

When $t_0/\tau = 0.1$ or $\omega t_0/2 = 0.1\pi$, the overestimation of the response caused by Equation 3.29 is only 1.7%. However, for $t_0/\tau = 0.25$ the error grows to 11%. Since Equation 3.30 is not much more difficult to use than Equation 3.29, there is little point in using the latter, in spite of it being quite popular in technical literature.

In the load case tabulations to follow, the term *short impulse* or *short pulse* is intended to represent a real impulse of magnitude $W_0 t_0$ and of duration t_0, which is so short that it can be treated as instantaneous. This is equivalent to imposing the initial velocity v_0 on mass M. If one begins with Equation 3.30 and assumes that a sine of a small angle can be replaced by an angle itself, Equation 3.29, or, in effect, a larger displacement is obtained. In this sense Equation 3.29 gives an upper bound of a result produced by Equation 3.30. This leads to the following rule in response calculations:

If a rectangular load pulse $W_0 t_0$ is replaced by an instantaneous one, equivalent to the initial velocity $v_0 = W_0 t_0/M$, the displacement response obtained in this way is the upper bound of the true result.

OSCILLATOR RESPONSE TO TRIANGULAR PULSES

There are three basic types of triangular pulses, as shown in Figure 3.6: symmetrical, increasing (with a sudden drop) and decreasing (after a sudden rise). The second one, of increasing magnitude, is typical of many impact situations. The third, a decreasing function, is often used as an approximation of the positive phase of a blast load. The symmetrical pulse displacement response, patterned after Harris and Crede [30], has three phases of motion separately described:

$$\left(\text{Notation:} \quad x = \frac{\omega t}{4} = \frac{\pi t}{2\tau}; \quad x_0 = \frac{\omega t_0}{4} = \frac{\pi t_0}{2\tau} \right)$$

$$\frac{u}{2u_{st}} = \frac{x}{x_0} - \frac{\sin(4x)}{4x_0} \quad \text{for } x \le \frac{x_0}{2} \text{ or } t \le \frac{t_0}{2} \tag{3.31a}$$

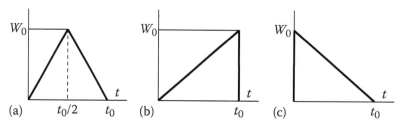

FIGURE 3.6   Basic triangular pulses: (a) symmetric, (b) increasing, and (c) decreasing.

$$\frac{u}{2u_{st}} = 1 - \frac{x}{x_0} - \frac{\sin(4x)}{4x_0} + \frac{\sin(4x - 2x_0)}{2x_0} \quad \text{for } \frac{x_0}{2} \le x \le x_0 \text{ or } \frac{t_0}{2} \le t \le t_0 \quad (3.31b)$$

$$\frac{u}{2u_{st}} = \frac{\sin^2 x_0}{x_0} \sin(4x - 2x_0) \quad \text{for } x \ge x_0 \text{ or } t \ge t_0 \quad (3.31c)$$

Static displacement, u_{st}, is calculated as if it was induced by the peak loading W_0. The response described by the above relationships is not only a function of time, but also depends on the ratio t_0/τ. It is easy to determine that for $t < t_0/2$ there is no extremum. For $t_0/2 < t < t_0$ one can obtain from Equation 3.31b the peak displacement of $u_m \approx 1.52 u_{st}$ when $t_0/\tau \approx 0.92$. Since two variables are involved, it makes the usual extremum search somewhat complex and necessitates the use of numerical methods. Equation 3.31c gives *residual response*, i.e., history of motion following the entire pulse application. It is a simple sinusoidal motion. The peak response here is $u/u_{st} = 1.45$, somewhat less than before, but that can be attained only if $t_0/\tau \approx 0.74$. A good graphical display of results can be viewed in TM5 [106].

The response to an increasing and a decreasing pulse is given in Case 3.3. One should remember that replacing any such shape with a symmetrical triangular pulse results in obtaining a higher u_m/u_{st} ratio. The exception is the decreasing pulse, which gives a higher response when $t_0 > 0.9\tau$. With regard to the latter shape, one should also keep in mind that when t_0 is extended far beyond the natural period, this loading begins to approximate a load step, rather than a distinct pulse. As the summary Table 3.1 shows, the peak responses to all three triangular pulses are similar in magnitude.

When pulses are short, i.e., when $t_0/\tau = 0.1$, the exact shape of the applied pulse is not very important, therefore an equivalent rectangular-shaped pulse (one that has the same impulse S) will give a good approximation of response. The distinctions between the intervals of time, as defined above, become meaningless and Equation 3.29 may be used. Here we have $S = W_0 t_0/2$, which leads to

$$\frac{u_m}{u_{st}} = \frac{\omega t_0}{2} \quad (3.32)$$

keeping in mind that $u_{st} = W_0/k$. For triangular pulses, the above expression, or its more accurate version, namely Equation 3.30, can be used up to $t_0/\tau = 0.4$.

TABLE 3.1
Single Pulse Parameters

	u_m/u_{st}	$S/(W_0 t_0)$
Rectangular	2.00	1.0
Triangular symmetric	1.52	0.5
Triangular increasing	1.59	0.5
Triangular decreasing	1.55	0.5
Half-sine	1.77	$2/\pi$
Versed sine	1.72	0.5

Note: For a triangular decreasing pulse the value of u_m is for $t_0 = \tau$.

SUMMARY OF RESPONSE TO SINGLE PULSES

The peak responses divided by static displacements (or DFs) are summarized in Table 3.1. The rectangular pulse must be at least as long as the half-period, i.e., $t_0 > \tau/2$ to achieve its peak response. For the triangular decreasing pulse, the longer the duration, the larger the peak value can be attained. If continued long enough, it tends to DF = 2.0, the value related to the step load. Since our interest is limited to relatively short pulses, the DF corresponding to $t_0/\tau = 1.0$ was selected for this pulse. The ratio of applied impulse to that associated with the rectangular pulse of the same duration is also given in Table 3.1.

Damped oscillator responses lead to more complex formulas. Step loading $W_0H(t)$ gives a response curve similar to what is seen in Figure 3.3, except that the amplitude is decreasing with every cycle. Like all cases of damped response, the peak spring and damper forces do not coincide in time. The details can be found in Case 3.9.

When a short pulse, resulting in the initial velocity v_0 is applied, the motion can be classified as "free-damped." This is a special case of Equation 3.13 with $u_0 = 0$. After differentiation with respect to time, velocity is obtained. The peak reaction may be larger than either the damper force (at the beginning) or the spring force (at maximum deflection). Refer to Case 3.9.

MULTIPLE SHOCK LOADS

Consider a torsional oscillator, as shown in Figure 3.1b, which is subjected to a series of pulses of torque with magnitude T_0. Each such load, when acting statically, produces α_{st} angle of rotation. As shown in Figure 3.7a, each pulse lasts for t_0 and then there is a pause of t_1. The objective here is to find a peak displacement of the system, when subjected to n such pulses.

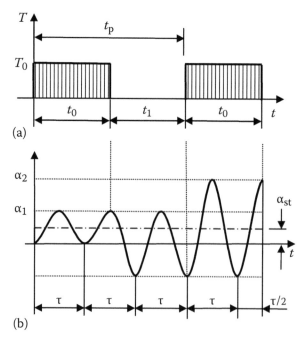

FIGURE 3.7 A periodic rectangular pulse (a) and the oscillator response (b).

On the basis of the section "Response of oscillator to step loading" it is known that the end of the first pulse must coincide with the odd number of natural half-periods $\tau/2$. As Figure 3.7b illustrates, we chose $t_0 = 3(\tau/2)$. After the load is removed, the amplitude of free vibrations becomes $2\alpha_1$ instead of α_1, as was the case under the load still being applied. After another $3(\tau/2)$ the torque applied again in the same direction as the shaft itself is driving the disk. To calculate the new shaft deflection, namely α_2 located on the opposite side of the neutral position ($\alpha = 0$), note that at the stagnation point (extreme deflection) the only energy component present is the strain energy of the spring. Equating the difference in strain energies with the work done by the torque, namely $T_0(\alpha_2 - \alpha_1)$, one finds

$$\alpha_2 = \alpha_1 + 2\alpha_{st} \tag{3.33a}$$

where

$$\alpha_{st} = T_0/K \tag{3.33b}$$

Continuing application of pulses in this manner one finds that each next one increases the deflection by $2\alpha_{st}$, therefore, after n load cycles the peak deflection becomes

$$\alpha_m = 2n\alpha_{st} \tag{3.34}$$

To illustrate the problem, we chose $t_0 = 3(\tau/2)$, but it could be any multiple of $\tau/2$, beginning with $\tau/2$ itself. But notice that the pause t_1 was also an odd multiple of $\tau/2$, which together made the period of the applied pulse t_p equal to an even multiple of $\tau/2$, or, as we may prefer to say, a multiple of the natural period τ. This result could be expected, because the objective was to maximize the response; consequently forcing must be tuned to natural period of vibration. This buildup of vibration leads to a condition known as *resonance*. It may be modified by onset of plasticity or terminated by a failure.

Consider a steady state, when a structure receives identical, periodic pulses, and oscillates with constant amplitude. This means that the work performed during one load period (which may be a multiple of the natural period) is offset by the energy loss due to damping. In basic vibration theory, the resonant condition is usually derived by considering sinusoidally varying force acting on the basic oscillator. Kaliski [44] develops an approximate equation, showing the amplitude growth during resonance when the applied force is $W = W_0 \sin \omega t$:

$$u \approx \frac{u_{st}}{2\zeta}(1 - \exp(-\zeta\omega t))\cos \omega t \tag{3.35a}$$

or

$$DF(u) \approx \frac{1}{2\zeta}(1 - \exp(-\zeta\omega t)) \tag{3.35b}$$

where
   $DF(u)$ is the dynamic factor for displacement
   u_{st} is the static displacement under the applied force ($u_{st} = W_0/k$) and the expression is
      written for a forcing frequency equal to the natural frequency ω

After some time the DF reaches $1/(2\zeta)$, a well-known relationship at the peak resonant response.

An approximate expression for the response after a number of arbitrarily shaped pulses can be found using as a reference the impulse applied by a sinusoidal force. In the latter case an impulse applied to the mass during one-half cycle is $(2/\pi)W_0(\tau/2)$. After one-half cycle the motion is reversed and so is the direction of the force. The total impulse in one cycle is

$$S_0 = \frac{2}{\pi}W_0\tau \qquad (3.36)$$

If there is a periodic pulse of a different magnitude S_1 applied, the response can be approximately found by multiplying the harmonic result, Equation 3.35, by the ratio S_1/S_0.

One more application of Equation 3.35a and b should be mentioned, when the exponent $\zeta\omega t$ is small, which happens for small damping or small number of cycles considered, or both. Noting that $e^{-x} \approx 1 - x$ when $x \ll 1$, one can write

$$u \approx \frac{\omega t}{2}u_{st}\cos\omega t = u_m\cos\omega t \qquad (3.37)$$

The accuracy of the above is checked in Example 3.7. The repeated pulse formulas are summarized in Cases 3.34 through 3.37.

Shock Isolation

Consider a damped oscillator, with some time-dependent force applied to the mass. The determination of the load applied to the ground by the spring–damper assembly is sometimes called an *isolation problem*, because that assembly is thought to isolate the ground from the dynamic action applied to the mass. This is called *active isolation*. A reverse problem would be to calculate the effect of the ground moving on the motion of the mass, also known as *passive isolation*. Only the latter type is discussed in this section. The active isolation response pertains to ground reactions and can be deduced from what was presented before. Only some simple types of isolators are considered in this section: a spring with or without a viscous damper.

Figure 3.8 shows the two versions of a damped oscillator. They do not differ from the oscillators previously presented, except that the base can perform a prescribed motion $u_b(t)$.

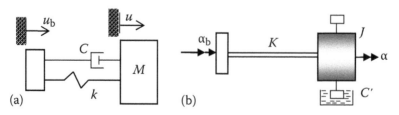

FIGURE 3.8 Damped isolators with a movable base.

The parameters defining what has now become an *isolator* (i.e., viscous constant C and the spring constant k) are so chosen in practical applications that the response of mass M is kept within the prescribed limits. Another way of putting it, in terms of acceleration, is to request that the overload factor n, defined as $n = a/g$, where g is the acceleration of gravity and a is the maximum acceleration to which the assembly is subjected, does not exceed the permissible value. (In other words, that the inertial loads stay within limits.)

Another concern is the strength of the isolator itself. In the case of a spring element, the applied load depends on the relative displacement of the base and the mass. For a viscous damper the relative velocity determines the applied load. In either case there is a limited magnitude of the load that can be safely applied.

To best visualize the effect of base or foundation motion in mass–spring system disregard the damper. The difference between the deflections of both ends of the spring results in the force $k(u_b - u)$ applied to the mass. The equation of motion is

$$M\ddot{u} = k(u_b - u) \quad \text{or} \quad M\ddot{u} + ku = ku_b \tag{3.38}$$

This is similar to Equation 3.10 except that the applied force is expressed in terms of foundation displacement. Thus we may say that prescribing a displacement $u_b(t)$ on the foundation gives the same result that would be obtained if the force ku_b were applied to the mass, while the foundation remained stationary.

Another manner of looking at the problem is this: if we are interested only in the *relative displacement* $u - u_b$ of the mass with respect to the base, we can introduce a new coordinate, $\bar{u} = u - u_b$. Then

$$u = \bar{u} + u_b \tag{3.39a}$$

and

$$\ddot{u} = \ddot{\bar{u}} + \ddot{u}_b \tag{3.39b}$$

while the equation of motion becomes

$$M\ddot{\bar{u}} + k\bar{u} = -M\ddot{u}_b \tag{3.40}$$

If the spring were rigid, the function on the right side of the equals sign would be the inertia force applied to the mass M. This is a known function of time and this equation can be solved in the same way as was done for an undamped oscillator. When $\bar{u}$ is calculated, the spring force $k\bar{u}$ also becomes known. The merit of using Equation 3.40 is that we can think of a mass connected to a fixed rather than moving foundation.

We can now readily use the solutions for a system with a grounded spring, with an appropriate force applied to the mass, rather than dealing explicitly with the moving support. The peak displacement $\bar{u}(t)$ can then be calculated from the equations developed for a configuration with a fixed base.

Similar reasoning applies when a damper is involved. The difference between the velocities of both ends of the damper results in the force $C(\dot{u}_b - \dot{u})$ applied to M. When both spring and damper are considered, one gets, in place of Equation 3.38:

$$M\ddot{u} + C\dot{u} + ku = ku_b + C\dot{u}_b \tag{3.41}$$

showing a combined effect of a spring and a damper. The previously introduced variable, $\bar{u} = u - u_b$ can also be employed. Substituting it in Equation 3.41 one gets the equation in terms of $\bar{u}$:

$$M\ddot{\bar{u}} + C\dot{\bar{u}} + k\bar{u} = -M\ddot{u}_b \tag{3.42}$$

This can be treated as an equation of motion of a mass attached to a fixed foundation and acted on by an external force $W = -M\ddot{u}_b$. If the isolator were rigid, W would be the inertia force applied to the mass as a result of the whole system moving with an acceleration $\ddot{u}_b$.

The essential formulas for the response of a system driven through its base are given in Cases 3.14 and 3.15.

Types of Damping

Various common types of damping will now be briefly discussed.

Viscous damping was introduced earlier in this chapter. A damper with a constant C generates a resistance force of magnitude $C\dot{u}$, where $\dot{u}$ is the velocity of one end of the damper with respect to the other. Suppose that a movement of the free end, as in Figure 3.9, is prescribed by a displacement, or the equivalent velocity:

$$u = u_0 \sin \Omega t \tag{3.43a}$$

or

$$v = v_0 \cos \Omega t \tag{3.43b}$$

where $u_0 = v_0/\Omega$. The driving force, needed to overcome the resistance is $P = Cv$ and the increment of work is $P\,du = Pv\,dt$. The work performed by the driving force and dissipated (lost) by the damper is

$$D = \int_0^t Cv^2\,dt = Cv_0^2\int_0^t (\cos \Omega t)^2\,dt \tag{3.44}$$

For a full cycle, $t = 2\pi/\Omega$:

$$D = \pi u_0^2 C\Omega = \pi P_m u_0 = \frac{\pi v_0^2 C}{\Omega} \tag{3.45}$$

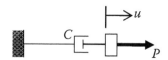

FIGURE 3.9 A viscous damper driven by a prescribed displacement or force.

where
$P_m = Cv_0$ is the peak force
v_0 is maximum velocity

In shock analyses a quarter-cycle is often of interest, which corresponds to $t = \pi/(2\Omega)$. For this time interval, one has the dissipated energy equal to 1/4 of the above.

Frictional damping, also called Coulomb damping, is the result of dry friction between two parts whose surfaces slide against each other. The energy lost in one cycle is

$$D = 4AF \tag{3.46}$$

where F is a constant friction force.

Structural damping. If a sample of material is subjected to a tensile or a compressive stress, which grows very slowly from zero to its maximum value, the characteristic is a straight line. After a few load cycles applied with some finite velocity, the characteristic forms a narrow loop. The area within the loop is equal to the energy dissipated by the sample in one cycle. The damping associated with this material behavior is called structural or material damping. It is thought to be caused by the internal friction within the volume of material. The energy dissipated in one cycle is assumed to be proportional to the square of displacement amplitude A:

$$D = sA^2 \tag{3.47}$$

where s is proportionality constant determined from an experiment.

Velocity-squared damping. This is often encountered in hydro-aerodynamical applications. It can be shown that when the resisting force is proportional to the square of velocity, the energy dissipated by a velocity-squared damper in one cycle of motion is

$$D_a = \frac{8}{3} C_a A^3 \omega^2 \tag{3.48}$$

where C_a is the proportionality constant.

All of these damping types are reducible to the equivalent viscous damping, as long as the resulting coefficient ζ is small, as demonstrated in Ref. [87]. This is convenient from the analytical viewpoint, as it allows to keep the equations of motion linear. If the condition of small ζ cannot be fulfilled, more elaborate methods of solution must be employed.

Having done this brief review of types of damping, one should keep in mind that the exact magnitude of damping is critical mainly for forced vibration, where, under harmonic forcing, the dynamic factor at resonance is approximately $1/(2\zeta)$. For most structures the range of values of ζ appears to be

$$0.5\% < \zeta < 5\%$$

This implies that including damping in a single-cycle event, such as shock or impact brings only a small reduction in calculated force.

One difficulty with damping is that it is usually known with lesser accuracy than other structural parameters. It was often found to vary strongly according to the prevailing stress. For example, Kroon [49] reports that certain turbine blades showed 80 times increase in the damping constant when their operating stress was changed from 20 to 20,000 psi. Another difficulty is that under extreme loads, where large distortions may take place, it may be quite difficult to separate damping proper from dissipation of energy via plastic deformation or by other means. In a real, MDOF structure analysis, carried out with the help of a finite-element program it is usual to employ *Rayleigh damping*. For each mode of vibration there is a modal damping coefficient ζ, such that

$$\zeta = \frac{\alpha}{2\omega} + \frac{\beta\omega}{2} \tag{3.49a}$$

$$\zeta = \frac{\alpha}{4\pi f} + \pi\beta f \tag{3.49b}$$

where $f = 2\pi\omega$ is the natural frequency. In the above formulation α is a mass-proportional coefficient and β is stiffness-proportional. While it is possible to use a combination of α and β in the same run, most users favor employing only β. This leaves only one term in Equation 3.49a and b to be considered. Once β is selected for a dominant frequency ω_n, then for every $\omega < \omega_n$ the modal damping ζ is less.

TABULATION OF CASES

CASE 3.1   ACCELERATING MASS M, INITIALLY AT REST

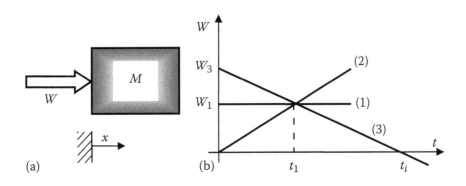

(a)   (b)

(1) Constant force $W = W_1$ applied.

$a = a_0 = W_1/M$; acceleration

$v = a_0 t$; velocity

$u = \dfrac{1}{2} a_0 t^2$; displacement

(2) Increasing force $W = W_1 \dfrac{t}{t_1}$ applied.

$a = \dfrac{W_1}{M} \dfrac{t}{t_1} = a_0 \dfrac{t}{t_1}$; acceleration

$v = \dfrac{a_0}{t_1} \dfrac{t^2}{2}$; velocity

$u = \dfrac{a_0 t^3}{6 t_1}$; displacement

(3) Decreasing force $W = W_3 \left(1 - \dfrac{t}{t_i}\right)$; $W = 0$ for $t = t_i$

$a = \dfrac{W_3}{M} \left(1 - \dfrac{t}{t_i}\right)$; acceleration, $a_\mathrm{m} = W_3/M$

$v = \dfrac{W_3}{M} t - \dfrac{W_3}{2M t_i} t^2$; velocity, $v_\mathrm{m} = \dfrac{W_3 t_i}{2M}$ at $t = t_i$; $v = 0$ for $t_\mathrm{m} = 2 t_i$

$u = \dfrac{W_3}{M} \dfrac{t^2}{2} - \dfrac{W_3}{M t_i} \dfrac{t^3}{6}$; displacement, $u_\mathrm{m} = \dfrac{W_3 t_\mathrm{m}^2}{6M}$ at t_m

Note: After u_m is reached in (3), a reverse cycle of motion begins.

CASE 3.2   BASIC OSCILLATOR, TRANSLATIONAL OR ROTATIONAL

(a) Increasing force W, as in Case 3.1 (1): $W = W_1 \dfrac{t}{t_1}$

$$u(t) = u_{st}\left(\frac{t}{t_1} - \frac{\sin \omega t}{\omega t_1}\right); \quad u_{st} = W_1/k$$

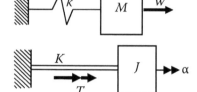

(b) Short impulse $S = W_0 t_0$, equivalent to initial velocity
$v_0 = S/M$.

$$u_m = v_0/\omega; \text{ at } t = \tau/4; \quad v_m = v_0; \quad R_m = v_0\sqrt{kM}$$

(c) Rectangular impulse $S = W_0 t_0$.

$$u_m = 2u_{st}\sin\frac{\omega t_0}{2} = 2u_{st}\sin\frac{\pi t_0}{\tau}; \quad u_{st} = W_0/k$$

$$u_m = 2u_{st} \quad \text{for } t_0 \geq \tau/2; \quad v_m = \omega u_m; \quad a_m = 2W_0/M$$

(d) Step load $W = W_0 H(t)$

$$u_{st} = W_0/k; \quad u_m = 2u_{st} \text{ at } t = \tau/2; \quad v_m = \omega u_{st}; \quad a_m = 2W_0/M$$

CASE 3.3   OSCILLATOR RESPONSE TO TRIANGULAR PULSE

Notation: $x = \dfrac{\omega t}{4} = \dfrac{\pi t}{2\tau}; \quad x_0 = \dfrac{\omega t_0}{4} = \dfrac{\pi t_0}{2\tau}$

(a) Symmetric triangular. Refer to text.
(b) Triangular increasing.

Maximum response takes place for $t > t_0$.
The response is then the same as for (c) below.

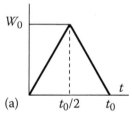

(a)

(c) Triangular decreasing.

$$\frac{u}{u_{st}} = 1 - \frac{x}{x_0} + \frac{\sin(4x)}{4x_0} - \cos(4x) \quad \text{for } x \leq x_0 \text{ or } t \leq t_0$$

The peak u_m/u_{st} grows asymptotically with t_0 up to $u_m/u_{st} = 2.0$.
(For $t_0 = \tau$, $u_m/u_{st} = 1.55$)

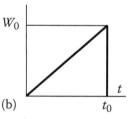

(b)

$$\frac{u_m}{u_{st}} = \left\{1 - \frac{\sin(4x_0)}{2x_0} + \left(\frac{\sin(2x_0)}{2x_0}\right)^2\right\}^{1/2} \quad \text{for } x \geq x_0 \text{ or } t \geq t_0$$

Peak response for $t > t_0$ is when $t_0/\tau = 0.65$, then $u_m/u_{st} = 1.26$.

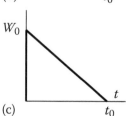

(c)

CASE 3.4   OSCILLATOR RESPONSE TO OTHER COMPACT PULSES

Notation: $x = \dfrac{\omega t}{4} = \dfrac{\pi t}{2\tau}; \ x_0 = \dfrac{\omega t_0}{4} = \dfrac{\pi t_0}{2\tau}$

(a) Half-sine

$$W(t) = W_0 \sin \frac{\pi t}{t_0} \quad \text{for } t < t_0$$

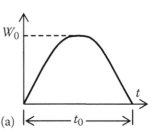

(a)

$$\frac{u}{u_{st}} = \frac{1}{1 - \left(\dfrac{\pi}{4x_0}\right)^2}\left[\sin\frac{\pi x}{x_0} - \pi - \frac{\sin(4x)}{4x_0}\right] \quad \text{for } x \le x_0 \text{ or } t \le t_0$$

$$\frac{u}{u_{st}} = \frac{\pi}{2x_0}\frac{\cos(2x_0)}{\left(\dfrac{\pi}{4x_0}\right)^2 - 1}[\sin(4x - 2x_0)] \quad \text{for } x > x_0 \text{ or } t > t_0$$

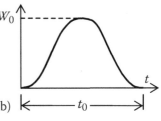

(b)

(b) Versed sine

$$W(t) = \frac{W_0}{2}\left(1 - \cos\frac{2\pi t}{t_0}\right) \quad \text{for } t < t_0$$

$$\frac{u}{u_{st}} = \frac{1}{2}\frac{1}{1 - (2x_0/\pi)^2}\left\{1 - \left(\frac{2x_0}{\pi}\right)^2\left[\cos\frac{2\pi x}{x_0} - 1\right] - \cos(4x)\right\} \quad \text{for } x \le x_0 \text{ or } t \le t_0$$

$$\frac{u}{u_{st}} = \frac{\sin(2x_0)}{1 - \left(\dfrac{2x_0}{\pi}\right)^2}[\sin(4x - 2x_0)] \quad \text{for } x > x_0 \text{ or } t > t_0 \text{ Ref.}[31]$$

CASE 3.5   OSCILLATOR RESPONSE TO AN INFINITE DURATION, EXPONENTIALLY DECAYING PULSE

$W = W_0 \exp\left(-\dfrac{t}{t_x}\right); \ u_{st} = W_0/k$

$S(t) = W_0 t_x\left\{1 - \exp\left(-\dfrac{t}{t_x}\right)\right\}; \ \text{current impulse}$

$S_m = W_0 t_x; \ \text{total impulse}$

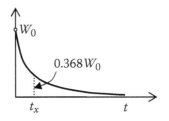

$$u(t) = \frac{u_{st}}{1 + (1/(\omega t_x)^2)}\left\{\frac{\sin\omega t}{\omega t_x} - \cos\omega t + \exp\left(-\frac{\omega t}{\omega t_x}\right)\right\}$$

1. Solution limit for $t_x \to \infty$: step loading with W_0; $u_m = 2u_{st}$ (Max. response)
2. Solution limit for $t_x \to 0$: motion with initial velocity $v_0 = \omega^2 u_{st} t_x$; $u_m = (\omega t_x) u_{st}$ (Max. response).

Use limit solutions for $\omega t_x > 40$ or $\omega t_x < 0.4$. For intermediate values use $\dfrac{u_m}{u_{st}} \approx \dfrac{2.222(\omega t_x)}{\omega t_x + 2}$; as approximation for u_m (accuracy per tabulation below):

Note: The first tabulated u_m/u_{st} is the true maximum value from the equation for $u(t)$.

ωt_x	u_m/u_{st}	$\approx u_m/u_{st}$
1	0.756	0.741
2	1.111	1.111
4	1.433	1.481

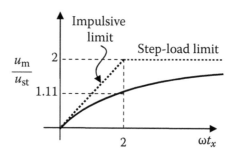

Note: For a further application of the concept of limiting load types see Chapter 13.

CASE 3.6   OSCILLATOR RESPONSE TO SLOPE-AND-STEP PULSE

$$W(t) = W_0 \frac{t}{t_0} \quad \text{for } t < t_0 \quad \text{and} \quad W(t) = W_0 \quad \text{for } t > t_0$$

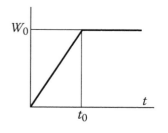

$$\frac{u}{u_{st}} = 1 - \frac{\tau}{\pi t_0} \sin \frac{\pi t_0}{\tau} \cos \frac{\pi(2t - t_0)}{\tau}$$

$$\frac{u_m}{u_{st}} = 1 + \frac{\tau}{\pi t_0} \left| \sin \frac{\pi t_0}{\tau} \right| \quad \text{Ref. [44].}$$

CASE 3.7   DAMPER-ONLY OSCILLATOR, TRANSLATIONAL
OR ROTATIONAL

Initial velocity v_0 or λ_0 prescribed

Translational:

$$u(t) = \frac{Mv_0}{C}\left[1 - \exp\left(-\frac{C}{M}t\right)\right]; \quad v(t) = v_0 \exp\left(-\frac{C}{M}t\right);$$

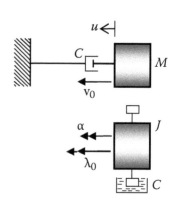

$$u(t_{05}) = \frac{Mv_0}{2C}; \quad u(\infty) = \frac{Mv_0}{C}$$

$$v(t_{05}) = \frac{v_0}{2}; \quad t_{05} = 0.6931\frac{M}{C}$$

$P_{dm} = P_0 = Cv_0$ (peak spring force)

$$u(t) = \frac{Mv_0}{C}\left[1 - \frac{v(t)}{v_0}\right] \quad \text{(displacement as a function of velocity)}$$

$$P_d(t) = Cv(t) = P_0 - \frac{C^2 u(t)}{M} \quad \text{(force as a function of displacement)}$$

Rotational:

$$\alpha(t) = \frac{J\lambda_0}{C'}\left[1 - \exp\left(-\frac{C'}{J}t\right)\right]; \quad \lambda(t) = \lambda_0 \exp\left(-\frac{C'}{J}t\right);$$

$$\alpha(t_{05}) = \frac{J\lambda_0}{2C'}; \quad \alpha(\infty) = \frac{J\lambda_0}{C'}$$

$$\lambda(t_{05}) = \frac{\lambda_0}{2}; \quad t_{05} = 0.6931\frac{J}{C'}$$

$T_{dm} = T_0 = C'\lambda_0$ (Peak torque)

$$\alpha(t) = \frac{J\lambda_0}{C'}\left[1 - \frac{\lambda(t)}{\lambda_0}\right] \quad \text{(Rotation as function of velocity)}$$

$$T_d(t) = C'\lambda(t) = T_0 - \frac{(C')^2 \alpha(t)}{J} \quad \text{(Torque as function of rotation)}$$

Note: t_{05} is the time needed for the initial velocity to drop by half.

CASE 3.8 DAMPER-ONLY OSCILLATOR. INITIAL BASE VELOCITY v_0 PRESCRIBED

$$u_b - u = \frac{Mv_0}{C}\left[1 - \exp\left(-\frac{C}{M}t\right)\right];$$

$$v_b - v = v_0 \exp\left(-\frac{C}{M}t\right);$$

$$u_b - u = \frac{Mv_0}{C} \quad \text{for } t \to \infty$$

$$P_{dm} = cv_0$$

$$v_b - v = 0 \quad \text{for } t \to \infty$$

Note: Refer to Case 3.7 to get analogous expressions for rotational case.

CASE 3.9   DAMPED OSCILLATOR, TRANSLATIONAL OR ROTATIONAL

(a) Short impulse: $S = W_0 t_0$. Equivalent to initial velocity v_0
   $= S/M$. Refer to Case 3.13.
(b) Step load $W = W_0 H(t)$

$$u_{st} = W_0/k;$$

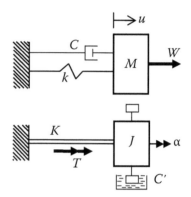

Moderate damping, $\zeta < 1.0$

$$\chi = \frac{\zeta}{\sqrt{1-\zeta^2}}; \text{ damping parameter}$$

$$u(t) = u_{st}[1-(\cos\omega_d t + \chi\sin\omega_d t)\exp(-\chi\omega_d t)]$$

$$v(t) = \omega_d u_{st}(1+\chi^2)\sin\omega_d t\exp(-\chi\omega_d t)$$

$$\bar{u}(t) = u_{st}\left[1+\sqrt{1+\zeta^2}\exp(-\chi\omega_d t)\right]; \text{ upper bound of displacement}$$

$$u_m = [1+\exp(-\pi\chi)]u_{st} \text{ at } t_1 = \pi/\omega_d;$$

$$v_m = \omega_d u_{st}(1+\chi^2)\sin\omega_d t_2\exp(-\chi\omega_d t_2); \text{ at } t_2,$$

$$\text{where } \tan(\omega_d t_2) = \frac{1}{\chi} P_{sm} = ku_m; P_{dm} = Cv_m$$

Small damping, $\zeta \ll 1.0$

$$u_m = [1+\exp(-\pi\zeta)]u_{st}; \text{ at } t_1 = \pi/\omega = \tau/2$$

$$v_m = \omega u_{st}\cdot\exp(-\pi\zeta/2); \text{ at } t_2 = \tau/4$$

$$P_{sm} = ku_m; \quad P_{dm} = Cv_m$$

CASE 3.10   DAMPER AND SPRING IN SERIES UNDER PRESCRIBED LOADING
(Maxwell element)

Generally: $u_2(t) = u_1(t) + \dfrac{P_0}{k}; \quad P = Cu$

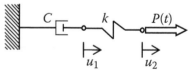

(a) Step load $P = P_0 H(t)$:

$$u_1(t) = \frac{P_0}{C}t; \quad u_2(t) = \frac{P_0}{k}\left[1+\frac{kt}{C}\right];$$

$$v_1 = v_2 = \frac{P_0}{C} = \text{const.}; \quad P_{dm} = P_{sm} = P_0$$

(b) Exponential decay loading $P = P_0 \exp(-bt)$:

$$u_1(t) = \frac{P_0}{Cb}(1 - e^{-bt}); \quad u_2(t) = \left(\frac{P_0}{k} - \frac{P_0}{Cb}\right)e^{-bt} + \frac{P_0}{Cb}$$

$$v_1(t) = \frac{P_0}{C}e^{-bt}; \quad v_2(t) = \left(\frac{P_0}{C} - \frac{bP_0}{k}\right)e^{-bt}; \quad P_{dm} = P_{sm} = P_0$$

(c) Sine loading by prescribed displacement: $u_2 = u_0 \sin \Omega t$

$$P(t) = \Omega C u_0 \left[\frac{\sin(\Omega t + \beta)}{\sin\beta} - \exp\left(-\frac{kt}{C}\right)\right]; \quad \sin\beta = \frac{1}{(1 + (\Omega C/k)^2)^{1/2}}$$

CASE 3.11   MASS RESTRAINED BY DAMPER AND SPRING IN SERIES. INITIAL VELOCITY v_0 APPLIED

When the spring relaxes during reverse motion, mass M separates with a rebound velocity V.

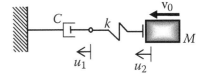

Upper bound of spring force: $P_{sm} = v_0\sqrt{kM}$ (when damper is rigid)

Upper bound of damper force: $P_{dm} = Cv_0$ (when spring is rigid)

Peak force in assembly: $P_m \approx (P_{sm}^{-1.15} + P_{dm}^{-1.15})^{-0.87}$

Mass velocity when P_m reached: $v_e = P_m/C$ (reversal point)

(*Note:* This is accurate to within a few percent over a large range of k, M, and C.)

Rebound velocity V and contact duration t_0:

Method A: $V = \sqrt{\dfrac{2}{M}(2E_e - E_0)}$; rebound velocity

$t_0 = \dfrac{\pi M v_0}{P_m}$; contact duration for a marked rebound

where $E_e = \dfrac{P_m^2}{2}\left(\dfrac{M}{C^2} + \dfrac{1}{k}\right)$ and $E_0 = \dfrac{1}{2}Mv_0^2$

Method B: Refer to Case 12.9. (A more general approach.)

Notes: Method A gives good results for a firm damper, when $v_e \ll v_0$. Otherwise, Method B may give better results. In those latter cases the rebound may be slower and, in effect, only the duration of the loading phase, lasting $t_0/2$, may be reasonably predicted by the above. Rebound velocities calculations are valid for a single-impact event only.

CASE 3.12   OSCILLATOR WITH INITIAL VELOCITY v_0 (OR λ_0) HAS ITS BASE SUDDENLY STOPPED

Equivalent to Giving the Fixed Oscillator in Case 3.2 a Short Impulse $S = Mv_0$ or $S = J\lambda_0$

Translational:

$$u_m = v_0/\omega; \text{ at } t = \tau/4; \quad v_m = v_0; \quad R_m = v_0\sqrt{kM}$$

Rotational:

$$\alpha_m = \lambda_0/\omega; \text{ at } t = \tau/4; \quad \lambda_m = \lambda_0; \quad T_m = \lambda_0\sqrt{KJ}$$

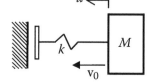

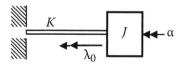

CASE 3.13   DAMPED OSCILLATOR WITH INITIAL VELOCITY v_0 (OR λ_0)

The base is suddenly stopped and held.

Moderate damping, $\zeta < 1.0$

$$\chi = \frac{\zeta}{\sqrt{1-\zeta^2}}; \text{ damping parameter}$$

$$u(t) = \frac{v_0}{\omega_d}\sin\omega_d t \exp(-\chi\omega_d t)$$

$$v(t) = v_0(\cos\omega_d t - \chi\sin\omega_d t)\exp(-\chi\omega_d t)$$

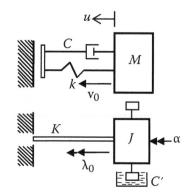

$$R(t) = k\frac{v_0}{\omega}\left\{\frac{1-2\zeta^2}{\sqrt{1-\zeta^2}}\sin\omega_d t + 2\zeta\cos\omega_d t\right\}\exp(-\chi\omega_d t);$$

the total base reaction as a sum of spring and damper forces. $u_m = \dfrac{v_0}{\omega_d}\sin\omega_d t_1 \exp(-\chi\omega_d t_1)$ at $\tan(\omega_d t_1) = \dfrac{1}{\chi}$; $v_m = v_0$ at $t = 0$; $P_{sm} = ku_m$; $P_{dm} = Cv_0$

Note: In general, the peak reaction may be larger than either the damper force (at the beginning) or the spring force (at maximum deflection).

Intermediate damping, $\zeta \leq 0.2$ $R_m = v_0\left(\dfrac{Mk}{1-\zeta^2}\right)^{1/2}\exp\left[\left(2\zeta - \dfrac{\pi}{2}\right)\zeta\right]$; peak base reaction.

Small damping, $\zeta \ll 1.0$

$$u_m = \frac{v_0}{\omega}\exp\left(\frac{-\pi\zeta}{2}\right) \quad \text{at } t_1 = \pi/(2\omega) = \tau/4$$

$$v_m = v_0 \text{ at } t = 0; \quad P_{sm} = ku_m; \quad P_{dm} = Cv_0$$

CASE 3.14 SIMPLE OSCILLATOR, PRESCRIBED BASE MOVEMENT

(a) Step acceleration a_0 applied to base, $a = a_0 H(t)$

$$\bar{u}_{st} = Ma_0/k; \quad \bar{u}_m = 2\bar{u}_{st} \text{ at } t = \tau/2;$$

$$\bar{v}_m = \omega\bar{u}_{st}; \quad P_m = Ma_0$$

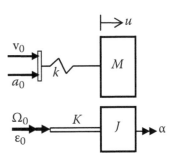

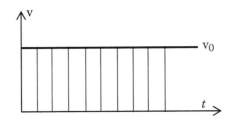

(b) Short acceleration pulse or velocity step

$$\bar{u}_{st} = \frac{v_0}{\omega} \text{ at } t = \tau/4 \quad \text{(peak relative displacement)}$$

$$P_m = v_0\sqrt{kM} \quad \text{(peak reaction)}$$

Note: Absolute values are shown.

CASE 3.15 DAMPED OSCILLATOR, PRESCRIBED BASE MOVEMENT u_b ($\bar{u} = u - u_b$)

(a) Acceleration step $a_b = a_{b0}H(t)$

Moderate damping, $\zeta < 1.0$

$\bar{u}_{st} = a_0 M/k$; $\bar{u}_m = [1 + \exp(-\pi\chi)]\,\bar{u}_{st}$ at $t_1 = \pi/\omega_d$ with

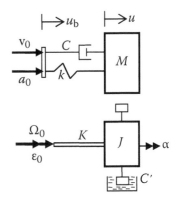

$$\chi = \frac{\zeta}{\sqrt{1-\zeta^2}}$$

$\bar{v}_m = \omega_d\bar{u}_{st}(1+\chi^2)\sin\omega_d t_2\exp(-\chi\omega_d t_2)$; at t_2,

where $\tan(\omega_d t_2) = \dfrac{1}{\chi}$ $P_{sm} = k\bar{u}_m$; $P_{dm} = C\bar{v}_m$.

Small damping, $\zeta \ll 1.0$

$\bar{u}_m = [1 + \exp(-\pi\zeta)]\bar{u}_{st}$; at $t_1 = \pi/\omega = \tau/2$

$\bar{v}_m = \omega\bar{u}_{st}\cdot\exp(-\pi\zeta/2)$ at $t_2 = \tau/4$; $P_{sm} = k\bar{u}_m$; $P_{dm} = C\bar{v}_m$

(b) Velocity step $v_b = v_{b0}H(t)$

Moderate damping, $\zeta < 1.0$

$$\bar{u}_m = \frac{v_{0b}}{\omega_d}\sin\omega_d t_1\exp(-\chi\omega_d t_1) \text{ at } \tan(\omega_d t_1) = \frac{1}{\chi} \text{ with } \chi = \frac{\zeta}{\sqrt{1-\zeta^2}}$$

at $t = 0$; $P_{sm} = k\bar{u}_m$; $P_{dm} = Cv_{0b}$

Small damping, $\zeta \ll 1.0$

$$\bar{u}_m = \frac{v_{0b}}{\omega}\exp\left(\frac{-\pi\zeta}{2}\right) \text{ at } t_1 = \pi/(2\omega) = \tau/4$$

$$P_{sm} = k\bar{u}_m; \quad P_{dm} = Cv_{0b}$$

Note: Absolute values are shown.

CASE 3.16 CANTILEVER BEAM AS A RIGID SEGMENT WITH AN ANGULAR SPRING

$$k = \frac{K}{L^2}; \quad \omega = \sqrt{\frac{3k}{M}}; \quad \tau = 2\pi\sqrt{\frac{M}{3k}}$$

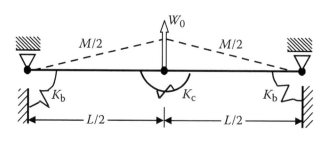

Equivalent tip mass: $M_e = M/3$; $u_{st} = W_0/k$

(a) Short impulse $S = W_0 t_0$, equivalent to initial velocity $v_0 = S/M_e$.

$$u_m = v_0/\omega; \text{ at } t = \tau/4; \; v_m = v_0; \; \mathcal{M}_m = \frac{Ku_m}{L} \text{ (maximum spring moment)}$$

(b) Rectangular impulse $S = W_0 t_0$.

$$u_m = 2u_{st}\sin\frac{\omega t_0}{2} \; (u_m = 2u_{st} \text{ for } t_0 \geq \tau/2)$$

$$v_m = \omega u_m; \quad a_m = W_0/M_e; \quad \mathcal{M}_m = \frac{Ku_m}{L}$$

(c) Step load $W = W_0 H(t)$

$$u_m = 2u_{st} \text{ at } t = \tau/2; \quad v_m = \omega u_m; \quad a_m = W_0/M_e; \quad \mathcal{M}_m = \frac{Ku_m}{L}$$

CASE 3.17 CLAMPED–CLAMPED BEAM AS TWO RIGID SEGMENTS WITH ANGULAR SPRINGS

$$k = \frac{8K_b + 16K_c}{L^2}; \quad \omega = \sqrt{\frac{3k}{M}}; \quad \tau = 2\pi\sqrt{\frac{M}{3k}}$$

Equivalent center mass: $M_e = M/3$; $u_{st} = W_0/k$

(a) Short impulse $S = W_0 t_0$, equivalent to initial velocity $v_0 = S/M_e$.

$u_m = v_0/\omega$; at $t = \tau/4$; $v_m = v_0$; $\mathcal{M}_b = \dfrac{2K_b u_m}{L}$ (maximum base spring moment)

$\mathcal{M}_c = \dfrac{4K_c u_m}{L}$ (maximum center spring moment)

(b) Rectangular impulse $S = W_0 t_0$.

$u_m = 2u_{st} \sin \dfrac{\omega t_0}{2}$ ($u_m = 2u_{st}$ for $t_0 \geq \tau/2$)

$v_m = \omega u_m$; $a_m = W_0/M_e$ (Moments expressed as above)

(c) Step load $W = W_0 H(t)$

$u_m = 2u_{st}$ at $t = \tau/2$; $v_m = \omega u_m$; $a_m = W_0/M$

CASE 3.18 SIMPLY SUPPORTED BEAM AS TWO RIGID SEGMENTS WITH ANGULAR SPRING

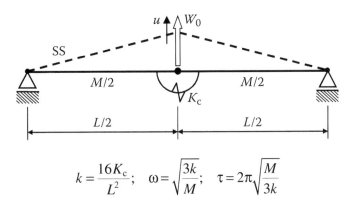

$$k = \frac{16K_c}{L^2}; \quad \omega = \sqrt{\frac{3k}{M}}; \quad \tau = 2\pi\sqrt{\frac{M}{3k}}$$

Equivalent center mass: $M_e = M/3$;

$$u_{st} = W_0/k$$

(a) Short impulse $S = W_0 t_0$, equivalent to initial velocity $v_0 = S/M_e$.

$u_m = v_0/\omega$; at $t = \tau/4$; $v_m = v_0$; $\mathcal{M}_m = \dfrac{4K_c u_m}{L}$ (maximum spring moment)

(b) Rectangular impulse $S = W_0 t_0$.

$$u_m = 2u_{st} \sin \frac{\omega t_0}{2} \quad (u_m = 2u_{st} \text{ for } t_0 \geq \tau/2)$$

$$v_m = \omega u_m; \quad a_m = W_0/M_e; \quad \mathcal{M}_m = \frac{4K_c u_m}{L}$$

(c) Step load $W = W_0 H(t)$

$$u_m = 2u_{st} \text{ at } t = \tau/2; \quad v_m = \omega u_m; \quad a_m = W_0/M_e; \quad \mathcal{M}_m = \frac{4K_c u_m}{L}$$

**CASE 3.19   SUPPORTED AND CLAMPED BEAM AS TWO RIGID SEGMENTS
WITH ANGULAR SPRINGS**

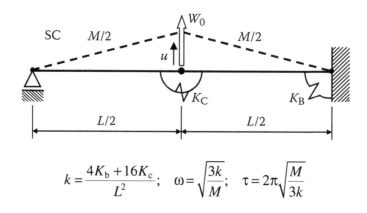

$$k = \frac{4K_b + 16K_c}{L^2}; \quad \omega = \sqrt{\frac{3k}{M}}; \quad \tau = 2\pi\sqrt{\frac{M}{3k}}$$

Equivalent center mass: $M_e = M/3$; $u_{st} = W_0/k$
The expressions for the response are the same as in Case 3.17.

**CASE 3.20   CLAMPED–CLAMPED BEAM WITH ONE END GUIDED PRESENTED AS A
RIGID SEGMENT WITH ANGULAR SPRINGS**

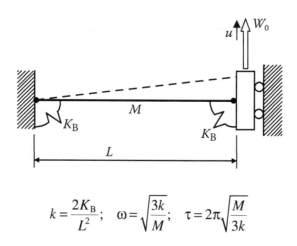

$$k = \frac{2K_B}{L^2}; \quad \omega = \sqrt{\frac{3k}{M}}; \quad \tau = 2\pi\sqrt{\frac{M}{3k}}$$

Equivalent end mass: $M_e = M/3$; $u_{st} = W_0/k$

(a) Short impulse $S = W_0 t_0$, equivalent to initial velocity $v_0 = S/M_e$.

$u_m = v_0/\omega$; at $t = \tau/4$; $v_m = v_0$; $\mathcal{M}_m = \dfrac{K_B u_m}{L}$ (maximum spring moment)

(b) Rectangular impulse $S = W_0 t_0$.

$$u_m = 2u_{st}\sin\frac{\omega t_0}{2} \quad (u_m = 2u_{st} \text{ for } t_0 \geq \tau/2)$$

$$v_m = \omega u_m; \quad a_m = W_0/M_e; \quad \mathcal{M}_m = \frac{K_B u_m}{L}$$

(c) Step load $W = W_0 H(t)$

$$u_m = 2u_{st} \text{ at } t = \tau/2; \quad v_m = \omega u_m; \quad a_m = W_0/M_e; \quad \mathcal{M}_m = \frac{K_B u_m}{L}$$

CASE 3.21   PENDULUM WITH A RIGID, MASSLESS ROD AND GRAVITY PRESENT

Stiffness: $k_{ef} = \dfrac{Mg}{L} + \dfrac{ke^2}{L^2}; \quad \omega = \sqrt{\dfrac{ke^2}{ML^2} + \dfrac{g}{L}}$

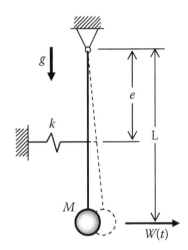

(a) Short impulse Mv_0:

$u_m = v_0/\omega$ (peak mass displacement at $t = \tau/4$)

$P_s = \dfrac{L}{e} k_{ef} u_m$ (peak spring force)

(b) Step load $W(t) = W_0 H(t)$:

$u_{st} = W_0/k_{ef}; \; u_m = 2u_{st} \text{ at } t = \tau/2;$

$P_s = \dfrac{L}{e} k_{ef} u_m$ (peak spring force)

CASE 3.22   MASSLESS CANTILEVER BEAM STRUCK BY MASS M WITH NO REBOUND

Stiffness, peak displacement, peak reaction, maximum bending, respectively:

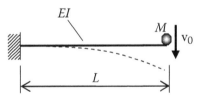

$$k = 3EI/L^3; \quad u_m = v_0\sqrt{\frac{ML^3}{3EI}}; \text{ (at } t = \tau/4)$$

$$R_m = ku_m = v_0\sqrt{kM}; \quad \mathcal{M}_m = R_m L$$

CASE 3.23   MASSLESS BEAM STRUCK BY MASS M WITH NO REBOUND

Stiffness, peak displacement, peak reaction, maximum bending, respectively:

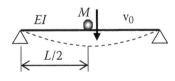

$$k = 48EI/L^3; \quad u_m = \frac{v_0}{4}\sqrt{\frac{ML^3}{3EI}}; \text{ (at } t = \tau/4);$$

$$R_m = ku_m = v_0\sqrt{kM}; \quad \mathcal{M}_m = R_m L/4$$

CASE 3.24 MASSLESS BEAM STRUCK BY MASS M WITH NO REBOUND

Stiffness, peak displacement, peak reaction, maximum bending, respectively, are

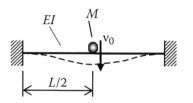

$$k = 192EI/L^3; \quad u_m = \frac{v_0}{8}\sqrt{\frac{ML^3}{3EI}}; \text{ (at } t = \tau/4);$$

$$R_m = ku_m = v_0\sqrt{kM}; \quad M_m = R_mL/8$$

CASE 3.25 CABLE PRELOADED WITH P_0 AND LATERAL LOAD APPLIED

(a) Short impulse $S = w_0t_0$

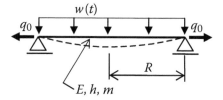

$$u_m = \frac{v_0}{\omega} = \frac{v_0L}{\pi}\left(\frac{m}{2P_0}\right)^{1/2} \text{ at } t = \tau/4; \, v_0 = S/m$$

(b) Step load, $w = w_0H(t)$

$$u_{st} = \frac{w_0L^2}{8P_0}; \quad u_m = 2u_{st} \text{ at } t = \tau/2$$

CASE 3.26 CIRCULAR MEMBRANE PRELOADED WITH q_0 (N/m) PER UNIT LENGTH OF EDGE

(a) Short impulse $S = w_0t_0$

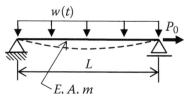

$$u_m = \frac{v_0}{\omega} = \frac{v_0R}{1.357}\left(\frac{m}{\pi q_0}\right)^{1/2} \text{ at } t = \tau/4; \, v_0 = S/m$$

(b) Step load, $w = w_0H(t)$

$$u_{st} = \frac{w_0R^2}{4q_0}; \quad u_m = 2u_{st} \text{ at } t = \tau/2$$

CASE 3.27 MASSLESS CIRCULAR PLATE STRUCK AT MIDPOINT BY MASS M WITH NO REBOUND

Stiffness, peak reaction, peak displacement, respectively:

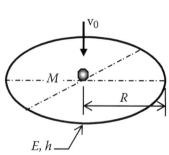

Edges supported: $k = \dfrac{16\pi D}{R^2}\dfrac{1+v}{3+v}$; $R_m = v_0\sqrt{kM}$; $u_m = R_m/k$; (at $t = \tau/4$)

Edges clamped: $k = \dfrac{16\pi D}{R^2}$; $R_m = v_0\sqrt{kM}$; $u_m = R_m/k$; (at $t = \tau/4$)

CASE 3.28 MASSLESS SQUARE PLATE STRUCK AT MIDPOINT BY MASS M WITH NO REBOUND. STIFFNESS, PEAK REACTION, PEAK DISPLACEMENT, RESPECTIVELY

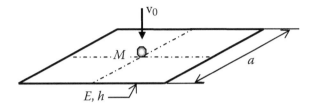

Edges supported: $k = \dfrac{86.18D}{a^2}$; $R_m = v_0\sqrt{kM}$; $u_m = R_m/k$; (at $t = \tau/4$)

Edges clamped: $k = \dfrac{178.7D}{a^2}$; $R_m = v_0\sqrt{kM}$; $u_m = R_m/k$; (at $t = \tau/4$)

CASE 3.29 CIRCULAR, MODERATELY THICK RING, PRESSURIZED

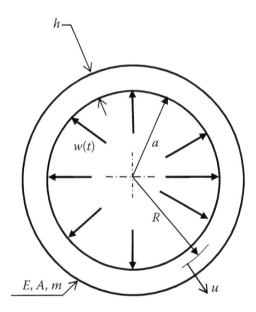

Static stress, strain, radial deflection (at R), and stretching work due to static load w_0 (N/m):

$$\sigma_{st} = \frac{w_0 a}{A}; \quad \varepsilon_{st} = \frac{w_0 a}{EA}; \quad u_{st} = \frac{w_0 aR}{EA}; \quad \mathcal{L} = \frac{\pi R}{EA}(w_0 a)^2$$

(a) Short impulse $S = w_0 t_0$:

$$u_m = \frac{v_0}{\omega} = v_0 R\left(\frac{m}{EA}\right)^{1/2} \text{ at } t = \tau/4$$

(b) Step load, $w = w_0 H(t)$

$$u_m = 2u_{st} \text{ at } t = \tau/2$$

CASE 3.30   CYLINDRICAL, MODERATELY THICK SHELL, SUDDENLY PRESSURIZED

(Refer to the illustration in Case 3.29)

Static stress, strain, radial deflection (at R), and stretching work due to static pressure p_0:

$$\sigma_{st} = \frac{p_0 a}{h}; \quad \varepsilon_{st} = \frac{p_0 a}{E''h}; \quad u_{st} = \frac{p_0 aR}{E''h}; \quad E'' = \frac{E}{1-v^2}; \quad \mathcal{L} = \frac{\pi R}{E''h}(p_0 a)^2$$

(a) Short impulse $S = p_0 t_0$: $u_m = \dfrac{v_0}{\omega} = v_0 R \left(\dfrac{m}{E''h}\right)^{1/2}$ at $t = \tau/4$

(b) Step load, $w = p_0 H(t)$: $u_m = 2u_{st}$ at $t = \tau/2$

Note: The equations pertain to a unit-long segment of shell.

CASE 3.31   SPHERICAL SHELL, MODERATELY THICK, PRESSURIZED

(Refer to the illustration in Case 3.29)

Static stress, strain, radial deflection (at R), and stretching work due to static pressure p_0:

$$\sigma_{st} = \frac{p_0 a}{2h}; \quad \varepsilon_{st} = \frac{p_0 a}{2E'h}; \quad u_{st} = \frac{p_0 aR}{2E'h}; \quad E' = \frac{E}{1-v}; \quad \mathcal{L} = \frac{\pi aR}{4E'h}(p_0 a)^2$$

(a) Short impulse $S = p_0 t_0$: $u_m = \dfrac{v_0}{\omega} = v_0 R \left(\dfrac{m}{2E'h}\right)^{1/2}$ at $t = \tau/4$

(b) Step load, $w = p_0 H(t)$: $u_m = 2u_{st}$ at $t = \tau/2$

CASE 3.32   CYLINDRICAL CAVITY, INFINITE ELASTIC BODY

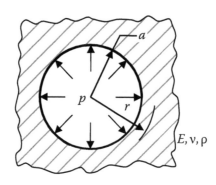

$$u_r = \frac{1+v}{E}\frac{pa^2}{r}; \text{ radial displacement at distance } r$$

$$u = \frac{1+v}{E} pa; \text{ displacement at } r = a$$

$$\sigma_r = -p\left(\frac{a}{r}\right)^2; \ \sigma_h = p\left(\frac{a}{r}\right)^2 \text{ radial and hoop stress, respectively, positive under tension.}$$

Note: Refer to Appendix D.

CASE 3.33   SPHERICAL CAVITY, INFINITE ELASTIC BODY

(Refer to the illustration in Case 3.32)

$$u_r = \frac{1+v}{E} \frac{pa^3}{2r^2}; \text{ radial displacement at distance } r$$

$$u_a = \frac{1+v}{2E} pa; \text{ displacement at } r = a$$

$$\sigma_r = -p\left(\frac{a}{r}\right)^3, \ \sigma_h = \frac{p}{2}\left(\frac{a}{r}\right)^3 \text{ radial and hoop stress, respectively, positive under tension}$$

Note: Refer to Appendix D.

CASE 3.34   ROTATIONAL OSCILLATOR, SUBJECTED TO PERIODIC FORCING
WITH REPEAT PERIOD OF t_p

Rectangular impulse $S = T_0 t_0$.

$\alpha_{st} = T_0/K$; static rotation under T_0

t_0 is an odd multiple of a $\tau/2$; t_p is a multiple of τ

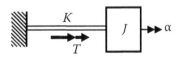

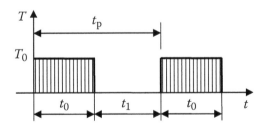

$\alpha_m = 2n\alpha_{st}$; peak rotation after n cycles (no damping)

$\alpha_m \approx \dfrac{\pi t_0}{4\tau\zeta} \alpha_{st}$; peak rotation after many cycles, in presence of damping.

Note: This is equally valid for a translational oscillator. See text for details and also Ref. [44].

CASE 3.35 ROTATIONAL OSCILLATOR, SUBJECTED TO CYCLIC FORCING WITH REPEAT PERIOD OF t_p

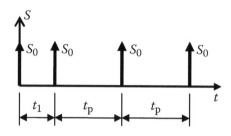

Short angular impulse S_0.

$\lambda_0 = \dfrac{S_0}{J}$; initial velocity induced by S_0

$\alpha_1 = \lambda_0/\omega$; peak rotation after first application of S_0

t_1 is a multiple of a $\tau + 3\tau/4$; t_p is a multiple of τ

$\alpha_m = \sqrt{n}\,\alpha_1$; peak rotation after n cycles (no damping) Ref. [44]

Note: This is equally valid for a translational oscillator. Time points t_1 and t_p are selected to maximize the response.

CASE 3.36 BASIC DAMPED OSCILLATOR, SUBJECTED TO SINUSOIDAL FORCING AT NATURAL FREQUENCY, $t_p = t$

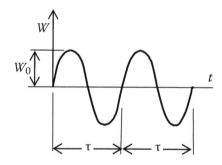

$u_{st} = W_0/k$; static deflection under W_0

$u_m = \dfrac{u_{st}}{2\zeta}$; upper deflection limit after large number of half-sine pulses

$u_t \approx u_m(1 - \exp(-\zeta\omega t))$; amplitude after p-pulses, or after $t = p(\tau/2)$ or $\omega t = \pi p$; p is odd number

$u \approx \dfrac{\omega t}{2} u_{st} = u_m$; approximation for a small number of applied cycles

$S_0 = \dfrac{2}{\pi} W_0 \tau$; total impulse in one full sine cycle for this harmonic forcing

S_1; impulse during one period applied by a periodic, nonsinusoidal pulse

$u_t \approx \dfrac{S_1}{S_0} u_m (1 - \exp(-\zeta \omega t))$; amplitude after p-pulses for the above impulse

Note: This is equally valid for a rotational oscillator. The above harmonic forcing pattern is regarded as a continuous sequence of sine impulses. The direction of force as well as of movement reverses after each half-period.

CASE 3.37 ROTATIONAL DAMPED OSCILLATOR, SUBJECTED TO PERIODIC, RECTANGULAR PULSES

The period of forcing $t_p = n\tau$ is a multiple of the natural period. Steady-state condition:

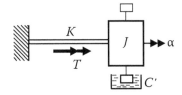

$$\omega^2 = \dfrac{K}{J}; \quad \zeta = \dfrac{C'}{C'_c}; \quad C'_c = 2\sqrt{KJ}$$

$t_0 = m(\tau/2)$; load duration, an odd multiple of half-period.

$t_1 = p(\tau/2) = t_p - t_0$; no load, an odd multiple of half-period.

$(m + p)/2 = n$

$\alpha_{st} = T_0/K$; static rotation under T_0

$S = T_0 t_0$; impulse

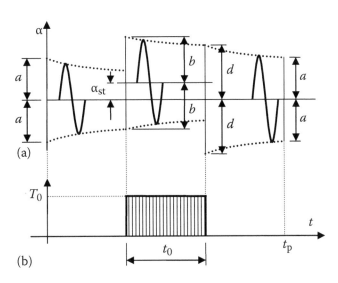

$b = \alpha_{st} \dfrac{1 + e^{-\pi\zeta p}}{1 - e^{-2\pi\zeta n}}; \quad d = \alpha_{st} \dfrac{1 + e^{-\pi\zeta m}}{1 - e^{-2\pi\zeta n}}$; amplitudes

$$\alpha_m = \frac{1}{2}(\alpha_{st} + b + d);\ \text{average maximum amplitude}$$

$$\alpha_m \approx \alpha_{st}\left(1 + \frac{1}{\pi\zeta n}\right);\ \text{close approximation}$$

Note: This is equally valid for a translational oscillator [49].

CASE 3.38   BASIC OSCILLATOR, INITIAL STATIC LOAD W_{st} FOLLOWED BY A SHORT IMPULSE

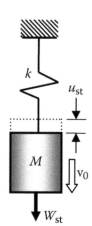

$u_{st} = W_{st}/k$

$S = W_0 t_0$, is a short impulse equivalent to the initial velocity $v_0 = S/M$

$u_m = u_{st} + \dfrac{v_0}{\omega};\ \text{peak displacement}$

$R_m = k u_m;\ \text{peak reaction}$

Note: The result does not depend on whether v_0 is up or down.

CASE 3.39   BASIC OSCILLATOR SUBJECTED SIMULTANEOUSLY TO AN INITIAL SHORT IMPULSE $S = W_1 t_1$ AND A RECTANGULAR IMPULSE $W_0 t_0$

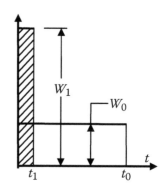

$v_1 = S/M = W_1 t_1/M;$ initial velocity from the short impulse
u_1 is the maximum displacement from this first impulse alone
$u_0 = W_0/k;$ static deflection from sustained load W_0 alone

(a) Step load $W_0 (t_0 > \tau/2)$

$$u(t) = u_0(1 - \cos\omega t) + \frac{v_1}{\omega}\sin\omega t;\ \text{resultant deflection}$$

$$u_m = \left(u_0^2 + \frac{M v_1^2}{k}\right)^{1/2} + u_0 = (u_0^2 + u_1^2)^{1/2} + u_0;\ \text{peak deflection}$$

(b) Initial impulse followed by a rectangular pulse

$$u_m \approx (u_p^2 + u_1^2)^{1/2};\ \text{where } u_p \text{ is the effect of the rectangular pulse acting alone.}$$

Note: The second equality in the first subcase is useful for more general systems acting in a SDOF mode. The term v_1/ω is the maximum displacement attained as a result of the initial impulse acting alone.

EXAMPLES

EXAMPLE 3.1 MASS M SUPPORTED BY TWO COLUMNS AND IMPACTED LATERALLY

Treating the beams as massless, find the peak deflection and bending stress for $W = 20\,$kN applied in the form of a rectangular pulse lasting longer than the natural period. The "universal beam" or the I-section is used, with the stronger direction opposing deflection. The depth of beam is 612 mm while $I = 986 \times 10^6\,$mm⁴, $H = 3,000\,$mm, and $E = 200,000\,$MPa.

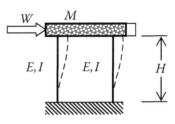

Case 3.2 shows that maximum deflection can be $u_m = 2u_{st}$ for an impulse lasting that long. To find stiffness, refer to Case 2.16:

$$k = \frac{24EI}{H^3} = \frac{24 \times 200,000 \times 986 \times 10^6}{3,000^3} = 175,290 \text{ N/mm}$$

Static deflection: $u_{st} = W/k = 20,000/175,290 = 0.1141$ mm
Peak dynamic deflection: $u_m = 2u_{st} = 0.2282$ mm

Both ends of the beam are fixed. It may be found in any reference text, that in this case the maximum moment is $M = QL/2$, where Q is the maximum lateral force per beam. The peak dynamic internal loading may be recovered from the peak deflection: $Q_m = ku_m/2 = 175,290 \times 0.2288/2 \approx 20,000\,$N (per beam)

(It is quite apparent here that the dynamic loading per beam would be twice the static loading. It is not always so clear in practical applications.)

$\mathcal{M}_m = 20,000 \times 3000/2 = 30 \times 10^6$ N-mm.

Peak bending stress: $\sigma = \mathcal{M}_m c/I = 30 \times 10^6 (612/2)/986 \times 10^6 = 9.31\,$MPa.

A shear check is desirable as well.

EXAMPLE 3.2 A TORQUE SUDDENLY APPLIED TO A DISK

The shaft-disk arrangement is as shown in Case 3.2. The torque is a triangular, decreasing pulse, as in Case 3.2d with the maximum of T_0. The torque duration is only 0.1τ. Estimate the peak torque experienced by the shaft, as a multiple of T_0.

Due to a short application time the pulse may be treated as rectangular with an impulse of $S = T_0 t_0/2$, equivalent to $T_0/2$ applied for t_0. From Case 3.2c, the maximum angle of twist is

$$\alpha_m = 2\alpha_{st} \sin\frac{\pi t_0}{\tau} = 2\alpha_{st} \sin(0.1\pi) = 0.618u_{st}$$

where u_{st} is the angle of rotation induced by the static torque $T_0/2$. The peak shaft torque can therefore be estimated as $T_m = 0.618T_0/2 = 0.309T_0$. The exact solution presented in Harris and Crede [30, pp. 8–34], clearly shows that for this ratio of t_0/τ the approximation is sufficiently accurate.

EXAMPLE 3.3   SUDDEN REMOVAL OF SUPPORT

Weight $W = 1000\,\mathrm{N}$ is suspended on a cable so that the spring below the weight is unstressed. If the cable suddenly breaks, what will be the maximum downward deflection and maximum velocity attained? How high will the weight go after the rebound? Use $k = 100\,\mathrm{N/mm}$.

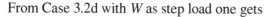

The mass is $W/g = 1000/9.81 = 101.94\,\mathrm{kg}$. Static displacement due to W is $u_{st} = 1000/100 = 10\,\mathrm{mm}$. The cable breaking is equivalent to a sudden application of the force of gravity.

From Case 3.2d with W as step load one gets

$$u_m = 2u_{st} = 20\,\mathrm{mm}$$

$$v_m = \omega u_{st} = \sqrt{\frac{k}{M}}\,u_{st} = \sqrt{\frac{100}{101{,}940}} \cdot 10 = 0.3132\ \mathrm{mm/ms};\ \text{attained at midpoint of downward}$$

travel.

After reaching peak downward deflection the weight returns to its original position, as may be inferred from the energy balance.

EXAMPLE 3.4   SUDDEN CHANGE OF DIRECTION OF MOVEMENT

The figure shows a single-axis trailer moving along the road, about to run over a discontinuity in the form of slope change. Assuming the vertical axis of the assembly to remain vertical after entering the slope, while forward velocity v_0 remains parallel to the surface, find the peak spring force and the peak damper force applied to the mass M:

$$v_0 = 12\,\mathrm{m/s}, \quad M = 700\,\mathrm{kg}, \quad k = 50\,\mathrm{N/mm}, \quad \zeta = 0.3.$$

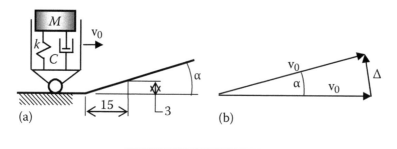

The velocity diagram in (b) shows that the change in the vertical direction is

$$\Delta v \approx v_0 \sin\alpha = 12 \times 0.1961 = 2.353\,\mathrm{m/s}$$

This velocity increase is acquired instantly and sustained, which means Case 3.15b, moderate damping, is applicable. (Except that v_0 in 3.15 is the same as Δv here). The dynamic parameters are

$$\omega = (50/700{,}000)^{1/2} = 8.452 \times 10^{-3}\,\text{kHz}$$

$C_c = 2M\omega = 11{,}833$; critical damping per Equation 3.3

$$C = \zeta C_c = 3550\,\text{N-ms/mm}; \quad \omega_d = \omega(1 - \zeta^2)^{1/2} = 8.063\,\text{Hz}$$

The largest damper force occurs at the beginning: $P_{dm} = C\Delta v = 3550 \times 2.353 = 8353\,\text{N}$

For $\chi = 0.3145$, $\tan(\omega_d t_1) = \dfrac{1}{0.3145}$, and $\omega_d t_1 = 72.54° = 1.266\,\text{rad}$, and $\bar{u}_m = \dfrac{v_{0b}}{\omega_d}$

$$\sin\omega_d t_1 \cdot \exp(-\chi\omega_d t_1) = \frac{2.353}{8.063 \times 10^{-3}}\sin 72.54° \cdot \exp(-0.3982) = 186.9\,\text{mm};$$

The largest spring force: $P_{sm} = 50 \times 186.9 = 9345\,\text{N}$.

EXAMPLE 3.5 PRELOADED SPRING AND DAMPER IN SERIES

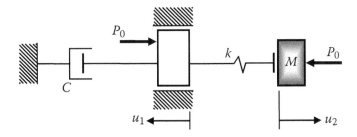

A weightless block separating the damper ($C = 47.54\,\text{N-s/mm}$) and the spring ($k = 100\,\text{N/mm}$) can slide without friction. The spring is preloaded and the system is motionless. At $t = 0$ both preloading forces $P_0 = 1164\,\text{N}$ are released and mass $M = 2.1\,\text{g}$ is pushed to the right. When the force compressing the spring drops to nil, the mass separates from the spring. Determine peak velocity v_f with which M flies off by assuming that one-half of the stored energy is lost due to damper action in the expansion phase.

An important thing to realize here is that after the release, the spring and the damper have the same force, as the connecting block is massless. Suppose the damper is rigid, i.e., the spring is grounded. Equating the strain energy in the spring, $P_0^2/(2k)$, with the kinetic energy of the moving mass, one has the fly-off velocity as

$$v_{fm}^2 = \frac{P_0^2}{kM} = \frac{1164^2}{100 \times 2.1}; \quad \text{then } v_{fm} = 80.32\,\text{m/s}$$

However, only one-half of the stored energy is recovered and converted to the kinetic energy because of the energy loss in the damper during expansion. In the equation above, the right side representing the strain energy has to be reduced by 1/2. Therefore, the fly-off velocity may be estimated as

$$v_{fm}^2 = \frac{1}{2}\frac{1164^2}{100\times 2.1} = 56.8\,\text{m/s}$$

EXAMPLE 3.6   EXPLOSIVE PRESSURE INSIDE SPHERICAL SHELL, ELASTIC RANGE

A shell with $a = 279.4\,\text{mm}$ and $h = 12.7\,\text{mm}$ has a charge exploding at its center. Two separate such events are recorded, each giving a pulse that can be simplified to a decreasing triangular shape. (Refer to Case 3.3c). The first event gave $p_0 = 3.64\,\text{MPa}$ and impulse $S = 72.41\,\text{kPa-ms}$, while for the second the quantities are 3.22 and 54.51, respectively. Determine the expected peak strain. The shell material data: $E = 82{,}760\,\text{MPa}$; $v = 0.3$ while density is uncertain. It should be recovered from the experimental observation that the natural frequency of radial motion is, on average, 2225 Hz. (The test results were reported by Johnson [40].)

Use Case 3.31:

$R = 279.4 + 12.7/2 = 285.8\,\text{mm}$; $E' = \dfrac{E}{1-v} = 118{,}230\,\text{MPa}$; mean radius and effective E

$u_{st} = \dfrac{p_0 aR}{2E'h} = \dfrac{3.64\times 279.4\times 285.8}{2\times 118{,}230\times 12.7} = 0.0968\,\text{mm}$; static deflection due to $p_0 = 3.64\,\text{MPa}$

Natural frequency $f = \omega/(2\pi)$ from Case 2.46:

$f = \dfrac{\omega}{2\pi} \approx \dfrac{1}{2\pi R}\sqrt{\dfrac{2E'}{\rho}} = \dfrac{1}{2\pi 285.8}\sqrt{\dfrac{2\times 118{,}230}{\rho}}$ or $2.225\,(\text{kHz}) = \dfrac{0.2708}{\sqrt{\rho}}$ from which

$\rho = 0.01481\,\text{g/mm}^3$. (Note that the thickness term in f was negligible). The density being so high (14.81 g/cm³) is indicative of some additional, nonstructural material lining the shell. The period is $\tau = 1/f = 0.4494\,\text{ms}$. To estimate impulse duration, assume it has a triangular decreasing shape (Chapter 13). Then $S = p_0 t_0/2$ or $t_0 = 2S/p_0$.

Here $t_0 = 2\times 0.07241/3.64 = 0.0398\,\text{ms}$. (Impulse must be expressed in MPa-ms here, to have consistent units.) Noting that $t_0 < 0.1\tau$, the impulse is decisively short, the shape of the pulse does not matter much and Case 3.2c may be invoked to find peak dynamic displacement:

$$u_m = 2u_{st}\sin\frac{\pi t_0}{\tau} = 2\times 0.0968\sin\frac{\pi 0.0398}{0.4494} = 0.05318\,\text{mm}$$

(This is smaller than static displacement under the peak load.) The peak hoop strain is $\varepsilon_h = u_m/R$

$$\varepsilon_h = 0.05318/285.8 = 186\times 10^{-6}.$$

Two test results are quoted in Ref. [40]; 168×10^{-6} and 187×10^{-6}, but it seems uncertain which one belongs to which event.

EXAMPLE 3.7 AMPLITUDE GROWTH DURING RESONANCE

The magnitude of force W_0 applied to an oscillator is such that it induces the static displacement $u_{st} = 10$ mm. If this force is applied with the natural frequency of the oscillator, find the amplitude of motion reached after two cycles and then after 20 cycles. Damping coefficient is $\zeta = 0.04$.

Apply Equation 3.35a, without the oscillating part: $u_m \approx \dfrac{u_{st}}{2\zeta}(1 - \exp(-\omega\zeta t))$ after two cycles

$t = 2\tau = 4\pi/\omega$, therefore $\omega t = 4\pi$ and after 20 cycles $\omega t = 40\pi$.

Peak displacement after two cycles: $u_m = \dfrac{u_{st}}{2 \times 0.04}(1 - \exp(-4\pi 0.04)) = 4.94 u_{st} = 49.4$ mm

Similarly, for 20 cycles: $u_{st} = 124.2$ mm.

Note that the upper limit of the amplitude is $u_m = \dfrac{u_{st}}{2\zeta} = \dfrac{10}{2 \times 0.04} = 125$ mm

To test the accuracy of a further approximation expressed by Equation 3.37, namely

$$u \approx \frac{\omega t}{2} u_{st} \cos \omega t$$

use the above for one half-cycle only, namely for a half-sine pulse. Then $\omega t = \pi$ and

$$|u_m| \approx \frac{\pi}{2} u_{st} = 1.571 u_{st}$$

According to Table 3.1, DF(u) for a half-sine pulse is 1.77, therefore the accuracy of the above approach for a one-half cycle is moderate.

EXAMPLE 3.8 RESPONSE TO A GROUND SHOCK

In a velocity record of an earthquake with a strong vertical component, the bell-shaped curve described by

$$v_b = \frac{v_0}{2}(1 - \cos \Omega t); \quad \text{with } \Omega t_0 = 2\pi \text{ when the pulse ends}$$

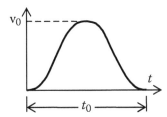

was identified in vertical ground movement. If a building is idealized to an oscillator, find the peak relative (ground-to-mass) acceleration that can take place.

Equation 3.40 tells us that if the relative motion only is of interest, then the ground accel-eration is the same as the relative mass-ground acceleration, except for the sign. When the expression for v_b is differentiated, the acceleration is obtained as

$$a_b = \frac{\Omega v_0}{2} \sin \Omega t; \quad \text{which has the peak of } \Omega v_0/2.$$

This represents a full sine wave during $\Omega t_0 = 2\pi$ with the force amplitude:

$$W_b = Ma_b = \frac{1}{2} M\Omega v_0$$

The relative static displacement, which this force can induce, is simply W_b/k. To maximize response, we set $\Omega = \omega$:

$$u_{st} = \frac{1}{2k} M\omega v_0 = \frac{v_0}{2\omega} \quad (\text{where } \omega^2 = k/M)$$

Because of small number of cycles involved (one full cycle or two half-sine impacts) one can use Equation 3.37, with $\omega t = 2\pi$, to obtain peak amplitude:

$$u_m = \frac{\omega t}{2} u_{st} = \frac{2\pi}{2} \frac{v_0}{2\omega} = \frac{\pi}{2} \frac{v_0}{\omega}$$

For the sake of comparison note that in Case 3.12, when a sudden application of velocity v_0 takes place, one obtains $u_m = v_0/\omega$. In this problem velocity is applied smoothly, but the pulse shape is such that it results in a larger response.

Example 3.9   A Steam Turbine with a Partial Admission

Such a turbine has its blades subjected to the driving steam force only on a portion of its circumference. The fact that during each rotation the blades become loaded and unloaded again may set up a resonant condition. Let a blade of such turbine have 40 vibratory cycles in one revolution. The steam is acting during just a little more than a half of the revolution. For the fundamental period, the damping level is $\zeta = 0.001$. The steam force is of such a magnitude, that when statically applied it induces a 0.1 mm deflection. Find the peak deflec-tion under the described conditions.

The above tells us that the steam forces act during over 20τ, say 20.5τ, while there are no forces during the remainder of one revolution, the latter being the period of forcing.

Referring to Case 3.37, which, while formulated for angles of rotation, applies to transla-tions as well. It tells us that the relevant multiples are: $n = 40$, $m = 41$, and $p = 39$. We have $u_{st} = 0.1$ mm and

$$\pi\zeta p = \pi 0.001 \times 39 = 0.1225, \quad \pi\zeta m = 0.1288, \quad \text{and} \quad 2\pi\zeta n = 0.2513$$

Substitution gives

$$b = u_{st} \frac{1+e^{-\pi \zeta p}}{1-e^{-2\pi \zeta n}} = u_{st} \frac{1+e^{-0.1225}}{1-e^{-0.2513}} = 8.482 u_{st}; \quad d = 8.457 u_{st}$$

$u_m = \frac{1}{2}(1+8.482+8.457)u_{st} = 8.97 u_{st}$; average maximum amplitude

$u_m \approx u_{st}\left(1+\frac{1}{\pi \zeta n}\right) = u_{st}\left(1+\frac{1}{\pi 0.001 \times 40}\right) = 8.958 u_{st}$; approximation of the above

The approximate expression gives a very close result.

4 Simple Nonlinear Systems

THEORETICAL OUTLINE

When the relationship between the external load and deflection experienced by a deformable system cannot be represented by a straight line, the system is said to be nonlinear. The natural frequency of such a system or structure is most often amplitude dependent, unlike the case of the perfectly linear system, where frequency remains constant regardless of amplitude. The usefulness of this, or the natural period concept in shock and impact analyses lies mainly in the fact that the deflection from the initial position to the final one can be regarded as taking place in one-half or one quarter of the natural period. For the system acted upon by a prescribed impulse, the main objectives are to determine the magnitude of that deflection (pseudo-amplitude) as well as duration of motion. This chapter is devoted to calculating these quantities.

GENERAL REMARKS ABOUT NONLINEARITIES

The equation of motion of the basic oscillator:

$$M\ddot{u} + C\dot{u} + ku = W(t) \tag{4.1}$$

is an example of a linear differential equation, because the unknown function $u(t)$ as well as its derivatives appears in the first power. When an equation of motion does not satisfy that condition, we call it a nonlinear equation and the system it describes is referred to as a nonlinear system. Most engineering objects show some deviation from linearity, but this is usually ignored for the practical reason of keeping computations simple. When deflection amplitudes are small in comparison with dimensions of structural elements, it is often quite realistic to assume proportionality between deflections and the elastic resistance. Yet, when those displacements significantly change the manner in which the external loads act on a structure, the nonlinearity of the system may not be ignored.

A good example of a nonlinear system is a body like a wing of an airplane, subjected to aerodynamic forces that are proportional to the square of velocity. Another large class of problems relates to systems with a general resisting force $R(u)$ so that we have

$$M\ddot{u} + C\dot{u} + R(u) = W(t) \tag{4.2}$$

instead of the previous equation. The function $R(u)$ is also referred to as a *characteristic* of a system. The first example of an element with a nonlinear characteristic is shown in Figure 4.1a. It is a cantilever that is slightly curved upward in its unstressed state. When subjected to a vertical downward force, the free length of the cantilever diminishes because the beam is gradually leaning against the rigid plane. In consequence, for every successive increment of displacement, a larger and larger increment of force W is needed.

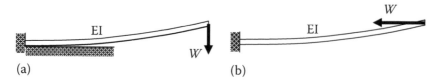

FIGURE 4.1   Hardening system (a) and softening system (b).

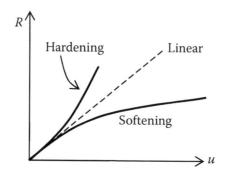

FIGURE 4.2   Characteristics of nonlinear systems.

Such elements in which the resistance grows faster than displacements are called *hardening* or *stiffening* elements.

Figure 4.1b gives an example of a *softening* element. This is the same cantilever as before, but loaded with a compressive force and having no additional restraint. As the deflection grows, the process of deforming becomes easier, at least up to a point. This is because the effective arm of force W with respect to the point of fixity is increasing. The characteristics of both types of element are plotted in Figure 4.2.

This example has illustrated an important principle, namely, that not only the element itself, but also its manner of loading and the magnitude of deflections may cause nonlinear effects to appear.

Most structural materials begin to yield at a certain stress level. Once yielding starts, it makes the characteristic of an element deviate from the initial slope. This is called a *material nonlinearity*, as opposed to *geometrical nonlinearity* caused by the change of shape or constraints during the loading process. The impact problem typically involves both types of nonlinearity, because the change in contact area may be accompanied by yielding.

Unlike the case of the linear problems, very few exact, closed-form solutions are available, and various approximations must be employed. That makes the field of nonlinear dynamics a diversified and difficult branch of engineering science.

An important computational aspect of a nonlinear system is that the principle of super-position does not in general apply. When a system is subject to several dynamic loads acting at the same time, the entire loading must be simultaneously considered.

ELASTIC, NONLINEAR SYSTEMS

A system or a body is called elastic if there is a 1 : 1 relationship between displacement and load. If in addition to that, the resistance–deflection curve is a straight line, the body is

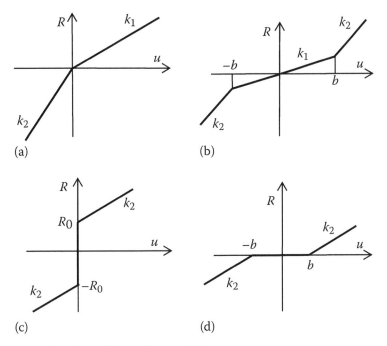

FIGURE 4.3 Some simple nonlinear characteristics.

called *linearly elastic*;* otherwise, it is *nonlinearly elastic*. The latter type of elements is described in this Section.

In practical situations, a linear characteristic is often inapplicable, so the first attempt is to replace it with straight line segments, making it piecewise-linear. A few such examples are shown in Figure 4.3. All are made of lines with two distinct slopes, but the term *bilinear* (BL) is usually applied only to the characteristic in Figure 4.3b. (The magnitude of slopes is indicated by k_1 or k_2.)

Elastic symmetry means that if the sense of displacement changes, so does the sense of the resisting force, but the magnitude of the latter remains unchanged. It is sufficient to define the resistance–deflection curve for positive values of displacement. Among the graphs in Figure 4.3, only the first one does not show the elastic symmetry.

Some simple structural arrangement can be used to obtain characteristics in Figure 4.3, such as the double-sided oscillator in Figure 4.4. First, consider the arrangement when there is no spring prestress and no gap. This may correspond to Figure 4.3a for springs of unequal

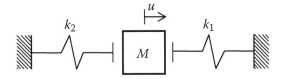

FIGURE 4.4 Double-sided oscillator.

---

stiffness. A discontinuity of slope, as in Figure 4.3c, may be caused by the springs being initially rigid and then beginning to yield at the R_0 level. A gap on both sides will give rise to a horizontal discontinuity, as in Figure 4.3d. Finally, to get the behavior like in Figure 4.3b, an additional set of linear springs is needed.

Unlike in a linear case, the natural frequency of a system is now dependent on the amplitude of vibrations the system is performing. The hardening system will exhibit a decrease of the natural period with the increase in amplitude.

MATERIAL NONLINEARITY, DUCTILE MATERIALS

Once the yield point of material is exceeded, the basic linear elastic (LE) model, used so far, is no longer satisfactory. Several simple, useful material models, approximating true stress–strain relationships will now be presented. The most popular approximations in nonlinear dynamics research appear to be the elastic-perfectly plastic (EPP) and rigid-plastic (RP) models. (Figure 4.5) The latter one represents a material, which is assumed to be undeformable up to a certain stress level σ_0, called here the *flow strength*.* When the level is reached, further elongation takes place with no increase of resistance. In EPP approximation, on the other hand, the elastic range is followed by deformation at the constant level of σ_0, the effective yield strength, reached when strain attains the value of ε_y. In general, the EPP model offers a more realistic representation because it has one parameter more. Based on strain magnitude alone, simplifying the EPP to a RP model seems almost natural. Yet, one must keep in mind that when bending is involved, a significant loss of accuracy may take place when EPP is replaced by a simpler RP model.

With reference to Figure 4.5 one may note that the curves are not drawn to scale, for the sake of clarity. The simplified graphs should be constructed in such a way that the

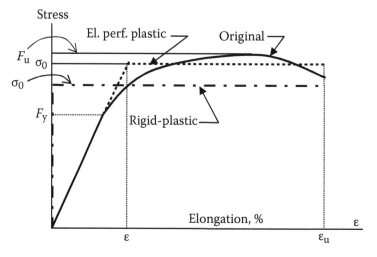

FIGURE 4.5   Original stress–strain curve and its two simple approximations.

---

* When reading technical papers, one of the items causing a frequent confusion is the word *stress*. Too often it is not known whether it means *the applied stress* or the *material strength*. Using the term *flow strength* makes it, beyond doubt, as the material property.

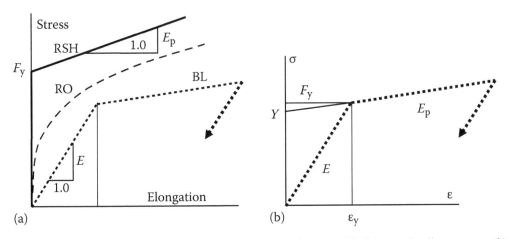

FIGURE 4.6   The remaining simple material models (a) and BL model with an unloading segment (b).

area under the actual curve is preserved. The yield strength is denoted by F_y in general and by σ_0 for materials with a constant yield stress level. There are two levels of σ_0, one for the elastic, perfectly plastic and one for a RP model.* When straining is large enough to reach a multiple of ε_y, in most cases those two levels coincide, or nearly so.

Several of the remaining models are shown in Figure 4.6. They are: BL, rigid, strain hardening (RSH), and Ramberg–Osgood (RO). Only for BL some details are shown: Young's modulus E, plastic modulus E_p, and unloading portion of the curve, which for steels has the same slope as the initial modulus E. One of the models that is not illustrated, is the power law (PL) material. When graphed, it is quite similar to RO. (When using RO form, $n > 1$. For PL, $n < 1$.) The equations, for the tensile stress are as follows:

$$\text{LE: } \sigma = E\varepsilon \tag{4.3a}$$

$$\text{RP: } \sigma = \sigma_0 \tag{4.3b}$$

$$\text{EPP: } \sigma = E\varepsilon \quad \text{for } \varepsilon \leq \varepsilon_y \quad \text{and}$$

$$\sigma = F_y \quad \text{for } \varepsilon > \varepsilon_y \tag{4.3c}$$

$$\text{BL: } \sigma = E\varepsilon \quad \text{for } \varepsilon \leq \varepsilon_y \quad \text{and}$$

$$\sigma - F_y = E_p(\varepsilon - \varepsilon_y) \quad \text{for } \varepsilon > \varepsilon_y \tag{4.3d}$$

$$\text{or} \quad \sigma = Y + E_p\varepsilon \quad \text{for } \varepsilon > \varepsilon_y$$

---

* There is a point called *proportionality limit*, at which the σ–ε begins to diverge from the straight line. Our simplifications ignore this fact and we act as if the curve was linear all the way to the yield point.

$$\text{RSH: } \sigma = F_y + E_p \varepsilon \tag{4.3e}$$

$$\text{RO: } \varepsilon = \frac{\sigma}{E} + \left(\frac{\sigma}{E_n}\right)^n \tag{4.3f}$$

$$\text{PL: } \sigma = B(C + \varepsilon)^n \tag{4.3g}$$

In the case of RO only it was more convenient to express strain in terms of stress. The model is very useful for steel working in high temperatures. One can often see a single-term stress–strain, relationships, written in a way similar to the second term of RO or PL above. The BL model is used quite frequently; two alternative expressions were given for the plastic range.

The area under the stress–strain curve is a measure of material ability to absorb strain energy and will be referred to as *material toughness*.* More precisely, when the strain reaches its peak ε_u, the toughness Ψ is also attained. This is equivalent to the following definition of toughness:

$$\Psi = \int_0^{\varepsilon_u} \sigma \, d\varepsilon \quad \text{(direct stress)} \tag{4.4a}$$

$$\Psi = \int_0^{\gamma_u} \tau \, d\gamma \quad \text{(shear)} \tag{4.4b}$$

For an RP material, where yielding takes place at σ_0 or at τ_0, the above expressions become very simple; $\Psi = \sigma_0 \varepsilon_u$ and $\Psi = \tau_0 \gamma_u$, where ε_u and γ_u are the ultimate strain and the distortion angle, respectively. In the listing of cases to follow, an expression for the strain energy Π is given as a function of the maximum strain reached, ε_m. When this is replaced with ultimate material strain ε_u, the equation for Ψ results.

To complete this description of ductile materials, one should include the stress–strain curve of a mild steel, a commonly used material with F_y between 200 and 350 MPa and illustrated in Figure 4.7. Apart from the plateau at the yield strength level F_y, the material also displays the upper yield point F_y'. The proportionality limit F_p is the stress where the characteristics shows a deviation from a straight line, as mentioned before.† The ultimate strain ε_u is the quantity specified by the manufacturer. There are many references in literature to the rupture strain, suggesting that this may refer to the strain associated with failure, as shown in the figure, but it appears that in many cases they really refer to ε_u. The falling part of the curve, between ε_u and ε_r is associated with the decrease in section area of the sample past the peak point.

---

* This is not to be confused with the toughness definition used in fracture mechanics.
† The knowledge of F_p is needed only in exceptional circumstances.

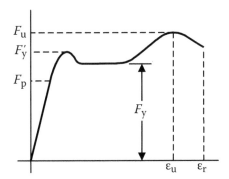

FIGURE 4.7   Stress–strain curve of a mild steel.

The discussion so far related to the results and simplifications of a tensile test. If a short sample is subjected to compression, the small-strain response is similar to that in tension. For larger strains there is swelling of the sample and other secondary effects. Presenting both tensile and the compressive stress in terms of logarithmic strain makes both curves more similar. When deformation is advanced, circumferential cracks appear in a compressed sample, which is not of great relevance with regard to compressive properties. It is not possible to induce a compressive failure, analogous to that in tension.

A selection of stress–strain curves was presented above, but it does mean that they are readily available. Instead, they must be constructed on the basis of experimental data. Typically, when working with metals in ambient temperature, one has only three numbers characterizing the material strength: F_y, F_u, and ε_u. It is natural to construct a BL characteristic based on these data, as well as E and v, unless we have a reason to expect a different shape. Also, if the material is simplified to resemble EPP, for example, then the yield point must be so selected that toughness Ψ is preserved.

SHEAR AND TWISTING OF SHAFTS

Material testing is usually done on tensile samples, but the same material often works in shear, so the question of material constants arises. As it is shown later, in Chapter 6, for metallic materials the yielding and the ultimate strength in pure shear, τ_y and τ_u, are related to the tensile properties as follows:

$$\tau_y = \frac{F_y}{\sqrt{3}} \tag{4.5a}$$

$$\tau_u = \frac{F_u}{\sqrt{3}} \tag{4.5b}$$

where the first equation results from Huber–Mises theory and the second is a customary assumption related to the ultimate strength.* A determination of stress–strain plot

---

* The reader should keep in mind that the above equations are based on small deflection theory.

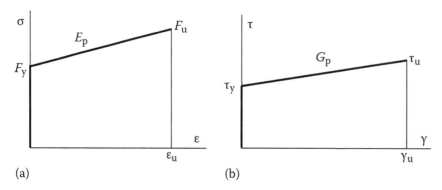

FIGURE 4.8   Direct stress–strain (a) and the corresponding shear stress vs. shear angle (b) for a RSH material.

parameters in shear is presented for a BL material as Case 4.8. When Equations 4.5a and b are augmented by putting a limit on shear strain, namely $\gamma_u = \sqrt{3}\varepsilon_u$ the relation between the tensile and the shear characteristics becomes complete. Its example for a RSH material is shown in Figure 4.8 While the above limit on shear strain has no theoretical justification, it gives a simple relationship between the moduli in plastic range, $G_p = E_p/3$, which is consistent with $\nu = 0.5$, generally assumed for this range. This ratio of G_p and E_p holds strictly true for the RHS material only, but under well-developed plasticity may be a good approximation for many real materials.

Twisting of shafts is detailed in Case 4.29. In general, this mode of deformation is more complex than tension, because of a nonuniformity in stress distribution over a cross section. The process of a gradual yielding of a solid, circular shaft begins in the outside fibers and continues until the near-center fibers reach yielding, in an asymptotic way. If an EPP model is used, the uniform shear stress distribution is a limiting condition. In practice it is approached only after the angle of twist is at least an order of magnitude larger than twisting associated with the initiation of yielding. In Case 4.29 the approximation is based on ignoring the angle of twist associated with the stress redistribution. Owing to this, a shaft made of a RSH material as in Figure 4.6a, can have its characteristic represented by a similar, trapezoidal shape. A thin-wall shaft is an exception. Its characteristic is simply a scaled material characteristic, as no redistribution takes place.

PROPERTIES OF CONCRETE

Concrete has a pronounced nonlinearity in its stress–strain curve, as Figure 4.9 illustrates. The nominal Young's modulus is taken as a secant modulus E_c, which may be related to its compressive strength as follows:

$$E_c = 5000\sqrt{F'_{cm}} \tag{4.6}$$

where $F'_c \sim$ MPa is the nominal compressive strength and $F'_{cm} \approx 1.1\,F'_c$ for most of industrial strength grades.* The σ–ε curve has a continuously decreasing slope with the maximum

---

* F'_c is the average 28 days compressive strength.

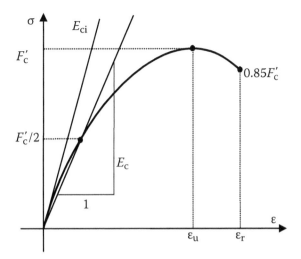

FIGURE 4.9   Typical, static stress–strain for concrete showing the initial E_{ci} and nominal E_c moduli.

tangent of E_{ci} at the origin. The strain at failure ε_u is usually assumed to be 0.002 for design purposes. The following approximation is also useful for cement-based materials as well as hard rocks:

$$E_c = 20 + 0.25F_c \ (E \sim \text{GPa} \quad \text{if } F_c \sim \text{MPa}) \tag{4.7}$$

A more detailed presentation of concrete stress–strain properties is given in Case 4.30. The strain-rate effect is also included there, but the situation is not quite clear with this variable. For a broader discussion of the subject the reader should refer to Chapter 6.

The above results describe the behavior of a cylindrical concrete sample under axial compression. There is also another test, called *triaxial compression*, which illustrates an important concrete property, namely sensitivity to confining pressure. The sample is placed in a device, which makes it possible to apply a uniform (hydrostatic) pressure to the side surface, in addition to axial compression. The larger the confining pressure, the larger the apparent axial strength. One of the simplest relationships was proposed by Reinhardt [71]:

$$F_{cc} = F_c' + 3p_c \tag{4.8}$$

where
F_c' is the nominal strength
F_{cc} is the confined strength under lateral hydrostatic pressure p_c

A similar relationship can be written for a hard rock.

When describing concrete, the attention is usually focused on its compressive properties. This material is weak in tension, as it can withstand only the stress level of

$$F_t = 0.6\sqrt{F_c'} \tag{4.9}$$

where F_c' is the nominal compressive strength, with both quantities in MPa.* Yet, the tensile strength is quite important, as it dictates the location of the first crack as well as subsequent breakup.

STRESS AND STRAIN IN A LARGE DEFORMATION RANGE

Consider a stretched bar in Figure 4.10. When extension is as large as the scale of the figure indicates, there is a need to distinguish between *engineering* strain, ε and the *natural* (or *logarithmic*) strain[†] $\bar{\varepsilon}$. The first relates to the change of length with respect to the initial length L and the second uses the current length l:

$$\varepsilon = \frac{l-L}{L} = \frac{u}{L} \tag{4.10a}$$

$$d\bar{\varepsilon} = \frac{dl}{l} \tag{4.10b}$$

The total natural strain $\bar{\varepsilon}$ is the sum of elementary increments:

$$\bar{\varepsilon} = \int_L^l \frac{dl}{l} = \ln\frac{l}{L} = \ln\left(1+\frac{u}{L}\right) \tag{4.11a}$$

and

$$\bar{\varepsilon} = \ln(1+\varepsilon) \tag{4.11b}$$

In this book, mostly, the strains are presumed small and there is no need to distinguish between $\bar{\varepsilon}$ and ε. For this reason strain is designated usually as ε, because $\bar{\varepsilon} \approx \varepsilon$, as long as

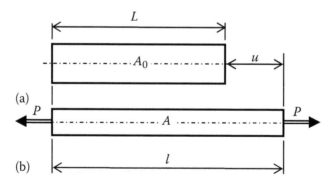

FIGURE 4.10   An axial bar, unstretched in (a) and stretched with a force P in (b).

---

* F_t as given above is the flexural strength, a typical application. Pure tension on concrete members is quite rare.
[†] There also is a concept of "true" strain, $(l-L)/l$. The same term is applied by some authors to the logarithmic strain defined above. Once we agree that specification of strain is only a matter of convention, then it is obvious the term "true strain" is a misnomer.

ε does not exceed a few percent.* For example, if $\varepsilon = 0.20$ then $\bar{\varepsilon} = 0.1823$, already a noticeable difference.

If deflections are as large as suggested by Figure 4.10, this means, for most materials, that they are well outside of the elastic range. It is usual to assume, in such a circumstance, that the volume of material remains invariant, i.e., $A_0 L = Al$, where A and l designate the current area and length. As stretching progresses, the section area decreases according to

$$A = \frac{L}{l} A_0 = \frac{A_0}{\exp \bar{\varepsilon}} \qquad (4.12a)$$

or

$$A = \frac{A_0}{1 + \varepsilon} \qquad (4.12b)$$

The tensile test results are usually prepared by plotting engineering stress, or stretching force divided by the original sample area, $\sigma = P/A_0$, vs. ε. If a *true stress* S is defined by dividing that force by the current area A, then

$$S = \frac{P}{A} = \frac{P}{A_0}(1 + \varepsilon) \qquad (4.13a)$$

or

$$S = \sigma(1 + \varepsilon) \qquad (4.13b)$$

A conventional σ–ε curve, which for metals is typically convex upward will become less convex (or may even become concave) when converted to the true strain. This is quite an important point when large distortions are involved.

There are several definitions of strain available. Which of them is best depends on the purpose at hand. If the objective of the investigation is to determine the load–deflection curve of a tensile member, then the engineering stress and strain are the most appropriate ones, because the basic tensile test uses these variables. There is nothing wrong, for example, with specifying true stress vs. engineering strain. While the true stress is a fundamental quantity, the expression of strain is a matter of convention.

One has to remember that the above distinctions have little relevance for typical design work. The metallic materials used in aerospace, for example, have maximum elongation typically not exceeding 15%, and some below 10%. The composites exhibit even smaller elongations. When considering unusual impact conditions, the maximum strain allowed will be, at best, only one-half of these figures, so that otherwise nonlinear investigations will be still in the small deflection range. Concrete can withstand only very small strains without cracking.

---

* Refer to approximations in Appendix B.

SOME RESPONSE FEATURES

The NL elements shown in tabulations are named after material characteristics of similar shapes: RP, EPP, BL, RSH, RO, and PL. There are two basic solutions provided for each of them: initial velocity response and step load response. The first solution of this type is always available, although the algebraic difficulties may exist. The solutions for the second have some limitations.

The maximum force in the structural element and maximum displacement u_m induced by a step load $W = W_0 H(t)$ can be found by applying the energy balance equation. We know from Chapter 3 that for a linear system under this type of loading the dynamic factor is 2.0, which means

$$u_m = 2u_{st} = 2W_0/k_1 \qquad (4.14)$$

This will also be true in case of EPP and BL as long as $W_0 < R_0/2$. When u_m is calculated according to this equation and is larger than u_y, this tells us that the deflection is in the plastic range. Equating the strain energy with the work of the external force W_0, we get the maximum deflection u_m.

A deflection cannot be determined in case of the RP system when the step load is applied. No deflection occurs for $W_0 < R_0$ and when $W_0 > R_0$ the displacement becomes unbounded. The latter merely means that the system is incapable of resisting the sustained applied load of this magnitude.

The plots of load–deflection characteristics shown in tabulations are patterned after stress–strain curves of various material models. They are more general relationships, but it is a simple matter to obtain from any of those the stress–strain curve for a particular material type. For any relationship characterized by two straight lines, one can do the following transformation:

$$R \rightarrow A\sigma, \quad u \rightarrow \varepsilon L, \quad k_1 \rightarrow EA/L, \quad k_2 \rightarrow E_p A/L \qquad (4.15)$$

resulting in a stress–strain curve of a corresponding shape. The value of "simple oscillator" type cases that are listed is not only in the material responses; often the entire structural elements respond according to such simple schemes. Consider two such approximations: Case 4.2 for the RP material and Case 4.5 for the EPP material. The peak displacements induced by the initial velocity are, respectively,

$$u_m = \frac{M v_0^2}{2R_0} \qquad (4.16a)$$

and

$$u_m = \frac{1}{2}\left(\frac{R_0}{k_1} + \frac{M v_0^2}{R_0}\right) \qquad (4.16b)$$

and the difference between the two is the $R_0/(2k_1)$ term, related to the elastic properties of the system. This is also the oscillatory term for the EPP, while the RP system determines only the permanent displacements. This difference between the two is the underestimate of the peak deflection if the RP material is used in place of EPP. The analogous difference in calculated response exists if a rectangular impulse of a prescribed force P_0 is applied.

DURATION OF FORWARD MOTION WITH THE INITIAL VELOCITY PRESCRIBED

Even with marked nonlinearities involved the forward motion of an impacted mass may resemble one quarter of harmonic period, especially when the initial velocity is prescribed. A good approximation may therefore be obtained by assuming that displacement is a harmonic function of time, per Equation 3.5 for an undamped oscillator:

$$u = u_m \sin \omega t \tag{4.17}$$

with the peak displacement obtained for $\omega t = \pi/2$, or the quarter of the natural period. The resisting function $R(u)$ can be of any shape, but the smoother it is, the better approximation is obtained. The equation of motion follows the general formula 4.2:

$$M\ddot{u} + R(u) = 0 \tag{4.18}$$

According to one variant of the Ritz method, described by Case 4.7, a pseudofrequency ω is calculated and then the loading phase duration is found from

$$t_m = \frac{\tau}{4} = \frac{\pi}{2\omega} \tag{4.19}$$

When the impulse S associated with the force R is prescribed, another possibility for approximating t_m exists. If deflection is a sinusoidal time function, so is the spring force or R in this event. It is easy to check, that for this one quarter of the period the impulse can be expressed as

$$S = \frac{2}{\pi} R_m t_m \tag{4.20a}$$

which gives

$$t_m = \frac{\pi S}{2 R_m} \tag{4.20b}$$

This indicates that if the impulse is established by kinetic means and peak reaction is known, the duration can be readily calculated.

One can also derive a simpler approximation from more elementary principles. Suppose that the peak deflection u_m attained during motion is known and so is the accompanying peak reaction R_m. From basic mechanics, the deflection can be written as $u_m = a_{av} t^2/2$, where a_{av} is the average acceleration. That average, however, relates to time domain, rather than to distance. For this reason the following expression for that average acceleration is adopted:

$$a_{av} = \frac{1}{2}\left(\frac{0.5R_0}{M} + \frac{1.5R_m}{M}\right) \qquad (4.21)$$

where the initial reaction R_0 is often nil. The associated time is $t_m = \sqrt{2u_m/a_{av}}$ or

$$t_m = \sqrt{\frac{8Mu_m}{R_0 + 3R_m}} \qquad (4.22)$$

Which of the approximate formulas is to be used depends on the problem and the accuracy desired. Example 4.1 illustrates several simple cases.

Duration of the loading phase of motion can also be estimated by using an *equivalent linear oscillator* concept. Consider a mass M attached to a nonlinear elastic spring and vibrating with amplitude A. Imagine an associated linear system with a spring of stiffness k^*, so chosen in which the maximum displacement $u_m = A$, the strain energy is the same in both springs (Figure 4.11). The natural period of that nonlinear system, which is characterized by $R(u)$, can now be approximated by

$$\tau = 2\pi\sqrt{\frac{M}{k^*}} \qquad (4.23)$$

The equivalent linear stiffness k^* is valid only for a given extreme displacement $u_m = A$. The deviation of curve $R(u)$ from a straight line must be small to make Equation 4.23 reasonably accurate.

It is not relevant whether the nonlinearity of system characteristic comes from material or geometric causes.

Keeping in mind that the duration is now equal to one quarter of the natural period, one can write

$$\omega^2 = \frac{R_m}{Mu_m} \quad \text{or} \quad t_m = \frac{\tau}{4} = \frac{\pi}{2}\sqrt{\frac{Mu_m}{R_m}} \qquad (4.24)$$

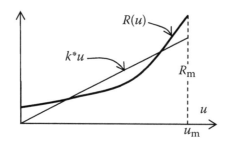

FIGURE 4.11   The characteristic $R(u)$ of the actual object and its equivalent linear oscillator with stiffness k^*.

This is evidently true for a linear element, as $R_m/u_m = k$, but for other systems it is merely a secant approximation of the characteristic. Yet, it can be a very effective formula if the ratio R_m/u_m, instead of being taken literally, is adjusted to correspond closer to k^*, as per Figure 4.11. Also, if the loading and unloading curves are the same, the validity of Equation 4.24 extends to cover the entire period. One should also keep in mind that if a step load is applied, then the loading phase lasts $\tau/2$.

CHARACTERISTICS OF A POLYTROPIC GAS

Consider a certain amount of gas with the initial volume V_1 and pressure p_1. When the volume decreases to V_2, pressure increases to p_2, but the product pV^n remains invariant:

$$p_1 V_1^n = p_2 V_2^n \tag{4.25}$$

The constant n is known as a polytropic exponent and is one of the characteristics of a gas. The energy content of the volume of gas is

$$Q = \frac{pV}{n-1} \tag{4.26}$$

The work performed on gas when compressing it from state 1 to state 2 is equal to the difference of energy contents and can be written using Equations 4.25 and 4.26:

$$\mathcal{L} = \frac{p_1 V_1}{n-1}\left(\left(\frac{V_1}{V_2}\right)^{n-1} - 1\right) \tag{4.27a}$$

or

$$\mathcal{L} = \frac{p_1 V_1}{n-1}\left(\left(\frac{p_2}{p_1}\right)^{(n-1)/n} - 1\right) \tag{4.27b}$$

The relation for mass density ρ is

$$\rho_2 = \rho_1\left(\frac{p_2}{p_1}\right)^{1/n} \tag{4.28}$$

For air, usually $n = 1.4$. Under normal conditions the atmospheric values are

$$\rho_0 = 1.225 \times 10^{-6} \text{ g/mm}^3; \quad p_0 = 0.1013 \text{ MPa}$$

The energy q per unit mass can be calculated from Equation 4.26 and, when related to some initial state 0, it is

$$q_0 = \frac{p_0}{\rho_0(n-1)} \tag{4.29}$$

The energy content per unit volume, e, related to the initial state is

$$e_0 = \frac{Q_0}{V_0} = q_0 \rho_0 \qquad (4.30a)$$

or

$$e_0 = \frac{p_0}{n-1} \qquad (4.30b)$$

For all those relationships, we have $n > 1$.

When gas is enclosed in a rigid container closed by a movable piston, as in the sketch for Case 4.9, then it acts as a uniaxial, nonlinear spring. For this case, the polytropic relation, in terms of pressure or the restraining force $R = Ap$ is

$$p_1 x_1^n = p_2 x_2^n \qquad (4.31a)$$

or

$$R_1 x_1^n = R_2 x_2^n \qquad (4.31b)$$

In the illustrated case going from state 1 to state 2 means expansion, or pressure drop. The work performed by gas when expanding from state 1 to state 2 is equal to the difference of energy contents and can be written using Equations 4.27 and 4.31:

$$\mathcal{L} = \frac{Ap_1 x_1}{n-1}\left[1 - \left(\frac{x_1}{x_2}\right)^{(n-1)}\right] \qquad (4.32)$$

To show the relation between a restraining force R and displacement u, use Equation 4.31b, choosing x_1 as the initial, or reference position. Then set $x_2 = x_1 + u$, obtaining

$$R = R_1 \left(\frac{x_1}{x_1 + u}\right)^n \qquad (4.33)$$

In this formula positive u decreases the restraining force R, exerted by gas.

LARGE DEFLECTIONS OF SIMPLE ELEMENTS

According to what was said in "Stress and strain in a large deformation range," there is a need to use true stress S and the logarithmic strain $\bar{\varepsilon}$ when large distortions are involved. However, to avoid unfamiliar notation, these will be designated by σ and ε, while their true meaning is kept in mind.

When *a bar* in Figure 4.10 is stretched, both strain and stress increase. There is the following gradient of the force $P = A\sigma$ with respect to strain:

$$\frac{dP}{d\varepsilon} = A\frac{d\sigma}{d\varepsilon} + \sigma\frac{dA}{d\varepsilon} = A\left(\frac{d\sigma}{d\varepsilon} - \sigma\right) \tag{4.34}$$

because one can conclude from Equations 4.12 and 4.13 that $dA/d\varepsilon = -A$. The maximum condition for P, $dP/d\varepsilon = 0$, becomes rather simple looking:

$$\frac{d\sigma}{d\varepsilon} = \sigma \tag{4.35}$$

The load carried by the bar is a function of current ε:

$$P(\varepsilon) = A\sigma = \frac{A_0\sigma}{\exp(\varepsilon)} \tag{4.36}$$

on the basis of Equation 4.12a. A specific material is needed to express $\sigma = \sigma(\varepsilon)$.

The growth of P, as elongation increases, is dictated by the slope of the σ–ε curve, so a simplified characteristic needs to retain some realism in this regard. A BL curve, which is quite adequate in many situations, may be used here only when a gentle change of slope of the original σ–ε curve is present. For a BL material: $\sigma = Y + E_p\varepsilon$ for $\varepsilon > \varepsilon_y$, consequently,

$$P = \frac{A_0(Y + E_p\varepsilon)}{\exp\varepsilon} \tag{4.37}$$

From Equation 4.35 the instability strain, stress, and peak force are

$$\varepsilon_{cr} = 1 - \frac{Y}{E_p}; \tag{4.38a}$$

$$\sigma_{cr} = E_p; \tag{4.38b}$$

$$P_m = \frac{A_0 E_p}{\exp\varepsilon_{cr}} \tag{4.38c}$$

For the PL material:

$$\sigma = B(C + \varepsilon)^n \tag{4.39a}$$

then

$$\frac{d\sigma}{d\varepsilon} = Bn(C + \varepsilon)^{n-1} \tag{4.39b}$$

The maximum condition becomes (from Equation 4.35)

$$\varepsilon_{cr} = n - C \tag{4.40a}$$

with

$$\sigma_{cr} = Bn^n \tag{4.40b}$$

From Equation 4.36

$$P_m = B(C + \varepsilon_{cr})^n \frac{A_0}{\exp\varepsilon_{cr}} = \frac{A_0 Bn^n}{\exp(n - C)} \tag{4.41}$$

This effect of the load capacity attaining a peak prior to material fracturing is referred to as a *tensile instability*. The term comes from two observations. The first is that past the critical point a further increase in deflection takes place with a gradually decreasing resisting force. Also, as soon as the critical tension is exceeded, there is a tendency in metals for necking to appear, which signals proximity of tensile failure.

The same method holds for a *circular ring* (R, A_0), shown in Case 4.22. This time the distributed load is w (N/m), for which the maximum is sought. It can be expressed as

$$w(\varepsilon) = \frac{P}{r} = \frac{A_0 \sigma}{R \exp(2\varepsilon)} \tag{4.42}$$

where the last inequality was written having Equation 4.12a in mind. Differentiating this with respect to ε (with σ as a function of ε) and setting the derivative equal to zero, the extremum condition is obtained as

$$\frac{d\sigma}{d\varepsilon} = 2\sigma \tag{4.43}$$

For a *spherical shell* (R, h_0), as described in Case 4.24, the loading is the internal pressure, which can be expressed as

$$p(\varepsilon) = \frac{2\sigma h}{r} = \frac{2h_0 \sigma}{R \exp(3\varepsilon)} \tag{4.44}$$

and the peak pressure condition is

$$\frac{d\sigma}{d\varepsilon} = 3\sigma \tag{4.45}$$

Sometimes the characteristic of material is such that the tensile instability cannot be attained. One can easily check that for $Y > E_p$, the stretching force is monotonically decreasing. The maximum value of P is therefore at the yield point, $P_m \approx P_y = A_0 F_y$. Equation 4.36 is used, as before, to calculate the current load P.

THICK RINGS AND SHELLS OF EPP MATERIAL

The method presented before is applicable to only moderately thick rings, say $b/a \leq 1.2$. When the ring is thicker, it is necessary to distinguish between the condition at the inner radius, where apart from the hoop stress there also is the applied pressure, and the outer radius, which is pressure-free. The analysis is usually carried out for the elastic state that ends when the limiting load w_y, associated with the onset of plasticity, is reached. For any larger loading there is a transitional, elastoplastic stage when the inner part of the ring is in plastic condition and the outer part is still elastic, as shown in Figure 4.12. This state ends when the load attains its ultimate value w_u coinciding with the thickness of ring in plastic condition.

According to Kaliszky [45], those critical load values, based on Huber–Mises theory, are

$$w_y = \left[1 - \left(\frac{a}{b} \right)^2 \right] \frac{B\sigma_0}{\sqrt{3}} \qquad (4.46a)$$

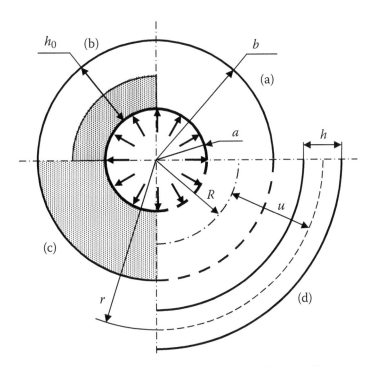

FIGURE 4.12 Stages of expansion due to internal pressure, as illustrated by four quadrants: Elastic (a), elastoplastic (b), totally plastic (c) with large deflections associated with it (d). (For most engineering materials displacements related to the fully plastic stage are very small.)

$$w_u = \frac{2B\sigma_0}{\sqrt{3}} \ln\left(\frac{b}{a}\right) \tag{4.46b}$$

$$u_u = \frac{\sqrt{3}\sigma_0}{2E} a \left(\frac{b}{a}\right)^2 \tag{4.47}$$

In the above u_u stands for the displacement at the inner radius, at the ultimate load w_u (fully plastic ring with $v = 0.5$) and B is the ring width. Using the expression

$$\frac{b}{a} = \frac{r+h/2}{r-h/2} = \frac{1+h/(2r)}{1-h/(2r)} \tag{4.48}$$

one can find, after transforming the above expression per Appendix B, that Equations 4.46a and b have the same form, in the limit, as the $w = B\sigma_0 h_0/R$ for thin rings, except for the effect of multiaxial stress coefficient, $2/\sqrt{3}$.

The conventional analysis usually ends here, because w_u is the largest load the ring can resist. To determine the strength at larger deflections, the assumption of the constant material volume is used as before. If the width B is invariant (plane strain), then the condition is $h_0 R = hr$. First of all, note from Equation 4.46b, that as R increases, h must decrease and so will the ratio b/a. This means the load needed to keep the ring at the deformed position decreases with increase in the mean radius. The other assumption is that Equation 4.46b holds not only for the initial, but also for the *current* values of the outer and inner radii.

If, in Equation 4.48, written for the current geometry, one designates $x = h/(2r)$, and if $x \ll 1$, which is usual, then $\ln(b/a) \approx 2x$ (in accordance with Appendix B) and the approximate expression for the current load is

$$w = \frac{2B\sigma_0}{\sqrt{3}} \ln\frac{1+\psi}{1-\psi} \approx \frac{2B\sigma_0}{\sqrt{3}} \frac{h_0 R}{r^2} \tag{4.49}$$

where volume invariance was used to replace hr with $h_0 R$, the latter product being a constant. If, for example, $b/a = 2$, which makes rather a thick shell, then using Equation 4.49 instead of Equation 4.46b with current values results in a 3.8% underestimate of w, which is often acceptable.

The equations developed for a ring also work for a cylindrical shell, if $B = 1$ and w is replaced by pressure p. A spherical shell needs new derivations, but the steps are exactly the same as before. The volume preservation condition is now $h_0 R^2 = hr^2$ (This is approximate, unlike the condition for a ring above). The formulas for the spherical shell are laid out in Case 4.20. One can notice that the decrease of load-carrying capacity is much steeper for a spherical shell than for a corresponding cylindrical shell, as evidenced by the higher power of the mean radius r in the denominator of the former.

Let us now determine the work needed to expand rings and shells of EPP material. The general expression for an incremental work performed by the load w (N/m) applied to the inside radius a of a ring is $(2\pi a)w(da)$, where $w=w(a)$ is a static load corresponding to the current radius a. The work performed between radii a_1 and a_2 is therefore

$$\mathcal{L} = 2\pi \int_{a_1}^{a_2} wa\,da \qquad (4.50)$$

Our main interest here is to determine this work in the plastic range, beginning with the point where fully plastic condition is attained. Using a three-point Simpson's integration formula, one gets, ignoring small elastic deflection

$$\mathcal{L}_{pl} = \frac{\pi}{3}(a_2 - a_0)(w_0 a_0 + 4w_1 a_1 + w_2 a_2) \qquad (4.51)$$

where indices 0, 1, and 2 refer to the initial, middle, and final expansion point with respect to the inner radius a. The initial point 0 coincides with the ultimate load w_u. For materials like steel, the displacements up to the point defined by w_u are quite small and so is the expansion work. We can therefore use an approximate expression for the elastoplastic range:

$$\mathcal{L}_{ep} \approx 2\pi a w_u u_u \qquad (4.52)$$

and the total expansion work is $\mathcal{L} = \mathcal{L}_{pl} + \mathcal{L}_{ep}$. The equations are applicable for a cylindrical shell, if $B=1$ and w is replaced by pressure p. For a spherical shell with internal pressure the incremental work is $(4\pi a^2)p(da)$, so the resulting expressions differ, as shown in Case 4.20.

For moderately thick rings, some simplifications are possible. By cutting a ring diametrically in half and showing the balancing forces, one finds $wa = A\sigma_0$. Using the volume invariance and noting that

$$da = \frac{da}{dr}dr = \left(1 + \frac{h}{2r}\right)dr \qquad (4.53)$$

Equation 4.50 can be written in terms of r rather than a. The simplification now is to assume $1 + h/(2r)$ to be constant, inserting the average values of h and r during the expansion process. The end result for the plastic work of expanding up to the current radius r is

$$\mathcal{L}_{pl} = 2\pi R(h_0\sigma_0)\left(1 + \frac{h_{av}}{2r_{av}}\right)\ln(r/R) \qquad (4.54)$$

Thin rings and shells are easier from the analytical viewpoint. For those, the hoop stress is assumed constant and no influence of radial stress is accounted for in the investigation of the yielding condition.

The distinction between the thickness grades is somewhat arbitrary. As our interests in this section revolve mostly around large deflections in a plastic state, it is appropriate to set the uniformity of the hoop stress in the fully plastic condition as a parameter. In this case a ring has a ratio of inner face to outer face hoop stress equal to $\eta = 1 - \ln(b/a)$. If one sets $b/a = 1.05$ (which gives $h/r \approx 0.05$) as a maximum radius ratio for a *thin* ring, then $\eta = 0.951$, quite a reasonable uniformity. For a *moderately thick* ring, $b/a = 1.2$ (equivalent to $h/r \approx 0.2$) gives $\eta = 0.818$. Anything with a larger b/a ratio will be called a *thick* ring. The same criteria apply to cylindrical shells.

A BODY OF A VARIABLE MASS

When a body mass M is constant, the usual way of expressing the Newton's law is

$$M\frac{dv}{dt} = P(t) \qquad (4.55)$$

If the mass M varies with time, i.e., $M = M(t)$, then a more general form is applicable:

$$\frac{d}{dt}(Mv) = P(t) \qquad (4.56)$$

Consider a special form of this equation, where $P = P_0$ is constant. The integration of the above, for the case of zero initial velocity, gives

$$Mv = P_0 t \qquad (4.57)$$

where the product of two functions of time, M and v, is proportional to time. For a pulse lasting t_0, Mv grows to $P_0 t_0$ and remains constant thereafter. The usefulness of this relation comes from the fact that at some instance either M or v may become known, which makes it possible to find the other one from Equation 4.57.

TABULATION OF CASES

CASE 4.1   SIMPLE OSCILLATOR WITH LINEAR ELASTIC (LE) SPRING

LE characteristic: $R = k_1 u$

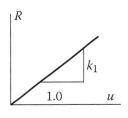

Strain energy: $\Pi = \dfrac{1}{2} k_1 u_m^2 \text{ or } \Pi = \dfrac{1}{2} \dfrac{R_m^2}{k_1}$

(a) Initial velocity $v_0 = \dfrac{S_0}{M} = \dfrac{W_0 t_0}{M}$

$$u_m = v_0 \sqrt{\dfrac{M}{k_1}} = \dfrac{S_0}{\sqrt{Mk_1}} \; ; \quad t_m = \dfrac{\pi}{2} \sqrt{\dfrac{M}{k_1}}$$

(b) Step load $W = W_0 H(t)$:

$$u_m = \dfrac{2W_0}{k_1} = 2u_{st} \quad t_m = \pi \sqrt{\dfrac{M}{k_1}}$$

Stress–strain relations: $\sigma = E\varepsilon$

$$\Pi = \dfrac{1}{2} E \varepsilon_m^2 \quad \text{or} \quad \Pi = \dfrac{1}{2} \dfrac{\sigma_m^2}{E} \quad \text{(energy content in a unit cube)}$$

CASE 4.2   SIMPLE OSCILLATOR WITH RP SPRING

RP characteristic: $R = R_0$
Strain energy: $\Pi = R_0 u_m$

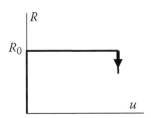

(a) Initial velocity $v_0 = \dfrac{S_0}{M} = \dfrac{W_0 t_0}{M}$; $W_0 > R_0$

$$u_m = \dfrac{Mv_0^2}{2R_0} = \dfrac{S_0^2}{2MR_0} \; ; \quad t_m = \dfrac{Mv_0}{R_0} \; ; \quad \text{peak displacement, at } t = t_m$$

(b) Rectangular pulse of W_0 during t_0:

$$v_m = \dfrac{W_0 - R_0}{M} t_0 ; \quad \text{peak velocity, at } t = t_0$$

$$u_m = \dfrac{1}{2} \dfrac{W_0}{M} t_0^2 \dfrac{W_0 - R_0}{R_0} \; ; \quad t_m = \dfrac{W_0 t_0}{R_0}$$

(c) Step load $W = W_0 H(t)$:

$$u_m = 0 \text{ for } W_0 < R_0 \quad \text{and} \quad u_m = \infty \text{ for } W_0 > R_0$$

(d) Exponential load $W = W_0(1 - t/t_0) \exp(-t/t_0)$

$$u_m = \frac{W_0 t_0^2}{M}\left[1 - \frac{R_0}{W_0}\left(1 + \ln\frac{W_0}{R_0} + \frac{1}{2}\ln^2\frac{W_0}{R_0}\right)\right]; \quad t_m = t_0 \ln\frac{W_0}{R_0}, \text{ Ref. [45]}.$$

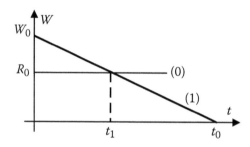

(e) Decreasing triangular pulse $W = W_0(1 - t/t_0)$

$$v_m = \frac{(W_0 - R_0)^2}{2MW_0}t_0; \quad \text{peak velocity, at } t = t_1$$

$$t_1 = \frac{W_0 - R_0}{W_0}t_0$$

$W_0/R_0 < 2$: $v = 0$ when $t = t_x = \dfrac{W_0 - R_0}{W_0}2t_0$

$$u_m = \frac{2}{3}\left(\frac{W_0 - R_0}{W_0}\right)^3\frac{W_0 t_0^2}{M}; \text{ valid for } W_0/R_0 < 2$$

$W_0/R_0 > 2$: $v_e = \left(\dfrac{W_0}{2} - R_0\right)\dfrac{t_0}{M}$; velocity at $t = t_0$

$$u_m = \left(W_0 - \frac{3}{2}R_0\right)\frac{t_0^2}{3M} + \frac{Mv_e^2}{2R_0}$$

Stress–strain relations: $\sigma = \sigma_0 = F_y$; $\Pi = \sigma_0 \varepsilon_m$ (energy content in a unit cube)

Note: This is also a description of the motion resisted by friction force R_0. In Case (e), when the ratio W_0/R_0 becomes large, but S_0 remains unchanged, u_m tends to Case (a) in the limit.

CASE 4.3   SIMPLE OSCILLATOR WITH BL SPRING

BL characteristic: $R = k_1 u$ for $u \leq u_y$ and

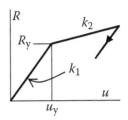

$$R - R_y = k_2(u - u_y) \quad \text{for } u > u_y$$

Strain energy: $\Pi = \dfrac{1}{2}k_2 u_m^2 + (k_1 - k_2)u_y u_m - \dfrac{1}{2}(k_1 - k_2)u_y^2$

(a) Initial velocity: $v_0 = \dfrac{S_0}{M} = \dfrac{W_0 t_0}{M}$

$$u_m = \left[(\rho^2 + \rho)u_y^2 + \frac{Mv_0^2}{k_2}\right]^{1/2} - \rho u_y \quad \text{with} \quad \rho = \frac{k_1}{k_2} - 1 \quad t_m \approx \sqrt{\frac{8Mu_m}{3R_m}}$$

(b) Step load $W = W_0 H(t)$:

$$u_m = \frac{W_0}{k_2} - \rho u_y + \left\{\left[\rho u_y - \frac{W_0}{k_2}\right]^2 + \rho u_y^2\right\}^{1/2} \qquad t_m \approx \sqrt{\frac{8Mu_m}{3R_m}}$$

Stress–strain relations: $\sigma = E\varepsilon$ for $\varepsilon \leq \varepsilon_y$ and $\sigma - F_y = E_p(\varepsilon - \varepsilon_y)$ for $\varepsilon > \varepsilon_y$

$$\Pi = \frac{1}{2}E_p\varepsilon_m^2 + (E - E_p)\varepsilon_y\varepsilon_m - \frac{1}{2}(E - E_p)\varepsilon_y^2 \quad \text{(energy content in a unit cube)}$$

CASE 4.4   SIMPLE OSCILLATOR WITH RSH SPRING

RSH characteristic: $R = R_y + k_2 u$

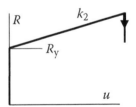

Strain energy: $\Pi = R_y u_m + \dfrac{1}{2}k_2 u_m^2$

(a) Initial velocity $v_0 = \dfrac{S_0}{M} = \dfrac{W_0 t_0}{M}$

$$u_m = \left[\frac{R_y^2}{k_2^2} + \frac{Mv_0^2}{k_2}\right]^{1/2} - \frac{R_y}{k_2} = \frac{R_y}{k_2}\left\{\left[1 + \frac{Mk_2 v_0^2}{R_y^2}\right]^{1/2} - 1\right\}; \quad t_m \approx \sqrt{\frac{8Mu_m}{R_y + 3R_m}}$$

(b) Step load $W = W_0 H(t)$:

$$u_m = 2\frac{W_0 - R_0}{k_2} \qquad t_m \approx \sqrt{\frac{8Mu_m}{R_0 + 3R_m}}$$

Stress–strain relations: $\sigma = F_y + E_p\varepsilon$

$\Pi = F_y\varepsilon_m + \dfrac{1}{2}E_p\varepsilon_m^2$ (energy content in a unit cube)

CASE 4.5   SIMPLE OSCILLATOR WITH EPP SPRING

EPP characteristic: $R = k_1 u$ for $u \le u_y$ and

$R = R_0$ for $u > u_y$

Strain energy: $\Pi = u_m R_0 - \dfrac{1}{2}\dfrac{R_0^2}{k_1}$

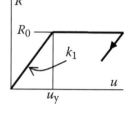

(a) Initial velocity $v_0 = \dfrac{S_0}{M} = \dfrac{W_0 t_0}{M}$

$$u_m = \frac{1}{2}\left(\frac{R_0}{k_1} + \frac{Mv_0^2}{R_0}\right) \quad t_m \approx \sqrt{\frac{8Mu_m}{3R_0}}$$

(b) Step load $W = W_0 H(t)$: $(R_0/2 < W_0 < R_0)$

$$\sin\omega t_y = \left\{1 - \left(\frac{u_y}{u_{st}} - 1\right)^2\right\}^{1/2} \; ; \text{ to find } t_y \text{ when } u_y \text{ is reached}$$

$$\omega t_m = \omega t_y + \frac{W_0}{R_0 - W_0}\sin\omega t_y$$

$$u_m = \frac{R_0^2}{2k_1(R_0 - W_0)} = \frac{u_y}{2(1 - W_0/R_0)} \quad t_m \approx \sqrt{\frac{8Mu_m}{3R_0}}$$

Stress–strain relations: $\sigma = E\varepsilon$ for $\varepsilon \le \varepsilon_y$ and $\sigma = \sigma_0$ for $\varepsilon > \varepsilon_y$

$\Pi = \varepsilon_m\sigma_0 - \dfrac{1}{2}\dfrac{\sigma_0^2}{E}$ (energy content in a unit cube)

Note: In case (a) when $u_m > 2u_y$, there are decaying oscillations about the permanent deflection equal to $u_m - u_y$.

CASE 4.6   SIMPLE OSCILLATOR WITH RO SPRING

RO characteristic: $u = \dfrac{R_m}{k_0} + \left(\dfrac{R_m}{k_n}\right)^n$

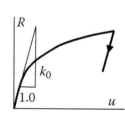

Strain energy: $\Pi = R_m\left(\dfrac{R_m}{2k_0} + \dfrac{n}{n+1}\left(\dfrac{R_m}{k_n}\right)^n\right)$

(a) Initial velocity $v_0 = \dfrac{S_0}{M} = \dfrac{W_0 t_0}{M}$

$$\frac{M v_0^2}{2} = R_{\mathrm{m}} \left(\frac{R_{\mathrm{m}}}{2k_0} + \frac{n}{n+1} \frac{R_{\mathrm{m}}^n}{k_n^n} \right) \quad t_{\mathrm{m}} \approx \sqrt{\frac{8 M u_{\mathrm{m}}}{3 R_{\mathrm{m}}}}$$

(b) Step load $W = W_0 H(t)$:

$$W_0 u_{\mathrm{m}} = R_{\mathrm{m}} \left(\frac{R_{\mathrm{m}}}{2k_0} + \frac{n}{n+1} \frac{R_{\mathrm{m}}^n}{k_n^n} \right) \quad t_{\mathrm{m}} \approx \sqrt{\frac{8 M u_{\mathrm{m}}}{3 R_{\mathrm{m}}}}$$

$$\text{Stress–strain relations: } \varepsilon = \frac{\sigma}{E} + \left(\frac{\sigma}{E_n} \right)^n ;$$

$$\Pi = \sigma_{\mathrm{m}} \left\{ \frac{\sigma_{\mathrm{m}}}{2E} + \frac{n}{n+1} \left(\frac{\sigma_{\mathrm{m}}}{E_n} \right)^n \right\} \quad \text{(energy content in a unit cube)}$$

CASE 4.7   NL OSCILLATOR WITH SMOOTHLY VARYING CHARACTERISTIC

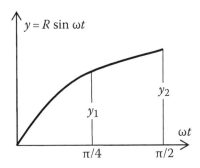

$y_1 = R_1 \sin(\pi/4); \; y_2 = R_2 \sin \pi/2 \equiv R_2$

$R_1 = R(0.7071 u_{\mathrm{m}})$; reaction at temporal midpoint

$R_2 = R(u_{\mathrm{m}})$; reaction at peak deflection

ω is pseudo-frequency (rad/s)

$$\omega^2 = \frac{4}{\pi M u_{\mathrm{m}}} \int_0^{\pi/2} R \cdot \sin \omega t \, \mathrm{d}(\omega t) \approx \frac{\pi}{12}(y_0 + 4y_1 + y_2)$$

$$\omega^2 = \frac{1}{3 M u_{\mathrm{m}}}(2.8284 R_1 + R_2); \text{ final expression}$$

$$t_{\mathrm{m}} = \frac{\tau}{4} = \frac{\pi}{2\omega}; \text{ duration of forward motion}$$

Note: The sketch shows a typical plot of an oscillator described. Simpson's integration formula is employed to find pseudo-frequency. The formulas are valid only for the loading phase of impact.

CASE 4.8   STRESS–STRAIN IN SHEAR WHEN THE TENSILE CURVE IS PRESCRIBED AS A BL MATERIAL

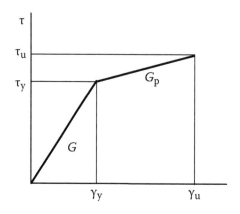

$$\tau = G\gamma \quad \text{for } \gamma < \gamma_y$$

$$\tau = \tau_y + G_p(\gamma - \gamma_y) \quad \text{for } \gamma > \gamma_y$$

$$\tau_y = \frac{F_y}{\sqrt{3}}; \quad \tau_u = \frac{F_u}{\sqrt{3}}$$

$$G = \frac{E}{2(1+\nu)}; \quad \gamma_y = \frac{\tau_y}{G}$$

$$\gamma_u = \sqrt{3}\varepsilon_u; \quad G_p \approx \frac{E_p}{E}G$$

$$\Psi_s = \int_0^{\gamma_u} \tau d\gamma; \text{ area under stress–strain curve or material toughness in shear}$$

For RSH material only:

$$G_p = E_p/3$$

$$\Psi_s = \frac{1}{2}(\tau_y + \tau_u)\gamma_u = \frac{1}{2}(F_y + F_u)\varepsilon_u; \text{ for RSH material}$$

For a rigid–plastic or RP material only: $G_p = 0$, $\gamma_y = 0$, $\tau_y \equiv \tau_0$

$$\Psi_s = \tau_0\gamma_u = \sigma_0\varepsilon_u; \text{ for RP material}$$

Note: The above is based on Huber–Mises theory along with the assumption of $\gamma_u = \sqrt{3}\varepsilon_u$. Material properties are usually provided in terms of a tensile test.

CASE 4.9   POLYTROPIC GAS SPRING

Balancing force $R = Ap$

$$R_1 x_1^n = R_2 x_2^n$$

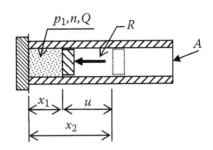

Energy content, for any x:

$$Q = \frac{Rx}{n-1}$$

The work performed on gas when compressing from
state 2 to state 1:

$$\mathcal{L} = \frac{R_1 x_1}{n-1}\left[1 - \left(\frac{x_1}{x_2}\right)^{(n-1)}\right]; \quad \text{also } R_2 = R_1\left(\frac{x_1}{x_1 + u}\right)^n$$

Note: It is assumed that piston motion is slow enough to be able to treat the volume of gas
in a quasistatic manner.

CASE 4.10   FRICTION-DAMPED OSCILLATOR, CONSTANT FRICTION RESISTANCE $F \sim N$

(a) Initial velocity $v_0 = \dfrac{S_0}{M} = \dfrac{W_0 t_0}{M}$

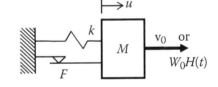

$$u_m = \left[\frac{F^2}{k^2} + \frac{Mv_0^2}{k}\right]^{1/2} - \frac{F}{k} \quad t_m \approx \sqrt{\frac{8Mu_m}{4F + 3ku_m}}$$

(b) Step load $W = W_0 H(t)$:

$$u_m = 2\frac{W_0 - F}{k} \quad t_m \approx \sqrt{\frac{8Mu_m}{4F + 3ku_m}}$$

CASE 4.11   OSCILLATOR WITH A RP SPRING, UNDER SUSTAINED STATIC LOAD W_{st} AND SUBJECTED TO A SHORT IMPULSE

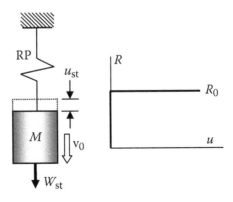

$W_{st} < R_0$

$S = W_0 t_0$, is a short impulse equivalent to the initial velocity $v_0 = S/M$.

Strain energy: $\Pi = R_0 u_m$

(a) Find u_m when S or kinetic energy is prescribed

$$u_m = \frac{M v_0^2}{2(R_0 - W_{st})} = \frac{S^2}{2M(R_0 - W_{st})}$$

(b) Find S or kinetic energy when u_m is prescribed

$$\frac{M v_0^2}{2} = \frac{S^2}{2M} = (R_0 - W_{st})u_m$$

Note: If v_0 is directed upwards, the unloading is elastic and so is the reloading to the original balanced position. The initial up-and-down movement is of no consequence and the result is the same as for v_0 directed down.

CASE 4.12   THIN CIRCULAR RING OF EPP MATERIAL, UNDER INTERNAL LOAD *w* (N/m), UNDERGOING LARGE STRAINS

$A_0 = B h_0$

σ_0 = flow strength ($b \le 1.05a$)

Width = B. Initial mean radius R and section area A_0. Later: $A = \dfrac{A_0 R}{r}$

$w_y = \dfrac{A_0 \sigma_0}{R}$; load at onset of yielding

$u_y = \dfrac{R \sigma_0}{E}$; radial displacement when w_y is reached.

$w = \dfrac{A_0 R}{r^2} \sigma_0$; large deflection

$\mathcal{L} = \mathcal{L}_{pl} + \mathcal{L}_e$; expansion work from initial (R) to final (r) condition.

$$\mathcal{L}_{pl} = 2\pi R(A_0 \sigma_0) \ln(r/R) \approx 2\pi R(A_0 \sigma_0)(u/R); \quad \mathcal{L}_e = \pi A_0 \sigma_0^2 / E$$

Note: For steel at ambient temperature the elastic component is small if the required expansion is large compared to u_y. There is no distinction between the yield load and the ultimate load for a thin ring of this material.

CASE 4.13   THIN CYLINDRICAL SHELL OF EPP MATERIAL, UNDER INTERNAL PRESSURE *p*, UNDERGOING LARGE STRAINS

σ_0 = flow strength.

Formulas as in Case 4.12, except $B = 1$ and load w is replaced by pressure p. All comments apply.

CASE 4.14   THIN SPHERICAL SHELL OF EPP MATERIAL, UNDER INTERNAL PRESSURE p, UNDERGOING LARGE STRAINS

σ_0 = flow strength ($b \leq 1.05a$).

Initial mean radius R and thickness h_0. Thickness changes as $h \approx \dfrac{R^2}{r^2} h_0$

$p_y = \dfrac{2h_0\sigma_0}{R}$; pressure at onset of yielding

$u_y = \dfrac{R\sigma_0}{E}(1-\nu)$; radial displacement when p_y is reached

$p = \dfrac{2h_0 R^2}{r^3}\sigma_0$; for large deflections

$\mathcal{L} = \mathcal{L}_{pl} + \mathcal{L}_e$; expansion work from initial (R) to final (r) condition

$$\mathcal{L}_{pl} \approx 8\pi R^2 (h_0\sigma_0)\ln(r/R) \approx 8\pi R^2(h_0\sigma_0)(u/R)$$

$$\mathcal{L}_e = 4\pi R^2 \frac{h_0\sigma_0^2}{E}(1-\nu)$$

Note: For steel at room temperature the elastic component $\mathcal{L}_e$ is small if the required expansion is large compared to u_u. There is no distinction between the yield pressure and the ultimate pressure for a thin ring of this material.

CASE 4.15   MODERATELY THICK CIRCULAR RING OF EPP MATERIAL, UNDER INTERNAL LOAD w (N/m), UNDERGOING LARGE STRAINS

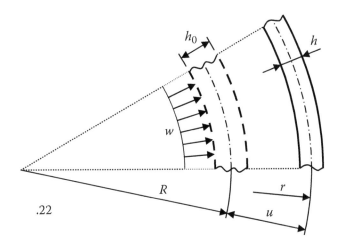

.22

$A_0 = Bh_0$

σ_0 = flow strength. ($b/a \leq 1.2$)

w_y, w_u, and u_u as for thick rings, Case 4.18. Area changes as $A = \dfrac{A_0 R}{r}$; B = constant (width)

$w = \dfrac{2Bh_0R}{r^2}\dfrac{\sigma_0}{\sqrt{3}}$ current load, for $u > u_\mathrm{u}$. $(r = R + u)$ $\mathcal{L} = \mathcal{L}_\mathrm{pl} + \mathcal{L}_\mathrm{ep}$; expansion work from initial (R) to current (r) condition.

$$\mathcal{L}_\mathrm{pl} = 2\pi R(A_0\sigma_0)\left(1 + \dfrac{A_\mathrm{av}}{2Br_\mathrm{av}}\right)\ln(r/R);$$

$$A_\mathrm{av} = (A_0 + A)/2, \quad r_\mathrm{av} = (R + r)/2; \quad \mathcal{L}_\mathrm{ep} \approx 2\pi a w_\mathrm{u} u_\mathrm{u}$$

$\mathcal{L}_\mathrm{pl} \approx 2\pi R A_0\sigma_0 u/R$; first approximation to the above.

Note: For steel at room temperature the elastoplastic component $\mathcal{L}_\mathrm{ep}$ is small if the required expansion is large compared to u_u.

CASE 4.16   MODERATELY THICK CYLINDRICAL SHELL OF EPP MATERIAL, UNDER INTERNAL PRESSURE *p*, UNDERGOING LARGE STRAINS

σ_0 = flow strength.
Formulas as in Case 4.15, except $B = 1$ and load w is replaced by pressure p.

CASE 4.17   MODERATELY THICK SPHERICAL SHELL OF EPP MATERIAL, UNDER INTERNAL PRESSURE *p*, UNDERGOING LARGE STRAINS

(Refer to the illustration in Case 4.15)

σ_0 = flow strength. $(b/a \leq 1.2)$

p_y, p_u, and u_u as for thick shells, Case 4.20. Thickness changes as $h \approx \dfrac{R^2}{r^2}h_0$

$p = \dfrac{2R^2h_0}{r^3}\sigma_0$ for large deflections

$\mathcal{L} = \mathcal{L}_\mathrm{pl} + \mathcal{L}_\mathrm{ep}$; expansion work from initial (R) to final (r) condition.

$\mathcal{L}_\mathrm{pl} \approx 8\pi R^2(h_0\sigma_0)\ln(r/R)$

$\mathcal{L}_\mathrm{ep} \approx 4\pi R^2 p_\mathrm{u} u_\mathrm{u}$

Note: For steel at room temperature the elastoplastic component $\mathcal{L}_\mathrm{ep}$ is small if the required expansion is large compared to u_u.

CASE 4.18   THICK CIRCULAR RING OF EPP MATERIAL, UNDER INTERNAL LOAD *w* (N/m), UNDERGOING LARGE STRAINS

σ_0 = flow strength. (Figure 4.12)
Width = B. Initial mean radius and area: R and A_0, respectively. Later: $Ar = A_0R$

$w_\mathrm{y} = \left[1 - \left(\dfrac{a}{b}\right)^2\right]\dfrac{B\sigma_0}{\sqrt{3}}$; load at onset of yielding

$$w_u = \frac{2B\sigma_0}{\sqrt{3}} \ln\left(\frac{b}{a}\right); \text{ load magnitude when ring becomes fully plastic}$$

$$u_u = \frac{\sqrt{3}\sigma_0}{2E} a\left(\frac{b}{a}\right)^2; \text{ displacement of inner surface when } w_u \text{ is reached}$$

$$w = \frac{2B\sigma_0}{\sqrt{3}} \ln\frac{1+\psi}{1-\psi} \approx \frac{2B\sigma_0}{\sqrt{3}} \frac{h_0 R}{r^2} \text{ [with } \psi = h/(2r)] \text{ for large deflections}$$

$\mathcal{L} = \mathcal{L}_{pl} + \mathcal{L}_{ep}$; expansion work from initial $(a = a_0)$ to final $(a = a_2)$ condition.

$\mathcal{L}_{pl} \approx \frac{\pi}{3}(a_2 - a_0)(w_0 a_0 + 4w_1 a_1 + w_2 a_2); \mathcal{L}_{ep} \approx 2\pi a w_u u_u$ with $a_1 = (a_0 + a_2)/2$ and w_0, w_1, and w_2 as corresponding equilibrium loads.

$\mathcal{L}_{pl} \approx 2\pi a A_0 \sigma_0(u/R)$; first approximation to the above (u is the *inside* radius displacement)

Note: For steel at room temperature the $\mathcal{L}_{ep}$ component is small if the required expansion is large compared to u_u. For expansion prescribed in terms of radii it is more convenient to use section area invariance as $b^2 - a^2 =$ constant, Symbol a stands for initial radius, $a = a_0$, Ref. [45].

CASE 4.19   THICK CYLINDRICAL SHELL OF EPP MATERIAL, UNDER INTERNAL PRESSURE p, UNDERGOING LARGE STRAINS

σ_0 = flow strength.

Formulas as in Case 4.18, except $B = 1$ and load w replaced by pressure p.

CASE 4.20   THICK SPHERICAL SHELL OF EPP MATERIAL, UNDER INTERNAL PRESSURE p, UNDERGOING LARGE STRAINS

(Refer Figure 4.12)

σ_0 = flow strength.

Initial radius and thickness: R and h_0, respectively. Later: $h \approx \frac{R^2}{r^2} h_0$; r is the current radius

$$p_y = \left[1 - \left(\frac{a}{b}\right)^3\right]\frac{2\sigma_0}{3}; \text{ pressure at onset of yielding}$$

$$p_u = 2\sigma_0 \ln\left(\frac{b}{a}\right); \text{ pressure when shell becomes fully plastic}$$

$$u_u = \frac{\sigma_0}{2E} a\left(\frac{b}{a}\right)^3; \text{ displacement of inner surface when } p_u \text{ is reached.}$$

$$p = 2\sigma_0 \ln\frac{1+\psi}{1-\psi} \approx \frac{h_0 R^2}{r^3} 2\sigma_0 \quad (\text{with } \psi = h/(2r)) \text{ for large deflections.}$$

$\mathcal{L} = \mathcal{L}_{pl} + \mathcal{L}_{ep}$; expansion work from initial $(a = a_0)$ to final $(a = a_2)$ condition.

$\mathcal{L}_{pl} \approx (2\pi/3)(a_2 - a_0)(p_0 a_0^2 + 4p_1 a_1^2 + p_2 a_2^2)$; $\mathcal{L}_{ep} \approx 4\pi a^2 p_u u_u$ with $a_1 = (a_0 + a_2)/2$ and p_0, p_1, and p_2 as corresponding (equilibrium) pressures.

$\mathcal{L}_{pl} \approx 8\pi a h \sigma_0 u_m$; first approximation of the above, with $u_m = r - R$.

Note: For steel at room temperature the elastoplastic component is small if the required expansion is large compared to u_u. For expansion prescribed in terms of radii it is more convenient to use volume invariance as $b^3 - a^3 = $ constant. Symbol a, when used without index means a_0, Ref. [45].

CASE 4.21   AXIAL BAR SUBJECTED TO AXIAL LOAD, LARGE TENSILE STRAINS RESULTING

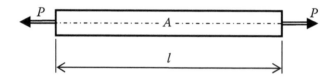

Initial length and section area:

L and A_0, respectively. Later: $A_0 L = Al$

$$P(\varepsilon) = A\sigma = \frac{A_0 \sigma}{\exp(\varepsilon)}; \text{ current bar load}$$

1. BL material: $\sigma = Y + E_p \varepsilon$: (for $\varepsilon > \varepsilon_y$)

    (a) $E_p > Y$, then P attains maximum P_m at $\varepsilon = \varepsilon_{cr}$ and $l = l_{cr} = L \exp(\varepsilon_{cr})$

$$\varepsilon_{cr} = 1 - \frac{Y}{E_p}; \quad \sigma_{cr} = E_p; \quad P_m = \frac{A_0 E_p}{\exp\varepsilon_{cr}}$$

    (b) $E_p \le Y$, then P attains maximum P_y at $\varepsilon = \varepsilon_y$

$$P_y = \frac{A_0 F_y}{\exp\varepsilon_y} \approx \frac{A_0 F_y}{1+\varepsilon_y} \approx A_0 F_y; \text{ approximation acceptable for } \varepsilon_y \ll 1$$

2. PL material: $\sigma = B(C + \varepsilon)^n$, $\sigma_0 = BC^n$

    (a) $n > C$, then P attains P_m at $\varepsilon = \varepsilon_{cr}$:

$$\varepsilon_{cr} = n - C; \quad \sigma_{cr} = Bn^n; \quad P_m = \frac{A_0 B n^n}{\exp(\varepsilon_{cr})}$$

(b) $n < C$, then P attains P_m at $\varepsilon = 0$:

$$P_0 = A_0 \sigma_0$$

Note: Material properties are specified as true stress vs. natural strain. BL material model can approximate only a gentle change of slope of σ–ε curve.

CASE 4.22 CIRCULAR RING SUBJECTED TO DISTRIBUTED LOAD w (N/m), LARGE STRAINS RESULTING

(Refer to the illustration in Case 4.15)

Initial radius and section area: R and A_0, respectively. Later: $A_0 R = Ar$. (Volume invariance)

$$w(\varepsilon) = \frac{P}{r} = \frac{A_0 \sigma}{R \exp(2\varepsilon)}; \text{ current distributed load. } P \text{ is the tensile force.}$$

1. BL material: $\sigma = Y + E_p \varepsilon$: (for $\varepsilon > \varepsilon_y$)

(a) $E_p > 2Y$, then $w(\varepsilon)$ attains maximum w_m at $\varepsilon = \varepsilon_{cr}$ and $\sigma = \sigma_{cr}$. Then $r = r_{cr} = R \exp(\varepsilon_{cr})$

$$\varepsilon_{cr} = \frac{1}{2} - \frac{Y}{E_p}; \quad \sigma_{cr} = E_p/2; \quad w_m = \frac{A_0 E_p}{2R \exp(2\varepsilon_{cr})}$$

(b) $E_p \leq 2Y$, then $w(\varepsilon)$ attains maximum w_y at $\varepsilon = \varepsilon_y$

$$w_y = \frac{A_0 F_y}{R \exp(2\varepsilon_y)} \approx \frac{A_0 F_y}{R(1 + 2\varepsilon_y)} \approx \frac{A_0 F_y}{R}; \text{ approximation acceptable for } 2\varepsilon_y \ll 1$$

2. PL material: $\sigma = B(C + \varepsilon)^n$, $\sigma_0 = BC^n$

(a) $n > 2C$, then w attains w_m at $\varepsilon = \varepsilon_{cr}$:

$$\varepsilon_{cr} = n/2 - C; \quad \sigma_{cr} = B\left(\frac{n}{2}\right)^n; \quad w_m = \frac{A_0 B(n/2)^n}{R \exp(2\varepsilon_{cr})}$$

(b) $n \leq 2C$, then w attains maximum w_0 at $\varepsilon = 0$:

$$w_0 = \frac{A_0 \sigma_0}{R}$$

Note: Case 4.21 notes are applicable. Also, the calculated w_m may be increased by $r_{cr}/(r_{cr} - h_{cr}/2)$. If the width, normal to page, is invariant, then the condition is $h_0 R = hr$.

CASE 4.23 CYLINDRICAL SHELL SUBJECTED TO PRESSURE p, LARGE STRAINS RESULTING

Formulas as in Case 4.22, except $B = 1$ and load w replaced by pressure p.

CASE 4.24　SPHERICAL SHELL SUBJECTED TO PRESSURE p, LARGE STRAINS RESULTING

(Refer to the illustration in Case 4.15)

Initial radius and thickness: R and h_0, respectively. Later: $h = \dfrac{R^2}{r^2} h_0$

$$p(\varepsilon) = \frac{2\sigma h}{r} = \frac{2h_0\sigma}{R\exp(3\varepsilon)}; \text{ current pressure}$$

1. BL material: $\sigma = Y + E_p\varepsilon$: (for $\varepsilon > \varepsilon_y$)

   (a) $E_p > 3Y$, then $p(\varepsilon)$ attains maximum p_m at $\varepsilon = \varepsilon_{cr}$ and $\sigma = \sigma_{cr}$. Then $r = r_{cr} = R\exp(\varepsilon_{cr})$

   $$\varepsilon_{cr} = \frac{1}{3} - \frac{Y}{E_p}; \quad \sigma_{cr} = B\left(\frac{n}{3}\right)^n; \quad p_m = \frac{2h_0 E_p}{3R\exp(3\varepsilon_{cr})}$$

   (b) $E_p \le 3Y$, then $p(\varepsilon)$ attains maximum p_y at $\varepsilon = \varepsilon_y$

   $$p_y = \frac{2h_0 F_y}{R\exp(3\varepsilon)} \approx \frac{2h_0 F_y}{R(1+3\varepsilon)} \approx \frac{2h_0 F_y}{R}; \text{ approximation acceptable for } 3\varepsilon_y \ll 1$$

2. PL material: $\sigma = B(C + \varepsilon)^n$, $\sigma_0 = BC^n$

   (a) $n > 3C$, then w attains p_m at $\varepsilon = \varepsilon_{cr}$:

   $$\varepsilon_{cr} = n/3 - C; \quad \sigma_{cr} = B\left(\frac{n}{3}\right)^n; \quad p_m = \frac{2h_0 B(n/3)^n}{R\exp(3\varepsilon_{cr})}$$

   (b) $n \le 3C$, then p attains maximum p_m at $\varepsilon = 0$:

   $$p_0 = \frac{2h_0\sigma_0}{R}$$

Note: Case 4.22 notes are applicable, but there is a different expression for the volume invariance here. Also, the calculated p_m may be increased by $r_{cr}/(r_{cr} - h_{cr}/2)$.

CASE 4.25　CYLINDRICAL CAVITY OF RADIUS a, INFINITE BODY. QUASISTATIC GAS EXPANSION

Polytropic gas initially at p_0 expanding to balance pressure p_b with radius growth by u.

$$\frac{u}{a}\left(1 + \frac{u}{a}\right)^{2n} = p_0\frac{1+v}{E}; \text{ equation to determine } u$$

$$p_b = \frac{Eu}{(1+\nu)a}$$

$\sigma_h = p_b$; hoop stress at the inside surface

CASE 4.26 SPHERICAL CAVITY OF RADIUS a, INFINITE BODY. QUASISTATIC GAS EXPANSION

Polytropic gas initially at p_0 expanding to balance pressure p_b with radius growth by u.

$$\frac{u}{a}\left(1+\frac{u}{a}\right)^{3n} = p_0 \frac{1+\nu}{2E}; \text{ equation to determine } u$$

$$p_b = \frac{2Eu}{(1+\nu)a}$$

$\sigma_h = p_b/2$; hoop stress at the loaded surface

CASE 4.27 PRESSURIZED CYLINDRICAL CAVITY, INFINITE BODY, EPP MATERIAL

Initiation of yield at $p = p_y = \sigma_0/\sqrt{3}$

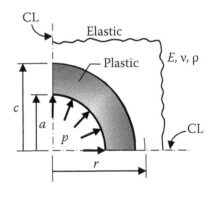

$$c = a\exp\left\{\frac{1}{2}\left(\frac{\sqrt{3}p}{\sigma_0} - 1\right)\right\}; \text{ radius of plastic zone}$$

$$u_r = \frac{1+\nu}{E}\frac{\sigma_0}{\sqrt{3}}\frac{c^2}{r}; \text{ radial displacement, elastic zone, } r \geq c$$

$$u_r = \frac{\sigma_0 a}{2\sqrt{3}E}\left\{(5-4\nu)\left(\frac{c}{a}\right)^2 - (1-2\nu)\left(3+6\ln\frac{c}{a}\right)\right\};$$

at $r = a$, inside face. If the cavity expands from zero to $u_r = a$, then $c \gg a$ and

$$c \approx a\sqrt{\frac{2\sqrt{3}}{\sigma_0}\frac{E}{5-4\nu}}$$

$$\sigma_r = -p\left(\frac{a}{r}\right)^2; \sigma_h = p\left(\frac{a}{r}\right)^2 \text{ radial and hoop stresses, respectively, elastic zone, } r \geq c$$

$$\sigma_r = -\frac{\sigma_0}{\sqrt{3}}\left(1 + 2\ln\frac{c}{r}\right); \text{ radial stress, plastic zone, } r \le c$$

$$\sigma_h = \frac{\sigma_0}{\sqrt{3}}\left(1 - 2\ln\frac{c}{r}\right); \text{ hoop stress, plastic zone, } r \le c, \text{ Ref. [45].}$$

Note: Cavity expansion from nil to a prescribed radius is described in Chapter 14.

CASE 4.28   PRESSURIZED SPHERICAL CAVITY, INFINITE BODY, EPP MATERIAL

(Refer to the illustration in Case 4.27)

Initiation of yield at $p = p_y = 2\sigma_0/3$

$$c = a\exp\left\{\frac{1}{3}\left(\frac{3p}{2\sigma_0} - 1\right)\right\}; \text{ radius of plastic zone}$$

$$u_r = \frac{1+\nu}{E}\frac{\sigma_0}{3}\frac{c^3}{r^2}; \text{ radial displacement, elastic zone, } r \ge c$$

$$u_r = \frac{\sigma_0 a}{E}\left\{(1-\nu)\left(\frac{c}{a}\right)^3 - \frac{2}{3}(1-2\nu)\left(1+3\ln\frac{c}{a}\right)\right\}; \text{ at } r = a, \text{ inside face}$$

$$c \approx a\left\{\frac{E}{(1-\nu)\sigma_0}\right\}^{1/3}; \text{ when the cavity expands from zero to } u_r = a, (c \gg a)$$

$$\sigma_r = -p\left(\frac{a}{r}\right)^3; \sigma_h = \frac{p}{2}\left(\frac{a}{r}\right)^2; \text{ radial and hoop stresses, respectively, elastic zone, } r \ge c$$

$$\sigma_r = -\frac{2\sigma_0}{3}\left(1+3\ln\frac{c}{r}\right); \text{ radial stress, plastic zone, } r \le c$$

$$\sigma_h = \frac{2\sigma_0}{3}\left(\frac{1}{2} - 3\ln\frac{c}{r}\right); \text{ hoop stress, plastic zone, } r \le c, \text{ Ref. [45].}$$

Note: Cavity expansion from nil to a prescribed radius is described in Chapter 14.

CASE 4.29   INELASTIC TWISTING OF CIRCULAR SHAFTS, TORQUE T APPLIED

L is the length of shaft

I_0 is the polar moment of inertia (m⁴)

τ_y is the shear stress at yielding

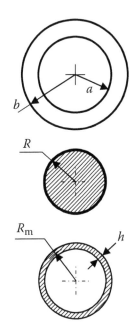

(a) Elastic range:

$$\tau = \frac{Tr}{I_0}; \text{ elastic stress at radius } r$$

$$\alpha = \frac{TL}{GI_0}; \text{ angle of twist}$$

$$I_0 = \frac{\pi}{2}(b^4 - a^4); \text{ hollow shaft with radii } a \text{ and } b$$

$$I_0 = \frac{\pi}{2}R^4; \text{ solid shaft}$$

$$I_0 = 2\pi h R_m^3; \text{ thin-wall shaft}$$

$$T_y = \frac{\tau_y I_0}{R}; \text{ onset of yield for section}$$

(b) Moment capacities for EPP material, $\tau_y \equiv \tau_0$

(1) Hollow shaft: $T_y' = \dfrac{2\pi}{3}\tau_y(b^3 - a^3)$

(2) Solid shaft: $T_y' = \dfrac{4}{3}T_y = \dfrac{2\pi}{3}\tau_y R^3 = \dfrac{4 I_0 \tau_y}{3R}$

(3) Thin-wall shaft: $T_y' = T_y = 2\pi\tau_y h R_m^2 = \dfrac{I_0 \tau_y}{R_m}$, Ref. [45].

(c) An approximate shaft characteristic for RSH material with τ_y and τ_u with G_p as modulus in plastic range.

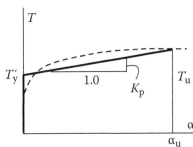

$\alpha = \dfrac{L}{R}\gamma$; angle of twist as a function of the material shear

angle γ (use R_m for a thin-wall shaft)

$$T_u = \left(\frac{\tau_y}{6} + \frac{\tau_u}{2}\right)\pi R^3; \text{ ultimate torque, solid shaft}$$

$$T_u = 2\pi\tau_u h R_m^2; \text{ ultimate torque, thin-wall shaft}$$

$$K_p = \frac{T_u - T_y'}{\alpha_u} \approx \frac{G_p I_0}{L}$$; shaft stiffness, N-m/rad. For a thin-wall shaft replace "$\approx$" with "$=$"

Note: The RSH material in shear is defined by Case 4.8. The above shaft characteristic ignores the fact that a transition from the yield to the ultimate condition, in case of a solid shaft, is more gradual, as suggested by the dashed line. For a thin-wall shaft the $T - \alpha$ characteristics is scaled $\tau - \gamma$ characteristics.

CASE 4.30   STRESS–STRAIN EQUATIONS FOR CONCRETE

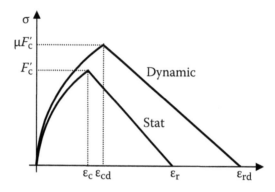

(a) Statics, compression (Modified Scott Model)

$$\sigma = \left[\frac{2\varepsilon}{\varepsilon_c} - \left(\frac{\varepsilon}{\varepsilon_c} \right)^2 \right] F_c' \quad \text{for } \varepsilon \leq \varepsilon_c$$

$$\sigma = F_c' - Z_s (\varepsilon - \varepsilon_c) \quad \text{for } \varepsilon \leq \varepsilon_c$$

$\varepsilon_r = \dfrac{F_c'}{Z_s} + \varepsilon_c$; strain at failure

$$Z_s = \frac{0.009 F_c' + 0.275}{(3 + 0.29 F_c' / (145 F_c' - 1{,}000)) - \varepsilon_c} ;$$

$\varepsilon_c = \dfrac{4.26}{(F_c')^{1/4}} \dfrac{F_c'}{E_c}$; strain at peak stress

$\Psi = \left(\dfrac{2\varepsilon_c}{3} + \dfrac{1}{2Z_s} \right) F_c'$; material toughness

(b) Statics, tension

$F_t = 0.6 \sqrt{F_c'}$; tensile strength ~ MPa, in flexure

where F_c' is the nominal compressive strength ~ MPa.

(c) Dynamic tension, prescribed strain-rate $\dot{\varepsilon}$ ($\mu = F_{tt}/F_t$, strengthening ratio)

$$\mu = \left(\frac{\dot{\varepsilon}}{\dot{\varepsilon}_{ts}}\right)^{1.016\beta} \quad \text{for } \dot{\varepsilon} \leq 30/\text{s}; \; \dot{\varepsilon}_{ts} = 30 \times 10^{-6} \; 1/\text{s}; \text{ assumed quasistatic strain rate}$$

$$\mu = \gamma\left(\frac{\dot{\varepsilon}}{\dot{\varepsilon}_{ts}}\right)^{1/3} \quad \text{for } \dot{\varepsilon} \geq 30/\text{s}; \; \beta = \frac{1}{10 + 0.6F_c'}; \; \gamma = \exp(7.11\beta - 2.33)$$

(d) Dynamic compression, prescribed strain-rate $\dot{\varepsilon}$ ($\mu = F_{cc}/F_c'$, strengthening ratio)

$$\mu = \left(\frac{\dot{\varepsilon}}{\dot{\varepsilon}_{cs}}\right)^{1.026\alpha} \quad \text{for } \dot{\varepsilon} \leq 30/\text{s}; \; \dot{\varepsilon}_{cs} = 30 \times 10^{-6} \; 1/\text{s}; \text{ assumed quasistatic strain rate}$$

$$\mu = \gamma(\dot{\varepsilon})^{1/3} \text{ for } \dot{\varepsilon} \geq 30/\text{s}; \; \alpha = \frac{1}{5 + 0.75F_c'}; \; \gamma = \exp(6.156\alpha - 0.49)$$

$\varepsilon_{cd} = \mu\varepsilon_c$; strain at peak dynamic strength, Ref. [120].

Note: The equations are valid up to the strain rate of 700/s, although the experience with strain rates above 300/s is quite limited. See text for further comments.

EXAMPLES

EXAMPLE 4.1 A RESTRAINED BLOCK, INITIAL VELOCITY APPLIED

An elastic rod, constraining a block with $M = 1,000\,g$ has $A = 10\,mm^2$, $E = 200,000\,MPa$, $L = 100\,mm$, and the yield stress $F_y = 250\,MPa$. Determine its response to the initial velocity of $1.0\,mm/ms$ for the elastic material, for the BL with $k_2 = k_1/4$, for the related material types RP, RSH, EPP, and then for RO with $k_0 = k_1$ and $n = 2$.

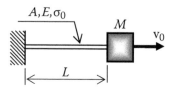

Follow Case 4.1, then Case 4.3. The yield strength of the rod is $W_y = AF_y = 10 \times 250 = 2,500\,N$.

Stiffness, up to yield: $k_1 = EA/L = 200,000 \times 10/100 = 20,000\,N/mm$

The deflection at yield: $u_y = W_y/k_1 = 2,500/20,000 = 0.125\,mm$

Inelastic $k_2 = 20,000/4 = 5,000$. Stiffness factor: $\rho = \dfrac{k_1}{k_2} - 1 = \dfrac{20,000}{5,000} - 1 = 3$

Maximum deflection:

$$u_m = \left[(\rho^2+\rho)u_y^2 + \frac{Mv_0^2}{k_2}\right]^{1/2} - \rho u_y = \left[(3^2+3)0.125^2 + \frac{1,000 \times 1^2}{5,000}\right]^{1/2} - 3 \times 0.125 = 0.2475\,mm$$

$R_m = k_2(u - u_y) + R_y = 5,000(0.2475 - 0.125) + 2,500 = 3,113\,N$; peak reaction

$$t_m \approx \sqrt{\frac{8Mu_m}{3R_m}} = \sqrt{\frac{8 \times 1,000 \times 0.2475}{3 \times 3,113}} = 0.46\,ms;\ \text{duration of loading phase.}$$

Alternatively, from Equation 4.20b: $t_m = \dfrac{\pi S}{2R_m} = \dfrac{\pi(1,000 \times 1)}{2 \times 3,113} = 0.505\,ms$

The above and the remaining characteristics are tabulated here.

Model Type	Model Type Description	u_m (mm)	R_m (N)	t_m (ms)	t_{m1} (ms)	t_{m2} (ms)	t_{m3} (ms)
LE	Linear elastic	0.224	4472	0.352	0.351	0.351	0.365
RP	Rigid plastic	0.200	2500	—	0.393	0.628	0.400
BL	Bilinear	0.248	3113	0.416	0.410	0.505	0.460
RSH	Rigid-strain hardening	0.171	3354	—	0.323	0.468	0.330
EPP	Elastic, perfectly plastic	0.263	2500	0.464	0.451	0.628	0.529
RO	Ramberg–Osgood	0.528	1531	—	0.879	1.026	0.959

The duration of forward motion in the t_m column was established using an implicit FEA program, but it was done only for some selected cases. The Ritz method, Case 4.7, gave t_{m1}. (The small difference between the two is due to a numerical damping in FEA, as seen in the LE formulation.) (The difference shows up only in duration, but not in amplitude.) The remaining two estimates, t_{m2} and t_{m3} come from Equations 4.20b and 4.22, respectively. It is seen that the last one, based on elementary considerations gives a sensible, although somewhat overestimated duration.

The t_{m1} column gives the best estimate.

EXAMPLE 4.2   A RESTRAINED BLOCK, STEP LOAD APPLIED

The rod is the same as described in Example 4.1, except the material model is now EPP with $\sigma_0 = 250\,\mathrm{MPa}$. Determine its response to the step load with the magnitude of $W_0 = 1{,}800\,\mathrm{N}$.

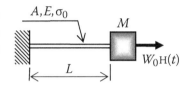

Circular frequency: $\omega = (k_1/M)^{1/2} = (20{,}000/1{,}000)^{1/2} = 4.472$ rad/ms.

Following Case 4.5, find t_y when u_y is reached:

$$u_{st} = W_0/k_1 = 1{,}800/20{,}000 = 0.09\,\mathrm{mm}$$

$$\sin \omega t_y = \left\{1 - \left(\frac{u_y}{u_{st}} - 1\right)^2\right\}^{1/2} = \left\{1 - \left(\frac{0.125}{0.090} - 1\right)^2\right\}^{1/2} = 0.9213$$

or $\omega t_y = 1.1714$ and $t_y = 0.2619\,\mathrm{ms}$, then

$$\omega t_m = \omega t_y + \frac{W_0}{R_0 - W_0}\sin \omega t_y = 1.1714 + \frac{1{,}800}{2{,}500 - 1{,}800}0.9213 = 3.54\,\mathrm{rad}$$

which gives $t_m = 3.54/4.472 = 0.7917\,\mathrm{ms}$ as the instant of peak deflection.

$$u_m = \frac{R_0^2}{2k_1(R_0 - W_0)} = \frac{2{,}500^2}{2 \times 20{,}000(700)} = 0.2232\,\mathrm{mm};\ \text{peak deflection}$$

$$t_m \approx \sqrt{\frac{8Mu_m}{3R_0}} = \sqrt{\frac{8 \times 1{,}000 \times 0.2232}{3 \times 2{,}500}} = 0.488\,\mathrm{ms};\ \text{duration of loading phase according to}$$

Equation 4.22

Using an FEA program, the following results were obtained:

$u_m = 0.2232\,\mathrm{mm}$ at $t_m = 0.9765\,\mathrm{ms}$, while $t_y = 0.445\,\mathrm{ms}$

One of the reasons for the discrepancy is the fact that the hand calculation ignored the rod mass, thereby artificially increasing the natural frequency.

EXAMPLE 4.3   A RIGID PLASTIC (RP) OSCILLATOR, TRIANGULAR DECREASING PULSE

Consider the same arrangement as in Example 4.2, except that the material is RP now. Determine its response to the pulse with the initial magnitude of $W_0 = 6A\sigma_0$ lasting 0.5 ms. Compare the displacement with that resulting from the equivalent instantaneous impulse.

$R_0 = A\sigma_0 = 10 \times 250 = 2,500\,\text{N}$; resisting force

$W_0 = 6A\sigma_0 = 6 \times 10 \times 250 = 15,000\,\text{N}$; the peak applied load

$S_0 = W_0 t_0/2 = 15,000 \times 0.5/2 = 3,750\,\text{N-ms}$; the associated impulse

Following Case 4.2e for $W_0/R_0 > 2$:

$$v_e = \left(\frac{W_0}{2} - R_0\right)\frac{t_0}{M} = \left(\frac{15,000}{2} - 2,500\right)\frac{0.5}{1,000} = 2.5\,\text{mm/ms}; \quad \text{velocity at } t = t_0$$

$$u_m = \left(W_0 - \frac{3}{2}R_0\right)\frac{t_0^2}{3M} + \frac{Mv_e^2}{2R_0} = (15,000 - 3,750)\frac{0.5^2}{3 \times 1,000} + \frac{1,000 \times 2.5^2}{2 \times 2,500} = 2.1875\,\text{mm}$$

From Case 4.2a (instantaneous impulse) using values calculated at the outset:

$$u_m = \frac{S_0^2}{2MR_0} = \frac{3750^2}{2 \times 1000 \times 2500} = 2.8125\,\text{mm}$$

In spite of the peak load W_0 being as large as $6R_0$, there is still a significant difference between the two answers. This difference will become smaller if the duration of the triangular impulse decreases while its magnitude is kept constant.

EXAMPLE 4.4   AN EQUIVALENT STRESS–STRAIN CURVE

On occasions, a simplified form of PL material is used, where the first term is absent. Johnson [40, p. 141] gives it as $\sigma = B\varepsilon^n$, where $B = 690\,\text{MPa}$ and $n = 0.25$ for a "typical" steel. Devise an equivalent BL material with the slope of $E = 200,000\,\text{MPa}$ at the origin, choosing the yield point σ_0 and the plastic modulus E_p in such a way, that the area under the σ–ε curve is retained and a reasonable approximation is obtained in the elongation range $\varepsilon = 0$ to $\varepsilon_u = 0.15$.

The prescribed curve will be designated as $\sigma_1 = 690\varepsilon^{0.25}$. For the general expression $\sigma = B\varepsilon^n$, the area under the stress–strain curve is

$$\Psi = \int_0^{\varepsilon_u} \sigma d\varepsilon = \frac{B\varepsilon_u^{n+1}}{n+1}; \quad \text{which gives } \Psi = \frac{690 \times 0.15^{1.25}}{1.25} = 51.53\,\text{MPa}$$

It seems that a RSH curve could be a good first try, with the $\sigma_0 = 200\,\text{MPa}$ being one reference point, while the other is at $\varepsilon = 0.10$, where $\sigma_1 = 388$. A line passing through these two points can be expressed as

$$\sigma_3 = 200 + \frac{388 - 200}{0.1}\varepsilon = 200 + 1880\varepsilon$$

Integrating (with the help of Mathcad) gives the area of 51.24 MPa, practically the same as before, confirming that the choice was quite good. To have an elastic range included in our $\sigma-\varepsilon$ curve, modify the above expression slightly:

$\sigma_s = 200,000\varepsilon$ up to $\sigma = 202\,\text{MPa}$
$\sigma_s = 202 + 1,880\varepsilon$ above $\sigma = 202\,\text{MPa}$

Integration yields 51.35 MPa area, or 0.5% below that of the target curve, which is quite acceptable for practical purposes. The original and the equivalent curves are illustrated below.

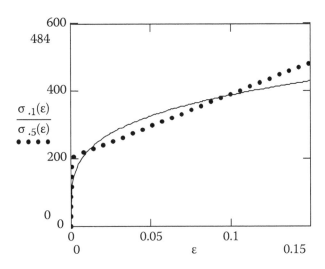

In retrospect, one can notice that an equivalent RSH curve would be just as effective as the BL curve developed. The former is much easier to describe and manipulate.

EXAMPLE 4.5   ENGINEERING VS. TRUE STRESS

Suppose that the BL material described in Example 4.4 was obtained by simplifying an engineering stress–strain curve. Consequently, the inelastic part of the curve should be written as $\sigma = 202 + 1880\varepsilon$. Assume this function to hold up to $\varepsilon = 0.50$. Convert this to a true stress vs. engineering strain, $S-\bar{\varepsilon}$ curve and graphically show difference between the two.

Using Equations 4.11 and 4.13 one could write a new equation. Instead, let us convert a single point from $\sigma-\varepsilon$ to $S-\bar{\varepsilon}$, to illustrate the procedure. Select the end point of the curve, $\varepsilon = 0.5$:

$\sigma = 202 + 1880 \times 0.5 = 1142\,\text{MPa}$; engineering stress
$S = \sigma(1+\varepsilon) = 1142(1+0.5) = 1713\,\text{MPa}$; true stress
$\bar{\varepsilon} = \ln(1+0.5) = 0.4055$ (A reference only)

and, repeating the operation for the remaining, reasonably spaced points, the following curves emerge.

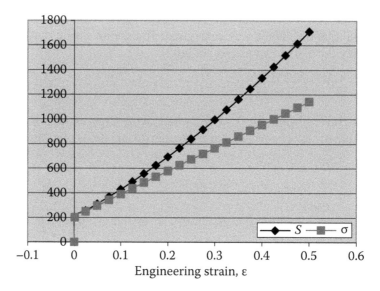

Note that if S is plotted vs. $\bar{\varepsilon}$, rather than ε, while the $\bar{\varepsilon}$-axis is overlaid on the ε-axis, the points of the upper curve shift to the left, showing a bigger discrepancy between the two. As mentioned before, σ is related to the load acting on the sample, rather than the applied stress.

EXAMPLE 4.6   POLYTROPIC SPRING APPLICATION

The system is as shown in Case 4.9 with the piston 5 mm thick having a surface area of 10 mm × 10 mm. The air is compressed to the pressure of 10 MPa, with the compressed air column length of 50 mm. The piston is then released to move freely in a long tube. Assuming the ambient air pressure $p_0 = 0.1$ MPa and the polytropic exponent $n = 1.4$, find the maximum speed attained by the piston v_m and the distance L it will travel before it stops. There is friction, exerting a resistance of about the same magnitude as the ambient air pressure. The dynamic air resistance applied to the piston is to be ignored. The piston is made of steel, $\rho = 0.00785$ g/mm³.

Calculate the resisting force applied to the piston by doubling the effect of p_0:

$$R_{ext} = 2 \times 10 \times 10 \times 0.1 = 20\,N.$$

The piston will accelerate as long as the driving gas force exceeds the above. This means that at the acceleration stage the air expands from 10 to 0.2 MPa. (Twice the ambient pressure.) The corresponding ratio of distances is, from Case 4.9:

$$\frac{x_2}{x_1} = \left(\frac{R_1}{R_2}\right)^{1/n} \quad \text{or} \quad \frac{x_2}{50} = \left(\frac{10}{0.2}\right)^{1/1.4} \quad \text{or} \quad x_2 = 817.6\,mm$$

(Piston area cancels out.) The work performed by the expanding gas is

$$\mathcal{L} = \frac{R_1 x_1}{n-1}\left[1 - \left(\frac{x_1}{x_2}\right)^{(n-1)}\right] = \frac{(10 \times 100)50}{0.4}\left[1 - \left(\frac{50}{817.6}\right)^{0.4}\right] = 84{,}122\,\text{N-mm}$$

This work, less resistance loss, is converted to a kinetic energy:

$$\frac{Mv_m^2}{2} = \mathcal{L} - R_{\text{ext}}(x_2 - x_1) \quad \text{or} \quad \frac{3.925v_m^2}{2} = 84{,}122 - 20(817.6 - 50) \quad \text{or} \quad v_m = 187.2\,\text{m/s}$$

where $M = 10 \times 10 \times 5 \times 0.00785 = 3.925\,\text{g}$.

The total energy content of the compressed gas, Equation 4.26, is

$$Q = \frac{pV}{n-1} = \frac{10(100 \times 50)}{1.4 - 1} = 125{,}000\,\text{N-mm}.$$

When this becomes equal to the work of the resistance, at some distance of travel L, the piston stops:

$$125{,}000 = R_{\text{ext}}L \quad \text{or} \quad L = 125{,}000/20 = 6{,}250\,\text{mm}$$

If the dynamic air resistance were included, both L and v_m would be smaller.

EXAMPLE 4.7   LARGE STRETCHING AND LOAD CAPACITY OF SIMPLE ELEMENTS

Johnson [40] quotes constants of a PL material curve for (half-hard) copper. After changing to metric units: $B = 42.85\,\text{MPa}$; $C = 0.114$; and $n = 0.3$.

Calculate the peak static load capacity for (a) axial bar $A = 66\,\text{mm}^2$, $L = 100\,\text{mm}$; (b) ring with $R = 45.8\,\text{mm}$, $h = 13.2\,\text{mm}$, $A = 66\,\text{mm}^2$; and (c) sphere with the same h_0 and R. The material can stretch only to $\varepsilon = 40\%$. (Refer to Cases 4.21, 4.22, and 4.24.)

$\sigma = B(C + \varepsilon)^n$ for all three shapes.

(a) Axial bar: $\varepsilon_{cr} = n - C = 0.3 - 0.114 = 0.186$; $\sigma_{cr} = Bn^n = 42.85 \times 0.3^{0.3} = 29.86\,\text{MPa}$

$$P_m = \frac{A_0 \sigma_{cr}}{\exp(\varepsilon_{cr})} = \frac{66 \times 29.86}{\exp(0.186)} = 1636\,\text{N}$$

(This simplified form will be used for the remaining shapes as well.)

(b) Ring: $\varepsilon_{cr} = n/2 - C = 0.036$; $\sigma_{cr} = B\left(\dfrac{n}{2}\right)^n = 24.25\,\text{MPa}$; $r_{cr} = R\exp(\varepsilon_{cr}) = 47.48\,\text{mm}$

$$w_m = \frac{A_0\sigma}{R\exp(2\varepsilon)} = \frac{66 \times 24.25}{45.8\exp(2 \times 0.036)} = 32.52\,\text{N/mm. This load may be increased by}$$

$r_{cr}/(r_{cr} - h_{cr}/2)$: $w'_m = 32.52\,\dfrac{47.48}{47.48 - 13.2/2} = 37.77\,\text{N/mm}$

(c) Sphere with the same thickness and radius: $\varepsilon_{cr} = n/3 - C = 0.3/3 - 0.114 = -0.014$

The negative value of ε_{cr} simply indicates that the material cannot reach the tensile instability, when in the spherical configuration. The peak load is attained at the outset.

$$\sigma_0 = BC^n = 42.85 \times 0.114^{0.3} = 22.34\,\text{MPa}$$

$$p_0 = \frac{2h_0\sigma_0}{R} = \frac{2 \times 13.2 \times 22.34}{45.8} = 12.88\,\text{MPa and the applied pressure at equilibrium must}$$

decrease thereafter.

EXAMPLE 4.8   LARGE STRETCHING OF THICK-WALL ELEMENTS

A tube with the section shown in Figure 4.12 has $a = 19$ and $b = 38\,\text{mm}$. It is made of steel with $E = 200,000\,\text{MPa}$, the yield stress $\sigma_0 = 500\,\text{MPa}$. Determine the critical pressure values as well as the pressure necessary to keep it in equilibrium when the outer radius doubles. Find the amount of work needed to achieve such a deformation. Repeat the calculation for a spherical shell of the same dimensions. Use the EPP material model.

Cylinder
Use Case 4.19 to find pressure at yield and the ultimate pressure:

$$p_y = \left[1 - \left(\frac{a}{b}\right)^2\right]\frac{\sigma_0}{\sqrt{3}} = \left[1 - \left(\frac{19}{38}\right)^2\right]\frac{500}{\sqrt{3}} = 216.5\,\text{MPa}$$

$$p_u = \frac{2\sigma_0}{\sqrt{3}}\ln\left(\frac{b}{a}\right) = \frac{2 \times 500}{\sqrt{3}}\ln\left(\frac{38}{19}\right) = 400.2\,\text{MPa}$$

one can also notice that the deflection u_u is only $0.1645\,\text{mm}$, which shows how small the deformation is, at the full yielded condition, compared with the values contemplated here. At peak displacement, $b = 2 \times 38 = 76\,\text{mm}$. The original section area must be preserved, therefore $38^2 - 19^2 = 76^2 - a^2$ gives $a = 68.51\,\text{mm}$ in the deformed condition.

The adjusted equation for p_u may now be used: $p_u = \dfrac{2 \times 500}{\sqrt{3}}\ln\left(\dfrac{76}{68.51}\right) = 59.9\,\text{MPa}$

The resistance against pressure when in deformed shape is therefore only a small fraction of what it was in the original shape. To calculate work needed to expand the shell, one needs to find the terms p_i for both conditions, designating them as initial (0), final (2), and middle (1). While 0 and 2 had been defined, the middle one still needs to be found.

Initial: $p_u = p_0 = 400.2\,\text{MPa}$, $a_0 \approx 19\,\text{mm}$

Final: $p_2 = 59.9\,\text{MPa}$, $a_2 = 68.51\,\text{mm}$

Midpoint: $a_1 = (19.0 + 68.51)/2 = 43.8\,\text{mm}$. Using the area invariance: $38^2 - 19^2 = b^2 - 43.8^2$ gives $b = 54.79\,\text{mm}$. Substituting into the expression for p_u gives $p_1 = 129.3\,\text{MPa}$. With all data points in hand:

$$\mathcal{L}_{pl} = \frac{\pi}{3}(68.51 - 19)(400.2 \times 19 + 4 \times 129.3 \times 43.8 + 59.9 \times 68.51) = 1.7815 \times 10^6\,\text{N-mm}$$

$$\mathcal{L}_{ep} = 2\pi a p_u u_u = 2\pi 19 \times 400.2 \times 0.1645 = 413.6; \quad \text{Total: } \mathcal{L} = 1.7819 \times 10^6\,\text{N-mm}$$

Check how big an error would be incurred if the approximate formulas for $\mathcal{L}_{pl}$ were used.

For a cylinder $R = (19 + 38)/2 = 28.5\,\text{mm}$, $u = 68.51 - 19 = 49.51\,\text{mm}$. Then

$$\mathcal{L}_{pl} \approx 2\pi a A_0 \sigma_0 u/R = 2\pi 19 \times 19 \times 500 \times 49.51/28.5 = 1.97 \times 10^6\,\text{N-mm, somewhat larger than}$$
the more accurate result.

Spherical shell
Repeating the operations for the per Case 4.20, one obtains

$$p_y = 291.7; \quad p_u = 693.1\,\text{MPa}$$

To preserve the original volume in the maximum deformed condition one can write $38^3 - 19^3 = 76^3 - a^3$, then $a = 73.12\,\text{mm}$.

$$p_u = 2\sigma_0 \ln\left(\frac{b}{a}\right) = 2 \times 500 \ln\left(\frac{76}{73.12}\right) = 38.63\,\text{MPa (small fraction of } p_u\text{)}$$

Thus $a_0 = 19$, $a_2 = 73.12\,\text{mm}$, $p_0 = 693.1\,\text{MPa}$, and $p_2 = 38.63\,\text{MPa}$

Then $a_1 = 46.06$, $b_1 = 52.62\,\text{mm}$, and $p_1 = 133.15\,\text{MPa}$

After substituting: $\mathcal{L}_{pl} = 179.86 \times 10^6\,\text{N-mm}$. A simpler formula is also available, but displacement of the mean radius, u_m is needed here. Initial $R = 28.5\,\text{mm}$, final $r = (73.12 + 76)/2 = 74.56\,\text{mm}$, then $u_m = 74.56 - 28.5 = 46.06\,\text{mm}$.

$$\mathcal{L}_{pl} = 8\pi a h \sigma_0 u_m = 8\pi 19 \times 19 \times 500 \times 46.06 = 208.95 \times 10^6\,\text{N-mm}.$$

This is some 16% more than the more accurate result.

EXAMPLE 4.9   LARGE STRETCHING OF THICK-WALL ELEMENTS, ANOTHER APPROACH

Use the same tube and spherical shell as in Example 4.8. In order to appreciate the difference in using moderately thick-shell expressions in place of those for a thick shell, apply Cases 4.16 and 4.17 to calculate the plastic component of expansion work under the same conditions as before.

Initial, final, and average thickness values, respectively:

$$h_0 = 19, \quad h_2 = 76 - 68.51 = 7.49, \quad \text{and} \quad h_{av} = 13.25 \, \text{mm}.$$

Initial, final, and average medium radii, respectively:

$$R = (19 + 38)/2 = 28.5, \quad r_2 = (68.51 + 76)/2 = 72.26, \quad \text{and} \quad r_{av} = 50.38 \, \text{mm}$$

Then, $h_{av}/(2r_{av}) = 0.1315$

$$\mathcal{L}_{pl} = 2\pi R(h_0\sigma_0)\left(1 + \frac{h_{av}}{2r_{av}}\right)\ln(r/R) = 2\pi 28.5(19 \times 500)1.1315 \ln(72.26/28.5) = 1.791 \times 10^6 \, \text{N-mm}$$

In spite of using this formula outside the intended range, the difference compared with the previous result is negligible.

Repeating the operations for the spherical shell, per Case 4.17, one obtains, for the maximum displacements:

$h = 76 - 73.12 = 2.88 \, \text{mm}, \, r = 76 - 2.88/2 = 74.56 \, \text{mm}, \text{and}$

$$\mathcal{L}_{pl} = 8\pi R^2(h_0\sigma_0)\ln(r/R) = 8\pi 28.5^2 (19 \times 500) \ln(74.56/28.5) = 186.51 \times 10^6 \, \text{N-mm}.$$

The answer here is nearly 4% higher than previously, quite acceptable deviation.

EXAMPLE 4.10   EQUIVALENT STRESS–STRAIN CURVE FOR CONCRETE

Create a simplified, EPP curve, as in Figure 4.6b, for $F_c' = 35 \, \text{MPa}$ concrete. The first, sloping part is to use the nominal E_c modulus and extend up to $F_c' = \sigma_0$. The extent of the flat part, up to $\varepsilon_u = \varepsilon_r$, is to be such that the material toughness, defined by the Modified Scott Model is to be retained.

$\sigma_0 = 35 \, \text{MPa}$, as given in the problem statement.

$E_c = 5,000\sqrt{F_{cm}'} = 5,000\sqrt{(1.1 \times 35)} = 31,020 \, \text{MPa}$; nominal Young's modulus per Equation 4.6.

$\varepsilon_y = \sigma_0/E_c = 35/31,020 = 0.00113$; strain at the onset of yield.

Follow Case 4.30: $\varepsilon_c = \dfrac{4.26}{(F_c')^{1/4}}\dfrac{F_c'}{E_c} = \dfrac{4.26}{(35)^{1/4}}\dfrac{35}{31,020} = \dfrac{1}{506}$; strain at peak stress

$$Z_s = \frac{0.009 \times 35 + 0.275}{((3 + 0.29 \times 35)/(145 \times 35 - 1000)) - (1/506)} = 471.7; \text{ post-yield coefficient.}$$

When the expression for the energy content from Case 4.5 is rewritten, one has

$$\Psi = \varepsilon_r\sigma_0 - \frac{1}{2} \times \frac{\sigma_0^2}{E} = 35\varepsilon_r - \frac{1}{2} \times \frac{35^2}{31,020} = 35\varepsilon_r - 0.0197 \, \text{MPa}$$

The analogous expression from Case 4.30:

$$\Psi = \left(\frac{2\varepsilon_c}{3} + \frac{1}{2Z_s}\right)F_c' = \left(\frac{2}{3}\times\frac{1}{506} + \frac{1}{2\times471.7}\right)35 = 0.0832\,\text{MPa}$$

Equating the two expressions gives $35\varepsilon_r - 0.0197 = 0.0832$, or $\varepsilon_r = 0.00294 \approx 0.003$

EXAMPLE 4.11   STRAIN-RATE EFFECT ON CONCRETE STRENGTH

Following Case 4.30 establish this effect for an $F_c' = 35\,\text{MPa}$ concrete under the strain-rate of 500/s.

$F_t = 0.6\sqrt{F_c'} = 0.6\sqrt{35} = 3.845\,\text{MPa}$; static tensile strength (in flexure)

Dynamic tension, prescribed strain-rate $\dot{\varepsilon} = 500$/s:

$$\beta = \frac{1}{10+0.6F_c'} = \frac{1}{10+0.6\times35} = \frac{1}{31}$$

$\gamma = \exp(7.11\beta - 2.33) = \exp(7.11/31 - 2.33) = \exp(-2.1006) = 0.1224$

$$\mu = 0.1224\left(\frac{500}{30\times10^{-6}}\right)^{1/3} = 31.26$$

$F_{tt} = \mu F_t = 31.26 \times 3.845 = 120.2\,\text{MPa}$; dynamic tensile strength

Dynamic compression:

$$\alpha = \frac{1}{5+0.75F_c'} = \frac{1}{5+0.75\times35} = 0.032$$

$\gamma = \exp(6.156\alpha - 0.49) = \exp(6.156 \times 0.032 - 0.49) = \exp(-0.293) = 0.746$

$\mu = \gamma(\dot{\varepsilon})^{1/3} = 0.746(500)^{1/3} = 5.921$

$F_{cc} = \mu F_c' = 5.921 \times 35 = 207.2\,\text{MPa}$; dynamic compressive strength

Note: Apart from the reservations against this approach, as expressed in Chapter 6, one must remember that failure may also be initiated when the strain-rate decreases faster than the applied load.

5 Wave Propagation

THEORETICAL OUTLINE

Sometimes it is important to learn about a deformed pattern and stress distribution in the vicinity of a suddenly applied force, just after that force is applied. This takes us to the topic of wave propagation. An elastic axial bar, in a small deflection regime and within the elastic range of stress, as considered below, is the best introduction to the subject. Other elements, relatively easy to describe, are quantified as well. The analysis involving a linear material is followed by an investigation of the same phenomena using a simple, bilinear material model. Also, the subjects of body waves and pulse transmission are outlined.

THE BASIC SOLUTION OF THE EQUATION OF MOTION

Deflection of a constant-section, slender bar is described by Equation 2.10 and written here in a more detailed form:

$$\frac{\partial^2 u}{\partial t^2} - c_0^2 \frac{\partial^2 u}{\partial x^2} = 0 \tag{5.1}$$

when no external loading is applied. One way to obtain a solution for $u(x,t)$ was previously discussed. There is, however, a more physically appealing way of interpreting the above relationship.

As one can easily check, when a function $u(x,t)$ satisfies Equation 5.1, another function, namely, $u(x - c_0 t)$ is a solution as well. The latter has exactly the same shape as the former, except is shifted by $c_0 t$ in the positive direction of the x-axis. Furthermore, the function $u(x + c_0 t)$, like the original, but shifted in negative direction, is also a solution. This tells us that the general solution of Equation 5.1 can be presented as

$$u(x,t) = u_1(x + c_0 t) + u_2(x - c_0 t) \tag{5.2a}$$

and that

$$u(x,0) = u_1(x,0) + u_2(x,0) \tag{5.2b}$$

The physical interpretation of the above is given in Figure 5.1. In Figure 5.1a, a fragment of a long, unconstrained bar is isolated from the rest by two rigid planes and stretched across two intermediate planes. This results in a distribution of axial forces (Figure 5.1b) and deflections (Figure 5.1c) in that isolated segment. The initial displacement condition is zero displacement and a nil force everywhere except the segment of interest. The initial velocity is zero everywhere. This is a status at $t = 0$.

147

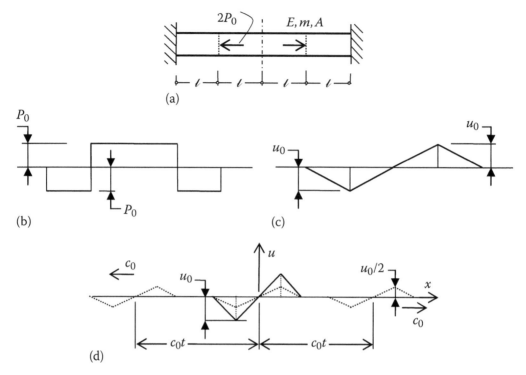

FIGURE 5.1 An example of generation of traveling waves: (a) initial state of an isolated segment, (b) axial force plot, (c) displacement plot, and (d) two opposite traveling waves after time t has elapsed from the release of constraints.

At that instant the constraints are removed and the initial deformed pattern, or the *initial disturbance* (as shown in Figure 5.1b and c), splits into two equal parts, moving away from each other with a velocity c_0 relative to the fixed coordinates' center. This is the reason why the constant c_0 in Equation 5.1 is called the *wave speed*. The general solution, Equation 5.2 shows two deformation waves traveling in opposite directions. In theory at least, those two *traveling waves* retain their shape and magnitudes regardless of how far they travel. The *wave front* is this end of a wave, which encounters an undisturbed observation point first. We can also speak of a *wave tail* that passes the observer last. Both wave front and wave tail are boundaries between disturbed and undisturbed medium. (A sharp boundary such as a wave front or tail is only sometimes encountered in practice; usually the transition is gradual rather than abrupt.)

When the equation of motion is linear, like Equation 5.1, a superposition of any two solutions is also a valid solution. Consider two opposite tensile waves traveling, this time, against one another, as in Figure 5.2a. When their fronts meet and then begin passing each other at point B, the superposition causes the tensile force to double in the overlapped segment (Figure 5.2b). After they pass each other, the magnitude of tension returns to the original value (Figure 5.2c). Now consider the displacements of the bar points coinciding momentarily with the wave fronts prior to their meeting. The front of the wave 1 is associated with the movement of a bar point to the left and the analogous point for wave 2 relates to movement with the opposite sense. When the fronts meet at point B, the motion cancels and point B remains motionless. This makes the point B fulfill the requirement of a fixed

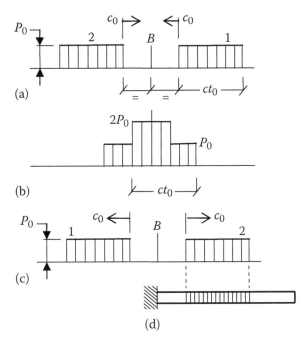

FIGURE 5.2 Superposition of two tensile waves: (a) two waves moving toward each other, (b) partially overlapping, (c) after passing each other, and (d) the above development seen as depicting a reflection of a single wave from the fixed end.

end of the bar. In Figure 5.2d, a bar to the right of point B is shown as fixed-ended. The described sequence of events may be reinterpreted as follows:

> **When a traveling wave rebounds from a fixed end, the magnitude of the axial force doubles during the rebound and, after completion of rebound, returns to the original magnitude. The sign of the force remains the same.**

During the time interval when the incoming wave overlaps with the reflected one the velocity of particles in the overlapped zone is nil. In the segment swept by the wave the sense of particle velocity is reversed after completion of rebound.

A different result arises if one of the waves in Figure 5.2a is tensile and the other compressive. When their fronts meet and then begin passing each other at point B, the superposition causes the forces to cancel at this location. The displacements associated with both fronts have the same sense with respect to the coordinate system, so they add up, or double when the fronts meet at point B. This absence of axial force makes point B meet the requirement of a free end of the bar. For the observer watching the events analogous to those in Figure 5.2d, the described sequence of events may be reinterpreted:

> **When a traveling wave rebounds from a free end, the magnitude of the axial force drops to nil during the rebound and, after completion of rebound, returns to the original magnitude. The sign of the force changes.**

During the time interval when the incoming wave overlaps with the reflected one, the magnitude of the force is nil, but the velocity of particles in the overlapped zone is doubled. The sense of particle velocity in the segment swept by the wave is unchanged after completion of rebound.

In order to clearly remember the relation between the sense of wave velocity and the particle velocity in a segment swept by a wave, the following rule is helpful:

In a compressive wave the sense of the wave speed is the same as that of particle velocity. In a tensile wave these senses are opposite.

The doubling of axial forces and velocities, for the respective cases, can extend over the entire length of a bar if the pulse duration is sufficiently long.

IMPOSITION OF END DISTURBANCE

This is imposed on an axial bar when a force is suddenly applied at the free end. In the example considered the force is constant, $P_0 H(t)$, as in Figure 5.3. After time t deformation spreads over a segment, whose length is $c_0 t$. The force in the segment is constant, equal to the applied force P_0. The static approach may be employed in calculating the end displacement:

$$u_e = \frac{P_0 c_0 t}{EA}$$

(5.3a)

with

$$c_0 = \sqrt{\frac{EA}{m}}$$

(5.3b)

When the only distributed mass is that of the bar itself, then $m = \rho A$ and

$$c_0 = \sqrt{\frac{E}{\rho}}$$

(5.4)

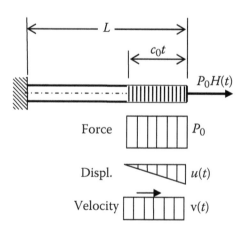

FIGURE 5.3   The status shortly after the end force is applied to an axial bar. In a tensile wave shown, the particle velocity is directed to the right, but the wave front moves to the left.

Differentiation of Equation 5.3a with respect to time gives the end velocity:

$$v_e = \frac{Pc_0}{EA} \qquad (5.5a)$$

or

$$\sigma_e = \frac{P}{A} = \rho c_0 v_e \qquad (5.5b)$$

where Equation 5.3b was used to obtain Equation 5.5b. The second of the above relationships seems to be the most popular wave formula. It relates the stress, σ_e, suddenly applied to the end of a bar to the velocity of that end section. This last quantity is also called the *particle velocity*, to distinguish it from the wave speed c_0. (Note that wave speed is independent of the magnitude of the applied stress, while the particle velocity is proportional to it.)

Equation 5.3a offers a simple way of calculating the end displacement u_e while the wave is sweeping the bar. At any time t, one regards the bar as fixed at $c_0 t$ from the end, where the wave has originated. The deflection is of the same magnitude as if a static force P_0 were stretching that segment of the bar. The result is so simple only because the load is invariant in time. The displacement of any intermediate point may be scaled, as Figure 5.3 indicates. This quasistatic relationship was taken here for granted, but it was not necessary to do so. One could deduce Equation 5.5 by a different approach and Equation 5.3 would result. This could be called a *quasistatic method* of finding deformation caused by a stress wave.

One should also note that the entire segment of length $c_0 t$ swept by the wave in Figure 5.3 has a particle velocity of v_e, not only the end section. A point of axis to the left of the incoming wave front is at rest, but as soon as the front passes, the point instantly acquires the velocity v_e.

There also is a counterpart of Equation 5.5b, which is more general in the sense that it can be applied to a bar that is already stressed and moving. In this case one speaks of stress increment $\Delta\sigma$ and (particle) velocity increment Δv. If $\Delta\sigma$ is applied during time t, then the wave sweeps a segment $c_0 t$ long, which acquires a velocity increment Δv. The impulse applied is therefore $(\Delta\sigma)t$ and the momentum change is $M(\Delta v) = (c_0 t)\rho\Delta v$. Equating the two one has

$$\Delta\sigma = \rho c_0 \Delta v \qquad (5.6a)$$

or

$$\Delta P = A\rho c_0 \Delta v \qquad (5.6b)$$

where the second equation is in terms of a force, rather than stress. The above can be applied when the additional particle velocity is additive with respect to one that was imposed before. When a reflection takes place, then

$$\sigma = \rho c_0 \Delta v \qquad (5.7a)$$

$$P = A\rho c_0 \Delta v \qquad (5.7b)$$

where
    σ is the total stress
    Δv is the total velocity change during rebound

For a reflection from a fixed end, for example, the incoming wave, with particle velocity v_0 and stress σ_0 is subjected to the total velocity increment of $2v_0$ and has the stress of $2\sigma_0$ during rebound.

TRAVELING WAVES, VIBRATION, AND A STANDING WAVE CONCEPT

When a disturbance is imposed on a bar of finite length, the periodicity motion becomes evident after some time. To illustrate the point, a bar in Figure 5.4 has a step load $P = 100H(t)$ imposed. (The result would not be much different if a finite pulse of duration t_0 were applied.) The above stress wave patterns result from successive reflection of the incoming wave from the ends, according to the rules discussed before. It can be noticed that after the wave traverses the bar four times, the motion is repeated, which means that what is shown in the figure is a single cycle of motion. It is instructive to compare this with the response of an oscillator to a step load, as discussed in Chapter 3. In both cases the

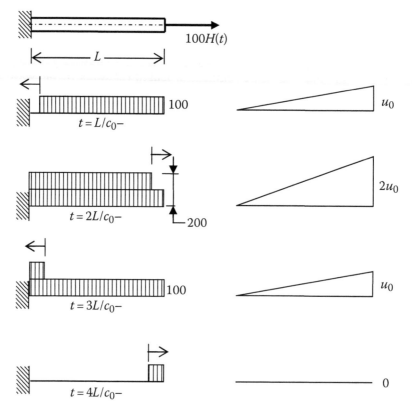

FIGURE 5.4 Changes in axial force and displaced pattern when a step load of magnitude $P_0 = 100\,\mathrm{N}$ is applied to the end. ($t = L/c_0-$ means just before $t = L/c_0$ has elapsed.). Static displacement from P_0 is designated by u_0. For simplicity, the displacement pattern is shown at the moment when the wave front arrives at an end.

displacement reaches its peak in the middle of the cycle and the system returns to the initial rest at the end. The quarter-cycle position, which corresponds here to static deflection of the rod, also corresponds to static deflection of an oscillator. The maximum excursion from the initial rest is twice the static displacement in both cases. One can easily note that the period of vibration here is $\tau = 4L/c_0$. Case 2.20b confirms this, even though the formula presented there was derived by a much different reasoning. Free vibration Case 2.20a, where the bar has both ends free, may be thought of as arising from a force, tensile at both ends, applied simultaneously. Here, the path that the wave has to traverse is only $L/2$, as the waves meet and rebound at the center, the latter acting like a fixed end. Consequently, the natural period is only one-half of what it was before. In Case 2.20c, free vibrations may result from a disturbance originating at the center. Just like in the previous case the path to be traversed is $4(L/2)$. Whatever was stated above with respect to the axial bar is also true for the three remaining members of the second-order group: a shear beam, a shaft, and a cable. (As for the latter, there is an obvious difficulty in handling and visualizing the response of an unconstrained end without invoking large deflections.) The impulsive action presented above leads to the conclusion that the bar will attain a vibratory pattern after some time. The same deformed shape can be arrived at in a different manner, by assuming a mode of vibration in a steady state. According to Timoshenko [110], an axial bar with ends fixed has a mode shape, which can be described by

$$X_n = \sin \frac{\pi x}{\Lambda_n} \tag{5.8}$$

where

$$\Lambda_n = \frac{L}{n} = \frac{\pi}{\omega_n}\left(\frac{EA}{m}\right)^{1/2} \tag{5.9}$$

in which ω_n is the natural frequency associated with the half-wave length Λ_n. This is depicted in Figure 5.5, showing the example for $n=2$. The deformation may be called a standing wave, because its shape does not change in time, but only the deflection magnitude does.

For every wave, one can write $u = ct$, where u is the distance traveled by a wave front moving with velocity c. If we put $u = 2\Lambda$, a distance corresponding to one period and time of a single period, $\tau = 2\pi/\omega$, then

$$2\Lambda = c\left(\frac{2\pi}{\omega}\right) \tag{5.10a}$$

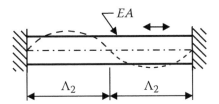

FIGURE 5.5   Lateral projection of displacement for a vibrating axial bar.

or

$$c = \frac{\Lambda\omega}{\pi} = \left(\frac{EA}{m}\right)^{1/2} \qquad (5.10b)$$

or, in other words, the nominal wave speed associated with a standing wave is the same, as that of a transient, traveling wave.

UNCONSTRAINED BAR PUSHED AT THE END

The motion of such a bar after the end impulse is applied can again be determined by using the wave propagation methodology. Let a short impulse $S = P_0 t_0$ be applied at one end. If the bar in Figure 5.6a is treated as a rigid body, the velocity gained is

$$v_r = \frac{S}{M} = \frac{P_0 t_0}{\rho A L} \qquad (5.11)$$

In Figure 5.6b, where the length of the pulse was for convenience taken as $c_0 t_0 = L/2$, the velocity history of the midpoint is shown. The axial force alternates between P_0 and $-P_0$ as the wave successively reflects from the free ends. The midpoint obtains an increment of velocity v_0 once the front of the wave passes it and then, at passing of the wave back, its velocity drops to nil again. The particle velocity in the bar has always the same sense, forward, regardless of the wave being compressive or tensile. This cycle repeats after every $2t_0$, because this is the time interval for the wave front to traverse the two half-lengths of the bar. According to Equation 5.5b the velocity initially applied is $v_0 = P_0/(\rho A c_0)$. If the average velocity for the typical interval is designated by $v_{av} = v_0/2$, one has

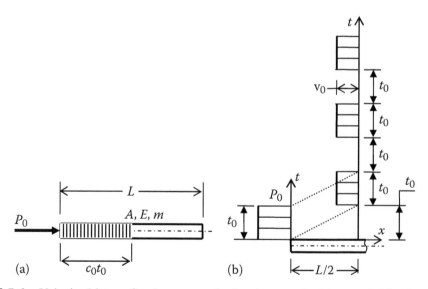

(a)   (b)

FIGURE 5.6 Velocity history for the center of a bar impacted with an axial load at the end: (a) loaded bar and (b) short pulse response.

$$\frac{v_r}{v_{av}} = \frac{P_0 t_0}{\rho AL} \frac{2}{v_0} = 1 \tag{5.12}$$

The average speed of midpoint is the same as if the bar were a rigid body. The actual speed, on the other hand, comes in discrete steps; the point either stands still or moves with $2v_r$. The shorter the end pulse is, the longer the periods of rest, although the time-averaged velocity is dictated by the magnitude of the end impulse only. One can also easily check that at the driven end (after the initial impulse) and at the far end the velocity jumps are twice as large, but they only last half as long as the initial impulse.

When a bar, moving with velocity v_0 impacts a rigid wall, Equation 5.5b is interpreted as follows. The change of speed of the impacted end is v_0, therefore the increment of stress (of the previously unstressed bar) is $\rho c_0 v_0$. Note that this is equivalent to a rigid wall impacting a stationary bar with the opposite, constant velocity. The details are given in Case 5.4.

COLLISION OF WAVES

In the section "The basic solution of the equation of motion" a superposition of two elastic waves, going in opposite directions was examined. Those waves were of the same sign, i.e., both were tensile or compressive. Another term for such an encounter of two waves is *collision*. As long as the material is elastic, the problem may be solved by superposition. For a more general material, a concept of *reflection from a moving interface* is more useful. This interface is an imaginary wall, moving with some velocity v_n, so that the resulting stress on each side of the interface is the same (Figure 5.7).

Suppose the interface velocity v_n is directed as shown in Figure 5.7. If the interface were fixed in space ($v_n = 0$) the right-going wave would experience the velocity change by $2v_1$. Taking v_n into account, however, results in the change of particle velocity being $2v_1 + v_n$. Repeating the reasoning for the left-going wave the following velocity changes, due to rebound from the interface, are obtained:

$$\Delta v_R = 2v_1 + v_n \text{ (right-going)} \tag{5.13a}$$

$$\Delta v_L = 2v_2 - v_n \text{ (left-going)} \tag{5.13b}$$

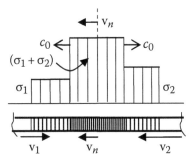

FIGURE 5.7 Left-moving and right-moving compressive waves collide and cause an increase in stress in the overlapping segment. Each of the waves can be thought of as reflecting from a rigid interface moving with a velocity v_n.

Associating these velocity increments with the interface stress and equating that stress on both sides of the interface results in finding v_n. For an elastic material stress is determined when the velocity increment is multiplied by impedance, as Equation 5.7 shows, which leads to

$$v_n = v_2 - v_1 \tag{5.14}$$

By substituting into Equation 5.13 one finds $\Delta v_R = \Delta v_L = v_1 + v_2$ and the total stress during rebound as $\sigma_1 + \sigma_2$, according to Equation 5.7a.

Transmission of Waves between Materials of Different Properties

Such transmission takes place when different materials are bonded at a junction, as illustrated in Figure 5.8. The stress in the incoming wave, Figure 5.8b, is σ_i. When the front encounters the interface, a reflected wave with stress σ_R appears at the same side and a transmitted wave, σ_T, emerges on the other side. All three waves are assumed to be compressive and in this case compression is treated as positive stress, which is usual when dealing with stress waves. The solution may again be based on the approach presented in the previous section, except that the interface location is quite specific now and the interface is expected to be moving to the right. Bar 2 is initially at rest, so its suddenly gained particle velocity can be written as $\Delta v_L = v_n$. The stress on both sides of the interface, σ_T, must be the same, therefore Equation 5.7a may be applied as follows:

$$\rho_1 c_1 (2v_1 - v_n) = \rho_2 c_2 v_n \tag{5.15}$$

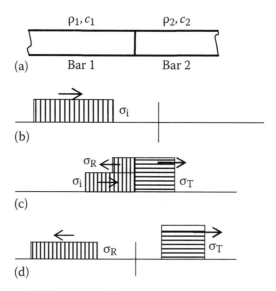

(a)   Bar 1   Bar 2

(b)

(c)

(d)

FIGURE 5.8   Joined bars of different materials: (a) configuration, (b) incoming wave, (c) interaction, and (d) after reflection/transmission.

where index 0 was dropped for simplicity. The product of ρ and c is called *specific impedance* of material. Using notation $i = \rho c$, the solution of the above for v_n is

$$v_n = \frac{2i_1}{i_1 + i_2} v_1 \qquad (5.16a)$$

or

$$\sigma_T = i_2 v_n = \frac{2i_2}{i_1 + i_2} \sigma_i \qquad (5.16b)$$

The stress on both sides of the interface must be equal, which means

$$\sigma_i + \sigma_R = \sigma_T \qquad (5.17)$$

from which the unknown σ_R can be found. In the end we have

$$\sigma_R = \frac{i_2 - i_1}{i_1 + i_2} \sigma_i; \qquad (5.18a)$$

$$\sigma_T = \frac{2i_2}{i_1 + i_2} \sigma_i \qquad (5.18b)$$

Several distinct conclusions from the above solution must be mentioned.

If the materials are identical, i.e., $i_1 = i_2$, then $\sigma_R = 0$ and $\sigma_T = \sigma_i$, which is rather obvious for a homogeneous bar, without any interface.
If the second bar is much more stiff and dense, i.e., $i_1 \ll i_2$, then $\sigma_R \approx \sigma_i$ and $\sigma_T \approx 2\sigma_i$. (Nearly fixed end of the first bar.)
If the first bar, containing the incoming wave, is much more stiff and dense, i.e., $i_1 \gg i_2$, then $\sigma_R \approx -\sigma_i$ (sign change) and $\sigma_T \approx 0$. (Nearly free end of the first bar.)

When the bars have the same section areas, as in this case, it is more convenient to work with stress, than a force. If they have different areas, A_1 on the left and A_2 on the right in Figure 5.8 and if the interface is capable of uniformly distributing stress over the cross sections, then it is more efficient to use the *impedance $I = A\rho c$*. Case 5.5 presents the details.

A brief review of uniaxial wave problems will perhaps help to clarify the logical development of the concepts involved so far. Imposition of a prescribed force at the end of the bar, such as shown in Figure 5.3 is equivalent to imposing the end velocity, with both quantities being related by Equations 5.5 through 5.7. Stopping the end of a moving bar is equivalent to applying the end velocity equal and opposite to the traveling speed of the bar. A reflection of a stress wave from a fixed or a free end, as initially discussed, is a more involved problem. In the former case the particle velocity of a traveling wave is first brought to nil by the fixity of the end. In the next step, that velocity is recovered, completing the rebound of a particle of the bar. The reflection from a free end involves doubling of particle velocity and a reversal of the sign of the wave stress.

WAVES IN BARS OF NONLINEAR MATERIAL

In the previous sections, the material studied was linearly elastic and so were the resulting waves. If a material is nonlinear, then the velocity of propagation depends on the stress level, i.e., instead of a constant c_0 one has $c(\sigma)$ [or, equivalently $c(\varepsilon)$] and Equation 5.1 becomes

$$\frac{\partial^2 u}{\partial t^2} - c^2(\sigma)\frac{\partial^2 u}{\partial x^2} = 0 \quad \text{with} \quad c^2(\sigma) = \frac{1}{\rho}\frac{\partial \sigma}{\partial \varepsilon} \tag{5.19}$$

For the sake of simplicity, further presentation will be limited to materials characterized by straight lines in the σ–ε plane or the bilinear characteristic. This is a fairly general, yet quite useful and frequently employed material model. (Refer to Figure 5.9.) The important feature now is to remember that unloading takes place along CD, which is parallel to the initial slope of OB. There is just one speed of propagation in the inelastic range, when the loading is increasing:

$$c_{pl} = \sqrt{\frac{E_p A}{m}} = \sqrt{\frac{E_p}{\rho}} \quad \text{where} \quad E_p = \frac{\partial \sigma}{\partial \varepsilon} = \text{const.} \tag{5.20}$$

as long as the stress level is above the yield point F_y. An example in Figure 5.9 shows the transmission of waves after loading the end of the bar. The applied stress increases linearly (in time) to the level of σ_m and remains constant thereafter ($\sigma_m > \sigma_0$).

The results are shown in Figure 5.9, in the *phase plane*, which means the x–t plane, as a function of time (t) for selected sections (x) along the bar. The first such section is the free

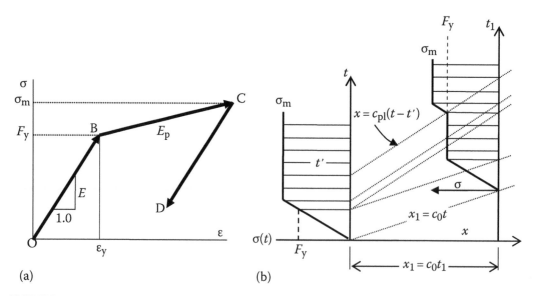

(a)   (b)

FIGURE 5.9   (a) Material with a bilinear stress–strain curve and (b) two stress histories shown in phase plane: the applied stress $\sigma(t)$ at the end at $x = 0$ and resulting stress at $x = x_1$.

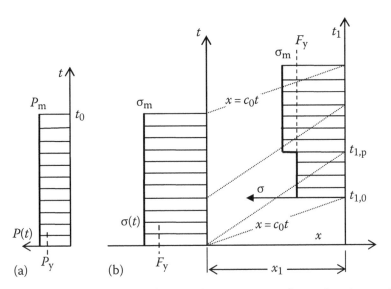

FIGURE 5.10   The applied load at $x = 0$ in (a) and stress patterns for $x = 0$ and $x = x_1$ in (b).

end, $x = 0$. The inclined lines connecting this end section "history" with $x = x_1$ are called characteristics. It can be demonstrated that the status (σ, v, and ε) at $x = 0$ is transferred unchanged to $x = x_1$, to the time point corresponding to that characteristics. As long as the applied stress is in the elastic range, the stress history at $x = x_1$ is essentially the same as at $x = 0$, although delayed by $t_1 = x_1/c_0$. Once the yield point F_y is reached, the speed of propagation suddenly drops to c_{pl}. This causes a delay in wave transmission, after which the increase of the applied stress resumes and continues until σ_m is reached. The reduction in speed of propagation causes the stress history at $x = x_1$ to be stretched in comparison with the original at $x = 0$.

Figure 5.10 shows stress distributions when a pulse with t_0 duration is applied at the end of a bar. (This time the prescribed end loading is shown in terms of a force, rather than stress, for a better clarity. Force P_y corresponds to yielding.) This may be regarded as a limit of what was presented before, when the rate of the applied load grows to infinity. At the beginning and until F_y is reached the wave front travels with the speed of c_0, but above F_y the speed drops to c_{0p}. This means there is an interval of time when the stress level remains at F_y at section x_1, just as in the case shown in Figure 5.9. This is a direct consequence of a sudden change of slope in σ–ε plot.

THE EXTENT OF A PLASTIC WAVE SPREADING

The extent can be assessed by considering a single, rectangular pulse of magnitude σ_m exceeding the yield level. A sudden removal of load at $t = t_0$ is transmitted as an elastic wave of opposite sign, which means that it spreads with velocity c_0. (Refer to Figure 5.9a for the unloading part of the curve.) The net effect is cutting off of the pulse at some x_1 location. In fact, when x_1 is chosen large enough, only the elastic component of the pulse will be transmitted to that location.

This is visualized in Figure 5.11, using a rectangular pulse of duration t_0 as an example. (All variables are the same as in Figure 5.10 except a new one, t_0, is introduced.) The length of the pulse t_0 is the same at x_1 as at the origin, but the full magnitude of σ_m is achieved

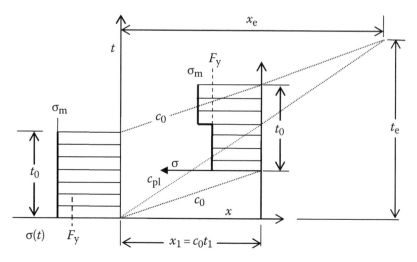

FIGURE 5.11 Rectangular pulse at the free end with the corresponding transmitted pulse at some distance away. The intersection of characteristics determines point (x_e, t_e) beyond which there is no inelastic action.

with some delay, as explained before. When thinking of unloading at x_1, it is helpful to refer to Figure 3.4, which shows a rectangular pulse as a sum of two infinite step loads, shifted with respect to one another and of the opposite sign. Alternatively, one might say that at t_0 unloading is achieved by applying an equal and opposite force, in addition to the force already existing. Unloading takes place along the original slope of $\sigma-\varepsilon$ for this material type and therefore the unloading pulse travels with the original, elastic speed. Once the characteristic c_0 emanating at the end of the pulse reaches x_1, the loading is simply cut off.

This illustrates that the plastic portion of the pulse becomes shorter, as we move the observation point (located on the axis of the rod) away from the origin. The intersection of characteristics (c_0 and c_{pl}) determines the position in space and time (x_e and t_e), beyond which there is no plastic action. Note that the following equations:

$$x_e = c_{pl} t_e \tag{5.21a}$$

and

$$t_e = t_0 + \frac{x_e}{c_0} \tag{5.21b}$$

solves the two unknowns:

$$x_e = \frac{c_{pl} c_0 t_0}{c_0 - c_{pl}} \tag{5.22a}$$

and

$$t_e = \frac{c_0 t_0}{c_0 - c_{pl}} \tag{5.22b}$$

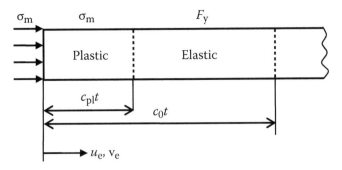

FIGURE 5.12   A bar of bilinear material suddenly loaded above the yield point.

The end velocity of a bar, pushed at the end by a load exceeding yielding is of practical interest. As Figure 5.12 shows, the first segment, near the origin, is loaded up to the external stress level σ_m. Consequently, its shortening is $\varepsilon_m(c_{pl}t)$. The remaining, elastic segment shortens by $\varepsilon_y(c_0t - c_{pl}t)$. The end displacement u_e at that time is equal to the sum of those two. The end velocity is therefore

$$v_e = \frac{u_e}{t} = c_{pl}(\varepsilon_m - \varepsilon_y) + c_0\varepsilon_y \tag{5.23}$$

If the above strains need to be replaced by stress, then usual bilinear equation applies. Noting that $\varepsilon_m - \varepsilon_y = (\sigma_m - F_y)/E_p$ and $\varepsilon_y = F_y/E$, the above equation can be written as

$$v_e = \frac{\sigma_m - F_y}{\rho c_{pl}} + \frac{F_y}{\rho c_0} \tag{5.24}$$

Note that when v_e is such that the yield point is reached, then $\sigma_m = F_y$ and the first term vanishes. The corresponding end velocity is the maximum value that can be applied without yielding the bar:

$$v_y = \frac{F_y}{\rho c_0} \tag{5.25}$$

For $v_e > v_y$, Equation 5.24 can be rewritten to

$$\sigma_m = F_y + \rho c_{pl}(v_e - v_y) \tag{5.26}$$

Finally, when the end velocity is suddenly decreased, from v_e to v_e', a partial unloading takes place along the elastic path. The new stress level is

$$\sigma_m' = \sigma_m - \rho c_0(v_e - v_e') \tag{5.27}$$

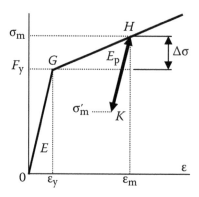

FIGURE 5.13   When the end is pushed faster than v_y, the maximum stress changes from point G to point H. Unloading will take it to point K. Reloading can take it back to point H and continue up along the plastic line.

which can be viewed as an alternative form of a bilinear stress–strain relation, illustrated in Figure 5.13.

When elongation at failure is large, say $\varepsilon_u = 30\%$ or more, we can speak of a *very* ductile material. Ignoring the terms relating to the elastic component of deflection, one can obtain a simple expression from Equation 5.24:

$$v_e \approx \frac{\sigma_m - F_y}{\rho c_{pl}} \approx c_{pl}\varepsilon_m \qquad (5.28)$$

and when v_e is so large that ε_m exceeds ε_u, failure eventuates. (This is applicable to tension, rather than to compression usually depicted in this chapter.) The reader can refer to the summary in Case 5.9 and the corresponding development for shearing in Case 5.10.

BAR IMPACTING RIGID WALL, INELASTIC RANGE

So far, the attention has been focused on the effect of applying a prescribed stress, or an equivalent velocity change to the end of a bar. When a bar, moving with the speed of v_0 impacts a rigid wall, the impacting end can be said to have experienced a velocity increment $\Delta v = -v_0$. Everything said in the preceding section about a sudden application of a constant end velocity v_e applies here as well. Regardless of the material involved, stopping the end of a bar moving to the left with v_0 has the same effect as applying the end velocity v_0, oriented to the right, to a stationary bar.[*] For example, Equation 5.26 can be used replacing symbol v_e by v_0.

The history of stress at any location along the axis will resemble, initially, that in Figure 5.10: The elastic wave appears first and is then followed by a slower, plastic wave, with both wave fronts shown in Figure 5.12. There is no cutoff due to limited pulse duration, however. Only after the elastic wave rebounds from the free end and returns to the impact point, there is a possibility of the latter separating from the wall. The situation is further

---

[*] In the first case the initial particle velocity of the bar is v_0, but in the second it is nil.

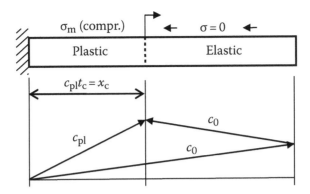

FIGURE 5.14 After the bar impacts a rigid wall, the plastic and the reflected elastic waves meet.

complicated by a collision of the returning elastic wave with the plastic wave still moving in the original direction. This is visualized, with the help of characteristics, in Figure 5.14. The wave fronts will meet at some point x_c, whose coordinate may be found as follows. The elastic wave will travel a distance $c_0 t = 2L - x_c$. At the same time, the plastic wave will traverse the distance of $c_{pl} t = x_c$. Eliminating time from these two relationships, one finds the location x_c and then the associated time point t_c:

$$x_c = \frac{2L}{(c_0/c_{pl}) + 1} \qquad (5.29a)$$

and

$$t_c = \frac{2L}{c_0 + c_{pl}} \qquad (5.29b)$$

After the returning elastic wave arrives at the collision point, as per Figure 5.14, the elastic part is stress-free and the part swept by the plastic wave has its stress σ_m determined by Equation 5.26. A detailed treatment of collision of the rebounding elastic wave with the progressing plastic wave can be found in Johnson [40]. The reasoning is similar to what was presented here for the interface of two bars of different materials, namely a compatibility of velocity and stress at the interface. For a moderate value of the impact velocity v_0, an elastic wave originates from the collision point, putting the elastic part in compression and unloading, to some extent, the plastic part of the bar. When the impact speed is as large as

$$v_0 = v_y \left(1 + \frac{2c_0}{c_0 + c_{pl}} \right) \qquad (5.30)$$

then, in addition to the elastic wave there also is a plastic wave spreading.

CONSTRAINED BAR PUSHED AT THE END

In Figure 5.2, there is an illustration of a wave rebounding from a fixed or constrained end, explained by means of a superposition of two pulses of the same kind, either tensile or

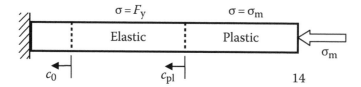

FIGURE 5.15   The initial wave pattern moving toward the fixed end of a bar.

compressive. Such a rebound results in doubling of the stress level when the incipient and the reflected waves overlap and the resulting stress does not exceed the yield limit. An alternative way of explaining the stress jump is this: the wave approaching the wall is defined by $\sigma_1 = \rho c_0 v_1$, which means it contains both the stress σ_1 and the particle velocity v_1. (This is another way of saying that it carries both strain and kinetic energies.) To cancel the particle velocity at the wall, one must apply σ_1 again, this time to the stressed bar section, thereby doubling the stress level.

In a situation where the applied stress σ_m exceeds F_y, as shown in Figure 5.15, the initial stress distribution does not differ from that in Figure 5.14, provided, of course, that the ends are reversed in the latter. The stress in the elastic wave approaching the wall is already at the yield point and the particles are traveling with the speed v_y determined by Equation 5.25, but it must undergo a sudden change of velocity by v_y, in the plastic range. The associated stress increase and the total reflected stress, respectively, are

$$\Delta\sigma = \rho c_{pl} v_y = \frac{c_{pl}}{c_0} F_y \qquad\qquad (5.31a)$$

and

$$\sigma_r = \left(\frac{c_{pl}}{c_0} + 1\right) F_y \qquad\qquad (5.31b)$$

The reflected plastic wave then returns to collide with one wave still moving against the wall. They rebound from each other as plastic waves as well. A similar problem, with a prescribed end velocity rather than a force, was solved in detail by Johnson [40] and its main results are presented as Case 5.13. After the precursor elastic wave rebounds, as described above, it is all plastic waves from that instant.

Two cases were discussed so far; a purely elastic one, where the incoming wave stress is doubled and one with an elastic wave at its upper limit F_y giving the reflected stress according to Equation 5.31b. However, there remains an intermediate range of the applied stress σ, namely $F_y/2 < \sigma < F_y$, which requires a slightly more general approach. The incoming wave velocity v is resolved into two components; $v_y/2$ and $\Delta v = v - v_y/2$. The first results in the rebound up to F_y, while Δv is associated with the plastic range. The stress increase from Δv and the total reflected stress, respectively, are

$$\Delta\sigma = 2\rho c_{pl}\left(v - \frac{v_y}{2}\right) \qquad\qquad (5.32a)$$

and

$$\sigma_r = F_y + 2\rho c_{pl}\left(v - \frac{v_y}{2}\right) \tag{5.32b}$$

it is easy to check that when $v = v_y$, then Equation 5.32 becomes identical with Equation 5.31.

TWISTING DEFORMATION OF SHAFTS

The analogy between translational and rotational motion was first presented in Chapter 1. Then, in Chapter 2, the equations of motion were developed, again displaying the analogous mathematical form for a bar and a shaft. The speed of a twisting wave c_1 is

$$c_1 = \left(\frac{GC}{\rho I_0}\right)^{1/2} = \left(\frac{G}{\rho}\right)^{1/2} \tag{5.33}$$

where the last equality, $C = I_0$ holds for the most common case, namely that of a shaft with a circular section. Acting in the manner that leads to writing Equation 5.5, we can show that the equations for a torque T_e, suddenly applied to a free end of a shaft is

$$T_e = \rho I_0 c_1 \lambda_e \tag{5.34}$$

where λ_e is the angular velocity enforced at that end. The dynamic relation between the end torque and the end twist is illustrated in Case 5.2. The analogy between tension and twisting, however, is clear only in the linear range, where the load and the stress are proportional. With the onset of plasticity the proportionality is lost due to redistribution of the shearing stress that follows.

To keep the calculations simple one may ignore that redistribution by pretending that the linear part of the torque–angle graph extends all the way to the full yielding of the section and at that point the whole section suddenly yields. (For example, Case 4.29 shows that for an elastic-perfectly plastic [EPP] material of a solid shaft the ultimate twisting moment, representative of the fully yielded condition, is 4/3 times the torque at the onset of yield.) In this chapter, we deal with bilinear material models, in order to avoid the situation when the plastic wave has zero speed. The relationship between the torque and the rotation angle can be made more manageable when plastic deformation is advanced, namely when the peak strain γ is several times larger than at the onset of yield taking place at $\gamma_y = \tau_y/G$. Under such a circumstance the torque–angle relation can be treated as trapezoidal, resulting from rigid strain hardening (RSH) stress–strain curve (Case 5.11). This method gives some underestimate of the twisting angle.

A thin-wall, circular shaft is free from this complication and its torque–angle characteristic can be assumed a rescaling of the stress–strain plot. The working of such a shaft, which is, in effect, a cylindrical shell, may be best imagined by cutting the generator to visualize the internal shear. An axial segment of the shell, when flattened, acts in the same way as a segment of a shear beam in Case 5.1.

STRONG SHOCK WAVES

When the applied action, in terms of end velocity or stress, is so intense that it results in stress being a number of times larger than the yield strength of material, one speaks of *shock waves* being transmitted. This is associated with a change of density of material and greatly complicates the analysis. One of the experimental equations applied in this field describes a relation between the speed of the wave front U and the particle velocity behind the front v:

$$U = c + sv \tag{5.35}$$

where coefficients c and s are specific for each material. This relationship is called a *Hugoniot* after one of the outstanding investigators in this field. The above version is valid for a material initially at rest, as illustrated in Figure 5.16.

The pressure–strain–velocity relationships for shock wave analysis are derived from hydrodynamic equations, which ignore material strength. An extensive treatment of the subject can be found in the books by Cooper [19] and Meyers [61]. For a material block, initially unstressed and at rest as in Figure 5.16, the pressure–velocity relation is

$$p = \rho_0 U v = \rho_0 (c + sv) v \tag{5.36}$$

where ρ_0 is the initial material density. One complication is obvious, compared with the elastic waves; the dependence of pressure on particle velocity is of the second order, unlike in Equation 5.5b, which is based on proportionality between the two variables. As a consequence, prescribing a constant pressure p_0 at the end of the bar makes it necessary to solve the quadratic Equation 5.36 for v.

As before for the elastic waves, it is the change of particle velocity rather than the velocity itself that determines the stress level. If the block moves with the speed v_1 and then the end speed is suddenly changed to v_2 (by $\Delta v = v_2 - v_1$) then Equation 5.36 may be written in a form analogous to Equation 5.7a:

$$p = \rho_0 [c(v_2 - v_1) + s(v_2 - v_1)^2] \tag{5.37a}$$

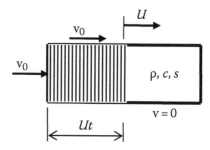

FIGURE 5.16 A block of material, whose end was subjected to a constant velocity v_0. After time t the wave had swept segment Ut and all particle velocities in this segment are v_0 now.

or

$$p = \rho_0[c(\Delta v) + s(\Delta v)^2] \qquad (5.37b)$$

After inspecting Equation 5.35 one could conclude that, when v becomes small, the wave speed U approaches c, and therefore c is the sonic velocity or the speed of propagation of elastic waves. This is not meant to be so, as c is different from the elastic wave speed. Equation 5.35 should therefore be seen only a good fit of experimental data.

A reflection of an elastic wave from a rigid wall was discussed in the section "The basic solution of the equation of motion." The doubling of stress was explained to be the effect of superposition of the two stress waves. The other way to present it was to note that the initial particle velocity was v_0, which, because of rebound was changed to $-v_0$. This means there was a net change of $2v_0$ in this velocity. Accordingly, Equation 5.6a would predict $\sigma = \rho c_0(2v_0)$. For a shock wave the procedure is similar, except that Equation 5.37a and b is used. The difference between the two is the presence of the quadratic term in the latter, which results in the reflected wave pressure being more than double the incipient pressure. This is detailed in Case 5.16. The collision of two shock waves is described in a manner analogous to that of an elastic wave, Case 5.15. Table 5.1 gives shock wave coefficients for various materials.

Transmission of a shock wave between two blocks of material is handled in a similar way as for elastic waves. The difference, of course, is the quadratic velocity term. Example 5.20 illustrates the procedure.

There is another type of a shock wave, which appears in media having a concave stress–strain curve, i.e., stiffening characteristic. During a dynamic loading process the wave speed increases, so that the waves emitted later are also faster, because of a larger $d\sigma/d\varepsilon$. This leads to an increase in the wave amplitude and may culminate in a strong shock wave. The formulas for a simple case, when the characteristic is bilinear, are provided in Case 5.18.

TABLE 5.1
Initial Density ρ_0 and Shock Wave Constants c and s

Material	ρ_0 (g/cm³)	c (km/s)	s
Calcium	1.55	3.60	0.95
Chromium	7.12	5.17	1.47
Copper	8.93	3.94	1.49
Iron	7.85	3.57	1.92
Magnesium	1.74	4.49	1.24
Nickel	8.87	4.60	1.44
Lead	11.35	2.05	1.46
Tungsten	19.22	4.03	1.24
Zinc	7.14	3.01	1.58
Sodium chloride	2.16	3.53	1.34
Al-6061	2.70	5.35	1.34
SS-304	7.90	4.57	1.49
Brass	8.45	3.73	1.43
Water	1.00	1.65	1.92

Source: After Meyers, M.A. *Dynamic Behavior of Materials*, Wiley, New York, 1994.

Body Waves in an Unbounded Elastic Medium

Equations of motion of an elementary volume of an unbounded elastic medium along with description of its deformation lead to the following conclusions. There are two distinct types of disturbances that can be propagated. The first type is a *pressure wave** or *P*-wave, which moves with the speed of c_p:

$$c_p = \sqrt{\frac{E_{ef}}{\rho}}$$

(5.38a)

where

$$E_{ef} = \frac{(1-\nu)E}{(1+\nu)(1-2\nu)}$$

(5.38b)

The second is the *shear wave,*[†] which has a speed of

$$c_s = \sqrt{\frac{G}{\rho}}$$

(5.39)

Any disturbance in the medium so described is a superposition of these two wave types. These expressions of wave velocities look familiar. In fact, when a comparison is made with the axial bar, which has no additional mass, so that A/m in Equation 5.3b becomes $1/\rho$, the only difference is E_{ef} now replacing E. If the axial bar is modified in such a way that no lateral movement is allowed, then, one can easily deduce from the generalized Hooke's law, that such a bar resists tension not with EA, but with $E_{ef}A$. The conclusion is the pressure wave speed c_p is the same, as a longitudinal wave speed in a bar, provided the bar is restrained against lateral motion.[‡]

The effective modulus, E_{ef}, does not greatly differ from E for typical Poisson's ratios. For $\nu = 0.15$, $E_{ef} = 1.056E$, and $c_p = 1.028c_0$. For $\nu = 0.3$, $E_{ef} = 1.346E$, and $c_p = 1.16c_0$. It is easy to determine that we always have $c_s < c_p$. It is useful to refer to Case 5.17, where the elastic constants are listed. Equation 5.39, on the other hand, is identical with Equation 2.22 for a shear beam, because, in absence of an added mass $A_s/m = 1/\rho$ which gives $c_s \equiv c_3$.

Consider a spherical cavity of radius a, existing in an infinite medium. If a uniform pressure is applied to the wall, it becomes a source of a pressure wave, moving away along the radius r. The wave displays spherical symmetry and all its quantities depend only on r and not on the remaining spatial coordinates. The important question is: If a pressure time step, $p = p_0H(t)$, is applied, how does the cavity surface respond? At this point one should recall that a mass of a simple oscillator, acted upon by such a load (Case 3.2) would move forward during the half-period and that it attains the peak displacement u_m of twice its static

---

* The term *compressional wave* seems to be the general favorite. The terms *dilatational, irrotational,* and *primary* are sometimes used.
[†] Also known as *equivoluminal, distortional,* or *secondary.*
[‡] Still, when one speaks about waves in bars, the lateral restraint is not implied, unless explicitly stated.

deflection [DF(u) = 2]. A similar solution is expected for a cavity in the sense that after some application time t_m the presence of the load is not relevant, because a rebound begins. A simplified solution for the cavity radius change is given by Case 5.19. It shows that both the u_m as well as the accompanying time t_m are linear functions of the cavity radius a. The DF is smaller for a cavity than for a linear oscillator.

At this point it is useful to review static results for a spherical cavity presented in Case 3.33. The radial displacement u, radial stress σ_r, and hoop stress σ_h are, respectively

$$u = \frac{1+\nu}{E} \frac{p_0 a^3}{2r^2} \tag{5.40}$$

$$\sigma_r = -p_0 \left(\frac{a}{r}\right)^3 \tag{5.41a}$$

$$\sigma_h = \frac{p_0}{2} \left(\frac{a}{r}\right)^3 \tag{5.41b}$$

where the minus sign indicates radial compression. An initial estimate of the dynamic stress experienced in the medium outside the cavity can be deduced from the following simple reasoning.

When a pressure wave with the shape of a rectangular pulse emanates from a cavity, then at some distance r_1 a layer of material, with a radial thickness of h, is compressed with by a stress of σ_1. At a larger distance, r_2, the stress amplitude will change to σ_2. One can assume that the strain energy associated with the wave remains constant and conclude that the amplitude decreases in proportion to $1/r$. The loss of amplitude due to the wave spreading in space is often called a *geometric attenuation*.

Unfortunately, the picture is not so simple, because the traveling pulse quickly loses its original shape. What initially is a compressive pulse begins to resemble, in time, a full sinusoid, which means that a tensile part develops. For this reason the decay is faster; and the proportionality is closer to $1/r^{1.5}$, rather than to $1/r$, as shown in Chapter 15.*

So, if pressure with a magnitude p_0 is applied statically in a spherical cavity, the stress decreases like $1/r^3$, as shown by Equation 5.41, but for the same pressure applied suddenly, the decrease in the peak value is proportional to $1/r^{1.5}$. This indicates that a pressure wave is a more efficient way of transmitting a mechanical effect than a static application of the same.

Consider now an infinitely long, *cylindrical* hole in the medium. When a uniform pressure is statically applied within that hole, a uniform, cylindrical stress field is induced. When viewed in two dimensions, in plane normal to the hole axis, it is the plane strain field. When the same pressure is suddenly applied, a *cylindrical wave* is emanated from the hole. (This wave may also be viewed as a superposition of spherical pressure waves emanating from points on the axis of the hole.) Using the same reasoning, as was applied for a spherical wave, one can demonstrate that the amplitude of such a wave is proportional to $1/\sqrt{r}$.

---

* This proportionality takes effect for r larger than several cavity radii of the hole and is valid for lightly damped media like a good quality rock.

The loss of pulse shape degrades the magnitude, but its retention is better than under a static load, where stress decreases in proportion to $1/r^2$.

Real mining boreholes, in which explosions take place, have only a finite length. Yet, in a part of the volume surrounding a hole a strong, cylindrical stress wave develops. Near the top and bottom of a borehole, static analysis would show local peaks of shear stress. Those areas are also sources of shear waves during an explosion.

The pressure waves discussed above bear a strong similarity to longitudinal waves in bars previously described. Yet, there are marked differences between the two. A rectangular pulse originating from an end of the bar, will maintain its amplitude and shape often for a substantial length of time. The same pulse emitted from a cylindrical cavity will lose its shape and have its magnitude diminished much faster, due to geometric attenuation. This effect is even more pronounced for a spherical cavity. Also, compressive and tensile stress components of a similar magnitude appear, even if the original pulse was only compressive.

In describing the above wave phenomena, the assumption was that in addition of being elastic, the medium was homogeneous and isotropic. The limitation of being "unbounded" should not, of course, be treated too literally. It only means "far enough from a boundary, so that within the time of interest the presence of boundary has no influence."

WAVES IN AN ELASTIC HALF-SPACE

When a boundary is a part of the picture, the equations of motion of an element have a third solution, known as *Rayleigh wave* or the R-wave in brief. The motion described by such a wave has its largest amplitude at the surface and quickly decays with depth. The speed of the R-wave c_R is slightly smaller than that of shear wave:

$$\text{for } \nu = 0, c_R = 0.874c_s; \quad \text{for } \nu = 0.5, c_R = 0.955c_s \tag{5.42}$$

For the intermediate values of ν the following relation, quoted by Persson et al. [66], can be used:

$$c_R = \frac{0.86 + 1.14\nu}{1 + \nu} c_s \tag{5.43}$$

The horizontal component of the R-wave motion is largest at the free surface. The same is nearly true for the vertical component, which peaks out a little below the surface. In this sense, the Rayleigh wave is more a surface wave than a body wave.

If there is a disturbance within the volume of a geological medium, an observer standing on its surface (be it a human observer or a geophone) will experience the following sequence of wave action:

1. P-wave, which arrives first, as the fastest one
2. S-wave, at some time afterward
3. R-wave, immediately following the S-wave

The distinction between 2 and 3 above may be at times difficult, due to varying properties of the strata forming the medium. The relative mechanical effect of the above components

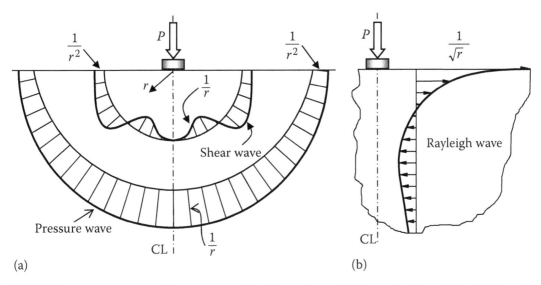

FIGURE 5.17 P-wave and S-wave spreading from the excitation point (a) and the horizontal component of the R-wave (b).

depends on the amount of energy associated with each. A study quoted by Richart et al. [72], carried out for a circular footing that is placed on an ideal medium surface and excited vertically, gives the following energy breakdown: 67% in R-wave, 26% in S-wave, and 7% in P-wave. This clearly shows that the effect of R-waves is dominating, which is true for man-made as well as for natural earthquakes. The main three waves spreading from an excitation point are illustrated in Figure 5.17 following the concept in Ref. [72]. The attenuation at the surface is considerably faster for P-wave and S-wave than the decay of the R-wave. (The vertical component of the R-wave is not shown, but it decays at about the same rate as the horizontal one.)

While seismic considerations are important, there is another area of applications of these concepts, namely (sub)surface blasting in the mining industry. The wave equation is used in the form $\Delta\sigma = \rho c_p(\Delta v)$ where the increment of longitudinal stress $\Delta\sigma$ is associated with the change of velocity Δv.*

It was demonstrated before that when a compressive wave reflects from the end of a bar, it changes to a tensile wave and vice versa. In a three-dimensional case as here, the effects are the same. When the pressure wave, originating far below the surface, reflects from that surface, the sign of the associated stress changes from compression to tension. Most media are weaker, or much weaker, in tension than in compression; therefore, the sign change may have a profound effect. This is investigated in detail in Chapter 15. If an interface with another medium is involved, the rules of reflection and transmission developed for axial bars also hold, except for a change in wave speeds.

When the source of a wave is close to the surface, the situation is greatly complicated. A portion of the emitted spherical wave impacts the surface at some angle. In addition to a reflected P-wave, a reflected S-wave also emerges. If, instead of a free surface, there is a boundary

---

* This is often quoted as σ rather than $\Delta\sigma$ here. The reason for distinguishing between the two is the presence of the pre-existing stress as well as hydrostatic stress in the ground.

TABLE 5.2

Wave Speeds for Common Materials

Material	E (GPa)	ρ (kg/m³)	ν	G (GPa)	c_0 (m/s)	c_s (m/s)	E_{ef} (GPa)	c_p (m/s)
Steel	200	7850	0.3	76.92	5047.5	3130.4	269.23	5856.4
Aluminum alloy	69	2700	0.35	25.56	5055.3	3076.5	110.74	6404.3
Copper	110	8970	0.3	42.31	3501.9	2171.8	148.08	4063.0
Brass	110	8500	0.35	40.74	3597.4	2189.3	176.54	4557.4
Lead	18	11300	0.45	6.21	1262.1	741.1	68.28	2458.1
Glass	70	2300	0.25	28.00	5516.8	3489.1	84.00	6043.3
Granite	65	2600	0.22	26.64	5000.0	3200.9	74.21	5342.5
Concrete	31	2400	0.2	12.92	3594.0	2319.9	34.44	3788.4
Hardwood	12	590	0	6.00	4509.9	3189.0	12.00	4509.9

Note: Wave speeds are designated as follows: c_0: uniaxial wave, in thin bar; c_s: shear wave, beam, or medium; c_p: pressure wave, confined bar or medium.

with another medium, one gets a reflected P-wave, reflected S-wave, refracted P-wave, and a refracted S-wave. A good, basic introduction to this aspect can be found in Johnson [40].

When the elastic medium has a layered structure, another type of disturbance can propagate, namely *Love waves*. An ideal case is presented by relatively flexible top layer supported by a more rigid, deep layer below it. If the S-wave speed of the top layer is c_{s1} and that of the layer below is c_{s2}, then, the condition $c_{s1} < c_{s2}$ makes the top layer a *wave trap*. This means that a disturbance originating within the top layer will reflect from the surface of the lower layer and return to the free surface. The repetition of this cycle makes the Love wave spread with the velocity c_L such that $c_{s1} < c_L < c_{s2}$. The wave energy is then confined to the top layer. The medium configuration mentioned above is quite common in geology, although perhaps not in such ideal form.

Nonreflecting Boundary

Nonreflecting boundary is a very useful device frequently used to limit the physical extent of finite element (FE) modeling. In Figure 5.3 it is evident that the wave issuing at the free end will, in time, reflect from the fixed end and return. In Figure 5.4 the effect of a repeated passage of a wave along a limited length of bar is illustrated. If one wants to study the wave generation and passage without the disturbance caused by an adjacent boundary, the model length must often be large. To eliminate the wave rebound altogether, one may cut off the model at a certain point and introduce an element, which would behave like a removed, semi-infinite part of the bar. Will a spring with a properly selected constant be such a boundary element? When a static displacement is imposed at the end of a semi-infinite bar, there is no reaction. However, in the presence of a spring there is a finite reaction, which is an inconsistency. The other possibility is to use a damper. When the end of the bar is pulled with a speed of v, the reaction is defined by Equation 5.5.[*] The damper reaction is therefore

---

[*] Even though the other equation was developed to give the effect of a suddenly applied force, it works in both ways, giving the resisting force when driving velocity is applied.

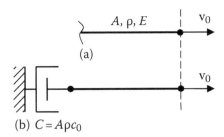

FIGURE 5.18 (a) A semi-infinite cable driven by v_0 and (b) the same but a finite-length cable with a nonreflecting boundary.

$$P_d = A\rho c_0 v \equiv Cv \qquad (5.44)$$

where C is a damper constant as defined in Chapter 3 and depicted in Figure 5.18. A damper is therefore an appropriate element to place where the rest of the bar is cut off. This concept is extended to infinite medium with loads symmetrical about a vertical midplane, as specified in Case 5.20. The main difference, here, compared with a uniaxial bar is that the medium offers a resistance to static loading and, therefore, both springs and dampers are needed to replace it. Those two types of elements are area-proportional. For this reason, it is easy to create a nonreflecting boundary around a cylinder. A sphere or an axisymmetric problem, on the other hand, is more laborious, as the tributary areas vary along the boundary.*

PULSE DECAY DUE TO MATERIAL DAMPING

A pulse of sinusoidal shape and of length b in time domain, as in Figure 5.19, is traveling through the medium with a sonic velocity c. A stress wave over segment b can be interpreted as one-half of the vibratory cycle undergone by that segment. (From zero to maximum and back to zero.) If a pulse travels distance L, this means the number of such imaginary cycles is

$$n = L/(2b) \qquad (5.45)$$

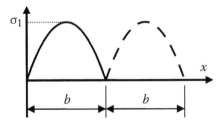

FIGURE 5.19 A pulse of length b traveling from an end of a bar. Solid line shows the initial position and dashed line shows the pulse after $t = b/c$.

---

* The procedure described is of benefit when a nonreflecting boundary is not available as a feature in a computer program being used.

From Chapter 3 it is known that after n cycles of motion, the ratio of the initial amplitude σ_1 to the current amplitude σ_n becomes

$$\frac{\sigma_1}{\sigma_n} = \exp(2\pi n\zeta) = \exp\left(\frac{\pi\zeta L}{b}\right) \tag{5.46}$$

provided ζ is small.* (Many authors use logarithmic decrement, $\Delta = 2\pi\zeta$ as the basic damping constant, instead of ζ.) The value of ζ is a property of the medium, through which the wave is transmitted. But regardless of what that value is, it must be within certain bounds for the sake of FE modeling. If ζ is too small, a numerical noise may cause some shape distortions. If is too large, the pulse will be decaying too rapidly. This will be made more specific with regard to an axial bar, with nodes uniformly spaced at distance l from one another. As Example 5.5 shows, the natural frequency of a continuous (real) bar will be preserved, when modeling is done correctly. Then, this frequency is

$$f = \frac{c}{2\Lambda} = \frac{c}{2l} \tag{5.47}$$

where l is a half-wave length involved. In this instance our interest is in the highest mode of vibrations that can be extracted from this model. A trial with an FE code will show that when the code operates on the basis of a consistent mass formulation, then the highest frequency, which can be extracted[†] is close to

$$f = \frac{0.55c}{l} \tag{5.48}$$

When the code employs only a stiffness-proportional damping (β damping), there is a simple relation, based on Equation 3.49:

$$\zeta = \frac{\beta\omega}{2} \tag{5.49a}$$

or

$$\zeta = \pi\beta f \tag{5.49b}$$

Inserting $f = 0.55c/l$ one finds the input constant β:

$$\beta = 0.5787\frac{l\zeta}{c} \tag{5.50}$$

---

[*] Strictly speaking, $n-1$ should be used in place of n, but this refinement is not necessary in view of other uncertainties as well as large number of cycles usually considered.
[†] If a lumped-mass formulation is used, the frequency results can be found in Case 2.19.

TABLE 5.3

Damping Properties of Selected Materials

Material	1/ζ	Deformation
Steel	3696	Flexure
Copper	1964	Flexure
Fused quartz	2417	Torsion
Lead glass	1496	Torsion
Soft glass	449	Torsion
Wood	233	Flexure
Polystyrene	131	Torsion
Charcoal granite	960	Direct
Basalt	504	Direct
Tennessee marble	378	Direct
Black slate	316	Direct

Note: *Direct* means tension or compression.

In a real, multi-degree system, there are a number of natural frequencies. If one uses a selected value of ζ in the above expression, the remaining modes, associated with lower frequencies, will exhibit lesser damping. The experience of this author is that when $\zeta = 0.4$ is used in the highest mode, the initial impulse shape appears regular and there is no excessive damping along the length.[*] This gives

$$\beta = 0.2315 \frac{l}{c} \tag{5.51}$$

For steel, from Table 5.2, $c = 5048$ mm/ms and the above results in $\beta = l/21{,}800$, when l is expressed in millimeters. Damping properties (ζ) in Table 5.3 are quoted along with the deformation mode, from which they were extracted. (The values are quoted after Kolsky [46] and Kutter and Fairhurst [50].)

It is quite clear that for most materials of interest ζ is a very small number. One reservation with regard to the last four entries in the table should be noted. The ζ values are given for a "competent" rock, i.e., a medium, which has only minor flaws. For a low quality rock, ζ can be several times larger.

SOME NOTES ON FINITE-ELEMENT MODELING OF WAVE MOTION

One of the fundamental questions facing every analyst setting out to create a dynamic model is the mesh density, i.e., how far should the nodes be from one another in order to adequately describe the event. If sustained vibrations are of interest, then the model must be

---

[*] One should not be alarmed with the high value of ζ recommended here. In a properly constructed FE model the contribution of the highest mode should be very small.

capable of responding to all frequencies of interest. In the case of a shock, however, such a criterion does not exist, as the number of relevant frequencies can be very large.

To make the explanation more specific, consider an axial bar where a rectangular load pulse lasting $t_0 = 0.5$ ms is applied. The bar has the sonic velocity of $c_0 = 2000$ m/s, so the pulse has the length of $L = c_0 t_0 = 1000$ mm, when measured along the bar. The question boils down to how many elements should be placed within the pulse length, so its shape can be reasonably accurately transmitted along the bar. The experience of this author indicates that placing 10–12 such elements is, in most cases, sufficient. In our example the nodes must be spaced at $l = L/10 = 100$ mm from one another, or closer. The rectangular pulse was used for simplicity of illustration only. The above applies to a less regular pulse as well, but a pronounced irregularity will demand more elements to be used.

Having the node spacing established, the next question is the time increment for integration, δt. The reference time is the interval needed for the wave to traverse the shortest element, l/c, where c is the speed of the wave of interest. The implicit or "typical" FEA codes (ANSYS, for example) need δt to be not larger than

$$\delta t = \frac{l}{2.5c} \tag{5.52}$$

Taking δt shorter than this would not do much for accuracy, but will increase the cost. Taking δt as a longer interval may decrease the accuracy. An implicit code (LS-Dyna for example) may have a default value of

$$\delta t = 0.9 \frac{l}{c} \tag{5.53}$$

This is quite sufficient for a moderate velocity impact or a sudden application of force. When the action of explosives is involved, it is recommended that the coefficient be reduced to 0.7, although a further reduction may be needed at times. For some special problems, like fragmentation, the reader may find that in order to achieve a good resolution the integration time increment needs to be substantially reduced, sometimes making the coefficient in Equation 5.53 as small as 0.1.

The β-damping criterion expressed by Equations 5.50 and 5.51 is a minimum requirement, related to computer simulation and allowing one to avoid artificially induced vibrations. The FE user should make certain that the damping coefficient β used in simulation is not less than what the actual material exhibits.

COMMENTS ON LIMITS OF SOME SOLUTIONS

In theory, a rectangular pulse emitted along an axial bar should continue moving along with the sonic velocity, while retaining its shape. In reality, the initial and the final part of the pulse gradually "spill" and, after a while, a finite-length pulse looks closer to a sine curve than to a rectangle.

The other question is how accurately a rectangular pulse can be produced. When experimental efforts are made, the beginning and the end of the pulse look like inclined rather than vertical lines and the top is more or less wavy, rather than straight. (The details are provided in Johnson [40].) Some of these effects can also be seen in FE simulations.

The subject of elastic wave propagation in an axial bar, as well as in other elements of similar mathematical description was one of the main topics in this chapter. In order that longitudinal waves may propagate in the manner described, the wave length must be significantly larger than the lateral dimension of the bar, in which the wave travels. If these two sizes are comparable, then some additional effects, arising out of the three-dimensionality of a bar take place and obscure the picture.

Closing Remarks

There are many terms associated with wave motion in solids. For example, one can speak of a wave of deformation in general, or about a stress or pressure or strain waves. Elastic waves in solids have a gradually rising front, while shock waves are distinguished by a steep, near vertical front.

A simplified theory for the NL range developed here and based on a bilinear stress–strain curve has some limitations. Even if the stress–strain seems like a good, overall idealization, the real material curves have a smoothly varying slope. This means that the wave speed changes gradually, rather than displaying a sudden decrease when going from an elastic to a plastic range. In real configurations, this may have some effect on failure predictions.

A quasistatic method of calculating displacements in a bar affected by a wave was discussed in the section "Imposition of end disturbance." This result holds true for any other nondispersive wave motion: lateral string deflection, twisting of a shaft and the pure shearing of a beam. One can say, in general, that even in highly transient wave motions statics remains a useful tool.* In later chapters some approximations will be developed along these lines for elements, which are dispersive by nature.

Achieving a complete fixity at an end of a bar is possible in practice only when this condition is associated with symmetry, which prevents some displacements. When a restraint is only an elastic body, it is difficult to preclude some displacements at the point of "fixity." While it is possible in theory and in virtual experiments, physical testing can rarely approach a perfect fixity. This matter is discussed further in chapters to follow.

---

* At the time of this writing this fact does not appear to be widely known.

TABULATION OF CASES

CASE 5.1   CONSTANT FORCE P_0 WITH A SHORT DURATION t_0 APPLIED TO
(1) AN END OF A LONG AXIAL BAR AND (2) FOR A SHEAR BEAM

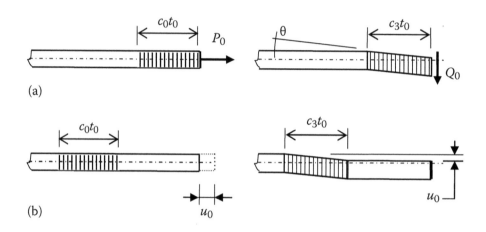

(a)

(b)

$$t \leq t_0 \text{ is shown in (a)}, \ t > 2t_0 \text{ in (b)}, \ m = A\rho$$

(1) Axial bar: $u_0 = \dfrac{P_0 c_0 t_0}{EA}$ with $c_0 = \sqrt{\dfrac{EA}{m}}$

$v_0 = \dfrac{u_0}{t} = \dfrac{P_0 c_0}{EA} = \dfrac{\sigma_0 c_0}{E}$; end velocity, equivalent to P_0 being applied

$P_0 = v_0 \sqrt{EAm}$; end load, equivalent to v_0

$\sigma_e = \rho c_0 v_0$; stress in wave-affected part

$\varepsilon = \dfrac{v_0}{c_0}$; strain in wave-affected part

(2) Shear beam: $u_0 = \theta c_3 t_0 = \dfrac{Q_0 c_3 t_0}{GA_s}$ with $c_3 = \left(\dfrac{GA_s}{m}\right)^{1/2} = \left(\dfrac{A_s}{A}\dfrac{G}{\rho}\right)^{1/2}$

$v_0 = \dfrac{u_0}{t} = \dfrac{Q_0 c_3}{GA_s} = \dfrac{\tau_0 c_3}{G}$; end velocity, equivalent to Q_0 being applied

$Q_0 = v_0\sqrt{GA_s m}$; end load, equivalent to v_0

$\tau_e = \rho c_3 v_0$; stress ($\tau_e = Q_0/A$)

$\gamma = \dfrac{v_0}{c_3}$; strain in wave-affected part

Note: The problem can be stated by specifying either end load or end velocity.

CASE 5.2 CONSTANT TORQUE T_m LASTING t_0 APPLIED TO AN END OF A SHAFT

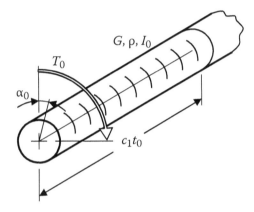

$c_1 = \left(\dfrac{GC}{\rho I_0}\right)^{1/2} = \left(\dfrac{G}{\rho}\right)^{1/2}$; speed of propagation of twisting wave

$\alpha_0 = \dfrac{T_m c_1 t_0}{GC}$; angle of twist after t_0

$\lambda_e = \dfrac{T_m c_1}{GC}$; angular velocity at the driven end

$\lambda_y = \dfrac{T_y' c_1}{GC}$; maximum λ_e possible without yielding

T_y' is the limit moment associated with the complete yielding (see Case 4.29)

$C = I_0$ for circular sections

$\tau_e = T_0 r/I_0 = \rho c_1 r \lambda_e = (G\rho)^{1/2} r \lambda_0$; shear stress at radius r

Note: See Case 4.29 for section properties.

CASE 5.3   CONSTANT FORCE P_0 WITH A SHORT DURATION t_0 APPLIED TO AN INTERMEDIATE POINT OF A SHEAR BEAM

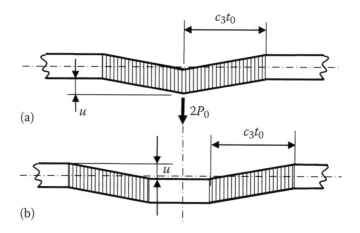

(a)

(b)

(a) Status at $t = t_0$
(b) Status at $t > t_0$

All formulas of Case 5.1 hold here.

Note: The problem can be stated by specifying either applied load or applied velocity. When axial loads are applied, the same approach holds.

CASE 5.4   AXIAL BAR IMPACTING RIGID WALL

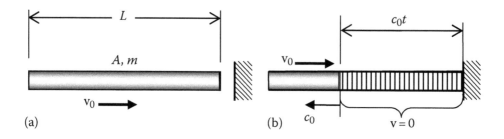

(a)

(b)

(a) Prior to impact
(b) Just after impact

Elastic wave originates at impact point.

Wave data:

$$c_0 = \sqrt{\frac{E}{\rho}} ;\text{ speed of wave front}$$

$\sigma_m = \rho c_0 v_0 = v_0 \sqrt{E\rho}$; stress behind wave front

$u_m = \dfrac{\sigma_m L}{E}$; maximum shortening

$t_0 = L/c$; time for the wave to traverse L

$t_d = 2L/c$; contact duration

Note: After $t = t_d$ the bar becomes unloaded and flies off with v_0. The bar impacting a rigid wall with v_0 is equivalent, in terms of stress, to the equal and opposite velocity being applied to the end.

CASE 5.5   WAVE TRANSMISSION BETWEEN TWO BONDED BARS

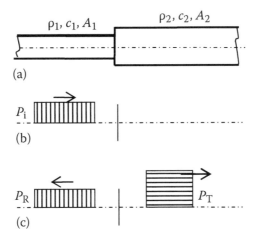

(a) Properties
(b) The incoming wave
(c) Status after reflection

$i_1 = \rho_1 c_1$; $i_2 = \rho_2 c_2$

$I_1 = A_1 \rho_1 c_1$; $I_2 = A_2 \rho_2 c_2$

$P_R = \dfrac{I_2 - I_1}{I_1 + I_2} P_i$; after reflection in bar 1

$P_T = \dfrac{2 I_2}{I_1 + I_2} P_i$; during reflection in both bars, after reflection in bar 2

When $A_1 = A_2$, and with $\sigma = P/A$:

$\sigma_R = \dfrac{i_2 - i_1}{i_1 + i_2} \sigma_i$; after reflection in bar 1

$$\sigma_T = \frac{2i_2}{i_1 + i_2}\sigma_i; \text{ during reflection in both bars, after reflection in bar 2}$$

Velocity of interface: $v_n = \dfrac{\sigma_T}{i_2}$ or $v_n = \dfrac{P_T}{I_2}$ or $v_n = \dfrac{2i_1}{i_1 + i_2}v_1$

Note: The interface is capable of uniformly distributing stress across sections.

CASE 5.6   WAVE TRANSMISSION BETWEEN TWO BONDED BLOCKS OF DIFFERENT MATERIALS

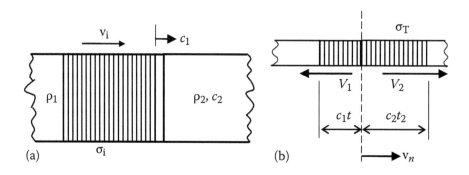

(a)   (b)

The approaching wave is in (a). Compressive waves resulting are shown in (b).

$i = \rho c = \sqrt{E_{ef}\rho}$; specific impedance

$$E_{ef} = \frac{(1-v)E}{(1+v)(1-2v)}; \text{ effective modulus}$$

$$\sigma_R = \frac{i_2 - i_1}{i_1 + i_2}\sigma_i; \text{ after reflection in block 1}$$

$$\sigma_T = \frac{2i_2}{i_1 + i_2}\sigma_i; \text{ during reflection in both blocks, after reflection in block 2}$$

$$v_i = \frac{\sigma_i}{i_1}; \text{ initial particle velocity in the left block}$$

$$v_n = \frac{2i_1}{i_1 + i_2}v_i = \frac{\sigma_T}{i_2}; \text{ velocity of interface during reflection}$$

$$V_1 = \frac{\sigma_R}{i_1}; \text{ final particle velocity, left block}$$

$$V_2 = \frac{\sigma_T}{i_2}; \text{ final particle velocity, right block}$$

Note: The contact area of blocks is assumed large enough in both directions, so that edge effects are minor. The above reduces to collision of bars when contact area is small (E_{ef} becomes E). The incoming wave length should be substantial to avoid 3D effects.

CASE 5.7   WAVE TRANSMISSION BETWEEN THREE BONDED BARS SHOWN AS CHANGES IN AXIAL FORCE *P*

$i_1 = \rho_1 c_1$; etc.

$I_1 = A_1 \rho_1 c_1$; etc.

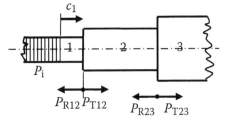

$$P_{R,1-2} = \frac{I_2 - I_1}{I_1 + I_2} P_i; \quad P_{T,1-2} = \frac{2I_2}{I_1 + I_2} P_i; \text{ at } 1\text{--}2$$

interface, initial impact

$$P_{R,2-3} = \frac{I_3 - I_2}{I_2 + I_3} P_{T,1-2}; \quad P_{T,2-3} = \frac{2I_3}{I_2 + I_3} P_{T,1-2}; \text{ at } 2\text{--}3$$

interface, initial impact

$$P'_{R,1-2} = \frac{I_1 - I_2}{I_1 + I_2} P_{R,2-3}; \quad P'_{T,1-2} = \frac{2I_1}{I_1 + I_2} P_{R,2-3}; \text{ at } 1\text{--}2 \text{ interface, secondary impact}$$

with reflection into bar 2, transmission into bar 1.

Note: The interfaces are capable of uniformly distributing stress across sections. The wave is short enough so that transmission and reflections at both interfaces do not interfere.

CASE 5.8   MASS *M* EMBEDDED BETWEEN TWO LONG BARS

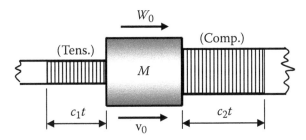

Either force W_0 or initial velocity v_0 is applied to the mass.

$$I_1 = A_1 \rho_1 c_1; \quad I_2 = A_2 \rho_2 c_2$$

$$C = I_1 + I_2$$

(a) Step load $W = W_0 H(t)$

$$v_0 = \frac{W_0}{C}; \text{ constant velocity attained}$$

(b) Initial velocity v_0

$$v(t) = v_0 \exp\left(-\frac{C}{M} t\right); \text{ asymptotically decreasing velocity}$$

(c) Rectangular pulse $W = W_0 t_0$

$$v = v_0 = \frac{W_0}{C}; \text{ while the pulse lasts, } t \leq t_0$$

$$v(t) = v_0 \exp\left[-\frac{C}{M}(t - t_0)\right] \quad \text{for } t \geq t_0$$

Note: The above may be seen as a special application of Case 3.7.

CASE 5.9   DIRECT STRESS $\sigma = \sigma_m H(t)$ APPLIED TO A FREE END OF A LONG BAR, INITIALLY AT REST

Bilinear material, yield point F_y is exceeded.
   Either σ or the equivalent end velocity v_e is prescribed.

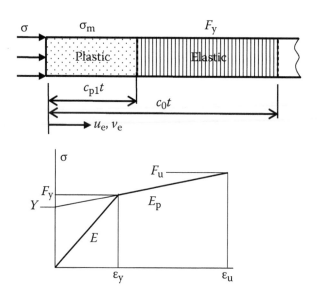

Elastic segment:

Wave speed: $c_0 = \sqrt{\dfrac{E}{\rho}}$; stress $\sigma = F_y$; end (particle) velocity $v = v_y = \dfrac{F_y}{\rho c_0} = \dfrac{F_y}{\sqrt{E\rho}}$; peak particle velocity (PPV) without yielding, $\sigma = F_y$.

Plastic segment:

Wave speed: $c_{pl} = \sqrt{\dfrac{E_p}{\rho}}$; stress $\sigma_m = F_y + \rho c_{pl}(v_e - v_y)$;

end velocity: $v_e = \dfrac{u_e}{t} = c_{pl}(\varepsilon - \varepsilon_y) + c_0\varepsilon_y = \dfrac{\sigma - F_y}{\rho c_{pl}} + v_y$; failure takes place for $\varepsilon = \varepsilon_u$ or $\sigma = \sigma_u$. For a very ductile material $\varepsilon_u \gg \varepsilon_y$: $v_e \approx c_{pl}\varepsilon_u$ produces failure (elastic contribution assumed negligible).

Note: The above was written for the case of the structural mass equal to total mass per unit length. If this is not the case, ρ must be adjusted.

CASE 5.10   SHEAR STRESS $\tau = \tau_M H(t)$ APPLIED TO A FREE END OF A LONG BAR, INITIALLY AT REST

Bilinear material, yield point τ_y is exceeded. Either τ or the equivalent end velocity v_e is prescribed.

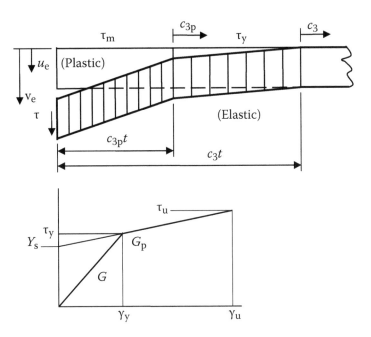

$G = \dfrac{E}{2(1+\nu)}$; shear modulus, elastic range; G_p in plastic range; τ_y = yield stress in shear.

Elastic segment:

Wave speed: $c_3 = \left(\dfrac{GA_s}{m}\right)^{1/2} = \left(\dfrac{GA_s}{\rho A}\right)^{1/2}$; peak stress $\tau = \tau_y$ over the entire segment: $v_e = v_y =$

$\dfrac{A_s}{A}\dfrac{\tau_y}{\rho c_3} = \dfrac{A_s}{A}\dfrac{\tau_y}{\sqrt{\rho G}}$; largest possible end particle velocity without yielding.

Plastic segment:

Wave speed: $c_{3p} = \left(\dfrac{G_p A_s}{m}\right)^{1/2}$; stress $\tau_m = \tau_y + \rho c_{3p}(v_e - v_y)\dfrac{A}{A_s}$;

end velocity: $v_e = c_{3p}(\gamma - \gamma_y) + c_3 \gamma_y = \dfrac{A_s}{A}\left\{\dfrac{\tau - \tau_y}{\rho c_{3p}} + \dfrac{\tau_y}{\rho c_3}\right\}$. Failure takes place for $\gamma = \gamma_u$ or

$\tau = \tau_u$. For a very ductile material, $\gamma_u \gg \gamma_y$: $v_e \approx c_{3p}\gamma_u$ produces failure (elastic contribution assumed negligible).

Note: The shear force applied is $Q_m = A_s\tau_m$. The above presentation is simplified in that the same shear area is used for stiffness and stress. Refer to Case 4.8 for the definitions of material shear properties. Structural mass is equal to total mass per unit length.

CASE 5.11   TWISTING MOMENT $T = T_M H(T)$ APPLIED TO A FREE END OF A LONG SHAFT, INITIALLY AT REST

Circular section. Bilinear material, yield point τ_y is exceeded (refer to figures in Case 5.10).
    Either T_m or the equivalent end velocity $\lambda = \lambda_e$ is prescribed.

Elastic segment: Refer to Case 5.2.

Plastic segment:

Wave speed: $c_{1p} = \sqrt{\dfrac{G_p}{\rho}}$; γ_u, τ_u; the ultimate shear angle and shear stress, respectively.

$C \equiv I_0$ for a circular shaft; $I_0 = \dfrac{\pi}{2}R^4$, solid shaft; $I_0 = 2\pi h R_m^3$, thin-wall shaft.

$$\tau_y = \frac{F_y}{\sqrt{3}}; \quad \tau_u = \frac{F_u}{\sqrt{3}}$$

$$G = \frac{E}{2(1+v)}; \quad \gamma_y = \frac{\tau_y}{G}$$

$$\gamma_u = \sqrt{3}\varepsilon_u; \quad G_p \approx \frac{E_p}{E}G$$

$\lambda_y = \dfrac{T_y' c_1}{GC}$,   maximum end velocity without yielding, alternatively

$\lambda_y = \dfrac{4}{3}\dfrac{c_1 \tau_y}{GR}$, solid shaft

$\lambda_y = \dfrac{c_1 \tau_y}{GR_m}$, thin-wall shaft

$\lambda_e = \lambda_y + \dfrac{c_{1p}}{R}\left(\dfrac{\tau_m - \tau_y}{G_p}\right)$, the end velocity (use R_m for a thin-wall shaft)

$\alpha = \lambda t$, angle of twist; $\gamma = \alpha\dfrac{R}{L}$, material shear angle. Failure takes place for $\tau = \tau_u$ or $\gamma = \gamma_u$.

For a very ductile material $\gamma_u \gg \gamma_y$, a bilinear material approaches the RSH model shown here.

Shear stress: $\tau_m \approx \tau_y + (\tau_u - \tau_y)\dfrac{\alpha}{\alpha_u}$

$$G_p \approx \frac{E_p}{3}$$

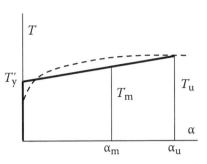

Note: The above equation is valid until (a) τ_u is reached, signaling the imminent failure or (b) until the plastic wave reaches the opposite end.

CASE 5.12 INELASTIC BAR IMPACTING RIGID WALL

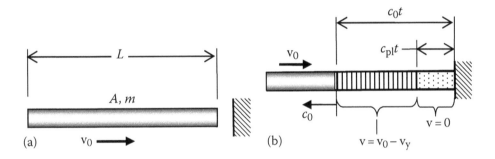

Elastic and plastic waves originate at impact point.

Elastic wave: Stress behind wave front and particle velocity, respectively: $\sigma = F_y$; $v = v_y = \dfrac{F_y}{\rho c_0}$.

Plastic wave: Speed, stress behind wave front and particle velocity, respectively: $c_{pl} = \sqrt{\dfrac{E_p}{\rho}}$; $\sigma_m = F_y + \rho c_{pl}(v_0 - v_y)$; $v = 0$.

Note: When the elastic wave rebounds from the free end and returns, it collides with the plastic wave. See text for details. The bar impacting a rigid wall with v_0 is equivalent, in terms of stress, to the equal and opposite velocity being applied to the end.

CASE 5.13 A SHORT BLOCK OF BILINEAR MATERIAL, RAPIDLY COMPRESSED

There is no friction at either end of the block. The driving velocity is larger than that necessary to initiate yielding:

$$v_0 > v_y = \frac{F_y}{\rho c_0}$$

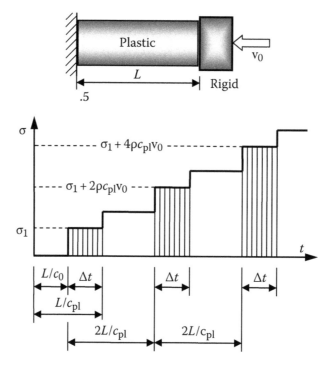

After reflection of the initial elastic wave from the stationary end the stress near this location is

$$\sigma_1 = \left(\frac{c_{pl}}{c_0} + 1\right) F_y$$

Afterward, there is an increase of stress by $\Delta\sigma = 2\rho c_{pl} v_0$ after every cycle lasting $2L/c_{pl}$, as shown (stationary end).

Note: The pattern will continue as long as v_0 is applied. The above condition is based on a small deflection approach.

CASE 5.14   END IMPACT WITH A STRONG SHOCK WAVE

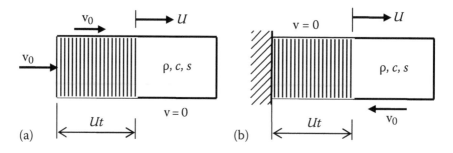

  (a) Sudden application of constant end velocity.
  (b) Stopping of the end against rigid wall.

$U = c + sv$; shock Hugoniot

U; wave speed

v; particle velocity (shown as constant v_0)

$p = \rho_0 U v_0 = \rho_0(c + sv_0)v_0$; end pressure developed due to impact with v_0

$p = \rho_0[c(\Delta v) + s(\Delta v)^2]$; pressure at the end of initially unstressed block when velocity changes $\Delta v = v_2 - v_1$ is experienced ($\Delta v = v_0$ here).

Note: The above two impacts are equivalent in terms of induced stress, when v_0 is the same.

CASE 5.15   WAVE COLLISION

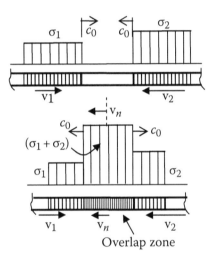

Two stress pulses, σ_1 and σ_2, start at opposite ends of the bar and meet at some intermediate point.

(1) Elastic material: $\sigma_1 = \rho c_0 v_1$ and $\sigma_2 = \rho c_0 v_2$

$v_n = v_2 - v_1$; interface velocity for $v_2 > v_1$

$\sigma = \sigma_1 + \sigma_2$; stress in overlap zone

(2) Hugoniot material; shock wave

$v_n = v_2 - v_1$; interface velocity for $v_2 > v_1$

$\Delta\sigma = \rho_0[c(\Delta v) + s(\Delta v)^2]$; stress increment in overlap zone

$\Delta v = v_1 + v_2$; velocity increment for either wave

$U = c + s(\Delta v)$; shock wave speed

Note: It is customary to use p rather than σ for shock wave stress, to be consistent with a hydrodynamic approach to the problem. When referring to shock waves, U replaces c_0 in the above sketch.

CASE 5.16   STRONG SHOCK REFLECTION FROM RIGID WALL

A pressure pulse p_1 becomes p_2, after reflection. Hugoniot material.

$U_1 = c + sv$; shock wave speed

$\Delta v = 2v_1$; velocity increment

$p_2 = \rho_0[c(\Delta v) + s(\Delta v)^2]$; stress in overlap zone

$U_2 = c + s(\Delta v)$

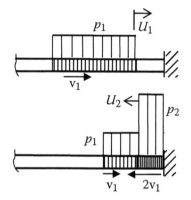

CASE 5.17   SOME CONSTANTS OF AN ELASTIC SPACE

E is Young's modulus and ν is the Poisson's ratio

$\varepsilon = \sigma/E$; uniaxial stress–strain relation

$G = \dfrac{E}{2(1+\nu)}$; shear modulus

$\gamma = \tau/G$; simple shear stress–shear strain relation

$K = \dfrac{E}{3(1-2\nu)}$; bulk modulus

$\varepsilon = -\dfrac{p}{K}$; volumetric strain, $\varepsilon = \varepsilon_x + \varepsilon_y + \varepsilon_z$, and pressure p

$E_{ef} = \dfrac{(1-\nu)E}{(1+\nu)(1-2\nu)}$; effective modulus of an axial bar, laterally constrained

$\sigma_x = \lambda\varepsilon + 2G\varepsilon_x$; $\sigma_y = \lambda\varepsilon + 2G\varepsilon_y$; $\sigma_z = \lambda\varepsilon + 2G\varepsilon_z$; principal stress–strain

$\lambda = \dfrac{\nu E}{(1+\nu)(1-2\nu)} = \dfrac{2\nu G}{1-2\nu}$

CASE 5.18   SHOCK WAVE IN SOIL (STIFFENING MEDIUM)

The applied pressure is $p = p_0 H(t)$, where $p_0 > \sigma^*$

Statics:

$\sigma = E\varepsilon$ for $\varepsilon \leq \varepsilon^*$ and

$\sigma = \sigma^* + E_s(\varepsilon - \varepsilon^*)$ for $\varepsilon > \varepsilon^*$

$\varepsilon = \dfrac{\sigma}{E}$ for $\varepsilon < \varepsilon^*$ and $\varepsilon = \varepsilon_s + \dfrac{\sigma}{E_s}$ for $\varepsilon > \varepsilon^*$

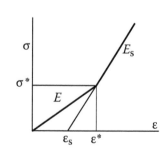

$$\varepsilon_s = \varepsilon^* - \frac{\sigma^*}{E_s}$$

$$\frac{1}{U^2} = \left(1 + \frac{E_s \varepsilon_s}{p_0}\right)\frac{\rho}{E_s}; \text{ to find the wave speed } U$$

$$v^2 = \frac{p_0}{\rho}\left(\varepsilon_s + \frac{p_0}{E_s}\right); \text{ particle velocity (squared) after the wave passage [44].}$$

Note: The medium is at rest prior to the wave passage.

CASE 5.19   DISPLACEMENT OF THE SPHERICAL VOID SURFACE UNDER THE ACTION OF A STEP PRESSURE PULSE $p = p_0 H(t)$

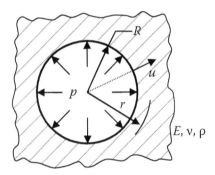

$$c_p = \sqrt{\frac{E_{ef}}{\rho}}; \text{ pressure wave speed}$$

$$E_{ef} = \frac{(1-v)E}{(1+v)(1-2v)}; \text{ effective modulus}$$

$$u_m = 0.6221\frac{p_0 R}{E}(1+v) - 0.002; \text{ peak displacement, mm}$$

$$t_m = 1.9718\frac{R}{c_p} + 0.0029; \text{ time when peak displacement is reached, ms}$$

Note: The radial movement, $u(t)$, resembles a damped sinusoid. The displacement quickly stabilizes at the level of u_{st}, or the static deflection due to p_0 (Case 3.33.) The numerical coefficients come from a FE studies. See text for further explanations.

CASE 5.20   A NONREFLECTING BOUNDARY AROUND A HALF-CYLINDER

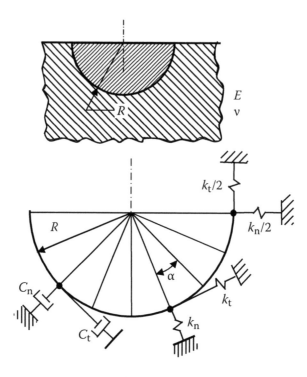

A half-cylinder, being a part of a continuous medium, is isolated from the medium by springs and dampers.

$$n = \frac{2\pi}{\alpha} = \frac{360^\circ}{\alpha}; \text{ angular distance between attachment points}$$

$$k_n = k_t = \frac{2\pi E}{n(1+\nu)} = \frac{4\pi G}{n}; \text{ stiffness of a typical spring, radial as well as tangential}$$

$$C_n = \frac{2\pi}{n} R\sqrt{E\rho}; \text{ damper constant, normal direction}$$

$$C_t = \frac{2\pi}{n} R\sqrt{G\rho}; \text{ damper constant, tangential direction}$$

Note: The above is based on circular symmetry of a cylindrical void in an elastic medium. For the two nodes at the surface only one-half of typical constants are used. Damper constant C is defined by $W = C\nu$, where ν is the velocity and W is the resisting force. If only short duration loading is applied and stress, rather than displacement, is of interest then the elastic springs need not be included in the model. This approach is suitable when the FE code has no such boundary capability.

EXAMPLES

EXAMPLE 5.1 LONGITUDINAL WAVE PASSAGE PARAMETERS FOR FE SIMULATION

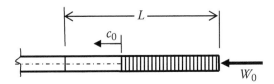

A long bar with section area $A = 1000\,mm^2$ has its free end impacted by a constant force W_0 lasting $t_0 = 0.8\,ms$. The force is such that the end is given the velocity $v_0 = 1\,m/s$. Calculate the induced stress, strain, and the time it takes the wave front to traverse the length $L = 6000\,mm$. Subdivide the bar into $l = 10\,mm$ elements for the purpose of FE simulation. Find a stiffness-proportional damping coefficient to obtain a stable solution. Find the time increment for an implicit FEA code simulation. Use aluminum data: $E = 69,000\,MPa$ and $\rho = 0.0027\,g/mm^3$.

Follow Case 5.1, but carry out the solution per unit cross section, i.e., $A = 1.0\,mm^2$.

$$m = A\rho = 0.0027\,g/mm; \quad c_0 = \sqrt{\frac{EA}{m}} = \sqrt{\frac{69,000 \times 1.0}{0.0027}} = 5,055\,mm/ms$$

$t_L = L/c_0 = 6000/5055 = 1.187\,ms$; transit time for the wave front

$\zeta = 0.4$; relative damping in highest mode (assumed)

$$\beta = 0.5787\,\frac{l\zeta}{c_0} = 0.5787\,\frac{10 \times 0.4}{5055} = 0.458 \times 10^{-3};\ \text{stiffness prop. damping (Equation 5.50)}$$

$i = \rho c_0 = 0.0027 \times 5055 = 13.649\,g/(mm^2\,s)$; specific impedance

$\sigma = iv_0 = 13.649\,MPa$; stress from unit velocity

$$\delta t = 0.4\,\frac{l}{c_0} = 0.4\,\frac{10}{5055} = 0.79 \times 10^{-3}\,ms;\ \text{maximum time increment for implicit analysis}$$

$b = c_0 t_0 = 5055 \times 0.8 = 4044\,mm$; pulse length

$4044/10 \approx 404$; number of elements in the pulse length

Note: This last number is more than adequate to accurately transmit the pulse.

EXAMPLE 5.2 SHEAR WAVE PASSAGE PARAMETERS FOR FE SIMULATION

A long bar with section area $A = 7.942\,mm^2$ and shear area $A_s = 7.148\,mm^2$ has its free end impacted by a constant force Q_0 lasting $t_0 = 0.8\,ms$. The force is such that the end is given the velocity $v_0 = 1\,m/s$. Calculate the induced stress, strain, and the time it takes the wave front to traverse the length $L = 1000\,mm$. Subdivide the bar into $l = 2\,mm$ elements for the purpose of FE

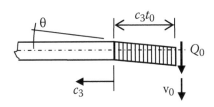

simulation. Find a stiffness-proportional damping coefficient to obtain a stable solution. Find the time increment for an implicit FEA code simulation. Use the same material data as in Example 5.1.

The calculation is quite similar to that in Example 5.1:

$$m = A\rho = 0.02144 \, \text{g/mm}; \quad c_3 = \sqrt{\frac{GA_s}{m}} = \sqrt{\frac{26,540 \times 7.148}{0.02144}} = 2,975 \, \text{mm/ms}$$

$t_L = L/c_3 = 1000/2975 = 0.3362 \, \text{ms}$; transit time for the wave front

$\zeta = 0.4$; relative damping in highest mode (assumed)

$$\beta = 0.5787 \, \frac{l\zeta}{c_3} = 0.5787 \frac{2 \times 0.4}{2975} = 0.156 \times 10^{-3}; \text{ stiffness-proportional damping}$$

$\tau = \rho c_3 v_0 = 0.0027 \times 2975 \times 1.0 = 8.033 \, \text{MPa}$; stress from unit velocity

$$\delta t = 0.4 \, \frac{l}{c_3} = 0.4 \, \frac{2}{2975} = 0.269 \times 10^{-3} \, \text{ms}; \text{ maximum time increment for implicit analysis}$$

Note: Shear beam is created by allowing only lateral displacements. In most practical cases of building design, for example, there is a much bigger difference between the axial area A and the shear area A_s. As in the previous example, the mesh density far exceeds the minimum requirements.

EXAMPLE 5.3   NATURAL FREQUENCY, STANDING WAVE, AND NODAL DENSITY OF AXIAL BARS

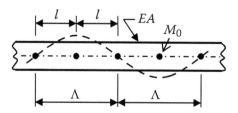

The sketch shows a bar vibrating in a natural, longitudinal mode in such a way, that the standing wave has a half-length Λ. What is the natural frequency associated with that wave length? Suppose a model is created in which the bar elements are weightless and the nodes are placed as shown with dots. What mass M_0 must be placed at each node so that the natural frequency is preserved? (Note that the dashed line shows the displacement magnitude, while the displacements are oriented along the bar.) What is the highest natural frequency of this model?

A segment of a vibrating bar is shown in Case 2.20c, which corresponds to the depicted mode shape or more precisely to one half-wave Λ, is associated with the following natural frequency:

$$\omega = \frac{\pi}{\Lambda}\sqrt{\frac{EA}{m}} = \frac{\pi}{\Lambda}\sqrt{\frac{E}{\rho}}$$

The postulated motion and node placing suggests that each segment Λ has two fixed nodes at the ends and a moving node with mass M_0 at center. The two halves of the segment work in parallel to resist the motion of the mass, i.e., their combined stiffness is

$$k = 2\frac{EA}{(\Lambda/2)} = \frac{4EA}{\Lambda}$$

which makes the natural frequency of the discrete model equal to $\omega = 2\sqrt{\dfrac{EA}{M_0\Lambda}}$. By equating both expressions for ω, the unknown M_0 may be found as

$$M_0 = \frac{4}{\pi^2}m\Lambda \approx 0.4053m\Lambda$$

which is somewhat less than the tributary mass $0.5m\Lambda$, which might be guessed on purely geometric basis. If the correct nodal mass is placed, then the natural frequency

$$f = \frac{c}{2\Lambda}$$

consistent with the standing wave concept, is retained. The highest frequency, expected from this model is when the deformation pattern depicted here is shrunk, so that $\Lambda = l$. Each point, midway between nodes is motionless and the two adjacent nodes move against one another. The whole reasoning displayed above is repeated with $\Lambda = l$ instead $\Lambda = 2l$. In effect

$$f = \frac{c}{2l}$$

is the highest frequency that can be detected.

EXAMPLE 5.4   ELASTIC BAR PUSHED AT ONE END, FREE AT THE OTHER

A bar of length L has a step load P_0 applied to its end. Calculate how the velocity of its center of gravity (CG) changes in time and compare the accelerations to those of a rigid bar.

In case of a rigid bar the acceleration is simply

$$a_r = \frac{P_0}{M} = \frac{\sigma_0}{\rho L}$$

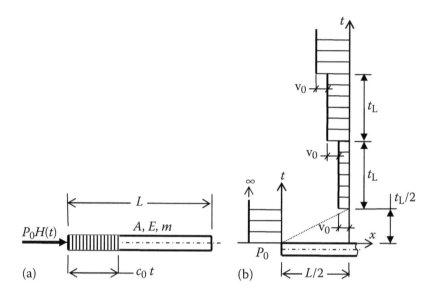

(a)

(b)

For an elastic bar, there will be no movement of CG until the stress wave gets there. When it does, after the time $L/(2c_0)$, the section at CG suddenly acquires $v_0 = \sigma_0/(\rho c_0)$, per Equation 5.5b. The wave continues then to the opposite end and on reaching it the particle velocity jumps from v_0 to $2v_0$. It must traverse another $L/2$ before it impacts the center, which then acquires another velocity increase. The initial history of CG movement is shown in the sketch. With the exception of the first half-length, the average velocity increase is steady: it is v_0 during the time needed to traverse the length L. The average acceleration is therefore

$$a_{av} = v_0/t_L = v_0 c_0/L$$

Comparing the two accelerations:

$$\frac{a_{av}}{a_r} = \frac{v_0 c_0}{L}\frac{\rho L}{\sigma_0} = \frac{\rho c_0 v_0}{\sigma_0} = 1$$

The average acceleration of the elastic bar is therefore the same as that of a corresponding rigid bar except during the initial interval, where the average acceleration of the elastic bar is twice as large. The average acceleration of the midpoint is the same as if the bar were a rigid body. The actual acceleration, on the other hand, comes in sharp pulses manifesting themselves as velocity steps.

EXAMPLE 5.5 TRIANGULAR PULSE REBOUND FROM A FREE END OF A BAR

A similar bar to that in Example 5.4 has a decreasing triangular pulse applied at one end. Illustrate how the impulse and the resulting stress change at the other free end.

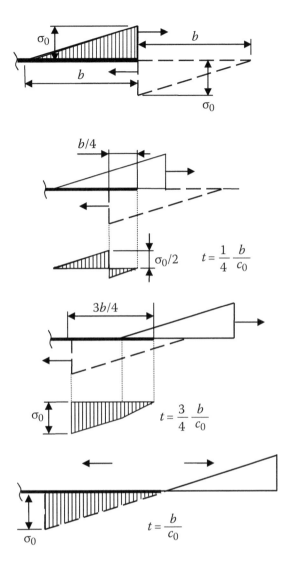

The first observation is that a pulse, decreasing in time will appear to be increasing when plotted traveling along a bar to the right. The free-end reflection is a superposition of two stress waves, one real and one imaginary, which are combined to obtain a stress-free end. The pulse positions are shown here for two selected time points in addition to the initial and the final instant. The resultant stress plots (hatched) are also displayed. The simplest way to quickly visualize a resultant stress is to rotate one of the two components about the bar axis and then to subtract the ordinates.

EXAMPLE 5.6 STOPPING ONE END OF A ROTATING SHAFT

A long shaft is spinning at 6000 rpm, when one end of it is suddenly stopped. Treating the other end as unrestrained find the peak-induced stress. The yield stress of material is 330 MPa. The outer diameter is 200 mm and wall thickness is 5 mm. Material data: typical steel, properties from Table 5.2.

Stopping the end shaft spinning with angular velocity λ_e is equivalent to suddenly applying said velocity to a stationary shaft. The shaft can be treated as a thin-wall tube. Referring to Case 5.2 and noting that $h = 5\,\text{mm}$ and $R_m = 97.5\,\text{mm}$, the twisting constant is

$$I_0 = 2\pi h R_m^3 = 2\pi 5 \times 97.5^3 = 29.12 \times 10^6 \text{ mm}^4$$

$$G = 76{,}920 \text{ MPa}; \quad \rho = 0.00785 \text{ g/mm}^3 \text{ (Table 5.2.)}$$

The angular velocity is $\lambda = 628.3\,\text{rad/s} = 0.6283\,\text{rad/ms}$. A sudden stopping of the end is equivalent to a sudden imposition of equal and opposite velocity λ, therefore Case 5.2 applies:

$$\tau_e = \rho c_1 R \lambda_e = (G\rho)^{1/2} R \lambda_e = (76{,}920 \times 0.00785)^{1/2} \, 100 \times 0.6283 = 1{,}544 \text{ MPa}$$

where R is the outer radius.

This is the peak calculated stress at the outer surface of the tube. It is obvious the shaft cannot be stopped in this manner without severe yielding or even failure, as inelastic analysis may reveal.

EXAMPLE 5.7   WAVE TRANSMISSION BETWEEN TWO BONDED BLOCKS OF DIFFERENT MATERIALS

A block of aluminum has its face bonded to a block of copper. A rectangular, compressive stress wave with the magnitude of 250 MPa and duration of 0.2 ms is traveling through aluminum toward the interface. Determine the magnitude of the reflected and the transmitted stress wave and show how their fronts move using time–distance $(t\text{–}x)$ diagram. All material constants to be taken from Table 5.2.

Using Case 5.6, with materials 1 and 2 being aluminum and copper, respectively:

$$i_1 = \sqrt{E_{ef}\rho} = \sqrt{110{,}740 \times 0.0027} = 17.29; \ i_2 = 36.45$$

$$\sigma_R = \frac{i_2 - i_1}{i_1 + i_2}\,\sigma_i = \frac{36.45 - 17.29}{17.29 + 36.45} \times 250 = 89.14 \text{ MPa}; \text{ in block 1}$$

$$\sigma_T = \frac{2i_2}{i_1 + i_2}\,\sigma_i = \frac{2 \times 36.45}{17.29 + 36.45} \times 250 = 339.14 \text{ MPa}; \text{ in block 2}$$

$$v_n = \frac{\sigma_T}{i_2} = \frac{339.14}{36.45} = 9.304 \text{ m/s}; \text{ velocity of interface during reflection}$$

$c_{p1} = 6404\,\text{m/s}$; $c_{p2} = 3502\,\text{m/s}$; wave front speeds in aluminum and copper, respectively, Table 5.2

The above is similar to a problem posed by Cooper [19], except that his purpose was to determine the effect of a strong shock wave, which is a more advanced subject. The sketch shows (no scale) location of the initial wave front (1) as well as that of the interface "i" during interaction (0.2 ms) between the incoming and the transmitted wave. The tangent of angle β is the velocity. R and T are the fronts of reflected and transmitted waves, respectively.

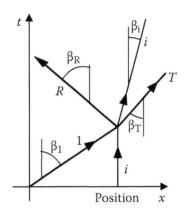

EXAMPLE 5.8   ELASTIC WAVE COLLISION

A rectangular, compressive stress wave with the magnitude of $\sigma_2 = 14$ MPa is applied at the right end of the bar and a similar wave with $\sigma_1 = 10$ MPa is applied at the other. Determine the magnitude of the combined pulse and the particle velocity when both waves overlap. The material is aluminum, see Example 5.12.

Refer to Case 5.15. The impedance is $i = \rho c_0 = 13.649$. The particle velocities in the waves: $v_2 = \sigma_2/i = 14/13.649 = 1.026$ m/s; $v_1 = 0.733$ m/s.

   Using superposition find the stress from collision of these compressive waves: $\sigma_1 + \sigma_2 = 24$ MPa and the interface velocity: $v_n = v_2 - v_1 = 0.293$ m/s (to the left). (The method of approach is explained in "Collision of Waves.")

EXAMPLE 5.9   END PULSE TRANSMISSION ALONG A BAR

A rock medium with $c = 6000$ mm/ms and damping ratio $\zeta = 0.01$ is subjected to single sinusoidal pulse, which lasts 0.068 ms, has amplitude of 82 MPa, and will travel 20 m. Calculate the amplitude reduction associated with this transmission.

The length of the pulse is $b = ct_0 = 6000 \times 0.068 = 408$ mm.
There will be 20,000/408 = 49.02 of such segments along the length.
According to Equation 5.45, there is $n = 49.02/2 \approx 25$ pseudo-periods during the travel.
The amplitude reduction ratio from Equation 5.46 is $\exp(2\pi n\zeta) = \exp(2\pi 25 \times 0.01) = 4.81$.
The reduced amplitude, after 20 m of travel, is 82/4.81 = 17 MPa.

EXAMPLE 5.10   STRESS WAVE TRANSMISSION THROUGH LAYERED MEDIUM

Consider a problem originally posed by Smith and Hetherington [80, p. 142]. A concrete structure is to be placed in a heavy clay soil and subjected to a ground shock in the form of a pressure wave of 60 kPa magnitude. Find the level of pressure transmitted into the concrete. Also, determine the effect of a layer of sand, having smaller density than clay, on the magnitude of transmitted wave. For simplicity, treat the problem as that of wave transmission between bars of unit sectional area and the incident wave as a rectangular pulse.

(1) Clay: $\rho = 0.0020\,\text{g/mm}^3$; $c = 2{,}100\,\text{mm/ms}$; $i_1 = \rho c = 4.2\,\text{g/(mm}^2\text{-ms)}$
(2) Concrete: $\rho = 0.0023$; $c = 3020$; $i_2 = 6.946$
(3) Sand: $\rho = 0.0016$; $c = 180$; $i_3 = 0.288$

For transmission from clay to concrete use Case 5.5:

$$\sigma_T = \frac{2i_2}{i_1 + i_2}\sigma_I \quad \text{or} \quad \sigma_T = \frac{2 \times 6.946}{4.2 + 6.946} \times 60 = 74.79\,\text{kPa}$$

The concrete target experiences a higher pressure level than so-called side pressure used to characterize the incident wave. If a layer of sand is placed between the two media, then two transmissions follow, clay to sand first and sand to concrete later:

$$\sigma_{T1} = \frac{2i_3}{i_1 + i_3}\sigma_I \quad \text{or} \quad \sigma_{T1} = \frac{2 \times 0.288}{4.2 + 0.288} \times 60 = 7.7\,\text{kPa}$$

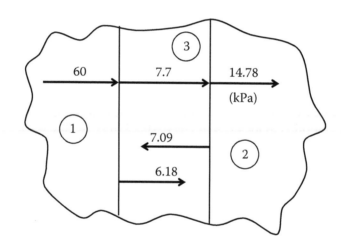

$$\sigma_{T2} = \frac{2i_2}{i_3 + i_2}\sigma_{T1} \quad \text{or} \quad \sigma_{T2} = \frac{2 \times 6.946}{0.288 + 6.946} \times 7.7 = 14.78\,\text{kPa}$$

The insertion of a soft sand layer has therefore decreased the pressure transmitted into concrete by the factor of about 5. This is not the whole story, as the referenced problem might suggest. There is a reflection back into sand as well with the magnitude of the stress wave being

$$\sigma_{R1} = \frac{i_2 - i_3}{i_2 + i_3}\sigma_{T1} \quad \text{or} \quad \sigma_{R1} = \frac{6.946 - 0.288}{6.946 + 0.288} \times 7.7 = 7.09\,\text{kPa}$$

and this wave now reflects from clay back into sand:

$$\sigma_{R2} = \frac{i_1 - i_3}{i_1 + i_3}\sigma_{R1} \quad \text{or} \quad \sigma_{R2} = \frac{4.2 - 0.288}{4.2 + 0.288} \times 7.09 = 6.18\,\text{kPa}$$

As the sketch illustrates, this second wave reaching concrete is only marginally less than the original one of 7.7 kPa and the transmitted wave resulting will also be proportionately less. The process of reflections is repeated, but with decreasing magnitude. The role of damping was neglected here, but damping makes successive reflections smaller and smaller. Still, for design purposes the original transmitted pressure of 14.78 MPa should be at least doubled. (Alternatively, one may leave the pressure unchanged but increase duration. It is the impulse, or the product of pressure and duration that does the damage.)

EXAMPLE 5.11   LONG WIRE STRETCHED BY A FALLING MASS

The material is patterned after a type of annealed copper quoted by Goldsmith [26]. It can be approximated by a bilinear curve:

$$\sigma = E\varepsilon \quad \text{for } \varepsilon \le \varepsilon_y \quad \text{and} \quad \sigma = Y + E_p\varepsilon \quad \text{for } \varepsilon > \varepsilon_y = 0.0038$$

$$E = 132{,}500 \text{ MPa}; \quad Y = 499 \text{ MPa}; \quad E_p = 1{,}082 \text{ MPa} \ (\rho = 0.00896 \text{ g/mm}^3).$$

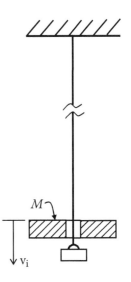

The ultimate strain is $\varepsilon_u = 0.10$ and the stress–strain curve should be assumed level past that point. The mass is quite large, therefore its velocity v_i may be assumed constant during and past the point of impact. Calculate maximum wire stress as a function of v_i.

We follow Case 5.9. The yield strength is $F_y = E\varepsilon_y = 503.5$ MPa. As long as the imposed velocity is small enough, the elastic relationships govern. The wave speed is $c_0 = \sqrt{\dfrac{E}{\rho}} = 3846$ m/s, and, according to Equation 5.5b:

$$\sigma_e = \rho c_0 v_e = 0.00896 \times 3846 v_e = 34.46 v_e$$

When the end speed is such that elastic stress is exceeded, i.e., $v_i \ge v_y = \dfrac{F_y}{34.46} = 14.61$ m/s, there is a plastic wave following its elastic precursor. The end velocity is then

$$v_e = \frac{\sigma_m - F_y}{\rho c_{pl}} + 14.61$$

In the plastic range we have $\rho c_{pl} = \sqrt{E_p \rho} = \sqrt{1082 \times 0.00896} = 3.114$ and the maximum stress is

$$\sigma_m = F_y + \rho c_{pl}(v_e - v_y) = 503.5 + 3.114(v_e - 14.61)$$

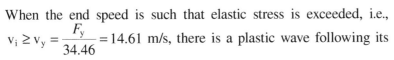

From the bilinear material curve, the ultimate strength is $F_u = Y + E_p \varepsilon_u = 499 + 1082 \times 0.1 = 607.2$ MPa. When the induced stress level grows to this point, the stress–strain curve becomes horizontal, which means that plastic wave speed becomes nil. This is synonymous with failure, as deformation becomes localized. The maximum attainable speed can be calculated from the above, by setting $\sigma_m = F_u$, which limits the impact to $v_e = 47.91$ m/s. The peak stress is a bilinear function of the applied impact velocity. The result shown in Ref. [26], obtained in a much more elaborate way, without simplifying the stress–strain curve, is close to $v_{im} = 54.2$ m/s.

EXAMPLE 5.12   INELASTIC IMPACT OF A BAR AGAINST RIGID WALL

An aluminum bar of length $L = 600$ mm is moving with the velocity of 11 m/s, when it impacts a rigid wall. Calculate the attributes of the elastic and the plastic component of the wave emanating from the impacted end. Describe the status of two parts of the bar just prior to collision of the plastic wave with the elastic wave, reflected from the opposite end. Material data: $E = 69,000$ MPa, $F_y = 100$ MPa, $E_p = 17,250$ MPa, and $\rho = 0.0027$ g/mm³. Perform the calculations per 1 mm² cross section.

Refer to Case 5.9:

$$c_0 = \sqrt{\frac{E}{\rho}} = \sqrt{\frac{69,000}{0.0027}} = 5,055 \text{ m/s}; \quad c_{pl} = \sqrt{\frac{E_p}{\rho}} = \sqrt{\frac{17,250}{0.0027}} = 2,528 \text{ m/s}$$

or, in terms of impedances: $i_0 = \rho c_0 = 13.649$ and $i_{pl} = \rho c_{pl} = 6.825$. Case 5.12 can now be used, with a minor change of symbols. The increment of end velocity needed to induce yield stress F_y is

$$\Delta v \equiv v_y = \frac{F_y}{\rho c_0} = \frac{100}{13.649} = 7.327 \text{ m/s}$$

The end velocity was changed by 11 m/s, which means that the remainder, $\Delta v = 11 - 7.327 = 3.673$ m/s is in the plastic range. From Equation 5.26:

$$\sigma_m = F_y + i_{pl}(\Delta v) = 100 + 6.825 \times 3.673 = 125.1 \text{ MPa}$$

This is the stress level at the end resulting from the impact, which had stopped the end of the rod. Soon after the impact, the status is as shown in Figure 5.12. The left part, swept by the plastic wave is stopped. The rightmost, unstressed part, still has the original velocity v_0. The segment, swept by the elastic wave only, has undergone the change of velocity consistent with Equation 5.25 in the text, so its resultant velocity is

$$v = v_0 - v_y = 11 - 7.327 = 3.673 \text{ m/s (oriented to the left)}$$

As was explained before, when the elastic wave rebounds from the free end, the particle velocity doubles, as long as the incoming and the reflected waves overlap, which is our case. This means that another velocity increment, $\Delta v = 7.327$, directed to the right again, will be experienced again. The resultant velocity of the segment swept by the reflected wave will therefore be

$$v = 3.673 - 7.327 = -3.673 \text{ m/s (oriented to the right)}$$

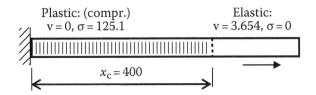

After reflection from the free end, the right segment is unstressed as shown below. The waves meet at x_c, as defined in the text.

EXAMPLE 5.13 INELASTIC IMPACT: WAVE INTERACTION

The bar, described in Example 5.12, undergoes an impact. Beginning with the instant when the plastic wave from the impacted end meets the elastic wave reflected from the opposite end, quantify the first collision, i.e., find the resulting stress levels and particle velocities.

Take the wave speeds and impedances from Example 5.12:

$$c_0 = 5055 \text{ m/s}; \quad c_{pl} = 2528 \text{ m/s}; \quad i_0 = \rho c_0 = 13.649; \quad \text{and} \quad i_{pl} = \rho c_{pl} = 6.825$$

The status of the bar just prior to wave collision was defined in Example 5.12 and it is shown on the sketch below.

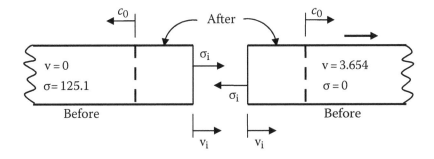

The fundamental observation is that stresses at the interface must be the same and that the interface has a specific velocity. These quantities are designated by σ_i and v_i in the sketch below, where the parts on each side of the interface are drawn separately. The signs were assumed on the expectation that the right part will pull the left, stationary part along its path. The wave spreading to the left is tensile; it will therefore be unloading the bar elastically. The resultant stress in this part is expressed as $\sigma_i = \sigma_1 - i_0(\Delta v) = 125.1 - 13.649\,v_i$ because v_i is the increment of velocity for this part. The right part, initially unstressed, is changing its velocity from $v_r = 3.654$ to v_i:

$$\sigma_i = 0 + i_0(v_i - v_r) = 13.649(v_i - 3.654)$$

After equating sides one finds $v_i = 6.41$ m/s, which leads to $\sigma_i = 37.62\,\text{MPa}$, using the second equation. The first equation gives $\sigma_i = 125.1 - 13.649 \times 6.41 = 37.61$.

EXAMPLE 5.14   INELASTIC WAVE REBOUND FROM A RIGID WALL

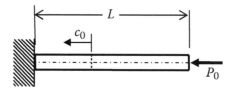

A bar of length $L = 600\,\text{mm}$ has one end fixed and the other one acted upon by a constant force $P_0 = 80\,\text{N}$ lasting $t_0 = 0.8\,\text{ms}$. Calculate the peak stress and the associated strain. The properties and areas are those in Example 5.12.

The wave speeds and impedances are $c_0 = 5055\,\text{m/s}$; $c_{pl} = 2528\,\text{m/s}$; $i_0 = \rho c_0 = 13.649$ and $i_{pl} = \rho c_{pl} = 6.825$. This an intermediate range of applied stress, where the result is given by Equation 5.32. The particle velocity in the incoming wave, from Equation 5.5b:
$v = \dfrac{\sigma}{i_0} = \dfrac{80}{13.649} = 5.861$ m/s. The change of end velocity needed to induce yield stress F_y is found from Equation 5.25:

$$\Delta v = \frac{\Delta \sigma}{i_0} = \frac{100}{13.649} = 7.327 \text{ m/s}$$

Only one-half of this is allowed prior to impact, if the reflected wave is to remain elastic: $v_y/2 = 3.664$ m/s. The remainder, namely, $\Delta v = 5.861 - 3.664 = 2.197$ m/s shows up as a plastic wave. From Equation 5.32a:

$$\Delta \sigma = 2i_{pl}(\Delta v) = 2 \times 6.825 \times 2.197 = 30 \text{ MPa}$$

The total maximum stress associated with the rebound of the wave is therefore

$$\sigma_r = 100 + 30 = 130 \text{ MPa}$$

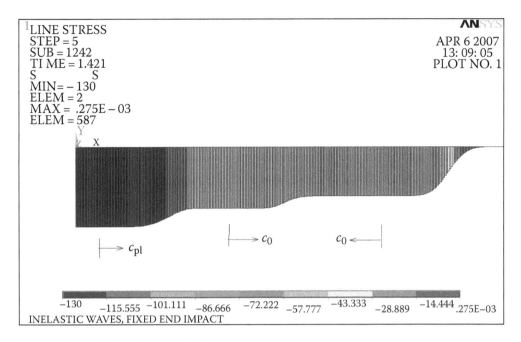

(See color insert following page 268.)

An ANSYS model was constructed using 600 line elements and the simulation of impact was made. The stress distribution after wave reflection is presented in the graph. Going from the left end, one sees the reflected plastic wave, then the elastic precursor, and then finally the original incoming wave. It is left as an exercise to the reader to see that, for example, when $E_p/E = 1/8$, the reflected stress is 121.2 MPa.

EXAMPLE 5.15   INELASTIC IMPACT OF A BAR AGAINST RIGID WALL; ALTERNATIVE APPROACH

The bar, treated in Examples 5.12 through 5.14 as being made of bilinear material, has a fairly complex pattern of behavior, due to interaction of elastic and plastic waves. This can be simplified by (a) ignoring the elastic strain and (b) assuming that the entire kinetic energy is converted to the work of plastic deformation. Following this approach find the maximum strain and the length of the bar swept by the plastic wave.

$$M = \rho L = 0.0027 \times 600 = 1.62 \text{ g}; \quad \text{mass, per unit cross section}$$

$$E_k = 1.62 \times 11^2/2 = 98.01 \text{ N-mm}; \quad \text{kinetic energy}$$

Ignoring the elastic displacements amounts to treating the material as rigid, strain hardening, or RSH. The total stress σ_m has an elastic and a plastic component. In Example 5.12 it was found that

$$\sigma_m = F_y + \sigma_{pl} = 100 + 25.1 = 125.1 \, (\text{MPa})$$

Designate the force at the impacted end by $R \equiv \sigma_m$ (unit area).
    Assuming R = constant makes it a motion with a constant acceleration of

$$a = R/M = 125.1/1.62 = 77.22 \, \text{mm}/(\text{ms})^2$$

The duration of motion, t_e, can be found from

$$v_i = at_e \text{ or } 11 = 77.22 t_e, \quad \text{then } t_e = 0.1424 \, \text{ms}$$

During this time the wave sweeps a segment $x = t_e c_{pl} = 0.1424 \times 2528 = 360.1$ mm.
    The plastic strain and the shortening of segment x are

$$\varepsilon_{pl} = \sigma_{pl}/E_p = 25.1/17,250 = 0.001455$$

$$\Delta x = x\varepsilon_{pl} = 360.1 \times 0.001455 = 0.524 \, \text{mm}$$

The method of finding duration t_e is equivalent to the momentum–impulse approach, as $a = \Delta v/\Delta t$. One can also equate the kinetic energy with the work supplied by R, by stating $E_k = R(\Delta x)$. This would result in the shortening of $\Delta x = 98.01/125.1 = 0.7835$ mm. Comparing this and the previous value shows us that the energy approach yields a higher value of Δx. The underlying physical reason for the discrepancy is that when spreading of the plastic wave ends, the entire bar is not at rest, as the latter method assumes.

EXAMPLE 5.16 AN AXIAL BAR, END-IMPACTED BY A LIGHT, HIGH-VELOCITY STRIKER

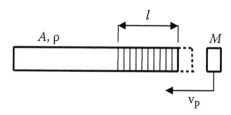

The same material ($\rho = 0.00896$ g/mm³) is used as in Example 5.11, but described as RSH with the following parameters:

$$\sigma = F_y + E_p\varepsilon; \quad F_y = 499; \quad E_p = 1,082 \, \text{MPa}$$

The bar has $A = 100$ mm² and is long compared to the striker, which has a length of 5 mm and which is made of the same material. The impact velocity is $v_p = 100$ m/s. Ignoring the elastic action, as appropriate for the RSH curve, determine the compressive force associated with the plastic wave and the maximum bar length deformed by the wave.

From Example 5.11 we have $c_0 = 3846$ m/s; sonic velocity. Then

$$c_{pl} = \sqrt{\frac{E_p}{\rho}} = 347.5 \text{ m/s}, \quad \rho c_{pl} = 3.114; \quad \text{the inelastic wave speed and impedance}$$

$$v_y = \frac{F_y}{\rho c_0} = \frac{499}{0.00896 \times 3846} = 14.48 \text{ m/s}; \quad \text{the limit velocity for an elastic wave}$$

Mass of the impactor: $M = 100 \times 5 \times 0.00896 = 4.48$ g.
   From Case 5.9, ignoring the elastic portion, find the peak-associated stress:

$$v_e \approx \frac{\sigma_m - F_y}{\rho c_{pl}} \quad \text{or} \quad 100 = \frac{\sigma_m - F_y}{3.114}; \text{ then } \sigma_m - 499 = 311.4 \quad \text{and} \quad \sigma_m = 810.4 \text{ MPa}$$

The peak compressive force is $P_m = A\sigma_m = 100 \times 810.4 = 81{,}040$ N. As the velocity of the impactor decreases, so does the force applied by it to the end of the bar. When this falls to F_y, the emission of the plastic wave stops. The average force in the plastic range is

$$R_{av} = (810.4 + 499)100/2 = 65{,}470 \text{ N}$$

The average acceleration of the impactor is $a = R_{av}/M = 65{,}470/4.48 = 14{,}614$ mm/(ms)^2.
   The velocity drop to the elastic level is $\Delta v = 100 - 14.48 = 85.52$ m/s. The time it takes may be found from

$$\Delta v = at_e \quad \text{or} \quad 85.52 = 14{,}614 t_e, \quad \text{then } t_e = 0.00585 \text{ ms}$$

During that time the plastic wave advances by $x = c_{pl}t_e = 347.5 \times 0.00585 = 2.03$ mm
   This is the permanently deformed length. The shortening of the bar from plastic strain is

$$\varepsilon_m = \frac{\sigma_m - F_y}{E_p} = \frac{810.4 - 499}{1082} = 0.2878$$

The affected portion of the bar undergoes shrinking by $u = x\varepsilon_m = 2.03 \times 0.2878 = 0.584$ mm.

EXAMPLE 5.17 END OF A BEAM DRIVEN WITH CONSTANT LATERAL VELOCITY

A cantilever with $L = 305$ mm and a 6.35 mm $\times 6.35$ mm cross section has its tip driven with a constant lateral velocity v_0. Find the value of v_0, which causes tip failure by exceeding the allowable shear strain, according to Case 5.10. The material is steel, treated as bilinear with $F_y = 400\,\text{MPa}$ and $F_u = 600\,\text{MPa}$. Do the calculation for two values of ε_u: 10% and 40%. (The data comes from the experiments by Parkes [65], except for material properties, which are assumed.)

For the square section, $m = \rho A = 0.00785 \times 6.35^2 = 0.3165\,\text{g/mm}$, $A = 40.32\,\text{mm}^2$.

G and c_s come from Table 5.2: $G = 76.92\,\text{GPa}$, $c_s = 3130\,\text{m/s}$. For a beam the shear wave speed is smaller, per Case 5.1:

$$c_3 = \left(\frac{A_s}{A}\frac{G}{\rho}\right)^{1/2} = c_s\left(\frac{A_s}{A}\right)^{1/2} = 3130\left(\frac{5}{6}\right)^{1/2} = 2857\,\text{m/s}$$

Determine shear properties for this bilinear material model with the help of Case 4.8. (The values resulting for $\varepsilon_u = 0.4$ are given in brackets.) Note that for a rectangular section, $A_s/A = 5/6$:

$$\tau_y = \frac{F_y}{\sqrt{3}} = \frac{400}{\sqrt{3}} = 230.9\,\text{MPa}; \quad \tau_u = \frac{F_y}{\sqrt{3}} = \frac{600}{\sqrt{3}} = 346.4\,\text{MPa}$$

$$\gamma_y = \tau_y/G = 230.9/76{,}920 = \frac{1}{333.1}; \text{ the same for both materials}$$

$$\gamma_u = \sqrt{3}\varepsilon_u = 0.1732 \text{ (or 0.6928)}$$

$$G_p = \frac{\tau_u - \tau_y}{\gamma_u - \gamma_y} = \frac{346.4 - 230.9}{0.1732 - 1/333.1} = 678.7\,\text{MPa (167.4 MPa)}$$

$$v_y = \frac{A_s}{A}\frac{\tau_y}{\sqrt{G\rho}} = \frac{5}{6}\frac{230.9}{\sqrt{76{,}920 \times 0.00785}} = 7.83\,\text{m/s}$$

Noting that $\rho c_3 = (G\rho)^{1/2} = (76{,}920 \times 0.00785)^{1/2} = 24.57$, (ignoring difference between A and A_s) and $\rho c_{3p} = (G_p\rho)^{1/2} = (678.7 \times 0.00785)^{1/2} = 2.308$ (1.146) one can use the following:

$$\tau_m = \tau_y + \rho c_{3p}(v_e - v_y)\frac{A}{A_s}$$

to find the breaking velocity v_e causing the maximum stress $\tau_m = \tau_u$:

$$361.8 = 230.9 + 2.308(v_e - 7.83)\frac{6}{5}; \quad \text{then } v_e = 55.09\,\text{m/s (103.0)}$$

This shows us that when ε_u grows, so does the peak speed, while all other factors remain the same.

EXAMPLE 5.18   END IMPACT WITH A STRONG SHOCK WAVE

A brass block moving with a velocity of 437 m/s collides with a hard obstacle moving at 100 m/s in the same direction. Determine the resulting shock wave pressure and the speed of the wave front using $\rho = 0.00845$ g/mm³ and the following Hugoniot data: $c = 3726$ m/s and $s = 1.434$.

In terms of Case 5.14, the velocity of impact is $\Delta v = 437 - 100 = 337$ m/s. (One can also say that the rigid obstacle approaches with the speed of 337 m/s.)

$$p = \rho_0[c(\Delta v) + s(\Delta v)^2] = 0.00845[3726 \times 337 + 1.434 \times 337^2]$$

$$= 11,990 \text{ MPa; pressure after impact}$$

$$U = c + sv = 3726 + 1.434 \times 337 = 4209 \text{ m/s; wave front speed.}$$

Note that this speed is measured with respect to the system moving with $v = 100$ m/s

EXAMPLE 5.19   SHOCK WAVE COLLISION

A rectangular, compressive stress wave with the magnitude of $\sigma_1 = 12$ GPa is applied at one end of a block and a similar wave with $\sigma_2 = 18$ MPa is applied at the other. Determine the magnitude of the combined pulse and the interface velocity when both waves overlap. The material is brass with $\rho = 0.00845$ g/mm³ and has the following Hugoniot data: $c = 3726$ m/s and $s = 1.434$.

Refer to Case 5.15. Calculate the initial particle velocity for each wave, per Case 5.14:

$$p = \rho_0(cv + sv^2) \quad \text{or} \quad 12,000 = 0.00845(3726 v_1 + 1.434 v_1^2)$$

which gives $v_1 = 337$ m/s. Similarly, for $p_2 = 18$ GPa, $v_2 = 482$ m/s

$$v_n = v_2 - v_1 = 482 - 337 = 145 \text{ m/s;} \quad \text{interface velocity}$$

$$\Delta v = v_1 + v_2 = 337 + 482 = 819 \text{ m/s;} \quad \text{increment in PV for either wave}$$

$p = \rho_0[c(\Delta v) + s(\Delta v)^2] = 0.00845[3,726 \times 819 + 1.434 \times 819^2] = 33,910$ MPa; peak pressure, applicable in overlap zone. This is *larger* than the sum of both wave pressures. $U = c + sv = 3,726 + 1.434 \times 819 = 4,900$ m/s; combined wave front speed.

The data in this problem are exactly the same as in Cooper [19] so that the reader can appreciate the difference in methodology.

EXAMPLE 5.20   SHOCK-WAVE TRANSMISSION BETWEEN DIFFERENT MATERIALS

An aluminum block on the left is bonded to the copper block on the right. A shock wave of magnitude 25 GPa is moving along the aluminum block and it impacts the interface.

(a) Determine the interface pressure and velocity. (b) Repeat the operation for this shock wave traveling the opposite way, from copper to aluminum.

$$\text{Aluminum 921-T: } \rho_1 = 0.002833 \text{ g/mm}^3; \quad c_1 = 5041 \text{ m/s}; \quad s_1 = 1.420$$

$$\text{Copper: } \rho_2 = 0.00893 \text{ g/mm}^3; \quad c_2 = 3940 \text{ m/s}; \quad s_1 = 1.489$$

Follow Case 5.15.

(a) The initial particle velocity for the wave:

$p = \rho_0(cv + sv^2)$ or $25,000 = 0.002833(5041v_1 + 1.42v_1^2)$; then $v_1 = 1,285$ m/s

Aluminum: $z_1 = \rho_1 c_1 = 14.28$ (this is not the impedance, hence different notation)

Copper: $z_2 = \rho_2 c_2 = 35.18$

$$v_n \approx \frac{2v_1}{1 + z_2/z_1} = \frac{2 \times 1,285}{1 + 35.18/14.28} = 742 \text{ m/s; first approximation of interface velocity}$$

(like in Case 5.6 for linear material).

Resulting pressures (based on velocity increments):

$$p_1 = \rho_1[c_1(2v_1 - v_n) + s_1(2v_1 - v_n)^2] = 14.28(2 \times 1,285 - 742)$$

$$+ 0.004023(2 \times 1,285 - 742)^2 = 39,550$$

$$p_2 = \rho_2[c_2(v_n) + s_2(v_n)^2] = 35.18 \times 742 + 0.0133 \times 742^2 = 33,430 \text{ MPa}$$

The value of v_n is too small, because it gives smaller pressure on the right block. Take the average of both values, $p = 36,490$ MPa, as the anticipated true value. With the linear term in the expression for p_2 dominating, estimate the new speed to be

$$v_n \approx \frac{36,490}{33,430} 742 = 810 \text{ m/s}$$

When new p_2 is recalculated, one finds $p_2 = 37,220$ MPa, sufficiently close to the anticipated value. The Cooper's solution, using the original method, gives $p_2 = 37,500$ MPa, and $v_n = 814$ m/s, so the differences are minor.

(b) For the opposite wave movement, one finds the initial PV to be $v_1 = 582$ m/s. (The indices of the materials are switched.) The approximate interface velocity:

$$v_n \approx \frac{2v_1}{1+z_2/z_1} = \frac{2 \times 582}{1+14.28/35.18} = 828 \text{ m/s;} \quad \text{then}$$

$$p_1 = 35.18(2 \times 582 - 828) + 0.0133(2 \times 582 - 828)^2 = 13{,}320 \text{ MPa}$$

$$p_2 = 14.28 \times 828 + 0.004023 \times 828^2 = 14{,}580 \text{ MPa}$$

The predicted v_n is too large, because it gives a larger pressure on the block on the left. The average of pressures is 13,950 MPa, probably close to the true value. Scaling v_n in the same manner as before, one gets 792 m/s. The answers obtained by Cooper were 14,200 MPa and 809 m/s, respectively.

EXAMPLE 5.21   STOPPING ONE END OF A ROTATING SHAFT—INELASTIC APPROACH

The shaft analyzed before, in Example 5.6 was treated as if made of linear material. The calculation showed that the peak stress attained is 1,544 MPa in shear. The steel properties for the plastic range are prescribed by $F_y = 340$ MPa, $F_u = 480$ MPa, and $\varepsilon_u = 0.20$. Treating the material as RSH, determine the extent of the plastic zone and the maximum stress. The shaft length is 2 m.

Much of the data used here will be taken from Example 5.6, but the NL properties will have to be generated.

$$E_p \approx \frac{F_u - F_y}{\varepsilon_u} = \frac{480 - 340}{0.20} = 700 \text{ MPa}$$

Calculate the shear parameters following Case 5.11:

$$\tau_y = \frac{F_y}{\sqrt{3}} = \frac{340}{\sqrt{3}} = 196.3 \text{ MPa;} \quad \tau_u = \frac{F_u}{\sqrt{3}} = \frac{480}{\sqrt{3}} = 277.1 \text{ MPa; yield and ultimate shear}$$

$G = 76{,}920$ MPa; $G_p = E_p/3 = 233.3$ MPa;

$$c_1 = \sqrt{\frac{G}{\rho}} = \sqrt{\frac{76{,}920}{0.00785}} = 3{,}130 \text{ m/s;} \quad c_{1p} = \sqrt{\frac{G_p}{\rho}} = \sqrt{\frac{233.3}{0.00785}} = 172.4 \text{ m/s; wave speeds}$$

Stopping the shaft rotating with $\lambda_e = 0.6283$ rad/ms is equivalent to a sudden spinning of its end to this angular velocity.

$$\lambda_y = \frac{c_1 \tau_y}{GR_m} = \frac{3{,}130 \times 196.3}{76{,}920 \times 97.5} = 0.0819 \text{ rad/ms; maximum velocity without yielding, thin-wall}$$

shaft

The remaining component is applied via plastic action:

$$\lambda_e - \lambda_y = \frac{c_{1p}}{R_m}\left(\frac{\tau_m - \tau_y}{G_p}\right) \text{ or } 0.6283 - 0.0819 = \frac{172.4}{97.5}\left(\frac{\tau_m - 196.3}{233.3}\right); \text{ then } \tau_m = 268.4\,\text{MPa}$$

With the strength of $\tau_u = 277.1$ MPa the shaft is very close to breaking.

It is easy to demonstrate that for any tubular element $J = \rho L I_0$, where J is mass-moment of inertia about the axis, L is length and I_0 is the polar moment of inertia of the cross section. Consequently, following Example 5.6, one has

$$J = \rho L I_0 = 0.00785 \times 2000 \times 29.12 \times 10^6 = 457.2 \times 10^6 \text{ g-mm}^2$$

The applied end moment associated with the τ_m as calculated above is assessed using Case 4.29:

$$T_m = 2\pi\tau_m h R_m^2 = 2\pi 268.4 \times 5 \times 97.5^2 = 80.16 \times 10^6 \text{ N-mm}$$

Assuming $T_m = $ const. makes it a motion with a constant (average) angular acceleration of

$$\breve{e} = T_m/J = 80.16 \times 10^6/457.2 \times 10^6 = 0.1753 \text{ rad/(ms)}^2 \text{ (see Table 1.1)}$$

The duration of motion, t_e, can be found from

$$\lambda_e = \breve{e}\,t_e \quad \text{or} \quad 0.6283 = 0.1753t_e, \quad \text{then } t_e = 3.584 \text{ ms}$$

During this time the wave sweeps a segment $x = t_e c_{1p} = 3.584 \times 172.4 = 617.8$ mm.

This is less than the shaft length, therefore the predicted breaking point is true. The reasoning is simplified in that it ignores the action of the elastic wave rebounding from the other end of the shaft and interfering with the plastic wave front.

EXAMPLE 5.22   SPHERICAL CAVITY UNDER A PRESSURE PULSE

The material surrounding the cavity is a hard rock with the following properties:

$$E = 97,200 \text{ MPa}, \quad v = 0.25, \quad \rho = 0.0027 \text{ g/mm}^3, \quad F_t = 21 \text{ MPa (tensile strength)}$$

The cavity radius is $R = 100$ mm and the pressure pulse is of a rectangular shape with a magnitude of $p_0 = 100$ MPa and duration of $t_0 = 0.16$ ms. Create a simple model consisting of one string of elements and examine the displacement response of the cavity surface.

Compare the peak displacement and the associated time with the results of Case 5.19. Also, determine the dynamic pressure level at which the rock begins to crack in tension thereby invalidating the elastic results.

Create a string of 50 brick elements in the form of a truncated pyramid with the small end at $R = 100$ mm and the large end at $r = 1100$ mm. In both circumferential directions the displacements are constrained making our model work like a fragment of a sphere. The only difference is that our elements are flat other than curved, but the error should be small for a small apex angle. Find the time t_t that will take the pressure wave to travel from inside to the outside surface, per case 5.19:

$$E_{ef} = \frac{(1-v)E}{(1+v)(1-2v)} = \frac{(1-0.25)97,200}{(1+0.25)(1-2 \times 0.25)} = 1.2E = 116,640 \text{ MPa; effective modulus}$$

$$c_p = \sqrt{\frac{E_{ef}}{\rho}} = \sqrt{\frac{116,640}{0.0027}} = 6,573 \text{ m/s; pressure wave speed}$$

$$t_t = L/c_p = (1,100-100)/6,573 = 0.1521 \text{ ms; traversing time}$$

It will take the emitted wave $2t_t$ to return to the cavity surface. By that time the applied load will be removed, therefore the boundary condition at the far end is not involved when monitoring the short-term response. (There is no need to implement a nonreflective boundary.) When static pressure p_0 was applied to the model so described, the displacement of 0.063 mm was obtained at the inside surface, a little less than 0.0643 mm from theory, Case 3.32. The rectangular pulse gave $u_m = 0.07923$ mm, which produces the dynamic factor $DF(u) = 0.07923/0.063 = 1.258$. The peak occurred at 0.033 ms from the beginning of load application. By comparison, the peak deflection u_m and time t_m from Case 5.19 are

$$u_m = 0.6221 \frac{p_0 R}{E}(1+v) - 0.002 = 0.6221 \frac{100 \times 100}{97,200}(1+0.25) - 0.002 = 0.078 \text{ mm}$$

$$t_m = 1.9718 \frac{R}{c_p} + 0.0029 = 1.9718 \frac{100}{6573} + 0.0029 = 0.0329 \text{ ms}$$

Case 3.33 gives the static hoop stress around the hole as $\sigma_h = p_0/2$. The DF = 1.258 reached here results in the dynamic stress of

$$\sigma_h = \frac{p_0}{2}(DF) = \frac{100}{2}(1.258) = 62.9 \text{ MPa}$$

However, only $\sigma_h = F_t$ is allowed, which means that the pulse magnitude must be limited to $p_0 = (21/62.9)100 = 33.4$ MPa to avoid tensile cracking.

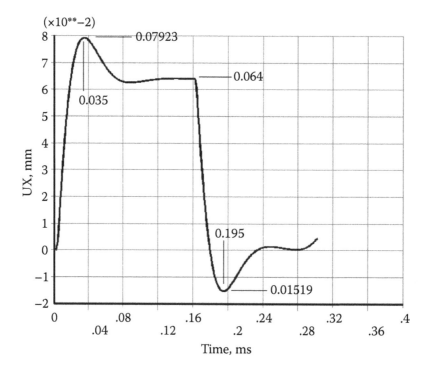

A graph of displacement–time is displayed above. It is fairly typical of elastic, semi-infinite space responses. After the initial spike due to dynamic action the displacement settles to a quasistatic level. The end of the pulse causes a sudden rebound and a spike of the same shape, just reversed.

6 Yield and Failure Criteria

THEORETICAL OUTLINE

BASIC THEORIES OF COMBINED STRESS

Consider a two-dimensional (2D) case of a plane stress, as described in Appendix A. Stress at a point is a tensor, which means that three components are needed to define it. Having those three components, σ_x, σ_y, and τ in some coordinate system is not always meaningful. For this reason, one often seeks the extreme values or three *stress invariants*, σ_{min}, σ_{max}, and τ_m. Having these three clarifies the state of stress, but does not answer the essential question: Will the material suffer permanent deformation or damage because of being stressed to that level?

Typically, materials are tested in a uniaxial fashion: steel in tension, concrete in compression, etc. Structural members, which are made of those materials, are usually subjected to more than one component of stress. To answer the questions related to the effect of a prescribed stress field, a designer must refer to *combined stress theories*.

The oldest of those is the *maximum stress theory*. It simply says that in deciding on the effects of the stress field one calculates the maximum direct stress and compares it with the allowable value. The above criterion was devised having brittle materials in mind. It is usually extended to accommodate the fact that tensile and compressive strengths are different. The resulting damage envelope, based on uniaxial compressive strength F_c and tensile strength F_t, is shown in Figure 6.1.

This theory works well for brittle materials, such as concrete, rock, and often soil. For those materials, damage usually means fracture. For ductile materials, the tensile yield strength F_y is the main material property, as far as yielding is concerned. The yield strength formulations are often extended to failure or fracture.

The *maximum shear stress theory* associated with the name of Tresca, is probably the most popular, as far as hand calculations of metallic structures are concerned. It says that yielding takes place when the maximum shear stress attains its critical value. The Mohr circle (Appendix A) tells us that in a simple tension of magnitude σ, the largest shear is $\sigma/2$ (at 45° to the direction of tension); consequently, a simple form of yield condition is to say that

$$\tau = \tau_m = F_y/2 \tag{6.1}$$

If the state of stress is three-dimensional (3D), then we have three principal stresses: σ_1, σ_2, and σ_3 and three Mohr circles representing three mutually perpendicular cutting planes. According to Tresca's theory, the material remains elastic if none of the principal stress difference exceeds F_y. This is equivalent to writing

$$|\sigma_1 - \sigma_2| < F_y; \quad |\sigma_2 - \sigma_3| < F_y; \quad \text{and} \quad |\sigma_3 - \sigma_1| < F_y \tag{6.2}$$

215

For a 2D state of stress $\sigma_3 = 0$, so the above equation is reduced to

$$|\sigma_1 - \sigma_2| < F_y; \quad |\sigma_2| < F_y; \quad \text{and} \quad |\sigma_1| < F_y \quad (6.3)$$

Equation 6.2 takes precedence when the principal components are of opposite signs, because then the largest Mohr circle is the one built on σ_1 and σ_2. In Figure 6.2 the envelope of yield points, according to this theory, is shown by the thinner lines for a 2D case.

A more recent theory, dating back to the beginning of twentieth century, is that of *distortion energy theory*, also referred to as Huber–Mises (HM) hypothesis, as mentioned before.* According to this concept, the state of stress can be resolved into a volumetric or hydrostatic component and a distortional or shear component. Yielding can only be caused by the latter, which leads to the following yield condition:

$$(\sigma_1 - \sigma_2)^2 + (\sigma_2 - \sigma_3)^2 + (\sigma_3 - \sigma_1)^2 \le 2F_y^2 \quad (6.4)$$

For a 2D state of stress, where $\sigma_3 = 0$, one has

$$\sigma_1^2 - \sigma_1\sigma_2 + \sigma_2^2 \le F_y^2 \quad (6.5)$$

When the above inequality is replaced by the equal sign, the yielding condition for a plane state of stress is prescribed. This is an ellipse, which is plotted in thicker line in Figure 6.2. A differentiation process gives us coordinates of points P and Q. It is interesting that one of the components of stress can exceed the yield strength F_y provided the other has a sufficiently large value and both components have the same sign. The ellipse in Figure 6.2 is an intersection of a cylinder in 3D spatial envelope (σ_1, σ_2, and σ_3) with the $\sigma_1 - \sigma_2$ plane.

Probably the simplest way of looking at the two theories is to use a concept of an *equivalent stress* σ_{eq} as combined measure of all six components being applied. The presence or absence of yielding as well

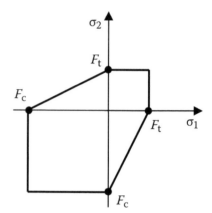

FIGURE 6.1 Damage envelope for brittle materials. For any combination of the applied principal stresses σ_1 and σ_2, there is no damage as long as the values fall within the envelope.

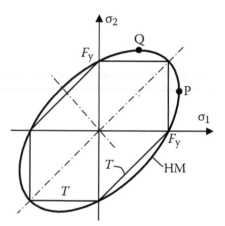

FIGURE 6.2 Interaction curves for Tresca's theory (*T*) and Huber–Mises (HM). The coordinates of point P are as follows:

$$\sigma_1 = \frac{2}{\sqrt{3}}F_y \approx 1.16F_y;$$

$$\sigma_2 = \frac{1}{\sqrt{3}}F_y \approx 0.58F_y$$

and, symmetrically, for point Q.

---

* Some authors point out that the idea of energy-of-distortion behind this theory belongs to the nineteenth century. While this is true, one should remember that the theory, in the way it is used today, was first formulated by Huber and then, 9 years later, independently by Mises.

as the margin of safety is found by comparing σ_{eq} with the yield stress F_y or an analogous allowable stress. The condition of nonyielding is satisfied when

$$\sigma_{eq} = |\sigma_1 - \sigma_2| \leq F_y \quad \text{(Tresca)} \tag{6.6}$$

provided, of course, the above equation represents the largest diameter of the Mohr circle. The other condition (HM) is in terms of principal stresses:

$$\sigma_{eq} = \left\{ \frac{1}{2} \left[(\sigma_1 - \sigma_2)^2 + (\sigma_2 - \sigma_3)^2 + (\sigma_3 - \sigma_1)^2 \right] \right\}^{1/2} \leq F_y \tag{6.7}$$

If all six independent stress components are present, the equivalent stress is found from

$$(\sigma_{eq})^2 = \frac{1}{2} \left[(\sigma_x - \sigma_y)^2 + (\sigma_y - \sigma_z)^2 + (\sigma_z - \sigma_x)^2 \right] + 3(\tau_{xy}^2 + \tau_{yz}^2 + \tau_{zx}^2) \tag{6.8}$$

The yield strength in pure shear can be found by noting that in such a state we have $\sigma_1 = \tau$ and $\sigma_2 = -\tau$, according to the related Mohr circle. Substituting into Equation 6.7 and observing that at the onset of yield $\sigma_{eq} = F_y$, one gets

$$\tau_y = \frac{1}{\sqrt{3}} F_y \approx 0.58 F_y \tag{6.9}$$

according to HM theory. A very common problem is that of direct stress σ and shear τ applied at right angles to the former as happens, for example, in a shaft that is bent and twisted. Using the Mohr circle and substituting into Equations 6.7 and 6.8, one can readily show that

$$\sigma_{eq} = \sqrt{\sigma^2 + 4\tau^2} \quad \text{(Tresca)} \tag{6.10}$$

$$\sigma_{eq} = \sqrt{\sigma^2 + 3\tau^2} \quad \text{(HM)} \tag{6.11}$$

The first of the above equations is slightly more conservative than the second. These two theories work well for ductile metals in that test results fall between the two, but usually are somewhat closer to the second one.

Strain criteria for failure can also be derived, using the similar concept of equivalent strain. Unfortunately, they were found not to agree well with the experiments. One exception is when a single component of strain predominates. Exceeding the ultimate material strain ε_u in this case will lead to failure.

FAILURE UNDER APPLIED LOAD

Failure is understood here as the initiation of a cracking process, which usually leads to a complete disintegration of a structural part. The subject in its entirety is quite complex, involving many variables. This brief discussion here is merely a simplified overview. At first, it is limited to ductile materials under quasistatic loads.

If a material sample yields at the stress of F_y and fails at F_u, then the ratio of load between two states is simply F_u/F_y. For elements, which are more complex than tensile samples, there is usually a stress redistribution, which one might call a self-adjustment of the stress field. The effect of the redistribution is such that the ratio of the failing load to that at the onset of yield is larger than F_u/F_y. The beneficial effect of such redistribution is one of the topics of the science of plasticity, on which a number of books were written.* The failure of parts in the plastic range is often synonymous with exceeding the ultimate strain ε_u.

The above is true for parts that have a smoothly varying geometry. If they have discontinuities, then the situation becomes less favorable, as there may be a marked growth of local strain near the notch. This leads not only to the reduction of load at failure, but may drastically reduce the permissible deformation of the part.

The effect of dynamics, which means rapid application of forces, does not make things any simpler. First of all, even a smooth material sample is often sensitive to the speed of load application, the so-called strain-rate effect. Typically, there is an increase in the apparent material strength, mostly in F_y, but also in F_u. This is, however, offset to some extent, by the decreased ductility. In real parts, there is also an effect of short duration of load growth, which prevents some of the benefit of plastic redistribution from taking place. When a load growth is very rapid, the resulting stress waves obscure the situation further.

Brittle materials, by definition, have only a small difference between the yield stress (or, perhaps more accurately, a stress where a marked deviation from linearity is encountered) and the stress at failure. The stress redistribution benefits are small. The loss of ductility due to a possible shape complication is, therefore, also small. Typically, a brittle part fails when the peak stress exceeds the material strength. The difference between tensile and compressive strengths can be very large, as Figure 6.1 implies.

The initiation of local cracking leads to failure, as long as a reasonable load level is maintained. There is some, although reduced, capacity of a part with an initial crack to resist the load. A situation of practical importance arises when a crack appears because of a past load history. Another, brief application of a moderate load will not break the part entirely, but will merely enlarge the crack. Quantification of such a development belongs to the science of fracture mechanics and is outside the scope of this book.

STRAIN-RATE EFFECTS ON SMOOTH MATERIAL SAMPLES

The theories described above are based on a very slow, quasistatic stretching of material samples. When one end of a sample is fixed and the other moves by u, the engineering strain is defined as $\varepsilon = u/L$, where L is the sample length. The strain rate, sometimes called the time rate of strain is

---

* One of the sources with many practical results is the book by Kaliszky [46].

$$\frac{d\varepsilon}{dt} = \frac{1}{L}\frac{du}{dt} \tag{6.12a}$$

or

$$\dot{\varepsilon} = \frac{\dot{u}}{L} \equiv \frac{v}{L} \tag{6.12b}$$

which is the end velocity divided by length. This quantity has an important influence on the measured material strength; in fact, most engineering materials appear stronger when the end of a sample is pulled faster. There are a number of formulas quantifying the strain-rate effect on the yield strength and among those the Cowper–Symonds relation appears to be the most popular one:

$$\sigma_{0n} = \left[1 + \left(\frac{\dot{\varepsilon}}{D}\right)^{1/q}\right]\sigma_{01} \tag{6.13}$$

where
  σ_{01} is the strength under quasistatic condition
  σ_{0n} is measured when the strain rate of $\dot{\varepsilon}$ is in effect

The quantities q and D are the material constants, which are specified in Table 6.1 for some materials.

Coefficient D represents a strain rate necessary to double the yield strength, regardless of the value of q. As Table 6.1 shows, it takes much higher rate to achieve such a growth in strength for aluminum compared with mild steel. For this reason, the former is regarded rate insensitive compared to the latter. Most authors use 1/s as a unit of strain rate, which makes the coefficient D 1000 × larger ($D = 40.4$ for mild steel then). It has not been established what the upper limit of applicability of the above expression is, with regard to the strain rate.

While there is an abundant literature on the subject of increase in strength of metals, when a sample is stretched at a higher than normal rate, there also is a related question, namely the effect of rapid straining on the ability of material to absorb energy prior to failure. This translates to simpler question: How does the increased strength influence the ability to stretch, also known as ductility?

There are several good texts dealing with the strain-rate effect, but the clues to answering the above question are few and far between. One of the exceptions is Stronge and Yu [84], who in his development relating to plastic hinges in beams, proposed that in order to account for decreased elongation at high strain rates, the product of the yield stress

TABLE 6.1
Strain-Rate Coefficients for Selected Materials

Material	D (1/ms)	q
Mild steel (1)	0.0404	5
Aluminum alloy	6.5	4
Stainless steel 304	0.1	10
High-strength steel	3.2	5

σ_0 and the corresponding ultimate elongation ε_u be kept invariant, regardless of the strain rate involved. In discussing a ballistic impact problem, Teng [107] quotes two sets of tested material properties. One of them was for Weldox 460E steel, which showed a decrease in ε_u when the strain rate was increasing and the other for 2024-T351 aluminum, which displayed an opposite trend.

There is some interesting experimental evidence available, which sheds light on the somewhat disorderly performance of materials under very high strain rates. One such long set of experiments was carried out by Lamborn [51] who worked on fragmentation of explosive-filled cylindrical shells. He compared dynamic ductility, or maximum strain prior to rupture, with static properties, as evidenced by standard tensile test and found little correlation in general. In fact, rapid stretching would improve ductility of more brittle materials and worsen the performance of those, which appeared quite ductile in a static test. There was some correlation between dynamic ductility and Charpy test energy, but it was a general tendency rather than 1:1 relationship.

Another reference is a little-known article by Belayev et al. [8], who describes high-speed stretching tests on samples of several metals. The experiments have demonstrated that as the strain rate grows, the ductility also increases, although slightly and then begins to decrease at some point. This decrease continues until a limit is reached where a brittle fracture is observed. For construction steels tested this limit was found ca. $\dot{\varepsilon} = 16,000/s$, which means a ballistic stretching rate.*

BALANCED MATERIAL CONCEPT

In Figure 6.3 a few stress–strain plots are presented, idealized to an elastic, perfectly plastic material model. All have the elastic part defined by the $\sigma = E\varepsilon$ curve, but the plastic strengths differ due to differences in the strain rates. Curve 1 is for static testing, and

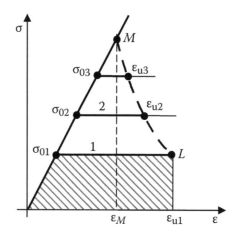

FIGURE 6.3 Elastic-perfectly plastic (EPP) material subjected to different strain rates of elongation.

---

* The reader should note that there is a significant difference between tensile samples and rings in that there is no stress wave problem for the latter except across thickness. The conclusions reached for tensile samples may not necessarily apply for rings or shells.

the hatched area under it represents the strain energy absorbed Π. When the process of stretching extends to nearly a breaking point, as in Figure 6.3, the hatched area is called the *material toughness* Ψ in accordance with Chapter 4. The remaining two curves will be discussed below.

The idea of a *balanced material* is that it retains the same toughness, regardless of the strain rate involved. This is to be applied to the most frequent situation in design, namely, when data regarding energy absorption at high strain rates are unknown or uncertain. A general expression for the strain energy content in a unit-size cube of elastic, perfectly plastic material is (Case 4.5):

$$\Psi = \varepsilon_{u1}\sigma_{01} - \frac{\sigma_{01}^2}{2E} \tag{6.14}$$

Index "1" in the above expression relates to the quasistatic curve 1. For a higher strain rate resulting in curve 2, the flow stress σ_{02} is calculated from Equation 6.13 and the limiting energy content $\Pi = \Psi$ is the same. The ultimate elongation is therefore

$$\varepsilon_{u2} = \frac{\Psi}{\sigma_{02}} + \frac{\sigma_{02}}{2E} \tag{6.15}$$

A similar calculation holds for curve 3, corresponding to a still higher strain rate. The dashed line in Figure 6.3 is the envelope of failure strain levels seen as a function of $\dot{\varepsilon}$. The failure strain ε_u decreases with the increasing yield strength σ_0, until it reaches the elastic line at σ_M. We then have

$$\Psi = \frac{\sigma_M^2}{2E} = \frac{1}{2}\sigma_M\varepsilon_M \tag{6.16}$$

The above explanations indicate that the dashed-line envelope in Figure 6.3 is drawn for a balanced material. In most test reports, however, there seems to be no clear trend with regard to the ultimate strain ε_u attained, when curves for several strain rates are drawn. The recommendation by Stronge, mentioned earlier, amounts to stating that

$$\sigma_{0n}\varepsilon_{un} = \sigma_{01}\varepsilon_{u1} \tag{6.17}$$

or that the product of the flow strength σ_0 and the corresponding ultimate elongation ε_u remains invariant, regardless of the strain rate involved. When the peak strain is an order of magnitude larger than the strain at the onset of yield (a typical situation), the above statement coincides with our balanced material concept.

STRENGTH OF NOTCHED ELEMENTS UNDER DYNAMIC LOADING

So far our attention has been focused on smooth elements, carefully prepared test samples subjected to tensile loads. Most structural parts, however, have strong discontinuities of shape, usually at ends but sometimes even in the interior. It has been known for a long time that such discontinuities cause stress concentrations and that they degrade the ductility of elements that contain them. This may manifest itself in a premature failure when such a

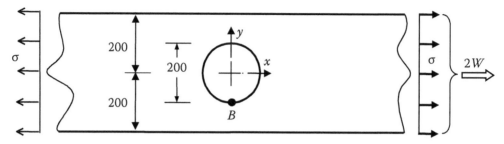

FIGURE 6.4   A structural member, 1 mm thick, with a hole. The load is applied as a step function of time.

part is subjected to shock loading. This is a very broad subject, which will be narrowed down in this section to the following conditions: (1) the material is ductile (2) there is a local weakening, or a local decrease in strength along a load path and (3) the load is abruptly applied and kept constant for a period of time. To make the discussion more meaningful, it will refer to a specific example shown in Figure 6.4.

Let $\sigma = W/A_{min}$ designate the average stress in the minimum section and let $\sigma_m = \sigma_0$ stand for the maximum (effective) stress at point B. The following variables are calculated:

$$K_\sigma = \sigma_m/\sigma \qquad (6.18a)$$

$$\varepsilon_m = \sigma_m/E \qquad (6.18b)$$

where
   K_σ is the stress concentration factor
   ε_m is the maximum strain associated with σ_m

(This second simple relation holds only for a free-edge stress, which, by definition is one-dimensional and which appears to be relevant for many stress concentrations.)

When the material model is EPP in brief, characterized by flow strength σ_0, the above relationships hold only up to $\sigma_m = \sigma_0$ or to $\varepsilon = \varepsilon_y$, the elongation at onset of yield at the stress concentration at point B. When the applied load grows to that point, it is designated as W_y, the load at the onset of yielding. When the load continues to grow, the ultimate strain ε_u is reached at the critical point, then tearing is about to begin. The load growth between the onset of yield (W_y) and beginning of tearing (W_u) belongs to the inelastic, nonlinear domain (Figure 6.5).

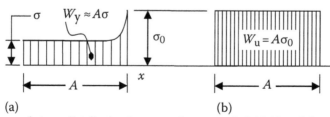

(a)   (b)

FIGURE 6.5   Change of stress distribution between the onset of yield (a) and the ultimate state (b) for a cross-section having a stress concentration point (EPP material).

When tearing is about to start, the important item is the average displacement of the loaded edge u_u, because this determines the absorbed strain energy. It is hypothesized here that the latter displacement, attained as a result of inelastic processes, may be expressed as follows:

$$\frac{u_u}{u_y} = \sqrt{\frac{\varepsilon_u}{\varepsilon_y} \frac{W_u}{W_y}} = \sqrt{\frac{\varepsilon_u}{\varepsilon_y} K_\sigma} \tag{6.19}$$

while the loading grows from W_y to the ultimate value of W_u. This hypothesis is based on a repeated observation that the growth of peak strain is always much faster than that of the overall deflection. In the last equality, it is assumed that plastic deformation is so extensive that, in the critical section, stress becomes uniformly distributed and attains the σ_0 level. We then have $W_u/W_y = \sigma_0/\sigma$, which, by definition, is the same as K_σ (Figure 6.5). From Equation 6.19, the displacement at the onset of breaking, u_u is found and then the strain energy absorbed:

$$\Pi \approx W_u u_u \tag{6.20}$$

The above equation disregards small elastic deflection and assumes the fully developed plastic load W_u from the outset. It ignores the additional energy absorbed during tearing. In a more general view, Equation 6.19 should be considered as an extrapolation of linear material results into a nonlinear material domain.* The above formulation is limited to a stress–strain curve, which is flat past the yield point. If there is strain hardening involved, or if the curve is of a more general character, the procedure of predicting stress and strain in the nonlinear range is not so straightforward, but it is not overly difficult. The procedure involves using Neuber's hypothesis and can be found in Fuchs and Stephens [25]. Also, when a more general stress–strain curve is used, the second equality in Equation 6.19 should not be used and FEA should be employed to determine W_u.

Let us look at the problem from a viewpoint of a structural designer. Suppose that a strip with a hole in Figure 6.4 is to be subjected to an infrequent load application with such a magnitude that the edge stress, ε is 150 MPa. As the net section area is one-half on the full section, the average applied stress is, therefore, 300 MPa at the weakest section. This is well below the yield point of 350 MPa, consequently there seems to be no problem. Without being specific, one knows that there will be a (net area) concentration factor of 2.0 or 3.0 (in case of a much wider element) because of the presence of the hole, but this will induce only a localized peak stress to be later smoothed out by plasticity. This reasoning is correct when only static loads are applied; in fact, for a long time, aircraft structures were designed using this approach. Correct, but not foolproof, as it focuses on statics alone. Ignoring the effect of discontinuities can be fatal in more than one way. The stress level can be limited, but strains are elevated, which may cause a fatigue failure. When shock/impact loading is applied, not only a certain nominal stress must be withstood, but also a pulse of energy is absorbed, which may not be possible with a limited ductility. This is illustrated, in greater detail, in Examples 6.1 and 6.2.

---

* It should also be treated as an unproven hypothesis and used with caution.

The above discussion is applicable to ductile materials, where most energy is absorbed prior to cracking. For those materials the initiation of visible cracks can, for practical purposes, be regarded as failure.

STRAIN RATE IN CONCRETE

There are many references to dynamic test results on material samples. Those are typically obtained using the split Hopinson bar and they demonstrate a visible growth of strength when the strain rate grows. This is especially pronounced above the rate of 100/s. The stress so obtained must be called "apparent," as it represents a dynamic force at the end of the sample divided by the sample area. One can ask how much of that increase in apparent stress is due to change of material properties and how much is due to side effects, such as the inertial resistance of the material being accelerated by the testing process. A thorough answer to this question is provided by Li and Meng [54], who determined that the dynamic increase factor for the apparent stress can be quantified as follows:

$$DFS = 1 + 0.03438(\log_{10}\dot{\varepsilon} + 3) \quad \text{for } \dot{\varepsilon} < 100/s \tag{6.21}$$

$$DFS = 8.5303 - 7.1372\log_{10}\dot{\varepsilon} + 1.729(\log_{10}\dot{\varepsilon})^2 \quad \text{for } \dot{\varepsilon} > 100/s \tag{6.22}$$

This is visualized in Figure 6.6. This effect is largely due to the radial confinement provided by the inertia of the sample material. To support that conclusion the authors demonstrate that the strength enhancement, of a similar magnitude is obtained in virtual testing with the strain-rate *insensitive* material. Other opinions on this subject differ and some authors demonstrate that the strengthening effect is partially due to the inertial confinement and partially to the inherent improvement in the material response. There appears to be a general agreement that for strain rates below 100/s the inertia effect is barely noticeable and for those above 1000/s it becomes very important. In view of the diversity of opinions and the fact that those opinions are based mainly on uniaxial testing, it seems prudent to allow only a small strain-rate enhancement.

By large, engineers involved in design use the strain-rate relationships, similar to those in Figure 6.6, as being the inherent concrete material property. If this procedure is applied

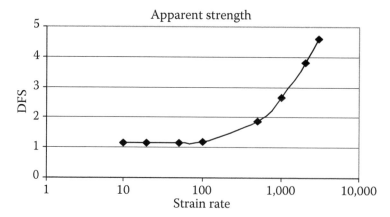

FIGURE 6.6   The strengthening factor for concrete in compression as a function of strain rate (1/s).

in the FEA environment, the configuration may be such that inertia effects are added on top of this, which leads to overstating of the material strength and renders this practice unsafe.

How much does this composition of the material resistance really matter? If certain strength was demonstrated by tests, is there a reason to argue with the findings? Yes, there is. While we can expect a cylinder of that material to perform in the field exactly as it does in test, this is not necessarily a typical stress pattern for critical concrete elements. A failure is often caused by bending combined with axial stress, when the extreme fibers do not have much benefit of the inertia of the surrounding material.

One should not lose sight of the fact that the confinement related to compressive testing is minor, as it is only the sample material itself, which is responsible for it. In other situations, where the stressed material is surrounded by a volume of concrete, the effect of confinement is indisputable.* The first approximation of that effect can be obtained from Equation 4.8, by setting the confinement pressure p_c as equal to the basic compressive strength F_c'.

CLOSING REMARKS

The coefficients of Equation 6.13, quantifying the strain-rate effect, are not known for a majority of materials. Also, the use of this effect complicates hand calculations and increases the cost of FE work. Additionally, in many cases the strain rate has only a minor effect on results.

With all this in mind, it often makes sense to adjust the material strength upwards and then treat it as strain-rate insensitive afterwards. One example of such analyses, which gave results close to experimental values, is offered by Smith [79], who used the following increase factors under moderate strain-rate conditions:

1.3 for mild steel
1.2 for steel with $F_y > 300\,\mathrm{MPa}$
1.1 for steel with $F_y > 500\,\mathrm{MPa}$

Most published material strength data are the minimum guaranteed values. In conventional design, their use increases the real safety margin. But such use may be inconsistent with the objective of dynamic analysis, where the purpose often is to find the most likely breaking load. This goal is better achieved by employing statistical mean values of F_y, F_u, and ε_u, which may often be established by inquiring with the manufacturers of the material in question.

The basic properties of concrete were presented in Chapter 4, along with often used expressions for the strain-rate effect. However, in view of the state of knowledge as described above, the optimum (and safe) procedure seems to be treating the material as strain-rate insensitive, but allowing a small strength enhancement, of an order of 10%–15%, due to strain rate when performing FE simulations. A simulation will exhibit the inertial component of the apparent strengthening anyway.

---

* Private communications with Leonard Schwer.

EXAMPLES

EXAMPLE 6.1 MATERIAL PERFORMANCE UNDER HIGH STRAIN RATE

The materials of interest have the following properties:

1. Mild steel.

   $E = 200,000$ MPa, static flow strength $\sigma_{01} = 250$ MPa, static ultimate strain $\varepsilon_{u1} = 0.40$.

2. Aluminum alloy.

   $E = 69,000$ MPa, $\sigma_{01} = 200$ MPa, $\varepsilon_{u1} = 0.08$

Find the material toughness Ψ as well as σ_M and ε_M for each material according to Figure 6.3.

----

1. Yielding starts at $\varepsilon_{y1} = \sigma_{01}/E = 0.00125$ and $\Psi = 99.84$ N/mm² from Equation 6.14.
   Then from Equation 6.16: $\sigma_M = 6320$ MPa and $\varepsilon_M = 0.0316$.
   Point M corresponds to brittle fracture of material caused by a very large strain rate of deformation, not attainable, practically, for this material.
   When the strain rate $\dot{\varepsilon} = D = 0.0404$/ms, the flow strength σ_0 is doubled. This, according to Equation 6.15, corresponds to nearly halving the ultimate strain so that $\varepsilon_{u2} = 0.201$. If a tested sample is $L = 60$ mm long, the velocity of the moving end is $v = \dot{\varepsilon}L = 2.424$ m/s.
2. Yielding starts at $\varepsilon_{y1} = 0.0029$ and $\Psi = 15.71$ N/mm², $\sigma_M = 1472$ MPa, and $\varepsilon_M = 0.02133$.
   When $\dot{\varepsilon} = D = 6.5$/ms, the flow strength σ_0 is doubled. This, according to Equation 6.15, corresponds to reducing the ultimate strain to $\varepsilon_{u2} = 0.04217$. If a tested sample is $L = 60$ mm long, the velocity of the moving end is $v = \dot{\varepsilon}L = 390$ m/s. This speed necessary to double σ_0 is two orders of magnitude larger than in the first example.

EXAMPLE 6.2 ENERGY ABSORPTION CAPACITY OF NOTCHED ELEMENTS

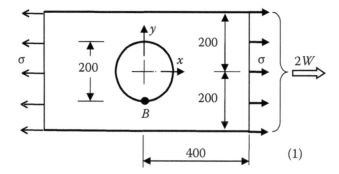

Estimate how much strain energy the elements shown here can absorb prior to failure when a step load is applied. The material is steel with $E = 200$ GPa, $v = 0.3$, and $\varepsilon_u = 0.16$. It is treated as EPP with $\sigma_0 = 350$ MPa. Thickness is $h = 1$ mm. The linear–elastic FEA of models showed the following, at the onset of yielding:

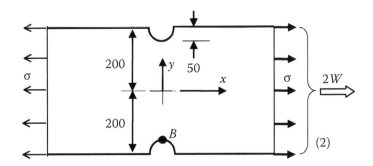

1. $\sigma_{ef} = 350$ MPa at point B, $W_y = 16{,}264$ N (load at the onset of yielding), and $u_y = 0.229$ mm (average displacement at loaded boundary)
2. $\sigma_{ef} = 350$ MPa at point B, $W_y = 25{,}880$ N, and $u_y = 0.2789$ mm

----

Strain at yield: $\varepsilon_y = 350/E = 0.00175$

1. Carry out the calculation for the upper half of the model only with deflections related to the length of 400 mm. Net section area $A_{min} = 100$ mm^2. Capacity load when σ_0 uniformly applied: $W_u = A_{min}\sigma_0 = 100 \times 350 = 35{,}000$ N.

   $K_\sigma = W_u/W_y = 35{,}000/16{,}264 = 2.152$; stress concentration factor. Then, from Equation 6.19: $\dfrac{u_u}{u_y} = \sqrt{\dfrac{\varepsilon_u}{\varepsilon_y}}\,K_\sigma = \sqrt{\dfrac{0.16}{0.00175}}\,2.152 = 14.03$; then $u_u = 14.03u_y = 3.213$ mm;

   elongation at failure. $\Psi \approx W_u u_u = 112{,}455$ N-mm (or 112.5 J); toughness of the member (Equation 6.20).

2. $A_{min} = 150$ mm^2 (upper half only and $L = 400$ mm).

   Capacity load: $W_u = 150 \times 350 = 52{,}500$ N.

   $K_\sigma = 52{,}500/25{,}880 = 2.029$; stress concentration factor $\dfrac{u_u}{u_y} = \sqrt{\dfrac{0.16}{0.00175}}\,2.029 =$ 13.62; then $u_u = 13.62u_y = 3.8$ mm.

   Also, $\Psi \approx W_u u_u = 199{,}500$ N-mm.

   It is interesting to compare the above to the toughness in absence of discontinuities. In (1), if not for the hole, the elongation at failure would be

$$u_u = L\varepsilon_u = 400 \times 0.16 = 64 \text{ mm}$$

With the same W_u, the magnitude of Ψ has dropped by $64/3.213 \approx 20$, while the load-carrying capacity fell only by the factor of 2.

7 Impact

THEORETICAL OUTLINE

Impact is defined here as a sudden contact of a moving body with a motionless barrier, or with a body of much larger size. This chapter analyzes impact of objects with linear, piece-wise-linear, or nonlinear characteristics. A distinction is made between local and overall deformation and their relative influence on response is compared. For compact bodies (all three dimensions of comparable magnitude), the local deformations are important and may predominate. The latter is the case when such a body strikes a barrier and the contact area is small. With respect to tangential contact, three surfaces are considered: *frictionless*, *rough*, and *no-slip*, with only the second one having a specified friction coefficient.

CENTRAL IMPACT AGAINST RIGID WALL

This is probably the simplest form of impact and as such, it provides a good illustration of the basic concepts involved. Let us consider a free fall and rebound of a ball against the ground, in the presence of gravity, and separate this event into distinct stages as shown in Figure 7.1.

The approach stage lasts from the moment the ball is released at height h_0 until it first touches the ground. At the end of this stage all particles of the falling body have velocity v, called the impact velocity. The instant of the first contact with the ground is the beginning of the *loading phase* of impact. During this phase, a deformation process takes place as a result of inertia forces acting on the ball. At the end of this stage all (or nearly all) particles have zero velocity, which means that the initial kinetic energy has been converted into strain and other forms of energy.

Then comes the *unloading phase* of impact, the recovery process during which the ball (nearly) returns to its original shape and gains the upward velocity *V*. (At the end of this phase the ball barely touches the ground.) The last, fly-off stage lasts from the time the ball leaves the ground until the peak point of rebound at height h_1 above the ground is reached. The duration of impact proper (loading and unloading) is usually very short in comparison with the free fall and fly-off.

The impulse applied to the ball by the ground is a vector quantity, which shows itself through the change of velocity of the impacted body. In general the impulse $S_{1,2}$ is

$$\vec{S}_{1,2} = M\vec{v}_2 - M\vec{v}_1 \tag{7.1}$$

where $\vec{v}_1$ and $\vec{v}_2$ are, respectively, the velocities before and after the impulse is applied to the body. In this particular case we choose the positive direction upward, along the z-axis and write the impulse of the loading phase as

$$S' = M \cdot 0 - (-Mv) = Mv \tag{7.2}$$

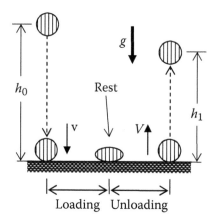

FIGURE 7.1   Stages of impact.

while at the unloading phase

$$S'' = MV - M \cdot 0 = MV \tag{7.3}$$

The ratio of the rebound velocity to the impact velocity is called the coefficient of restitution:

$$\kappa = \frac{V}{v} = \sqrt{\frac{h_1}{h_0}} \tag{7.4}$$

The second equality is written on the basis of the free fall equations:

$$V = \sqrt{2gh_1} \quad \text{and} \quad v = \sqrt{2gh_0} \tag{7.5}$$

The total impulse applied to the ball has two components, S' which stops the ball and S'' which makes it rebound:

$$S = S' + S'' = Mv + MV = Mv(1 + \kappa) \tag{7.6}$$

When the kinetic energy is changed into strain energy, and vice versa, during the process of deformation of the ball, some energy losses take place, which may be measured by the loss of velocity:

$$\Delta E_k = \frac{1}{2} Mv^2 - \frac{1}{2} MV^2 = \frac{1}{2} Mv^2 (1 - \kappa^2) \tag{7.7}$$

When the velocities of rebound and impact are the same, the coefficient of restitution $\kappa = 1.0$ and there are no energy losses. This ideal event is called a perfectly elastic impact and, needless to say, it cannot be attained in practice. The other extreme is the perfectly plastic impact, for which $\kappa = 0$ and the entire kinetic energy is lost. This can be quite well approximated in the real world if the impacting body is made of easily and permanently deformable material, e.g., a bag of sand. Most likely is an intermediate type of impact, for which $0 < \kappa < 1$.

As far as the impact against a rigid surface is considered, the coefficient of restitution characterizes the impacting body. For a deformable surface, the above formulas would not change; however, the coefficient would then also depend on the properties of the wall material.

The term *central impact* is used when the velocity of the center of gravity (CG) of the impacting body moves along the same line as the reaction at the impacted surface.

PARTICLE IMPACT AGAINST RIGID SURFACE

The term *particle* implies a body so small that its dimensions need not be considered. The impact of a particle against a smooth (no friction) surface is shown in Figure 7.2. The normal velocities and impulses are related as before, namely,

$$V = \kappa v_n \tag{7.8a}$$

and

$$S = M v_n (1 + \kappa) \tag{7.8b}$$

while the tangential component remains unchanged due to lack of friction. If there is friction present, the tangential component v_t becomes reduced to V_t:

$$M V_t = M v_t - \mu S \tag{7.9}$$

where
  μ is the coefficient of friction
  S is the normal impulse

The movement of the particle has two components: normal, which is treated in exactly the same manner as a central impact and the tangential one, which may exhibit slowing down due to friction.

BODY IMPACT AGAINST FRICTIONLESS SURFACE

In general, the velocity of the CG is not aligned with the normal to the surface at the impact point, which means the impact is *eccentric*. The solution of the kinetic problem

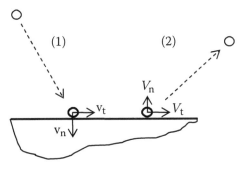

FIGURE 7.2 Particle impact against rigid surface. Approach stage (1) and rebound (2).

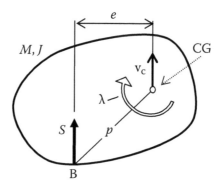

FIGURE 7.3   Unconstrained body subjected to impulse S.

for such a body becomes much simpler if one knows the effect of a specified eccentric impulse, as shown in Figure 7.3.

When a force is applied at some point B, it may be resolved, relative to the direction B-CG (segment p), into a component along this direction and one across it. The first induces translation of CG and the other one translation plus rotation about CG. The same is true for impulses, in a translational and a rotational direction. The net effect of such a resolution is

$$v_c = \frac{S}{M} \tag{7.10a}$$

and

$$\lambda = \frac{Se}{J} \tag{7.10b}$$

where λ is the angular velocity induced by the moment of impulse Se acting on a body possessing a moment of inertia J about an axis passing through the CG and normal to the page. There is another relevant question here, which leads to a concept of mass reduced to the impact point, M_r. After application of impulse S the impact point will acquire some velocity v_s, in the direction of impulse:

$$v_s = v_c + \lambda e = \frac{J + Me^2}{JM} S \tag{7.11}$$

But this can also be written as $v_s = S/M_r$, where M_r is the reduced mass being sought, so

$$M_r = \frac{JM}{J + Me^2} \tag{7.12}$$

This will now be applied to the impact against a frictionless surface, as illustrated in Figure 7.4a, showing a body in a translational motion with the initial velocity components v_n and v_t impacting the surface at point B.

The impulse S of the contact force is directed along the normal at point B. Figure 7.4b shows the kinematic condition just after the impact ends. Initially, there was only a

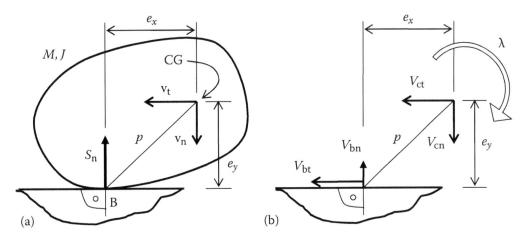

FIGURE 7.4 Eccentric impact, frictionless surface: (a) impacting body and velocities before impact (plus applied impulse) and (b) velocities after impact.

translational motion defined by v_n and v_t, while afterwards the angular velocity λ appeared. The approach to this type of impact is similar to that used in the section "Central impact against rigid wall" and described by Equation 7.8b, except that the velocity jump is related to the point of contact rather than to the CG. The coefficient of restitution κ again determines the ratio of velocities after and before the time of contact, $V_n = \kappa v_n$. Due to eccentricity of impact, however, the impulse is related to the reduced, rather than the total mass:

$$S_n = M_r v_n (1 + \kappa) \tag{7.13a,b}$$

Based on the angular impulse–momentum preservation one can write the equation for the angular velocity:

$$\lambda = \frac{M_r}{J}(1 + \kappa)v_n e_x \tag{7.14}$$

The rebound velocity of CG attains the following value:

$$V_{cn} = v_n - \frac{S_n}{M} \tag{7.15}$$

The horizontal velocity of CG remains unchanged, $V_{ct} = v_t$ because there is no friction and consequently, no tangential force. Note that in the more general setting, as depicted by Figure 7.4a and Equation 7.13a, the coordinate e_x appears, rather than e as in Equation 7.12.

BODY IMPACT AGAINST NO-SLIP SURFACE

A surface is regarded as *no-slip* when the friction force is so large that no slippage along the contact area is possible. In physical world this is likely to happen when a sharp corner presses on the surface, as Figure 7.5a suggests. The treatment is now more complicated, as a clear distinction must be made between the loading and unloading phase of impact. During the approach stage, the body has only a normal velocity component $v_n = v_0$. At the loading stage the contact point B remains motionless and the body acquires the angular velocity λ', with segment p rotating about point B. (The CG then moves with the velocity $\lambda' p$, depicted

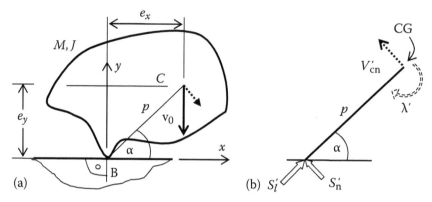

FIGURE 7.5   Impact against rough surface.

by a dotted line.) Using the preservation principle, the moment of the body momentum about point B, namely Mv_0e_x, is equated with the angular momentum $J_B\lambda'$, and the angular velocity is found:

$$\lambda' = Mv_0e_x/J_B \tag{7.16}$$

In which $J_B = J + Mp^2$. The impulse applied to the body and associated with this change of motion can be thought of as a sum of a longitudinal component S_l' and normal, S_n'. The first stops the velocity component along the p-axis, while the second produces translation of CG and rotation about it, both shown with dashed lines in Figure 7.5b:

$$S_l' = Mv_0 \sin\alpha \tag{7.17a}$$

$$V_{cn}' = S_n'/M \tag{7.17b}$$

$$\lambda' = pS_n'/J \tag{7.17c}$$

The combination of V_{cn}' and λ' must give zero normal velocity at point B, which leads to

$$S_n' = M_r'v_0 \cos\alpha \quad \text{with } M_r' = \frac{MJ}{J + Mp^2} \tag{7.18}$$

Assuming that the same restitution coefficient applies to both normal and longitudinal components, the total impulse values are $1+\kappa$ of what is shown above. The impulses so obtained can also be resolved into components in the xy coordinate frame. For example, the surface-normal impulse of the loading phase can be expressed as

$$S_y' = (M\sin^2\alpha + M_r'\cos^2\alpha)v_0 \tag{7.19}$$

while the total of the two phases, stopping and rebound, is

$$S_y = S_y'(1+\kappa) \tag{7.20a}$$

$$S_t = 0 \tag{7.20b}$$

BODY OF REVOLUTION IMPACTING A SURFACE

Consider an oblique impact in Figure 7.6. If the surface is frictionless, the problem is simple in that horizontal velocity will continue after contact terminates while the vertical rebound velocity will equal to κv_n. For a no-slip surface, however, the character of motion will change drastically as the contact point of the body becomes momentarily locked in horizontal direction.

This happens at the beginning of the loading phase. The tangential impulse applied at contact point is

$$S_t' = M_r v_t \tag{7.21}$$

with M_r according to Equation 7.12. The loading phase in Figure 7.6b is a virtual rotation of the body about the base point. The center velocity at the end of the loading phase is

$$v_t' = v_t - \frac{S_t'}{M} \tag{7.22}$$

which can also be found from preservation of momentum. The contact point rebounds with a tangential velocity κv_t, as seen in Figure 7.6c and the speed of the center is further decreased. The same κv_t defines the final angular velocity λ.

When a body spins with an angular velocity Ω while impacting a no-slip surface, the approach to defining post-impact parameters is the same as above, in that the contact point becomes momentarily locked and then rebounds, both tangentially and normally. The initial circumferential velocity is made up of the algebraic sum of ΩR and whatever translational tangential velocity is prescribed. The normal impact accompanied by the initial spinning produces an oblique rebound.

The impact treated in the manner described is referred to as a *stereomechanical impact*, which is concerned with rigid bodies and exchange of impulses between them resulting in sudden changes of velocity. It has been approached here in a simplified way: The same coefficient of restitution was assumed for the normal and the tangential component and the character of contact is treated as unchanging during the process. A more refined methodology can provide more accurate results, but that can rarely be successful due to lack of information on the contact area properties.

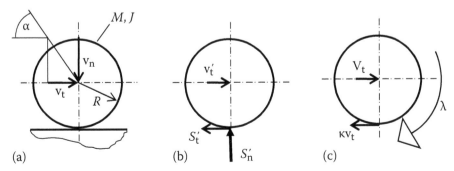

FIGURE 7.6 Oblique impact on no-slip surface with no initial angular velocity. (a) Approach stage, (b) status just after loading phase, and (c) horizontal velocity distribution after rebound.

Two extreme cases of tangential interaction were presented so far: unlimited slippage on a frictionless surface and no slipping. An intermediate condition, namely a rough surface with a prescribed friction μ is given as Case 7.6. In practical situations μ is often unknown and the bounds of a solution are provided by considering the two extremes.

PEAK IMPACT FORCE AND DURATION OF IMPACT

So far, only the kinetic effects of impact were considered, namely the change of velocities and associated impulses. The interaction of the impactor with the surface was expressed by the velocity restitution coefficient κ, while no specific references to elasticity involved were made. In order to calculate the remaining parameters like duration of impact and the peak contact force, the force–deflection characteristic of the impact area is needed. To visualize the combined deflections of both bodies one can think of a fictitious contact spring, as in Figure 7.7. (The length of the spring usually does not enter the calculations.) Introducing that contact spring makes it possible to treat both the impactor and the surface as rigid.

Suppose the contact spring is linear and has stiffness k_1. With some change of notation in Case 3.2b, the maximum deflection δ_m is obtained along with the peak force W_m:

$$\delta_m = v_0 \sqrt{\frac{M}{k_1}} \tag{7.23a}$$

and

$$W_m = v_0 \sqrt{k_1 M} \tag{7.23b}$$

From a simple observation that peak displacement results from integration of velocity over a quarter-period one can find the loading phase duration:

$$t_1 = \frac{\pi u_m}{2 v_0} \tag{7.24}$$

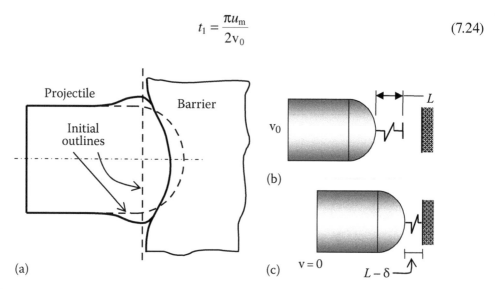

FIGURE 7.7 Actual contact deformation and equivalent "bumper" spring. Initial and deformed shapes in (a), equivalent spring, uncompressed in (b), and compressed in (c).

Unfortunately, in most real situation the characteristics of "bumper springs" are NL. However, many NL problems can be closely approximated as described in Chapter 4, where t_m, or duration of "forward motion" is the same as the loading phase of impact. This phase lasts from the first instant of contact until the impactor reaches the stationary point. The reaction can usually be determined from energy equivalence. The loading phase duration can be written as a quarter-period when the contact is linear or, more generally, using Equation 4.20b:

$$t_m = \tau/4 \tag{7.25a}$$

or

$$t_m = \frac{\pi S_n}{2W_m} = \frac{\pi M v_0}{2W_m} \tag{7.25b}$$

In calculating the above equation only the normal component of impact, S_n is used. One of the simplest situations exists when each of the two phases can be approximated as linear, but each of the lines has different slopes, as depicted in Figure 7.8. In this event the loading and the unloading phase last one-quarter period each, except that the second (unloading) phase is shorter if k_1 is smaller than k_2.

The impact ends when the contact force reaches zero value during the unloading phase. This corresponds to a point with coordinates $W = 0$ and $\delta = \delta_p$ in Figure 7.8. One of the results of the spring being stiffer during unloading, $k_2 > k_1$, is the permanent deformation δ_p. The shaded area represents the kinetic energy ΔE_k lost during impact:

$$\Delta E_k = \frac{1}{2} k_1 \delta_m^2 \left(1 - \frac{k_1}{k_2}\right) \tag{7.26}$$

This energy loss may also be expressed in terms of the coefficient of restitution κ (Equation 7.7). Since $M v^2 = k_1 \delta_m^2$ we have

$$\kappa^2 = \frac{k_1}{k_2} \tag{7.27}$$

For the basic oscillator with a damper the coefficient of restitution is defined in Case 7.14.

LOCAL DEFORMATION FORMULAS FOR ELASTIC CONTACT

Consider two bodies with spherical surfaces that are in contact. The theory of elasticity demonstrates (see Timoshenko [108]) that it is mainly the immediate vicinity of the contact area that contributes to an overall deflection. For this reason, we can apply a load through two reference planes,

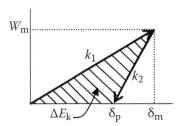

FIGURE 7.8 A simple piece-wise-linear characteristic.

as in Figure 7.9. The planes are rigid in vertical direction, and are frictionless. As we slowly increase the load W, the displacement δ of the moving plane a–a increases according to the Hertz formula:

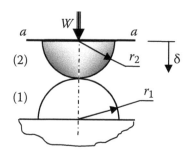

FIGURE 7.9 Contact of two curved surfaces.

$$\delta = \left[\frac{3}{4}\left(\frac{1-v_1^2}{E_1} + \frac{1-v_2^2}{E_2}\right)\sqrt{\frac{1}{r_1} + \frac{1}{r_2}}\right]^{2/3} W^{2/3} \qquad (7.28)$$

This above equation is practically the same as the relative movement of centers two spheres colliding with each other. A few important special cases may be deduced from this equation. When r_1, or r_2 is infinitely large, that is, when $1/r_1$, or $1/r_2$ is zero, we deal with a sphere pressed into an elastic half-space. When the larger radius (r_1 in this case) is negative, it means that the upper sphere is pressing on the bottom of a spherical cavity having the radius r_1. If either of the two bodies is assumed to be rigid, we set the appropriate E as infinitely large.

Although Equation 7.28 is based on elastic material properties, the relation between the resistance force W and local deflection δ is nonlinear. This deflection is treated as an "approach," or shortening of the distance between the body centers. A briefer form is obtained when the multiplier of W in Equation 7.28 is renamed to $1/k_0$:

$$\delta = \left(\frac{W}{k_0}\right)^{2/3} \qquad (7.29a)$$

or

$$W = k_0 \delta^{3/2} \qquad (7.29b)$$

where

$$\frac{1}{k_0} = \frac{3}{4}\left(\frac{1-v_1^2}{E_1} + \frac{1-v_2^2}{E_2}\right)\sqrt{\frac{1}{r^*}} \qquad (7.30a)$$

$$\frac{1}{r^*} = \frac{1}{r_1} + \frac{1}{r_2} \qquad (7.30b)$$

as illustrated later in Figure 7.10. The increase of contact area during the loading process is responsible for this geometric nonlinearity. The maximum contact radius a is given by

$$a = \left[\frac{3}{4}\left(\frac{1-v_1^2}{E_1} + \frac{1-v_2^2}{E_2}\right)Wr^*\right]^{1/3} \qquad (7.31a)$$

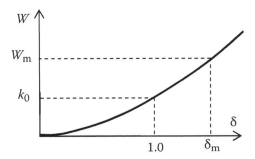

FIGURE 7.10 Resistance–deflection plot for local deformation.

or

$$a = \left(\frac{W}{k_0}\right)^{1/3} \sqrt{r^*} \qquad (7.31b)$$

For two spherical surfaces, the compressive stress, or pressure, applied to the circular contact area varies along the radius of that area and attains its peak in the middle. According to Timoshenko [108], one can show the peak stress to be

$$\sigma_c = \frac{3W}{2\pi a^2} = \frac{3}{2\pi} \frac{k_0^{2/3} W^{1/3}}{r^*} \qquad (7.32)$$

When a rigid rod of radius r_2 is pressed axially into an elastic half-space, the displacement is

$$\delta = \frac{1-v^2}{2Er_2} W \qquad (7.33)$$

When a uniformly distributed pressure with the same resultant W is applied to a circular area within r_2, the average displacement is only 1.08× larger. Under impact condition, the flexibility of a short cylinder in the axial direction may be approximated by the flexibility of one-half of its length. Grybos [27] derives a more general formula, which holds for the same rod axially pressed into a larger radius rod (r_1):

$$\delta = \frac{1-v^2}{2Er_2}\left(1-\frac{r_2}{r_1}\right) W \qquad (7.34)$$

PARAMETERS OF IMPACT ACCORDING TO HERTZ THEORY

The characteristic in Figure 7.10 is a plot of Equation 7.29b. The area under the curve W–δ represents the strain energy absorbed, so it follows that

$$\Pi_m = \int_0^{\delta_m} W d\delta = \int_0^{\delta_m} k_0 \delta^{3/2} d\delta = \frac{2}{5} k_0 \delta_m^{5/2} \qquad (7.35)$$

As demonstrated later in Chapter 8, the energy flow during a collision can be seen as having two stages: one of them is the approach, where the kinetic energy of the relative motion is converted into strain energy of the contact spring and a rebound, where some of this energy is regained. The kinetic energy loss during the approach, also called here the loading phase, can be expressed by

$$E_k = \frac{1}{2} M^* v_r^2 \tag{7.36a}$$

with

$$\frac{1}{M^*} = \frac{1}{M_1} + \frac{1}{M_2} \tag{7.36b}$$

where v_r is the relative speed. Equating the strain and the kinetic energies, as defined above, gives the maximum relative movement and the peak force of impact:

$$\delta_m = \left(\frac{5 M^* v_r^2}{4 k_0} \right)^{0.4} \tag{7.37a}$$

and

$$W_m = \left(\frac{5}{4} M^* v_r^2 \right)^{0.6} k_0^{0.4} \tag{7.37b}$$

Occasionally, it may be convenient to get an equivalent spring constant k_{ef} in Hertz-type impact. The strain energy of such a spring is $k_{ef} \delta_m^2 / 2$. When equated with the expression for Π_m above, one obtains

$$k_{ef} = \frac{4}{5} k_0^{2/3} W^{1/3} \tag{7.38a}$$

or

$$k_{ef} = \frac{4}{5} k_0 \delta_m^{1/2} \tag{7.38b}$$

The above was written for collision of two objects. When considering an impact of a single body with M_2, moving with a velocity v_0 against a half-space, which by definition has an infinite mass, one gets $M^* = M_2$ and $v_r = v_0$. For two identical spheres about to collide, each with mass M and the speed relative to a fixed system, $v_r = 2v_0$ and $M^* = M/2$. The duration of impact (including rebound) is found by integrating the equation of motion. As a result

$$t_0 = \frac{3.214}{v_r^{0.2}} \left(\frac{M^*}{k_0} \right)^{0.4} \tag{7.39}$$

The Hertz theory of impact is based on the use of local deformation only. The stress induced by impact (Equation 7.32) should remain within the elastic range.

The critical location within the volume of material depends, to some extent, on the material properties. For elastic/ductile materials, like steel, in a multiaxial state of stress, Tresca's yield criterion is the most convenient rule. The critical location is just below the contact surface where the magnitude of resultant shear reaches about $0.31\sigma_c$ for $\nu = 0.3$ (Equation 7.32). Keeping in mind that the yield stress in shear is $F_y/\sqrt{3}$, where F_y is the tensile yield, we can state that the onset of yield takes place when the peak surface pressure reaches $1.862F_y$. (Another way is to relate this to the average surface pressure, σ_{av}, which, according to Hertz theory, is 2/3 of the peak pressure value. The onset of yield therefore takes place at $\sigma_{av} = 1.241F_y$.)

For elastic/brittle materials, like concrete, maximum tension decides. The peak tension appears near the boundary of the contact area and attains

$$\sigma_t = \frac{1 - 2\nu}{3}\sigma_c \qquad (7.40)$$

When the applied stress exceeds the yield stress, one speaks of *onset of yielding*, even though in the case of brittle materials it is not always applicable.

Quantifying the onset of yield for a cylinder, axially pressing against a surface, is, strictly speaking, not possible. Assuming that the existence of sharp corners leads to a singularity, i.e., infinite stress level reached at the surface where the corners impinge. But then, in real cylinders, the corners are never sharp and moreover, the small-scale local yielding is inconsequential. In absence of better data the value of $0.5\sigma_{cav}$ is suggested for both peak shear and peak tension where σ_{cav} is the average value of contact stress at peak load.

ELASTOPLASTIC CONTACT AND PERFECTLY PLASTIC CONTACT

The elastic phase of impact ends when yielding of the material is initiated. As demonstrated above, ductile material yielding begins when the average contact pressure σ_{av} between spherical surfaces (force divided by projected contact area) reaches $1.241F_y$.*

Stronge and Yu [84] present a detailed picture of transition between different phases of contact of ductile spherical surfaces. The beginning of elastoplastic phase, according to the above, is at $\sigma_{av} \approx 1.1F_y$ and the resultant load is designated by W_y. When the contact load grows, the plastic zone below the surface spreads and reaches the surface. The condition becomes fully plastic when $\sigma_{av} = 2.8F_y$. At this point the resultant is $W_p \approx 424W_y$ and further increase in contact load is due only to the increase in contact area, while the surface stress remains equal to $2.8F_y$. Many authors take the value of $3F_y$ as the beginning of plastic phase of deformation. In practice, the transition is a somewhat academic problem and requires very exacting FEA studies. The assumption of a fully plastic interaction, from the beginning to the end of the loading phase, is much more manageable, from an engineering viewpoint at least.

More often, an engineer performing an impact analysis will find the resulting stress to be above the yield point of material. The bigger the exceedence, the less accurate the Hertz

---

* Some authors are more conservative in this regard and use Tresca's criterion to calculate the equivalent yield in shear, which is $\tau_y = 0.5F_y$. This results in the onset of yield at $1.075F_y$.

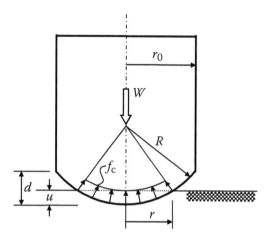

FIGURE 7.11   A striker indenting a surface.

formulas become. In the elastoplastic range, described above, some rules of finding maximum deflection and peak force may be devised, but they will hold only for specific combination of materials. The situation becomes easier to handle when the impact is assumed to be fully plastic.

When a striker is of a compact or pointed shape, it may create a dent or crater in the impacted surface. This is even more likely to happen when a striker is made of a stronger material than the target surface. The simplest approach in such a case is to assume (1) that as a result of impact the striker penetrates the impacted body by u_m and (2) that the contact stress level f_c remains constant during impact. The geometry of the event is presented in Figure 7.11, for a curved-surface striker, which is assumed to be rigid. The increase in contact force, as the striker travels downwards, is ascribed to the enlargement of the contact area. To keep the picture simple, the impacted surface just outside the contact area is treated as undeformable; it does not sink or swell.

A spherical surface striker of radius R pressed into a flat deformable surface is used as an example. At the penetration depth u, the radius of contact r is given by

$$r^2 = 2Ru - u^2 \tag{7.41}$$

and the contact force, according to the rules of hydrostatics, becomes

$$W = 2\pi R f_c u \left(1 - \frac{u}{2R}\right) \tag{7.42}$$

This is linearized by ignoring the second term in the above brackets. The energy absorbed by this local compression is then

$$\Pi = \pi R f_c u^2 \tag{7.43}$$

If one thinks about the deformability of the surface in terms of an equivalent spring with a stiffness k_e, as in Figure 7.7, the strain energy stored in that spring is $k_e u^2 / 2$, therefore

$$k_e = 2\pi R f_c \tag{7.44}$$

The above equation allows us to treat this impact as pseudolinear, because the effective spring stiffness is constant. (Refer to Case 7.26.) The kinetic energy may be expressed either by the drop height h or the equivalent initial velocity, $v_0 = \sqrt{2gh}$. Maximum deflection u_m is obtained by equating the energy of the striker, $Mgh = Mv_0^2/2$ with the strain energy:

$$u_m = \sqrt{\frac{Mgh}{\pi R f_c}} = v_0 \sqrt{\frac{M}{2\pi R f_c}} \tag{7.45}$$

while the corresponding maximum force is

$$W_m = v_0 (2M\pi R f_c)^{1/2} \tag{7.46}$$

The linearized equations (Equations 7.43 through 7.46) are quite accurate for a typical case of u/R being a small fraction. These expressions serve only to a point when the depth of penetration becomes equal to the depth of the spherical segment, $u = d$ in Figure 7.11, or when $r = r_0$. In this case the maximum attainable force of contact becomes

$$W_p = \pi r_0^2 f_c \tag{7.47}$$

and then remains constant with a further increase in displacement. When a striker is hemispherical with a radius a, then $W_0 = \pi a^2 f_c$ can be reached.

The duration of the loading phase can be found by equating the energy at u_m from Equation 7.43 to the initial kinetic energy. The assumption of harmonic motion gives Equation 7.24 again. Similar relations may be found in Johnson [41].

The procedure for a cylindrical surface striker of radius R is quite similar and is presented as Case 7.27. The force is now proportional to $\sqrt{u}$. The true characteristic when deflections are not necessarily small is illustrated in Figure 7.12. In this case an attempt to linearize the characteristic based on equal energy is much less accurate.

The above equations involve only the deflection component resulting from local deformation of the surface. Their main application is to provide preliminary estimates, but they can also be useful when creating a computer model in establishing a local stiffness at the point of impact.

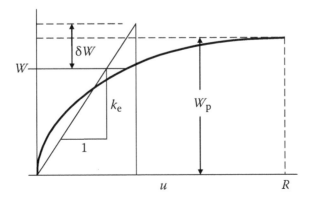

FIGURE 7.12   The true characteristic for a cylindrical striker.

These relationships were presented by this author [94], as a part of a study of steel strikers impacting heavy concrete beams. The range of impact velocities was between 1 and 10 m/s. There was a question of how to select the value of f_c to best match the test data. This author anticipated the following relationship: $f_c = nF_c'$, where n is a multiplier and F_c' is the conventional compressive strength of concrete. On the basis of these tests as well as other experiments in low-velocity rock mechanics, $n = 3$ was selected. This method gave much better results, in terms of the calculated impact force, than Hertz formulas. The reference to a similar approach, although in a more general setting, is given by Goldsmith [26]. Exactly the same result for the spherical striker is given by Johnson [40], who did not interpret the meaning of f_c correctly in his otherwise excellent book.

RIGID MASS IMPACTING AN AXIAL BAR

Figure 5.3 illustrates what happens, when a prescribed force is applied to a free end of a bar. The magnitude of force in those examples was constant, but it is not necessary this be so. If the force were decreasing in time, then the force diagram in Figure 5.3 would be decreasing, when going from left to right. Whatever the magnitude of the end force, the corresponding end velocity is defined by Equation 5.5. If there is an object of mass M attached to a bar, the bar itself acts as a velocity-proportional damper. In Case 3.7, which describes the movement of a mass M with initial velocity v_0, the movement resisted by a damper with a constant C, we had

$$v(t) = v_0 \exp\left(-\frac{Ct}{M}\right) \tag{7.48}$$

Figure 7.13 shows a bar impacted at the end, for which the corresponding damping constant is $C = A\rho c_0$, as per Equation 5.5. The end velocity varies according to

$$v(t) = v_0 \exp\left(-\frac{mc_0}{M}t\right) \quad \text{with } m = A\rho \tag{7.49}$$

Next, in Figure 7.13b the initial wave is shown, with the front of magnitude of $P_0 = A\rho c_0 v_0$ and the current force P at the tail, corresponding to the current velocity $v(t)$. At $t_1 = L/c_0$

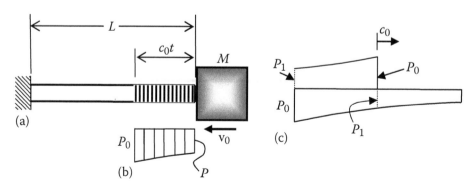

(a)

(b)

(c)

FIGURE 7.13 (a) Impacted bar; (b) axial force distribution at time t; (c) the first pass of the wave (below horizontal line) and after reflection (above).

this wave will reflect from the fixed end, with the same magnitude P_0 (causing a temporary doubling of force to $2P_0$ at that end) and travel to the impacted end. The status in Figure 7.13c is shown for time $t \approx 1.5L/c_0$.

If the motion of the impacted end is prescribed, like in this case, this end is not free any more. Because of this, the incoming wave will reflect from it as if from a fixed end, raising the total force to

$$P_2 = 2P_0 + A\rho c_0 v_2 \tag{7.50a}$$

with

$$v_2 = v_0 \exp\left(-\frac{2mL}{M}\right) \tag{7.50b}$$

or

$$P_2 = P_0 \left[2 + \exp\left(-\frac{2mL}{M}\right)\right] \tag{7.51}$$

The above happens at $t_2 = 2L/c_0$ and velocity v_2 is calculated from Equation 7.49. As detailed investigations show (Timoshenko [108] and Goldsmith [26]) further development depends on the mass ratio, $\alpha = A\rho L/M$. Unless there appears a loss of contact between M and the bar (accompanied by impacted end force dropping to nil) the compressive wave will reflect from the fixed end and impact the other end again. The actual peak force reached at the impacted end will vary depending on α, but P_2 may be used as a practical estimate of that peak. The largest compressive force appears at the fixed end and is estimated in Case 7.24.

This is merely one of the examples where the wave action is not consistent with basic engineering intuition. First of all, a bar, whose end is driven in a prescribed manner, reacts like a damper, rather than a spring. (The simple relationship is, of course, later upset by the reflection from the fixed end.) Also, one would expect that the contact force to be dependent on the impacting mass M. Although there is some relationship between the contact force and the mass ratio, at least in the initial phase, it is solely the impact speed v_0 that matters.

When the striking mass is very small, i.e., $M \ll A\rho L$, then the second term in square brackets of Equation 7.51 becomes negligible and the peak force in the bar is close to $2P_0$. When the reverse is true, namely, $M \gg A\rho L$, the second term tends to unity.

This latter possibility, $M \gg A\rho L$, brings up the matter of a substantial difference between results derived here and what is obtained when the basic stereomechanical approach is employed. If the mass of the bar is small, it is natural to treat the bar as a massless spring and use Case 3.2 to find the peak impact force of $R_m = v_0 \sqrt{kM}$. This can be rewritten by noting $k = EA/L$ and in the end we find $R_m = A\rho c_0 v_0$, which is identical with P_0, as defined here.

Thus, if Case 3.2 is used to derive the peak contact force by disregarding the bar mass and treating it as a spring, a substantial underestimate is obtained for small ratios $M/(mL)$. The difference, of course, is not as large as implied here, because of damping, which had not been accounted for and which will reduce the peak values. With this approach the contact between the bodies lasts for one-half of the natural period, calculated in a usual way, namely,

$$t_d = \pi\sqrt{\frac{M}{k}} = \pi\sqrt{\frac{ML}{EA}} \tag{7.52}$$

and the contact force is calculated per Case 3.2 with $k=EA/L$. According to what was said before, one should increase the peak force by the following P_0 to obtain a better approximation:

$$P_0 = A\rho c_0 v_0 = A v_0 \sqrt{E\rho} \tag{7.53}$$

AXIAL BAR WITH THE END SPRING

This configuration needs a different analytical approach. As Figure 7.14 shows, the initial velocity applied to the bar is equivalent to the constant velocity applied to the rigid base, but the second option is easier to deal handle. Again assuming the compressive force to be positive, one can write the displacements and velocities for the spring ends as follows:

$$P = k(u_1 - u_2) \quad \text{and} \quad \dot{P} = k(\dot{u}_1 - \dot{u}_2) \tag{7.54}$$

At the driven end $\dot{u}_1 = v_0$, while at the other Equation 5.5b holds. The second of the above expressions can be written as

$$\dot{P} + \frac{k}{A\rho c_0}P = kv_0 \tag{7.55}$$

This first-order differential equation for P can be easily solved by separation of variables with the initial condition of $P(0) = 0$:

$$P = P_0\left(1 - \exp\left(\frac{-kt}{A\rho c_0}\right)\right) \quad \text{with } P_0 = A\rho c_0 v_0 \tag{7.56}$$

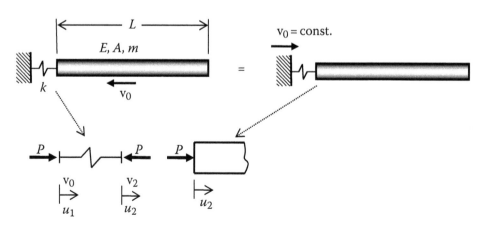

FIGURE 7.14 Axial bar impacting wall through a spring. Giving initial velocity to the bar is equivalent to applying constant velocity v_0 to the wall. Internal forces are shown below.

This equation is valid only until $t_m = 2L/c$, i.e., until the initially compressive wave comes back as tensile, after reflecting from the free end. Prior to that time

$$P_m = P_0 \left(1 - \exp\left(\frac{-2kL}{EA} \right) \right)$$ (7.57)

was, for all practical reasons, the maximum compressive force. Note that as the ratio $k/(EA/L)$ becomes bigger, so does P_m, which may come close to P_0, but not to exceed it. The spring may also be used to represent an element acting in accordance with the Hertz theory, provided that said element is linearized according to the principles described before, giving the constant k_{ef} expressed by Equation 7.38a and b. The peak load W_m is unknown, causing the need for successive approximations, but a reasonable starting point makes the process rapidly converge.

CLOSING REMARKS

A number of simplifications have been used in this chapter to make a description of physical events more manageable. In the stereomechanical impact, for example, the nature of contact between an impacting solid and the rigid base remained invariable throughout the event, excluding a possibility of intermittent slippage. The coefficient of restitution was defined solely for the normal velocity component. Relaxing these limitations may occasionally lead to a better insight into physics, but also makes the analysis more complex and less transparent. (Some examples of more refined analyses can be found in Goldsmith [26].)

A fair amount of space was devoted to determination of a local flexibility, which is often dominant for bodies of a compact shape. When deformability of an axial bar was investigated, it was found that the axial bar responds to a mass impact as a damper, rather than a spring, at least in the initial phase of contact. The reaction of a beam to a lateral impact will be presented in Chapter 10, while a plate impact will be analyzed in Chapter 12.

When an elastic body impacts an obstacle, intense vibration takes place. This makes the contact force oscillate and go through a number of successive peaks and valleys. Our simplified analyses can predict only one peak, which should be regarded as the highest one.

TABULATION OF CASES

COMMENTS

Case 7.15 specifies the applied force and impulse necessary for overturning a standing object. The force must be large enough to lift one side of the object against gravity. The impulse must be sufficient for the body to acquire a kinetic energy to raise the position of the CG so that it reaches a point directly above the corner, about which overturning takes place. This formulation is applicable for relatively short impulses, so that the impulse (or its main part) ends while the rotation angle is still small.

Cases 7.19 through 7.22 are coarse approximations of mass-impacted beams when the kinetic energy of the impacting mass is large enough to cause a pronounced plastic action. The approximation is reasonable when the impacting mass is larger than the beam mass.

CASE 7.1   PARTICLE REBOUNDING FROM RIGID SURFACE

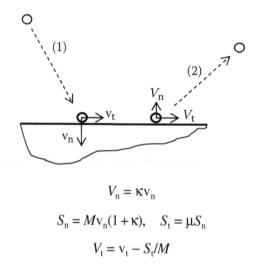

$$V_n = \kappa v_n$$

$$S_n = M v_n (1 + \kappa), \quad S_t = \mu S_n$$

$$V_t = v_t - S_t/M$$

For the case with friction the tangential velocity will not reverse, i.e., $V_t \geq 0$. If viscous damping is involved, $\kappa = \exp(-\pi\chi)$ with $\chi = \dfrac{\zeta}{\sqrt{1-\zeta^2}}$. For small damping: $\kappa \approx 1 - \pi\zeta$.

CASE 7.2   THE INVERSE PROBLEM FOR A PARTICLE REBOUNDING FROM SURFACE (AS IN CASE 7.1)

(Pre-impact status from post-impact data.)

$$v_n = V_n/\kappa$$

$$S_n = M V_n \frac{1+\kappa}{\kappa}$$

$$S_t = \mu S_n$$

$$v_t = V_t + \mu V_n \frac{1+\kappa}{\kappa}$$

If viscous damping is involved: $\kappa = \exp(-\pi\chi)$ with $\chi = \dfrac{\zeta}{\sqrt{1-\zeta^2}}$. For small damping: $\kappa \approx 1 - \pi\zeta$.

CASE 7.3   BODY WITH A TRANSLATIONAL VELOCITY IMPACTING A FRICTIONLESS SURFACE

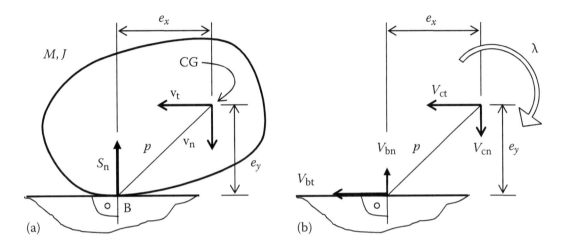

$$V_{bn} = \kappa v_n$$

$$M_r = \frac{JM}{J + Me_x^2}; \text{ reduced mass}$$

$$S_n = M_r v_n (1 + \kappa); \text{ normal impulse}$$

$$\lambda = \frac{S_n e_x}{J}$$

$$V_{cn} = v_n - \frac{S_n}{M}$$

$$V_{ct} = v_t$$

$$V_{bt} = v_t + \lambda e_y$$

Note: As the surface is frictionless, only the normal component of impulse, S_n, is present.

CASE 7.4   A ROD IMPACTED NORMALLY WITH IMPULSE S, OFF CENTER, IN PRESENCE OF GRAVITY

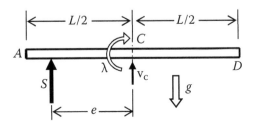

$$M = mL; \quad J = ML^2/12$$

$$v_c = \frac{S - Mgt_0}{M}; \text{ velocity of CG}$$

$$\lambda = \frac{Se}{J}; \text{ angular velocity}$$

$$v_A = v_c + \lambda L/2; \text{ at end A}$$

$$v_D = v_c - \lambda L/2; \text{ at end D, up when positive}$$

Special case: $e = L/2$, no gravity: $v_c = \dfrac{S}{M}; \lambda = \dfrac{6S}{ML}; v_a = \dfrac{4S}{M}; v_b = -\dfrac{2S}{M}$

Note: Duration t_0 must be large enough to justify inclusion of gravity effect.

CASE 7.5   THE INVERSE PROBLEM FOR A BODY WITH TRANSLATIONAL VELOCITY IMPACTING ON FRICTIONLESS SURFACE (AS IN CASE 7.3)

(Pre-impact status from post-impact data.)

$$v_n = V_{bn}/\kappa$$

$$M_r = \frac{M(v_n - V_{cn})}{v_n(1+\kappa)}$$

$$S_n = M_r V_{bn} \frac{1+\kappa}{\kappa}$$

$$e_x = \frac{\lambda J}{S_n}$$

$$v_t = V_{ct}$$

$$e_y = \frac{V_{bt} - v_t}{\lambda}$$

CASE 7.6   BODY WITH TRANSLATIONAL VELOCITY IMPACTING ON A ROUGH SURFACE

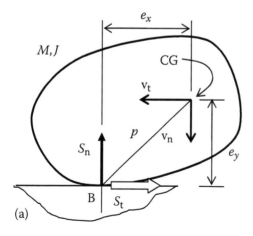

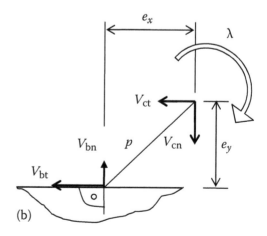

(a)   (b)

(Friction coefficient is μ)

$$V_{bn} = \kappa v_n; \text{ rebound velocity at B}$$

$$S_n = M_{rn} v_n (1 + \kappa)$$

$$S_t = \mu S_n$$

$\dfrac{1}{M_{rx}} = \dfrac{1}{M} - \dfrac{\tilde{e}e_y}{J\mu}; \dfrac{1}{M_{rn}} = \dfrac{1}{M} + \dfrac{\tilde{e}e_x}{J};$ reduced masses of normal and tangential impulse, respectively

$$\lambda = \frac{S_n \tilde{e}}{J} \equiv \frac{S_n}{J}(e_x - \mu e_y) \text{ with } \tilde{e} = e_x - \mu e_y$$

$$V_{cn} = v_n - \frac{S_n}{M}$$

$$V_{ct} = v_t - \frac{S_t}{M}$$

$$V_{bt} = V_{ct} + \lambda e_y$$

Note: The interaction between the impacting body and the target surface is of unchanging character during the event.

CASE 7.7 THE INVERSE PROBLEM FOR A BODY WITH TRANSLATIONAL VELOCITY IMPACTING ON A ROUGH SURFACE (AS IN CASE 7.6)

(Pre-impact status from post-impact data.)

$$v_n = V_{bn}/\kappa$$

$$S_n = M(v_n - V_{cn}); \quad S_t = \mu S_n$$

$$M_{rn} = \frac{S_n}{v_n(1 + \kappa)}$$

$$\tilde{e} = \frac{\lambda J}{S_n}$$

$$e_y = \frac{V_{bt} - V_{ct}}{\lambda}$$

$$v_t = V_{ct} + \frac{S_t}{M}$$

$$e_x = \tilde{e} + \mu e_y$$

CASE 7.8 BODY IMPACTING ON A NO-SLIP SURFACE WITH A NORMAL VELOCITY v_0

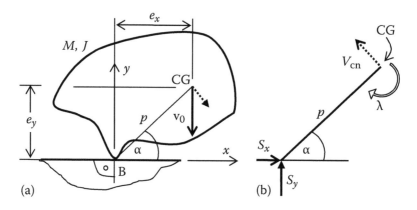

(a)   (b)

$$M'_r = \frac{JM}{J + Mp^2}$$

$$V_{Bl} = \kappa v_0 \sin \alpha; \quad V_{Bn} = \kappa v_0 \cos \alpha$$

longitudinal and normal relative to B-CG, at B

$$V_{cl} = \kappa v_0 \sin \alpha; \quad V_{cn} = \left(1 - \frac{J(1+\kappa)}{J + Mp^2}\right) v_0 \cos \alpha$$

longitudinal and normal rebound velocities, at CG

$$S_l = M(1+\kappa)v_0 \sin \alpha; \quad S_n = M'_r(1+\kappa)v_0 \cos \alpha$$

longitudinal and normal impulses, relative to B-CG, applied at B

$$S_x = (M - M'_r)\sin \alpha \cos \alpha (1+\kappa)v_0$$

$$S_y = (M \sin^2 \alpha + M'_r \cos^2 \alpha)(1+\kappa)v_0$$

$$\lambda = \frac{Mp(1+\kappa)v_0 \cos \alpha}{J + Mp^2}$$

Note: Refer to text. λ is the angular velocity after rebound. S_x and S_y are total impulses applied to the body by impact.

CASE 7.9 SPHERICAL STRIKER IMPACTING A SURFACE OF THE SAME MATERIAL

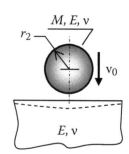

$\kappa = 1 - \dfrac{X}{1 + (Y/v_0)}$; restitution coefficient for a sphere rebounding from a thick plate.

Note: These are experimental results after Ref. [27].

Material	X	Y
Construction steel	0.427	1.57
Rail steel	0.550	1.48
Nickel steel	0.575	1.28
Aluminum alloy 6061-T6	0.715	0.19
Aluminum 1100F	0.780	0.15
Lead	0.885	0.041
Brass	0.560	0.36

CASE 7.10   RESTITUTION COEFFICIENTS FOR NATURAL ROCK

Restitution coefficient range when rock falls on concrete surface:

"Square" rock: $\kappa = 0.09–0.34$
"Round" rock: $\kappa = 0.08–0.10$

Friction coefficient range for several surface types:

Note: These are the results of the experimental work by Masuya et al. [60]. "Natural" means "as found."

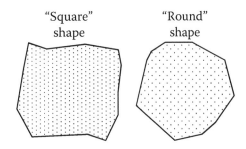

	μ_{min}	μ_{max}
Concrete slab	0.30	0.90
Gravel road	0.25	1.03
Soil	0.29	1.29

CASE 7.11   RESTITUTION COEFFICIENTS FOR A FALLING BEAM

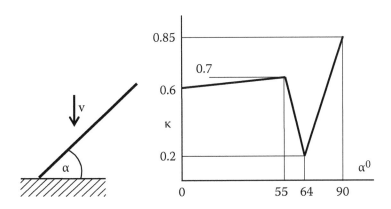

The steel beam is falling on a steel base with a small friction coefficient ($\mu \leq 0.075$).

Note: The κ values graphically presented come from tests and FEA work by Low and Zhang [58]. As usual with a deformable body impact, there was a loss and regaining of contact between surfaces during the impact duration.

CASE 7.12   OBLIQUE IMPACT OF BODY OF REVOLUTION AGAINST NO-SLIP SURFACE

Initial velocities in (a) and final in (b).

$$M_r' = \frac{JM}{J + MR^2}$$

$$V_n = \kappa v_n; \quad V_t = \left(1 - \frac{J(1+\kappa)}{J + MR^2}\right)v_t$$

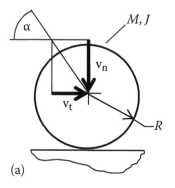

(a)

longitudinal and normal rebound velocities, at CG

$$S_t = M_r'(1+\kappa)v_t; \quad S_n = M(1+\kappa)v_n$$

longitudinal and normal impulses, applied at base

$$\lambda = \frac{MR(1+\kappa)v_t}{J + MR^2}; \text{ angular velocity.}$$

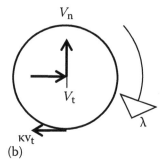

(b)

CASE 7.13   NORMAL IMPACT OF A SPINNING BODY OF REVOLUTION AGAINST NO-SLIP SURFACE

Initial velocities in (a) and final in (b).

$$M_r' = \frac{JM}{J + MR^2}$$

$$V_n = \kappa v_n; \quad V_h = \frac{J\Omega R(1+\kappa)}{J + MR^2}$$

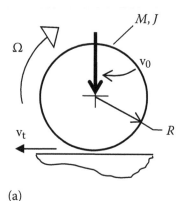

(a)

normal and tangential rebound velocities, at CG

$$S_t = M_r'(1+\kappa)\Omega R; \quad S_n = M(1+\kappa)v_0$$

tangential and normal impulses, respectively, applied at base

$$\lambda = \frac{J - \kappa MR^2}{J + MR^2}\Omega$$

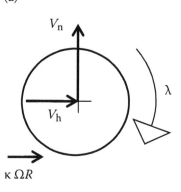

(b)

CASE 7.14   VISCOUS DAMPING AND COEFFICIENT OF RESTITUTION

v_0 is the impact velocity

$V = \kappa v_0$; rebound velocity

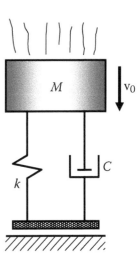

$$\kappa = \exp\left(-\frac{\pi\zeta}{\sqrt{1-\zeta^2}}\right) \approx \exp(-\pi\zeta) \approx 1 - \pi\zeta\,;\text{ coefficient of restitution}$$

The first value should be used for a relatively large damping, when $0.5 < \zeta < 1.0$. The last value is applicable for small damping, when ζ is only a few percent.

Note: Refer to Case 3.13, where the solution to the equation of motion is given.

CASE 7.15   IMPULSE S TO OVERTURN A HEAVY BLOCK STANDING ON A NO-SLIP SURFACE

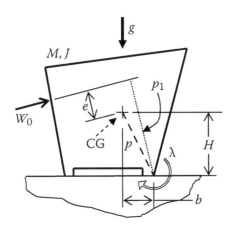

$(S = W_0 t_0$ when t_0 is short$)$

$W_0 > Mgb/p_1$ peak force needed

$S > \dfrac{1}{p_1}\left[2MgJ_B(p - H)\right]^{1/2}$; impulse required

$\lambda = Sp_1/J_B$; angular velocity gained, $J_B = J + Mp^2$

Note: J is the moment of inertia about the centroidal axis, normal to paper. For the overturning to take place, both inequalities must be satisfied.

CASE 7.16 SPHERICAL STRIKER IMPACTING RIGID SURFACE

(Equivalent to two identical spheres colliding.)

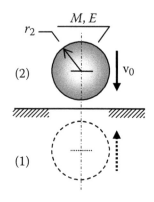

$$\frac{1}{k_0} = \frac{3}{2}\left(\frac{1-v^2}{E}\right)\sqrt{\frac{2}{r_2}}; \quad \delta_m = \left(\frac{5Mv_0^2}{2k_0}\right)^{0.4}$$

$$W_m = \left(\frac{5}{2}Mv_0^2\right)^{0.6}k_0^{0.4}; \quad t_0 = \frac{3.214}{(2v_0)^{0.2}}\left(\frac{M}{2k_0}\right)^{0.4}$$

$$a = \left(\frac{5Mv_0^2}{2k_0}\right)^{0.2}\sqrt{r_2}; \text{ contact radius}$$

CASE 7.17 SPHERICAL STRIKER IMPACTING AN ELASTIC SURFACE

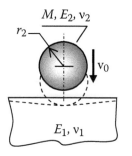

$$\frac{1}{k_0} = \frac{3}{4}\left(\frac{1-v_1^2}{E_1} + \frac{1-v_2^2}{E_2}\right)\frac{1}{\sqrt{r_2}}; \quad \delta_m = \left(\frac{5Mv_0^2}{4k_0}\right)^{0.4}$$

$$W_m = \left(\frac{5}{4}Mv_0^2\right)^{0.6}k_0^{0.4}; \quad t_0 = \frac{3.214}{v_0^{0.2}}\left(\frac{M}{k_0}\right)^{0.4}$$

$$\sigma_{cm} = 0.499\frac{M^{0.2}}{r_2}k_0^{0.8}v_0^{0.4}; \text{ peak surface contact stress}$$

$$\tau_m = 0.31\sigma_{cm}; \text{ peak shear underneath the contact surface}$$

$$\sigma_{tm} = \frac{1-2v}{3}\sigma_{cm}; \text{ peak surface tension.}$$

CASE 7.18 CYLINDRICAL STRIKER IMPACTING AN ELASTIC SURFACE

$$(H/r_2 \le 4)$$

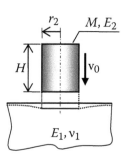

$$\frac{1}{k} = \frac{1}{2r_2E_1}\left(1 - v_1^2 + \frac{E_1 H}{\pi E_2 r_2}\right); \quad \delta_m = \frac{W_m}{k}$$

$$W_m = v_0\sqrt{kM}; \quad t_0 = \frac{\pi}{2}\sqrt{\frac{M}{k}}$$

CASE 7.19   A RIGID BEAM IMPACTED BY MASS M

Angular spring with RPP characteristic and $\mathcal{M}_0$ capacity at the supported end $(w_0L^2/2 > \mathcal{M}_0)$. Plastic impact; mass and beam travel together.

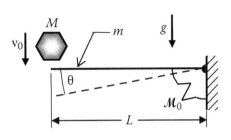

$$E_{k1} = Mv_0^2/2 \text{ before contact.}$$

$$E_{k2} = \frac{M}{M_{ef}} E_{k1} \text{ with } M_{ef} = M + mL/3 \text{ (after)}$$

(a) Without gravity: $\theta_m = \dfrac{M}{M_{ef}} \dfrac{Mv_0^2}{2\mathcal{M}_0}$; $u_m = \theta_m L$; peak deflection

(b) With gravity: $\theta_m = \dfrac{M}{M_{ef}} \dfrac{Mv_0^2}{2(\mathcal{M}_0 - MgL)}$

Note: The decrease in kinetic energy comes from preservation of angular momentum.

CASE 7.20   A RIGID BEAM ASSEMBLY IMPACTED AT CENTER BY MASS M

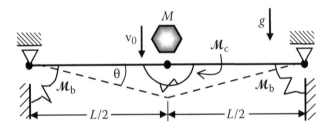

The angular springs have RP characteristics and capacities $\mathcal{M}_b$ and $\mathcal{M}_c$, respectively. Plastic impact; mass and beams travel together.

$$E_{k1} = Mv_0^2/2; \text{ before contact.}$$

$$E_{k2} = E_{k2} = \frac{M}{M_{ef}} E_{k1} \text{ with } M_{ef} = M + mL/3; \text{ after contact}$$

(a) Without gravity: $\theta_m = \dfrac{M}{M_{ef}} \dfrac{Mv_0^2}{4(M_b + M_c)}$; $u_m = \theta_m L/2$; peak deflection

(b) With gravity: $\theta_m = \dfrac{M}{M_{ef}} \dfrac{Mv_0^2}{2(2M_b + 2M_c - MgL/2)}$; u_m as above

Note: The decrease in kinetic energy comes from preservation of angular momentum.

CASE 7.21   A RIGID BEAM ASSEMBLY IMPACTED AT CENTER BY MASS M

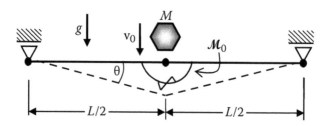

(A simply supported beam). The angular spring has RP characteristic and moment capacity M_0. Plastic impact; mass and beams travel together.

$$E_{k1} = Mv_0^2/2 \text{ before contact}$$

$$E_{k2} = \frac{M}{M_{ef}} E_{k1} \text{ with } M_{ef} = M + mL/3; \text{ after contact}$$

(a) Without gravity: $\theta_m = \dfrac{M}{M_{ef}} \dfrac{Mv_0^2}{4M_0}$; $u_m = \theta_m L/2$; peak deflection

(b) With gravity: $\theta_m = \dfrac{M}{M_{ef}} \dfrac{Mv_0^2}{2(2M_0 - MgL/2)}$; u_m as above

Note: The decrease in kinetic energy comes from preservation of angular momentum.

CASE 7.22   A RIGID BEAM ASSEMBLY IMPACTED AT CENTER BY MASS M

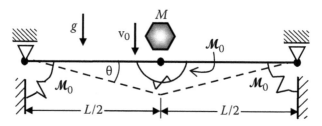

(A clamped–clamped beam) The angular RP springs have M_0 capacity. Plastic impact; mass and beams travel together.

$$E_{k1} = Mv_0^2/20 \text{ before contact.}$$

$$E_{k2} = \frac{M}{M_{ef}} E_{k1} \text{ with } M_{ef} = M + mL/3; \text{ after contact}$$

(a) Without gravity: $\theta_m = \dfrac{M}{M_{ef}} \dfrac{Mv_0^2}{8M_0}$; $u_m = \theta_m L/2$; peak deflection

(b) With gravity: $\theta_m = \dfrac{M}{M_{ef}} \dfrac{Mv_0^2}{2(4M_0 - MgL/2)}$; u_m as above

Note: The decrease in kinetic energy comes from preservation of angular momentum.

Case 7.23 Mass Impacting an Unconstrained Elastic Bar

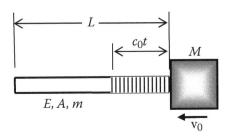

$P_0 = A\rho c_0 v_0 = A v_0 \sqrt{E\rho}$; initial impact force: $(P = A\sigma)$

After rebound from free end and return, at $t_r = 2L/c_0$:

$$P = -P_0 + P_0 \exp\left(-\frac{2mL}{M}\right) \text{ (tensile prevails)}$$

$$v(t) = v_0 \exp\left(-\frac{mc_0}{M}t\right); \text{ impacted end velocity, for } t < t_r \text{ with } m = A\rho$$

$$v_M = v_0 \exp\left(-\frac{2mL}{M}\right); \text{ mass velocity at } t_r$$

$$v_e = v_0\left[1 + \exp\left(-\frac{2mL}{M}\right)\right]; \text{ impacted bar end at } t_r \text{ directed left}$$

Note: The resultant is a tensile force, which means separation; end of contact. (Tension is shown as negative here.)

Case 7.24 Mass Impacting Constrained Elastic Bar

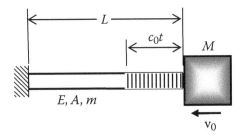

$P = A\sigma$

$P_0 = A\rho c_0 v_0 = A v_0 \sqrt{E\rho}$; initial impact force

$$P_2 \approx \left[1 + \exp\left(-\frac{2mL}{M}\right)\right]2P_0; \text{ estimated peak value for the constrained end}$$

$$P_1 = \left[2 + \exp\left(-\frac{2mL}{M}\right)\right]P_0; \text{ force at the impacted end, just after } t = t_{tr}$$

$t_{tr} = 2L/c_0 = L/c_0$ (time for the wave to traverse L and return)

$$P_\mathrm{m} \approx P_0 \left(\sqrt{\dfrac{M}{mL}} + 1 \right); \text{ approximation for } M > 5mL$$

Contact duration for the same: $t_\mathrm{d} = \pi \sqrt{\dfrac{ML}{EA}}$

Note: Contact duration varies from $3.07t_\mathrm{tr}$ for $M/(mL) = 1$ to $7.42t_\mathrm{tr}$ for $M/(mL) = 6$. In the last equation t_d is half-period, $\tau/2$, from Equation 1.4 with the bar mass ignored. The equations for P_1 and P_2 give the approximate bounds for the peak compressive force in the bar, Ref. [108].

CASE 7.25   PILE DRIVING—A SIMPLIFIED THEORY

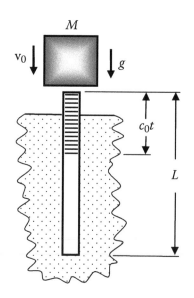

Hammer M impacts the top end of the pile with velocity v_0. The pile is submerged in soft soil. Gravity assists in driving the pile.

$P_0 = A\rho c_0 v_0 = Av_0\sqrt{E\rho}$; initial contact force

$t_\mathrm{d} = 2L/c$; contact duration

$$v(t) = \left(v_0 - \dfrac{Mg}{mc_0} \right) \exp\left(-\dfrac{mc_0}{M} t \right) + \dfrac{Mg}{mc_0}; \text{ velocity of top end}$$

for $t < t_\mathrm{d}$

$P(t) = A\rho c_0 v(t)$; contact force at the top end for $t < t_\mathrm{d}$

$P' \approx A\rho c_0 v_\mathrm{d}$; estimated force of second impact, where $v_\mathrm{d} = v(t_\mathrm{d})$

Note: In this simplified theory the presence of soil is essentially neglected, except when assuming the top end to be stationary before the second impact. At $t = t_\mathrm{d}$ the top end of pile has an abrupt velocity increase by v_0, downward.

CASE 7.26   A HARD STRIKER WITH A SPHERICAL CONTACT FACE IMPACTING A DEFORMABLE SURFACE

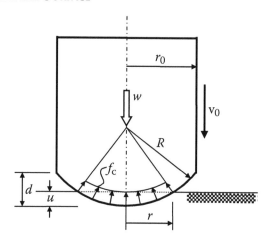

(Plastic impact)

For initial velocity v_0 or the drop height h, with $v_0 = \sqrt{2gh}$:

$$u_m = \sqrt{\frac{Mgh}{\pi R f_c}} = v_0 \sqrt{\frac{M}{2\pi R f_c}}$$

$W_m = v_0 (2M\pi R f_c)^{1/2}$; peak contact load

$W_m = v_0 (k_e M)^{1/2}$; peak contact load

$W_m = W_p \dfrac{r^2}{r_0^2} \approx W_p \dfrac{u}{d}$; maximum contact force

$k_e = 2\pi R f_c$; effective stiffness, then

$t_m = \sqrt{\dfrac{8Mu_m}{3W_m}}$; duration of loading phase

$\Pi = \pi R f_c u^2$; strain energy absorbed

For the case when $u = d$: $W_p = \pi r_0^2 f_c$

$W_0 = \pi R^2 f_c$ (Maximum attainable force in case of half-spherical striker, $u = R$)
$W_m = 2W_0 x_m (1 - x_m/2)$; peak contact force (large defl.)

Note: Relations valid for $u \ll R$. f_c is the dynamic bearing strength; for concrete $f_c \approx 3F_c'$, where F_c' is the nominal compressive strength and the impact speed is between 1 and 10 m/s. A more general observation: When mass M, moving with a velocity v, impacts a solid wall via a linear spring of stiffness k, the peak contact force is given by $W = v\sqrt{kM}$. The analogous form here is $W_m = v_0 (k_e M)^{1/2}$.

CASE 7.27 A HARD STRIKER WITH A CYLINDRICAL CONTACT FACE IMPACTING A DEFORMABLE SURFACE

(Refer to the illustration in Case 7.26)
(Plastic impact)

For initial velocity v_0 or the drop height h, with $v_0 = \sqrt{2gh}$:

$$x_m = \left(\frac{3Mgh}{2\sqrt{2}RW_0} \right)^{2/3} = \left(\frac{3Mv_0^2}{4\sqrt{2}RW_0} \right)^{2/3} \quad \text{with } x_m = u_m/R$$

u_m is the peak displacement
$W_m = 2bR f_c (2x_m)^{1/2}$; peak contact force

$t_m = \sqrt{\dfrac{8Mu_m}{3W_m}}$; duration of loading phase

$W_0 = 2bR f_c$ (maximum attainable force in case of half-cylinder striker, $u = R$)

$\Pi = \dfrac{2}{3}\sqrt{2} R x_m^{3/2} W_0$; strain energy absorbed

$W_m = W_0 (2x_m - x_m^2)^{1/2}$; peak contact force (large defl.)

Note: The striker length b (normal to paper) must be sufficiently large so that the end effects do not matter. In addition, notes for Case 7.26 are valid here.

CASE 7.28 TAYLOR CYLINDER TEST: A SHORT BAR IMPACTING A RIGID WALL

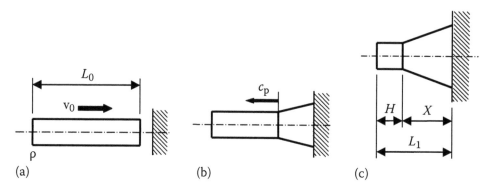

(a)   (b)   (c)

Purpose of test: determine the dynamic yield strength F_{yd} and the plastic wave speed c_p.
 Sequence as shown:

(a) Initial shape
(b) During deformation
(c) Final deformed shape

$$F_{yd} = \frac{\rho v_0^2}{2} \frac{L_0 - H}{L_0 - L_1} \frac{1}{\ln(L_0/H)}; \text{ dynamic yield strength}$$

$$c_p = \frac{v_0}{2} \frac{X}{L_0 - L_1}; \text{ plastic wave speed}$$

$$c_p = \frac{X}{t_d}; \text{ alternative expression, where } t_d \text{ is from first contact to rest}$$

Note: The mushrooming of the head is a large-strain phenomenon, consequently the above is a simplified theory. v_0 must be less than c_p in order to avoid shock waves not considered in deriving the above equations, Ref. [121].

CASE 7.29 LIGHT AIRCRAFT IMPACT AGAINST GROUND

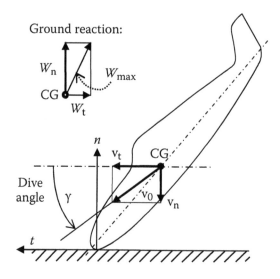

v_0 is the impact speed

W_{max} is the peak dynamic load applied to the aircraft during impact. The smaller sketch shows ground reaction at CG.

$n = \dfrac{W_{max}}{Mg}$; overload factor, to be assumed between $n = 17$ and $n = 33$, average $n = 25$

$W_{max} = Mgn$

$W_t = 0.4W_n$; peak horizontal load component

$W_n = \dfrac{W_{max}}{1.077}$; vertical component

$v_n = v_0 \sin \gamma$; vertical impact speed

$S_n = Mv_n$; vertical impulse applied to aircraft

$t_0 = \dfrac{2Mv_n}{W_n}$; impact duration

Note: The above assumes friction $\mu = 0.4$ and a triangular plot of impact force vs. time. The sketch shows peak forces applied to the centroid. Refer to Example 7.18 for additional comments.

CASE 7.30   A LIQUID BODY IMPACTING RIGID SURFACE

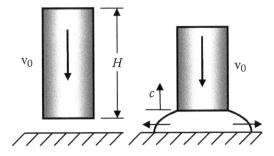

ρ, c, and K are density, sound velocity, and bulk modulus of liquid, respectively; M is the impactor mass:

$c = \sqrt{\dfrac{K}{\rho}}$

$p_m = \rho c v_0$; initial impact pressure

$W_m = A p_m$; initial impact force for a body with a constant section A, parallel to rigid surface

$W_f = A \rho v_0^2$; hydraulic impact force, in effect shortly after the first contact

$S = Mv_0$; total impulse applied to the surface

Note: W_m is applied during the first instant only. The hydraulic force W_f is applied as long as the initial body shape can be distinguished, or for nearly $t_d = H/v_0$. Water at 15°: $\rho = 999\,kg/m^3$ and $K = 2140\,MPa$. Zukas et al. [121] quotes a study of a water drop impact showing that p_m can be exceeded by 20%.

CASE 7.31   CYLINDRICAL, METALLIC STRIKER IMPACTING LIQUID SURFACE

Contact stress on the flat face at the beginning of impact

$$\sigma = \frac{v_0}{(1/\rho_w c_w) + (1/\rho_c c_c)} \quad \text{Ref. [40]}$$

$$v_w = \frac{\sigma}{\rho_w c_w}; \text{ interface velocity}$$

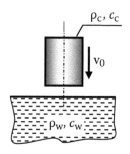

CASE 7.32   SPHERICAL STRIKER IMPACTING A BODY OF WATER

M_w is mass of water contained in a sphere of radius R.

$$a = \frac{3}{8} C_D \frac{v_0^2}{R} \frac{M_w}{M}; \text{ deceleration with the drag coefficient } C_D \text{ variable in}$$
time, $C_D \approx 1.03$ (max.)

$$\Delta v = v_0 \left(1 - \frac{1}{\exp(M_w/4M)} \right); \text{ velocity loss on submerging; Ref. [6]}$$

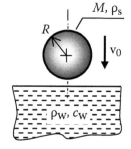

Note: If M is non-homogeneous, the average ρ_s should be used.

CASE 7.33   SUDDEN VALVE CLOSING IN A PIPELINE WHERE LIQUID IS
FLOWING—WATER HAMMER

R, h, and E are pipe radius, wall thickness, and Young's modulus, respectively

p_0 and v_0 are static pressure and velocity of flowing liquid, respectively

$$t_d = 2L/c$$

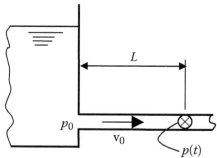

$\Delta p = \rho c v_0$; pressure jump due to valve closing

$$c = \sqrt{\frac{K_{ef}}{\rho}} \quad \text{where} \quad \frac{1}{K_{ef}} = \frac{1}{K} + \frac{2R}{Eh}$$

Note: If closing of the valve is gradual, so is the pressure increase. Pressure must not drop below atmospheric p_a for the process to develop as described. The real event shows pressure attenuating, rather than periodically repeating as illustrated here. The resultant pressure just upstream of the valve is shown. For more details see Tullis [112].

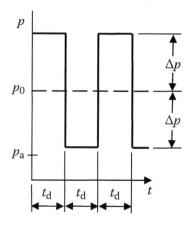

EXAMPLES

EXAMPLE 7.1   ECCENTRIC IMPACT ON A BEAM

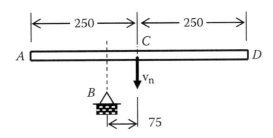

A 500 mm long rod with mass of 2500 g is falling with a vertical velocity of 4 m/s when it hits a stop. The restitution coefficient is 0.6. Calculate the velocity distribution just after the impact. Then, solve the reverse problem. From the knowledge of kinematics just after the rebound find the impact velocity, the impulse, as well as the offset from the CG.

Formulas of Case 7.3 apply, except that there is no horizontal component. The inertia constants are

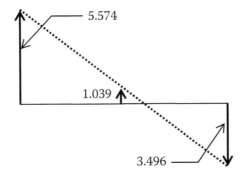

$M = 2500$ g, $J = ML^2/12 = 52.083 \times 10^6$ g/mm², $e_x = 75$ mm, $M_r = 1968.5$ g

Rebound velocity at impact point B and the impulse S are

$V_{bn} = \kappa v_n = 0.6 \times 4 = 2.4$ mm/ms; rebound velocity at B

$S = M_r v_n (1 + \kappa) = 1968.5 \times 4 \times 1.6 = 12{,}598$ N-ms/g; impulse

$\lambda = \dfrac{S_n e_x}{J} = 12{,}598 \times 75/J = 0.01814$ rad/ms; angular velocity

$V_{cn} = v_n - \dfrac{S_n}{M} = 4 - 12{,}598/2{,}500 = -1.039$ (up); velocity at CG

$V_{an} = -1.039 - 250\lambda = -5.574$; $V_{dn} = -1.039 + 250\lambda = 3.496$ mm/ms; velocity at ends

Next, using Case 7.5, find the pre-impact quantities. Note that $e_y = 0$, $V_{ct} = V_{bt} = 0$.

$$v_n = V_{bn}/\kappa = 2.4/0.6 = 4.0 \text{ m/s}$$

$$M_r = \frac{2500(v_n - V_{cn})}{v_n(1+\kappa)} = \frac{2500(4+1.039)}{4(1+0.6)} = 1968.4 \text{ g}$$

$$S_n = M_r V_{bn}\frac{1+\kappa}{\kappa} = 1968.4 \times 2.4\frac{1.6}{0.6} = 12{,}598 \text{ N-ms/g}$$

$$e_x = \frac{\lambda J}{S_n} = \frac{0.01814 \times 52.083 \times 10^6}{12{,}598} = 75 \text{ mm}; \quad v_t = 0; \quad e_y = \frac{V_{bt} - v_t}{\lambda} = 0$$

The answers are consistent with the status before impact.

EXAMPLE 7.2   CONTACT FORCE AND DURATION OF IMPACT

Using a linear spring with $k = 2000$ N/mm, calculate peak reaction and duration of loading phase during impact described in Example 7.1.

When M is replaced by the reduced mass M_r, the peak displacement can be found from Equation 7.23:

$$\delta_m = v_0\sqrt{\frac{M_r}{k_1}} = 4\sqrt{\frac{1968.5}{2000}} = 3.968 \text{ mm} \quad \text{and} \quad R_m = k\delta_m = 2000 \times 3.968 = 7937 \text{ N}$$

Duration from Equation 7.24: $t_1 = \dfrac{\pi\delta_m}{2v_0} = \dfrac{3.968\pi}{2 \times 4.0} = 1.558 \text{ ms}$

The reader can check that the more general formula (Equation 7.25b) gives the same result, provided M_r is used rather than M.

EXAMPLE 7.3   FALLING ROD IMPACTING A NO-SLIP SURFACE

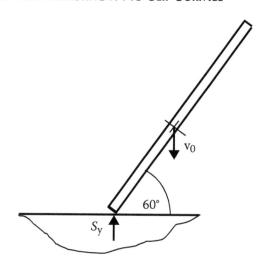

The same rod, as used in Example 7.1 is falling with a translational vertical velocity of 4 m/s when it impacts a no-slip surface. The restitution coefficient is also 0.6. Calculate surface-normal impulse and the final angular velocity. Compare these with the values obtained for frictionless surface.

Formulas of Case 7.8 apply for a no-slip surface. Note that for a slender rod, with $p = L/2$

$$\frac{J}{J + Mp^2} = 0.25$$

$M = 2500\,\text{g}; J = ML^2/12 = 52.083 \times 10^6\,\text{g/mm}^2$

$$M'_r = \frac{JM}{J + Mp^2} = 0.25 \times 2500 = 625\,\text{g}$$

$S_y = (0.75M + 0.25M'_r)(1 + 0.6)4.0 = 13,000$ N-ms; surface-normal impulse

$$\lambda = \frac{2500 \times 250(1.6)4 \times 0.5}{52.083 \times 10^6 + 2500 \times 250^2} = 9.6 \times 10^{-3}\ \text{rad/ms}\ ;\ \text{angular velocity}$$

As an exercise, one can find that when Case 7.3 is used for a frictionless surface we have

$$S = 9143 \quad \text{and} \quad \lambda = 21.94 \times 10^{-3}$$

In spite of vastly differing friction coefficients the differences in the surface-normal impulse are not very large, but the rod is spinning much faster after impact on the frictionless surface.

EXAMPLE 7.4   CONTACT FORCE AND DURATION OF A ROD IMPACT

Using a spring of rigid, strain-hardening (RSH) type, defined by constants $R_0 = 2,500$ N and $k_2 = 1000$ N/mm, calculate peak reaction component, normal to surface, and duration of loading phase during impact described in Example 7.3.

Case 4.4a applies, but M is replaced by the reduced mass $M_r = 625$ g.

$$u_m = \sqrt{\frac{2500^2}{1000^2} + \frac{625 \times 4^2}{1000} - \frac{2500}{1000}} = 1.531\,\text{mm} \quad \text{and} \quad R_m = 2500 + 1000 \times 1.531 = 4031\,\text{N}$$

The impulse associated with the loading phase is smaller than the entire impulse calculated in the example:

$$S'_n = (1/(1 + \kappa))S_y = (1.0/1.6)S_y = 13,000/1.6 = 8,125\ \text{N-ms.}$$

$$t_m \approx \sqrt{\frac{8 \times 625 \times 1.531}{2500 + 3 \times 4031}} = 0.724\ \text{ms}\ ;\ \text{duration per Equation 4.22.}$$

The reader can check that the alternative formula (Equation 7.25b), with M_r, gives 0.974 ms, which is a less accurate result. (Note that using S'_n would cause a serious error.) However, the basic approximation (Equation 7.24) gives 0.601 ms, a closer result.

EXAMPLE 7.5   INCLINED FALLING ROD IMPACTING ROUGH SURFACE

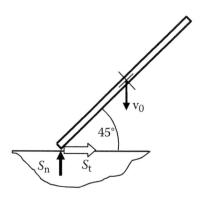

A rod is falling with a vertical velocity of 1 m/s when it impacts a surface with a friction coefficient of $\mu = 0.45$. The restitution coefficient is $\kappa = 0.5$. Calculate post-impact velocities at CG, at impact point B, as well as the angular velocity. $M = 1\,\text{kg}$, $L = 1\,\text{m}$.

Formulas of Case 7.6 apply.

$$J = ML^2/12 = 1000 \times 1000^2/12 = 83.33 \times 10^6\,\text{g/mm}^2$$

$$V_{bn} = \kappa v_n = 0.5 \times 1.0 = 0.5\,\text{m/s}$$

$$e_x = e_y = (L/2)\cos 45° = 353.6\,\text{mm}$$

$$\tilde{e} = e_x - \mu e_y = 353.6(1 - 0.45) = 194.5\,\text{mm}$$

$$\frac{1}{M_m} = \frac{1}{M} + \frac{\tilde{e}e_x}{J} = \frac{1}{1000} + \frac{194.5 \times 353.6}{83.33 \times 10^6}; \text{ then } M_{rn} = 547.9\,\text{g; reduced mass}$$

$S_n = M_{rn}v_n(1+\kappa) = 547.9 \times 1.0(1+0.5) = 821.8$
$S_t = \mu S_n = 0.45 \times 821.8 = 369.8$; N-ms, normal and tangential impulse,
respectively

$$\lambda = \frac{S_n\tilde{e}}{J} = \frac{821.8 \times 194.5}{83.33 \times 10^6} = \frac{1}{521.3}\,\text{rad/ms}$$

$$V_{cn} = v_n - \frac{S_n}{M} = 1 - \frac{821.8}{1000} = 0.1782\,\text{m/s}$$

$$V_{ct} = v_t - \frac{S_t}{M} = 0 - \frac{369.8}{1000} = -0.3698 \text{ (to the right)}$$

$$V_{bt} = V_{ct} + \lambda e_y = -0.3698 + 353.6/521.3 = 0.3085\,\text{m/s}$$

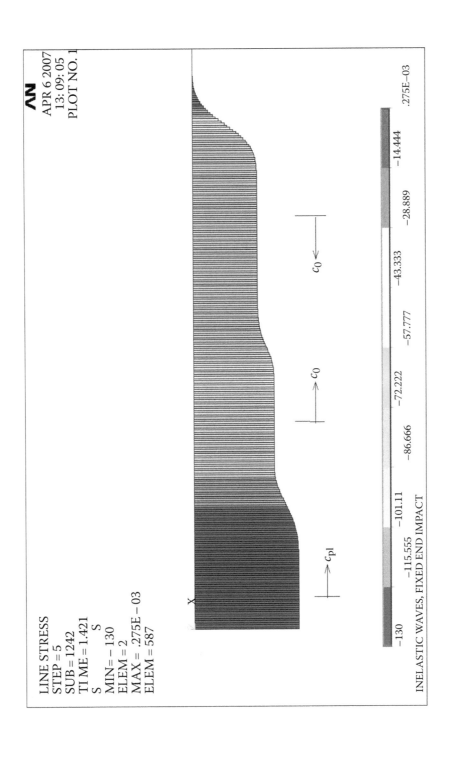

LINE STRESS
STEP = 5
SUB = 1242
TI ME = 1.421
S   S
MIN = − 130
ELEM = 2
MAX = .275E − 03
ELEM = 587

APR 6 2007
13: 09: 05
PLOT NO. 1

X

c_{pl}

c_0

c_0

−130   −115.555   −101.11   −86.666   −72.222   −57.777   −43.333   −28.889   −14.444   .275E − 03

INELASTIC WAVES, FIXED END IMPACT

The following will serve as an indirect check. The rebound velocity $V_{bn} = 0.5\,\text{m/s}$ comes directly from our formulation. It can also be calculated from the overall motion after rebound:

$$V_{bn} = \lambda e_x - V_{cn} = \frac{353.6}{521.3} - 0.1782 = 0.5 \text{ m/s}$$

The input data used here was exactly the same as in one of Brach's [12] problems, so that solutions employing different formulations can be compared. There is a good agreement, except for the round-off errors.

EXAMPLE 7.6   THE REVERSE PROBLEM FOR THE INCLINED FALLING ROD IMPACTING ROUGH SURFACE

(Example 7.5.) Calculate pre-impact kinematics from post-impact data from the referenced Example.

Case 7.7 applies: $v_n = V_{bn}/\kappa = 0.5/0.5 = 1.0\,\text{m/s}$

$$S_n = M(v_n - V_{cn}) = 1000(1 - 0.1782) = 821.8; \quad S_t = \mu S_n = 0.45 \times 821.8 = 369.8$$

$$M_{rn} = \frac{S_n}{v_n(1+\kappa)} = \frac{821.8}{1.0(1.5)} = 547.9$$

$$\tilde{e} = \frac{\lambda J}{S_n} = \frac{83.33 \times 10^6}{521.3 \times 821.8} = 194.5$$

$$e_y = \frac{V_{bt} - V_{ct}}{\lambda} = \frac{0.3085 + 0.3698}{1/521.3} = 353.6$$

$$v_t = V_{ct} + \frac{S_t}{M} = -0.3698 + 369.8/1000 = 0$$

$$e_x = \tilde{e} + \mu e_y = 194.5 + 0.45 \times 353.6 = 353.6$$

All values agree with the initial data for Example 7.5.

EXAMPLE 7.7   SPHERICAL STRIKER IMPACTING A THICK PLATE

A solid steel ball with 20 mm diameter is being dropped from the height of 300 mm onto a thick plate made from the same steel. Determine the maximum local shear stress in the plate resulting from that event and also the velocity of impact without triggering the onset of yielding. Repeat the calculation for a stronger plate, as specified below.

$E = 200,000\,\text{MPa}; \nu = 0.3; \rho = 0.00785\,\text{g/mm}^3.$

$F_y = 250\,\text{MPa}$ for the first plate and $F_y = 1000$ for the second.

The impact speed: $v_0 = \sqrt{2gh} = \sqrt{2 \times 0.00981 \times 300} = 2.426$ mm/ms.

For $r_2 = 10$ mm, $M = 32.88$ g

Case 7.17 applies. For identical materials:

$$\frac{1}{k_0} = \frac{3}{2}\left(\frac{1-0.3^2}{200,000}\right)\frac{1}{\sqrt{10}} = \frac{1}{463,340}$$

$$\sigma_{cm} = 0.499\frac{32.88^{0.2}}{10} \, 463,340^{0.8} \times 2.426^{0.4} = 4,877 \text{ MPa} \text{ ; average contact stress}$$

$$\tau_m = 0.31 \times 4,877 = 1512 \text{ MPa}$$

This is far in excess of the shear yield strength, $\sim 0.6 \times 250 = 150$ MPa. Permanent deformation is to be expected. For the second plate, 4× stronger, the degree of yielding is much smaller. In order to find the limiting impact speed v_{01}, note that the above can be written as

$$\sigma_{cm} = 3421.4v^{0.4}$$

which, along with the allowable $\sigma_{cm} = 0.6F_y/0.31 = 1.935F_y$ limits the impact velocity to

$$v = 7.522 \times 10^{-3} \text{ mm/ms} \quad \text{for } F_y = 250 \text{ MPa}$$

and

$$v = 0.241 \quad \text{for } F_y = 1000$$

These velocities are very small indeed, even for very strong steel.

EXAMPLE 7.8   AXIAL BAR IMPACTING WALL

A steel rod axially impacts an obstacle with $v_0 = 2.591$ m/s. Calculate the peak stress for the following configurations: (a) flat impacting end against rigid wall; (b) hemispherical end against a thick steel plate of the same material. For the sake of simplicity the flexibility of the hemispherical end should be linearized. $L = 3048$ mm and $D = 9.525$ mm for the rod.
     Steel properties: $E = 200,000$ MPa; $v = 0.3$ and $\rho = 0.00785$ g/mm³.

(a) According to Equation 7.53: $\sigma_0 = \rho c_0 v_0 = v_0\sqrt{E\rho} = 2.591\sqrt{200,000 \times 0.00785} = 102.66$ MPa

(b) The reduction of the peak compressive force can be accomplished by a "bumper" spring inserted between the rod and the wall. The hemispherical ending is, in principle, such a bumper.

When one of the two spherical surfaces is a plane, we have

$$\left(\frac{1}{k_0}\right)^{2/3} = \left[\frac{3}{4}\left(\frac{1-v_1^2}{E_1}+\frac{1-v_2^2}{E_2}\right)\sqrt{\frac{1}{r_2}}\right]^{2/3} = \left[\frac{3}{2}\left(\frac{1-v^2}{E}\right)\sqrt{\frac{1}{r_2}}\right]^{2/3}$$

in accordance with Equations 7.30a and b. (The first of the two similar terms reflects the additional deflection of the plate, while the second is due to the hemispherical end.) This local flexibility can now be linearized according to Equation 7.38a, where the first approximation of the peak contact force W_m needs to be input. Let us use

$$W_m \approx P_0 = A\sigma_0 = 71.26 \times 102.66 = 7315\,\text{N}$$

Employing $r_2 = 4.763\,\text{mm}$ gives us $k_0 = 639,506$ and the linearized stiffness as

$$k^* = \frac{4}{5}k_0^{2/3}W_m^{1/3} = \frac{4}{5}639,506^{2/3}7,315^{1/3} = 115,271\,\text{N/mm}$$

The corrective term in Equation 7.57, which defines the reduction in peak load is

$$\exp\left(\frac{-2k^*L}{EA}\right) = \exp\left(\frac{-2\times115,271\times3,048}{200,000\times71.26}\right) \cong 0$$

which means that the softening is insufficient to visibly reduce the peak load of P_0 or the peak stress of σ_0.

The above example relates to the result of a physical test quoted by Goldsmith [26]. Two such rods were involved, moving with a relative velocity twice larger than defined here. The stress wave in a rod was reported to be roughly rectangular, with the peak stress of 15,000 psi (103.5 MPa). Considering irregularities always involved in such testing (as well as rounding off of units) the agreement is very good.

EXAMPLE 7.9   SPHERICAL SHAPE IMPACTING A STATIONARY SPHERE

This is a reinterpretation of a physical test reported by Goldsmith [26]. A solid steel impactor with a 3.969 mm diameter strikes a beam with a $v_0 = 1$ m/s velocity. On the surface of the beam at the impact point there is welded hemisphere of 6.35 mm diameter, made of the same material.

The beam is steel, 25.4 mm × 25.4 mm solid section, 2591 mm long. As the beam section has its size several times larger than the impactor, it is permissible to treat the impact as against a fixed elastic object. Acting in this manner, determine the peak contact force. For steel use $E = 200,000\,\text{MPa}$, $\rho = 0.00785\,\text{g/mm}^3$, and $v = 0.3$.

The impactor mass is $M = 0.257\,g$. Use Equation 7.30a and b to extract constant k_0:

$$\frac{1}{k_0} = \frac{3}{4}\left(2 \times \frac{1-0.3^2}{200,000}\right)\sqrt{\frac{1}{1.984} + \frac{1}{3.175}} \quad \text{or} \quad k_0 = 161,904$$

The peak force may be found from Equation 7.37b, where M replaces M^* and v_0 replaces v_r:

$$W_m = \left(\frac{5}{4} \times 0.257 \times 1.0^2\right)^{0.6} 161,904^{0.4} = 61.35\,N$$

This force was measured to be $69.4\,N$ (15.6 lbf).

EXAMPLE 7.10 SOLID BALL DROPPED ON A THICK PLATE, IMPACT DURATION

A solid steel ball with 50 mm diameter is being dropped with the speeds of 1.2, 2.4, and 3.6 ft/s onto a thick plate made from the same steel. Determine the contact duration for these impacts and the peak contact stress according to Hertz theory and compare those with experimental results. Given: $E = 210,000\,MPa$; $v = 0.3$; $\rho = 0.00785\,g/mm^3$.

Follow Case 7.17. Show details for the 2.4 ft/s = 0.7315 m/s. For identical materials:

$$\frac{1}{k_0} = \frac{3}{2}\left(\frac{1-0.3^2}{210,000}\right)\frac{1}{\sqrt{25}} = \frac{1}{769,230} \; ; \text{mass } M = 513.8\,g$$

$$t_0 = \frac{3.214}{v_0^{0.2}}\left(\frac{M}{k_0}\right)^{0.4} = \frac{3.214}{0.7315^{0.2}}\left(\frac{513.8}{769,230}\right)^{0.4} = 0.1837\,ms; \text{duration}$$

$$\sigma_{cm} = 0.499\frac{513.8^{0.2}}{25}769,230^{0.8} \times 0.7315^{0.4} = 3,140\,MPa \; ; \text{peak surface stress}$$

The average surface stress is only 2/3 of the above, but is still well in excess of the yield strength of most steels. Consequently, longer duration is expected from an experiment. The table below shows the calculated as well as the experimental t_0' according to Goldsmith [26], p. 266.

v (ft/s)	v (m/s)	t_0 (ms)	σ_m	t_0' (ms)
1.2	0.3658	0.2111	2379.3	0.244
2.4	0.7315	0.1837	3139.5	0.200
3.6	1.0973	0.1694	3692.3	0.178

EXAMPLE 7.11 SOLID BALL DROPPED ON A THICK PLATE, CONTACT RADIUS

The problem description is the same as in Example 7.10.

The impact speeds to be considered are 1, 2, 3, and 4 ft/s. Determine the contact radius according to Hertz theory and compare it with the experimental results provided below. Also, use the perfectly plastic impact approach. Determine the dynamic flow stress f_c in such a way that it fits Hertz contact load for 2 ft/s impact and compare other results.

The details will be shown for the 2.0 ft/s = 0.6096 m/s.

$k_0 = 769{,}230$; $M = 513.8$ g; as before.

$$W_m = \left(\frac{5}{4}Mv_0^2\right)^{0.6} k_0^{0.4} = \left(\frac{5}{4}513.8\times 0.6096^2\right)^{0.6} 769{,}230^{0.4} = 6{,}041 \text{ N}$$

Contact radius: $a = \left[\frac{3}{4}\left(\frac{1-v_1^2}{E_1}+\frac{1-v_2^2}{E_2}\right)Wr*\right]^{1/3}$ with $\frac{1}{r*}\equiv\frac{1}{r_2}=\frac{1}{25}$

$$a = \left(\frac{W}{6153.8}\right)^{1/3} = \left(\frac{6041}{6153.8}\right)^{1/3} = 0.9938 \text{ mm; contact radius}$$

According to Goldsmith [26], the experimental contact radius was 1.019 mm. As the first guess of the appropriate value of f_c take 250 MPa, or close to the yield point of a mild steel. The peak load for plastic impact (Equation 7.46):

$$W_m = v_0(2M\pi Rf_c)^{1/2} = 0.6096(2\times 513.8\times\pi\times 25\times 250)^{1/2} = 2738.2 \text{ N}$$

This is much less than calculated from Hertz formula. To make both values agree, one needs to take $f_c = 1217$ MPa. The contact radius r_0 based on the applied average stress f_c:

$$r_0 = \left(\frac{W}{\pi f_c}\right)^{1/2} = \left(\frac{6041}{1217\pi}\right)^{1/2} = 1.257 \text{ mm}$$

When the peak loads match, the contact radius predicted by plastic impact method is larger. The table shows the results for several velocities of impact.

v (ft/s)	v (m/s)	W (N)	a (mm)	a (exp)	W' (N)	a' (mm)	a'/a
1	0.3048	2629.4	0.7532	0.7657	3020.9	0.8889	1.180
2	0.6096	6041	0.9939	1.019	6041.7	1.2571	1.265
3	0.9144	9826.7	1.1688	1.1948	9062.6	1.5397	1.317
4	1.2192	13,878	1.3114	1.4021	12083.5	1.7778	1.356

a = radius of contact according to Hertz formula

a(exp) = experimentally obtained radius

a' = radius calculated for a perfectly plastic impact

W = peak contact force according to Hertz

W' = force in perfectly plastic impact with $f_c = 1217$ MPa, per Equation 7.46.

EXAMPLE 7.12 SPHERICAL STEEL STRIKER IMPACTING CONCRETE BLOCK

Treating steel as undeformable, establish the impact parameters for the following, as per Case 7.26.

Drop height $h = 1835$ mm, $R = 1407$ mm, $d = 25$ mm, and $M = 200$ kg for the striker

$$f_c = 3\times 27.9 = 83.7 \text{ MPa}$$

$$u_m = \sqrt{\frac{Mgh}{\pi R f_c}} = \sqrt{\frac{200{,}000 \times 0.00981 \times 1{,}835}{\pi 1407 \times 83.7}} = 3.12 \text{ mm} \quad (\text{less than } d = 25\text{mm, formulas hold})$$

$$v_0 = \sqrt{2gh} = \sqrt{2 \times 0.00981 \times 1835} = 6 \text{ m/s}$$

$$W_m = v_0 (2M\pi R f_c)^{1/2} = 6(2 \times 200{,}000\pi 1407 \times 83.7)^{1/2} = 2.308 \times 10^6 \text{ N}$$

$$t_1 = \frac{\pi u_m}{2v_0} = \frac{\pi 3.12}{2 \times 6} = 0.817 \text{ ms}$$

EXAMPLE 7.13   HERTZIAN CONTACT VS. PLASTIC CONTACT FORMULATION

Reference [81] presents a set of test results obtained by dropping steel spheres of several diameters on a lucite plate, $h = 12.7$ mm thick. The drop height was 635 mm resulting in impact velocity of 3.53 m/s. The spheres of 6.35, 12.7, and 17.5 mm in diameter will be taken here as representative of the tested range. The spheres may be treated as undeformable. The mechanical properties of lucite were

$$E = 2500 \text{ MPa}, \quad \rho = 0.00118 \text{ g/mm}^3, \quad v = 0.3, \quad G = 961.5 \text{ MPa}$$

Typical tensile strength $F_t = 70$ MPa

For the plate $m = \rho h = 0.015$ g/mm^2.

For a small contact area, the equivalent insert diameter may be assumed as $h/2 = 6.35$ mm.

The shear thickness is $h_s = h/1.2 = 10.58$ mm.

The detailed calculations will be shown for 12.7 mm diameter only.

$M = \pi d^3 \rho / 6 = 8.419$ g

For Hertzian contact, use Case 7.17:

$$k_0 = 1831.5\sqrt{r} = 1831.5\sqrt{12.7/2} = 4615.2$$

$$W = \left(\frac{5}{4}Mv_i^2\right)^{0.6} k_0^{0.4} = \left(\frac{5}{4}M \cdot 3.53^2\right)^{0.6} 4615.2^{0.4} = 544.9 \text{ N}$$

For a perfectly plastic contact use Case 7.26. To calculate contact stiffness k_e assume, as the first try, $f_c = F_t = 70$ MPa:

$$k_e = 2\pi R_s f_c = 2\pi(12.7/2)70 = 2793 \text{ N/mm}$$

Peak of contact force: $W = v_0 \sqrt{k_e M} = 3.53\sqrt{2793 \times 8.419} = 541.3 \text{ N}$.

The summary of calculation is shown in the table. The closeness of W and W' is quite remarkable.

Hertz peak load (W) and the perfectly plastic value (W')

D (mm)	M (g)	k_0	W (N)	k_e (N/mm)	W' (N)
6.35	1.052	3263.46	136.22	1396.37	135.32
12.7	8.419	4615.23	544.89	2792.73	541.28
17.5	22.028	5417.65	1034.62	3848.25	1027.76

EXAMPLE 7.14   AXIAL BAR FIXED AT ONE END, IMPACTED AT THE OTHER

A bar with section area $A = 1,000\,\text{mm}^2$ and length $L = 400\,\text{mm}$ has its free end impacted by a mass $M = 2,160\,\text{g}$, moving with velocity $v_0 = 20\,\text{m/s}$. Calculate the induced peak stress. Use aluminum data: $E = 69,000\,\text{MPa}$ and $\rho = 0.0027\,\text{g/mm}^3$. Repeat the calculation for $M = 5,400\,\text{g}$.

Case 7.24 applies.

$m = A\rho = 1000 \times 0.0027 = 2.7\,\text{g/mm}$; distributed mass, $mL = 1080\,\text{g}$; bar mass

$$c_0 = \sqrt{\frac{EA}{m}} = \sqrt{\frac{69,000 \times 1.0}{0.0027}} = 5,055 \text{ mm/ms; sonic velocity}$$

$t_L = L/c_0 = 400/5055 = 0.0791$ ms; transit time for the wave front

$\sigma_0 = \rho c_0 v_0 = 0.0027 \times 5055 \times 10 = 136.5\,\text{MPa}$; initial impact stress

$P_0 = A\sigma_0 = 136,500\,\text{N}$; initial impact force

$$P_2 = \left[1 + \exp\left(-\frac{2mL}{M}\right)\right]2P_0 = \left[1 + \exp\left(-\frac{2 \times 1,080}{2,160}\right)\right]2 \times 136,500 = 373,430 \text{ N; upper}$$

bound

for the constrained end. To evaluate the peak force by the basic engineering method, find the axial stiffness: $k = EA/L = 69,000 \times 1,000/400 = 172,500\,\text{N/mm}$; then, from Case 3.2b

$$P_m = v_0\sqrt{kM} = 10\sqrt{172,500 \times 2,160} = 193,040 \text{ N}$$

This is only about one-half of the previous value. When P_0 is added to it, as recommended in the text, the total becomes 329,540 N, a much better estimate.

For $M = 5400\,\text{g}$ (which is $5\,\text{mL} = 5 \times 1080$), the approximate formula (Case 7.24) gives

$$P_m \approx P_0\left(\sqrt{\frac{M}{mL}} + 1\right) = 136,500\left(\sqrt{\frac{5,400}{1,080}} + 1\right) = 441,720 \text{ N}$$

while the engineering method gives $P_m = 305,220\,\text{N}$ only.

EXAMPLE 7.15   REACTION OF AN AXIAL BAR TO THE IMPACT OF MASS M

It is said that a bar, when stricken by a mass M flying with a velocity v_0, reacts like a damper. What is the constant C of such a damper?

Case 3.7 tells us that the velocity of the mass impacting a damper changes according to

$$v(t) = v_0 \exp\left(-\frac{C}{M}t\right)$$

When a bar is impacted, one has, from Case 7.23:

$$v(t) = v_0 \exp\left(-\frac{mc_0}{M}t\right)$$

Comparing the two, we simply have $C = mc_0 = A\rho c_0 = \sqrt{EAm}$

EXAMPLE 7.16   PILE DRIVING

The pile has a rectangular, 200 mm × 200 mm section, is 5 m long and is made of hardwood:

$$E = 12{,}000 \text{ MPa}; \quad v = 0 \quad \text{and} \quad \rho = 0.00059 \text{ g/mm}^3$$

A solid hammer, with mass M = 200 kg, regarded as undeformable, is dropped from the height of 5 m onto the upper end of the pile, which is partially submerged in soft soil. Using a simplified theory of pile driving, as described in Case 7.25, determine the contact duration of the main impact and the peak force. Assuming that the hammer and the pile meet again after separation, the pile end is at rest, estimate the force of the second impact.

The distributed and the total mass of the pile is, respectively: $m = A\rho = 200 \times 200 \times 0.00059 = 23.6$ g/mm. $M = mL = 118{,}000$ g. Longitudinal wave speed $c_0 = \sqrt{12{,}000/0.00059} = 4{,}510$ m/s. Anticipated main contact duration: $t_d = 2L/c_0 = 2 \times 5000/4510 = 2.217$ ms.

The impact velocity: $v_0 = \sqrt{2gh} = \sqrt{2 \times 0.00981 \times 5000} = 9.9$ m/s

Initial contact force: $P_0 = A\rho c_0 v_0 = mc_0 v_0 = 23.6 \times 4510 \times 9.9 = 1.054 \times 10^6$ N ($\sigma_0 = 26.34$ MPa)

To calculate the top end velocity at $t = t_d$, find

$$\frac{Mg}{mc_0} = \frac{118{,}000 \times 0.00981}{23.6 \times 4{,}510} = 0.01088 \quad \text{and} \quad \frac{mc_0}{M} = \frac{23.6 \times 4{,}510}{118{,}000} = 0.902$$

$$v = \left(v_0 - \frac{Mg}{mc_0}\right)\exp\left(-\frac{mc_0}{M}t_d\right) + \frac{Mg}{mc_0} = (9.9 - 0.01088)\exp(-0.902 \times 2.217) + 0.01088$$

$$= 1.35 \text{ m/s.}$$

The distance between the separated surfaces will be small, so the gain of velocity of the hammer between the impacts is negligible. Using the last value as the impact velocity, one gets $P' = A\rho c_0 v = 23.6 \times 4{,}510 \times 1.35 = 143{,}690\,\text{N}$, or about 14% of the initial impact force.

Example 7.17   Car Impacting a Fixed Barrier; Stiffness Properties

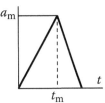

Johnson [40] quotes the following results of crash testing of British cars, as reported in 1,970. A typical deceleration response for the impact speeds between 25 and 45 mph (11.2–20.12 m/s) is approximately triangular, as shown in the graph. The peak of $a_m = 40g$ is attained at $t_m = 0.05$ s.

Interpret this result by (a) treating the vehicle as an inelastic bar impacting a rigid wall find the speed of the plastic wave traveling along the structure and (b) treating the vehicle as undergoing a basic stereomechanical impact find the natural frequency and the impacting spring stiffness. Assume the length $L = 4$ m and mass M to be 1100 kg.

Obviously, the impact is much more complex than either of the two simple interpretations suggest, therefore we may only find some indicators from the above results.

(a) The plastic wave spreads with the speed of c_{pl} and is associated with some maximum stress σ_m. If we assume that the peak reaction takes place when the plastic wave reaches the opposite end, then $c_{pl} = L/t_m = 4000/50 = 80\,\text{mm/ms}$.

(b) In Case 3.2b (where the motion is harmonic), one has the maximum reaction of the impacted wall as $R_m = v_0\sqrt{kM}$. Dividing both sides by M, one gets

$$a_m = \frac{R_m}{M} = v_0\sqrt{\frac{k}{M}} = \omega v_0,$$ which is a well-known relation for a harmonic motion.

The whole range of velocities is reported here. The average is $v_0 = 15.66$ m/s. The corresponding peak acceleration, in time, is 40 g or 0.3924 mm/ms^2, as per the test data. Substituting

$$0.3924 = 15.66\omega \quad \text{or} \quad \omega = 0.02506 \text{ kHz} \quad \text{or} \quad \omega = 25.06 \text{ Hz}$$

Then $k = \omega^2 M = (25.06 \times 10^{-3})^2 1.1 \times 10^6 = 690.8\,\text{N/mm}$

While such interpretations are inaccurate, they still give the analyst some parameters to predict performance of similar structures, under different conditions.

EXAMPLE 7.18   LIGHT AIRCRAFT IMPACT AGAINST GROUND

An aircraft with $M = 2700\,\text{kg}$, and speed $v_0 = 28.6\,\text{m/s}$ at the dive angle of $\gamma = 47.5°$ collides with the ground. Measured impact duration is $t_0 = 174\,\text{ms}$. The observed restitution coefficient $\kappa = 0$. Assuming that the contact force is triangular, determine the peak acceleration.

This example will explain some of the details of Case 7.29.

The velocities and the impulse components needed to stop the aircraft are

$$v_n = v_0 \sin\gamma = 28.6 \ \sin 47.5° = 21.09 \ \text{mm/s};$$

$$S_n = Mv_n = 2.7\times10^6 \times 21.09 = 56.94\times10^6\,\text{g/mm/s}$$

$$v_t = v_0 \cos\gamma = 19.32 \ \text{mm/s}; \quad S_t = 52.16\times10^6\,\text{g/mm/s}$$

If the normal contact force is triangular with a peak value of W_n, then it applies the impulse of $0.5W_n t_0 = S_n$ or $0.5W_n 174 = 56.94 \times 10^6$; then $W_n = 654{,}480\,\text{N}$.

With the friction of $\mu = 0.4$, the horizontal component is $0.4W_n = 261{,}790\,\text{N}$

The resultant force acting on the aircraft is therefore $W_m = \sqrt{1.0^2 + 0.4^2}\,W_n = 1.077W_n = 704{,}900\,\text{N}$.

The overload factor: $n = \dfrac{W_m}{Mg} = \dfrac{704{,}900}{2.7\times10^6 \times 0.00981} = 26.61\ \text{g}$

The experimentally measured factor was $21.85\,\text{g}$, which indicates that the triangular assumption is not very accurate for this event. This factor is a measure of fuselage strength and therefore it should be in a similar range for aircraft that are alike, although it may be influenced by the impact angle. This example comes from a set of experiments carried out by NASA and compiled by Brach [12]. The range of overload factors was from 16.9 to 33.3 with the average value of 25.1. This range of n values is the basis of estimation presented by Case 7.29.

8 Collision

THEORETICAL OUTLINE

This chapter is a continuation of Chapter 7 into problems involving impacts between movable objects. The word *collision* used in the title implies a sudden contact of two or more bodies with masses of the same order of magnitude.

Although there will be more degrees of freedom involved now, the response is computed using similar methods. Both slender and compact bodies are featured with either linear or nonlinear contact characteristics.

The concept of impulse and impulsive loading was initially defined in Chapter 3. In general, the impulse applied by a force is an integral of the force–time diagram. A rectangular impulse is merely a product of the magnitude of a force and its duration. The concept is used in the same manner as in Chapter 7, but here the exchange of impulse between colliding bodies is of interest.

CENTRAL COLLISION OF BODIES

This type of interaction takes place when the contact point of the colliding objects is located on the same line along which the centers of gravity are moving. Like a similar, but simpler phenomenon described in the section "Central impact against rigid wall" of Chapter 7, the collision is separated into several distinct stages illustrated in Figure 8.1.

At the beginning, we have ball 1 approaching ball 2, $v_1 > v_2$. Stage 1 lasts until the balls touch each other. At this moment, they still have their original velocities. During Stage 2, the first ball tries to accelerate the second one by means of its own inertia, and the resulting contact force deforms both bodies. Stage 2 ends at that instant when both balls travel with the same velocity v_0. Stage 3 is the recovery process during which the balls return to their original shape (if the collision is elastic) and their relative velocities change sign. (At the end of this stage, they barely touch each other.) At Stage 4, the balls continue with their new velocities, V_1, and V_2. The time of the collision contact (Stages 2 and 3) is very short in comparison with entire process described.

Figure 8.2 illustrates another way of looking at the collision of two bodies, as originally presented by this author in Ref. [87]. If an observer is moving with velocity v_0, he sees the balls approaching each other, and then going apart following the collision. In fact, he can think of each ball rebounding from a rigid wall that appears stationary to him. This interpretation helps us to treat collision as a form of impact against a rigid barrier that is moving with the previously defined velocity v_0. The common velocity v_0, which is in effect at that instant when the balls are stationary, relative to one another, may be found from the principle of preservation of momentum: $M_1 v_1 + M_2 v_2 = (M_1 + M_2) v_0$:

$$v_0 = \frac{M_1 v_1 + M_2 v_2}{M_1 + M_2} \tag{8.1}$$

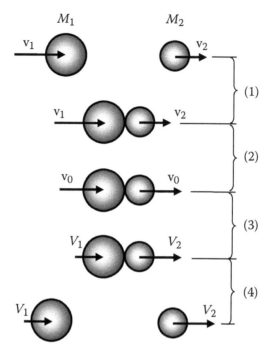

FIGURE 8.1 Four stages of collision.

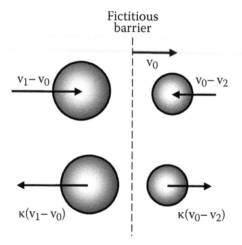

FIGURE 8.2 Collision treated as a rebound from a rigid barrier.

To find the velocities after the collision, we return for a moment to Figure 7.1, which shows us that the final velocity V with respect to the barrier is $-\kappa v$, where κ is the restitution coefficient and v is the impact velocity. The analogous change of sign and magnitude is shown in Figure 8.2. Taking into account that the entire system is moving with the velocity v_0, we have

$$V_1 = v_0 - \kappa(v_1 - v_0) \tag{8.2a}$$

$$V_2 = v_0 + \kappa(v_0 - v_2) \tag{8.2b}$$

A positive velocity is directed to the right. By subtracting both sides of these equations, we can eliminate v_0 and obtain the relationship between the relative velocities:

$$V_2 - V_1 = \kappa(v_1 - v_2) \tag{8.3a}$$

or

$$V_r = \kappa v_r \tag{8.3b}$$

The energy loss during collision is expressed by

$$\Delta E_k = \frac{1}{2} M_1 (v_1^2 - V_1^2) + \frac{1}{2} M_2 (v_2^2 - V_2^2) \tag{8.4a}$$

or

$$\Delta E_k = \frac{1}{2} M^* \, v_r^2 (1 - \kappa^2) \tag{8.4b}$$

where M^* is an equivalent mass calculated from

$$\frac{1}{M^*} = \frac{1}{M_1} + \frac{1}{M_2} \tag{8.5}$$

while v_r is the initial relative velocity, which was used in Equations 8.3a and b. The collision may be perfectly elastic ($\kappa = 1.0$), inelastic ($0 < \kappa < 1.0$), or perfectly plastic ($\kappa = 0$). The meaning of these terms is the same as described in the section "Central impact against rigid wall" of Chapter 7. There is also another way of looking at Equation 8.4b, if a concept of a *kinetic energy of relative motion* is used. The expression $M^* \, v_r^2 / 2$ is such an energy, completely lost in the loading phase, as the relative motion stops. But then, in the rebound phase, the kinetic energy of $\kappa^2 M^* \, v_r^2 / 2$ is regained. The impulse experienced by both bodies during the collision is

$$S_n = M_1 (v_1 - V_1) = M_2 (V_2 - v_2) = M^* v_r (1 + \kappa) \tag{8.6}$$

The above developments were carried out using the techniques of the rigid-body mechanics. The deformability was accounted for in an indirect way, by means of the restitution coefficient. To establish such important parameters as the duration of contact and the magnitude of contact force, some information is needed with regard to the resistance–deflection properties of the colliding bodies.

COLLISION OF PARTICLES IN A PLANE

Collision of particles in plane is considered when there are both normal and tangential velocity components at the impact zone. The normal components are treated in exactly the same manner as in a central impact. The tangential one causes a change of tangential velocities due

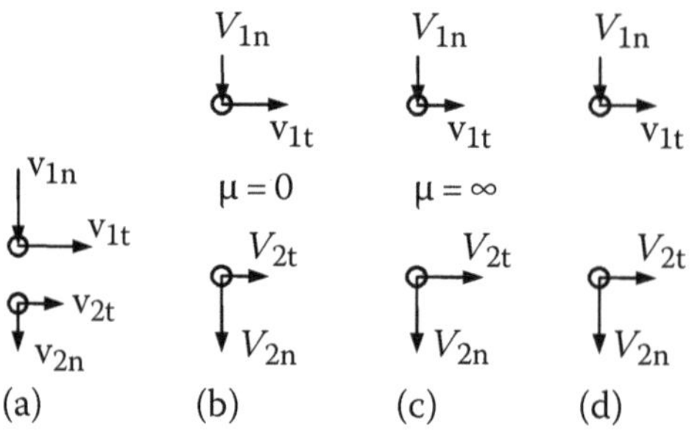

FIGURE 8.3 Two particles before collision (a), after frictionless collision (b), after no-slip collision (c), and after rough (moderate friction) collision (d).

to friction. The tangential impulse is defined as $S_t = \mu S_n$, where μ is the friction coefficient. As shown in Figure 8.3, there are three important cases related to the latter components:

(1) The particles are frictionless, (a) and (b). The tangential velocity components are the same before and after impact.
(2) There is no slippage during contact, (c). The collision is considered independently in both directions.
(3) The particles are rough, i.e., there is some change in tangential velocity as dictated by the friction coefficient μ (d).

The normal velocities and impulses are related as in the central impact, i.e., Equations 8.1 through 8.3 hold, with an added index n for normal direction. The exchange of impulse in the tangential direction depends on which of the above cases is involved. We have, for a frictionless contact

$$V_{2t} = v_{1t} \tag{8.7}$$

For no-slip contact, the tangential impact is analogous to that for the normal direction:

$$v_{t0} = \frac{M_1 v_{t1} + M_2 v_{t2}}{M_1 + M_2} \tag{8.8}$$

$$V_{t1} = v_{t0} - \kappa(v_{t1} - v_{t0}) \tag{8.9a}$$

$$V_{t2} = v_{t0} - \kappa(v_{t0} - v_{t2}) \tag{8.9b}$$

In the last, intermediate case, the faster particle is retarded while the slower one is accelerated by the tangential impulse $S_t = \mu S_n$ applied during contact. The details are given in Case 8.2.

COLLISION OF TWO BODIES AND THE INFLUENCE OF FRICTION

The simplest derivation is available when the surfaces are frictionless. Following the terminology of Chapter 7, this is the case when no tangential interaction exists between the colliding bodies. The equations similar to those of Chapter 7 apply here as well. The

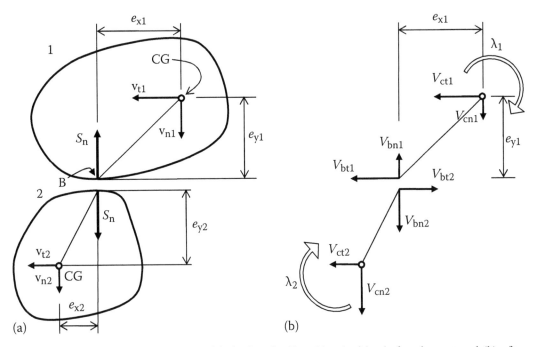

FIGURE 8.4 Eccentric collision of frictionless bodies: (a) velocities before impact and (b) after impact. (A small separation is shown for clarity.)

kinematic state before and after collision is shown in Figure 8.4. The contact takes place at point B with the impulse S_n of the interaction force and is directed along the joint normal at point B.

Initially, there is only a translational motion defined by v_n and v_t for each body, while afterwards the angular velocities λ_1 and λ_2 appear. The velocity jump is related to the point of contact B. The coefficient of restitution κ again determines the ratio of relative velocities after and before the collision. Due to eccentricity of impact, the reduced masses rather than the actual masses are involved. Following Equations 7.12 through 7.14 and keeping Equation 8.5 in mind

$$V_{nr} = \kappa V_{nr} \quad \text{or} \quad V_{bn2} - V_{bn1} = \kappa(V_{n1} - V_{n2}) \tag{8.10}$$

$$S_n = M_r^* v_{rn}(1+\kappa) \tag{8.11a}$$

$$M_{r1} = \frac{J_1 M_1}{J_1 + M_1 e_{x1}^2} \tag{8.11b}$$

$$M_{r2} = \frac{J_2 M_2}{J_2 + M_2 e_{x2}^2} \tag{8.11c}$$

$$\frac{1}{M_r^*} = \frac{1}{M_{r1}} + \frac{1}{M_{r2}} \tag{8.11d}$$

The moments of impulse about the center of masses are converted into angular velocities:

$$\lambda_1 = \frac{S_n e_{x1}}{J_1} \quad \text{and} \quad \lambda_2 = \frac{S_n e_{x2}}{J_2} \tag{8.12}$$

Velocities at the CGs change their magnitudes by S_n/M_i in the normal direction, but remain constant in the tangential direction. At the contact point B the tangential speeds change due to acquired angular velocities.

When friction is included, the additional impulse tangent to the contact plane appears. It is designated by S_t in Cases 8.2 and 8.3, which exemplify the application of rough contact surfaces. When no-slip surfaces are involved ($\mu = \infty$), the formulation further complicates, especially in the most general case. Still, when the geometry is simple the calculation can be easily done. (Example 8.3 may be viewed as relevant). It is the impression of this author that the presence of friction, at least for finite values of μ, does not drastically influence the results in most practical applications, therefore an assumption of a frictionless collision is a valid first estimate.

IMPACT AGAINST BODY FREE TO ROTATE ABOUT FIXED AXIS

A particular type of this impact is considered, when the direction of the contact force is perpendicular to the line determined by the pivot point and the center of gravity, Figure 8.5.

Similarly to what was done in "Central collision of bodies," we refer to an instant when the impacting object has velocity v_0 that is the same as the velocity of the contact point of the rotating body. Since the angular momentum of the system is preserved, we can write

$$M_1 v_1 d = M_1 v_0 d + J_0 v_0 / d \tag{8.13a}$$

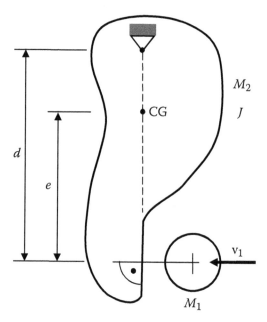

FIGURE 8.5   Impact against body free to rotate about fixed axis.

or

$$V_0 = \frac{M_1 v_1}{M_1 + J_0/d^2} \quad \text{with} \quad J_0 = J + M_2(d-e)^2 \tag{8.13b}$$

where J_0 is the mass moment of inertia of the body about the axis of rotation, while J is about the center of gravity. Note that if we put $M_2 = J_0/d^2$, then Equation 8.13a becomes a particular case of Equation 8.1 when the impacted mass M_2 is in the state of initial rest, $v_2 = 0$. With this substitution we can use the formulas described in "Central collision of bodies" section. In particular, the velocities after the impact are, by Equation 8.2a:

$$V_1 = v_0 - \kappa(v_1 - v_0) \tag{8.14}$$

in which V_1 is the post-impact velocity of M_1, positive in the same direction as v_1, while the angular velocity of the constrained body is

$$\lambda = V_2/d = v_0(1+\kappa)/d \tag{8.15}$$

PARAMETERS OF COLLISION WITH LINEAR CHARACTERISTIC

Two objects that are to come in contact are shown in the sketch for Case 8.7. The midpoint of impact (with respect to time) is characterized by the common velocity v_0 of both bodies. At this moment, the compression of the springs reaches its maximum. The linearity of spring and lack of damping result in $\kappa = 1$, or no energy loss. The post-collision velocities may be found from Equations 8.2. A comparison of kinetic plus strain energy at the beginning and the end of the loading phase of contact gives

$$\frac{1}{2}M_1 v_1^2 + \frac{1}{2}M_2 v_2^2 = \frac{1}{2}(M_1 + M_2)v_0^2 + \frac{1}{2}k_1\delta_1^2 + \frac{1}{2}k_2\delta_2^2 \tag{8.16}$$

in which δ_1 and δ_2 are the maximum spring deflections. The maximum contact force resulting from this energy balance is

$$R_{\mathrm{m}} = v_{\mathrm{r}}\sqrt{M^* k^*} \tag{8.17a}$$

with

$$v_{\mathrm{r}} = v_1 - v_2 \tag{8.17b}$$

where v_{r} is the relative velocity before the contact, as used in Equation 8.3. Symbol M^* stands for the effective mass and k^* is the effective spring stiffness:

$$\frac{1}{k^*} = \frac{1}{k_1} + \frac{1}{k_2} \tag{8.18a}$$

and

$$\frac{1}{M^*} = \frac{1}{M_1} + \frac{1}{M_2}$$

(8.18b)

As long as the springs are in contact, their flexibility is additive, so Equation 2.29 for the series connection of springs holds here as well. Note that Equations 8.17a and 8.18b, developed from energy considerations, are also consistent with Case 2.2, which gives the same effective mass for two unconstrained masses connected by a spring. Deflection of any of the two springs is found from

$$R_m = k_1 \delta_1 = k_2 \delta_2$$

(8.19)

Let t_1 be the duration of the loading phase of contact, which is depicted as Stage 2 in Figure 8.1. This interval may be identified with one-quarter of the natural period of a system consisting of two masses and a spring of stiffness k^* joining them. The natural frequency of such an arrangement is given in Case 2.2. The duration of the loading phase is

$$t_1 = \frac{\pi}{2} \sqrt{\frac{M^*}{k^*}}$$

(8.20)

If the collision is indeed linear–elastic, the entire time of contact is $2t_1$. Otherwise, the time of the unloading phase t_2 may be calculated using the unloading stiffness of the springs.

PARAMETERS OF COLLISION WITH NONLINEAR CHARACTERISTIC

The colliding bodies may be treated as rigid after putting their entire flexibility into a "bumper" spring separating them. A general form of a force–deflection relationship for such a spring is written as

$$R(\delta) = k_0 f(\delta)$$

(8.21)

where
  δ is the relative movement of the bodies
  k_0 is a constant coefficient, which sometimes has a meaning of initial stiffness. (If the properties of springs are specified separately for each body as $R_1(\delta)$ and $R_2(\delta)$, those two springs must be joined in series so that an expression in the form of Equation 8.21 may be derived.)

This corresponds to Figure 8.6, except that there is now only a single, NL spring. The kinetic energy lost in the loading phase of collision changes into strain energy Π:

$$\Pi = \frac{1}{2} M_1 v_1^2 + \frac{1}{2} M_2 v_2^2 - \frac{1}{2}(M_1 + M_2) v_0^2$$

(8.22)

A much simpler looking expression can also be derived from the above:

$$\Pi = \frac{1}{2} M^* v_r^2 = \frac{1}{2} M^* (v_1 - v_2)^2$$

(8.23)

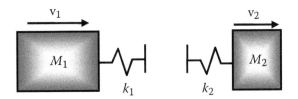

FIGURE 8.6 Two-body collision.

With the value of Π and the knowledge of the resistance–deflection curve, one is able to calculate the maximum reaction R_m and the associated deflection δ_m.

The time of the loading phase t_1 is the same as a quarter of the period of free vibration. Some of the methods of finding a natural frequency of single-mass nonlinear systems were presented in Chapter 4 (see "Duration of forward motion with the initial velocity prescribed" section). They may be adapted by changing M to M^* and u_m to δ_m. If the time-dependence of reaction can be approximated by a sinusoid, the simplest relation comes from Equation 7.24:

$$t_1 \approx \frac{\pi \delta_m}{2 v_r} \tag{8.24}$$

A detailed presentation of Hertz-type contact was made in Chapter 7, where a discussion of interaction of two deformable bodies was also conducted. An equivalent linear spring constant k_{ef} was also derived, so that it can replace the deformability of the two contacting bodies. Such a constant is obviously load dependent, so an iterative procedure is needed to correctly assess its value. For compact bodies, however, it is not necessary to derive this constant in order to determine the peak contact load W_m and the relative approach δ_m of the contacting bodies, when acting in accordance with the energy method.

SHOCK LOAD ON UNCONSTRAINED BODIES AND DYNAMIC EQUILIBRIUM CONCEPT

Using d'Alembert's principle, we can look at an accelerating body as being in a special state of equilibrium where the active, applied loads are balanced by the inertia forces. An example in Figure 8.7 shows an axial bar with lumped masses, subjected to a force W at one end. When we think of the assembly as a rigid body, the magnitude of the inertia force applied to mass M_i, is

$$R_i = a M_i \tag{8.25}$$

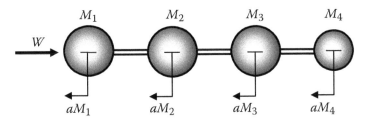

FIGURE 8.7 Dynamic load applied to unconstrained body.

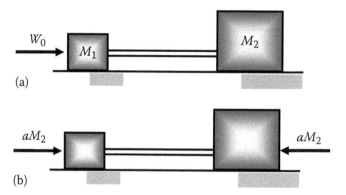

FIGURE 8.8   (a) Load applied to unconstrained body; (b) compressive component of that load.

where $a = W/(M_1 + M_2 + M_3 + M_4)$ is the acceleration of the body. The load W is a function of time, and the inertia forces are also time dependent. The rigid-body type of distribution just presented is only one possibility, but an important one, since it provides a reference level for calculation of internal forces.

At this point, it is useful to refer to "Unconstrained bar pushed at the end" section in Chapter 5, where the analysis of an axial bar pushed at one end is conducted. As a result of a sustained load $W = W_0 H(t)$ applied to one end, the internal force is oscillating between 0 and W_0. That suddenly applied load W_0 is the largest internal force, which is present at every section at some instant of time.

The dynamic equilibrium concept suggests that also in less regular configurations, such as one shown in Figure 8.7, where the continuous bar is interspersed with lumped masses, the reasoning based on inertia alone may be used to determine an upper bound of the internal force. A detailed analysis of a system in Figure 8.8, called a "two-mass oscillator," presented in Chapter 2 is continued here, although in a simplistic way.

The external load (suddenly applied W_0) is resolved into a rigid-body component (forces aM_1 and aM_2, not shown in Figure 8.8) and the compressive component (forces aM_2 and $-aM_2$). The first component applies a uniform acceleration. The second one compresses two single oscillators, which, in effect, are making up this system, as previously demonstrated. The motion of masses in time will be synchronous, because the natural frequencies of both components are the same. The dynamic factor of 2.0 gives the following peak compression in the elastic link:

$$|P|_{max} = 2aM_2 = \frac{M_2}{M_1 + M_2}(2W_0) \qquad (8.26)$$

When this procedure is extended to larger systems, like that in Figure 8.7, the results should be treated as approximations, which may, at places, considerably deviate from the true values. The reason, of course, is the presence of lumped masses causing partial reflections of the stress waves. A more general treatment of the problem in Figure 8.8 is given by Case 8.10.

AXIAL COLLISION OF TWO BARS WITH IDENTICAL PROPERTIES

The axial collision of two bars with identical properties, except possibly the length, is easier to follow, if their velocities are resolved as in Figure 8.9.

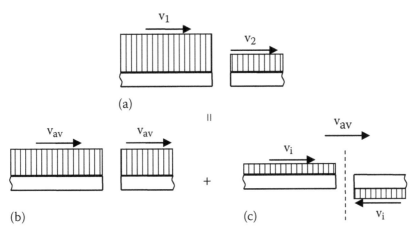

FIGURE 8.9 Bar velocities in (a) resolved into two components: pure translation with v_{av} (b) and impact v_i (c).

Pure translational velocity v_{av} and impact velocity v_i are defined as follows:

$$v_{av} = \frac{1}{2}(v_1 + v_2) \qquad (8.27a)$$

and

$$v_i = \frac{1}{2}(v_1 - v_2) \qquad (8.27b)$$

If an observer is moving with velocity v_{av}, he sees the bar faces approaching each other, in a symmetric fashion, with regard to the observation plane, marked in the figure by a dashed line. This approach helps us to treat this collision as a form of impact against a rigid barrier, each bar impacting with a velocity v_i. Interestingly, the length of bars is not involved, at least in the initial phase of the event. The symmetry of the arrangement in Figure 8.9c continues, after contact is initiated, by the compressive wave spreading with equal speed, away from the plane of symmetry. This resolution of velocities is equivalent to considering the motion of bars in a coordinate system moving with velocity v_{av}, the same as that of the observation plane.

   If there is a spring of stiffness k, softening the impact between bars, one can treat it as two springs in series, each with a stiffness of $2k$, thereby achieving symmetry with respect to the impact plane, marked with a dashed line in Figure 8.9c. The problem of finding the time-dependence of the contact force is essentially the same as a one-sided impact shown before in Figure 7.14, except that the spring is twice as stiff. This is quantified in Case 8.13.

   Only after the wave reflects from the free end and comes back to the impacted end is the symmetry upset by unequal length of bars. If the lengths are the same then the situation in Figure 7.14 is replicated here for each of the bars. (In the description of the effects, v_0 in that figure is identical with v_i as defined here.) In fact, Figure 7.14 in conjunction with Figure 8.9 above can be used to determine the post-impact velocities, V_1 and V_2 of the two colliding identical bars. It is sufficient to say that the collision speed v_i, which is reversed following the impact, was defined as a part of a broader picture, namely Figure 8.9a. The total post-impact velocities are therefore obtained by adding v_{av} to the result, which is that $V_1 = v_2$ and $V_2 = v_1$, i.e., the velocity reversal takes place. Examples 8.6 and 8.7 deal with bars that are of unequal length.

AXIAL COLLISION OF TWO BARS WITH DIFFERENT PROPERTIES

This type of collision of two bars, having initial velocities as in Figure 8.9a requires a slightly more complex approach. Suppose that after contact the interface velocity is v_n. That means the face of bar 1 experienced a change of speed $v_1 - v_n$ to the left while the face of bar 2 changed its speed by $v_n - v_2$ to the left. The contact force $P = A\rho c(\Delta v)$ is the same for both bars. (This is the same as Equation 5.4b, except that velocity increment is used because the initial velocity is not zero.) Writing the expression for the contact force to be equal, one has

$$A_1\rho_1 c_1(v_1 - v_n) = A_2\rho_2 c_2(v_n - v_2) \tag{8.28}$$

noting that the speed of the second bar relative to the moving wall is $v_n - v_2$. From this equation, with $I = A\rho c$, one has

$$v_n = \frac{I_1 v_1 + I_2 v_2}{I_1 + I_2} \tag{8.29}$$

Note the similarity of this expression to Equation 8.1. Figure 8.10 illustrates a similar way of looking at the collision of two bars. If an observer is moving with velocity v_n, he sees the bars approaching each other, and then remaining stationary in contact. Compressive waves, originating at the moment of impact, move away from the interface, each with its

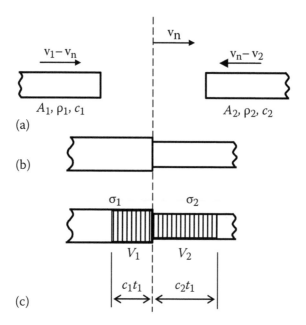

FIGURE 8.10   Collision of two different bars presented as rebound from a moving wall.

own speed. This interpretation helps us to treat this collision as a form of impact against a rigid barrier that is itself moving with v_n. The contact force

$$P = I_1(v_1 - v_n) = I_2(v_n - v_2) \tag{8.30}$$

remains in effect until the reflections from free ends cause the bars to separate.

COLLISION METHOD FOR A MASS–BEAM IMPACT PROBLEM

This is a rather simple engineering approach, where a collision between the impactor of mass M and a portion of the beam, M_b is considered. The problem is illustrated in Case 8.18, using an example of a concrete beam and a nondeformable striker. If the contact can be represented as linear with a stiffness k, the peak reaction and the energy lost are, respectively

$$R_m = v_0 \sqrt{kM^*} \tag{8.31a}$$

and

$$\Delta E_k = \frac{1}{2} M^* v_0^2 \tag{8.31b}$$

where both equations are specific forms of Equations 8.17a and 8.5, respectively. In NL contact problems, the determination of the peak contact force is less direct.

A beam may be built of material, which makes it necessary to use a large cross section. This usually results in a three-dimensional problem, at least locally. Case 8.18 is an example of such a configuration. If a beam is made from solid metal, a relatively longer segment may be expected to interact. One can try to determine M_b in such a way that the natural frequency of the actual beam is preserved. This suggests lumping of the beam inertia into a modal mass, as defined in Chapter 2. For example, a simply supported beam, struck at midpoint, would have $M_b = 0.493\,mL$ according to Table 2.1.*

From the contact viewpoint, the determination of the peak force and, possibly, the duration of the loading phase, is the solution. With regard to the impacted beam, the peak response is of interest. This usually takes place at some time after the initial contact. In NL problems, it is common to assume no rebound takes place. This leads to the following statement of the problem for a beam: The initial velocity

$$v_i = \frac{Mv_0}{M_b + M} \tag{8.32}$$

is suddenly applied to the part of the beam designated as M_b as well as to the mass M of the attached impactor. This is an adaptation of Equation 8.1, as applicable to this arrangement. The formulation is correct as long as the duration of the loading phase of collision is much less than the natural half-period of the beam.

---

* An approach, which assumes that the deflected line of an impacted beam has the same shape as under static loading is presented in detail by Goldsmith [26], who arrives at (17/35) mL, not much different from our number.

This methodology is effective only for relatively small impact velocities. One can antici-pate quite small values of M_b when velocities are large and the deformed shape is far from its static counterpart. The beam response calculations are presented in Chapter 10.

COLLISION ACCOMPANIED BY STRONG SHOCK WAVES

As discussed in Chapter 5, a very large contact stress induces strong shock waves originat-ing at the interface. The quantification of this phenomenon is approximated by means of hydrodynamic analysis, in which material strength is ignored. Specifically, a collision of two blocks with different properties is approached in the same manner as that of two axial bars, presented before in Figure 8.10. One of the unknown variables is the interface velocity v_n, which must result in the same pressure being applied to the faces of the blocks in con-tact. In this discussion a special case will be considered, when the block on the left is moving with the speed of v_1 and the other block is stationary prior to collision, Figure 8.11. The first one is then moving, relative to the interface, with the speed of $v_1 - v_n$ (relative to that imaginary moving wall). When it reaches the wall, its front face stops, therefore its change of speed is $v_1 - v_n$. The other is moving towards the interface with $-v_n$ and its change of speed on contact is $|v_n|$. The Hugoniot will be written for each block in the form of Equation 5.37b, while suppressing the 0 index in the initial density ρ:

$$p_1 = \rho_1[c_1(v_1 - v_n) + s_1(v_1 - v_n)^2] \tag{8.33}$$

$$p_2 = \rho_2[c_2(v_n) + s_2(v_n)^2] \tag{8.34}$$

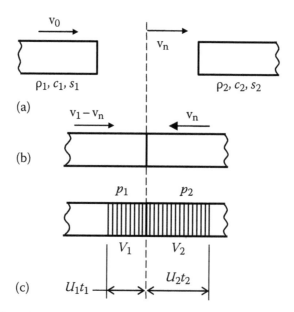

FIGURE 8.11 Shock collision of two different bars presented as rebound from a moving wall. Note that v_n calculated here is different from v_n in Figure 8.10.

By definition, the interface stress is the same for both blocks, i.e., $p_1 = p_2$. The constants c_1, c_2, s_1, and s_2 are the appropriate terms in the basic Hugoniot, namely $U = c + sv$. Equating the above two expressions gives a quadratic equation for the unknown v_n. The solution is somewhat uncomfortable in that each coefficient has several terms. The following approximation of the problem can be found useful (Figure 8.11).

Calculate quantities $z_1 = \rho_1 c_1$ and $z_2 = \rho_2 c_2$, which might be called *notional imped-ances*, as c_1 and c_2 do not have a precise physical meaning. In the first approximation, omit the square terms in Equations 8.33 and 8.34; this makes the problem statement similar to that of a collision of elastic blocks. Equation 8.29 can now be used to find an approximate interface speed:

$$v_n \approx \frac{z_1 v_1}{z_1 + z_2} = \frac{v_1}{1 + z_2/z_1} \tag{8.35}$$

Because the relation is approximate, the pressures on each side of interface, as calculated by Equations 8.33 and 8.34 will not be the same. One of the ways to execute the second approximation is illustrated in Example 8.8.

TABULATION OF CASES

COMMENTS

Cases 8.16 and 8.17 relate to a situation where two masses, subjected to the force of gravity, are capable of colliding. The collision is caused by failure of supports, which may be brittle or ductile. When one of the masses loses its support, for whatever reason, motion begins and a collision results.

CASE 8.1   CENTRAL COLLISION OF TWO MASSES WITH COEFFICIENT OF RESTITUTION κ

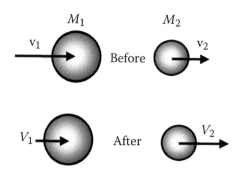

$$v_0 = \frac{M_1 v_1 + M_2 v_2}{M_1 + M_2}; \text{ inst. common velocity}$$

$$V_1 = v_0 - \kappa(v_1 - v_0); \text{ post-impact velocity}$$

$$V_2 = v_0 + \kappa(v_0 - v_2)$$

$$v_r = v_1 - v_2$$

$$\Delta E_k = \frac{1}{2} M^* v_r^2 (1 - \kappa^2) \text{ (kinetic energy loss)}$$

$$\frac{1}{M^*} = \frac{1}{M_1} + \frac{1}{M_2}; \text{ to find the effective mass}$$

CASE 8.2   COLLISION OF PARTICLES IN PLANE, ROUGH CONTACT (FRICTION COEFFICIENT, μ)

(a) Before contact; (b) after contact

$$v_{n0} = \frac{M_1 v_{n1} + M_2 v_{n2}}{M_1 + M_2}; \text{ inst. common normal velocity}$$

$$V_{n1} = v_{n0} - \kappa(v_{n1} - v_{n0}); \text{ normal velocities after rebound}$$

$$V_{n2} = v_{n0} + \kappa(v_{n0} - v_{n2})$$

$$S_n = M_1(v_{n1} - V_{n1}) = M_2(V_{2n} - v_{2n}); \text{ normal impulse}$$

$$S_t = \mu S_n; \text{ tangential impulse}$$

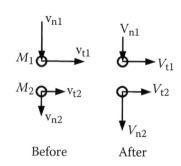

Before   After

$V_{t1} = (M_1 v_{t1} - \mu S_n)/M_1$; tangential velocities after rebound

$V_{t2} = (M_2 v_{t2} + \mu S_n)/M_2$

$S_{tl} = M v_{t1}/\mu$; upper limit of tangential impulse

Note: S_t is limited to the value, which brings V_{t1} to nil.

CASE 8.3 THE REVERSE PROBLEM OF COLLISION OF PARTICLES IN PLANE, ROUGH CONTACT

As in Case 8.2 (pre-impact status from post-impact data).

$v_{n0} = \dfrac{M_1 V_{n1} + M_2 V_{n2}}{M_1 + M_2}$; inst. common normal velocity

$v_{n1} = v_{n0} + (v_{n0} - V_{n1})/\kappa$; normal velocities before rebound

$v_{n2} = v_{n0} + (v_{n0} - V_{n2})/\kappa$

$S_n = M_1(v_{n1} - V_{n1}) = M_2(V_{2n} - v_{2n})$; normal impulse

$S_t = \mu S_n$; tangential impulse

$v_{t1} = (M_1 V_{t1} + \mu S_n)/M_1$; tangential velocities before rebound

$v_{t2} = (M_2 V_{t2} - \mu S_n)/M_2$

$S_{tl} = M v_{t1}/\mu$; upper limit of tangential impulse

CASE 8.4 COLLISION OF TWO BODIES WITH A ROUGH SURFACE CONTACT

(Friction Coefficient is μ)

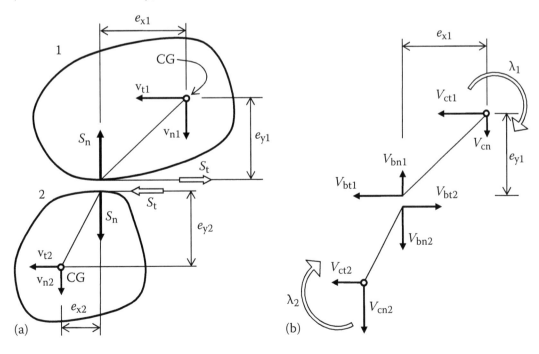

(a)   (b)

$V_{nr} = \kappa v_{nr}$ or $V_{bn2} - V_{bn1} = \kappa(v_{n1} - v_{n2})$; relative velocity rebound at contact point

$$S_n = M_r^* v_{rn}(1+\kappa); \quad S_t = \mu S_n$$

$$M_{r1} = \frac{J_1 M_1}{J_1 + M_1 e_{x1}^2}; \quad M_{r2} = \frac{J_2 M_2}{J_2 + M_2 e_{x2}^2}; \text{ component reduced masses}$$

$$\frac{1}{M_r^*} \approx \frac{1}{M_{r1}} + \frac{1}{M_{r2}}; \text{ effective reduced mass}$$

The following are for Body 1, with index "1" omitted for clarity

$$\lambda = \frac{S_n \tilde{e}}{J} \equiv \frac{S_n}{J}(e_x - \mu e_y) \quad \text{with} \quad \tilde{e} = e_x - \mu e_y$$

$$V_{cn} = v_n - \frac{S_n}{M}$$

$$V_{ct} = v_t - \frac{S_t}{M}$$

$$V_{bt} = V_{ct} + \lambda e_y$$

For Body 2: $\tilde{e}_2 = e_{x2} - \mu e_{y2}$

Note: Analogous equations hold for Body 2. Relative magnitude of components in the above sketch will help to adjust signs. For the sake of simplicity, the combined reduced mass M_r^* is taken as if the contact were frictionless.

CASE 8.5   THE INVERSE PROBLEM FOR A COLLISION OF TWO BODIES WITH A ROUGH SURFACE CONTACT AS IN CASE 8.4

(Pre-impact status from post-impact data)

$$M_{r1} = \frac{J_1 M_1}{J_1 + M_1 e_{x1}^2}; \quad M_{r2} = \frac{J_2 M_2}{J_2 + M_2 e_{x2}^2}; \text{ component reduced masses}$$

$$\frac{1}{M_r^*} \approx \frac{1}{M_{r1}} + \frac{1}{M_{r2}}; \text{ effective reduced mass (approximate)}$$

$$v_{nr} = V_{nr}/\kappa \quad \text{or} \quad v_{n1} - v_{n2} = (V_{bn2} - V_{bn1})/\kappa$$

$$S_n = M_r^* v_{rn}(1+\kappa); \quad S_t = \mu S_n$$

$$v_{t1} = V_{ct1} + \frac{S_t}{M_1}; \quad v_{n1} = V_{cn1} + \frac{S_n}{M_1}$$

$$\tilde{e}_1 = \frac{\lambda_1 J_1}{S_n} \quad \text{where} \quad \tilde{e}_1 = e_{x1} - \mu e_{y1}$$

$$e_{y1} = \frac{V_{bt1} - V_{ct1}}{\lambda_1}$$

$e_{x1} = \tilde{e}_1 + \mu e_{y1}$; determination of eccentricities from other variables

CASE 8.6 CENTRAL COLLISION OF TWO MASSES WITH SPRING AND DAMPER ASSEMBLY

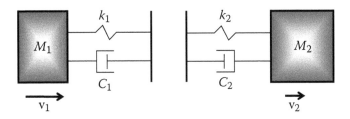

$v_0 = \dfrac{M_1 v_1 + M_2 v_2}{M_1 + M_2}$; "moving wall" velocity

$$V_1 = v_0 - \kappa(v_1 - v_0)$$
$$V_2 = v_0 + \kappa(v_0 - v_2)$$

are velocities after impact

$v_r = v_1 - v_2$; relative velocity

$\kappa \approx 1 - \pi\zeta$; effective restitution coefficient

$R_m = v_r \sqrt{\dfrac{M^* k^*}{1 - \zeta^2}} \exp\left[\left(2\zeta - \dfrac{\pi}{2}\right)\zeta\right]$; peak contact force

$\zeta = \dfrac{C^*}{2\sqrt{k^* M^*}}$; effective damping

$$\frac{1}{k^*} = \frac{1}{k_1} + \frac{1}{k_2}; \quad \frac{1}{M^*} = \frac{1}{M_1} + \frac{1}{M_2}; \quad \frac{1}{C^*} = \frac{1}{C_1} + \frac{1}{C_2}$$

$t_1 = \dfrac{\pi}{2}\sqrt{\dfrac{M^*}{k^*(1 - \zeta^2)}}$ (duration of loading phase)

Note: This is a good approximation for the peak force up to $\zeta = 0.2$.

CASE 8.7 CENTRAL COLLISION OF TWO MASSES WITH SPRING CONSTANTS SPECIFIED

$V_1 = 2v_0 - v_1$

$V_2 = 2v_0 - v_2$; velocities after impact

$v_r = v_1 - v_2$; relative velocity

$R_m = v_r \sqrt{M^*k^*}$; peak contact force

$t_1 = \dfrac{\pi}{2}\sqrt{\dfrac{M^*}{k^*}}$; duration of the loading phase

Note: Other symbols as in Case 8.6. This is elastic collision, with $\kappa = 1$. There is no kinetic energy loss.

CASE 8.8   CENTRAL COLLISION OF TWO MASSES WITH A NL SPRING CONSTANT

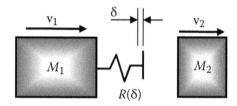

$R(\delta) = k_0 f(\delta)$; prescribed NL characteristic

$$\frac{1}{M^*} = \frac{1}{M_1} + \frac{1}{M_2}; \quad v_r = v_1 - v_2$$

$\Delta E_k = \dfrac{1}{2} M^* v_r^2$; kinetic energy loss of the relative condition

$$R_m = k_0 f(\delta_m)$$

$t_1 \approx \dfrac{\pi \delta_m}{2 v_r}$; duration of the loading phase

Note: This is an inelastic collision. The relative kinetic energy loss is equated to the work of resisting force, which gives peak displacement δ_m.

CASE 8.9   MOVING MASS IMPACTING A SPRING-SUPPORTED MASS

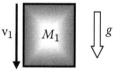

$V_2 = \dfrac{M_1 v_1 (1 + \kappa)}{M_1 + M_2}$; post-impact velocity of M_2

$u_m = u_{st} + \left(u_{st}^2 + \dfrac{M_1}{M_1 + M_2} \dfrac{2E_k}{k} \right)^{1/2}$; maximum displacement of M_2

$$u_{st} = \frac{M_1 g}{k} \quad \text{and} \quad E_k = \frac{1}{2} M_1 v_1^2$$

Note: A fully plastic impact is assumed.

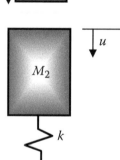

CASE 8.10   TWO-MASS OSCILLATOR UNDER TWO STEP LOADS: $W_1H(t)$ AND $W_2H(t)$

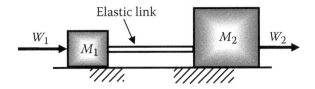

This is resolved into accelerating force

$W_1 + W_2$ giving uniform acceleration: $a = \dfrac{W_1 + W_2}{M_1 + M_2}$

and link-compressing force: $W_c = \dfrac{M_2W_1 - M_1W_2}{M_1 + M_2}$

For a linear-elastic link the peak compression attained is $2W_c$.

CASE 8.11   COLLISION OF TWO BLOCKS OF DIFFERENT MATERIALS

Presented as rebound from a rigid wall moving with the interface velocity v_n

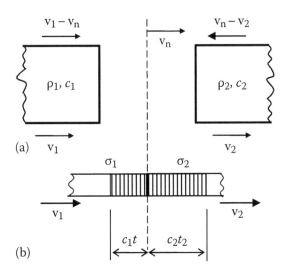

The initial velocities of blocks relative to a fixed system are v_1 and v_2, as shown in (a) compressive waves resulting are shown in (b).

$i_1 = \rho_1 c_1 = \sqrt{E_{ef1}\rho_1}$; specific impedance

$E_{ef1} = \dfrac{(1 - v_1)E_1}{(1 + v_1)(1 - 2v_1)}$; effective modulus

and similar for bar 2.

$v_n = \dfrac{i_1v_1 + i_2v_2}{i_1 + i_2}$; interface velocity following contact

$\sigma = \rho_1 c_1 (v_1 - v_n)$; contact stress

The shaded areas in (b) move with velocity v_n, while the original speeds remain outside of the shaded areas.

Note: The motion of blocks may be interpreted as relative to a frame moving with interface velocity v_n. The contact area of the blocks is assumed large enough in both directions, so that edge effects are minor. The above reduces to collision of bars when contact area is small (E_{ef} becomes E).

CASE 8.12   A SHORT BAR IMPACTING A STATIONARY LONG BAR

(The same bar diameter.)

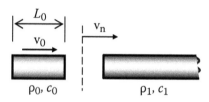

$i_1 = \rho_1 c_1 = \sqrt{E_1 \rho_1}$; specific impedance; similar for bar 0.

$v_n = \dfrac{i_0 v_0}{i_0 + i_1}$; interface velocity following contact

$\sigma = \rho_1 c_1 v_n$; contact stress, valid until $t = 2L_0/c_0$

Special case, when materials are the same:

$$v_n = v_0/2; \quad \sigma = \rho_0 c_0 v_0/2$$

CASE 8.13   COLLISION OF TWO BARS OF DIFFERENT LENGTHS WITH A SPRING IN BETWEEN

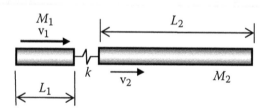

$v_{av} = \dfrac{1}{2}(v_1 + v_2)$; average velocity

$v_i = \dfrac{1}{2}(v_1 - v_2)$; impact velocity

$v_r = v_1 - v_2$; relative speed

(a) Rigid-bodies-with-a-spring approach

$$\frac{1}{M^*} = \frac{1}{M_1} + \frac{1}{M_2}; \quad R_m = v_r \sqrt{kM^*}; \text{ peak reaction}$$

(b) Wave approach

$$P = P_0 \left(1 - \exp\left(\frac{-2kt}{A\rho c_0}\right)\right); \text{ contact force, with } P_0 = A\rho c_0 v_i \text{ and } c_0 = \sqrt{\frac{E}{\rho}}$$

This equation is valid only until $t_m = 2L_1/c_0$, i.e., until the initially compressive wave comes back after reflection from the free end of the shorter bar. (It was assumed that bar 1 is shorter.)

$$P_m = P_0\left(1 - \exp\left(\frac{-4kL_1}{EA}\right)\right); \text{ peak contact force}$$

Note: One could also introduce a more advanced engineering approach in place of (a). The bars can have their masses at the respective CGs. The contact spring would consist of three springs in series, where the first one represents the flexibility of one-half of bar 1 and the third spring relates to one-half of bar 2 in the same way.

CASE 8.14   COLLISION OF TWO ROUND BARS WITH SPHERICALLY ROUNDED ENDS

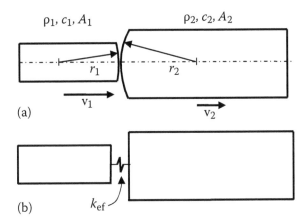

(a) The original bars in contact
(b) The equivalent assembly with an intermediate spring.

$$I_1 = A_1\rho_1 c_1; \quad I_2 = A_2\rho_2 c_2$$

$$v_n = \frac{I_1 v_1 + I_2 v_2}{I_1 + I_2}; \text{ notional interface speed}$$

$$P_0 = I_1(v_1 - v_n); \text{ the upper bound of the contact force}$$

$$P_m \approx P_0\left\{1 - \frac{1}{2}\left[\exp\left(\frac{-2k_{ef}L_1}{E_1 A_1} + \frac{-2k_{ef}L_1}{E_2 A_2}\right)\right]\right\}; \text{ approximate peak force}$$

L_1 stands for the length of a shorter bar.

$$\frac{1}{k_0} = \frac{3}{4}\left(\frac{1-v_1^2}{E_1} + \frac{1-v_2^2}{E_2}\right)\sqrt{\frac{1}{r^*}}; \quad \frac{1}{r^*} = \frac{1}{r_1} + \frac{1}{r_2}$$

$$k_{ef} = \frac{4}{5}k_0^{2/3}P_m^{1/3}$$

Note: The upper bound P_0 is determined as if the ends were flat. The expressions for k_0 and k_{ef} were developed in Chapter 7. The first approximation for k_{ef} is found by putting P_0 in place of P_m in the last formula.

CASE 8.15   SPLIT HOPKINSON PRESSURE BAR

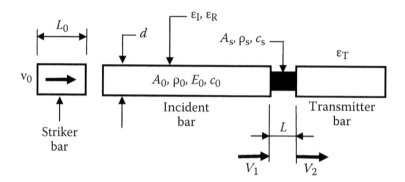

All bars have the same diameter. Strains in the bars are measured by strain gauges: incipient ε_I, reflected ε_R, and transmitted ε_T.

$\sigma = \rho_0 c_0 v_0 / 2$; stress due to striker bar impact

$\varepsilon_I = \sigma / E_0$; incipient strain

$V_1(t) = c_0(\varepsilon_I - \varepsilon_R)$, $V_2(t) = c_0 \varepsilon_T$; velocities of sample ends

$$\frac{d\varepsilon}{dt} = \frac{1}{L}\left[V_1(t) - V_2(t)\right] = \frac{c_0}{L}\left[\varepsilon_I - \varepsilon_R - \varepsilon_T\right];$$ strain rate for the sample

$P_1(t) = E_0 A_0(\varepsilon_I + \varepsilon_R)$; $P_2(t) = E_0 A_0 \varepsilon_T$; forces at sample ends

$$\sigma_s = \frac{1}{2A_s}\left[P_1(t) + P_2(t)\right] = \frac{E_0 A_0}{2A_s}\left[\varepsilon_I + \varepsilon_R + \varepsilon_T\right];$$ average stress in the sample

When a sample is assumed to be in equilibrium:

$$P_1(t) = P_2(t) \quad \text{or} \quad \varepsilon_I + \varepsilon_R = \varepsilon_T$$

$\dot{\varepsilon}(t) = \dfrac{-2c_0}{L}\varepsilon_R$; strain rate of the sample

$$\sigma_s = \frac{P_2(t)}{A_s} = \frac{E_0 A_0}{A_s}\varepsilon_T(t); \quad \varepsilon(t) = \frac{-2c_0}{L}\int_0^t \varepsilon_R \, dt$$

Note: The theory of slender bars is used to interpret this experimental procedure. However, the samples are not, typically, slender [61].

CASE 8.16 COLLISION RESULTING FROM A COLLAPSE OF FRANGIBLE SUPPORTS

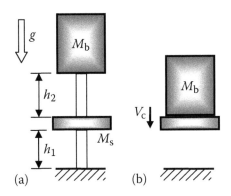

The static condition prior to the event is shown in (a) and motion after supports collapse in (b)

(1) The upper support (under M_b) collapses first.

$v_0 = \sqrt{2gh_2}$; velocity of M_b prior to impacting mass M_s

On impact the lower support collapses.

$V_c = \dfrac{M_b v_0}{M_b + M_s}$; post-impact velocity of assembly,

no rebound, sketch (b)

$\Delta v = v_0 - V_c = \dfrac{M_s v_0}{M_b + M_s}$; decrease of speed of M_b after impacting mass M_s

$\Delta E_k = \dfrac{1}{2} M^* v_0^2 = \dfrac{1}{2} \left(\dfrac{M_s M_b}{M_s + M_b} \right) v_0^2$; kinetic energy loss due to the blocks colliding

$E_g = (M_b h_2 + M_b h_1 + M_s h_1)g - \Delta E_k$; energy of the assembly when impacting ground

$v_f = \sqrt{2gh_1 + V_c^2}$; final impact velocity of the assembly

$S = (M_b + M_s)v_f$; impulse applied to ground by the falling assembly

$t_f = \sqrt{\dfrac{2h_2}{g}} + \dfrac{2h_1}{v_f + V_c}$; event duration

(2) The lower support (under M_s) collapses first.

$v_s = \sqrt{2gh_1}$; velocity of M_s and M_b prior to M_s impacting ground

On impact the upper support collapses. Afterwards

$v_f = \sqrt{2g(h_1 + h_2)}$; velocity of M_b prior to impacting ground

$S_1 = M_s v_s$ and $S_2 = M_b v_f$; two separate impulses are applied to ground

$t_f = \sqrt{\dfrac{2(h_1 + h_2)}{g}}$; event duration

Note: The supports are assumed to instantly collapse, with no energy absorption. The initial failure (weakening to the point when gravity cannot be resisted) is triggered by external means, such as fire or an explosive action.

CASE 8.17 COLLISION RESULTING FROM A COLLAPSE OF ELASTOPLASTIC SUPPORTS

(Refer to the illustration in Case 8.16)

$\Pi_2 = R_2 h_2$; energy absorption capacity of the upper support equal to the average resistance multiplied by the squashing height.

$\Pi_1 = R_1 h_1$; analogous for the lower support

(1) The upper support (under M_b) weakens, initiating collapse.

$M_b g h_2 > \Pi_2$; energy condition for M_b to impact M_s

$$v_0 = \sqrt{2\left(\frac{g h_2 - \Pi_2}{M_b}\right)}; \text{ velocity of } M_b \text{ prior to impacting mass } M_s$$

$$V_c = \frac{M_b v_0}{M_b + M_s}; \text{ post-impact velocity of assembly, no rebound}$$

$$\Delta v = v_0 - V_c = \frac{M_s v_0}{M_b + M_s}; \text{ decrease of speed of } M_b \text{ after impacting mass } M_s$$

$$\Delta E_k = \frac{1}{2}M^* v_0^2 = \frac{1}{2}\left(\frac{M_s M_b}{M_s + M_b}\right)v_0^2; \text{ kinetic energy loss due to the blocks colliding}$$

$$\frac{(M_b + M_s)V_c^2}{2} + (M_b + M_s)g h_1 > \Pi_1; \text{ energy condition for } M_b \text{ and } M_s \text{ to impact ground}$$

$$v_f = \sqrt{\frac{2g h_1 + V_c^2 - 2\Pi_1}{M_b + M_s}}; \text{ impact velocity of the assembly}$$

$S = (M_b + M_s)v_f$; impulse applied to ground by the falling assembly

$E_g = (M_b h_2 + M_b h_1 + M_s h_1)g - \Delta E_k - \Pi_1 - \Pi_2$; energy of the assembly when impacting ground

$$t_f \approx \frac{2h_2}{v_0} + \frac{2h_1}{v_f + V_c}; \text{ event duration}$$

(2) The lower support (under M_s) weakens, initiating collapse.

$(M_b + M_s)g h_1 > \Pi_1$; energy condition for M_b and M_s to impact ground

$v_1 = \sqrt{2g h_1 - 2\Pi_1/(M_b + M_s)}$; velocity of M_s and M_b prior to M_s impacting ground

On impact, the upper support begins to yield.

$M_b(v_1^2/2 + g h_2) > \Pi_2$; energy condition for M_b to impact ground

$v_f = \sqrt{2g h_2 + v_1^2 - 2\Pi_2/M_b}$; impact velocity of M_b

Note: The first weakening or local collapse is triggered by external means, such as fire. If the energy condition is not fulfilled, motion will stop at an intermediate point.

CASE 8.18 COLLISION METHOD FOR A MASS-BEAM IMPACT

(a) Metallic beam, relatively slender, with a compact cross section.
The interacting beam mass M_b may be taken as the modal mass M_m, the latter given in Table 2.1 before.

$$\frac{1}{M^*} = \frac{1}{M} + \frac{1}{M_b}; \text{ to find the effective mass } M^* \text{ in collision.}$$

When there is a linear contact stiffness k, then $R_m = v_0\sqrt{kM^*}$ is the peak contact reaction. If the contact is NL, then use

$$\Delta E_k = \frac{1}{2}M^*v_0^2; \text{ as the energy loss is associated with straining of the NL spring.}$$

$$v_i = \frac{Mv_0}{M_b + M}; \text{ the velocity of the impactor and } M_b \text{ travelling together. This is the initial}$$
velocity to calculate the global beam response.

(b) Typical concrete beam under localized impact from a heavy striker. Rectangular section beam, width B normal to page.

$$M_b = \rho BH(s + H)$$

interacting beam mass during collision.

$s = \sqrt{A}$; where A is the maximum contact area. Formulas for M^*, R_m, ΔE_k, and v_i as above.

Note: For NL cases like (b) above, first estimate of s may be obtained with the help of Cases 7.17 and 7.18, treating the beam as rigidly supported along the bottom surface. Then, after finding M_b and M^*, the value of ΔE_k is used to determine the impact force. Case type (b) can be expected to yield a better prediction than (a).

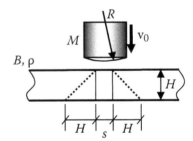

EXAMPLES

EXAMPLE 8.1 COLLISION OF BILLIARD BALLS

One of the two identical billiard balls has an inclined velocity vector and the other is stationary, as indicated by "Before" in the sketch. For $v_1 = 4$ m/s calculate the post-impact velocities for the following cases: (1) the balls are treated as rough particles and the friction coefficient is 0.04, (2) the balls are treated as frictionless, and (3) the impact is assumed to be without slippage. In each case the restitution coefficient is 0.9.

----

(1) Resolving v_1 into components, get $v_{1n} = 3.464$ and $v_{1t} = 2.0$ m/s. Using Case 8.2 get the velocity of the rebound plane and then the post-impact velocities of balls:

$$V_{0n} = \frac{Mv_{1n} + M \cdot 0}{M + M} = v_{1n}/2 = 1.73$$

$$V_{1n} = 1.732 - 0.9(3.464 - 1.732) = 0.1732; \quad V_{2n} = 3.2908 \text{ m/s}$$

The normal and tangential impulse, respectively: $S_n = MV_{2n}$; $S_t = 0.04MV_{2n}$.

If mass M of each object is the same, it cancels out:

$$V_{t1} = 2 - 0.04 \times 3.2908 = 1.8684 \text{ m/s}$$

$$V_{t2} = 0 + 0.04 \times 3.2908 = 0.1316 \text{ m/s}$$

This shows that with such a small friction the tangential change of velocities is negligible. It is left as an exercise for the reader to determine that $M^* = M/2$ and that Equation 8.2b gives the same answer for V_{2n}.

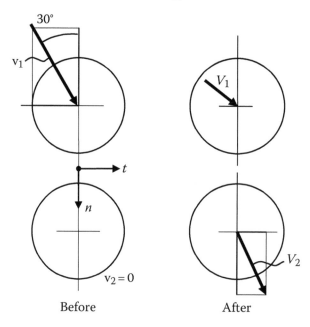

Before   After

(2) The vertical component of collision remains unchanged. There is no change of horizontal components because of lack of friction. The post-impact velocity diagram is the same as before, save for the horizontal components.

(3) Use Equations 8.8 and 8.9.

EXAMPLE 8.2   THREE MASSES ALONG A ROD

A rod as in Figure 8.7, but with three masses only is subjected to a step load $W = 120H(t)$. Determine the extreme axial force by the dynamic equilibrium method outlined earlier. Next, construct a simple FE model and calculate peak forces more precisely that way. The masses are $M_1 = 10$ g, $M_2 = 20$ g, and $M_3 = 30$ g. The rod is made of aluminum alloy, $E = 69,000$ MPa, with section $A = 1.0$ mm². The length is 100 mm. Disregard the mass of the rod itself, as it is quite small in relation to the attached masses.

Following the method outlined in the text: $a = 120/(10 + 20 + 30) = 2.0$. With this, the rigid-body diagram can be constructed indicating the quasistatic axial forces.

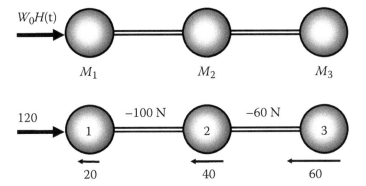

One can therefore expect a maximum compression of $2 \times 100 = 200$ N. After the FE model was constructed and run for about two periods (2.2 ms) the peak axial forces were: −193.4 in segment 1–2 and −172.4 in segment 2–3. The peak compression estimate from hand calculations is therefore quite satisfactory.

EXAMPLE 8.3   SHOCK LOAD IN GEARS

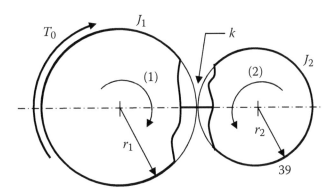

Two mating gears are capable of freely rotating about their respective centers. The flexibility of the contacting teeth is symbolically shown as a short beam with a stiffness k. If a torque $T_0 = 170$ N-m is applied to the larger gear as a step load in time, what is the maximum circumferential contact force between the teeth resulting from this shock load? Use $r_1 = 125$ mm, $r_2 = 75$ mm, $J_1 = 80 \times 10^6$ g/mm², and $J_2 = 10 \times 10^6$ g/mm². (J is the moment of inertia about the axis of rotation.)

The methodology is the same as in Case 8.10, except that the angular accelerations are not identical, but linked with a kinematic condition. Consider a rigid-body rotation when T_0 is applied. The gears then rotate with their respective angular accelerations $\breve{e}_1$ and $\breve{e}_2$. This also means that the smaller gear is subjected to a torque of $J_2\breve{e}_2$ or, alternatively, to a tangential interaction force of $J_2\breve{e}_2/r_2$. This force is applied by the action of the large gear, which needs a driving torque of

$$J_2\breve{e}_2 \frac{r_1}{r_2} = J_2\breve{e}_1 \left(\frac{r_1}{r_2}\right)^2 ; \text{ since } \breve{e}_1 r_1 = \breve{e}_2 r_2 \text{ results from kinematics.}$$

But the large gear also needs a torque of $J_1\breve{e}_1$ to overcome its own inertial resistance, so the total that must be externally applied to it is

$$J_1\breve{e}_1 + J_2\breve{e}_1 \left(\frac{r_1}{r_2}\right)^2 = 80 \times 10^6 \breve{e}_1 + 10 \times 10^6 \breve{e}_1 \left(\frac{125}{75}\right)^2 = 107.78 \times 10^6 \breve{e}_1 \text{ (N-mm)}$$

Equating this with the prescribed torque gives the large gear acceleration as

$$\breve{e}_1 = 170,000/107.78 \times 10^6 = 1.577 \times 10^{-3} \text{ rad/(ms)}^2$$

For the small gear, the corresponding acceleration and the torque are

$$\breve{e}_2 = \breve{e}_1 r_1/r_2; \quad T_2 = J_2\breve{e}_2 = J_2\breve{e}_2 r_1/r_2 = 10 \times 10^6 \times 1.577 \times 10^{-3} \times 125/75 = 26,283 \text{ N-mm}$$

This results solely from the action of the contact force. The above calculation was carried out in a quasistatic manner. To account for the dynamic nature of the event, the interaction force W must be multiplied by the dynamic factor 2.0:

$$W = 2.0 \frac{T_2}{r_2} = 2.0 \frac{26,283}{75} = 701 \text{ N}$$

EXAMPLE 8.4 ECCENTRIC COLLISION OF TWO RODS

A 500 mm long rod with mass $M = 2500$ g is falling with vertical velocity of 4 m/s when it hits a stationary rod of the same properties. The restitution coefficient is 0.6. Calculate velocities of the CGs and the angular velocities of both rods just after the impact. The "bumper" of a negligible mass on the lower rod positions the contact point. (This is the same rod as used in Example 7.1 which was then impacting an eccentric fixed bumper.)

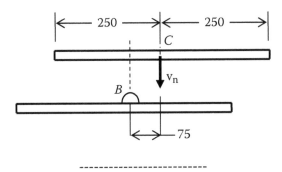

Follow text, Equations 8.10 through 8.12. The basic properties for a single rod hold for both now, but the offsets and reduced masses are different: $M_1 = M_2 = 2500\,\text{g}$, $J_1 = J_2 = ML^2/12 = 52.083 \times 10^6\,\text{g/mm}^2$, $e_{x1} = 75\,\text{mm}$, $e_{x2} = 0$, $M_{r1} = 1968.5\,\text{g}$, $M_{r2} = M = 2500\,\text{g}$

$$\frac{1}{M_r^*} = \frac{1}{M_{r1}} + \frac{1}{M_{r2}} = \frac{1}{1968.5} + \frac{1}{2500}; \quad M_r^* = 1101\,\text{g}$$

$S_n = M_r^* v_m (1+\kappa) = 1101 \times 4 \times 1.6 = 7046.4\,\text{g/mm/ms}$. The moment of this impulse about CG results in angular velocity: $\lambda_1 = 7046.4 \times 75/J = 0.01015\,\text{rad/ms}$; $\lambda_2 = 0$. The normal impulse causes CG velocity to change: $V_{c1} = 4 - 7046.4/2500 = 1.1814$ (down); $V_{c1} = 7046.4/2500 = 2.8186\,\text{m/s}$ (down).

EXAMPLE 8.5   A SIMPLE TWO-CAR COLLISION

Vehicle 1, with $M_1 = 2223\,\text{kg}$ and $v_1 = 13.4\,\text{m/s}$ impacts a stationary vehicle, $M_2 = 1023\,\text{kg}$.

The impact is central, i.e., gravity centers are aligned. It was observed that the duration of the contact lasted 150 ms. There is no visible rebound, therefore $\kappa = 0$. Assuming the force–time plot to be a triangle, find the peak contact force and the average accelerations of both vehicles.

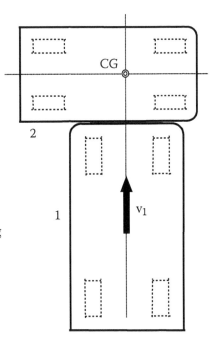

The first task is to estimate the effective mass. From Case 8.4:

$$\frac{1}{M^*} = \frac{1}{M_1} + \frac{1}{M_2} = \frac{1}{2223} + \frac{1}{1023} \quad \text{or} \quad M^* = 700.6\,\text{kg}$$

The common velocity, at the peak of contact:

$$v_0 = \frac{M_1 v_1 + M_2 v_2}{M_1 + M_2} = \frac{2223 \times 13.4}{2223 + 1023} = 9.177\,\text{m/s}$$

$V_2 = v_0$; post-impact velocity of either vehicle. $S_2 = M_2 v_0 = 1.023 \times 10^6 \times 9.177 = 9.388 \times 10^6$ g/mm/s; impulse applied to Vehicle 2. $S_2 = R_m t_0/2$; impulse expressed by duration t_0. Equating both expressions obtain

$$R_m = \frac{2S_2}{t_0} = \frac{2 \times 9.388 \times 10^6}{150} = 125,170 \text{ N}; \quad \text{peak impact force}$$

$$a_1 = \frac{R_m}{M_1 g} = \frac{125,170}{2223 \times 10^3 \times 0.00981} = 5.74\,\text{g}; \quad a_2 = 12.47\,\text{g}$$

As usual during collisions, the lighter body suffers a larger acceleration. The data for this problem come from Brach [12], who uses a different formulation, but obtains the same results. The assumption of a triangular graph of a contact force during the entire contact duration gives only about 2/3 of the actual peak force that was experienced in that crash. Assuming a rectangular graph during 1/3 of the crash duration only gives a reasonably close peak prediction.

EXAMPLE 8.6   COLLISION OF TWO BARS, SAME PROPERTIES, UNEQUAL LENGTH

Examine the changes of stress and velocity when two bars, one 3x as long as the other, collide with equal but opposite initial velocities.

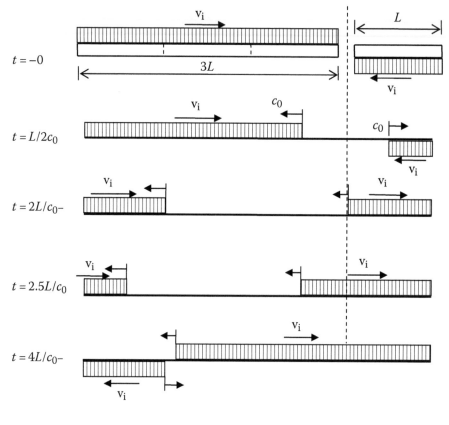

In the picture above the longer bar is subdivided for visual purpose only. Symbol v_i designates current velocity (long arrows) while c_0 stands for the wave speed. (Minus sign means "just before.") After the initial collision the bars are depicted by a solid line only.

The initial position is shown just before impact initiating at $t=0$. The location of the initial contact is marked by a dashed line, fixed in space.

Both bars experience a compressive stress wave, which results in a velocity increment of v_i, directed to the right for the short bar. That increment annihilates the initial velocity, so that after $t = L/c_0$ the short bar is stopped. The rules for the particle velocity changes, which are spelled out in Chapter 5 state that, in effect, the same increment of v_0 is applied by the wave on rebounding from the free end. Therefore, after $t = 2L/c_0$ the entire short bar is moving to the right. At that instant the interface, previously loaded with σ_0 by the compressive wave, becomes unloaded. But this is equivalent to applying a tensile wave to a longer bar. (In summary of what happens at the interface one might say that the tensile wave, originating at the free end of the short bar, just passes through the interface into the long bar.) The interfacing ends move with the same speed.

The history of the long bar follows a similar pattern, but with some complications. The initial compressive wave forces the particle motion to the left, thus cancelling the original motion to the right. It would take $t = 3L/c_0$ to make this bar stand still for an instant, if not for the earlier action at the right end. The tensile wave originating at $t = 2L/c_0$ forces the increment of velocity by v_i, the increment being directed to the right. At $t = 4L_0$ it meets the tensile wave, the latter resulting from the reflection from the left end at $t = 3L/c_0$. The latter wave is imposing the particle velocity to be directed to the left. The overlap of both waves result in $v = 0$ and doubling of tensile stress to $2\sigma_0$. At $t = 6L/c_0$ the right end of the long bar becomes motionless and the bars separate.

From the viewpoint of the short bar, the event does not differ much from impacting against a rigid wall. The rebound velocity is the same and so is the maximum stress. (The interfacing ends were travelling together after the short bar rebound, but they were stress-free.)

The velocity of the long bar at the instant of separation is nonuniform. The left 1/3 will be moving with v_i, directed to the right. The remainder will be at rest, which means the averaged velocity, from momentum viewpoint, is $v_i/3$. (The momentum, before and after collision, remains the same.)

EXAMPLE 8.7   COLLISION OF TWO BARS SEPARATED BY A SPRING

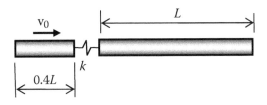

The colliding bars, one moving with $v_0 = 12$ m/s, the other stationary, are made of aluminum alloy. Each has $A = 50$ mm², while $L = 100$ mm.

First, determine the peak contact force treating the bars as rigid, connected by a spring of $k = 345,000$ N/mm². Then, use the wave approach and compare results.

(a) Rigid-bodies-with-spring approach, Case 8.13a:

$$M_1 = A\rho L_1 = 50 \times 0.0027 \times 0.4 \times 100 = 5.4\,g, \quad M_2 = 13.5\,g$$

$$\frac{1}{M^*} = \frac{1}{M_1} + \frac{1}{M_2} = \frac{1}{5.4} + \frac{1}{13.5}, \text{ then } M^* = 3.857\,g; \text{ effective mass}$$

$$v_r = v_0$$

$$R_m = v_0\sqrt{kM^*} = 12\sqrt{345,000 \times 3.857} = 13,842\,N; \text{ peak reaction}$$

(b) Wave approach, Case 8.7. Use Aluminum alloy properties from Table 5.2.

$$v_i = \frac{1}{2}(v_1 - v_2) = \frac{1}{2}(12-0) = 6\,m/s$$

$$P_0 = A\rho c_0 v_i = 50 \times 0.0027 \times 5055 \times 6 = 4095\,N$$

$$P_m = P_0\left(1 - \exp\left(\frac{-4kL_1}{EA}\right)\right) = 4095\left(1 - \exp\left(\frac{-4 \times 345,000 \times 40}{69,000 \times 50}\right)\right) \approx 4,095\,N; \text{ peak contact force}$$

This last equation is written for $t_m = 2L_1/c_0$, i.e., for the instant when the unloading wave comes back after reflecting from the free end of bar 1. Using the wave approach results in a much smaller impact force because a larger effective flexibility is involved.

EXAMPLE 8.8   A SHOCK-WAVE COLLISION OF TWO BLOCKS

An aluminum block moving with the speed of 1800 m/s impacts a stationary block of stainless steel. The following properties are used:

Aluminum 2024: $\rho_1 = 0.002785\,g/mm^3$; $c_1 = 5328\,m/s$; $s_1 = 1.338$

Stainless steel 304: $\rho_2 = 0.007896\,g/mm^3$; $c_1 = 4569\,m/s$; $s_1 = 1.490$

Determine the interface pressure and velocity as well as the shock wave speeds for the materials involved.

Find the notional impedances first:

Aluminum: $z_1 = \rho_1 c_1 = 14.84$

Steel: $z_2 = \rho_2 c_2 = 36.08$

Then use Equations 8.33 through 8.35 to calculate the following:

$$v_n \approx \frac{v_1}{1 + z_2/z_1} = \frac{1800}{1 + 36.08/14.84} = 524.6\,m/s; \text{ first approximation of interface velocity}$$

Pressures resulting from the above:

$$p_1 = \rho_1[c_1(v_1 - v_n) + s_1(v_1 - v_n)^2] = 14.84(1,800 - 524.6) + 0.00373(1,800 - 524.6)^2 = 24,990\,MPa$$

$$p_2 = \rho_2[c_2(v_n) + s_2(v_n)^2] = 36.08 \times 524.6 + 0.01177 \times 524.6^2 = 22,170\,MPa$$

The value of v_n is too small, because it gives a smaller pressure in the driven block. Take the average of both, $p = 23,580\,MPa$, as the anticipated true value. With the linear term in the expression for p_2 dominating, estimate the new speed to be

$$v_n \approx \frac{23,580}{22,170}524.6 = 558\,\text{m/s}$$

When new p_2 is recalculated, one finds $p_2 = 24,185\,MPa$, sufficiently close to the trial value. The particle velocity change for aluminum is $1,800 - 558 = 1,242\,m/s$. The shock wave speeds are

$$U_1 = 5328 + 1.338 \times 1242 = 6990 \text{ m/s}$$

$$U_2 = 4569 + 1.490 \times 558 = 5400 \text{ m/s}$$

The solution of Cooper [20], who originally posed this problem but handled it differently, results in $p = 24,000\,MPa$, then 563, 6080, and 5,410 m/s, respectively. The differences are negligible except for U_1. If $v_n = 563$ is used, then one should be getting $U_1 = 6983\,m/s$, so our number appears correct.

9 Cables and Strings

THEORETICAL OUTLINE

From an engineering viewpoint, a *cable* or a *string* is a very simple structural element, which resists loads applied only along its axis. In a common parlance,* there are some differences between the two objects, but here both names are used interchangeably, to designate elements, which have no bending stiffness.

The above description of string properties should be augmented by the following: If a string is preloaded to some level of tension, it will also resist lateral forces, owing to the geometric stiffness, as described in Chapter 2. This is how musical string instruments work, although they were invented long before the terms used here were conceived.

A structural beam offers two components of resistance to deformation: bending and stretching. The first of those is related to what we understand as a "beam proper" while the second is associated with a cable-like action. Under some circumstances, the second component becomes predominant, which enables us to simplify such a beam to a cable. This makes the following study of string-like action so much more important.

STATICS OF CABLES

While a string is a simple structural element, the lack of lateral stiffness puts it into a moderate deflection range, which makes the description of its response a nonlinear procedure.

Consider a cable, initially in the form of a straight line, spanning the distance between two points. Two types of distributed load may be involved, as Figure 9.1 illustrates. The first one is the action of forces, like gravity, which retain their original direction as in Figure 9.1a while the second is a uniform pressure, changing direction according to the deflected position (Figure 9.1b). The difference between Figure 9.1a and b is not significant until deflections are quite visible.

It is easy to show that, for pressure-like loading in Figure 9.1c, the cable becomes curved with a constant radius R, so that the tensile force P is given by

$$P = w_0 R \tag{9.1}$$

The force P can be found from moment equilibrium with respect to the support point in Figure 9.1c:

$$P \approx \frac{w_0 L^2}{8\delta} \tag{9.2}$$

---

* A *string* is usually made of fiber strands and can be bent by hand without much resistance. A *cable*, on the other hand, suggests at least some metallic fibers, which can offer a substantial resistance to local bending.

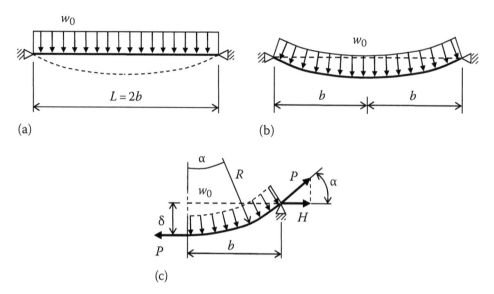

FIGURE 9.1   Initially straight cable, with gravity load applied (a), with pressure load (b), and the equilibrium diagram for the latter (c).

while ignoring the small horizontal component of the distributed loading. The above relationship, arising from equilibrium, is valid for all material models. If, for example, an RPP material is involved, the situation is much simpler in that the cable force is constant, equal to the yield value; $P_y = A\sigma_0$. Specifying w_0 gives R from Equation 9.1 and δ from Equation 9.2. In general, the relationship between δ and R can be written as

$$2R\delta - \delta^2 = b^2 \tag{9.3}$$

where $b = L/2$. More details of statics under distributed loading can be found in Case 9.1.

Geometry of a cable under a point load is shown in Figure 9.2. (Gravity is not involved here, as may be the case with a horizontally placed cable.) The equilibrium equation of forces at the application point gives

$$W = 2P \sin \alpha \tag{9.4}$$

When e designates the elongation of a half-length, the strain becomes

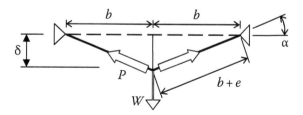

FIGURE 9.2   Static equilibrium of a point-loaded cable ($L = 2b$).

$$\varepsilon = \frac{e}{b} = \left(1 + \frac{\delta^2}{b^2}\right)^{1/2} - 1 \qquad (9.5)$$

This is true for any material as long as the shape is maintained. For moderate deflections, the second term in brackets is much smaller than the first, so the equation can be simplified, as done in Case 9.2. When the cable, along with the load W, is cut in half, the moment equation about a support point gives

$$P \approx \frac{WL}{4\delta} \qquad (9.6)$$

Like Equation 9.2, this is only an approximate relation, which assumes $\cos \alpha \approx 1$ and can be used for moderate deflections only. Refer to Case 9.2 for more details. If an RPP material is involved, the constant cable force $A\sigma_0$ is in effect.

STRAIN ENERGY AND WORK

Consider an elastic cable fixed at both ends. Its elongation $\lambda = 2e$, is the difference between the deformed length and the initial length L. Timoshenko [110] shows that

$$\lambda = \frac{1}{2} \int_0^L \left(\frac{du}{dx}\right)^2 dx \qquad (9.7)$$

Although this is derived for small deflections, it will hold very well for moderate deflection as long as the ends are fixed. If the peak lateral displacement δ is known, then, assuming a deformed shape makes it possible to get the approximate elongation. If the deflected curve resembles a sinusoid,* $u = \delta \sin \frac{\pi x}{L}$ as often assumed, then the integration gives

$$\lambda_1 = \frac{\pi^2 \delta^2}{4L} \qquad (9.8)$$

This deformed pattern will be referred to as a *sine mode*. In general, the elastic strain energy is $EA\lambda^2/(2L)$, so here we have

$$\Pi_1 = \frac{\pi^4}{32} \frac{EA\delta^4}{L^3} \qquad (9.9)$$

The distributed load, $w = w_0 H(t)$, performs the work proportional to the area between the original axis and the maximum deflected line. Using the sine mode approximation

$$\mathcal{L}_1 = w \int_0^L \delta \sin \frac{\pi x}{L} dx = \frac{2}{\pi} w \delta L \qquad (9.10)$$

---

* There is not much difference between a flat circular arc and a flat sinusoid.

can be readily found from Appendix B. The kinetic energy concept is useful in finding a response, when a short impulse lasting t_0 is applied. For a distributed load w, the initial velocity is then $v_0 = wt_0/m$ and a kinetic energy of one-half of a symmetric cable is

$$E_k = \frac{1}{2}\left(m\frac{L}{2}\right)v_0^2 = \frac{1}{2}M_h v_0^2 \tag{9.11}$$

where M_h is the mass of one-half of the cable. In a linear elastic case the strain energy of a half-cable is $\Pi = EAe^2/L$. Equating this to E_k gives the elongation e of a half-cable and the associated peak force P_m:

$$e = \frac{1}{2}\frac{w_0 t_0 L}{\sqrt{EAm}} \tag{9.12a}$$

$$P_m = \frac{2EA}{L}e \tag{9.12b}$$

General expressions for strain energy and work of external load of any elastic string can be written, respectively, as

$$\Pi = \Psi\frac{EA\delta^4}{L^3} \tag{9.13a}$$

$$\mathcal{L} = \Theta L w\delta \tag{9.13b}$$

where Ψ and Θ are the coefficients to be determined. If the step load $w = w_0 H(t)$ is applied, then the work and the energy can be equated, at the extreme deflection, to obtain

$$\delta = L\left(\frac{\Theta L w_0}{\Psi EA}\right)^{1/3} \tag{9.14}$$

The above expressions are written having in mind a distributed loading $w(t)$. The *kink mode* equations are given by Case 9.3. The assumption of two halves of the cable remaining straight during elongation results in somewhat smaller energy input:

$$E_k = \frac{3}{8}M_h v_0^2 \tag{9.15}$$

as compared with the sine mode before. This is the direct result of our assumption that in the sine mode there is a minimal obstruction of the initial cable motion by the end supports. (Note that in the kink mode the velocity $v_0 = wt_0/m$ is merely a reference value.) When a point load $W_0 H(t)$ is applied at mid-length, the expressions developed for sine mode are adjusted by a change in coefficients and by replacing wL in Equation 9.10 by W_0.

For various reasons, cables are often prestressed to σ_i or preloaded to $P_i = A\sigma_i$. A procedure to determine their response is similar to the methodology used so far, except that instead of equating work or energy input to that of strain energy, we equate it with the change of that energy from the preloaded to the current position. When prestretching, the initial elongation may often be quite sizeable, for example, in various nonmetallic fibers used for musical instruments. The change of section area must then be taken into account. Case 9.4 presents some formulas for prestressed cables.

TRANSIENT CABLE MOTION*

There are two components of cable motion relative to its axis: longitudinal and lateral. The first is analogous to that of an axial bar, discussed at length in Chapter 5. For the analogy to be complete, the cable must remain straight, that is, not allowed to sag. The other condition is that the cable remains in tension, which can be realized if the cable is preloaded. In this case an apparent compression of limited magnitude can take place, said compression being merely a decrease in tension.

A lateral component of motion was briefly explained in Chapter 2. As long as there is axial tension present, there also is a resistance to lateral motion. The mathematical description of lateral motion in a linearly elastic cable is similar to that of the longitudinal component, except the difference in constants. A general treatment of extensible strings is given by Cristescu [18]. He demonstrates that two types of elastic waves are involved: a longitudinal one with a velocity c_0, which propagates elongations, without changing the shape of the cable and a transverse wave, which affects the shape of the string only, without changing its longitudinal strain. Still, both waves influence one another. Their velocities, consistent with our notation, are

$$c_0 = \left(\frac{EA}{m} \right)^{1/2} = \left(\frac{E}{\rho} \right)^{1/2} \quad \text{(longitudinal)} \tag{9.16}$$

$$c_2 = \left(\frac{P}{m} \right)^{1/2} = \left(\frac{\sigma}{\rho} \right)^{1/2} \quad \text{(transverse)} \tag{9.17}$$

where the second equality is valid only if the structural area is such that we can write $\rho A = m$, that is, there is no additional inertia. The form of equations presented above is for a linear case only, which means not only the material nonlinearity is ignored, but also changes in cross-sectional area due to stretching are not taken into account. Still, in spite of these simplifications a transient cable motion is less than simple. Consider the case in Figure 9.2, with the cable initially straight, but not preloaded. This means that at the outset there is no resistance against the lateral load W. As soon as the motion starts, the applied force is resisted by the inertia of the cable. This also means that the transverse wave may now propagate due to appearance of the axial load spreading along cable.

In Figure 9.3 a part of a long cable is shown, with the point being pulled laterally with a constant velocity v_0. The whole segment of the initial length $x = c_2 t$ acquires the lateral speed of v_0. Angle β remains constant, as long as velocity v_0 does. The increase in length u of the activated segment is ascribed to elongation caused by the longitudinal wave, which has traversed segment $c_0 t$ after time t. From simple geometry, one can write $(x + u)^2 = x^2 + y^2$ or

$$u^2 + 2xu = y^2 \tag{9.18}$$

---

* It is recommended that the reader refreshes his memory of Chapter 5 prior to reviewing this part.

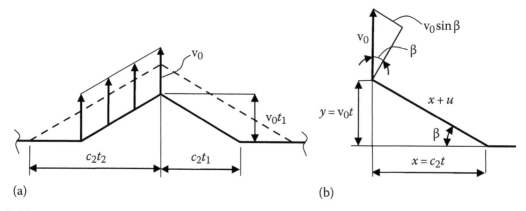

FIGURE 9.3 (a) An intermittent point of a cable is driven with velocity v_0. (b) One-half of this cable.

where the components of the above are

$$u = \frac{P(c_0 t)}{EA} \qquad (9.19a)$$

$$x = c_2 t \qquad (9.19b)$$

$$y = v_0 t \qquad (9.19c)$$

In this set of equations there are two unknowns: c_2 and P. Substituting the latter into the former and taking advantage of Equations 9.16 and 9.17, one obtains

$$q^2 + 2\sqrt{c_0} q^{3/2} = v_0^2 \qquad (9.20a)$$

with

$$q = \frac{P c_0}{EA} \qquad (9.20b)$$

Solving for q gives, in effect, the stretching force P and, this, in turn, allows one to find the transverse wave velocity from Equation 9.17, which determines cable motion. The associated driving force W_0, applied at the apex, can be found from a simple equilibrium:

$$W_0 = 2P \sin \beta \qquad (9.21)$$

with $W_0/2$ applied when a semi-infinite* cable is involved, as in Figure 9.4. This is valid, of course, as long as the transverse wave, spreading with velocity c_2, does not encounter a support or another obstacle so that the reflected pulse may come back and blur the picture.

---

* This term simply means that the cable is so long, when measured from the end under consideration, that the presence of the other end is not felt.

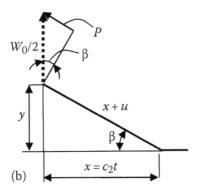

FIGURE 9.4 Semi-infinite cable driven by a prescribed force $W_0/2$.

One should note that when a cable is driven as described above, the velocity component along the axis is $v_0 \sin \beta$, so one can write, using the basic relation between the end force and the end velocity:

$$\sigma = \rho c_0 v_0 \sin \beta \qquad (9.22a)$$

or

$$P = m c_0 v_0 \sin \beta \qquad (9.22b)$$

When the driving force W_0 is prescribed, as in Figure 9.4, the procedure is somewhat different. Solving Equations 9.20 gives P and also determines c_2:

$$\tan \beta = \frac{v_0}{c_2} \approx \sin \beta \qquad (9.23)$$

for small-to-moderate angles. Equation 9.22 then gives

$$W_0 \approx 2 \frac{v_0}{c_2} P \qquad (9.24)$$

which makes it possible to find W_0 if P is known. However, if W_0 is the starting point, an iterative procedure is necessary.

The lateral deformation spreads with the velocity c_2 and the latter depends, in turn on the axial load P. The strong nonlinearity of the above equations comes from said load being nil at the beginning of the process and acquiring a finite value dependent on the external loading. If, on the other hand, a certain preload P_0 is imposed at the beginning and then the lateral action is applied, the situation simplifies. If, in addition, the lateral action is small enough, so that it does not cause a significant change in P, then c_2 is practically constant and the string behaves in a linear fashion, as described in Chapter 2. The deflection due to

a lateral step loading can be calculated in a quasistatic manner first presented in Chapter 5.*
Examples 9.4 and 9.6 present numerical illustrations of an initially stress-free and a preloaded
cable, respectively.

One may note that Equation 9.18 allows a simple formulation of the breaking condition
for a cable, whose point becomes driven with a constant velocity. If one writes $\varepsilon = u/x$ and
also puts $y/x = \tan \beta = v_0/c_2$, this results in a formula given in Case 9.8, provided the cable
remains elastic until failure.

The considerations so far are related to an elastic cable. When *rigid-plastic* material is
assumed, with the cable preloaded to the flow strength σ_0, the problem of constant velocity
driving, analogous to what is shown in Figure 9.3, becomes quite simple and is presented
as Case 9.11. The cable force P remains constant and so does the speed of the lateral deformation wave front c_2. One should note, however, that the force needed to drive the cable
increases with the increasing angle.

INELASTIC CABLE

Inelastic cable is often conveniently approximated with a rigid-plastic material. The cable
is preloaded to the flow strength σ_0, said preload determining the speed of propagation of
lateral deflections by defining c_2 according to Equation 9.17, where $P = P_0 = A\sigma_0$. As shown
in Figure 9.5, at time t the deformation wave has swept the distance $\xi = c_2t$ on each side of
the point of force application.

A simple motion results if the application point is driven with a constant velocity v_0, as illustrated by Case 9.11, consistent with Figure 9.3. The mobilized part of the string remains straight,
as there is no acceleration present. The angle β remains constant and so does driving force W_d.

A cable can rest on a continuous foundation, which acts like the medium described in
Chapters 2 and 10. The difference now is that it is not an elastic foundation, but a rigid-plastic
form of it. This means that when the cable deflects, the foundation reacts with a constant,
distributed load w_y (N/m) independent of the deflection magnitude.

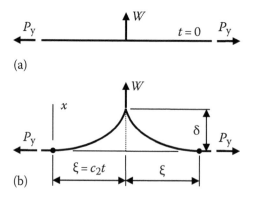

FIGURE 9.5   Rigid-plastic cable. (a) Initial position and (b) after time t.

---

* Another nonlinearity appears when angle β in Figure 9.4 becomes significant, but a quasistatic calculation
  due to step loading should still give a good approximation.

A quasistatic solution of a point-loaded string, including the presence of a foundation, is quite easy to obtain. The inertia of cable is ignored and time is used as a parameter merely indicating how far the deformation has spread. When Equation 2.26 is invoked, with the inertia term omitted and with $w = -w_y$, one has

$$u''(x) = \frac{w_y}{P_y} \qquad (9.25)$$

which means a deflected line of constant curvature. Assuming u and the slope u' to vanish at $x = 0$, according to Figure 9.5, one obtains, after integrating twice

$$u' \equiv \theta = \frac{w_y}{P_y} x \qquad (9.26a)$$

and

$$u = \frac{w_y x^2}{2P_y} \qquad (9.26b)$$

with peak deflection u designated by δ in Figure 9.5. Further details can be found in Case 9.12.* Whenever one considers a problem without including the end conditions, as here, it is implied that those ends are far enough not to interfere with the event. After obtaining a solution, one must check if this indeed is the case.

PROJECTILE IMPACT

Projectile impact against an elastic cable is quantified by Cases 9.5 and 9.6. Those summaries are applicable when a projectile mass is larger than the cable mass. When the material is rigid-plastic, some closed-form solutions are available for the transient motion without the above limitation.

When a mass M impacts with velocity v_0, as in Figure 9.6a, the deformed shape after time t is illustrated in Figure 9.6b. The projectile travels with the cable and the deformation wave sweeps distance $\xi = c_2 t$ on each side of the impact point. For the loading stage of motion, during which there is no reversal of velocity, the simplifying assumption is to treat the entire velocity field to be constant, at any time-point t. The principle of momentum conservation gives us

$$Mv_0 = Mv + mDv + 2m\xi v \qquad (9.27)$$

Differentiating both sides with respect to time gives the following equation of motion:

$$(M + mD + 2m\xi)\frac{dv}{dt} + 2mcv = 0 \qquad (9.28)$$

---

* One should note that the position of origin, $x = 0$, changes with time, as the wave spreads.

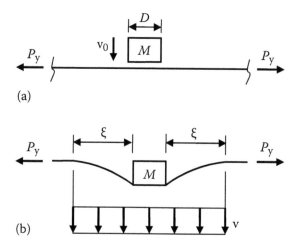

(a)

(b)

FIGURE 9.6   Cable impacted by a mass of finite width D. (a) Prior to impact and (b) deflection and velocity after impact.

According to Teng [107], the solution for the velocity is

$$v = \frac{v_0}{1 + \beta + 2\beta ct/D} \quad \text{with } \beta = \frac{mD}{M} \tag{9.29}$$

and the time counts from the instant the segment of length D had acquired its initial velocity $v_0/(1+\beta)$. (Refer to Case 9.13.)

When a cable rests on a continuous foundation, as shown in Figure 9.7, the procedure is essentially the same. The momentum conservation statement has the same form as Equation 9.27, except the term mDv is missing, because of a point contact considered now.

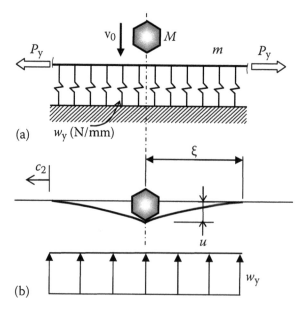

(a) w_y (N/mm)

(b)

FIGURE 9.7   (a) A projectile approaching a rigid-plastic cable on a rigid-plastic foundation. (b) deflected shape after time t and the foundation reaction w_0.

When the cable deflects, the foundation reacts with a load, whose resultant is $2\xi w_y$, independent of deflection magnitude.

Differentiating the momentum with respect to time gives the externally applied force, so here one gets

$$(M + 2m\xi)\frac{dv}{dt} + 2mcv = -2\xi w_y \tag{9.30}$$

This is a differential equation for v with time t hidden in $\xi = c_2 t$. An in-depth analysis of this problem given by Wierzbicki and Hoo Fatt [116] gives the following solution:

$$v = \frac{1 - (t/t_f)^2}{1 + \mu(t/t_f)} v_0 \quad \text{with } t_f = \sqrt{\frac{Mv_0}{c_2 w_y}} \quad \text{and} \quad \mu = \frac{2mc_2 t_f}{M} \tag{9.31}$$

This satisfies the initial condition, $v = v_0$ for $t = 0$. The expression for v shows that at $t = t_f$ one has $v = 0$, that is, the mass stops. At that time $\xi_f = c_2 t_f$, which makes it evident that μ is the ratio of mass of mobilized string to the impacting mass, at the end of motion.

A special case of the above is when there is no foundation, that is, $w_y = 0$. The momentum expression is the same as before and it gives us

$$v = \frac{v_0}{1 + (2mc_2 t/M)} \tag{9.32}$$

In this case, the velocity of the impacted point decreases asymptotically with time. The above solutions imply that, for the sake of convenience, the cable had been preloaded to the yield stress σ_0.

END DISTURBANCE IN CABLE MOTION

It is instructive to first consider the cable movement without the presence of end supports, merely as a free body. The distributed loading, constant in time, applies the following acceleration, velocity, and displacement to a laterally unconstrained cable element:

$$a_0 = w_0/m \tag{9.33a}$$

$$v_e = a_0 t \tag{9.33b}$$

$$u_e = a_0 t^2/2 \tag{9.33c}$$

where $m = A\rho$ is the mass per unit length. This motion by itself does not induce any tension in the cable. It is the supports that are the sources of disturbance spreading from their respective locations. Consider a very short pulse of loading, equivalent to initial velocity v_0. In the early stage of motion, one can replace Figure 9.8a with b, where the central part of the cable is stationary, but there is a fictitious motion of the ends, in the direction opposite to that of the load. Both states of motion may be treated as equivalent, as far as cable loading is concerned. (Figure 9.8) (It is obvious that such a reversal makes no difference only until the lateral movement has spread to the midpoint of cable.)

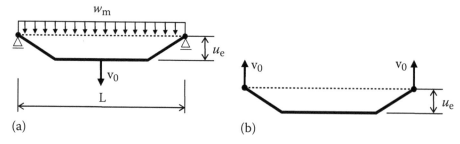

FIGURE 9.8 Deflected shape after the initial velocity is applied (a) and the same shape seen as a result of a stationary cable affected by the prescribed movement of ends (b).

Once the waves issued from both ends meet at the center, they rebound and return intensified toward the ends. Then they rebound from the ends, intensified again by the order of magnitude of 2.0 and travel toward the center. This bouncing back and forth is also accompanied by up-and-down motion of the cable. Example 9.4 illustrates some aspects of this phenomenon.

CLOSING REMARKS

The subject of the safety factor (SF) is not one of the primary topics of this book, nevertheless it always deserves some attention. In NL problems this factor is often not explicitly stated, but the engineer conducting the work should be quite sure that he is on the safe side of the equation. A typical situation with cables is that an increase in external loading is accompanied by a smaller relative increase of cable stress, as described in detail by Irvine [38]. For this reason it is safer to formulate the SF, often assumed as 2.5 for cables, with respect to the resulting cable stress, rather than to the imposed load.

If a loaded cable becomes strongly curved during its deformation process, local bending may become important. The combined action of bending and stretching in a moderate deflection range can lead to substantial analytical complexities. Only in exceptional cases a simple treatment may be available. The practical way of going about it is to assign beam-like properties to at least parts of the cable and perform a FE simulation.

There also is an important aspect to consider when specifying the end conditions. Fixing the ends longitudinally means removing the freedom of movement of the ends along the axis. But, in practice, it is not easy to provide a perfect axial constraint, so there is likely to always be some flexibility at an end. Empirically measured deflection will be larger than one calculated using perfect axial constraint as a basis. Also, loads from dynamic action will be larger due to wave reflection. Lateral flexibility at the ends, if any, is less of a problem, as the corrections are usually minor. A good modeling of any real problem should therefore include at least the axial flexibility of ends. This is a common concern in the analysis of slender elements and it will also be discussed in the chapters to follow.

There often is a price to be paid for attempting to give simple solutions to complex problems. Most frequently, it shows itself in the loss of accuracy or, perhaps in the lack of consistency, when dealing with various aspects of the same event. The example may be found in Case 9.2, where two different equivalent (lumped) masses were used.

TABULATION OF CASES

CASE 9.1   CABLE BETWEEN TWO SUPPORTS, INITIALLY STRAIGHT, UNDER DISTRIBUTED LOAD

Length L, section area A, distributed mass m, load $w \sim \text{N/m}$, one-half shown. e is the elongation of one-half of the cable.

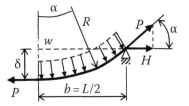

(a) Statics

$R = P/w$; curvature radius

$$\delta \approx \frac{L}{4}\left(\frac{3wL}{EA}\right)^{1/3}; \quad \alpha \approx \frac{4\delta}{L}; \quad R \approx \frac{L^2}{8\delta}$$

$$P \approx \frac{wL^2}{8\delta}; \text{ stretching force}$$

$$\varepsilon = \frac{2e}{L} = \frac{P}{EA} = \frac{\pi^2 \delta^2}{4L^2}; \text{ strain, } \Pi = \frac{\pi^4}{32}\frac{EA\delta^4}{L^3}; \text{ strain energy}$$

H is the horizontal component of the stretching force. When the vertical gravity load mg is applied to the cable, one has $H \approx \dfrac{mgL^2}{8\delta}$

(b) Impulsive load w_0 lasting t_0

Then initial velocity at midpoint $v_0 = w_0 t_0 / m$.

$$e = \frac{1}{2}\frac{w_0 t_0 L}{\sqrt{EAm}}; \text{ elongation of one-half}$$

$$\delta = e\left(1 + \frac{L}{e}\right)^{1/2} \approx \sqrt{Le}; \text{ maximum deflection}$$

$$P_{\text{m}} = \frac{2EA}{L}e; \text{ peak cable force}$$

(c) Step load $w = H(t)w_0$

$$\delta = L\left(\frac{\Theta L w_0}{\Psi EA}\right)^{1/3}; \quad \Psi = \frac{\pi^4}{32} = 3.044; \quad \Theta = 2/\pi$$

$$\lambda = \frac{\pi^2 \delta^2}{4L}; \text{ total elongation}$$

$$P_{\text{m}} = \frac{EA}{L}\lambda; \text{ peak cable force}$$

Note: The static expressions for δ and P are for moderate deflections. When the angle $\alpha = 40°$, then $c/\delta = 0.364$ and the error of the first equation is within 4.1% of the true value. For small angles, there is no need to distinguish between loads applied to the initial and to the deformed shape.

CASE 9.2   CABLE BETWEEN TWO SUPPORTS, INITIALLY STRAIGHT, UNDER POINT LOAD $W \sim N$

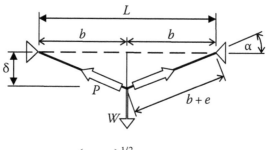

$$\delta = e\left(1 + \frac{L}{e}\right)^{1/2} \quad \text{or} \quad \delta \approx \sqrt{Le}$$

(a) Statics

$$\frac{W}{2EA} = \tan\alpha - \sin\alpha \approx \frac{\alpha^3}{2}$$

$$\delta \approx \frac{L}{2}\left(\frac{W}{EA}\right)^{1/3}; P = \frac{W}{2\sin\alpha} \approx \frac{WL}{4\delta}; \text{stretching force}$$

$$\varepsilon = \frac{2e}{L} = \frac{P}{EA} = \frac{W}{2EA\sin\alpha} \approx \frac{2\delta^2}{L^2}; \text{strain}$$

$$\Pi_2 = 2\frac{EA\delta^4}{L^3}; \text{strain energy}$$

(b) Impulsive load W_0 lasting t_0

$M_{\text{ef}} = 0.15(mL)$; effective lumped mass for peak load estimate

$$e = \sqrt{\frac{b}{EA}}\frac{W_0 t_0}{\sqrt{2M_{\text{ef}}}}; \text{elongation of one-half}$$

$$P_{\text{m}} = \frac{2EA}{L}e; \text{peak cable force}$$

$M'_{\text{ef}} = (mL)/3$; effective lumped mass for peak deflection estimate

$$e' = \sqrt{\frac{b}{EA}}\frac{W_0 t_0}{\sqrt{2M'_{\text{ef}}}}; \text{elongation of one-half}$$

$\delta \approx \sqrt{Le'}$; maximum deflection

(c) Step load $W = W_0 H(t)$

$\mathcal{L} = W_0 \delta$; work of applied load

$$\delta = L \left(\frac{W_0}{2EA} \right)^{1/3}; \text{ maximum deflection}$$

$e = \delta^2/L$; elongation of one-half

$$P_{\mathrm{m}} = 2EA \left(\frac{\delta}{L} \right)^2; \text{ peak cable force}$$

Note: The static expressions for δ and P are for moderate deflections. When the angle $\alpha = 20°$, the error of simplifying δ is within 3% of the true value.

CASE 9.3   CABLE BETWEEN TWO SUPPORTS, INITIALLY STRAIGHT, DEFLECTING IN KINK MODE

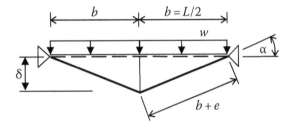

Distributed load $w \sim$ N/mm.

$$\delta = e \left(1 + \frac{L}{e} \right)^{1/2} \text{ or } \delta \approx \sqrt{Le}$$

$$\Pi = 2 \frac{EA\delta^4}{L^3}; \text{ strain energy}$$

$$P_{\mathrm{m}} = \frac{2EA}{L} e; \text{ peak cable force (for all elastic cases below)}$$

(a) Impulsive load w_0 lasting t_0

$$v_0 = \frac{3}{4} \frac{w_0 t_0}{m}; \text{ initial velocity at center, after the impulse of } w_0$$

$$E_{\mathrm{k}} = \frac{3}{8} \frac{(wt_0)^2 b}{m}; \text{ kinetic energy of one-half}$$

$$\Pi = \frac{EAe^2}{L}; \text{ strain energy of one-half}$$

$$e = \frac{\sqrt{3}}{4} \frac{w_0 t_0 L}{\sqrt{EAm}}; \text{ resulting elongation of one-half}$$

(b) Step load $w = w_0 H(t)$

$$\mathcal{L} = \frac{1}{2} w \delta L; \text{ work of applied load}$$

$$\delta = L \left(\frac{L w_0}{4EA} \right)^{1/3}; \quad e = \frac{\delta^2}{L}$$

CASE 9.4 AXIALLY PRELOADED CABLE, UNDER DISTRIBUTED IMPULSIVE LOAD $w_0 \sim$ N/mm, LASTING t_0

The prestress is $\sigma_i < \sigma_0$.

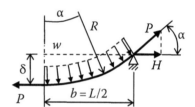

$$e_i = \frac{P_i L}{2EA} = \frac{\sigma_i L}{2E}; \text{ the initial prestretch of half-cable}$$

$$\delta = e \left(1 + \frac{L}{e} \right)^{1/2}; \text{ peak deflection}$$

(a) Elastic cable:

$$\frac{e^2 - e_i^2}{L} EA = \frac{1}{4} \frac{w_0^2 t_0^2 L}{m}; \text{ work–energy equivalence}$$

$$e^2 = \frac{1}{4} \frac{(L w_0 t_0)^2}{EAm} + e_i^2; \text{ maximum stretch of half-cable}$$

$$P_m = \frac{2EA}{L} e; \text{ peak stretching force}$$

(b) Inelastic cable: it is conveniently assumed that preloading is close to $P_y = A\sigma_0$

$$P_y e - \frac{1}{2} P_y e_i = \frac{1}{4} \frac{w_0^2 t_0^2 L}{m}; \text{ work–energy equivalence}$$

$$e = \frac{1}{4} \frac{L w_0^2 t_0^2}{P_y m} + \frac{e_i}{2}; \text{ maximum stretch of half-cable}$$

Note: When peak deflection is a multiple of that attained at the onset of yielding, the effect of the preload is insignificant.

CASE 9.5 STRAIGHT CABLE SUBJECTED TO MASS IMPACT

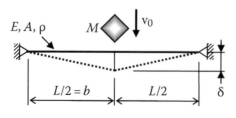

$M > Lm$

$$\delta = \left(\frac{ML^3 v_0^2}{4EA} \right)^{1/4}$$

$e = \dfrac{\delta^2}{L}; \, \varepsilon_m = \dfrac{2\delta^2}{L^2} = 2v_0 \left(\dfrac{M}{4EAL} \right)^{1/2}$; elongation and maximum strain, respectively, of

one-half.

$$P_m = \frac{2EA}{L} e = EA\varepsilon_m; \text{ peak cable force}$$

$$\Pi = \frac{bP^2}{2EA}; \text{ strain energy in one-half of the cable}$$

Note: Refer to Case 9.2 for a static loading with force W at center. This case does not address the possibility of a failure near the impact point. As M decreases to less than mL, the accuracy of the above formulas also decreases.

CASE 9.6   A CABLE WITH ELASTICALLY RESTRAINED ENDS, SUBJECTED TO MASS IMPACT

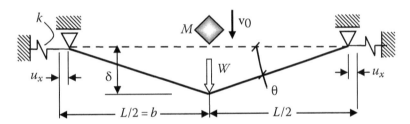

$M > Lm$

b measured along the initial, undeflected position.

(a) Statics, under an applied force W

$$W = F(\tan\theta - \sin\theta) \approx F\frac{\theta^3}{2}; \; F = \frac{2bk}{1 + bk/EA}$$

$$\theta^3 \approx \frac{2W}{F}; \; P = \frac{W}{2\sin\theta}; \text{ cable force}$$

$$H = P\cos\theta = \frac{W}{2\tan\theta}; \text{ spring stretching force}$$

$u_x = H/k$; spring elongation

$$\Pi = \frac{W^2}{8\sin^2\theta} \left(\frac{\cos^2\theta}{k} + \frac{1}{K} \right); \text{ with } K = \frac{EA}{b}; \text{ accumulated strain energy, one-half}$$

$$\cos\theta = \frac{b - u_x}{b + e}; \text{ checking equation, where } e = \frac{Pb}{EA}; \text{ elongation of one-half}$$

$$\delta = (b + e)\sin\theta = \left(1 + \frac{P}{EA}\right)b\sin\theta$$

(b) Mass impact with kinetic energy $E_k = Mv_0^2/2$

$$\frac{E_k}{2} = \frac{P^2}{2}\left(\frac{\cos^2\theta}{k} + \frac{1}{K}\right); \quad ') \text{ equating input kinetic energy with strain energy at the}$$

reversal point

$$P = k*b\left(\frac{1}{\cos\theta} - 1\right); \quad '') \text{ with } \frac{1}{k*} = \frac{1}{k} + \frac{1}{K}$$

$$P \approx v_0\sqrt{MK/2}; \text{ peak cable load when } K \ll k, \text{ usual situation}$$

$$\frac{1}{\cos\theta} = 1 + \frac{P}{k*b}; \text{ to determine } \theta$$

$H = P\cos\theta$; force applied to the restraining spring

Note: Equations ') and '') are exact.

CASE 9.7   LONG CABLE, INITIALLY STRAIGHT AND STRESS-FREE, ONE POINT DRIVEN LATERALLY WITH PRESCRIBED, CONSTANT VELOCITY v_0

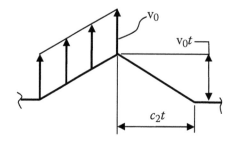

$$c_0 = \left(\frac{EA}{m}\right)^{1/2}; \quad c_2 = \left(\frac{P}{m}\right)^{1/2}; \text{ longitudinal and lateral wave velocities, respectively}$$

$$q^2 + 2\sqrt{c_0}\,q^{3/2} = v_0^2; \text{ to find factor } q$$

$$P = \frac{EA}{c_0}q; \text{ force in cable}$$

$$W_0 = 2P\sin\beta; \text{ associated driving force}$$

Note: The validity of the above was tested only for $v_0 < c_2$, the latter found from the above calculation. A support or any other discontinuity of the cable must be far enough, so that the head of the longitudinal wave traveling with c_0 speed cannot return before the observation time ends.

CASE 9.8 BREAKING OF A LONG CABLE, INITIALLY STRAIGHT, ONE POINT DRIVEN LATERALLY WITH CONSTANT VELOCITY V_0

(Refer to the illustration in Case 9.7)

Failure when strain reaches ε_u.

P_u is peak cable force associated with ε_u.

$$v_{br} = \left(\frac{P_u}{m}(2\varepsilon_u + \varepsilon_u^2) \right)^{1/2}$$

Note: This is for a cable assumed to remain elastic until failure.

CASE 9.9 INELASTIC CABLE BETWEEN TWO SUPPORTS, INITIALLY STRAIGHT, UNDER DISTRIBUTED LOAD $w \sim N/mm$

(Refer to the illustration in Case 9.1)

Rigid-plastic material with σ_0.

(a) Statics $P_y = A\sigma_0$; stretching force, remains constant.

$$\delta \approx \frac{wL^2}{8P_0}; \text{ maximum deflection; } \alpha \approx \frac{4\delta}{L}$$

$$R = \frac{P_y}{w}; \text{ curvature radius}$$

$$e = 2R(\alpha - \sin\alpha); \text{ total elongation}$$

(b) Impulsive load w_0 **lasting** t_0. Initial velocity at midpoint $v_0 = wt_0/m$.

$$P_y = A\sigma_0$$

$$e = \frac{1}{4}\frac{w_0^2 t_0^2 L}{P_0 m}; \text{ elongation of one-half}$$

$$\delta = e\left(1 + \frac{L}{e}\right)^{1/2} \approx \sqrt{Le}; \text{ maximum deflection}$$

Note: Comments for Case 9.1 apply. Sine mode of deflection was employed.

CASE 9.10   INELASTIC CABLE BETWEEN TWO SUPPORTS, INITIALLY STRAIGHT UNDER POINT LOAD $W \sim N$

Rigid-plastic material with σ_0.

(Refer to the illustration in Case 9.2)

(a) Statics

$P_0 = A\sigma_0$; constant stretching force.

$\delta \approx \dfrac{WL}{4P_0}$; maximum deflection

$e = \dfrac{\delta^2}{L}$; elongation of one-half

(b) Impulsive load W_0 lasting t_0

$P_y = A\sigma_0$

$M_{ef} = 0.15(mL)$; effective center mass

$e = \dfrac{(W_0 t_0)^2}{4M_{ef}P_y}$; elongation of one-half

$\delta \approx \sqrt{Le}$; maximum deflection

Note: Comments for Case 9.2 apply. Kink mode of deflection was employed.

CASE 9.11   RIGID-PLASTIC LONG CABLE, INITIALLY STRAIGHT, ONE POINT DRIVEN LATERALLY WITH PRESCRIBED, CONSTANT VELOCITY v_0

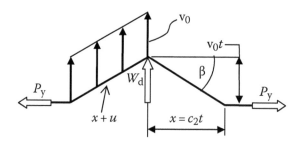

The preload P_y is equal to yield load, $P_y = A\sigma_0$

$c_2 = \left(\dfrac{P_y}{m}\right)^{1/2} = \left(\dfrac{\sigma_0}{\rho}\right)^{1/2}$; lateral wave velocity

$\tan\beta = \dfrac{v_0}{c_2}$

$W_d = 2P_y \sin\beta$; associated driving force

$$\varepsilon = \frac{u}{x} = \frac{1}{\cos\beta} - 1; \text{ strain } (u \text{ is the elongation of one-half of cable})$$

Note: A support or any other discontinuity of the cable must be far enough, so that the head of the lateral wave traveling with c_2 speed cannot return before the observation time ends.

CASE 9.12   RIGID-PLASTIC CABLE, INITIALLY STRAIGHT, PLACED ON RIGID-PLASTIC FOUNDATION

One point driven laterally with a force W. The preload P_y is equal to yield load. This is a quasistatic approximation of the problem. (One-half of cable is shown.)

$$c_2 = \left(\frac{P_y}{m}\right)^{1/2} = \left(\frac{\sigma_0}{\rho}\right)^{1/2}; \text{ lateral wave velocity}$$

$$u' \equiv \theta = \frac{w_y}{P_y} x \text{ and } u = \frac{w_y x^2}{2P_y}; \text{ slope and deflection,}$$
respectively

$\xi = c_2 t$; half-length of deformed cable

$$\delta = \frac{W^2}{8P_0 w_0} = \frac{w_y t^2}{2m}; \text{ peak deflection, } W = \sqrt{8P_y w_y \delta}; \text{ driving force}$$

$$\lambda = \left(\frac{w_y}{P_y}\right)^2 \frac{\xi^3}{6} = \frac{W^3}{48P_y^2 w_y}; \text{ elongation of a half-cable} \sim m, \text{ Ref. [116].}$$

Note: Here the time t is a parameter merely indicating how far the deformation had spread.

CASE 9.13   IMPACT OF THE PROJECTILE M ON A RIGID-PLASTIC CABLE, INITIALLY STRAIGHT

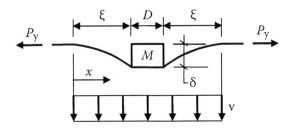

The preload P_y is equal to yield load, $P_y = A\sigma_0$.

$$c_2 = \left(\frac{P_y}{m}\right)^{1/2} = \left(\frac{\sigma_0}{\rho}\right)^{1/2}; \text{ lateral wave velocity}$$

$\xi_f = c_2 t_f$; half-length of deformed cable at the end of impact

$$v(t) = \frac{v_0}{1 + \beta + 2\beta c_2 t / D} \quad \text{with } \beta = \frac{mD}{M}; \text{ velocity (the same for all points) and mass ratio}$$

(Time t counts from the instant the segment of length D has acquired its initial velocity $v_0/(1 + \beta)$.)

$$a(t) = \frac{-2\beta c_2 v_0}{D(1 + \beta + 2\beta c_2 t / D)^2}; \text{ acceleration, } a_m = \frac{2\beta c_2 v_0}{D(1 + \beta)^2}; \text{ peak acceleration}$$

$$\delta(t) = \frac{Dv_0}{2\beta c_2} \ln\left(1 + \frac{2\beta}{1 + \beta} \frac{c_2 t}{D}\right); \text{ center deflection}$$

$$\theta_m = \frac{v_0}{c_2} \frac{1}{1 + \beta}; \text{ peak slope (near } M)$$

$$\varepsilon_m = \frac{1}{2}\left(\frac{v_0}{c_2}\right)^2 \frac{1}{(1 + \beta)^2}; \text{ peak strain (near } M\text{), Ref. [107]}.$$

Note: The peak strain is based on the local slope calculation. When point contact is involved, that is, $D = 0$, replace β/D with m/M and set $\beta = 0$.

CASE 9.14   IMPACT OF THE PROJECTILE M ON A RIGID-PLASTIC CABLE, INITIALLY STRAIGHT, PLACED ON A RIGID-PLASTIC FOUNDATION

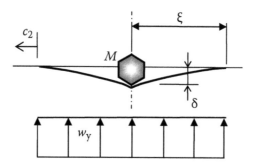

The preload P_y is equal to yield load.

$$c_2 = \left(\frac{P_y}{m}\right)^{1/2} = \left(\frac{\sigma_0}{\rho}\right)^{1/2}; \text{ lateral wave velocity}$$

$\xi_f = c_2 t_f$; half-length of deformed cable at the end of impact

$$t_f = \sqrt{\frac{Mv_0}{c_2 w_y}}; \text{ duration of motion, } \mu = \frac{2m\xi_f}{M}; \text{ mass ratio}$$

$$v = \frac{1 - t^2/t_f^2}{1 + \mu(t/t_f)} v_0; \text{ velocity of } M \text{ as function of time}$$

$$\delta_f = \frac{v_0 t_f}{\mu} \left[\frac{\mu^2 - 1}{\mu^2} \ln(1 + \mu) + \frac{1}{\mu} - \frac{1}{2} \right]; \text{ final peak deflection}$$

$$\delta_f = \frac{2}{3} v_0 t_f; \text{ deflection limit for } \mu \to 0, \text{ that is, a very large } M \text{ or } v_0$$

$$\delta_f = \frac{1}{2} v_0 t_f; \text{ deflection limit for } \mu = 1$$

$$W_d \approx \sqrt{8 P_y w_y \delta_f}; \text{ peak contact force between cable and } M$$

$$\lambda \approx \left(\frac{w_y}{P_y} \right)^2 \frac{\xi_f^3}{6}; \text{ elongation of cable (m)}$$

Note: The expressions for W_d and λ come from quasistatic approximation, Case 9.12. Those results are valid for $\mu = 1$ and are reasonably accurate for $0.5 < \mu < 5.0$, Ref. [116].

EXAMPLES

EXAMPLE 9.1 NATURAL FREQUENCY OF A PRE-STRETCHED CABLE

In Case 2.7 the natural period τ of a cable with length L, pre-stretched with a force P_0 was found from the equation of motion. In the section "Traveling waves, vibration, and a standing wave concept" of Chapter 5, the relations between those quantities were presented. Check, if that relation produces the same result as given by Case 2.7.

If the lateral motion results from a disturbance spreading from the center, then the path the disturbance must traverse is $L/2$. When the wave speed is c_2, the traversing time is

$$t(L/2) = \frac{L}{2c_2} = \frac{L}{2}\sqrt{\frac{\rho}{\sigma_0}} = \frac{L}{2}\sqrt{\frac{m}{P_0}}$$

because $m = A\sigma$ and $P_0 = A\sigma_0$. By definition, this time is one-quarter of the natural period of lateral vibrations as given by Case 2.7.

EXAMPLE 9.2 CABLE UNDER UNIFORMLY DISTRIBUTED LOAD

The material is aluminum with a round solid section, $\phi = 3.18\,\text{mm}$, $L = 100\,\text{mm}$, and the following properties: $E = 69,000\,\text{MPa}$, $\rho = 0.0027\,\text{g/mm}^3$, structural $A = 7.942\,\text{mm}^2$, and the total area twice as large, so that $m = 2A\rho = 0.04288$.

(a) Apply uniform loading $w_0 = 100\,\text{N/mm}$ lasting $t_0 = 0.05\,\text{ms}$ for the material as described and
(b) to the same cable made of elastic-perfectly plastic (EPP) material, with $\sigma_0 = 200\,\text{MPa}$.

Determine the elongation, the center deflection, and the peak cable axial force.

Note that $EA = 69,000 \times 7.942 = 548,000$.

(a) Use Case 9.2. The elongation of one-half is

$$e = \frac{1}{2}\frac{w_0 t_0 L}{\sqrt{EAm}} = \frac{1}{2}\frac{100 \times 0.05 \times 100}{\sqrt{548,000 \times 0.04288}} = 1.6309\,\text{mm}$$

$$\delta = e\left(1 + \frac{L}{e}\right)^{1/2} = 12.87\,\text{mm; center deflection}$$

$$P_{\text{m}} = \frac{2EA}{L}e = \frac{2 \times 548,000}{100}1.6309 = 17,875\,\text{N; peak cable force}$$

(b) Use Case 9.9. The constant force in cable: $P_y = A\sigma_0 = 1588\,\text{N}$.

$$e = \frac{1}{4}\frac{w_0^2 t_0^2 L}{P_y m} = \frac{1}{4}\frac{(100 \times 0.05)^2 100}{1588 \times 0.04288} = 9.179\,\text{mm}$$

$$\delta = e\left(1 + \frac{L}{e}\right)^{1/2} = 31.66\,\text{mm}$$

The above problems were also solved using LS-Dyna models made up of 100 segments and the results are given in brackets below. The energy method gives reasonably accurate results.

	Linear Elastic Material	EPP Material
δ_m (mm)	12.87 (12.124)	31.66 (30.51)
P_m (N)	17,875 (17,930)	—

EXAMPLE 9.3   CABLE UNDER A POINT LOAD AT CENTER

The same cable to be used as in Example 9.2.

(a) Apply $W_0 = 5000\,\text{N}$ lasting $t_0 = 0.05\,\text{ms}$ to the cable made of linear elastic material and

(b) $W_0 = 2000\,\text{N}$ lasting $t_0 = 0.05\,\text{ms}$ to EPP material.

Determine: Elongation, peak cable force, and maximum center deflection.

(a) Use Case 9.2. $EA = 548{,}000$, as before.

$M_{ef} = 0.15(mL) = 0.15 \times 0.04288 \times 100 = 0.6432\text{g}$; effective mass for cable load

$$e = \sqrt{\frac{b}{EA}}\frac{W_0 t_0}{\sqrt{2M_{ef}}} = \sqrt{\frac{50}{548{,}000}}\frac{5000 \times 0.05}{\sqrt{2 \times 0.6432}} = 2.105\,\text{mm};\ \text{elongation of one-half}$$

$$P_m = \frac{2EA}{L}e = \frac{2 \times 548{,}000}{100}2.105 = 23{,}075\ \text{N};\ \text{peak cable force}$$

$M_{ef}' = mL/3 = 1.4293\ \text{g}$; effective mass for peak deflection

$$e' = \sqrt{\frac{b}{EA}}\frac{W_0 t_0}{\sqrt{2M_{ef}'}} = 1.4124\,\text{mm}$$

$\delta \approx \sqrt{Le'} = 11.88\,\text{mm}$; maximum center deflection

(b) Use Case 9.10. The constant force in cable: $P_y = A\sigma_0 = 1588\,N$.

$$e = \frac{(W_0 t_0)^2}{4 M_{ef} P_y} = \frac{(2000 \times 0.05)^2}{4 \times 0.6432 \times 1588} = 2.448 \text{ mm}$$

$$\delta = e\left(1 + \frac{L}{e}\right)^{1/2} = 15.84 \text{ mm}$$

The above problems were also solved using LS-Dyna models made up of 100 segments and the results are given in brackets below. In the elastic case, this simplified method gives a good approximation of the stretching force, but overestimates the deflection. This is due to the fact that the response is not closely related to a simple, assumed deflection pattern.

	Linear Elastic Material	EPP Material
δ_m (mm)	11.88 (8.258)	15.84 (15.67)
P_m (N)	23,075 (22,650)	—

EXAMPLE 9.4   SEMI-INFINITE CABLE UNDER PRESCRIBED TRANSVERSE VELOCITY AT THE END

The cable is the same as in Example 9.2, except lighter, with only the structural area present.
   For geometry refer to Figure 9.3b. The driving velocity is $v_0 = 100\,m/s$. Determine displacement and stress at $t = 0.07$, 0.14, and 0.21 ms after start. Construct an FEA model to indirectly verify the results.

Use Case 9.7. The longitudinal wave speed and mass density, respectively:

$$c_0 = \left(\frac{E}{\rho}\right)^{1/2} = \left(\frac{69,000}{0.0027}\right)^{1/2} = 5,055 \text{ mm/ms}; \quad m = A\rho = 0.02144 \text{ g/mm}$$

$q^2 + 2\sqrt{c_0} q^{3/2} = v_0^2$; from which $q = 16.72$. Noting that $EA = 548,000$

$$P = \frac{EA}{c_0} q = \frac{548,000}{5,055} 16.72 = 1,813 \text{ N}$$

This gives the speed of the transverse wave as $c_2 = \left(\frac{P}{m}\right)^{1/2} = \left(\frac{1813}{0.02144}\right)^{1/2} = 290.8 \text{ m/s}$

The horizontal position of the kink is determined by

$$u_x = c_2 t = 290.8t$$

and it is tabulated below. Also, $\tan \beta = v_0/c_2 = 100/290.8$, then $\sin \beta = 0.3252$ and $W_0 = 2P \sin \beta = 2 \times 1813 \times 0.3252 = 1179\,\mathrm{N}$. (Only one-half of this drives the half-cable.)

The vertical kink position, for constant vertical velocity is simply $u_y = v_0 t$.

To verify this result numerically, an FE model was constructed and executed using LS-Dyna code. The cable was 100 mm long and had 100 segments. The left end was driven vertically and restrained horizontally. The right end was constructed as a nonreflecting boundary, which, according to Equation 5.44, involves a damper with a constant C of

$$C = A\rho c_0 = 7.942 \times 0.0027 \times 5055 = 108.4\,\mathrm{N\ ms/mm}$$

This assures non-reflectivity only for longitudinal waves, because the observation time is too short to experience any lateral wave reflection. The motion of the left end was maintained for 0.25 ms. The output axial force was 1883 N (nearly 4% above the calculated result), almost constant during the first 0.25 ms, except for a short, 0.01 ms duration of the initial growth. The position of a kink, or rather intersection of the deflected and undeflected lines is tabulated as $u(D)$. It can be seen that the motion from FEA is somewhat faster than that from the analytic solution, although the relative difference decreases with time. ($u(D)$ was read for t + 0.01 ms to compensate for the initial speeding up from rest. If the results were read without that lag, there would be a little better agreement between both sets of numbers.)

t (ms)	u_x (mm)	u (D)
0.07	20.36	22.7
0.14	40.71	42.6
0.21	61.07	62.4

EXAMPLE 9.5 INITIAL END REACTION OF A CABLE UNDER IMPULSIVE, DISTRIBUTED LOAD

Use the same cable as in Example 9.2, but for the elastic material only. Apply the same impulse, but more abruptly: $w_0 = 1000\,\mathrm{N/mm}$ lasting $t_0 = 0.005\,\mathrm{ms}$. Treating the cable as put in motion by the initial velocity, calculate the initial load level in the cable in accordance with Case 9.7.

The speed of sound in the cable, which has the effective density of $2 \times 0.0027 = 0.0054\,\mathrm{g/mm^3}$ is

$$c_0 = \left(\frac{E}{\rho}\right)^{1/2} = \left(\frac{69,000}{0.0054}\right)^{1/2} = 3,575\ \mathrm{m/s}$$

$$v_0 = w_0 t_0/m = 1000 \times 0.005/0.04228 = 116.6\ \mathrm{m/s}$$

To use Case 9.7 find $2\sqrt{c_0} = 119.6$ and substitute

$q^2 + 119.6q^{3/2} = 116.6^2$; which results in $q = 22.86$; then

$$P = \frac{EA}{c_0}q = \frac{548,000}{3,575}22.86 = 3,504\ \mathrm{N}$$

When the FEA was done for this problem, the first load plateau was attained at 3660 N, 4% more than using the analytical approach. Then, in four jumps the load peaked at 18,603 N. The interval between the jumps was noted to be close to 0.03 ms. This interval is explained by the traversing time for the head of the wave, from the end to the center (where the opposing waves meet), and back, $L/c_0 = 0.028$ ms ≈ 0.03 ms.

EXAMPLE 9.6   LONG PRESTRESSED CABLE UNDER PRESCRIBED TRANSVERSE FORCE AT A POINT

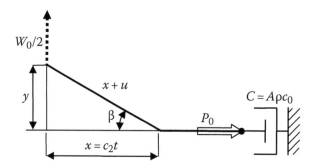

(One-half shown)

The cable is the same as in Example 9.4, namely $E = 69,000$ MPa, $\rho = 0.0027$ g/mm³, $L = 100$ mm, $A = 7.942$ mm² (structural area only.) The preload is $P_0 = 5000$ N and the applied step load is $W_0/2 = 1000$ N.

Determine the transient displacement as a function of time. Construct an FEA model to indirectly verify the results.

Taking data from Example 9.4:

$$c_0 = 5,055 \text{ mm/ms}; \quad m = A\rho = 0.02144 \text{ g/mm}; \quad EA = 548,000 \text{ N}$$

Our preload gives the following speed of the transverse wave:

$$c_2 = \left(\frac{P}{m}\right)^{1/2} = \left(\frac{5000}{0.02144}\right)^{1/2} = 482.9 \text{ m/s}$$

The horizontal position of the kink is determined by $x = c_2 t = 482.9t$. The same FE model, as described in Example 9.4, was used except for addition of P_0. The preload was applied first and, after the axial force stabilized at P_0, the lateral step $W_0/2 = 1000$ N load was applied at 0.15 ms. The deflection snapshot showed a straight line, as on this sketch. At time $t = 0.26$ ms, the output was $x \approx 56.76$ mm, $y = 9.84$ mm. According to what was derived above, $x = 482.9(0.26 - 0.15) = 53.12$ mm. Using the quasistatic approach to find the lateral deflection, one puts $L/2 = x = 53.12$ mm as cable length and $W_0 = 2000$ N as the applied load to use in the static formula of Case 9.2:

$$\delta \approx \frac{L}{2}\left(\frac{W_0}{EA}\right)^{1/3} = 53.12\left(\frac{2,000}{548,000}\right)^{1/3} = 8.18 \text{ mm}$$

The reason for the discrepancy was implied in the text. The ratio P_0/W_0 is not sufficiently large to cause only a negligible increase in the axial load. In fact, the latter was observed to jump to 5850 N in the deformed part and persist near that level. This caused an increase in the length swept by the wave of deformation as well as in the calculated δ.

EXAMPLE 9.7   PRELOADED CABLE UNDER UNIFORMLY DISTRIBUTED LOAD

The same cable with the same load will be used as in Example 9.2. However, the cable is now preloaded to 150 MPa. Determine peak deflections for the elastic and inelastic materials.

Use Case 9.4: Initial elongation is $e_i = \dfrac{\sigma_i L}{2E} = \dfrac{150 \times 100}{2 \times 69,000} = 0.1087$

(a) *Elastic cable*:

$$e^2 - e_i^2 = \frac{1}{4}\frac{w_0^2 t_0^2 L^2}{EAm} = \frac{1}{4}\frac{(100 \times 0.05 \times 100)^2}{548,000 \times 0.04288} = 2.66 \text{ mm; Hence } e = 1.6345 \text{ mm}$$

$$\delta = e\left(1 + \frac{L}{e}\right)^{1/2} = 12.89 \text{ mm; center deflection}$$

$$P_m = \frac{2EA}{L}e = \frac{2 \times 548,000}{100}1.6345 = 17,914 \text{ N peak cable force}$$

(b) *Inelastic cable*: The constant stretching force: $P_y = A\sigma_0 = 1588$ N.

$$e = \frac{1}{4}\frac{w_0^2 t_0^2 L}{P_y m} + \frac{e_i}{2} = \frac{1}{4}\frac{(100 \times 0.05)^2 100}{1588 \times 0.04288} + \frac{0.1087}{2} = 9.233 \text{ mm}$$

$$\delta = e\left(1 + \frac{L}{e}\right)^{1/2} = 31.76 \text{ mm}$$

In both cases there is only a small difference between the above results and those of Ex. 9.2.

EXAMPLE 9.8   CABLE IMPACTED BY A FLYING MASS

An initially straight cable with $L = 12$ m, $E = 150,000$ MPa, $A = 60$ mm^2 is impacted, mid-length, by a mass of 600 kg flying with a velocity of 75 m/s. Determine maximum deflection and strain as well as the equivalent static loading W to produce the same deflection.

Use Case 9.5. The center deflection is

$$\delta = \left(\frac{ML^3 v_0^2}{4EA}\right)^{1/4} = \left(\frac{600,000 \times 12,000^3 \times 75^2}{4 \times 150,000 \times 60}\right)^{1/4} = 3,568 \text{ mm}$$

$$\varepsilon_m = \frac{2\delta^2}{L^2} = \frac{2 \times 3568^2}{12,000^2} = 0.1768; \quad P_m = EA\varepsilon_m = 150,000 \times 60 \times 0.1768 = 1,591,330 \text{ N}$$

To get a static load W corresponding to this displacement, rewrite an equation in Case 9.2:

$$W \approx 4P\delta/L = 4 \times 1.591 \times 10^6 \times 3,568/12,000 = 1.892 \times 10^6 \text{ N}$$

EXAMPLE 9.9   ELASTICALLY RESTRAINED CABLE IMPACTED BY A FLYING MASS

A cable described in Example 9.8 has a horizontal restraining spring at each end with the stiffness of $k = 15,000\,\text{N/mm}$. Determine maximum deflection and strain from the same flying mass.

Use Case 9.6. One-half kinetic energy is $E_k/2 = 600,000 \times 75^2/4 = 0.8438 \times 10^9$ N mm.

The half-cable stiffness is $K = EA/b = 150,000 \times 60/6,000 = 1,500\,\text{N/mm}$.

The combined stiffness k^* is found from a series connection formula:

$$\frac{1}{k^*} = \frac{1}{k} + \frac{b}{EA} = \frac{1}{15,000} + \frac{1}{1,500}; \quad \text{then } k^* = 1363.6\,\text{N/mm}$$

Noting that $k \gg K$, find P from the approximate formula:

$$P \approx v_0\sqrt{MK/2} = 75\sqrt{600,000 \times 1,500/2} = 1,591,000 \text{ N}$$

The maximum angle is found from

$$\frac{1}{\cos\theta} = 1 + \frac{P}{k^* b} = 1 + \frac{1.591 \times 10^6}{1363.6 \times 6000}, \text{ then } \theta = 33.16°$$

The corrected value of P is then obtained from the expression for the absorbed energy:

$$\Pi = \frac{P^2}{2}\left(\frac{\cos^2\theta}{k} + \frac{1}{K}\right) \quad \text{or} \quad 0.8438 \times 10^9 = \frac{P^2}{2}\left(\frac{\cos^2 33.16°}{15,000} + \frac{1}{1,500}\right)$$

which gives $P \approx 1,538,000\,\text{N}$, marginally smaller result than from the initial approximation:

$$\delta = \left(1 + \frac{P}{EA}\right)b\sin\theta = \left(1 + \frac{1.538 \times 10^6}{150,000 \times 60}\right)6,000\sin 33.16 = 3,843 \text{ mn}$$

The equivalent static load W to induce P of this magnitude is found from

$$W = 2P\sin\theta = 2\times 1,538,000\times\sin 33.16° = 1.683\times 106\ \mathrm{N}$$

$$\varepsilon = \frac{P}{EA} = \frac{1.538\times 10^6}{150,000\times 60} = 0.1709;\ \text{cable elongation}$$

The deflections are larger and cable stretching is smaller, compared to the previous example, but the differences are minor due to a relatively large end-spring stiffness. If that stiffness was much smaller, another iteration, beginning with the energy balance, would be needed.

10 Beams

THEORETICAL OUTLINE

A beam is a structural element in which the dimension of length is much larger than the width and depth. When viewed in three dimensions, there are six components of internal beam forces: axial load, two bending moments, two shears, and a torque. The most frequent application is that of plane deformation, where only three components exist. Bending is thought to be a typical mode of deformation, which is especially true for small deflections. The shearing stress, particularly near the ends, or near the point-load application, can be quite important, even more so in dynamics than under static loading. The circumstances where the axial load is of primary concern are quantified in Chapter 11.

BASIC RELATIONSHIPS

Pure Bending

A fragment of a bent beam is shown in Figure 10.1. The usual assumptions are made: plane sections remain plane and longitudinal fibers do not interact. The relations between strain ε and curvature radius ρ are as follows:

$$\varepsilon = \frac{y}{\rho}; \quad \varepsilon_1 = \frac{h_1}{\rho}; \quad \varepsilon_2 = -\frac{h_2}{\rho} \tag{10.1}$$

A moment about the neutral axis originating from an incremental area of width b is $(b\mathrm{d}y)$ σy; consequently, the entire bending moment is

$$\mathcal{M} = \int_{-h_1}^{h_2} \sigma b y \mathrm{d}y \tag{10.2}$$

The linear material, for which Hooke's Law, $\sigma = E\varepsilon$ applies, is the easiest to quantify. Substitution into the above gives

$$\mathcal{M} = \frac{E}{\rho} \int_{-h_1}^{h_2} y^2 b \mathrm{d}y \quad \text{or} \quad \frac{1}{\rho} = \frac{\mathcal{M}}{EI} \quad \text{where} \quad I = \int_{-h_1}^{h_2} y^2 b \mathrm{d}y \tag{10.3a,b,c}$$

Section constant I is known as the *second area moment* or the *section moment of inertia*. When Equation 10.3b is combined with Equation 10.1, the elastic bending stress equation results:

$$\sigma = \frac{\mathcal{M}y}{I} \tag{10.4a}$$

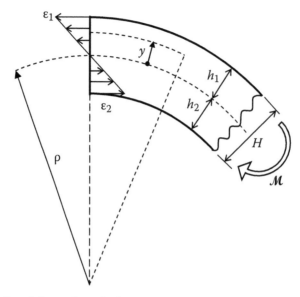

FIGURE 10.1 Bending deformation of a beam.

or

$$\varepsilon = \frac{\mathcal{M}y}{EI} \tag{10.4b}$$

This is valid only up to the nominal yield stress of F_y. The elastic moment associated with this stress level at the extreme fiber $y = c$ marks the beginning, or onset, of yielding:

$$\mathcal{M}_y = \frac{IF_y}{c} \tag{10.5}$$

The other important material model is the rigid plastic (RP) or rigid-perfectly plastic, as described in Chapter 4. The stress is either σ_0 or $-\sigma_0$, (flow strength), so the integration of Equation 10.2 gives

$$\mathcal{M}_0 = \sigma_0 \int_{-h_1}^{h_2} bydy \quad \text{or} \quad \mathcal{M}_0 = Z\sigma_0 \quad \text{where} \quad Z = \int_{-h_1}^{h_2} bydy \tag{10.6a,b,c}$$

The constant Z is usually called the *first area moment*. The described case is an ideal situation where, upon attaining $\mathcal{M}_0$, the bent section does not increase its resistance with the increase of strain. $\mathcal{M}_0$ is referred to as the fully plastic moment, or the plastic moment capacity of a section. Adoption of this material leads to the concept of a plastic joint, to be discussed later.

The geometries of some of selected common sections are shown in Figure 10.2. Their properties are defined in Table 10.1.

The relation between $\mathcal{M}$ and σ, expressed by Equations 10.4, is very simple only for a linear material. The two-flange sections in Table 10.1, are exceptional in this regard, because their moment–stress relations are statically determinate. From equilibrium of the symmetrical,

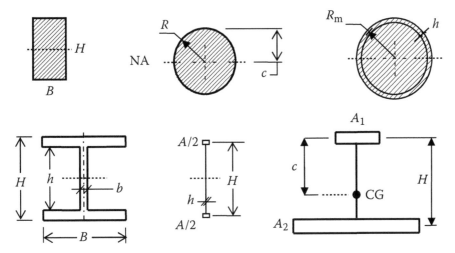

FIGURE 10.2 Some common sections are shown with the neutral axis of bending marked on each section.

TABLE 10.1
Properties of Some Common Sections

Sections		A	I	Z	α	ψ
	Rectangle	BH	$\dfrac{BH^3}{12}$	$\dfrac{BH^2}{4}$	3/2	6/5
	Solid circle	πR^2	$\dfrac{\pi}{4}R^4$	$\dfrac{4}{3}R^3$	4/3	10/9
	Tube	$2\pi R_m h$	$\pi R_m^3 h$	$4R_m^2 h$	2	2
	I—section	$BH - (B-b)h$	$\dfrac{1}{12}\left[BH^3 - (B-b)h^3\right]$	$\dfrac{1}{4}\left[BH^2 - (B-b)h^2\right]$	$\dfrac{A}{bH}$	$\dfrac{A}{bH}$
	TFS	A	$\left(\dfrac{H}{2}\right)^2 A$	$\dfrac{AH}{2}$	1.0	1.0
	TFUS	$A_1+A_2 = A_1(1+p)$	$\dfrac{A_2 H^2}{1+p}$	$\dfrac{2HA_2}{1+p}$	1.0	1.0

two-flange section, one can write $\mathcal{M} = (\sigma A/2)H$, thereby demonstrating the proportionality between stress and moment for *any* material. Once the stress σ is established, the corresponding strain ε can be found from the appropriate σ–ε curve. The yield and the ultimate bending capacities, respectively, can be determined for this section from the following:

$$\mathcal{M}_y = \frac{1}{2}AHF_y \quad \text{and} \quad \mathcal{M}_m = \frac{1}{2}AHF_u \tag{10.7}$$

where F_u is the ultimate strength of the material. The section fails when the ultimate moment $\mathcal{M}_m$ is reached, as dictated by the breaking (or buckling) strain $\varepsilon = \varepsilon_m$.

For a bilinear material, the expressions relating to a rectangular section (Case 10.1) are still manageable, but for other material models they become much more involved, as was demonstrated by this author [85]. Yet, in the same reference it is shown that assuming a bilinear moment–curvature relation provides a reasonably accurate solution.

When the applied load is large enough to cause the applied strain to be an order of magnitude larger than the yield strain ε_y, the material can be treated as rigid strain hardening (RHS) and the bending moment has two distinct components (which becomes evident when stress distribution is sketched):

$$M = (\sigma - F_y)\frac{I}{c} + ZF_y \tag{10.8}$$

where F_y is the yield strength. This is true for all sections listed here. When σ reaches F_u or the ultimate material strength, then the moment reaches its maximum as well, $\mathcal{M} = \mathcal{M}_m$.

Section Constants

Our attention will be focused on the most common sections, which are symmetrical about the neutral axis, but with one exception.

For the last section, $p = A_2/A_1$ and $c = Hp/(1+p)$, measured as in Figure 10.2. The constant Z, the second area moment, which is much less often mentioned than I, can be found, for example, in Szuladziński [86]. The last two coefficients in Table 11.1 relate to the peak shear stress (α) and shear stiffness (ψ).

The uniform stress due to a tensile force P is simply expressed as

$$\sigma = P/A \tag{10.9}$$

while the same uniformity does not usually hold for shear stress τ* due to the shear force Q, where the maximum stress is larger than the average:

$$\tau_m = (Q/A)\alpha \tag{10.10}$$

The other shear coefficient is related to stiffness. To calculate shear deflection one uses shear area A_s, instead of the section area A:

$$A_s = A/\psi \tag{10.11}$$

For the two *I*-sections, defining the coefficients α and ψ as was done in the Table 10.1 is equivalent to saying that the entire web should be used as a shear area. (This is only an approximate formula; for more details see Young [118].)

The symmetric "two-flange" section is idealized: it consists of two flanges resisting bending and a web working in shear only. (In spite of this limitation the results obtained using this beam

---

* It is customary to use τ as the symbol for shear stress in relevant textbooks. Although it is identical to the natural period designation introduced before, the possibility of confusion is small.

differ only a little from those of an industrial *I*-beam.) The shear coefficients are stated as 1.0 with the understanding that Q is applied only to the web.

Static and Dynamic Properties of Elastic Beams

The following end conditions will be considered: CF or clamped-free (cantilever), SS or simply supported, SC or simply supported and clamped, CC or clamped–clamped, and CG or clamped-guided; all of these illustrated in Figure 10.3. (Clamping means removing all motion, except that at least one end of a beam must be unconstrained in the axial direction.) The beams are under a distributed load w_0 or under a point load W_0. (The point load is considered in the next section, but it is convenient to make a general summary now.) The maximum static deflections can be written as the sum of the flexural and shear components:

$$u = u_b + u_s \qquad (10.12a)$$

or

$$u = b\frac{w_0 L^4}{EI} + s\frac{w_0 L^2}{GA_s} \qquad (10.12b)$$

or

$$u = b\frac{W_0 L^3}{EI} + s\frac{W_0 L}{GA_s} \qquad (10.12c)$$

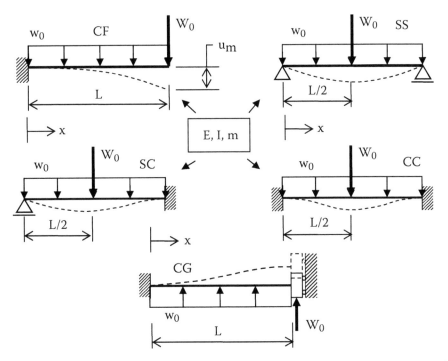

FIGURE 10.3 End conditions for the beams under consideration. (Note that the point load W and the distributed load w are not usually applied simultaneously in this chapter.)

TABLE 10.2

Deflection, Shear, and Moment Coefficients

(End Conditions in Figure 10.3)

Type/Load	b	s	Q	$\mathcal{M}$
CF/w_0	1/8	1/2	$w_0 L$	$(1/2)w_0 L^2$
SS/w_0	5/384	1/8	$(1/2)w_0 L$	$(1/8)w_0 L^2$
SC/w_0	1/185	0.1953	$(5/8)w_0 L$	$(1/8)w_0 L^2$
CC/w_0	1/384	1/8	$(1/2)w_0 L$	$(1/12)w_0 L^2$
CG/w_0	1/24	1/2	$w_0 L$	$(1/3)w_0 L^2$
CF/W_0	1/3	1	W_0	$W_0 L$
SS/W_0	1/48	1/4	$(1/2)W_0$	$(1/4)W_0 L$
SC/W_0	0.00932	11/32	$(11/16)W_0$	$(3/16)W_0 L$
CC/W_0	1/192	1/4	$(1/2)W_0$	$(1/8)W_0 L$
CG/W_0	1/12	1	W_0	$(1/2)W_0 L$

Notes: For SC beam, the maximum shear force is near the clamped end in both cases and so is the maximum (negative) moment quoted above. The same situation exists for the CC beam under distributed loading. (For the latter support case and the point load W_0, $\mathcal{M}$ is equal, in absolute values, at the loaded point and at a support.)

where the coefficients b and s are shown in Table 10.2 along with the largest shear force Q and the bending moment $\mathcal{M}$. (Bending deflections, shear forces, and moments come from Ref. [118], while shear deflections were calculated by this author.) The symbol w_0 refers to distributed loading and W_0 to point loads.

The stiffness k of any of the beams under consideration is defined as the total load divided by the maximum deflection, $w_0 L/u$ or W_0/u. For slender beams of compact cross sections, one can ignore the contribution of shear to the overall deflections. In this case, on the basis of Equation 10.12, one has

$$k = \frac{1}{b}\frac{EI}{L^3} \tag{10.13}$$

The natural frequency f or the circular frequency $\omega = f/(2\pi)$ is an important reference property, but the natural period $\tau = 2\pi/\omega$ is more useful in impact studies. Using the expressions developed in Chapter 2 one can write, in the absence of shear deformation,

$$\tau = \varepsilon L^2 \sqrt{\frac{m}{EI}} \tag{10.14}$$

where the coefficient ε is shown in Table 10.3.

TABLE 10.3
The Coefficient ε for Natural
Period Calculation

Type	ε
CF	1.787
SS	$2/\pi$
SC	1/2.454
CC	1/3.561
CG	1.1234

DISTRIBUTED LOADING, ELASTIC RANGE

Estimation of Response to Rectangular Load Pulses

Several time functions of load will be considered, as illustrated in Figure 10.4. The first is a rectangular pulse, characterized by an abrupt application of force, which is then suddenly removed after t_0. If the duration is infinitely long, then the function is called a step load. Another pulse, of triangular shape and increasing magnitude, is typical of many impact situations. The decreasing triangular function can often be used as an approximation of the positive phase of a blast load. Finally, there is a symmetrical triangular pulse, again representative of impact situations. This approach here is based on small deflections and linear material properties; therefore the response is proportional to the load magnitude, for each individual pulse shape. While all of the pulse shapes presented above have some importance, the attention will be focused on the first two.

In order to make the analysis of a beam simpler, it is often useful to treat it, from a dynamic viewpoint, as an SDOF, or an oscillator having mass M supported by a spring with stiffness k. This follows from the basic assumption that all points of the beam move in unison, although with different amplitudes, and therefore only a single time function is needed to describe the motion. A further assumption is that the beam deflects as in a fundamental mode and that the mode is similar to the static deflection under uniform loading. (In reality, it takes some time after load application for the first mode to become dominant. For this reason the accuracy decreases with the decrease of load duration. This will be demonstrated later.)

According to Equation 3.25, the motion of an oscillator, after a rectangular impulse lasting t_0, has an amplitude of

$$u_d = 2u_{st}\sin(\pi t_0/\tau) = 2u_{st}\sin(\omega t_0/2) \tag{10.15}$$

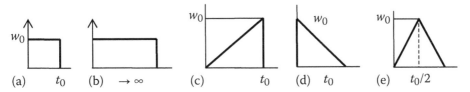

FIGURE 10.4 Some load pulses under consideration: (a) rectangular, (b) step load, (c) triangular increasing, (d) triangular decreasing, and (e) triangular symmetric.

This reaches maximum for $t_0 = \tau/2$, or after a half-period, as well as every odd multiple of the half-period. (Any shorter t_0 makes the peak deflection smaller than $2u_{st}$.) Once the peak deflection u_d is calculated, the peak shear and moment are approximated as

$$Q_d = \frac{u_d}{u_{st}} Q_{st} \tag{10.16a}$$

$$\mathcal{M}_d = \frac{u_d}{u_{st}} \mathcal{M}_{st} \tag{10.16b}$$

When the impulse duration is very short, one can simplify it to an instantaneous impulse S, which imparts the initial velocity v_0 to mass M according to $S = Mv_0$. The peak displacement then becomes

$$u_d = S\omega/k \tag{10.17a}$$

or

$$u_d = 2u_{st}(\omega t_0/2) \tag{10.17b}$$

as found by substituting $S = P_0 t_0$, which is the approximate value of Equation 10.15 for small angles. (At this point it is useful to review the remarks in the section "Response of oscillator to impulsive load.")

The above methods can be compared to another one, in which three coefficients have to be calculated for a particular support case (see, e.g., TM-855-1 [105] or Biggs [9]). Those coefficients (load, mass, and load–mass factors) are tabulated for a few basic cases, but for any nonstandard beam the analyst has to carry out lengthy integrations. The method presented here relies on the natural frequency and a static deflection instead, and in a nonstandard case any mathematics or complications involved relate to finding those quantities only.

Alternative Estimates of Peak Responses

Equations 10.16 often result in underestimates of shear and bending, especially if t_0 is a small fraction of the natural period. An alternative method for the peak shear determination is documented in Ref. [105] and is also used by Biggs [9] and other authors. It is developed using somewhat convoluted reasoning, but it works very well in many cases, especially for very short load durations. Here is the way it is applied. Let $W_0 = w_0 L$ designate the original, short lasting load magnitude. Let W_e be the dynamic load calculated as

$$W_e = (u_d/u_{st})W_0 \tag{10.18}$$

where W_0 is the static resultant. The shear force near a support can be then approximated by

$$Q'_d = cW_e + dW_0 = \left(c\frac{u_d}{u_{st}} + d \right)W_0 \tag{10.19}$$

where the coefficients c and d depend on the end conditions as shown in Table 10.4.

TABLE 10.4
Peak Shear Coefficients, for Distributed Load w and
Point Load W

Beam	c	d	Beam	c	d	Application Point
CF, w	0.68	0.32	CF, W	1.35	−0.35	Tip
SS, w	0.39	0.11	SS, W	0.78	−0.28	Center
SC, w	0.43	0.19	SC, W	0.54	0.14	Center
CC, w	0.36	0.14	CC, W	0.71	−0.21	Center
CG, w	0.78	0.22	CG, W	1.42	−0.42	Guided end

For SC, CC, and CG the shear force so calculated relates to a clamped end. The results from the references quoted above are available for cases SS, SC, and CC. Some slight deviations in comparison with the above values are due to differences in the assumed deflected shapes. Refer to Figure 10.3 for orientation.

Treating a beam support as a discontinuity permits an estimate of the peak bending in a dynamically loaded beam. This approach was already explained in Chapter 9 in relation to a cable. When the applied load is uniform and suddenly imposed on a beam, its supports become a source of discontinuity spreading as a bending stress wave. The estimate of a peak moment $\mathcal{M}$ may be related to the development shown later is the section "Transient beam response due to a suddenly applied point load" by writing

$$\mathcal{M} = qW_0l \tag{10.20}$$

where
q is a coefficient depending on the manner of support
W_0 is the resultant load applied
l is either l_1 or l_2 as given by Equations 10.24 and the relevant time is t_0 or load duration

Specific forms of Equation 10.20 are given in Table 10.5 with the coefficients based on FEA results. If a pulse is of a general shape, its duration t_0 should be taken as that of an equivalent rectangular pulse.

The upper bound of the peak bending moment is the larger of the two estimates, one being Equation 10.16b and the other presented in Table 10.5. The reader should refer to Example 10.2.

TABLE 10.5
Estimates of Peak Bending
Moments

Beam	$\mathcal{M}$
CF	$0.47W_0l_2$
SS	$0.60W_0l_1$
SC	$0.27W_0l_2$
CC	$0.20W_0l_2$

Response to Other Impulse Shapes

A step load (Figure 10.4b) is like a rectangular pulse with $t_0 = \infty$, but one should keep in mind that $t_0 = \tau/2$ is sufficient to reach an extreme response. (See section "Response of oscillator to step loading.") The oscillator has a peak displacement u equal to twice the static one. The other way to say it is that the dynamic factor is

$$DF(u) = 2.0 \tag{10.21}$$

While the same factor is also true for the peak spring force of an oscillator, the spring force does not always relate well to the parameters of the actual system that the oscillator is to represent. As detailed calculations show, anticipating twice the static response not only for u, but also for Q and $\mathcal{M}$, gives a good approximation for uniformly distributed loading.

The oscillator response to a triangular pulse is discussed in detail in Chapter 3. Depending on the shape of a pulse the magnitude of response will vary, but those variations are relatively small and of no great significance for an approximate analysis. The expression for the peak response of an oscillator to a symmetrical pulse, given by Equation 3.31c, is

$$u_{\mathrm{d}} = \frac{\sin^2 x}{x}(2u_{\mathrm{st}}) \tag{10.22}$$

where $x = \pi t_0/(2\tau) = \omega t_0/4$ and u_{st} is calculated as if it was induced by the static loading w_0.

This equation gives the peak of a *residual* response, i.e., history of motion following the entire pulse application. The numerical studies conducted by this author indicate that this compact expression gives a good approximation for all three triangular, relatively short pulses. (The peak attainable $DF(u) = u_{\mathrm{d}}/u_{\mathrm{st}}$ from Equation 10.22 is 1.42. The absolute maximum that can be reached during the pulse is $DF(u) = 1.52$.)

The Accuracy of the SDOF Approach

This can best be appreciated by an FEA study of each case of support. Example 16.1 shows such a study conducted for a rectangular pulse. The accuracy of these predictions strongly depends on the ratio of impulse duration t_0 to the natural period τ. Probably the best approximation is obtained when $t_0/\tau = 0.1$ or thereabouts. For shorter duration impulse the approximation underpredicts, with the biggest errors being for Q, smaller for $\mathcal{M}$, and the smallest for u. (Note the degradation of shear results for very short relative durations applied to a CF beam.) For longer duration times, $0.5 > t_0/\tau > 0.1$, the largest overprediction is for Q, a much smaller for $\mathcal{M}$, and slight underprediction for u. From a practical viewpoint, the concern is mainly for very short durations and the resulting underestimation of the shear force.

As for the peak shear, the alternative method would typically improve accuracy for very short pulses (as per the above considerations), although it could cause overpredictions for longer pulses.

A similar study was conducted by this author for triangular pulses. The increasing pulse was applied to the SS and CC beams and a decreasing pulse to CF and SC. The best approximation was obtained when $t_0/\tau \approx 0.1$, up to 0.25. For a shorter time duration the approximation underpredicts, with the biggest errors being for Q, smaller for $\mathcal{M}$, and the smallest for u. For longer duration times, $0.5 > t_0/\tau > 0.25$, there was a similar pattern, except that overpredictions took place. With regard to shear, the alternative method gave an overestimate, while the oscillator method resulted in an underestimate. The best result was obtained by taking the average of values predicted by both methods.

The step loading would generate a dynamic factor of about 2.0, as mentioned before. A sample run on a model of SS beam, as described in Example 10.2, but with duration of $\tau/2$ gave the following factors, for displacement, moment, and shear, respectively,

$$DF(u) = 2.015; \quad DF(\mathcal{M}) = 2.115; \quad DF(Q) = 1.819$$

POINT LOADS, ELASTIC RANGE

Transient Beam Response due to a Suddenly Applied Point Load

When the load is applied, the deflection gradually spreads over time until the entire beam length is affected. The early deflected shape does not resemble the fundamental mode of vibration. The normal-mode superposition, briefly described in Chapter 2, is made awkward by the fact that the number of modal components necessary to get a reasonably accurate answer increases as the time of interest decreases. Even the "simplified" method, as developed by Goldsmith [26] (who, *nota bene*, goes through lengthy developments in detail) is tedious enough to render it impractical for engineering estimates.

For elastic beams, the final maximum moment response is attained at some time between $\tau/4$ and $\tau/2$, where τ is the natural period of vibration of a beam. It is usually quite easy to estimate this peak response, especially when there is a resemblance to the fundamental mode. However, when the load magnitude is large, so that the evaluation renders the response far in excess of moment capacity of the beam, one needs to consider what happens much earlier. In this situation, excessive straining or failure may take place at a small fraction of τ and at a different location from what is normally expected.

Based on physical reasoning as well as observation of deflected shapes emerging out of FEA simulations, this author proposed a concept of a *prime flexural wave*, or a portion of a beam significantly deformed by bending at the time following impact [87]. In Figure 10.5 such *flex waves* (in brief) have the lengths of l_1 (semi-infinite beam) and l_2, (infinite-length beam) respectively. As the deflection grows, so do these lengths.

Before quantifying the response, let us review what is known about the simple, nondispersive axial wave motion, such as resulting in a cantilever, which is suddenly loaded by an axial force at the free end. In section "Traveling waves, vibration and a standing wave concept" it was explained that it takes one-quarter of the natural period before the wave, originating at the free end sweeps the entire length and reaches the fixed end.

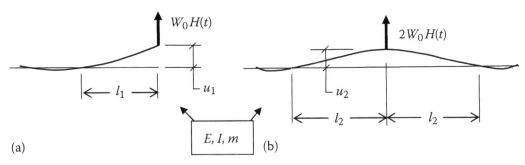

(a)   (b)

FIGURE 10.5 Flex waves in a semi-infinite beam (a) and in an infinite-length beam (b) resulting from application of a step load.

Returning to our main interest, i.e., lateral response, we therefore conclude that if the waves so defined are indeed the predominant forms of early motion, the time needed by the wave to sweep the length of the beam* should be close to one-quarter of the natural period of lateral vibrations. Such quarter-periods are, from Chapter 2:

$$\frac{\tau}{4} = 0.4468L^2 \sqrt{\frac{m}{EI}}$$
(10.23a)

and

$$\frac{\tau}{4} = \frac{L^2}{3.561} \sqrt{\frac{m}{EI}}$$
(10.23b)

respectively, for Figure 10.5a and b, keeping in mind that in case (b) the vibrating beam, if clamped at both ends, is $2L$ long. Assuming that that the growth of the wave (or the position of the intercept point) is proportional to time, one can generalize the above by replacing L by the flex wave length l and $\tau/4$ by the elapsed time obtaining the following:

$$l_1 = 1.496 \left(\frac{EI}{m} \right)^{1/4} \sqrt{t}$$
(10.24a)

and

$$l_2 = 1.887 \left(\frac{EI}{m} \right)^{1/4} \sqrt{t}$$
(10.24b)

The above formulas are the same as those developed by this author [87], except for a minor difference in coefficients. Once the time point is selected and the corresponding length of the wave is known, the deflections can be approximated by static relationships, i.e., customary beam equations (Figure 10.6)[†]:

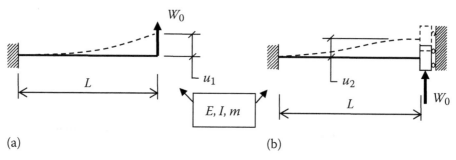

(a)   (b)

FIGURE 10.6   The same deflected patterns as in Figure 10.5 applied to finite-length beams.

---

* This is the time needed for l_1 or l_2 to grow to L.
[†] For nondispersive waves, like one spreading along an axial bar, calculating displacements as if they were caused by a static load applied to a deformed portion of the structural member gives exact results, as was found in Chapter 5. (The same can be better visualized using a shear beam rather than an axial bar concept.) For dispersive cases, like here, using a quasistatic method is merely an approximation.

$$u_1 = \frac{W_0 l_1{}^3}{3EI}$$

(10.25a)

and

$$u_2 = \frac{W_0 l_2{}^3}{12EI}$$

(10.25b)

The intercept point, defined by Equations 10.24, moves in proportion to $t^{1/2}$, which is not a physically intuitive result.

In retrospect, the above procedure is applicable to infinitely long beams. In our case the observation time to is limited to $\tau/4$, corresponding to a beam of length L under consideration.

One should also keep in mind that the flexural deflection is a result of several bending waves, each with different amplitude and a different speed of propagation. The presence of constraint points makes the situation more difficult, as reflections of those waves as well as interference take place along the length. Obviously, the approach presented above is only an approximation. Still, Equations 10.24 and 10.25 are accurate up to $\tau/4$ and even somewhat beyond.

Bending moments as functions of time are specified in Cases 10.6 and 10.7. The peak moment for the cantilever appears at about $l_1/2$, while for the guided beam it is always located at the loaded point. The larger the applied forces, the smaller intervals of time are of interest. This elastic analysis can indicate the location of a plastic "joint," although the indication is not very accurate. (See Examples 10.4 and 10.5.)

Global Response of an Elastic Beam to a Point–Load Impact

This discussion is limited to short, rectangular pulses of a force W_0 acting over an interval of t_0. In the previous section the response was determined for a period immediately after the load application. If no tendency to overload the beam is discovered in that early stage, the peak global response becomes of interest then. It often takes place between $t = \tau/4$ and $t = \tau/2$, where τ is the fundamental period of vibrations.

The peak response is most readily found by using the (SDOF) approach, already summarized by Equations 10.15 and 10.16. The only difference is that now u_{st} is the static deflection attained due to the point load W_0 rather than the distributed w_0. The SDOF approach implies that the maximum dynamic factor, i.e., $DF(u) = u_d/u_{st}$ does not exceed 2.0, which is usually a sufficient accuracy for deflections.

The peak bending moment is often underestimated when SDOF method is used. This is especially true for t_0 much shorter than the natural period. The other rough estimate is

$$\mathcal{M}_d = W_0 l_1$$

(10.26a)

or

$$\mathcal{M}_d = \frac{W_0 l_2}{2}$$

(10.26b)

TABLE 10.6

Shear Coefficients for Dynamic Response, Point Loads

Condition	Q_{st}	c	d
CF	W_0	6.013	1.415
SS	$(1/2)W_0$	0.176	0.502
SC	$(11/16)W_0$	0.431	0.678
CC	$(1/2)W_0$	0.726	0.541
CG	W_0	2.583	1.005

Note: The end condition designation is in Figure 10.3.

where Equation 10.26a is for a cantilever only, while Equation 10.26b holds for the remaining four cases under consideration. Symbols l_1 and l_2 stand for the distances traveled by a flex wave and are defined by Equations 10.24. The peak bending moment is expected to be the larger value of the two estimates given by Equations 10.16b and 10.26. The approach gives close results for displacement throughout the entire period τ. For bending it gives a sensible approximation provided the length of a calculated flex wave l does not go past the boundary. (In most cases, this means $t < \tau/4$.) When calculated l is larger, one can expect overprediction of moments.

With regard to shear, Equation 10.19 will be employed, as for distributed load, but it has to be used with caution. One method gave the coefficients in Table 10.4, but unfortunately the accuracy of results is poor. It is difficult to create a two-term equation, which would be accurate in the entire range of interest, from very short impulses up to those lasting $\tau/2$. Table 10.6, on the other hand, results from an alternative method of determining c and d. They are calculated based on a known, accurate solution for two time points. It gives reasonable results for $t_0/\tau = 0.1$, overpredicts for shorter times and underpredicts for longer durations.

A detailed example (Example 16.2) shows that the dynamic factors for moments, DF(M), came close to 2.5 for all support cases. There is much more dispersion of results for shears, with DF(Q) resulting between 3.5 and 4.9. One should remember that the factors obtained there are merely a guide in that they depend, to some degree, on the length of the beam. (For a selected set of section properties the peak responses will vary somewhat when the beam length is varied.)

The Influence of Shear Deformation and Sustained Axial Force

In practical applications the shear deflection component is often negligible, but in some cases it may not be ignored. Including this deflection in accordance with Equation 10.12 increases flexibility and therefore decreases the natural frequency. The following approximation works well provided that shear component is smaller than the flexural one:

$$\frac{\omega}{\omega_b} = \sqrt{\frac{u_b}{u_b + u_s}}$$

(10.27)

where
  ω is the true value of the frequency
  ω_b is this quantity found with the help of formulas in Chapter 2, which do not include shear deflections

When there is an axial force P_0 present in a beam subjected to a lateral load, one speaks of that beam as of a *beam–column*. The exact expressions for deflections are available in sources such as Young [118], but they are usually too involved to be used in initial estimates. A simpler way is to employ an approximate expression

$$u_{bc} = \frac{u_b}{1 + \dfrac{P_0}{P_{cr}}} \tag{10.28}$$

which is analogous to Equation 2.38 and which says that the deflections of a stretched beam–column u_{bc} are smaller than beam deflections in absence of axial loading, u_b. (Positive value of P_0 corresponds to tension.) The approximate expression for the natural frequency, in presence of tension, is

$$\omega_{bc} = \omega_b \sqrt{1 + \frac{P_0}{P_{cr}}} \tag{10.29}$$

where
  ω_b is the frequency calculated in absence of an axial force
  P_{cr} is the elastic buckling force calculated for (approximately) the same deflected shape as used for deriving the natural frequency value (this is consistent with Equation 2.39)

The tensile load makes the beam apparently stiffer, thereby increasing its natural frequency while compression has the opposite effect. In case of a beam with simple supports, for example, P_{cr} is the familiar Euler force

$$P_{cr} = P_e = \pi^2 EI/L^2 \tag{10.30}$$

When there is a marked influence of a sustained axial force, one must include a beam–column correction for the response components. When the influence of the rest of the structure is such that the axial load grows in time, the validity of the above small-deflection approach is limited to the early phase of motion. Everything stated in this section applies to both distributed loads, point loads, and every combination of lateral loading.

MASS–BEAM IMPACT

The Lumped-Parameter Methods for Mass–Beam Impact Problem

The lumped-parameter methods for mass–beam impact problem represent simple idealizations of the interaction between the striker and the beam. In the *basic method*, a beam is treated as a massless, flexible member. The elementary form of this approach was already presented in Cases 3.22 through 3.24, where applying the energy balance equation gives

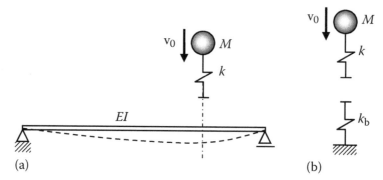

FIGURE 10.7   Actual configuration (a) and the equivalent oscillator (b).

the peak impact force. The next step in making the problem statement a little more realistic is to introduce a contact flexibility $1/k$, which is usually not negligible. (Even if the striker itself is very hard, there is a local deformation of the beam section, which is then represented by a spring with a stiffness k.) In Figure 10.7 the original problem is shown, as well as its simplified representation, equivalent to an oscillator with an effective stiffness k^* calculated by Equation 2.29. The mass is assumed to travel with the beam, without a rebound, until the peak displacement is reached. This becomes the initial velocity problem for such an oscillator whose solution was given before by Case 3.2. The peak contact force is

$$R_m = v_0 \sqrt{k^* M} \qquad (10.31)$$

Once this force is established, the beam response (deflection, moment, and shear) is found as if it arose from static application of R_m. Needless to say, the beam flexibility is related to the actual contact point. (As Table 10.2 shows, a SC beam, for example, struck at midpoint would have $k_b = 48EI/L^3$.) If the contact point is close to a support, the shear component of deflection must be included; this decreases the effective stiffness k^*. As Example 10.6 shows, the method gives a sensible first approximation for the peak bending moment, but is deficient with regard to other parameters.

Another step in analytical refinement is the *collision method*, which considers the interaction between a striker of mass M and a portion of a beam adjoining the impact point, M_b, as described in Chapter 8 and summarized in Case 8.18. The formulation involves a contact flexibility $1/k$, which is understood to be a result of a local deformation. In other words, the collision method deals with a contact spring between two moving masses and determines the peak contact force according to the rules for short-lasting collisions.

Application of a Shear Beam Concept

This needs to be considered before another solution to mass–beam impact may be introduced. Lateral deflection of a two-dimensional Timoshenko beam has two components; bending and shear. The first is due to curving of the axis, while the second results from distortion of elements across the axis. When only the latter is accounted for, we speak of a *shear beam* as depicted in Figure 10.8 and previously analyzed in Chapters 2 and 5.

The tip static deflection of the beam in Figure 10.8 is

$$u = \frac{W_0 L}{GA_s} \qquad (10.32)$$

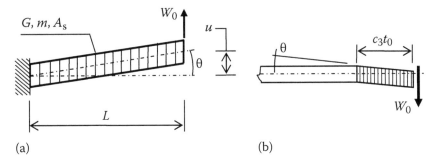

FIGURE 10.8   Shear beam loaded statically (a) and loaded by a short-duration pulse (b).

where
  G is the shear modulus
  A_s is the shear area

When a load is applied in the form of a short-duration pulse lasting t_0, as in Figure 10.8b, the tip deflection u_0 and the speed of propagation of a shear disturbance c_3 are, respectively, per Case 5.1:

$$u_0 = \frac{W_0 c_3 t_0}{GA_s} \tag{10.33a}$$

and

$$c_3 = \left(\frac{GA_s}{m} \right)^{1/2} \tag{10.33b}$$

Note that as long as the driving force remains constant, so is the lateral velocity of the tip, u_0/t_0. If the force is decreasing, the deflected shape becomes curved and the smallest curvature is at the tip. At the termination of the pulse for $t = t_0$, the deflected shape moves along the beam and the tip deflection does not change until the pulse, reflected from the fixed end comes back. The above equations are also applicable to a beam loaded at an intermediate point, rather than at the tip, provided W_0 is replaced by $2W_0$.

A Mass–Beam Impact Approach Involving Transient Beam Deformation

This approach can be called *a stress-wave method*, as it considers the deformation spreading from the impact point while the impact is in progress.

A beam stricken at a point undergoes both shearing and flexural deformation. The second component of deflection is dominant most of the time. However, at the first instant of contact, the shear component prevails. As the contact duration is very short and characterized by a rapid change of the magnitude of the contact force, our formulation is simplified in that only shear deflection is included in the beam flexibility. The equivalent system to be used in analysis is shown in Figure 10.9. The damper represents the deformability of the beam and M_m represents its inertia, or a portion of beam mass.

Returning to Equation 10.33, related to Figure 10.8, one can notice that as long as the driving force W remains constant, equal to W_0, the tip velocity of the beam also remains constant. Denoting the latter by v_e, one has

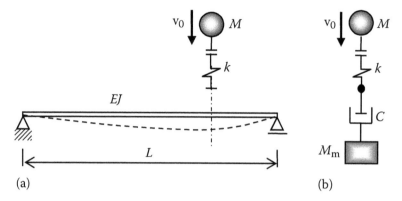

FIGURE 10.9   Mass-impacted beam (a) and the equivalent system (b).

$$v_e = \frac{u}{t} = W_0 \left(\frac{1}{GA_s m} \right)^{1/2} \tag{10.34}$$

or, simply $W_0 = v_e C$,   where $C = \sqrt{GA_s m}$ \qquad\qquad (10.35a,b)

The last statement simply tells us that a point on a beam, when pressed, moves like an end of a viscous damper with a constant C. This is the reason for placing a damper at the end of the contact spring in Figure 10.9b. This damper constant was derived for a beam impacted at the tip. If the impact is well inside the length of a beam, the constant should be doubled. Mass M_m, placed at the lower end, indicates a portion of the beam mass included in the simplified model.

The mass in Figure 10.9b is the same as the modal mass M_m defined in Table 2.1. The modal mass is treated here as a limit of an inertial resistance of an impacted beam. Depending on circumstances, M_m may or may not be explicitly used in the calculation.

As evident from Figure 10.9, the problem is reduced to that of a two-mass collision cushioned by a contact stiffness k and contact viscosity C in series. An attempt to solve the governing equations of motion and recover the peak contact force leads to somewhat complex expressions. For this reason a simple, engineering approach is proposed, based on the following reasoning.

From elementary dynamics, the peak contact force of a mass–spring system impacting a rigid wall with a speed v_i is $R = v_i \sqrt{kM}$ (Chapter 3). If two masses, M and M_b, collide with a relative velocity v_i, with only a spring k between them, one finds (Equation 8.17) that the formula for the peak force remains the same, except that M is replaced with a reduced mass M^*, which is smaller than either of the two colliding masses. In our case, where two elements provide cushioning, the most effective approach seems to be as follows: mass M^* is set to be equal to the striker mass M, but not larger than the modal mass M_m. (Although this is not strictly consistent with Equation 8.17, it agrees with the principle that the effective mass is not larger than either of the two components.) The estimated peak force is

$$R_{ms} = v_i \sqrt{kM^*} \tag{10.36}$$

If those two colliding masses are connected by a damper only, then the peak damper force is Cv, where v is the peak relative velocity, which is the same as the impact velocity v_i. However, in our case, the damper is squeezed not by a suddenly applied velocity, but by a gradually growing force. For this reason, a good guess, based on observation of FEA results, of the largest likely force to be developed by a damper is

$$R_{md} = Cv_i/2 \tag{10.37}$$

Returning now to Figure 10.9b, one can approximate the peak force as follows. Suppose the damper is rigid. The peak contact force is then given by Equation 10.36. If the spring is assumed rigid, the peak contact force is given by Equation 10.37. The largest impact force R_m can now estimated from

$$\frac{1}{R_m} = \frac{1}{R_{ms}} + \frac{1}{R_{md}} \tag{10.38}$$

as appropriate for two elements in a series connection. (This approach was FE-tested by the author for the impactor mass not exceeding the mass of the beam.)

Duration of contact t_0 can be calculated from momentum considerations. If the rebound velocity is negligible and if the force–time plot resembles a triangle, then the momentum–impulse relation is

$$R_m t_0/2 = Mv_0 \tag{10.39a}$$

or

$$t_0 = \frac{2Mv_0}{R_m} \tag{10.39b}$$

If, however, the plot is sinusoidal, then the coefficient of 2 is replaced by $\pi/2$. (Johnson [41] demonstrated that this type of contact is between a sine and a triangle.) In case of a perfectly elastic rebound, which means doubling of the impulse, the calculated t_0 will double, too. (This is an upper bound of t_0 for our model).

Equation 10.36 implied that the contact spring is linear. In practical applications, however, most springs are not. One way to solve a problem is to linearize a spring, but it often leads to complications and loss of accuracy. It is not necessary to do so if the force–displacement relation for the contact can be analytically expressed. In case of a Hertz-type contact, the peak force can be found from Equation 7.37b.

This methodology described is valid within reasonable limits. One of those is the size ratio of the striker and the beam cross section. If the (equivalent) diameter of the cross section is several times that of the striker and if they are made of similar materials, then it becomes, for all practical reasons, a case of striker impacting a solid, elastic obstacle. In this event the contact force is found simply from Equation 10.36 with striker mass M replacing the reduced mass M^*. (A configuration like this is presented by Example 7.9.)

In many practical cases the contact stiffness is larger than that of the beam itself, i.e., $k > k_b$. One should note that a more complete picture is obtained if one includes both bending and shear components of deflection. In cases of a hard impact, however, where the fall of the contact force from its peak value is quite rapid, the shear deformation dominates.* In fact, the impact must be hard for the shear-wave method to be applicable. An advantage of being able to ignore the bending component is that shear is simpler to analytically describe.

Elastic Strain Energy and Approximate Deflected Shapes

When a beam is subjected to a 3D load system including axial force P, bending moment M, shear force Q, and torque T, the strain energy can be expressed as follows:

$$\Pi = \int \frac{P^2 dx}{2EA} + \int \frac{Q^2 dx}{2GA_s} + \int \frac{M^2 dx}{2EI} + \int \frac{T^2 dx}{2GC} \qquad (10.40)$$

where integration extends over the whole length. For the bending term, which is most frequently used, one can employ the small-deflection equality, $EIu'' = M$ to express bending energy Π_b:

$$\Pi_b = \frac{EI}{2} \int (u'')^2 dx \qquad (10.41)$$

This formula becomes useful when it is necessary to express the bending energy stored by the maximum deflection δ. If that deflection is caused by a uniformly distributed load w, and, in addition, an approximation by trigonometric functions is acceptable, then, for the end conditions of interest one has

$$u = \delta\left(1 - \cos\frac{\pi x}{2L}\right) \quad \text{for CF} \qquad (10.42a)$$

$$u = \delta \sin\frac{\pi x}{L} \quad \text{for SS} \qquad (10.42b)$$

$$u = 3.858\delta(3\eta^3 - 2\eta^4 - \eta) \quad \text{with } \eta = x/L \text{ for SC} \qquad (10.42c)$$

$$u = \frac{\delta}{2}\left(1 - \cos\frac{2\pi x}{L}\right) \quad \text{for CC} \qquad (10.42d)$$

$$u = \frac{\delta}{2}\left(1 - \cos\frac{\pi x}{L}\right) \quad \text{for CG} \qquad (10.42e)$$

where δ is the same as maximum displacement u_m mentioned before. After performing the integration, one finds

$$\Pi_b = \Gamma EI \frac{\delta^2}{L^3} \qquad (10.43)$$

where coefficients Γ are case-dependent. These coefficients are tabulated in Chapter 11.

---

* "Rapid" is a qualitative term. Another way to put it would be to say that the impact is hard when the contact force is reduced to a fraction of its original value well before the wave of deformation reaches the nearby support.

If the deflected shape due to point load W_0 is assumed, the work done or the accumulated strain energy is simply $W_0\delta/2$. In this case it is helpful to first notice that Equation 10.12, limited to bending deflections, can be written in one of two ways:

$$\delta = b\frac{W_0 L^3}{EI} \qquad (10.44a)$$

or

$$W_0 = \frac{EI}{L^3}\frac{\delta}{b} \qquad (10.44b)$$

The strain energy can now be written in terms of δ or W_0:

$$\Pi_b = \frac{b}{2}\frac{W_0^2 L^3}{EI} \qquad (10.45a)$$

or

$$\Pi_b = \frac{EI}{2b}\frac{\delta^2}{L^3} \qquad (10.45b)$$

where coefficients b can be found from Table 10.2. In spite of being related to point loads, coefficients $1/(2b)$ are not much different from coefficients Γ shown above.

Equations 10.44 and 10.45 are useful in approximate calculations, which involve the use of strain energy. Table 10.7 below lists the absolute values of the peak deflection u_m, slope $(u')_m$, and the curvature $(u'')_m$ according to the approximate Equation 10.42.

The shapes described above may dominate the deflected line, but one has to keep in mind that this simple picture often requires an adjustment. Consider, for example, a beam on two supports with the load concentrated over a small portion of the length. Apart from bending deflection, there is always some deformation of the beam cross section. For a slender beam with a solid section, the latter may usually be ignored when compared with the former; that is, the *local deformation* is negligible in comparison with the *global deflections*. When a beam is relatively short, however, and the cross section is hollow, the preceding statement will not be true. It is a rule that for compact bodies (all three dimensions of comparable magnitude), the local deformations are important and may even predominate. The latter is the case when a compact body strikes a barrier and the contact area is small.

DISTRIBUTED LOADING, INELASTIC RANGE

Assumptions and Procedures

When a dynamic load is applied to a beam, with a very large magnitude, but also with short duration, the first step is to check its response assuming that it is still in the elastic range. If the stress so calculated is much larger than the yield strength of material, this does not necessarily imply failure, but merely makes the analyst change his approach.

TABLE 10.7

Parameters of a Deflected Shape

End Conditions	$\|u\|_m$	$\|u'\|_m$	$\|u''\|_m$	Loc. $\|u''\|_m$
CF	δ	$\dfrac{\pi\delta}{2L}$	$\left(\dfrac{\pi}{2L}\right)^2\delta$	$x=0$
SS	δ	$\dfrac{\pi\delta}{L}$	$\left(\dfrac{\pi}{L}\right)^2\delta$	$x=L/2$
CG	δ	$\dfrac{\pi\delta}{2L}$	$\left(\dfrac{\pi}{L}\right)^2\dfrac{\delta}{2}$	$x=0$
SC	δ	$3.858\dfrac{\delta}{L}$	$\dfrac{23.148}{L^2}\delta$	$x=L$
CC	δ	$\dfrac{\pi\delta}{L}$	$\left(\dfrac{2\pi}{L}\right)^2\dfrac{\delta}{2}$	$x=0$

Note: For SC, u_m is at $x/L = 0.4215$. The coordinate system is shown in Figure 10.3.

The concept of a *plastic joint*, which makes such analyses easier, is well established. When bending moment attains a certain limiting value in a particular section, we assume the yielding starts abruptly at this location and a hinge forms there. Regardless of how large the deflection is, the bending moment stays within its limit. The counterpart of this concept in shear, namely, the *shear slide* is a relatively new concept, attributed to Nonaka [63].

In order to keep the formulation simple, it is necessary to replace an actual stress–strain curve of material with a simplified shape. The most popular approximations, when large strains are involved, are the elastic-perfectly plastic (EPP) and the RP. The latter one represents a material, which is assumed undeformable up to a certain stress level σ_0, the flow strength. When this level is reached, further curving of the axis takes place with no increase of resistance. Based on strain magnitude alone, simplifying EPP to RP model seems almost natural, at least in tension. In bending, however, the consequences are more serious. While RP makes calculations easier, it is definitely less realistic than the EPP model.

A detailed solution for a SC beam, with a continuously distributed loading, was developed in Ref. [63], which presented five initial modes of motion, depending on beam properties and load magnitude. A similar formulation was carried out for a clamped–clamped beam by Li and Jones [53]. A simpler, approximate approach presented below offers only two modes of motion. All of the above are based on RP model.

Limit Values of Bending Moment and Shear Force

In pure bending the yielding begins when bending moment M_y (Equation 10.5) is first attained. The bending moment capacity for fully developed plastic stress distribution is M_0, per Equation 10.6b. The flow strength σ_0 is the nominal yield strength, which should be selected to give the actual moment capacity. The yield, or flow strength in shear is

$\tau_0 = \sigma_0/\sqrt{3}$, in accordance with Huber–Mises theory. The corresponding onset of yield in shear and the shear capacity are, respectively,

$$Q_y = (A/\alpha)\tau_0 \tag{10.46a}$$

and

$$Q_0 = A_{ef}\tau_0 \tag{10.46b}$$

where A_{ef} is the section area effective for the limiting shear condition, shown in Table 10.8 as a fraction of the total cross-sectional area A. The ratio of peak elastic shear stress to the average shear stress, designated by α, can be found Table 10.1. One should also note that for material models other than RP and EPP, as defined in Chapter 4, the ultimate strength of material F_u is usually specified. The corresponding shear strength is then $\tau_u = F_u/\sqrt{3}$ and the shear strength of the section becomes

$$Q_u = \tau_u A_{ef} \tag{10.47}$$

The L_{cr} column below defines the length of a beam (with both ends clamped) below which the beam is more critical in shear than in bending, under static conditions. It is expressed by the section constants as

$$L_{cr} = \sqrt{3}\frac{8Z}{A_{ef}} \tag{10.48}$$

Static Collapse Values for Distributed Load

These values will now be determined, keeping in mind that there is no movement until the load attains its limiting value. The values of shear forces and bending moments resulting from a uniformly distributed load w are taken from Table 10.2.

The simplest case is for a cantilever beam, as in Figure 10.10. Two independent modes of deformation are possible: shear sliding and joint rotation. The corresponding limiting loads are designated as w_s and w_b:

$$w_s = Q_0/L \tag{10.49a}$$

TABLE 10.8

Additional Constants for Plastic Bending

Section Type	A_{ef}	$\mathcal{M}_0/\mathcal{M}_y$	L_{cr}
Solid rectangle, $B \times H$	A	1.5	$3.464h$
Solid circle, R	$0.9A$	1.698	$6.534R$
Thin tube, $R_m \times h$	$(2/\pi)A$	1.273	$13.86R_m$
I-beam, B, H, b, h	$A_w = bH$	Zc/I	$13.86Z/(Hb)$
Two-flange, A_f, h, H	$A_w = hH$	1.0	$6.928(A_f/A_w)H$

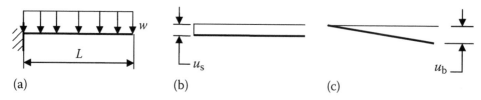

FIGURE 10.10 Original CF beam (a), shear slide (b), and bending motion (c). (From Szuladziński, G., *J. Eng. Mech.*, 133, March 2007.)

and

$$w_b = 2\mathcal{M}_0/L^2 \qquad (10.49b)$$

If $w_s < w_b$, shear will govern, but for most practical configurations $w_s > w_b$ and bending, or joint rotation will take place. The length of a CC beam, above which bending is critical is designated by L_{cr} in Table 10.8. In Figure 10.11, depicting other end conditions, only the bending modes are illustrated. Movement is assumed to start whenever bending moment at any location attains its limit. At this instant another joint is activated to make the movement possible and the same limit moment is assumed at the second joint. (This is applicable to SC beam. Three joints are needed for CC.)

The determination of limiting load values is quite straightforward, except, perhaps, for SC beam, which is both unsymmetric and redundant. One joint appears at the clamped end, where bending is maximum. The second joint is needed to create a mechanism, as per Figure 10.11c. A free-body diagram in Figure 10.12 visualizes the internal moment. (Note the absence of shear force at the intermediate joint. This is based on a basic beam theory, which says that if a moment reaches a local maximum, shear force becomes zero. The support points are possible exceptions to this.)

The relation of lengths l_1 and l_2 making up the total L is found by equating moments $\mathcal{M}_0$ needed to balance each of the parts separately:

$$\frac{l_2}{l_1} = \sqrt{2} \qquad (10.50a)$$

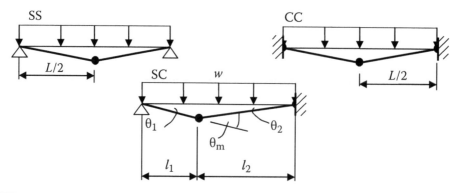

FIGURE 10.11 Remaining cases of end support. (From Szuladziński, G., *J. Eng. Mech.*, 133, March 2007.)

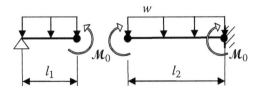

FIGURE 10.12 Rigid-body diagram for SC case. (From Szuladziñski, G., *J. Eng. Mech.*, 133, March 2007.)

and

$$l_1 = \frac{L}{1+\sqrt{2}}$$ (10.50b)

From the equilibrium of the left segment, for example, the limiting distributed load w_b can be found, while the shearing limit w_s is determined from the right segment:

$$w_b = 11.66 \mathcal{M}_0 / L^2$$ (10.51a)

and

$$w_s = 1.707 Q_0 / L$$ (10.51b)

While the above is derived from a consistent deformation pattern, one should note, that the shear force of $(5/8)wL$ at the clamped end is capable of creating a shear slide, which is a mechanism. This reduces the coefficient in Equation 10.51b from 1.707 to 1.6. In general, the load limits can be expressed as follows:

$$w_b = \xi \mathcal{M}_0 / L^2$$ (10.52a)

and

$$w_s = \eta Q_0 / L$$ (10.52b)

where the coefficients ξ and η are found in Table 10.9.

Simplified Estimates of Dynamic Response

The dynamic load of interest has a form of a rectangular pulse; a constant magnitude w_m with a duration of t_0. The impulse applied to the beam is $S = Lw_m t_0$. If t_0 is very short, which for a prescribed impulse makes w_m large, this type of loading is called *impulsive*. (For the practical purpose of distinguishing impulsive loading, one may say that w_m must be at least 10 times as large as the static limit load.) In the following description, impulsive loading is assumed to be in effect.

TABLE 10.9

Response Coefficients for Classical End Conditions, Distributed Load

	ξ (Equation 10.52a)	η (Equation 10.52b)	γ (Equation 10.62a)	θ_m/θ_e	β (Equation 10.62b)
CF	2	1	3/8	1	3/8
SS	8	2	3/16	2	3/32
SC	11.66	1.6	0.1553	1.707	1/15.55
CC	16	2	3/32	2	3/64

CF in Figure 10.10 presents the simplest scheme. The first case is shear deformation or translational motion. The impulse applied to the beam with a mass $M = Lm$ can be equated with the momentum gained to find the initial velocity v_0:

$$S = Lw_m t_0 = Mv_0 \qquad (10.53)$$

The maximum deflection is estimated by comparing the initial kinetic energy with the strain energy of shear deformation:

$$E_k = \Pi \quad \text{or} \quad (mL)v_0^2/2 = Q_0 u_s \qquad (10.54)$$

Substituting v_0 from Equation 10.53 into Equation 10.54 one obtains the maximum sliding deflection as

$$u_s = \frac{S^2}{2dMQ_0} \qquad (10.55)$$

Coefficient $d = 1$ for this cantilever, but for all other cases of support one has $d = 2$, because the resisting shear force is then present at both ends, rather than at one end only. (Note that motion starts here concurrently with load application. This is made possible by the assumption that $w_0 \gg w_s$.)

A similar approach is used for the case of joint rotation simulating bending motion, except that the moment of impulse is equated to the angular momentum gained:

$$(Lw_0)\frac{L}{2}t_0 = \frac{ML^2}{3}\lambda \qquad (10.56)$$

where

$ML^2/3$ is the moment of inertia of a stick with mass M and length L rotating about its end

λ is the angular velocity

The latter is found and inserted in the energy-balance equation below

$$\frac{1}{2}\left(\frac{ML^2}{3}\right)\lambda^2 = \mathcal{M}_0\theta \tag{10.57}$$

where the right side is the strain energy absorbed in angular deformation with the resisting moment $\mathcal{M}_0$. This gives the maximum rotation θ as

$$\theta = \frac{3S^2}{8M\mathcal{M}_0} \tag{10.58}$$

The associated tip translation is $u_b = L\theta$. For the SC beam the reasoning is the same, except the process has to be carried out separately for each segment and the results added. In addition, the continuity relations are used:

$$\theta_1 l_1 = \theta_2 l_2 \tag{10.59a}$$

and

$$\lambda_1 l_1 = \lambda_2 l_2 \tag{10.59b}$$

When the sum of moments of impulse for each part is calculated and equated to the sum of the angular momenta gained, one obtains the relation from which the unknown λ_1 can be found:

$$M_1 l_1^2 \lambda_1 = (3/2)w_0 t_0 l_1^2 \tag{10.60}$$

The total kinetic energy is set equal to the work of resisting moments at joints

$$\frac{1+\sqrt{2}}{6} M_1 l_1^2 \lambda_1^2 = \left(1+\sqrt{2}\right)\mathcal{M}_0\theta_1 \tag{10.61}$$

keeping in mind that the resisting end joint rotates by θ_2 and the middle one by $\theta_1+\theta_2$. When λ_1 is inserted in the above, one finds the same equation type describing the final angle of rotation, as before:

$$\theta_e = \frac{\gamma S^2}{mL\mathcal{M}_0} \tag{10.62a}$$

or

$$u_m = \frac{\beta S^2}{m\mathcal{M}_0} \tag{10.62b}$$

but now the coefficient is $\gamma = 0.1553$. Here θ_e is the left end rotation, the distinction being important for SC beam only. The displacement of the intermediate joint is u_m and obviously $\beta = \gamma l_1$. The remaining two cases of support are simpler to calculate and the corresponding

---

TABLE 10.10

Peak Shear Coefficients, for Distributed Load *w* and Midpoint Load *W*

Beam	*c*	*d*	Beam	*c*	*d*
Plastic range of deformation					
SS, *w*	0.38	0.12	SS, *W*	0.75	−0.25
SC, *w*	0.38	0.12[a]	SC, *W*	0.75	−0.25[a]
CC, *w*	0.38	0.12	CC, *W*	0.75	−0.25

[a] Add the third term, $\pm M_0/L$, to maximize the shear result.

---

coefficients are displayed in Table 10.9. Angle θ_m designates the discontinuity between both parts of the beam, which is $\theta_1 + \theta_2$ for SC and $2\theta_e$ otherwise.

The peak values of shear at the supports can be found using the same methodology as previously described for elastic beams. The same Equation 10.19 is used, with the coefficients shown in Table 10.10 for both the distributed as well as the concentrated loads, according to Biggs [9].

The Influence of Material Model and the Assumed Shape

As mentioned before, exact solutions exist for some end conditions of beams dynamically loaded well into the inelastic range. Compared with those, our simplified solutions neglect the resistance of material while it is acting as a rigid body. This, by itself, points to the overestimate of peak deflections, when simplified solutions are employed. Another variable is material modeling. The exact solutions are based on RP material, which ignores the existence of the elastic range. When the material model is upgraded from RP to a more realistic EPP model, the magnitude of computed deflections should increase. When a computer simulation is conducted, using EPP material, one finds that the simplified approach gives an underestimate, but the exact formulas underestimate even more. (This problem is described in detail by Szuladziński [101], as well is illustrated by Example 10.14.)

The increase in deflection when changing the material model does not nearly account for the difference between the expected and the actual results of the FEA. The main problem lies in the energy input level. We have used a consistent approach, treating the beams as rigid throughout the process, as it is normally done. In reality, however, when a distributed load is applied, it affects nearly the whole beam with the exception of a small segment near the end. (This was illustrated for a cable in Chapter 9.) Our approach, as expressed by Equations 10.56 and 10.57, gave the kinetic energy of

$$E_k = \frac{3}{8}(mL)v_0^2 = \frac{3}{8}(mL)\left(\frac{w_0 t_0}{m}\right)^2$$

while, if the entire beam was regarded as receiving the impulse, the above coefficient would be 1/2. If the latter assumption is more realistic, then the energy input in the above developments was only 3/4 of what it should have been. Incorporating the change would increase the hand-calculated values of plastic displacements by the factor of 4/3. This is left as an option for the reader.

There also is a qualitative difference between the results obtained by using these two materials. The change of bending moment, when viewed at some instant along the length, is much smoother when the material is upgraded to EPP. It shows itself quite clearly in computer-generated plots as disappearance of such sharp transitions as predicted by the RP material.

POINT LOADS, INELASTIC RANGE

When dynamic loads become large enough to significantly exceed the yield strength, a concept of a plastic hinge, or joint, is often invoked, in conjunction with the RP material model to quantify the collapse mechanism. The first step, as before, is to determine the corresponding static load W_b capable of inducing the initiation of collapse. Depending on the problem type and objectives of the analysis, either a *stationary* or a *traveling* plastic joint is involved.

Basic Response to Point Loads

This development closely follows what was done for distributed loads before including the predetermined position of plastic hinges. The general expressions for bending and shear collapse, respectively, are

$$W_b = \xi_m \mathcal{M}_0 / L \tag{10.63a}$$

and

$$W_s = \eta Q_0 \tag{10.63b}$$

as a counterpart of Equations 10.52. The values in Table 10.11 come from a simple assumption that both the plastic hinges as well as plastic slides occur at the same locations as the corresponding peak responses in the elastic range. When a point load W is applied, the impulse

TABLE 10.11
Response Coefficients for Classical End Conditions, Point Loads

	ξ_m (Equation 10.63a)	η (Equation 10.63b)	γ_1 (Equation 10.65)
CF	1	1	1/2
SS	4	2	1/2
SC	16/3	16/11	0.3536
CC	8	2	1/4
CG	2	1	1/4

is $S = Wt_0$, where t_0 is the load duration. For large, short impulses, far exceeding the plastic resistance moment M_0, the kinetic energy E_k and the effective mass M_{ef} can be respectively expressed as

$$E_k = \frac{S^2}{2M_{ef}} \tag{10.64a}$$

$$M_{ef} = \frac{mL}{3} \tag{10.64b}$$

for all support conditions of interest. The strain energy Π can be expressed as a function of the discontinuity angle θ_m. Equating both energy components one gets

$$\theta_m = \frac{\gamma_1 S^2}{M_{ef} M_0} \tag{10.65}$$

As before, θ_m stands for the angle of discontinuity. Only for CF and CG this angle is the same as the end rotation.

Infinite or Semi-Infinite Beams, Stationary Joints, and Traveling Joints

When a cantilever is subjected to a tip load large enough, it is usual to see it yield or break at the base. However, if the load is large and rapidly applied, yielding may take place at some intermediate location. One can postulate an appearance of a *plastic joint* at some distance η from the impacted tip, as shown in Figure 10.13a, where the load $W = W_0 H(t)$ is suddenly applied. Such a hinge is expected to occur only when bending moment reaches its local maximum, which means there is no shear force present at that location. This mechanism is quantified in conjunction with the RP material. The distributed inertia force, whose resultant is $R = \eta_1 ma_t/2$ and the lumped (tip) inertia force, Ma_t, must be in balance with the applied load. The equation of the vertical equilibrium is

$$W_0 - Ma_t - \frac{1}{2}\eta_1 ma_t = 0 \tag{10.66}$$

Both parts of the beam outside the hinge remain rigid; therefore linear distribution of motion governs the moving part. The moment equilibrium about the tip

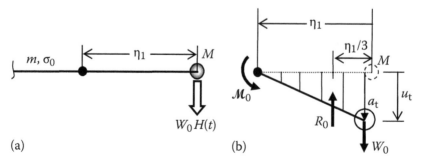

(a)   (b)

FIGURE 10.13   A long beam with a tip mass (a) and a dynamic balance of rotating segment (b).

$$\left(\frac{1}{2}\eta_1 ma_t\right)\frac{\eta_1}{3} - \mathcal{M}_0 = 0 \tag{10.67}$$

makes it possible to eliminate the tip acceleration a_t from Equation 10.66. Solving the resulting equations for η_1 and the tip acceleration a_t gives

$$\eta_1 = \frac{3}{2}\frac{\mathcal{M}_0}{W_0}\left[1 + \left(1 + \frac{8}{3}\frac{W_0 M}{\mathcal{M}_0 m}\right)^{1/2}\right] \tag{10.68}$$

$$a_t = \frac{W_0}{M + (\eta_1 m/2)} \tag{10.69}$$

Once this is determined, the velocity and displacement after some time t can be found from elementary relationships:

$$v_t = a_t t \tag{10.70a}$$

$$u_t = \frac{a_t t^2}{2} \tag{10.70b}$$

The work $\mathcal{L}_t$ performed by the applied load W_0 and the strain energy Π absorbed in the joint during some time t are, respectively,

$$\mathcal{L}_t = W_0 u_t \tag{10.71a}$$

$$\Pi = \mathcal{M}_0 \theta_t \tag{10.71b}$$

where
  $\mathcal{M}_0$ is the resisting moment
  θ_t is the angle of rotation

The existence of a stationary joint ends when the applied force is removed, at some time t_0. At that instant the segment η_1 is still rotating and the kinetic energy remaining is

$$E_k = \mathcal{L}_t - \Pi \tag{10.72}$$

Another important configuration is a long beam impacted at an intermediate point, with a lumped mass present. This is shown as Case 10.30.

The relationships for an infinite or semi-infinite beam without a lumped mass may be obtained by setting $M = 0$ in the above, or by applying simpler derivations from the outset. The results are given in Cases 10.27 through 10.28. It is interesting to note that in absence of lumped mass the strain energy of the stationary joint is 1/3 of the work of external load and that this observation holds both for the tip load as well as an interior point loading.

When the beam is of finite length L (measured from the joint), which is the usual case, and calculations show that $\eta > L$, then the position of the hinge is not an issue any more, as the hinge will form at the end of a beam. The equations of the angular momentum and the energy expressions must then be written with respect to L rather than η. (See Cases 10.23 through 10.26.)

Another way of determining whether a hinge can form at an intermediate point is to consider the ratio of applied load to that inducing static collapse. Note from Equation 10.68 that when $M = 0$, or no lumped mass is present, one has

$$\eta_1 = \frac{3\mathcal{M}_0}{W_0} \tag{10.73a}$$

or

$$W_0 = \frac{3\mathcal{M}_0}{L} \tag{10.73b}$$

in the limit, when $\eta_1 = L$. This is the limiting value of W_0 (equal to $3W_b$, according to Table 10.11), which is still associated with the hinge forming at the base. Thus, a cantilever will develop an intermediate hinge if the tip load, suddenly applied, is larger than 3× the static collapse load.* For a load application at an intermediate point the answer is not so straightforward, in terms of load magnitude, as one can see in Example 10.16.

So far the impact of a constant force was considered, applied to a beam having a lumped mass at the point of impact. A different problem, namely, that of a striker with mass M_0 and a velocity v_0 can be handled with the help of the *traveling plastic hinge* concept, an alternative approach to what is presented above. Some of the fundamental work was done by Parkes [65], who performed experiments and carried out calculations. The latter were further developed and presented by Johnson [40] and Stronge [84]. The beam in Figure 10.14 has the same properties as before, in Figure 10.13. When it is struck transversely at the tip, mass M becomes bonded to the beam. A plastic hinge forms and moves toward the base. Two parts of the beam connected by the joint behave as rigid sticks. Based on small deflection theory, the formulas given in Case 10.33 were derived, along with those in Case 10.34

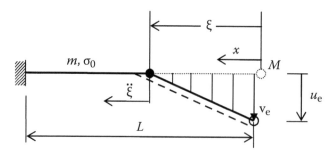

FIGURE 10.14   A cantilever is struck with a moving mass M and a traveling plastic hinge forms.

---

* Stronge [85] uses the term *moderate dynamic load* when $W_0 < 3W_b$ and *intense load* when $W_0 \geq 3W_b$ at the tip of a cantilever.

for a midpoint impacted beam. Interestingly, and in early phases of motion the position of the joint is approximately proportional to $t^{1/2}$.

The approach taken by the authors mentioned above was to investigate in detail two distinct groups of impacts: the heavy strikers, namely, when $M \gg mL$ and the light strikers, $M \ll mL$. As for the first group, the action of the traveling plastic joint is not very relevant with regard to the final, residual deflection reached. More accurate deflection figures are obtained from a simple energy balance based on preservation of angular momentum. This may also include the effect of gravity, as shown before in Case 7.18. A more interesting outcome of this theory eventuated with respect to the light strikers. They create a characteristic final deformation pattern, whereby a cantilever, like that in Figure 10.18 later, has the largest curvature at the tip, gradually decreasing toward the end of the beam. While the extent of deformation is closely predicted by the theory (Cases 10.33 through 10.34), the magnitude of deflection is substantially overpredicted. (Example 10.18.) A simpler way of finding the end deflection based on energy approach is summarized by Case 10.35 and illustrated by Example 10.19. One should also note that according to this theory, the initial contact force between the striker and the beam is infinitely large. The magnitude of that force may sometimes be important, as in the event where the integrity of the striker itself is of interest. If this is the case, the mass–beam impact analysis presented in the earlier part of this chapter should provide a more reasonable estimate.

One should note here that a traveling plastic joint is somewhat of an enigma. First, in any realistic material there is nothing like a joint in a kinematic sense, there is only a zone of marked plastic deformation, with the curvature larger than that of a surrounding beam. Secondly, it is often very difficult to tell if it is a traveling joint, or merely an expansion of a yielded zone. Some sightings of the traveling joints have been reported, but the extent of deformation remains vague. It appears that moving joints arising out of experiments with beams clamped at both ends, subjected to uniform dynamic loading, are more convincing than those with cantilevers.

Finite-Length Unconstrained Beams, Step Load, Stationary Joints

A beam in Figure 10.15a is unconstrained, so when a step load $W = 2W_0 H(t)$ is applied in the middle, the dynamic equilibrium concept helps to determine the internal forces. The distributed inertia force, $w_0 = W_0/L$, which has a uniform intensity for a rigid beam, results in a bending moment $W_0 L/2$ at the center. As soon as the moment reaches the bending capacity $\mathcal{M}_0$, a plastic joint forms and both halves begin to rotate about it, as shown in Figure 10.15b. The minimum value of the driving force to induce such a hinge is therefore

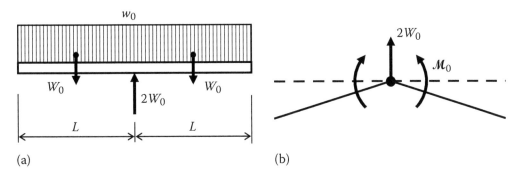

(a)   (b)

FIGURE 10.15 Dynamic equilibrium of a free beam (a) and the joint forming (b).

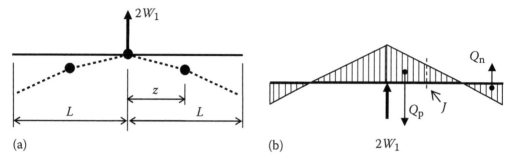

FIGURE 10.16 Postulated secondary joint pair (a) and necessary shape of acceleration pattern (b).

$$W_0 = \frac{2\mathcal{M}_0}{L} \tag{10.74}$$

After joint rotation begins, the center of the beam keeps accelerating. The next question that comes to mind is this: how big must W_0 be in order to induce formation of a pair of hinges, somewhere within the beam length, as shown in Figure 10.16a? This time, providing an answer requires more steps. An intermediate hinge requires that the bending moment attains a local peak, or that the shear force is zero at that location. To realize this, the inertia loading caused by acceleration must change its sign within length L, as shown in Figure 10.16b. (In the instance shown, the shear is zero at point J, which means a joint can appear at point J.) In addition to location as described, the magnitude of the applied force must be such that, going from the end, the resultant moment about J from Q_n and part of Q_p must equal $\mathcal{M}_0$. The detailed derivation is presented by Johnson in [40]. The magnitude of W_0 and the joint location are given in Case 10.32.

Everything, that was described above for a beam impacted with $2W_0H(t)$ is also true for an unconstrained (translation-wise) cantilever of length L, impacted at the base with $W_0H(t)$. This is illustrated by Case 10.31.

Some Cases of Strain Distribution along a Beam

A plastic hinge is a convenient abstraction; an axis point where the slope changes suddenly. In reality, the change of slope must take place over a certain length of the beam. Once a rotation at a joint is found from analysis of displacements, this rotation must be related to strain that it induces. To achieve this, one needs to know the length ℓ_h of a joint, as shown in Figure 10.17.

It is not easy to decide what joint length to use, as several authors mention different multiples of H. The experimental results of large-deformation impacts, such as presented by Liu and Jones [56], for example, indicate that the change of slope can be quite abrupt and takes place over a short segment. For this reason one can use $\mathcal{L}_h = H$ (or less) or the joint at the end of a beam, as in Figure 10.17. Then, for joints that form at intermediate points, one would take $\ell_h = 2H$. With the joint rotation prescribed as θ, we have $\theta = \ell_h/\rho$ or $\rho = \ell_h/\theta$, which defines the outer fiber strain, per Equation 10.1:

$$\varepsilon_1 = \frac{h_1}{\rho} = \frac{h_1\theta}{\ell_h} \tag{10.75}$$

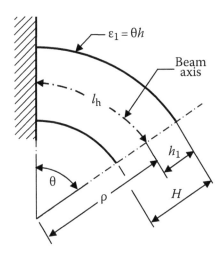

FIGURE 10.17   Deformation of a plastic joint of a finite length, undergoing rotation θ. Outside the joint the beam is assumed to be straight.

Suppose that a rotation near the fixed point, $\theta = 36°$, is prescribed and that the section is a rectangle, $h_1/H = 1/2$. Assuming $\ell_h = H$ gives the corresponding strain as $\theta/2 = 0.6283/2 = 0.314$.

A thorough discussion of plastic hinges is given by Alves and Jones [1]. One possible definition of a hinge length is a distance between a point on the beam axis where onset of yielding takes place and a point where the full value of plastic moment is attained. The same method of approach can be used for shear loading and the associated hinge (slide) lengths. This is simple for basic cases of support, where static moment distributions are used to determine those locations. It is not usually practical to do so for impact loading. It seems that $H/2$ as a hinge length may often be a reasonable choice, for both bending joints and shear slides. However, when deflection grows in presence of a large axial force, this assumption may not be adequate, as further discussed in Chapter 11.

There are two major reasons for constructing models involving plastic joints. One is for the determination of the peak deflection attained as a result of applied load and the other is to establish a load limit or deflection limit associated with failure. It appears that different joint lengths have to be used for those purposes. More details on this subject are provided in Chapter 15.

A plastic joint is an extreme case of deflections (and permanent strains) localized in one area. A different situation is presented by a deformation pattern of a cantilever tip impacted by a light striker, Figure 10.18. The strain and deflection are smoothly varying along the beam reaching their peaks at the tip. The dashed curve shown here is the experimental

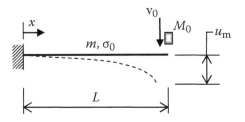

FIGURE 10.18   A RP cantilever tip impacted by a high-velocity striker.

result of the impact. A realistic picture of this deformed shape is obtained by assuming that the outer fiber strain varies, for example, as

$$\varepsilon = \varepsilon_y + \frac{x^2}{L^2}\varepsilon_e \tag{10.76}$$

along the beam. If a beam segment of length l is bent with a constant bending moment $\mathcal{M}_0$, the work done by the moment is

$$\mathcal{L} = \mathcal{M}_0\theta = \mathcal{M}_0\frac{l}{\rho} = \mathcal{M}_0 l\frac{2\varepsilon}{H} \tag{10.77}$$

The last equality was written on the basis of Equation 10.1 and is valid for sections symmetric about the bending axis, a rectangle in this case. When ε varies continuously, l is replaced by dx and the incremental work is integrated along the whole length. When this is equated with the kinetic energy of the striker, one can find the peak strain ε_e. (See Case 10.35.) By integrating the curvature and then the slope one finds the tip deflection as

$$u_m = \frac{L^2}{H}\left(\varepsilon_y + \frac{1}{6}\varepsilon_e\right) \tag{10.78}$$

The above approach appears to give more faithful results when compared with graphical descriptions of the tests in Ref. [40] than what was obtained from the moving joint theory, as tried in Example 10.17.

The partition of kinetic energy of the projectile depends on the mass ratio $\gamma = M/(mL)$, as presented in Case 10.33. For light strikers, $\gamma \ll 1$, almost all energy is lost in the pattern shown in Figure 10.18. For heavy strikers, $\gamma \gg 1$, nearly all energy is dispersed in root section bending, in Figure 10.17. The energy is partitioned equally, when $\gamma = 0.694 \approx 0.7$. If only the tip deflection is of interest, then the root-bending pattern, as in Case 7.19, is a good approximation provided $\gamma \geq 1$. For $\gamma = 1$, only about 60% of the strain energy is absorbed by root bending, so the prediction of peak strain based on this mode would be conservative.

When beams are rapidly loaded, the strain-rate effect is likely to increase the effective magnitude of the moment capacity $\mathcal{M}_0$. This is due to the increase in yield strength, previously quantified by Equation 6.13. Stronge [84] quotes some results on mild steel beams to which loading was applied laterally by magnetomotive means. The largest increase in yield strength noted was 1.6, relative to quasistatic conditions. It seems unlikely that even under extreme loading conditions the relative increase will exceed the factor of 2.0 or 3.0.

The presence of strain-rate effects can greatly complicate hand calculations. One of the effective, relatively simple approaches is to assess the average strain rate in the most critical location, based on elastic action alone. The increased yield strength is then calculated from Equation 6.13 and used for subsequent analyses treating the material as rate-insensitive.

Deformation History When Plastic Joints Are Involved

The discussion will relate to the basic case of a cantilever beam struck at the tip, but, with minor modifications, it also applies to the effect of force at an interior point of a beam. So far the stationary and the traveling joint concepts were treated as two different approaches. However, the two are inseparable when the applied load has a form of a finite-duration pulse. Once the pulse terminates the joint ceases to be stationary and begins to move towards the root of a beam. After arriving at the root, the joint becomes stationary again and the whole beam begins to rotate about that joint until it comes to a stop, when its kinetic energy is exhausted. When the process is complete, the deformed pattern is as follows. There is a kink at the first and the last joint, a gently deformed length of beam between the two.

A creation of a stationary hinge, as in Figure 10.13, and further developments are discussed in detail by Stronge [84]. The energy input is $W_0 u_t$, where u_t is found from Equation 10.70b. One-third of this energy is dissipated during the stationary joint formation. The part that is lost in the final phase, namely, rotation about the base, depends on the ratio of the applied force and the collapse load, W_0/W_b. The larger the ratio, the less deformation takes place in the final phase. From an engineering viewpoint two locations are of possible concern: the initial plastic joint and the base joint. Both must be checked whether straining is within permissible limits.

A continuously traveling joint, as presented in section "Infinite or semi-infinite beam, stationary joints, and traveling joints," is typical effect of a large applied point load, which then rapidly decreases. One of the most important examples is a hard mass–beam impact. Two phases of motion may be then distinguished: the joint travel as the initial phase and base rotation as the final. The initial phase is characterized by a gradually changing beam curvature with the largest straining taking place near the tip. How much energy is absorbed in each phase is a function of the mass ratio, $\gamma = M_0/(mL)$ and was given previously. In summary, a fast, small mass can break (or badly bend) a cantilever near the tip, but a large mass will break it near the base.

When a FEA of an impact is carried out using an EPP material, the results should be a step closer to physical reality and, one would think, not that much different from the RP model. However, further complications arise. An impacted end emanates two types of waves: an elastic and a plastic one, the latter in the form of a hinge. The first wave travels much faster, so it rebounds from the fixed end and interferes with the incoming plastic hinge. The results of such collisions may vary between a disappearance of the hinge and a permanent kink forming somewhere about the mid-length of cantilever.

SPECIAL TOPICS

High-Speed Collision of Two Beams

The intensity of the impact is assumed to be large enough for the beams to achieve moderate deflections. The setting of the problem, originally presented by Teng [107], is in Figure 10.19. The beams are at right angles to each other and one of them stationary, while the other is moving with a velocity v_2. The end conditions are not stated, which is the other way of saying that time interval of interest is so short, that those conditions are not relevant. When the contact takes place, the overlapping rectangles $(B_1 \times B_2)$ acquire a common velocity V_0.

A simplifying assumption is also made that bending rigidity can be ignored compared with string-like action; therefore the deformed beam segment can be treated as a cable.

Additionally, the material is modeled as RP with the flow strength of σ_0. Some simple solutions for cables were presented in Chapter 9 and the same methodology will be employed here. For each beam the mass involved in momentum conservation will be $Bm + 2\xi m$, where

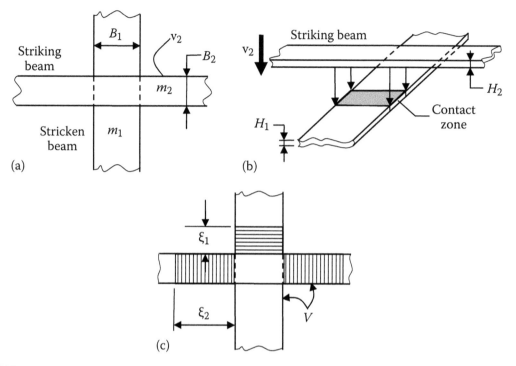

FIGURE 10.19   The beams shown in (a) plan view and (b) isometric view before collision. In (c) the view is after collision, with lateral deformation spreading.

B is the width of the center segment and ξ is the mobilized length per side, $\xi = ct$. Symbol c used here stands for the speed of propagation of lateral deflections* of a cable preloaded with the yield load $BH\sigma_0$. The velocity of the striking beam decreases from v_2 to V, while the other one gains a velocity V. Equating the change of momenta one writes

$$(m_2 B_1 + 2m_2\xi_2)\,(v_2 - V) = (m_1 B_2 + 2m_1\xi_1)V \tag{10.79}$$

Differentiating both sides with respect to time gives the following equation of motion:

$$(m_1 B_2 + m_2 B_1 + 2m_1\xi_1 + 2m_2\xi_2)\frac{dV}{dt} + (2m_1 c_1 + 2m_2 c_2)V = 2m_2 c_2 v_2 \tag{10.80}$$

in which time is hidden in variables $\xi_1 = c_1 t$ and $\xi_2 = c_2 t$. According to Teng [107], the solution for velocity is

$$V(t) = \frac{v_2}{1 + \dfrac{m_1(B_2 + 2c_1 t)}{m_2(B_1 + 2c_2 t)}} \tag{10.81}$$

---

*This speed is designated by c_2 throughout this book, but this index is dropped here for the sake of clarity. Consequently, c_1 and c_2 refer to the first and the second beam.

The ratio in the denominator is simply that of mobilized masses of the colliding beams. At $t = 0$ one has

$$V_0 = \frac{v_2}{1 + \frac{m_1 B_2}{m_2 B_1}}$$

(10.82)

because of the initial collision of the overlapping rectangles. (This is consistent with a more general Equation 8.1.) One should note that an exceptionally simple form of Equation 10.81 exists when the beam width is negligible; the common velocity V is then independent of time. A detailed study in Ref. [107] shows that the deflected shape of either beam following impact can be concave, convex, or straight. Either of the two beams, or both can fail due to excessive strain. The break point can be near the contact zone or at some intermediate location. While the exact theory is quite complex, some simplifications are presented in Case 10.38. They are mainly based on the fact that $V(t)$ changes with time rather slowly and that the duration of interest cannot be too long, because of the finite length of the real beams involved.

One should be aware of the fact that in problems like this, where resistance to deformation comes from inertia of the beam itself, rather than from the supports, relatively large impact speeds, or, more importantly, large accelerations on impact are needed to cause rupture.

Beams on Elastic Foundation

An elastic foundation works as a dense row of springs with stiffness k_f per unit length of beam. The static differential equation of the deflected line of a flexural beam is obtained from Equation 2.26 by ignoring the inertia term:

$$EIu^{IV} + k_f u = w(x)$$

(10.83)

When a lateral force W is applied to the tip of a semi-infinite beam, as in Figure 10.20b, the expressions for displacement, rotation, bending moment, and shear force can be written, according to Hetényi [33]:

$$u(x) = \frac{2W\lambda}{k_f} e^{-\lambda x} \cos \lambda x \quad \text{with } \lambda^4 = \frac{k_f}{4EI}$$

(10.84a)

$$\theta(x) = \frac{2W\lambda^2}{k_f}(\cos \lambda x + \sin \lambda x)e^{-\lambda x}$$

(10.84b)

$$\mathcal{M}(x) = \frac{W}{\lambda} e^{-\lambda x} \sin \lambda x$$

(10.84c)

$$Q(x) = W(\cos \lambda x - \sin \lambda x)e^{-\lambda x}$$

(10.84d)

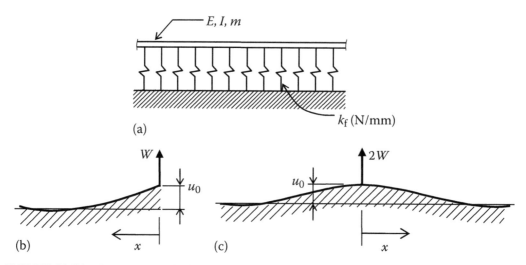

(a)

(b) x (c) x

FIGURE 10.20   Beam on elastic foundation (a), semi-infinite beam with end load (b) and infinite with a point load (c).

When the force $2W$ is applied at some point of an infinite beam (Figure 10.20c), the relations are

$$u(x) = \frac{W\lambda}{k_f}(\cos\lambda x + \sin\lambda x)e^{-\lambda x}$$ (10.85a)

$$\theta(x) = \frac{2W\lambda^2}{k_f}e^{-\lambda x}\sin\lambda x$$ (10.85b)

$$M(x) = \frac{W}{2\lambda}(\cos\lambda x - \sin\lambda x)e^{-\lambda x}$$ (10.85c)

$$Q(x) = We^{-\lambda x}\cos\lambda x$$ (10.85d)

Sandwich Beams

Sandwich beams have outer facing plates and a solid, usually a lightweight, weaker core. The plates are intended to carry axial loads and bending, while the core resists shearing. In accordance with Figure 10.21, the effective axial area and the neutral axis location are, respectively,

$$A = B(h_1 + h_2)$$ (10.86a)

$$c = \frac{Hh_2}{h_1 + h_2}$$ (10.86b)

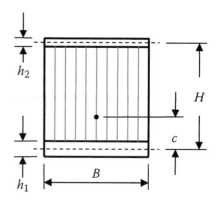

FIGURE 10.21   Section of a sandwich beam.

while in bending the second area moment is

$$I = \frac{BH^2 h_1 h_2}{h_1 + h_2} \qquad (10.87)$$

as long as the facing plates are thin. If the sheet thicknesses are the same, i.e., $h_1 = h_2 = h$, then

$$A = 2Bh \qquad (10.88a)$$

$$c = H/2 \qquad (10.88b)$$

$$I = BH^2 h/2 \qquad (10.88c)$$

The shear stiffness is $GA_s \approx G_c BH$ and the shear stress: $\tau = Q/A_s$, where G_c stands for the shear modulus of core material. This construction is very efficient in resisting bending, but the negative feature is the decreased shear stiffness, when compared with a monolithic construction. Also, a number of failure modes exist, which appears not only in a sandwich, but in composite structures in general. Those include buckling of facing plates with a possible separation from the core, crushing of the core, and shear failure between the core and a plate.

Laminated Beams

Laminated beams are beams made up of layers, or *laminae*. The purpose of such construction is structural efficiency usually achieved by placing the stiffest and the strongest layers on the outer surfaces. In what follows, the treatment will be limited to beams of doubly symmetric section, not only with respect to geometry, but also in regard to mechanical properties. The first and obvious problem is a determination of the effective stiffness properties of such a layered beam, so that, at least for some part of the analysis, one can treat the section as homogeneous. The postulate of the plane sections remaining plane is preserved here as well.

The axial stiffness parameter $\bar{E}A$ of the beam is formulated here as a product of the equivalent tensile modulus $\bar{E}$ and the full section area A. The definition implies that $\bar{E}A$ is a sum of rigidities of the component layers. In accordance with Figure 10.22:

$$\bar{E}A = 2\sum E_i A_i = 2B\sum E_i \delta_i = 2B(E_1\delta_1 + E_2\delta_2 + \cdots + E_n\delta_n) \qquad (10.89)$$

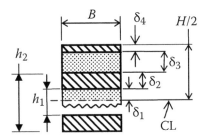

FIGURE 10.22   A symmetrical, laminated beam. The half-section, above the centerline, is shown as well as one layer only below the centerline.

where summation extends to all layers above the centerline. A similar procedure holds for the bending stiffness, where $\breve{E}$ is the equivalent bending modulus. Each layer k contributes $BE_k(h_k^3 - h_{k-1}^3)/12$ to the overall rigidity. The total equivalent bending stiffness therefore is

$$\breve{E}I = \frac{B}{12}\sum E_i(h_i^3 - h_{i-1}^3) = (B/12)\left[E_1(h_1^3 - 0) + E_2(h_2^3 - h_1^3) + \cdots + E_n(h_n^3 - h_{n-1}^3) \right] \quad (10.90)$$

The composite beam may now be treated, as far as the response is concerned, in the same manner as a homogeneous beam, provided that $\bar{E}A$ and $\breve{E}I$ are used in place of appropriate axial and bending terms. When the loads are known, the stress levels can be recovered. If the axial strain is $\varepsilon_a = P/(\bar{E}A)$ then the stress level in layer k is

$$\sigma_k = E_k \frac{P}{\bar{E}A} = \frac{E_k}{\bar{E}}\sigma_a \quad (10.91)$$

where $\sigma_a = P/A$ is the average section stress. The same result is obtained if one uses a computer program, where the beam is treated as homogeneous and then the output stress σ_a is corrected by the $E_k/\bar{E}$ ratio. The obvious result of the variation in stiffness along the depth is the variation of axial stress, while strain remains constant.

A similar approach holds in bending. With $\bar{E}$ as an equivalent bending modulus, the strain and stress in the kth layer are, respectively,

$$\varepsilon_k = \frac{My_k}{\breve{E}I} \quad (10.92a)$$

$$\sigma_k = \frac{E_k}{\bar{E}}\frac{My_k}{I} \quad (10.92b)$$

where the first equation is written on the basis of Equation 10.4, replacing c with a coordinate y_k.

The calculation of *interlaminar shear*, or the shear stress in a horizontal section, acting parallel to the beam axis, is somewhat more complicated. First, consider the equation for the shear in a homogeneous beam section, at the level y above the neutral axis:

$$\tau_{xy} = \frac{QZ_y^c}{BI} \quad \text{with} \quad Z_y^c = B(c - y)y_m \quad (10.93)$$

where

Z_y^c is the second area moment of the part of the section above y-coordinate, as indicated in Figure 10.23a

y_m is the center coordinate of that part

(This can be found in Popov [69], for example, and adjusted for the case of a constant width B.) Q stands, as usual, for the shear force, resulting from the distributed load $w(x)$. This can be rewritten to visualize the change of axial force:

$$B\tau(\Delta x) = \Delta P = \frac{QZ_y^c}{I}\Delta x \quad (10.94)$$

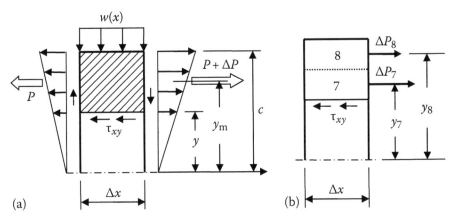

FIGURE 10.23 Increment of axial force balanced by axial shear. Homogeneous beam (a) and laminated beam (b).

If this is the top layer k of a composite beam, with a modulus E_k, the only difference now is the expression for bending stress. Following Equation 10.92b, this leads to

$$B\tau_k(\Delta x) = \Delta P_k = \frac{E_k}{\tilde{E}}\frac{QZ_{yk}^c}{I}\Delta x \tag{10.95}$$

For the sake of clarity, a more general expression will now be written for a specific case of a beam with eight layers above the neutral axis. The variable of interest is the shear stress below layer 7, as per Figure 10.23b. Equation 10.95, applied for the force increments ΔP_7 and ΔP_8 gives, in effect:

$$\tau_{xy} = \frac{Q}{B\tilde{E}I}(E_7\delta_7 y_7 + E_8\delta_8 y_8) \tag{10.96}$$

Example 10.11 illustrates the flow of calculation. One should be aware that the structural efficiency of the laminar beams is achieved with respect to bending. At the same time sensitivity to shear is increased, especially in dynamic applications.

Dynamic Strength of a Circular Ring in Radial Motion

The equation of motion may be constructed by considering a fragment of the ring as shown in Case 2.43, in particular a unit length of such a ring as measured along the median line. It has a mass ρA and its static stiffness is $k_{st} = w/u = EA/(Rr_i)$. The equation of motion corresponds to a simple oscillator:

$$\rho h \ddot{u} + ku = \frac{r_i}{R}w(t) \tag{10.97}$$

where the distributed load w is reduced because of being applied to the inside, rather than to the median radius R. The more convenient form is

$$\ddot{u} + \omega^2 u = \frac{r_i}{R}\frac{w(t)}{\rho h} \tag{10.98}$$

$$\omega^2 = E/(\rho R^2) \tag{10.99}$$

Natural frequencies of rings can be found in Chapter 2.

Closing Remarks

A set of beams with classical end conditions is presented in Figure 10.3. Those conditions are often mentioned and, one might say, they form the backbone of our numerical results. Some thought is therefore needed on the relevance of those conditions to the physical world in which the constraints are unlikely to act in a rigid or nearly rigid way. Beginning with an SS beam, a fair comment would be that the flexibility of the supports, if significant, can be subtracted from a total measured deflection and the remainder would be the beam component, as usually defined. When we progress to another case, namely a SS beam with axially restrained ends, the situation becomes more involved. It is not easy to provide a near-perfect axial constraint. Empirically measured deflection will be larger than one calculated using a perfect axial constraint as a basis. Again, the additional deflection may be subtracted from the total, netting a beam and/or cable component, but with much greater difficulty.

When there is an end clamping in addition to a lateral restraint, the situation further complicates, especially in the presence of an axial constraint. "Clamping" normally implies a zero end slope, but this is very difficult to attain in physical experiments. Such clamping is usually imperfect, which means another variable (support flexibility) distorts the anticipated deflection pattern. In total, the magnitude of deflection will be larger and the end bending moment smaller, when flexibility of constraints is included in the picture. This problem will be further discussed in Chapter 11.

One of the realistic experimental setups, where these problems are alleviated or removed, is made possible by symmetry. A beam with a plane of symmetry across a middle support, for example, offers zero slope and a perfect axial constraint over that support.

If a beam has a hollow cross section, the additional deflections caused by the deformation of that section must be included. If the impact, be it by means of a prescribed force or a flying mass, is localized, then the local deflections may overshadow the global component. This will render inapplicable some of the cases presented here, or at least there will be a need for adjustments.

The plastic joints theory as herein presented arose based on results of experiments on metallic elements. It has only a limited application to brittle structures, as reinforced concrete. Still, when the applied bending moment exceeds the initial cracking moment, one can anticipate a local damage and a joint-like behavior at some distance from the impact point.

In order to assess the onset of yielding or, sometimes, a safety factor, one has to combine the influence of the load components. This topic is presented in Chapter 11, where the need for such combinations is more pronounced.

TABULATION OF CASES

CASE 10.1   PURE BENDING OF RECTANGULAR SECTION, NONLINEAR MATERIALS

Onset of yielding when $\mathcal{M} = \mathcal{M}_y = \dfrac{IF_y}{c} = \dfrac{BH^2}{6} F_y$

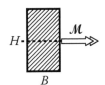

(a) Beam material: bilinear

$$\frac{\mathcal{M}}{\mathcal{M}_y} = \frac{3}{2}\left(1 - \frac{1}{3e^2}\right)\left(1 - \frac{E_p}{E}\right) + \frac{E_p}{E} e \quad \text{with } e = \frac{\varepsilon}{\varepsilon_y} > 1; \text{ Ref. [85].}$$

(b) Beam material: RSH

$$\mathcal{M} = (\sigma - F_y)\frac{BH^2}{6} + \frac{BH^2}{4} F_y \quad \text{with } \sigma > F_y$$

(c) Beam material: EPP; $\mathcal{M}_y \equiv 2\mathcal{M}_0 / 3$; $F_y = \sigma_0$

$$\frac{\mathcal{M}}{\mathcal{M}_y} = \frac{3}{2}\left(1 - \frac{1}{3e^2}\right) \quad \text{with } e = \frac{\varepsilon}{\varepsilon_y} > 1$$

(d) Beam material: PL (power law: $\sigma = B\varepsilon^n$)

$$\frac{\mathcal{M}}{\mathcal{M}_y} = \frac{3}{2+n}\left\{e^n - \frac{1-n}{3e^2}\right\} \quad \text{with } e = \frac{\varepsilon}{\varepsilon_y} > 1. \text{ Ref. [84].}$$

Note: For material definitions see Chapter 4.

CASE 10.2   PURE BENDING OF A SYMMETRIC, TWO-FLANGE SECTION, NONLINEAR MATERIALS

Onset of yielding when $\mathcal{M} = \mathcal{M}_y = \dfrac{AH}{2} F_y$

(a) Beam material: bilinear

$$\mathcal{M} = \mathcal{M}_y + \frac{AH}{2} E_p(\varepsilon - \varepsilon_y); \quad \text{with } \varepsilon > \varepsilon_y$$

(b) Beam material: RSH; $M = M_y + \dfrac{AH}{2} E_p \varepsilon$

(c) Beam material: EPP; $M_y = M_0 = \dfrac{AH}{2} \sigma_0$; with $\varepsilon > \varepsilon_y$

(d) Beam material: RP; $M = M_0 = \dfrac{AH}{2} \sigma_0$

(e) Beam material: RO (Ramberg–Osgood)

$$\frac{1}{\rho} = \frac{M}{EI} + \left(\frac{M}{E_n I_n} \right)^n ; \quad I_n = A \left(\frac{H}{2} \right)^{1 + \frac{1}{n}}$$

Note: For material definitions see Chapter 4. In definition of RO the strain–stress curve replaces the usual stress–strain. Ref. [86].

CASE 10.3   PURE BENDING OF A ROUND SECTION, NONLINEAR MATERIALS

Onset of yielding when $M = M_y = \dfrac{I F_y}{c} = \dfrac{\pi R^3}{4} F_y$

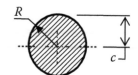

(a) Beam material: RSH

$$M = (\sigma - F_y) \frac{\pi R^3}{4} + \frac{4 R^3}{3} F_y \quad \text{with } \sigma > F_y$$

(b) Beam material: EPP; $F_y \equiv \sigma_0$

$$\frac{M}{M_y} = \frac{2}{3\pi} \left(5 - \frac{2}{e^2} \right) \sqrt{1 - \frac{1}{e^2}} + (3e) \arcsin \left(\frac{1}{e} \right) \quad \text{with } e = \frac{\varepsilon}{\varepsilon_y} > 1$$

Note: For material definitions see Chapter 4. Ref. [84].

CASE 10.4   ELASTIC CANTILEVER (CF BEAM) UNDER A RECTANGULAR PULSE LASTING t_0

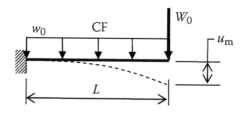

(a) Distributed load w_0

$$u_{st} = \frac{1}{8}\frac{w_0 L^4}{EI} + \frac{1}{2}\frac{w_0 L^2}{GA_s}; \; M_{st} = w_0 L^2/2; \; Q_{st} = w_0 L$$

$$u_d = 2u_{st}\sin(\pi t_0/\tau);$$

$$M_d = 2M_{st}\sin(\omega t_0/2); \text{ or } M_d = 0.47 W_0 l_2; \text{ whichever larger, } l_2 = 1.887\left(\frac{EI}{m}\right)^{1/4}\sqrt{t}$$

$$Q_d = 2Q_{st}\sin(\pi t_0/\tau), \text{ or}$$

$$Q_d' = \left(0.68\frac{u_d}{u_{st}} + 0.32\right)w_0 L \text{ (for } t_0 < 0.1\tau)$$

(b) Point load W_0

$$u_{st} = \frac{1}{3}\frac{W_0 L^3}{EI} + \frac{W_0 L}{GA_s}; \; M_{st} = W_0 L; \; Q_{st} = W_0$$

$$u_m = u_d = 2u_{st}\sin(\pi t_0/\tau); \text{ peak dynamic deflection}$$

$$DF(u) = u_m/u_{st}; \text{ dynamic factor (for displacements)}$$

$$M_d = W_0 l_1 \text{ or } M_d = \frac{u_d}{u_{st}}M_{st}; \text{ whichever larger; } l_1 = 1.496\left(\frac{EI}{m}\right)^{1/4}\sqrt{t}$$

$$Q_d = \left(6.013\frac{u_d}{u_{st}} + 1.415\right)W_0; \text{ but } Q_d \le 5Q_{st} \text{ (*)}$$

Note: These peak dynamic response formulas are approximations only. For coefficients related to other support conditions refer to the text. The coefficients in (*) are intended for $t_0 \approx 0.1\tau$.

CASE 10.5   PEAK RESPONSE OF A DAMPED BEAM RESULTING FROM A SHORT PULSE $w_0 t_0$

(Damping $\zeta = 0.05$). Response components:

$$u_m = \frac{(w_0 t_0)L^2 d}{\sqrt{EIm}}; \quad M_m = (w_0 t_0)d\sqrt{\frac{EI}{m}}; \quad Q_m = \frac{(w_0 t_0)d}{L}\sqrt{\frac{EI}{m}}$$

Beam Supports	Component	Maximum d	Minimum d
	$u(L)$	0.426	−0.348
	$M(0)$	−1.665	1.421
	$Q(0)$	3.420	−2.569

(continued)

Beam Supports	Component	Maximum d	Minimum d
SS	$u(L/2)$	0.119	−0.102
	$M(L/2)$	1.381	−1.028
	$Q(0)$	6.315	−3.653
SC	$u(0.4L)$	0.0810	−0.0675
	$M(0.4L)$	1.296	−0.967
	$M(L)$	−2.093	1.409
	$Q(0)$	5.104	−4.834
	$Q(L)$	−6.128	4.272
CC	$u(L/2)$	0.0533	−0.0469
	$M(L/2)$	1.132	−0.872
	$M(0)$	−1.997	1.425
	$Q(0)$	12.988	−7.487
CG	$u(L)$	0.2132	−0.1876
	$M(0)$	−1.997	1.425
	$Q(0)$	6.494	−3.744

Note: The first four cases come from Henrych [32] and the fifth is scaled from the fourth. Positive displacement u is in the direction of applied load. Signs of moments and shears are according to general convention. "Maximum" means the largest absolute value of the coefficient. The results are based on an Euler beam with no shear flexibility.

CASE 10.6 PEAK TRANSIENT BEAM RESPONSE DUE TO FREE-END POINT LOAD

$$l_1 = 1.496\left(\frac{EI}{m}\right)^{1/4}\sqrt{t}; \quad u_1 = \frac{W_0 l_1^3}{3EI}$$

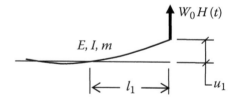

For a cantilever of length L

$$\frac{M}{M_{st}} = 0.05 + 3.24\frac{t}{\tau} \quad (\text{For } l_1 < L)$$

M_{st} is the peak static moment, τ is natural period.

For very small values of time, $\dfrac{M}{M_{st}} = 0.837\sqrt{\dfrac{t}{\tau}}$ $(t < \tau/170)$

Note: The peak moment appears at about $l_1/2$. Ref. [87].

CASE 10.7   PEAK TRANSIENT BEAM RESPONSE DUE TO A POINT LOAD APPLIED FAR FROM AN END

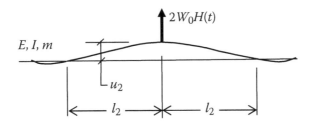

$$l_2 = 1.887 \left(\frac{EI}{m} \right)^{1/4} \sqrt{t}; \quad u_2 = \frac{W_0 l_2{}^3}{12EI}$$

For a beam of length $2L$

$$\frac{\mathcal{M}}{\mathcal{M}_{st}} = 0.05 + \frac{1}{2\pi} \left(4\omega t + \sin(4\omega t) \right); \quad (l_2 < L)$$

$\mathcal{M}_{st} = W_0 L/2$ is the peak static moment, ω is the circular frequency.

For very small values of time, $\dfrac{\mathcal{M}}{\mathcal{M}_{st}} = 1.677 \sqrt{\dfrac{t}{\tau}} \ (t < \tau/100)$

Note: The peak moment location is at the loaded point. The results are also valid for a beam with a guided end, l_2 long and loaded with W_0. Ref. [87].

CASE 10.8   TRANSIENT BEAM RESPONSE DUE TO A CONSTANT DRIVING VELOCITY v_0 APPLIED FAR FROM AN END

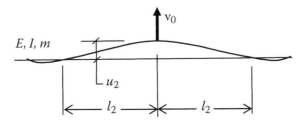

$u_2 = v_0 t$

$\mathcal{M} = v_0 \sqrt{EIm}$; peak moment, at driven point

Note: Deformed length l_2 can be assumed as for Case 10.7, for the same u_2. The results are also valid for a guided beam, l_2 long, one end driven with v_0. Ref. [26].

CASE 10.9   REGID-PLASTIC BEAM WITH DISTRIBUTED DEFORMATION, CF

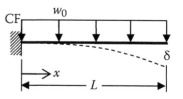

$S_0 = w_0 L t_0$; short impulse applied

$$E_k = \frac{M v_0^2}{2} = \frac{S_0^2}{2M}; \text{ kinetic energy supplied}$$

$$\Pi = \frac{\pi \mathcal{M}_0 \delta}{2L}; \text{ strain energy}$$

$$\delta = \frac{L S_0^2}{\pi M \mathcal{M}_0}; \text{ peak deflection}$$

$$\varepsilon_m = \left(\frac{\pi}{2L}\right)^2 c\delta; \text{ peak strain}$$

For a rectangular section $B \times H$: $\mathcal{M}_0 = BH^2\sigma_0/4$; $M = mL = \rho BHL$

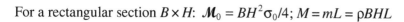

$$\Pi = \frac{\pi BH^2 \delta \sigma_0}{8L}; \quad \delta = \frac{4S_0^2}{\pi B^2 H^3 \rho \sigma_0}; \quad \varepsilon_m = \left(\frac{\pi}{2L}\right)^2 \frac{H}{2}\delta$$

CASE 10.10   REGID-PLASTIC BEAM WITH DISTRIBUTED DEFORMATION, SS

(Variables identical with Case 10.9 not repeated here.)

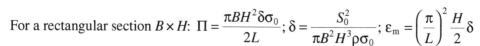

$$\Pi = \frac{2\pi \mathcal{M}_0 \delta}{L}; \text{ strain energy}$$

$$\delta = \frac{L S_0^2}{4\pi M \mathcal{M}_0}; \text{ peak deflection}$$

$$\varepsilon_m = \left(\frac{\pi}{L}\right)^2 c\delta; \text{ peak strain}$$

For a rectangular section $B \times H$: $\Pi = \frac{\pi BH^2 \delta \sigma_0}{2L}; \delta = \frac{S_0^2}{\pi B^2 H^3 \rho \sigma_0}; \varepsilon_m = \left(\frac{\pi}{L}\right)^2 \frac{H}{2}\delta$

CASE 10.11   RIGID-PLASTIC BEAM WITH DISTRIBUTED DEFORMATION, CC

(Variables identical with Case 10.9 not repeated here.)

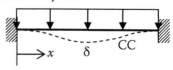

$$\Pi = \frac{4\pi \mathcal{M}_0 \delta}{L}; \text{ strain energy}$$

$$\delta = \frac{L S_0^2}{8\pi M \mathcal{M}_0}; \text{ peak deflection}$$

$$\varepsilon_{\mathrm{m}} = \left(\frac{2\pi}{L}\right)^2 \frac{c\delta}{2}; \text{ peak strain}$$

For a rectangular section $B \times H$

$$\Pi = \frac{\pi B H^2 \delta \sigma_0}{L}; \quad \delta = \frac{S_0^2}{2\pi B^2 H^3 \rho \sigma_0}; \quad \varepsilon_{\mathrm{m}} = \left(\frac{2\pi}{L}\right)^2 \frac{H\delta}{4}$$

CASE 10.12 ELASTIC CANTILEVER (CF BEAM) SUBJECTED SIMULTANEOUSLY TO THE INITIAL SHORT IMPULSE $s = w_1 t_1$ AND A RECTANGULAR IMPULSE $w_0 t_0$

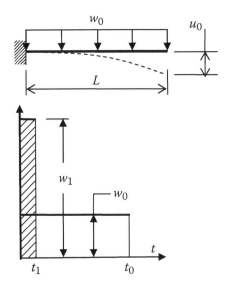

$v_i = s/m = w_1 t_1/m$; initial velocity from the short impulse

u_1 is the maximum displacement from this first impulse alone

$$u_0 = \frac{1}{8}\frac{w_0 L^4}{EI} + \frac{1}{2}\frac{w_0 L^2}{GA_{\mathrm{s}}}; \text{ static deflection due to } w_0 \text{ alone}$$

(a) Initial impulse followed by a step load $w = w_0 H(t)$; $(t_0 > \tau/2)$

$u(t) = u_0(1 - \cos\omega t) + \dfrac{v_i}{\omega}\sin\omega t$; resultant tip deflection

$u_{\mathrm{m}} = (u_0^2 + u_i^2)^{1/2} + u_0$; peak deflection from the combined action

$u_i = \dfrac{v_i}{\omega}$; deflection from the initial impulse alone

(b) Initial impulse followed by a rectangular pulse

$u_{\mathrm{m}} \approx (u_{\mathrm{p}}^2 + u_i^2)^{1/2}$; where u_{p} is the effect of the longer pulse acting alone

Note: These expressions for peak dynamic response holds for other end conditions as well, as long as SDOF is a reasonable approximation.

CASE 10.13   MASS–BEAM IMPACT, SHEAR WAVE METHOD

Beam inertia is represented by M_m and its deformability by damper C. M is the striker mass.

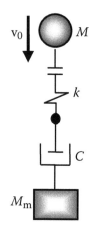

$R_{ms} = v_0\sqrt{kM^*}$; peak force when damper C is rigid

M^* is equal to the striker mass M, but not larger than the modal mass M_m. (See Table 2.1 for modal masses)

$R_{md} = Cv_0/2$; peak force when spring k is rigid

$C = \sqrt{GA_s m}$; for a cantilever impacted at the tip

$C = 2\sqrt{GA_s m}$; for an intermediate point impact

$\dfrac{1}{R_m} \approx \dfrac{1}{R_{ms}} + \dfrac{1}{R_{md}}$; to calculate the actual peak force R_m

$$t_0 = \frac{2Mv_0}{R_m}; \text{ duration of the loading phase of impacts}$$

Note: Duration of contact t_0 is calculated from momentum considerations. The formula is given for nil rebound velocity and the force–time plot resembling a triangle. For a sinusoidal force–time plot, use $\pi/2$ instead of 2. In case of a perfectly elastic contact the duration is $2t_0$. See text for further comments.

CASE 10.14   PORTAL FRAME WITH HEAVY AND STIFF CAPPING MEMBER

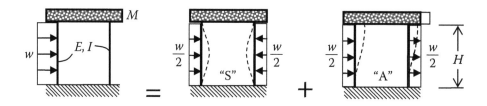

A short rectangular pulse of load w (N/mm) applied during t_0.

Natural frequencies of both modes are given by Case 2.29. The responses in different points at each mode are patterned after Case 10.4, i.e., using Tables 10.2 and 10.4.

Symmetric mode "S": CC beam under $w/2$

Antisymmetric mode "A": CG beam under $w/2$

Note: Modal responses are superposed. Superposition is strictly correct only in the elastic range.

CASE 10.15   A RIGID BEAM (CF) ACTED UPON BY IMPULSE S_0 APPLIED
BY A DISTRIBUTED LOAD w_0

Angular spring with RP characteristic and $\mathcal{M}_0$ capacity is at the base.

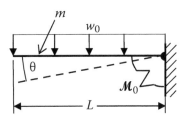

Shear capacity: $Q_0 = A_{ef}\tau_0$

$w_b = 2\mathcal{M}_0/L^2$; static collapse load in bending

$w_s = Q_0/L$; static collapse load in shear

$M_{ef} = \dfrac{mL}{3}$; effective mass at the tip

$w_0 > w_b$ or $w_0 > w_s$; the smaller of the two will govern

Short impulse $S_0 = Lw_0t_0/2$:

$v_0 = \dfrac{S_0}{M_{ef}}$; initial velocity, tip

$E_k = \dfrac{S_0{}^2}{2M_{ef}}$; kinetic energy

$\theta_m = \dfrac{E_k}{\mathcal{M}_0} = \dfrac{S_0{}^2}{2\mathcal{M}_0 M_{ef}} = \dfrac{3(w_0t_0)^2 L}{8\mathcal{M}_0 m}$; peak rotation; $u_m = L\theta_m$; peak displacement

Note: For tip loading with W_0 set $W_0 = w_0L/2$. Response is written for bending failure, for shear use Equation 10.55.

CASE 10.16   A RIGID BEAM (CF) WITH A TIP MASS, ACTED UPON BY IMPULSE S_0,
APPLIED BY DISTRIBUTED LOAD w_0

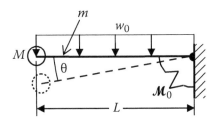

Angular spring with RP characteristic and $\mathcal{M}_0$ capacity is at the base.

$$M_{ef} = \dfrac{mL}{3} + M$$

The response formulas are the same as in Case 10.15, but with a different M_{ef}.

Note: For tip loading with W_0 set $W_0 = w_0L/2$ and $M_{ef} = \dfrac{mL}{3} + M$ and refer to Case 10.24.

CASE 10.17   A RIGID BEAM, (SS) ACTED UPON BY IMPULSE S_0, APPLIED BY DISTRIBUTED LOAD w_0

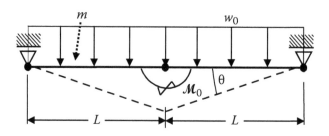

Angular spring with RP characteristic and $\mathcal{M}_0$ capacity at center.

$S_0 = w_0 L t_0$; equivalent center impulse

$Q_0 = A_{ef} \tau_0$; shear capacity, each end

$w_b = 2\mathcal{M}_0/L^2$; static collapse load in bending

$w_s = Q_0/L$; static collapse load in shear

$M_{ef} = \dfrac{2mL}{3}$; effective mass at center

$w_0 > w_b$ or $w_0 > w_s$; the smaller of the two will govern

Short impulse S_0: $v_0 = \dfrac{S_0}{M_{ef}}$; initial velocity, center

$E_k = \dfrac{S_0^2}{2M_{ef}}$; kinetic energy, $\theta_m = \dfrac{E_k}{2\mathcal{M}_0} = \dfrac{S_0^2}{4\mathcal{M}_0 M_{ef}}$; peak rotation

$u_m = L\theta_m$; peak displacement

Note: Response is written for bending failure, for shear use Equation 10.55

CASE 10.18   A RIGID BEAM, (SS) ACTED UPON BY IMPULSE S_0, APPLIED BY DISTRIBUTED LOAD w_0

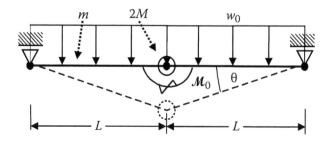

Angular spring with RP characteristic and $\mathcal{M}_0$ capacity at center.

$M_{ef} = \dfrac{2mL}{3} + 2M$; effective mass

The response formulas are the same as in 10.17, but with a different M_{ef}.

CASE 10.19 A RIGID BEAM, (SS) DESCRIBED AS IN CASE 10.17 WITH AN ADDITIONAL, SUSTAINED STATIC LOAD w_{st}

$w_b = 2M_0/L^2$; static collapse load in bending

$S_0 = w_0 L t_0$; equivalent center impulse

$M_{ef} = \dfrac{2mL}{3}$; effective mass at center

Short impulse S_0: $v_0 = \dfrac{S_0}{M_{ef}}$; initial velocity, center

$E_k = \dfrac{S_0^2}{2M_{ef}}$; kinetic energy,

(a) Find θ_m when S_0 or E_k prescribed

$$\theta_m = \frac{E_k}{2M_0(1 - M_{st}/M_0)} = \frac{S_0^2}{4M_{ef}M_0(1 - M_{st}/M_0)}; \text{ peak rotation}$$

(b) Find S_0 or E_k when θ_m prescribed

$$\frac{M_{ef}v_0^2}{2} = \frac{S_0^2}{2M_{ef}} = 2M_0(1 - M_{st}/M_0)\theta_m$$

Note: The response is written for a bending failure.

CASE 10.20 BEAM OF RP MATERIAL (SC) UNDER A RECTANGULAR PULSE LASTING t_0

Moment capacity: $M_0 = Z\sigma_0$, Shear capacity: $Q_0 = A_{ef}\tau_0$

Applied short impulse: $S = Lw_0 t_0$; $M = mL$

$w_b = 11.66 M_0/L^2$; static collapse load in bending

$w_s = 1.6 Q_0/L$; static collapse load in shear

$w_0 > w_b$ or $w_0 > w_s$; the smaller of the two will govern

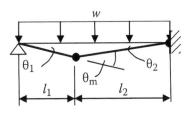

$l_1 = \dfrac{L}{1+\sqrt{2}}$; $l_2 = \sqrt{2}l_1$; $v_0 = \dfrac{w_0 t_0}{m}$; initial velocity

$u_s = \dfrac{S^2}{4MQ_0}$; maximum shear deflection at shear failure mode

$\theta_e = \dfrac{0.1553S^2}{MM_0}$; is the maximum θ_1; $\theta_m = 1.707\theta_e$; kink angle at bending failure

$u_m = l_1\theta_e$; peak displacement

Note: The bending mode of failure is illustrated, taking place when $w_b < w_s$. For coefficients related to other support conditions refer to the text. Ref. [101].

CASE 10.21 A RIGID BEAM (CC) ACTED UPON BY IMPULSE S_0 APPLIED BY DISTRIBUTED LOAD w_0

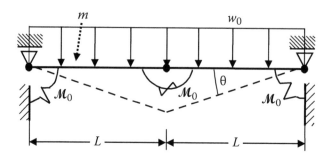

Angular springs with RP characteristic and $\mathcal{M}_0$ capacity at center and at ends.

$S_0 = w_0 L t_0$; equivalent center impulse

$Q_0 = A_{ef}\tau_0$; shear capacity of each end

$w_b = 4\mathcal{M}_0/L^2$; static collapse load in bending

$w_s = Q_0/L$; static collapse load in shear

$$M_{ef} = \frac{2mL}{3}$$

$w_0 > w_b$ or $w_0 > w_s$; the smaller of the two will govern

Short impulse S_0:

$v_0 = \dfrac{S_0}{M_{ef}}$; initial velocity, center

$E_k = \dfrac{S_0^2}{2M_{ef}}$; kinetic energy

$\theta_m = \dfrac{E_k}{4\mathcal{M}_0} = \dfrac{S_0^2}{8\mathcal{M}_0 M_{ef}}$; peak rotation; $u_m = L\theta_m$; peak displacement

Note: For other pulses take results from Case 10.26 by setting $W_0 = w_0 L/2$. The response is written for bending failure, for shear use Equation 10.55.

CASE 10.22 A RIGID BEAM, WITH A CENTER LUMPED MASS, ACTED UPON BY IMPULSE S_0 APPLIED BY DISTRIBUTED LOAD w_0

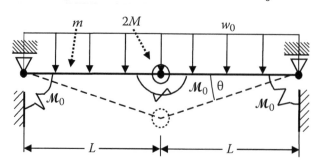

Angular springs with RP characteristic and M_0 capacity at center and at the ends.

$$M_{ef} = \frac{2mL}{3} + 2M$$

The response formulas are the same as in Case 10.21, but with a different M_{ef}.

CASE 10.23   A RIGID BEAM (CF) ACTED UPON BY IMPULSE $S = W_0 t_0$ APPLIED BY TIP FORCE W_0

Angular spring with RP characteristic and M_0 capacity is at the base.

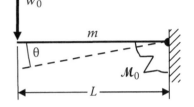

$W_0 L \gg M_0$

$u_m = L\theta_m$; peak displacement for the cases below

(a) Short impulse $S_0 = W_0 t_0$

$$v_0 = \frac{3S_0}{mL} \text{ (initial velocity);} \quad E_k = \frac{3(S_0)^2}{2mL} \quad \text{and} \quad \theta_m = \frac{3(S_0)^2}{2M_0 mL}$$

(b) Rectangular impulse $W_0 t_0$

$$\theta_m = \frac{3W_0 t_0^2}{2L^2 m}\left(\frac{W_0 L}{M_0} - 1\right); \quad \text{at } t_m = \frac{LW_0 t_0}{M_0}$$

(c) Triangular decreasing pulse $W = W_0(1 - t/t_0)$

$$\frac{W_0 L}{M_0} < 2; \ \theta_m = \frac{2W_0 t_0^2}{mL^2}\left(1 - \frac{M_0}{W_0 L}\right)^3; v = 0 \text{ when } t = t_x = \left(1 - \frac{M_0}{W_0 L}\right)$$

$$\frac{W_0 L}{M_0} > 2; \ \theta_m = \frac{W_0 t_0^2}{mL^2}\left(1 - \frac{3M_0}{2W_0 L}\right)^2 + \frac{mLv_e^2}{6M_0}$$

$$v_e = \frac{3W_0 t_0}{Lm}\left(\frac{1}{2} - \frac{M_0}{W_0 L}\right); \quad \text{velocity at } t = t_0$$

Note: When $W_0 \gg 3M_0/L$, a joint will appear at an interior point (refer to Case 10.27). If this is the case, the above response is an upper bound.

CASE 10.24   A RIGID BEAM, WITH A TIP MASS, ACTED UPON BY IMPULSE $S = W_0 t_0$, APPLIED BY POINT LOAD W_0

Angular spring with RP characteristic and M_0 capacity is at the base.

$$M_{ef} = \frac{mL}{3} + M$$

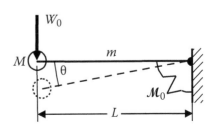

(a) Short impulse, $S_0 = W_0 t_0$. $(3\mathcal{M}_0 \gg W_0 L \gg \mathcal{M}_0)$

$$v_0 = \frac{S_0}{M_{ef}} \text{ (initial velocity)}; \quad E_k = \frac{S_0^2}{2M_{ef}} \quad \text{and} \quad \theta_m = \frac{E_k}{\mathcal{M}_0}$$

(b) Rectangular impulse $W_0 t_0$

$$\theta_m = \frac{W_0 t_0^2}{2L M_{ef}}\left(\frac{W_0 L}{\mathcal{M}_0} - 1\right); \quad \text{at } t_m = \frac{L W_0 t_0}{\mathcal{M}_0}$$

(c) Triangular decreasing pulse $W = W_0(1 - t/t_0)$

$$\frac{W_0 L}{\mathcal{M}_0} < 2; \quad \theta_m = \frac{2 W_0 t_0^2}{3 M_{ef} L}\left(1 - \frac{\mathcal{M}_0}{W_0 L}\right)^3; \quad v = 0 \text{ when } t = t_x = \left(1 - \frac{\mathcal{M}_0}{W_0 L}\right) 2 t_0$$

$$\frac{W_0 L}{\mathcal{M}_0} > 2; \quad \theta_m = \frac{W_0 t_0^2}{3 M_{ef} L}\left(1 - \frac{3\mathcal{M}_0}{2 W_0 L}\right)^2 + \frac{M_{ef} v_e^2}{2\mathcal{M}_0}$$

$$v_e = \frac{W_0 t_0}{M_{ef}}\left(\frac{1}{2} - \frac{\mathcal{M}_0}{W_0 L}\right); \quad \text{velocity at } t = t_0$$

Note: When W_0 is large enough to cause $\eta_1 < L$, a joint to appears at an interior point (refer to Case 10.29). If this is the case, the above response is an upper bound.

CASE 10.25   A RIGID BEAM (CC) ACTED UPON BY IMPULSE $S = 2 W_0 t_0$

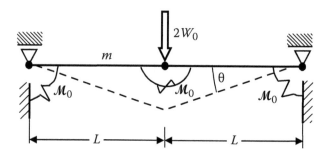

Angular springs with RP characteristic and $\mathcal{M}_0$ capacity at center and at ends.

$$M_{ef} = \frac{2mL}{3}$$

The response formulas are the same as in Case 10.26, but with a different M_{ef}.

Note: When $W_0 \gg 6\mathcal{M}_0/L$, this will cause a joint to appear at an intermediate point (refer to Case 10.28). In this event, the above response is an upper bound.

CASE 10.26 A RIGID BEAM, WITH A CENTER MASS, ACTED UPON BY IMPULSE $S = 2W_0t_0$

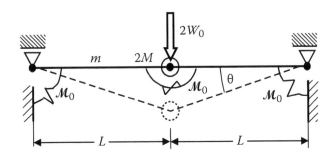

Angular springs with RP characteristic and M_0 capacity at center and at the ends.

$$M_{ef} = \frac{2mL}{3} + 2M$$

$u_m = L\theta_m$; peak displacement

All cases are mentioned below

(a) Short impulse $S_0 = 2W_0t_0$.

$$(W_0L \gg 2M_0)$$

$$v_0 = \frac{S_0}{M_{ef}} \text{ (initial velocity); } \quad E_k = \frac{1}{2}\frac{(S_0)^2}{M_{ef}} \quad \text{and} \quad \theta_m = \frac{E_k}{4M_0}$$

(b) Rectangular impulse W_0t_0

$$\theta_m = \frac{W_0t_0^2}{M_{ef}}\left(\frac{W_0L}{2M_0}-1\right); \quad \text{at } t_m = \frac{LW_0t_0}{2M_0}$$

(c) Triangular decreasing pulse $W = W_0(1-t/t_0)$

$$\frac{W_0L}{M_0} < 4: \theta_m = \frac{4W_0t_0^2}{3M_{ef}L}\left(1-\frac{2M_0}{W_0L}\right)^3; \quad v = 0 \text{ when } t = t_x = \left(1-\frac{2M_0}{W_0L}\right)2t_0$$

$$\frac{W_0L}{M_0} > 4: \theta_m = \frac{2W_0t_0^2}{3M_{ef}L}\left(1-\frac{3M_0}{W_0L}\right)+\frac{M_{ef}v_e^2}{8M_0}$$

$$v_e = \frac{2W_0t_0}{M_{ef}}\left(\frac{1}{2}-\frac{2M_0}{W_0L}\right); \text{ velocity at } t = t_0$$

Note: When W_0 is large enough to cause $\eta_2 < L$, a joint appears at an intermediate point (refer to Case 10.30). If this is the case, the above response is an upper bound.

CASE 10.27 STATIONARY PLASTIC JOINT APPEARING IN A LONG BEAM WITH MOMENT
CAPACITY $\mathcal{M}_0$, STRUCK AT THE TIP WITH A CONSTANT LOAD LASTING t_0

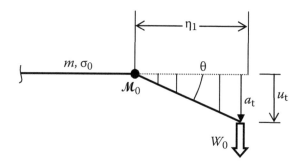

$$\eta_1 = \frac{3\mathcal{M}_0}{W_0}; \quad a_t = \frac{2W_0}{m\eta_1}$$

After time t_0

$$v_t = \frac{2W_0 t_0}{m\eta_1} = \frac{2W_0^2 t_0}{3\mathcal{M}_0 m}; \text{ tip speed}$$

$$u_t = \frac{1}{2} a_t t_0^2 = \frac{(W_0 t_0)^2}{3\mathcal{M}_0 m}; \text{ tip displacement}; \quad \theta_t = \frac{u_t}{\eta_1} = \frac{W_0^3 t_0^2}{9\mathcal{M}_0^2 m}; \text{ segment rotation}$$

$$\mathcal{L}_t = \frac{W_0}{2} a_t t_0^2 = \frac{W_0^3 t_0^2}{3\mathcal{M}_0 m}; \text{ work done by applied load}$$

$$\omega = \frac{\mathcal{M}_0 \theta_t}{\mathcal{L}_t} = \frac{1}{3}; \text{ fraction of external work absorbed by the stationary joint}$$

Note: After time t_0 the joint begins to move towards the other end of the beam.

CASE 10.28 STATIONARY PLASTIC JOINTS APPEARING IN A LONG BEAM
WITH MOMENT CAPACITY $\mathcal{M}_0$, STRUCK AT AN INTERIOR
POINT WITH A CONSTANT LOAD LASTING t_0

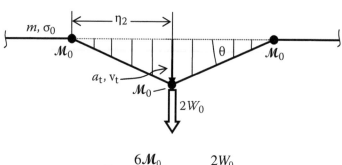

$$\eta_2 = \frac{6\mathcal{M}_0}{W_0}; \quad a_t = \frac{2W_0}{m\eta_2}$$

After time t_0

$$v_t = \frac{2W_0 t_0}{m\eta_2} = \frac{W_0^2 t_0}{3\mathcal{M}_0 m}; \quad u_t = \frac{1}{2}a_t t_0^2 = \frac{(W_0 t_0)^2}{6\mathcal{M}m}; \quad \theta_t = \frac{u_t}{\eta_2} = \frac{W_0^3 t_0^{\,2}}{36\mathcal{M}_0^2 m}$$

$$\mathcal{L}_t = W_0 a_t t_0^2 = \frac{W_0^3 t_0^2}{3\mathcal{M}_0 m}$$

$$\omega = \frac{\mathcal{M}_0 \theta_t}{W_0 u_t} = \frac{1}{6}; \text{ fraction of external work absorbed by the center stationary joint}$$

Note: After time t_0 the outer joints begin to move towards the other ends of the beam.

**CASE 10.29   STATIONARY PLASTIC JOINT APPEARING IN A LONG BEAM WITH MOMENT
CAPACITY $\mathcal{M}_0$, STRUCK AT THE TIP WITH A CONSTANT LOAD LASTING t_0**

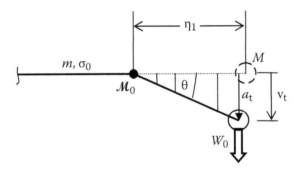

Mass M placed at the tip

$$\eta_1 = \frac{3}{2}\frac{\mathcal{M}_0}{W_0}\left[1+\left(1+\frac{8}{3}\frac{W_0 M}{\mathcal{M}_0 m}\right)^{1/2}\right]$$

$$a_t = \frac{W_0}{M + \dfrac{\eta_1 m}{2}}$$

After time t_0

$$v_t = \frac{W_0 t_0}{\dfrac{m\eta_1}{2}+M}; \quad u_t = \frac{v_t t_0}{2}; \quad \theta_t = \frac{v_t t_0}{2\eta_1}; \quad \mathcal{L}_t = W_0 u_t$$

$$\omega = \frac{\mathcal{M}_0}{W_0 \eta_1}; \text{ fraction of external work absorbed by the stationary joint}$$

Note: At $t = t_0$, $\omega < 1/3$ for $M > 0$. After time t_0 the joint begins to move towards the other end of the beam.

CASE 10.30   STATIONARY PLASTIC JOINTS APPEARING IN A LONG BEAM WITH MOMENT CAPACITY $\mathcal{M}_0$, STRUCK WITH A CONSTANT LOAD LASTING t_0

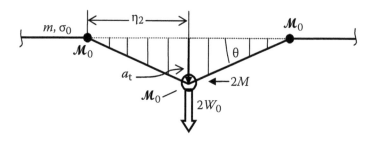

Mass $2M$ is placed at the impact point.

$$a_t = \frac{W_0}{M + \dfrac{\eta_2 m}{2}}; \quad \eta_2 = 3\frac{\mathcal{M}_0}{W_0}\left[1 + \left(1 + \frac{4}{3}\frac{W_0 M}{\mathcal{M}_0 m}\right)^{1/2}\right]$$

After time t_0: $v_t = \dfrac{W_0 t_0}{\dfrac{m\eta_2}{3} + M}$; $u_t = \dfrac{v_t t_0}{2}$; $\theta_t = \dfrac{v_t t_0}{2\eta_2}$; $\mathcal{L}_t = 2W_0 u_t$ (total)

$\omega = \dfrac{\mathcal{M}_0}{W_0 \eta_2}$; fraction of external work absorbed by the center stationary joint

Note: At $t = t_0$, $\omega < 1/6$ for $M > 0$. After time t_0 the outer joints begin to move towards the other ends of the beam.

CASE 10.31   PLASTIC JOINTS APPEARING IN UNCONSTRAINED BEAM WITH MOMENT CAPACITY $\mathcal{M}_0$, STRUCK AT THE BASE WITH A STEP LOAD W_0

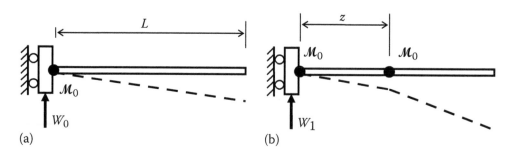

(a)   (b)

Pattern (a) when $\dfrac{2\mathcal{M}_0}{L} \le W_0 < \dfrac{11.44\mathcal{M}_0}{L}$

Pattern (b) when $W_0 = W_1 \ge \dfrac{11.44\mathcal{M}_0}{L}$; $z = 0.404L$. Ref. [40].

CASE 10.32 PLASTIC JOINTS APPEARING IN UNCONSTRAINED BEAM WITH MOMENT
CAPACITY $\mathcal{M}_0$, STRUCK AT THE CENTER WITH A STEP LOAD $2W_0$

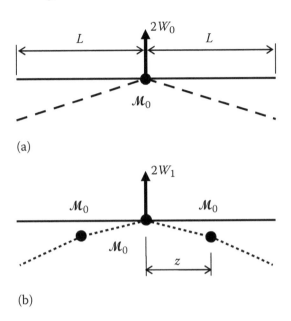

(a)

(b)

Pattern (a) when

$$\frac{2\mathcal{M}_0}{L} \leq W_0 < \frac{11.44\mathcal{M}_0}{L}$$

Pattern (b) when

$$W_0 \geq \frac{11.44\mathcal{M}_0}{L}; \quad z = 0.404L. \text{ Ref. [40]}.$$

Note: If W_0 is larger than prescribed by the above equation, the outer pair of joints shifts inward, towards the center point.

CASE 10.33 TRAVELING PLASTIC JOINT IN A BEAM IMPACTED AT THE FREE END
BY MASS M

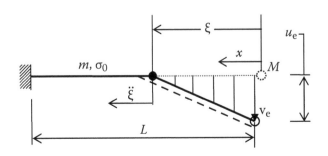

Impact speed v_0. Bending capacity $\mathcal{M}_0$.

$E_{k0} = Mv_0^2/2$; initial kinetic energy

$\gamma = \dfrac{M}{mL}$; mass ratio

$v_e = \dfrac{v_0}{1 + \dfrac{m\xi}{2M}}$; the speed of impacted tip

$t(\xi) = \dfrac{mv_0}{6\mathcal{M}_0} \dfrac{\xi^2}{\left(1 + \dfrac{m\xi}{2M}\right)}$; time for the joint to travel distance ξ

$\dfrac{1}{\rho} = \dfrac{2v_0^2 mL\gamma^2}{3\mathcal{M}_0} \dfrac{(x+4\gamma)}{(x+2\gamma)^3}$; curvature along x; $\dfrac{1}{\rho} = \dfrac{v_0^2 m}{3\mathcal{M}_0}$ for $x = 0$

$u_m = \dfrac{(Mv_0)^2}{3\mathcal{M}_0 m}\left(\dfrac{1}{1+2\gamma} + 2\ln\left(1 + \dfrac{1}{2\gamma}\right)\right)$; permanent deflection, after the base is bent.

$u_m = \dfrac{MLv_0^2}{2\mathcal{M}_0}$; permanent deflection, for $\gamma \gg 1$ (heavy striker)

$\Pi_2 = \dfrac{4\gamma(1+3\gamma)}{3(1+2\gamma)^2}E_{k0}$; part of the initial energy absorbed in root joint bending

$c_M = \dfrac{6\mathcal{M}_0}{mLv_0}\left(1 + \dfrac{1}{2\gamma}\right)$; average speed of plastic wave: $c_M = L/t(L)$

Note: Deflection u_m describes the process until $\xi = L$ and then for some time to allow for the rest of energy to be dissipated by the stationary end joint. See text for further comments. Refs. [43,84].

CASE 10.34   TRAVELING PLASTIC JOINT IN A BEAM IMPACTED AT AN INTERMEDIATE POINT BY MASS M

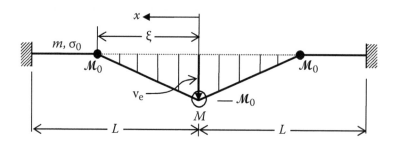

Impact speed v_0. Bending capacity $\mathcal{M}_0$.

$$v_e = \frac{v_0}{1 + \dfrac{m\xi}{M}}; \quad \gamma = \frac{M}{mL}$$

$$t(\xi) = \frac{mv_0}{12\mathcal{M}_0} \frac{\xi^2}{\left(1 + \dfrac{m\xi}{M}\right)}; \text{ time to travel distance } \xi$$

$$u_m = \frac{(Mv_0)^2}{24\mathcal{M}_0 m}\left(\frac{1}{1+\gamma} + 2\ln\left(1 + \frac{1}{\gamma}\right)\right); \text{ permanent deflection, after the base is bent.}$$

$$c_M = \frac{12\mathcal{M}_0}{mLv_0}\left(1 + \frac{mL}{M}\right); \text{ average speed of plastic wave: } c_M = L/t(L)$$

Note: All of the above describes the process until $\xi = L$ and then for some time to allow for the rest of energy to be dissipated by the stationary end joints. See text for further comments. Ref. [43].

CASE 10.35   DEFORMATION OF A CANTILEVER TIP-IMPACTED BY A LIGHT STRIKER M_0

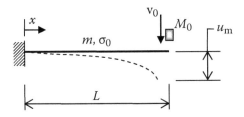

Rectangular section of depth $= H$. RP beam material with limit moment $\mathcal{M}_0 = BH^2\sigma_0/4$.

$$\varepsilon = \varepsilon_y + \frac{x^2}{L^2}\varepsilon_e; \text{ assumed strain distribution}$$

$$\varepsilon_y = \sigma_0/E$$

$$\frac{1}{2}M_0 v_0^2 = \frac{2\mathcal{M}_0}{H}L\left(\varepsilon_y + \frac{1}{3}\varepsilon_e\right); \text{ work–energy equivalence to find } \varepsilon_e$$

$$u_m = \frac{L^2}{H}\left(\varepsilon_y + \frac{1}{6}\varepsilon_e\right); \text{ tip deflection}$$

Note: The assumed strain distribution is based on residual deflections obtained from tests. The moment capacity $\mathcal{M}_0$ must be appropriate for the strain rate involved. Refer to Example 10.19.

CASE 10.36 EQUIVALENT RECTANGULAR PULSE FOR ELEMENTS MADE OF RP MATERIAL, IMPULSIVELY LOADED

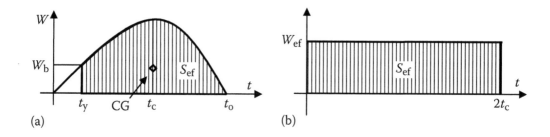

(a)							(b)

Static collapse load W_b is reached at $t = t_y$.

Original pulse shape (a) and equivalent rectangular pulse (b).

$$S_{ef} = \int_{t_y}^{t_0} W(t)\,dt;\ \text{effective impulse magnitude (*)}$$

t_c; centroid of a hatched area in (a)

$$W_{ef} = \frac{S_{ef}}{2t_c};\ \text{effective applied force. Ref. [84].}$$

Note: This is intended for a beam under an intense load, so that a plastic joint can appear at an intermediate point. Equation (*) is approximate, as the integration should be only up to t_f. The latter can be found using $S_{ef} = W_b(t_f - t_y)$.

CASE 10.37 TRIANGULAR DECREASING PULSE FOR A TIP-LOADED CANTILEVER OF CASE 10.23, MADE OF RP MATERIAL

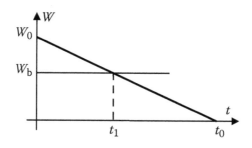

Traveling hinge and subsequent rotation at the base. $W_b = M_0/L$

(a) $3M_0 > W_0L > M_0$; no hinge, Case 10.23

(b) $W_0L > 3M_0$

Plastic hinge appears at $\eta_1 = \dfrac{3M_0}{W_0}$ and travels towards the base.

(c) For the case $W_0L = 6M_0$

$$u_i = \frac{11}{18} \frac{(W_0 t_0)^2}{M_0 Lm}; \text{ tip deflection, initial phase}$$

$$u_f = \frac{(W_0 t_0)^2}{M_0 Lm} \left[\frac{23}{18} + \frac{2}{3} \ln\left(\frac{W_0 L}{6 M_0} \right) \right]; \text{ final tip deflection}$$

$$\theta_f \approx \frac{u_f - u_i}{L}; \text{ an upper bound of base rotation. Ref. [84].}$$

Note: Initial phase is when the plastic hinge travels along the beam. Final deflection when all input energy is exhausted.

CASE 10.38   HIGH-SPEED COLLISION OF TWO BEAMS OF RP MATERIAL AND WITH A NEGLIGIBLE BENDING STIFFNESS

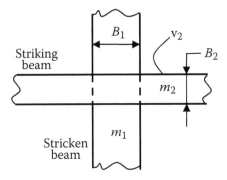

$P_0 = A\sigma_0$; the axial load in the deformed part equal to the yield load. (See Figure 10.19.) The beams may have different strengths: σ_{01} and σ_{02}.

v_2 is the initial velocity of the striking Beam 2

$\varepsilon_{1u}, \varepsilon_{2u}$ are breaking strains for Beam 1 and 2, respectively

$$c_1 = \left(\frac{P_{01}}{m_1} \right)^{1/2} = \left(\frac{\sigma_{01}}{\rho_1} \right)^{1/2}; \text{ lateral wave velocity for Beam 1}$$

$\xi_1 = c_1 t$; half-length of the deformed Beam 1 at time t, outside contact zone.

$m_1 = A\rho_1$ distributed mass of Beam 1

$$V(t) = \frac{v_2}{1 + \dfrac{m_1(B_2 + 2c_1 t)}{m_2(B_1 + 2c_2 t)}}; \text{ current speed of the contact zone and the mobilized mass outside,}$$

common for both beams

$$V_0 = \frac{v_2}{1 + \dfrac{m_1 B_2}{m_2 B_1}}; \text{ initial velocity immediately after the contact is attained}$$

$$V_i = \frac{v_2}{1 + \dfrac{m_1 c_1}{m_2 c_2}}; \text{ limit velocity when } t \to \infty. \text{ Also, common constant speed when } B_1 \approx 0 \text{ and}$$

$B_2 \approx 0$

$$V_{av} \approx \frac{1}{2}(V_0 + V_i); \text{ average velocity}$$

$$V_{1s} = \frac{4}{\pi} c_1 \sqrt{\varepsilon_{1u}}; \text{ safe } average \text{ speed for Beam 1, no break when } V_{av} < V_{1s}$$

$\delta = u(t) \approx V_{av} t$; displacement of contact zone at time t

$$\lambda_1 = \frac{\pi^2 \delta^2}{4 L_1}; \text{ total elongation for Beam 1 with } L_1 = B_2 + 2\xi_1$$

$$\varepsilon_1 = \frac{\lambda_1}{L_1}; \text{ strain of Beam 1}$$

$$\frac{1}{t_1} = \frac{1}{B_2}\left(\frac{\pi V_{av}}{2\sqrt{\varepsilon_{1u}}} - 2c_1 \right); \text{ to calculate time to breakage, } t_1, \text{ for Beam 1}$$

$\xi_{1b} = c_1 t_1$; the extent of plastic wave spreading at t_1. Ref. [107].

Note: The expressions are given for Beam 1 and those for Beam 2 are quite similar, obtained by switching index 1 to 2 and vice versa. For the results to be valid, ξ must be smaller than the distance to a boundary or to a free end.

EXAMPLES

EXAMPLE 10.1   LOAD CAPACITIES OF A CIRCULAR CROSS SECTION

Determine the onset of yield and load capacities for bending and shear for a beam specified below. The following material models, as defined in Chapter 4, are to be used: LE, RP, EPP, BL, and RSH. For the elastic range of material properties:

$$E = 69{,}000\,\text{MPa}, \quad \nu = 0.3, \quad G = 26{,}540\,\text{MPa}$$

while for the inelastic range:

The yield strength: $F_y = 200\,\text{MPa}$, plastic modulus: $E_p = 20\,\text{MPa}$, and breaking strain: $\varepsilon_u = 0.12$. Shear yield strength: $\tau_y = 200/\sqrt{3} = 115.5\,\text{MPa}$.

The beam is a cantilever (CF), a slender aluminum rod with a round solid section and with following properties: $L = 100$ mm, $\phi 3.18$ mm

$$\rho = 0.0027\,\text{g/mm}^3, \quad A = 7.942\,\text{mm}^2, \quad I = 5.02\,\text{mm}^4, \quad m = \rho A = 0.02144\,\text{g/mm}$$

$$(EI/L^3 = 0.34638)$$

Remaining section properties, per Tables 10.1 and 10.8:

$$A_s = 0.9A = 7.148\,\text{mm}^2, \quad A_{ef} = 0.9A = 7.148\,\text{mm}^2, \quad Z = 5.36\,\text{mm}^3$$

The onset of yielding takes place at the tensile stress F_y or the tensile strain $\varepsilon_y = 200/69{,}000 = 0.0029$. Section capacities:

$$\mathcal{M}_y = \frac{IF_y}{c} = \frac{5.02 \times 200}{1.59} = 631.4\,\text{N-mm (Equation 10.5)}$$

$$Q_y = (A/\alpha)\tau_0 = (3 \times 7.942/4)115.5 = 688\,\text{N (Equation 10.46a)}$$

Fully plastic values are associated with RP model, and those are also limits for the EPP model: (Equations 10.6 and 10.46b):

$$\mathcal{M}_0 = Z\sigma_0 = 5.36 \times 200 = 1072\,\text{N-mm (Equation 10.6b)}$$

$$Q_0 = A_{ef}\tau_y = 7.148 \times 115.5 = 825.6\,\text{N (Equation 10.46b)}$$

The ultimate strength F_u per BL model is related to the ultimate elongation ε_u as follows:

$$F_u = F_y + E_p(\varepsilon_u - \varepsilon_y) = 200 + 20(0.12 - 0.0029) = 202.3\,\text{MPa}$$

(For RHS it would come out practically the same, 202.4 MPa)
    For RSH model use Equation 10.8:

$$\mathcal{M}_m = (F_u - F_y)\frac{I}{c} + \mathcal{M}_y = (202.3 - 200)\frac{5.02}{1.59} + 1072 = 1079.3\ \text{N-mm}$$

The shear strength (for BL and RSH materials) is $\tau_u = F_u/\sqrt{3} = 116.8\,\text{MPa}$.
From Equation 10.47:

$$Q_u = \tau_u A_{ef} = 116.8 \times 7.148 = 835\,\text{N}$$

The RSH model gives an upper bound of section capacity for the BL model and it is usually only slightly above the true value. The same relation exists between RP and EPP models. (Needless to say that maximum strains attained must be at least several times larger than ε_y for these statements to be true.)

One could notice here that section capacities should not be strongly dependent on the material model assumed, even for sections going much further into plasticity than in this example. A reference is made here to the comment made in Chapter 4, where it is stated that the material constants are dependent on which model is to approximate the real stress–strain curve. For this reason F_y, for example, could be a larger number if material EPP is selected than it would be for BL representation (Figure 4.5).

EXAMPLE 10.2   A BEAM UNDER SUDDENLY APPLIED, DISTRIBUTED LOAD

Estimate the dynamic response to load $w_0 = 0.3\,\text{N/mm}$, applied as a decreasing triangular impulse over $t_0 = 0.4446\,\text{ms}$. The beam is a cantilever (CF), a slender aluminum rod, $L = 100\,\text{mm}$, with a round solid section, $\phi 3.18\,\text{mm}$. The section properties are as specified in Example 10.1, except that now $A_s = 0.5A$ and the material is used in elastic range.

Refer to Case 10.4.

$$u_{st} = \frac{1}{8}\frac{0.3 \times 100^4}{69,000 \times 5.02} = 10.826\,\text{mm; static deflection}$$

The natural period from Equation 10.14:

$$\tau = \varepsilon L^2 \sqrt{\frac{m}{EI}} = 1.787 \times 100^2 \sqrt{\frac{0.02144}{69,000 \times 5.02}} = 4.446\ \text{ms};\ \text{or}\ \omega = 1.4132\,\text{rad/ms}$$

$W_0 = 0.3 \times 100 = 30\,\text{N}$; the resultant force

$Q_{st} = W_0 = 30\,\text{N}$; static shear at base

$M_{st} = 0.5 w_0 L^2 = 1500\,\text{N-mm}$; static bending moments at the base

Because this is the triangular, decreasing pulse, employ Equation 10.22:

$$x = \omega t_0/4 = 1.4132 \times 0.4446/4 = 0.1571$$

$$u_d = \frac{\sin^2 x}{x}(2u_{st}) = \frac{\sin^2 0.1571}{0.1571}(2 \times 10.826) = 3.374\,\text{mm};\quad DF(u) = u_d/u_{st} = 0.3117$$

Estimate the peak shear and bending response by employing the above ratio (Equations 10.16):

$$Q_d = 30(u_d/u_{st}) = 9.35\,N$$

$$M_d = 1500(u_d/u_{st}) = 467.6\,N\text{-mm}$$

$$Q'_d = (0.680 \times 0.3117 + 0.32)30.0 = 15.96\,N \text{ (Alternative estimate by Equation 10.19)}$$

To get an upper-bound value of bending, find the flex wave length at $t = t_0$, according to Equation 10.24a. Taking t_0 as the length of an equivalent rectangular pulse, namely,

$$t_0 = 0.4446/2 = 0.2223: \quad l_2 = 1.887\left(\frac{69,000 \times 5.02}{0.02144}\right)^{1/4}\sqrt{t_0} \text{ or } l_2 = 119.6\sqrt{0.2223} = 56.4\,mm.$$

From Table 10.5: $M_d = 0.47 \times 30 \times 56.4 = 795\,N\text{-mm}$.
(This larger value should be used in absence of more reliable data.)

To check on results, an LS-Dyna model of the beam, with the properties as described above, made up of 100 elements, was constructed. The following peak values were obtained:

$$u_d = 3.558; \quad Q = 13.51; \quad M = 528.1$$

The approximate deflection results are close, 5% below these. The shear is closer to the calculated upper bound and bending closer to the lower bound.

EXAMPLE 10.3   A BEAM UNDER DISTRIBUTED LOAD, ALTERNATIVE APPROACH

Find the dynamic response to load $w_0 = 0.15\,N/mm$, applied as a rectangular impulse over $t_0 = 0.35$ ms. The beam is a cantilever (CF), as described in Example 10.1. Treat the load as a short impulse and apply the results of Case 10.5.

Find the constants: $w_0 t_0 = 0.15 \times 0.35 = 0.0525\,N\text{-s/mm}$

$$\frac{1}{\sqrt{EIm}} = \frac{1}{\sqrt{69,000 \times 5.02 \times 0.02144}} = \frac{1}{86.18}$$

$$\sqrt{\frac{EI}{m}} = 4019.4$$

From Example 10.2: $\tau = 4.446$ ms, therefore $t_0 = 0.35$ ms makes it a short impulse.

$$u_m = \frac{(w_0 t_0)L^2 d}{\sqrt{EIm}} = \frac{(0.0525)100^2 \times 0.426}{\sqrt{EIm}} = 2.595\,mm$$

$$M_m = (w_0 t_0)d\sqrt{\frac{EI}{m}} = 0.0525 \times 1.665 \times 4019.4 = 351.3\,N\text{-mm}$$

$$Q_{\mathrm{m}} = \frac{(w_0 t_0)d}{L}\sqrt{\frac{EI}{m}} = \frac{(0.0525)3.42}{100}\,4019.4 = 7.216\,\mathrm{N}$$

The above results are only slightly smaller than what would be obtained using the basic method. A FEA simulation gave the following extreme values, respectively:

$$u_{\mathrm{m}} = 2.813; \quad \mathcal{M} = 443.9, \quad \text{and} \quad Q = 11.367$$

EXAMPLE 10.4   RESPONSE DUE TO A SUDDENLY APPLIED POINT LOAD

A cantilever, as described in Example 10.1, has a point load of 100 N applied at the tip, in the form of a time step. Determine the natural periods of motion for (a) the flexural component and (b) the shear component. From this estimate the natural period of vibrations. Show the magnitude of bending and shear deflection components in early motion, when the deformation waves do not yet reach the fixed end.

From Example 10.2, the bending frequency is $\omega_b = 1.4132$ rad/ms.

From Case 2.21b, the shear frequency is

$$\omega_s = \frac{\pi}{2L}\sqrt{\frac{GA_s}{m}} = \frac{\pi}{2\times100}\sqrt{\frac{26{,}540\times0.5\times7.942}{0.02144}} = 34.826 \text{ rad/ms}$$

These are the components of the same motion, which means flexibilities are additive and therefore Equation 2.45 should be a good approximation:

$$\frac{1}{\omega^2} = \frac{1}{\omega_b^2} + \frac{1}{\omega_s^2}; \quad \text{which gives } \omega_b = 1.412 \text{ rad/ms.}$$

As anticipated for this shape of a beam, the shear component produces only a minor adjustment to the vibratory frequency. The quarter-period, i.e., the anticipated arrival time at the fixed end, is much shorter for shear, namely,

$$\frac{\tau_s}{4} = \frac{\pi}{2\omega_s} = \frac{\pi}{2\times34.826} = \frac{1}{22.17}\,\mathrm{ms}$$

The comparison of component deflections is meaningful only within this time interval.

The spreading of the prime flexural wave and growth of deflection at the impact point is given by Equations 10.24a and 10.25a, respectively:

$$l_b = 1.496\left(\frac{EI}{m}\right)^{1/4}\sqrt{t} = 1.496\left(\frac{69{,}000\times5.02}{0.02144}\right)^{1/4}\sqrt{t} = 94.845\sqrt{t}$$

$$u_b = \frac{W_0 l_1^3}{3EI} = \frac{100\times\left(94.845\sqrt{t}\right)^3}{3\times69{,}000\times5.02} = 82.104 t^{3/2}$$

The wave speed and the corresponding quantities for the shear wave from Equations 10.33:

$$c_3 = \left(\frac{GA_s}{m}\right)^{1/2} = \sqrt{\frac{26{,}540 \times 0.5 \times 7.942}{0.02144}} = 2217.1 \text{ m/s}; \quad l_s = 2217.1t$$

$$u_s = \frac{W_0 c_3 t}{GA_s} = \frac{100 \times 2217.1t}{26{,}540 \times 0.5 \times 7.942} = 2.1037t$$

When the shear wave reaches the clamped end, at $t = 1/22.17\,\text{ms}$, the above equations give

$$l_b = 20.14\,\text{mm}; \quad u_b = 0.787\,\text{mm}$$

$$l_s = 100\,\text{mm}; \quad u_s = 0.095\,\text{mm}$$

At this time the flex wave has engulfed a smaller part of the beam, yet it caused a larger deflection than the shear wave. By equating l_b and l_s as functions of time one finds that the deflections are equal for $t = 0.657 \times 10^{-3}$ ms and prior to that the shear component predominates.

The segment deformed by the shear wave is then 1.456 mm long. After rebounding from the clamped end, and returning to the application point (at $t = \tau_s/2$) the shear deflection reaches its maximum.

Another significant aspect of early motion is also the initial velocity, du/dt, for each component. For shear it is constant, proportional to the magnitude of the applied force. By differentiating the above expression for u_b one finds that at $t = 0$ the velocity is nil, regardless of the magnitude of forcing.

Another important observation is with regard to geometry of this beam, which implies a small shear component of the overall deflection. It is obvious that beams used in the construction industry are, typically, shaped for a better bending efficiency, which almost invariably implies a larger shear deflection component. A shorter length/depth ratio will also cause an increase in importance of shear deflections. In this sense the above example is "bending-friendly."

EXAMPLE 10.5   POINT LOADING AND LOCAL, TRANSIENT EFFECTS

Consider a beam with clamped ends and the same properties as in Example 10.1. Apply a point load $W_0 = 200\,\text{N}$, as a step function in time, at midpoint of the beam. Calculate peak deflection after 0.05 ms at midpoint and peak bending at that instance. Evaluate the magnitude of the load needed for the stationary joints to appear at $x = 20\,\text{mm}$ from the midpoint assuming the material to be rigid-perfectly plastic with $\sigma_0 = 250\,\text{MPa}$.

Follow Case 10.7. The flex wave will spread by

$$l_2 = 1.887\left(\frac{69{,}000 \times 5.02}{0.02144}\right)^{1/4}\sqrt{0.05} = 26.75\,\text{mm}$$

$$u_2 = \frac{200 \times 26.75^3}{12 \times 69{,}000 \times 5.02} = 0.921 \text{ mm, the center deflection}$$

The static peak moment is found from Table 10.2: $\mathcal{M}_{st} = 200 \times 100/8 = 2500\,\text{N-mm}$.

The period, from Equation 10.14, is 0.699 ms, then $\omega = 8.989$ kHz.

The peak moment at this instant, for $\omega t = 8.989 \times 0.05 = 0.4494$ is

$$\frac{M}{M_{st}} = 0.05 + \frac{1}{2\pi}\left(4 \times 0.4494 + \sin(4 \times 0.4494)\right) = 0.4912;$$

$$M = 2500 \times 0.4912 = 1228 \text{ N-mm.}$$

The moment capacity, per Equation 10.6b and Table 10.1 is

$$M_0 = 4R^3\sigma_0/3 = 4 \times 1.59^3 \times 250/3 = 1340 \text{ N-mm.}$$

To establish a joint at 20 mm from impact point, one may use Case 10.28, with $\eta_2 = 20$ mm:

$$W_0 = \frac{6M_0}{\eta_2} = \frac{6 \times 1340}{20} = 402 \text{ N}$$

According to our approach this means that $2W_0$ or 804 N must be suddenly applied.

EXAMPLE 10.6   MASS–BEAM IMPACT. THE BASIC METHOD

A cantilever is impacted at the tip. It is a slender aluminum rod with a round solid section and properties as in Example 10.1. A solid striker with mass M and velocity $v_0 = 10$ mm/ms (=10 m/s) impacts the tip laterally. There is a contact spring with stiffness k between the striker and the tip. Find the peak impact force and the beam response for $k = 10$, 100, and 10,000 N/mm and for $M = 0.021$, 0.21, and 2.1 g.

It is easy to calculate that the beam mass is $M_b = 2.144$ g, so the selected striker masses are about $0.01M_b$, $0.1M_b$, and M_b. A detailed calculation will be shown for $M = 0.21$ and $k = 100$.

From Case 10.1, one can find the cantilever stiffness, referred to the tip force:

$$k_b = 3EI/L^3 = 3 \times 0.34638 = 1.039$$

From Equation 2.29: $\dfrac{1}{k^*} = \dfrac{1}{k_b} + \dfrac{1}{k} = \dfrac{1}{1.039} + \dfrac{1}{100}$ or $k^* = 1.0283$

From Equation 10.31: $R_m = v_0\sqrt{k^*M} = 10\sqrt{1.0283 \times 0.21} = 4.647$ N

Peak displacement: $u_m = R_m/k_b = 4.647/1.039 = 4.47$ mm

Peak bending moment: $M = R_m L = 4.647 \times 100 = 464.7$ N-mm

This and the remaining cases are shown in the table below. The contact stiffness is so large compared to the beam itself that using one or the other value has little effect on the resulting force.

Hand-Calculated vs. FEA Values of Beam Bending Moments

Case	k	M	k*	R_m	u_m	$\mathcal{M}$	Ra($\mathcal{M}$)
A1	10	0.021	0.9412	1.405894	1.35	140.59	0.981
1	100	0.021	1.0283	1.469511	1.41	146.95	1.155
2	10,000	0.021	1.0389	1.477049	1.42	147.70	1.121
A3	10	0.21	0.9412	4.445827	4.28	444.58	0.614
3	100	0.21	1.0283	4.647002	4.47	464.70	0.736
4	10,000	0.21	1.0389	4.670839	4.50	467.08	0.907
A5	10	2.1	0.9412	14.05894	13.53	1405.89	0.888
5	100	2.1	1.0283	14.69511	14.14	1469.51	1.310
6	10,000	2.1	1.0389	14.77049	14.22	1477.05	1.193

To have a set of reference values for a comparison with the above, an ANSYS [2] model of the beam was constructed. It consisted of 100 elements of BEAM3 type (2D elastic beam). The material and section properties were as described above.

The detailed comparison was made only in reference to bending moment. [The ratios of our results to those based on the FEA calculations are given by symbol Ra($\mathcal{M}$)]. While the agreement is far from perfect, it gives a reasonable first estimate. The deflections calculated using this method are excessive, compared to FEA. The calculated value of R_m is much smaller than that of the actual contact force between the mass and the beam, which renders it useless for peak shear determination (refer to Example 10.7).

EXAMPLE 10.7 MASS–BEAM IMPACT. COLLISION METHOD

Using the data from Example 10.6 determine the peak contact load. Solve for $M = 0.21$ g and $k = 100$.

Follow Case 8.18. The first task is to estimate the interacting beam mass. For a cantilever, from Table 2.1, $M_b = 0.243$, $mL = 0.243 \times 0.02144 \times 100 = 0.521$ g.

$$\frac{1}{M^*} = \frac{1}{M_b} + \frac{1}{M} = \frac{1}{0.521} + \frac{1}{0.21} \text{ or } M^* = 0.15 \text{ g; effective mass at collision}$$

$$R_{mc} = v_i \sqrt{kM^*} = 10\sqrt{100 \times 0.15} = 38.73 \text{ N ; the peak contact force}$$

The peak contact force was seriously underestimated in Example 10.6, where the value of 4.65 N was found. Still, the basic method is more accurate with respect to peak bending. As mentioned before, this method is not very accurate for slender beams as this one.

EXAMPLE 10.8 AN ELASTIC BEAM IMPACTED BY MASS M

The beam is a cantilever with the same properties as in Example 10.2, impacted laterally at the tip. Find the peak impact force for several permutations of impact parameters: $v_i = 1$, 10, and 100 m/s; $k = 10, 100$, and

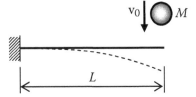

10,000 N/mm; and $M = 0.021$, 0.21, and 2.1 g. To accentuate the influence of shear, use $A_s = 0.5A$. Employ a FEA program to verify the answers.

Follow Case 10.13. Show detailed calculations only for $M = 0.21$ g (about 1/10 of the beam mass), contact stiffness $k = 100$ N/mm and $v_0 = 10$ m/s.

$M_m = 0.243(mL) = 0.243 \times 0.02144 \times 100 = 0.521$ g; modal mass

Use $M^* = M = 0.21$ g

$R_{ms} = v_0 \sqrt{kM^*} = 10\sqrt{100 \times 0.21} = 45.83$ N; spring component

$A_s = 0.5A = 0.5 \times 7.942 = 3.971$ mm²; assumed shear area

$C = \sqrt{GA_s m} = \sqrt{26,540 \times 3.971 \times 0.02144} = 47.53$ N/(mm/s); damper constant

$R_{md} = Cv_0/2 = 47.53 \times 10/2 = 237.65$ N; damper component of force

$\dfrac{1}{R_m} = \dfrac{1}{45.83} + \dfrac{1}{237.65}$ or $R_m = 38.42$ N; the impact force

The natural period of the first mode is $\tau = 4.446$ ms (Example 10.2).

The impact duration, for a case of a negligible rebound:

$$t_0 = \frac{2Mv_0}{R_m} = \frac{2 \times 0.21 \times 10}{38.42} = 0.1093 \text{ ms (it is only 1/41 of } \tau)$$

To check on our results indirectly, a FEA model of the beam, with the properties as described above, made up of 100 elements, was constructed and run. (The code was Ansys, v10 [2]. The elements were specified as BEAM3 elastic beams.) The simulation gave the peak force of 27.86 N and duration of 0.118 ms. The peak force, while not accurately calculated, is a reasonable approximation. The force and velocity plots for the mass M are given in Example Figure 10.1.

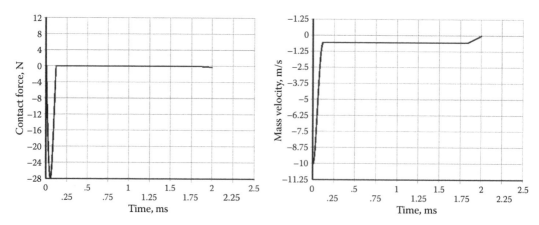

EXAMPLE FIGURE 10.1   Impact force history (a) and velocity history (b).

Many other combinations of contact stiffness k, impacting mass M, and impact velocity v_0 were run, as shown in the Example Table 10.1. The overall agreement with what was found by our approximate method (Example Table 10.2) is quite good, considering the nature of the contact force.

EXAMPLE 10.9   ELASTIC BEAMS WITH VARIOUS END CONDITIONS, IMPACTED BY MASS *M*

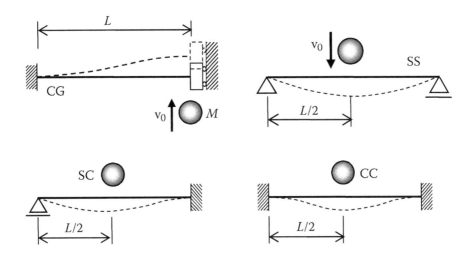

As in Example 10.8, determine the peak contact force for the following support conditions: CG, SS, SC, and CC, with $M = 0.21$ g, $k = 100$ N/mm, and $v_i = 10$ m/s. Compare the results with those of FEA.

In case of CG, the damper constant is the same as in Example 10.8, so the answers should also be the same (38.42 N). For the three remaining ones it is easy to rework a part of the example, doubling the damper constant and finding $R_m = 41.8$ N and $t_0 = 0.1005$ ms. (The impact is in the middle of the beam, consequently the damper constant doubles in comparison with what was used before.) The duration t_0 was calculated assuming no rebound. For the rebound velocity, $v_r = 6$, as found from FEA, the impulse transferred increases by a factor of 1.6 and the calculated t_0 is 0.1608 ms.

The FEA results, tabulated below, are the same for all cases except the first. This is an indirect confirmation that the impact itself is a local phenomenon. The global beam responses are likely to be different, of course.

Peak force, duration, and rebound velocity from FEA:

Type	R_m	t_0	v_r
CG	34.25	0.1317	3.05
SS	39.1	0.1407	5.904
SC	39.1	0.1401	5.943
CC	39.1	0.1401	5.982

EXAMPLE TABLE 10.1

Ansys-Computed Contact Force (N) for Different Values of M and k (Cantilever)

v_0	$M = 0.021$	$M = 0.021$	$M = 0.021$	$M = 0.21$	$M = 0.21$	$M = 0.21$	$M = 2.1$	$M = 2.1$	$M = 2.1$
	$k = 10$	$k = 100$	$k = 10,000$	$k = 10$	$k = 100$	$k = 10,000$	$k = 10$	$k = 100$	$k = 10,000$
(m/s)									
1	—	1.286	9.52	—	2.786	15.33	—	3.687	16.77
10	4.284	12.86	97.06	10.57	27.86	154	27.03	36.87	168.7
100	—	128.6	982.4	—	278.6	1544	—	368.7	1690

EXAMPLE TABLE 10.2

Hand-Calculated Contact Force (N) for the Same Parameters as in Example Table 10.1

1	0.45	1.37	9.00	1.37	3.84	15.65	2.08	5.53	17.88
10	4.50	13.66	90.03	13.66	38.42	156.52	20.81	55.33	178.77
100	44.96	136.59	900.28	136.59	384.19	1565.15	208.07	553.27	1787.71

Note that the deflected shape, as marked in the above sketches, takes effect long after the impact is over.

EXAMPLE 10.10   SIMPLY SUPPORTED BEAM IMPACTED BY MASS M

The configuration of interest was thoroughly analyzed by Goldsmith [26]. A steel beam, 12.7 mm × 12.7 mm section, 762 mm long, simply supported, was impacted by a 12.7 mm diameter steel sphere, with a velocity of 45.72 m/s. Determine the peak contact force.

$$E = 200,000\,\text{MPa}, \quad \rho = 0.00785\,\text{g/mm}^3, \quad \nu = 0.3, \quad \text{and} \quad G = 76,920\,\text{MPa}$$

Locally, this is the case of a sphere impacting a plane; therefore the local stiffness may be found from Case 7.17:

$$\frac{1}{k_0} = \frac{3}{2}\frac{1-\nu^2}{E}\frac{1}{\sqrt{r_2}} = \frac{3}{2}\cdot\frac{0.91}{200,000}\cdot\frac{1}{\sqrt{6.35}} = \frac{1}{369,220}$$

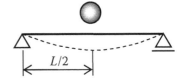

Impacting mass $M = 8.419\,\text{g}$. Section area: $A = 161.3\,\text{mm}^2$, $A_s = A/1.2 = 134.4\,\text{mm}^2$, $m = A\rho = 1.266\,\text{g/mm}$

Determine the peak contact force using Case 10.13:

$$C = 2\sqrt{GA_s m} = 2\sqrt{76,920 \times 134.4 \times 1.266} = 7235.5\,\text{N/(mm/s)}$$

$$R_{\text{md}} = Cv_i/2 = 7235.5 \times 45.72/2 = 165,404\,\text{N}$$

The modal mass from Table 2.1 is $M_m = 0.493 \times 1.266 \times 762 = 475.6\,\text{g}$. It is larger than the impacting mass M, so that $M^* = M = 8.419\,\text{g}$ is set. This is a Hertz-type impact; therefore Equation 7.37b can be used:

$$R_{\text{ms}} \equiv W_m = \left(\frac{5}{4}M^* v_r^2\right)^{0.6} k_0^{0.4} = \left(\frac{5}{4} \times 8.419 \times 45.72^2\right)^{0.6} 369,220^{0.4} = 67,974\,\text{N}$$

The impact force is

$$\frac{1}{R_m} = \frac{1}{R_{\text{ms}}} + \frac{1}{R_{\text{md}}} = \frac{1}{165,404} + \frac{1}{67,974} \quad \text{or} \quad R_m = 48,176\,\text{N}; \quad \text{according to our stress wave}$$

method.

In Goldsmith [26, p. 119], this is solved including both bending and shear deformation and employing 65 normal modes to calculate the contact force of 55,845 N (12,550 lbf). One can also note that an upper bound of our calculation will be obtained by suppressing the reduction by 2 in Equation 10.37. If we do so, the calculated force becomes 56,388 N.

EXAMPLE 10.11   STATIC LOADING OF A COMPOSITE BEAM

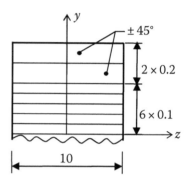

The beam is made up of graphite epoxy layers. Each of the two top layers has two plies of fibers, one at +45° to the beam axis and the other at −45°. For such a layer the relevant modulus is $E_2 = 91.3$ GPa, as calculated in Appendix B. In the rest of the beam, the six plies have fibers running along the beam axis, with $E_1 = 148$ GPa. The beam is 50 mm long, simply supported. A pressure of 0.1 MPa is applied at the top surface. Calculate the bending stress and interlaminar shear at critical locations.

The first six plies above the z-axis are treated as combined layer 1 and the remainder as layer 2.

For the whole section: $I = 10 \times 2^3/12 = 6.667$ mm^4

From Table 10.2, the maximum moment and shear are, respectively,

$$Q = wL/2 = (0.1 \times 10)50/2 = 25 \text{ N.}$$

$$\mathcal{M} = wL^2/8 = (0.1 \times 10)50^2/8 = 312.5 \text{ N-mm.}$$

By our convention, $h_1 = 2 \times 6 \times 0.1 = 1.2$ mm and $h_2 = 2(0.6 + 0.4) = 2$ mm. According to Equation 10.90, the combined bending stiffness is

$$\breve{E}I = (B/12)\left[E_1(h_1^3 - 0) + E_2(h_2^3 - h_1^3)\right] = (10/12)\left[148 \times 1.2^3 + 91.3 \times (2^3 - 1.2^3)\right]10^3$$

$$= 690,310 \text{ N-mm}^2$$

Then $\breve{E} = \breve{E}I/I = 690,310/6.667 = 103,550$ MPa. From Equation 10.92b

$$\sigma_k = \frac{E_k}{\breve{E}} \frac{\mathcal{M}y_k}{I} = \frac{E_k}{103,550} \frac{312.5y_k}{6.667} = \frac{E_k y_k}{2,209}$$

At the top and bottom of the upper layer the stress is, respectively,

$$\sigma_{2t} = \frac{91,300 \times 1.0}{2,209} = 41.33 \text{ MPa} \quad \text{and} \quad \sigma_{2b} = 0.6\sigma_{2t} = 24.8 \text{ MPa}$$

while at the top of the lower layer $\sigma_{1t} = \dfrac{148,000 \times 0.6}{2,209} = 40.2$ MPa

It is therefore seen that there is a stress discontinuity where the layers meet. The shear stress from Equation 10.95 has to be checked at the junction of layers, for which the area moment is

$$Z_{0.6}^c = B(c-y)y_{\mathrm{m}} = 10(1.0 - 0.6)0.8 = 3.2\,\mathrm{mm}^3$$

$$\tau = \frac{E_2}{\tilde{E}}\frac{QZ_{0.6}^c}{BI} = \frac{91.3}{103.55}\frac{25 \times 3.2}{10 \times 6.667} = 1.058\,\mathrm{MPa}$$

The shear is expected to peak at $y = 0$, as for a homogeneous beam. The area moment for layer 1 is $Z_0^{0.6} = 10 \times 0.6 \times (0.6/2) = 1.8\,\mathrm{mm}^2$. The summation of the axial force differentials, as per Equation 10.96, leads to

$$\tau_0 = \frac{Q}{BI\tilde{E}}(E_1 Z_0^{0.6} + E_2 Z_{0.6}^c) = \frac{25}{10 \times 6.667 \times 103.55}(148 \times 1.8 + 91.3 \times 3.2) = 2.023\,\mathrm{MPa}$$

If the free ends of the beam are not prevented from warping along the depth, the situation complicates and a numerical treatment is necessary.

Example 10.12   Static Response of a Sandwich Beam

This symmetrical section of the beam has two graphite epoxy face sheets and the core is a hexagonal honeycomb. The beam is 400 mm long and clamped at both ends. The loading is uniform, $w = 10\,\mathrm{N/mm}$. The mechanical properties are

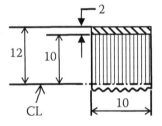

Face sheets: $E_f = 148\,\mathrm{GPa}$ (along axis); Core: $G_c = 500\,\mathrm{MPa}$

Calculate deflections taking shear distortion into account. Also find the peak bending stress and shear.

According to Figure 10.21: $B = 10$, $H = 2(10+1) = 22$, $h = 2\,\mathrm{mm}$.

$$A = 2Bh = 440\,\mathrm{mm}^2; \quad c = H/2 = 11\,\mathrm{mm}; \quad I = BH^2h/2 = 4840\,\mathrm{mm}^4$$

The deflection is a sum of bending and shear components as per Equation 10.12b. The coefficients come from Table 10.2:

$$u = \frac{1}{384}\frac{w_0 L^4}{EI} + \frac{1}{8}\frac{w_0 L^2}{G_c BH} = \frac{1}{384}\frac{10 \times 400^4}{148{,}000 \times 4{,}840} + \frac{1}{8}\frac{10 \times 400^2}{500 \times 10 \times 22}$$

$$= 0.931 + 1.818 = 2.749\,\mathrm{mm}$$

Interestingly, the shear component is larger than the bending component. The peak shear force is

$$Q = wL/2 = 10 \times 400/2 = 2000\ \mathrm{N}.$$

The shear stress: $\tau = Q/A_s = 2000/(10 \times 22) = 9.09\,\mathrm{MPa}$.

The strength of the shear bond between the plates and the core must be larger than this number.

Also, $M = wL^2/12 = 10 \times 400^2/12 = 133,333$ N-mm.

Bending stress: $\sigma = Mc/I = 133,333 \times 11/4,840 = 303$ MPa.

EXAMPLE 10.13   ELASTIC, PERFECTLY PLASTIC BENDING OF A BEAM SEGMENT

Consider a cantilever with a rectangular section, $b = 1$, $h = 30$, and length $L = 600$ mm, bent about the stiffer axis. A static bending moment M is applied at the free end. Calculate the magnitude of deflection that is sustained after initiation of yielding, but prior to the development of a fully plastic moment. Use $E = 200,000$ MPa and $\sigma_0 = 200$ MPa.

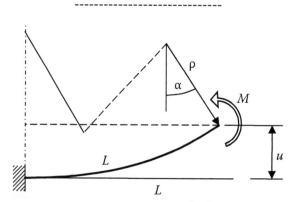

The onset of yielding takes place for $\sigma_0/E = \varepsilon_y = 0.001$. If we prescribe the largest strain in the section, ε_m, it also sets the radius of curvature, according to Equation 10.1:

$$\varepsilon_m = \frac{h_1}{\rho} = \frac{15}{\rho}; \quad \text{or} \quad \rho = \frac{15}{\varepsilon_m}$$

The relation between bending moment and strain is given in Case 10.1c, rewritten here as

$M = M_0 \left(1 - \dfrac{1}{3e^2} \right)$ where $e = \dfrac{\varepsilon_m}{\varepsilon_y} >$; and $M_0 = 1.5 M_y$ where M_y is the moment at the onset of yield and M_0 is the limit moment for the EPP material. The calculation is presented in the table. The peak strain ε_m is selected first, the corresponding moment found next, and then the radius of curvature. The latter is constant, which makes the axis assume the shape of a circular arc. It is then a simple matter, using the above sketch, to calculate angle α and the corresponding lateral deflection u as

$$\alpha = L/\rho \quad \text{and} \quad u = \rho(1 - \cos \alpha)$$

ε_m	M/M_0	ρ (mm)	α	u (mm)
0.001	2/3	15,000	0.04	12.0
0.005	0.987	3,000	0.20	59.8
0.020	0.999	750	0.80	227.5

The first obvious conclusion from the above is that it takes a relatively small strain (in terms what metallic structures can typically withstand) to come close to the moment capacity $\mathcal{M}_0$. The second observation is that quite significant deflections take place in the elastoplastic range considered above; for the strain as small as 0.005 the deflection is nearly 10% of the length of beam. As said before, a RP analysis implies no deflection until $\mathcal{M}_0$ is reached. This is one of the main reasons why RP analyses tend to underestimate deflections. As a side observation note that this manner of determining displacements is valid in the large deflection range, because it amounts merely to projecting the curved axis on the vertical reference line. The fact that the radius of curvature is constant makes the calculation much easier.

EXAMPLE 10.14   DISTRIBUTED LOAD IMPULSE APPLIED TO A BEAM. INELASTIC RANGE

Consider a thin tube with outer diameter of 609.6 mm and wall thickness of 24.6 mm ($R_m = 292.5$ mm). With the help of Tables 10.1 and 10.8:

$$A = 45{,}211 \text{ mm}^2, \quad A_{ef} = 28{,}782 \text{ mm}^2, \quad I = 1{,}934 \times 10^6 \text{ mm}^4, \quad Z = 8.419 \times 10^6 \text{ mm}^3$$

The available material properties of this steel tube are

$$E = 177{,}240 \text{ MPa}, \quad \rho = 0.00785 \text{ g/mm}^3, \quad G = 68{,}170 \text{ MPa}, \quad F_y = 197 \text{ MPa}, \quad F_u = 455 \text{ MPa}$$

with maximum elongation of 22%. (10% increase in F_y and F_u had been allowed in the above due to dynamic loading. No strain-rate data was available for this material, but some adjustment seemed appropriate.) The beam is clamped at both ends (CC). Two cases are under consideration: $L = 2.5$ m and $L = 5$ m. Calculate peak deflections due to an applied uniform loading of 50,000 N/mm over a time of 0.2 ms.

Since the stress–strain curve is not available, assume $\sigma_0 = 0.5(F_y + F_u) = 326$ MPa.

Shear strength per Equation 10.46b: $Q_0 = 28{,}782 \times 326/\sqrt{3} = 5.417 \times 10^6$ N

Bending strength per Equation 10.6b: $\mathcal{M}_0 = 8.419 \times 10^6 \times 326 = 2744.5 \times 10^6$ N-mm

Distributed mass $m = A\rho = 354.9$ g/mm

Determine static strength first. (The results for the longer beam are given parenthetically.)

Using Equations 10.52 with the coefficients $\xi = 16$ and $\eta = 2$ from Table 10.9 gives

$$w_b = 7026 \text{ N/mm } (1756.5) \quad \text{and} \quad w_s = 4334 \text{ N/mm } (2167)$$

which shows that the shorter beam is weaker in shear and the longer one in bending. (In fact, Table 10.8 indicates that $L_{cr} = 4{,}054$ mm.) Apply a uniform loading of 50,000 N/mm over a time of 0.2 ms, which results in a global impulse of $S = 25 \times 10^6$ N-mm (50×10^6). The shorter beam will undergo shear sliding with displacement given by Equation 10.55, namely, $u_s = 32.51$ mm. The longer beam will act in bending mode with the deflection prescribed by Equation 10.62b and the coefficients in Table 10.9, $u_m = 120.3$ mm.

In order to verify these results, a model was constructed with the properties as described, using LSDYNA code [59]. The elements were specified as Hughes–Liu beams with cross-section integration. Each of the two models had 100 elements. The EPP material model employed (with $\sigma_0 = 326$ MPa and zero plastic slope) was designated as mat_plastic_kinematic in the code. The key input data and the deflection results (u_d) are shown in the table.

Deflections by FEA (u_d) and by simplified equations (u_m):

L	w_b	w_s	w_m	t_0	u_d	u_m
2,500	7,026	4,334	50,000	0.20	35.1	32.51
5,000	1,757	2,167	50,000	0.20	134.7	120.3

Note: The short beam has a significant shear component of deflection.

The hand-calculated values are underestimates. This is quite typical of an analysis with these assumptions.

EXAMPLE 10.15 PLASTIC JOINT IN CANTILEVER STRUCK AT THE TIP

A long cantilever beam with the tip mass $M = 368$ kg, distributed mass $m = 461$ kg/m, and plastic moment capacity $M_0 = 1.246 \times 10^6$ N-m is subjected to a suddenly applied, sustained tip load $W_0 = 1.53 \times 10^6$ N. Calculate the distance η_1 at which a plastic hinge will form assuming the material to be RP.

Case 10.29 can be used for this purpose. Data is input using g-mm units.

$$\eta_1 = \frac{3}{2}\frac{M_0}{W_0}\left[1+\left(1+\frac{8}{3}\frac{W_0 M}{M_0 m}\right)^{1/2}\right] = \frac{3}{2}\frac{1.246\times10^9}{1.53\times10^6}\left[1+\left(1+\frac{8}{3}\frac{1.53\times10^6\times368,000}{1.246\times10^9\times461}\right)^{1/2}\right]$$

$\eta_1 = 3,544$ mm

If the beam is shorter than this, the hinge will form at the base.

EXAMPLE 10.16 PLASTIC JOINT AT AN INTERMITTENT LOCATION

A beam is struck at an interior point, as in Figure 10.16, with the force of $2W_0$ and no lumped mass present. Consider two support conditions: SS and CC with length L on each side of impact point. Treating the material as RP, find the limiting load for each condition when a plastic joint can be formed at a distance shorter than L.

Case 10.28 can be used for this purpose. The general location of the joint is at

$$\eta_2 = \frac{6M_0}{W_0} \text{ or in the limit } L = \frac{6M_0}{W_0}; \text{ or } 2W_0 = \frac{12M_0}{L}$$

From Table 10.11 one finds that the static collapse load for an SS beam, L-long, is $W_b = 4M_0/L$. Our beam is $2L$ long, however, so it needs only $W_b/2$ at the center. In Case 10.28,

$2W_0$ is applied, so $2W_0 = 2\mathcal{M}_0/L$. A comparison with the previously calculated limit tells us that that we need 6× the static collapse load or more at center to induce a joint in an interior point, $2W_0 = 6W_b$.

For the CC condition, the minimum load to create an interior joint is the same as before:

$$2W_0 = \frac{12\mathcal{M}_0}{L}$$

The static collapse load is twice as large as before, so $W_b/2 = 4\mathcal{M}_0/L$ needs to be applied. This time we need 3× (static collapse load) at the center. In fact, the absolute value of the point load is the same in both cases, while it is only the collapse loads that differ. For this reason it is not very instructive, in general, to state the magnitude of a joint-inducing load as a multiple of the static collapse force.

EXAMPLE 10.17 MOVING PLASTIC JOINT IN A CANTILEVER, TIP IMPACTED BY A LIGHT, FAST STRIKER

A cantilever beam $L = 305$ mm and a 6.35 mm × 6.35 mm cross section has its tip impacted by a striker of mass $M_0 = 2.27$ g. The impact velocity, at right angle to the beam, is $v_0 = 481.6$ m/s. Assume that a plastic joint will form and travel towards the base. Calculate the final tip deflection, the average joint speed, and the peak strain near the tip. The material is steel, treated as RP with $\sigma_0 = 400$ MPa. (The data comes from the experiments by Parkes [65], except σ_0, which is assumed.)

Case 10.33 refers to this situation. For the square section

$$m = \rho A = 0.00785 \times 6.35^2 = 0.3165 \text{ g/mm}$$

$$2\gamma = \frac{2M_0}{mL} = \frac{2 \times 2.27}{0.3165 \times 305} = \frac{1}{21.26} g < 0.2 \text{, this is definitely a light striker.}$$

$$\mathcal{M}_0 = Z\sigma_0 = (H^3/4)\sigma_0 = (6.35^3/4)\,400 = 25{,}605 \text{ N-mm}$$

$$u_m = \frac{(Mv_0)^2}{3\mathcal{M}_0 m}\left(\frac{1}{1+2\gamma} + 2\ln\left(1 + \frac{1}{2\gamma}\right)\right) = \frac{(2.27 \times 481.6)^2}{3 \times 25{,}605 \times 0.3165}(0.9551 + 2\ln 22.26)$$

$$= 352 \text{ mm}$$

$$c_M = \frac{6\mathcal{M}_0}{mLv_0}\left(1 + \frac{mL}{2M_0}\right) = \frac{6 \times 25{,}605}{0.3165 \times 305 \times 481.6}(1 + 21.26) = 73.56 \text{ m/s}$$

The magnitude of the deflection predicted is unreasonably large and inconsistent with the small deflection approach. To appreciate the relative importance of the two phases of motion, find the energy fraction expended in the final phase:

$\dfrac{4\gamma(1+3\gamma)}{3(1+2\gamma)^2} = 0.031$; this is so little, that the initial phase (bending along the beam) dominates.

$$\frac{1}{\rho} = \frac{v_0^2 mL}{3M_0} = \frac{481.6^2 \times 0.3165}{3 \times 26{,}605} = 0.92$$

$\varepsilon_1 = \dfrac{h_1}{\rho} = \left(\dfrac{6.35}{2}\right) 0.92 = 2.92$; this is unrealistically large

The tip deflection, as reported by Parkes, appears to be less than 1/3 of what was calculated here. The attempts to adjust the results of this theory were made along the lines of increasing M_0 due to strain hardening. The influence of the latter would have to be very large indeed.

EXAMPLE 10.18   ELASTIC AND PLASTIC WAVES

Consider the same cantilever as in Example 10.17. Determine the elastic wave speeds: shear c_s and bending c_b. What would the impact velocity have to be to achieve the same plastic joint speed c_M as that of the elastic bending wave c_b?

In a rectangular section the shear-effective area is nearly as big as the total area (the ratio is 5/6, per Table 10.1). The wave speed in Table 5.2, $c_s = 3130$ m/s, will therefore serve as a good approximation. To calculate the flex wave speed, as defined in section "Transient beam response due to a suddenly applied point load," one first needs to find the natural period from Equation 10.14:

$$\tau = \varepsilon L^2 \sqrt{\frac{m}{EI}} = 1.787 \times 305^2 \sqrt{\frac{0.3165}{200{,}000 \times 135.5}} = 17.965\,\text{ms}$$

where $I = 6.35^4/12 = 135.5\,\text{mm}^4$. The average flex wave speed is then

$$c_b = \frac{L}{(\tau/4)} = \frac{305}{(17.965/4)} = 67.9 \text{ mm/ms}$$

The plastic wave speed, or the joint velocity, from Example 10.17, is $c_M = 73.56$ mm/ms.
   This shows a disparity in speeds between the plastic wave and the elastic bending wave. c_M is inversely proportional to impact speed v_0, per Case 10.33. To reduce c_M, the impact speed would have to be increased to

$$v_e = 481.6(73.56/67.9) = 521.7\text{m/s}$$

Most industrial applications would involve much smaller impact speeds than the ballistic value of 481.6 m/s used in Example 10.17. One should keep in mind that when c_M and c_b, based on different material models, are of a similar magnitude, the response will not necessarily conform closely to either an elastic wave or the moving joint theory.

EXAMPLE 10.19   DEFORMATION OF A CANTILEVER, TIP IMPACTED BY A FAST, LIGHT STRIKER

The beam and impactor were described in Example 10.17. The dashed curve shown in Case 10.35 is a typical experimental result of a tip impact. Assuming that the outer fiber strain varies like $\varepsilon = \varepsilon_y + \dfrac{x^2}{L^2}\varepsilon_e$ along the beam, find the unknown total strain ε_e and the deflection at the tip of the beam.

Case 10.35 describes this approach. The yield strain is $\varepsilon_y = 400/200{,}000 = 0.002$.

The moment capacity: $M_0 = 25{,}605\,\text{N-mm}$

The work–energy equivalence gives us the peak plastic strain ε_e:

$$\frac{2 \times 25{,}605}{6.35}\,305\left(0.002 + \frac{1}{3}\varepsilon_e\right) = \frac{1}{2}\,2.27 \times 481.6^2 \quad \text{or} \quad \varepsilon_e = 0.315$$

The total strain at the tip is therefore $\varepsilon_m = 0.002 + 0.315 = 0.317$

The beam deflection is

$$u_m = \frac{L^2}{H}\left(\varepsilon_y + \frac{1}{6}\varepsilon_e\right) = \frac{100^2}{6.35}\left(0.002 + \frac{1}{6}0.315\right) = 85.82\,\text{mm}$$

The numerical deflection resulting from the referenced texts are not quoted, but the above figure is more consistent with the graphical descriptions of the test in Ref. [65] than what was obtained in Example 10.17.

EXAMPLE 10.20   CANTILEVER, TIP IMPACTED BY A HEAVY STRIKER

Consider the same cantilever as in Example 10.17. The striker has the speed of $v_0 = 5.5\,\text{m/s}$ and mass $M = 454\,\text{g}$. Determine the amount of plastic straining resulting from impact.

Taking necessary data from Example 10.17, the beam mass is $mL = 0.3165 \times 305 = 96.5\,\text{g}$. The mass ratio is $\gamma = M/(mL) = 454/96.5 = 4.7$. For this ratio most energy will go into root-bending mode, as can be easily found from Case 10.33. The easiest way to find an approximate tip deflection seems to be the application of Case 7.19:

$M_0 = 25{,}605\,\text{N-mm}$; moment capacity

$$M_{ef} = M + mL/3 = 454 + 96.5/3 = 486.2\,\text{g}$$

$$\theta_m = \frac{M}{M_{ef}} \frac{Mv_0^2}{2\mathcal{M}_0} = \left(\frac{454}{486.2}\right)\frac{454 \times 5.5^2}{2 \times 25,605} = 0.2504 = 14.35°; \text{ base rotation}$$

$u_m = L \sin \theta_m = 75.6\,\text{mm}; \text{ tip deflection}$

The same variable from Case 10.33:

$$u_m = \frac{(Mv_0)^2}{3\mathcal{M}_0 m}\left(\frac{1}{1+2\gamma} + 2\ln\left(1+\frac{1}{2\gamma}\right)\right) = \frac{(454 \times 5.5)^2}{3 \times 25,605 \times 0.3165}(0.0962 + 2\ln 1.106)$$

$$= 76.53\,\text{mm}$$

The first result for u_m is from a formula where the loss of energy comes from converting a rectilinear motion of the impactor itself to rotational motion of the assembly. In the second one, the loss results from straining of the beam itself. As the numbers demonstrate, for a sizeable ratio γ both results are close.

The strain at the base joint can be evaluated by Equation 10.75 with $l_h = H$:

$$\varepsilon_1 = \frac{h_1}{H}\theta = \frac{\theta}{2} = \frac{0.2504}{2} = 0.1252$$

This is a minor strain, when mild steel is used. However, a marked stress concentration at the root section would require careful checking to make certain a rupture does not take place.

The above configuration is one of those examined by Parkes [65]. When compared with his results, one finds that bending deformation is indeed limited to the root area only. The experimental magnitude of the tip deflection appears significantly smaller than that calculated here.

This is probably the result of not including contact losses of kinetic energy between the striker and the beam.

EXAMPLE 10.21 CLAMPED–CLAMPED BEAM IMPACTED AT CENTER BY A HEAVY STRIKER

Solve the same problem as is Example 10.20, except that the beam is now 610 mm long, clamped at both ends. The striker is the same, $M = 454\,\text{g}$. Determine the amount of plastic straining resulting from impact.

The beam mass is now 96.5 g per each half. One can anticipate deflection to be less, because the same mass is now impacting the equivalent of two beams working in parallel. Let us assume the analogous mode of deformation: plastic joints at the center and at the ends. Use Case 7.22 with $L = 610\,\text{mm}$:

$\mathcal{M}_0 = 25,605\,\text{N-mm}; \text{ moment capacity}$

$$M_{ef} = M + 0.3165 \times 610/3 = 454 + 96.5/3 = 518.4\,\text{g}$$

$$\theta_m = \frac{M}{M_{ef}} \frac{Mv_0^2}{8\mathcal{M}_0} = \left(\frac{454}{518.4}\right) \frac{454 \times 5.5^2}{8 \times 25,605} = 0.05872\, \text{rad} = 3.364°$$

$u_m \approx (L/2)\theta_m = 305 \times 0.05872 = 17.91$ mm; tip deflection

The deflection from Case 10.34: The mass ratio is $\gamma = 4.7$, the same as before.

$$u_m = \frac{(Mv_0)^2}{24\mathcal{M}_0 m}\left(\frac{1}{1+\gamma} + 2\ln\left(1+\frac{1}{\gamma}\right)\right) = \frac{(454 \times 5.5)^2}{24 \times 25,605 \times 0.3165}(0.1754 + 2\ln 1.2128)$$

$$= 17.99\, \text{mm}$$

The difference between the two answers is minimal.

The strain at the base joint can be evaluated by Equation 10.75:

$$\varepsilon_1 = \frac{h_1}{H}\theta = \frac{\theta}{2} = \frac{0.05872}{2} = 0.02936$$

Compared with Equation 10.20, we now have two component beams rather than one. Additionally, each of these has twice the resistance of the single beam in the previous example. The experimental results are not available now, but probably the calculated deflection is too large because contact losses between the striker and the beam are again not included.

EXAMPLE 10.22 HIGH-SPEED COLLISION OF TWO BEAMS

This is the same input data as in the problem posed in Ref. [107] except for the reduced material strength and ductility. The impact velocity is $v_2 = 240$ m/s and the beam properties are

1. Aluminum alloy, $\sigma_{01} = 250$ MPa, $\rho_1 = 0.0027$ g/mm³, $\varepsilon_{u1} = 0.20$, $B_1 = 10$ mm, $H_1 = 2$ mm
2. Steel, $\sigma_{02} = 780$ MPa, $\rho_2 = 0.00785$ g/mm³, $\varepsilon_{u2} = 0.40$, $B_2 = 10$ mm, $H_2 = 2$ mm

Determine duration of impact until one of the beams breaks.

Follow Case 10.38:

$$c_1 = \left(\frac{\sigma_{01}}{\rho_1}\right)^{1/2} = \left(\frac{250}{0.0027}\right)^{1/2} = 304.3;\ c_2 = 315.2\, \text{m/s; lateral wave velocities}$$

$m_1 = \rho_1 B_1 H_1 = 0.0027 \times 10 \times 2 = 0.054;\ m_2 = 0.157$ g/mm

$$V_0 = \frac{V_2}{1 + \frac{m_1 B_2}{m_2 B_1}} = \frac{240}{1 + \frac{0.054 \times 2}{0.157 \times 2}} = 178.6\,\text{m/s; initial velocity of overlapping area}$$

$$V_i = \frac{V_2}{1 + \frac{m_1 c_1}{m_2 c_2}} = \frac{240}{1 + \frac{0.054 \times 304.3}{0.157 \times 315.2}} = 180.2\,\text{m/s; limit velocity when } t \to \infty.$$

$$V_{av} \approx \frac{1}{2}(V_0 + V_i) = \frac{1}{2}(178.6 + 180.2) = 179.4\,\text{m/s; average velocity between contact and}$$

rupture

$$V_{1s} = \frac{4}{\pi} c_1 \sqrt{\varepsilon_{1u}} = \frac{4}{\pi} 304.3 \sqrt{0.2} = 173.3\,\text{m/s; safe average speed for Beam 1}$$

Note that $V_{1s} < V_{av} \approx 179.4\,\text{m/s}$; this beam will break. For Beam 2: $V_{2s} = 253.8\,\text{m/s}$, therefore Beam 2 will not break. Time needed for Beam 1 to fail is

$$\frac{1}{t_1} = \frac{1}{B_2}\left(\frac{\pi V_{av}}{2\sqrt{\varepsilon_{1u}}} - 2c_1\right) = \frac{1}{10}\left(\frac{\pi 179.4}{2\sqrt{0.2}} - 2 \times 304.3\right); \text{ then } t_1 = 0.465\,\text{ms}$$

At this point, one may attempt to improve the accuracy by finding the velocity at time t_1 and then calculating the new average speed

$$V(t_1) = \frac{V_2}{1 + \frac{m_1(B_2 + 2c_1 t)}{m_2(B_1 + 2c_2 t)}} = \frac{240}{1 + \frac{0.054(2 + 2 \times 304.3 \times 0.465)}{0.157(2 + 2 \times 315.2 \times 0.465)}} = 180.2\,\text{m/s}$$

This gives a minor change in t_1, but the gain in accuracy is not worth the effort.

$\xi_1 = c_1 t_1 = 304.3 \times 0.465 = 141.5\,\text{mm}$ is the half-length of the deformed Beam 1 at time t_1, outside contact zone. For the results to be valid, ξ_1 must be larger than the distance to the support.

$\delta = u(t) \approx V_{av} t_1 = 179.4 \times 0.465 = 83.4\,\text{mm}$; displacement of contact zone at time t_1.

11 | Columns and Beam–Columns

THEORETICAL OUTLINE

This chapter continues the analysis of beams/columns with particular attention to the effects of the axial load, be it sustained or dynamically applied. Such problems are, typically, non-linear, but an array of means exists to simplify analytical formulations.

The shape imperfections or deviations from a straight line feature prominently in this chapter. If a magnitude of deviation is not explicitly known, it must be assumed in accordance with the standards of the industry, from which the column originates.

Axially Compressed Beams

When axial compression predominates, such beams are usually referred to as columns or beam-columns. The differential equation of the deflected line of an elastic column, as shown in Figure 11.1a, with no inertia forces involved, is

$$EIu^{IV} + Pu'' = w; \quad \text{where } u^{IV} \equiv \frac{\partial^4 u}{\partial x^4} \quad \text{and} \quad u'' \equiv \frac{\partial^2 u}{\partial x^2} \tag{11.1}$$

This expression can be found, for example, in Timoshenko [110] or it can be derived from Equations 2.24 and 2.25 by noting that in our sign convention one has

$$Q = \frac{d\mathcal{M}}{dx} + P\frac{du}{dx} \tag{11.2}$$

where the compressive force is regarded as positive. If no lateral loading is applied, that is, $w(x) = 0$, then a simpler form of Equation 11.1 may be used:

$$EIu'' = -\mathcal{M} \tag{11.3}$$

with $\mathcal{M} = Pu$ in our case. This equation may be solved for deflection $u(x)$ and the associated force P for a simply supported beam in Figure 11.1. As presented in Ref. [110], the solution yields the following family of forces:

$$P_{cr} = n^2 \frac{\pi^2 EI}{L^2} = n^2 P_e \tag{11.4}$$

where $n = 1, 2, 3 \ldots$, and the associated shapes $u(x)$ are sinusoidal half-waves. Two of these shapes, related to $n=1$ and $n=3$, are presented in Figure 11.1b. The forces defined by Equation 11.4 are referred to as *critical*, because each is capable, at least in principle, to hold the beam in equilibrium. The smallest of these forces, for $n=1$, is the *buckling force*

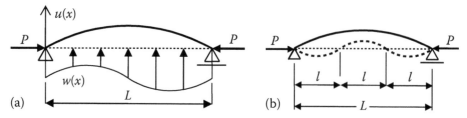

FIGURE 11.1   (a) A deformed beam with an axial force and lateral loading. (b) The same beam, initially straight, in a buckled shape.

and it is the only one that has physical meaning in the static domain.* However, when the dynamic compressive force is applied, the *higher modes* of buckling, corresponding to $n > 1$, can be induced. This will be demonstrated later.

Symbol P_e in Equation 11.4 stands for the Euler force, defined in Chapter 2. A buckling force for any other support condition shown in Figure 11.2 can be compactly written as

$$P_{cr} = \mu \frac{\pi^2 EI}{L^2} = \mu P_e \tag{11.5}$$

where P_{cr} stands for the smallest critical force, or a buckling force in the static sense. The coefficients μ are given in Table 11.1. If a customary concept of radius of inertia i is invoked, then *slenderness* Λ is defined as

$$\Lambda = \frac{L}{i\sqrt{\mu}} \quad \text{with} \quad i = \sqrt{\frac{I}{A}} \tag{11.6}$$

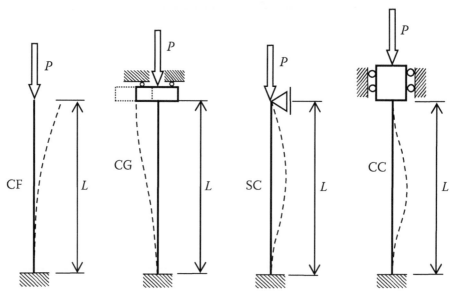

FIGURE 11.2   The remaining end conditions under consideration for buckling.

---

* The remaining shapes, associated with larger critical forces, can be realized only if the intermediate supports are introduced.

TABLE 11.1
Various Column Parameters

End Conditions	$\mu = P_{cr}/P_e$	Γ	$L\lambda/\delta^2$	Ψ	Θ
Cantilever (CF)	1/4	$\pi^4/64$	$\pi^2/16$	—	—
Simply supported (SS)	1	$\pi^4/4$	$\pi^2/4$	$\pi^4/32$	$2/\pi$
Clamped-guided (CG)	1	$\pi^4/16$	$\pi^2/16$	—	—
Supported-clamped (SC)	2.047	53	2.552	—	—
Clamped (CC)	4	π^4	$\pi^2/4$	$\pi^4/32$	1/2

while the buckling stress σ_{cr}, according to this linear buckling theory is*

$$\sigma_{cr} = \frac{P_{cr}}{A} = \frac{\pi^2 E}{\Lambda^2} \tag{11.7}$$

Equation 11.7 is valid only in the elastic range, $\sigma_{cr} \leq F_y$, or for $\Lambda > \Lambda_y$:

$$\Lambda_y = \pi\sqrt{\frac{E}{F_y}} \tag{11.8}$$

where Λ_y so defined is the limiting value of slenderness. One should notice here that most industrial columns are connected to other members, which may posses a substantial flexibility. For this reason, the perfect angular fixity will rarely be approached. (Coefficient μ will rarely come close to 4.)

STRAIN ENERGY IN AN ELASTIC COLUMN

This variable can be found using the approximate deflected shapes introduced in Chapter 10, where δ is designated as the largest lateral deflection. Equations 10.43 through 10.45 serve to calculate the numerical value of coefficient Γ in the bending energy expression:

$$\Pi_b = \Gamma EI \frac{\delta^2}{L^3} \tag{11.9}$$

The results of the integration, in the form of the Γ coefficient, are presented in Table 11.1. Another important variable is the axial shrinking of the column due to bending alone (with no contribution of compression). This quantity is identical, except for the sign, with λ that was introduced in Chapter 9 to determine the difference in length between a deformed and undeformed cable extending between two fixed points. For small deflections, it can be shown to be

$$\lambda = \frac{1}{2}\int_0^L \left(\frac{du}{dx}\right)^2 dx \tag{11.10}$$

* Some authors merely state $\Lambda = L/i$, but this is in reference only to a simply supported condition, or equivalent.

Using the approximate shape definitions according to section "Elastic strain energy and approximate deflected shapes" again, one finds the nondimensional shrinking, $(L/\delta^2)\lambda$ to be as in Table 11.1. The definition of this shortening makes it possible to derive the buckling load by the energy method. The column in Figure 11.1 is first compressed up to the load P, while lateral displacement is prevented. In the second step, with P kept at a constant level, the lateral restraints are slowly removed, thereby allowing the column to bend. In this step the axial shortening λ is caused entirely by flexure and defined by Equation 11.10. The work of the compressive load P is therefore $P\lambda$. This can be equated with the bending energy defined by Equation 11.9, to obtain

$$P_{cr}\lambda = \Gamma EI \frac{\delta^2}{L^3} \tag{11.11}$$

The value of P, which ensures this equality, is the buckling load P_{cr}. The Γ coefficient as well as the multiple of λ in Table 11.1 are based on approximate deflected shapes. When the exact values of buckling loads are known for the end conditions of interest, Equation 11.11 provides an opportunity to check on their consistency of those shape approximations. Typically, the agreement is perfect in that P_{cr} calculated by substituting the Table 11.1 values for λ and Γ into Equation 11.11 gives the result in the μ column of the Table 11.1. The only exception is the "irregular" SC condition, where $\mu = 2.127$ rather than 2.047.

BUCKLING WHEN PEAK STRESS EXCEEDS THE YIELD POINT

The above developments were presented for an elastic material. If the critical stress from Equation 11.7 exceeds the proportionality limit (which we assume to coincide with the yield point F_y) a formula, accounting for the material nonlinearity must be employed. The simplest, and quite realistic at the same time, is the tangent modulus theory, which says that above the yield point one must merely replace the elastic modulus E by the tangent modulus E_p in Equation 11.7:

$$\sigma_{crp} = \frac{P_{crp}}{A} = \frac{\pi^2 E_p}{\Lambda^2} \quad \text{with} \quad E_p = \frac{d\sigma}{d\varepsilon} \tag{11.12}$$

The tangent modulus varies from point to point in the inelastic part of the stress–strain curve. These formulas suggest the following procedure for the determination of the inelastic buckling stress. Beginning with the stress–strain curve, find E_p for each point, then calculate $\Lambda = \pi\sqrt{E_p/\sigma}$. The resulting curve is sketched in Figure 11.3 for $\Lambda < \Lambda_y$. The curve ends at the transition point where $\sigma_{crp} = \sigma_{cr} = F_y$. The elastic formula governs for $\Lambda > \Lambda_y$.

In many applications, the σ–ε curve is treated as a straight line above $\sigma = F_y$. If such a simplification is made, in accordance with Chapter 4, then only a single value of E_p is available. In a typical situation, where the σ–ε curve is not available but the variables F_y, F_u, and ε_u are, the secant value of E_p can be determined using these variables along with a bilinear material model. Unfortunately, E_p found in this manner is unsuitable for use in Equation 11.12. (See Example 11.1.) Lacking better data, one can employ the following estimate for structural materials:

$$\sigma_{cr} = F_u - (F_u - F_y)\left(\frac{\Lambda}{\Lambda_y}\right) \tag{11.13}$$

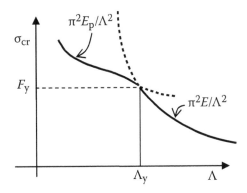

FIGURE 11.3 Average buckling stress resulting from stress–strain curve properties.

The above is merely an interpolation between $\Lambda = 0$, where σ_{cr} is assumed as F_u and Λ_y, where it is equal to F_y.

One could bring up the following argument against the tangent modulus concept. When a column is compressed, some eccentricity is always present, due to either geometrical imperfections or the lateral loading. As a result, the total stress resulting from superposition of bending and compression is different on each side of the column. One side may be in the plastic range, while the other is still likely to remain in the elastic state. If so, applying the same material modulus to both sides of the column is not correct. While there is some rationale in the above, there are also some complicating factors in column response to increasing compressive loading. The experimental evidence is strongly in favor of the tangent modulus theory as compared with its more complex competitors.

Another comment with regard to material properties is necessary. In the above description, it was assumed that the σ–ε curve has a smoothly varying and monotonically decreasing slope once it deviates from the Hooke's law. This can be expected in the majority of engineering materials, but there are exceptions. Some types of mild steel, which have a pronounced level segment of the σ–ε curve is then followed by material hardening as the strain increases. If such a plateau exists, it has to be regarded as an "absolute" yielding and the upper limit of the buckling stress. In this event, F_u in Equation 11.13 must be replaced by F_y and the latter, in turn, by the proportionality limit F_p.*

TWO-FLANGE SYMMETRICAL COLUMN

This section type or TFS in brief, has a straightforward description, because of its geometrical simplicity. The external force P and moment $\mathcal{M}$ shown in Figure 11.4 induce the flange loads R_l and R_u:

$$R_l = \frac{P}{2} + \frac{\mathcal{M}}{H} \tag{11.14a}$$

---

* Ignoring the distinction between the proportionality limit and the yield point, as is done in most of this book, is usually of no consequence. This is one of the exceptions where it must be mentioned.

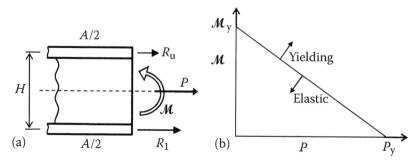

FIGURE 11.4   Two-flange beam forces (a) and moment–force interaction diagram (b).

$$R_u = \frac{P}{2} - \frac{M}{H} \qquad\qquad (11.14b)$$

The above equations ignore, of course, the contribution of web, consistent with the assumptions made earlier. For the yield stress of F_y, the yield flange forces and the resulting load capacities become

$$R_y = AF_y/2 \qquad\qquad (11.15a)$$

$$P_y = AF_y \qquad\qquad (11.15b)$$

$$\mathcal{M}_y = HAF_y/2 \qquad\qquad (11.15c)$$

The equation for R_1 may be rewritten in order to determine the onset of yield when both loading components are interacting:

$$\frac{P}{P_y} + \frac{M}{\mathcal{M}_y} = 1 \qquad\qquad (11.16)$$

as is visualized in Figure 11.4b. The section loads may also be expressed by strains. From Equation 11.14 one obtains

$$\mathcal{M} = \frac{AH}{4}(\sigma_1 - \sigma_u) \qquad\qquad (11.17a)$$

$$P = \frac{A}{2}(\sigma_1 + \sigma_u) \qquad\qquad (11.17b)$$

All of the above relations come from equilibrium only and, as such, are material-independent. For elastic material, $\varepsilon = \sigma/E$, one has

$$\mathcal{M} = \frac{EAH}{4}(\varepsilon_1 - \varepsilon_u) \qquad\qquad (11.18a)$$

$$P = \frac{EA}{2}(\varepsilon_l + \varepsilon_u) \tag{11.18b}$$

which tells us that the moment is associated with the extreme strain difference and the axial load is related to the average strain. One can now use Equations 10.1 to notice that since $\varepsilon_l = y_1/\rho$ and $\varepsilon_u = -y_2/\rho$, then

$$\varepsilon_l - \varepsilon_u = \frac{H}{\rho} \tag{11.19a}$$

and

$$M = \frac{EI}{\rho} \tag{11.19b}$$

where $I = AH^2/4$. The first of the above relations was written considering bending only, but one should note, however, that the difference of strains remains the same after addition of an axial load or axial strain. The second relationship was derived in Chapter 10.

Next, consider a pure bending of the TFS past the yield point. A general bilinear stress–strain curve of the material is shown in Figure 11.5a. As already noted in Chapter 10, the moment–curvature relation is merely a rescaling of a stress–strain curve, in absence of an axial force:

$$\sigma = \frac{2M}{AH} \tag{11.20a}$$

or

$$M = \frac{AH}{2}\sigma = \frac{AH}{2}(Y + E_p\varepsilon) \tag{11.20b}$$

consistent with Chapter 4. Noting that for a section in bending $\varepsilon = H/(2\rho)$, one obtains

$$M - M'_y = \frac{E_pI}{\rho} \quad \text{where} \quad M'_y = M_y\frac{Y}{F_y} \tag{11.21}$$

The above expressions, valid past the onset of yield, are illustrated in Figure 11.5b. Equation 11.21a is quite similar to the elastic relation given by Equation 11.19b, with the plastic

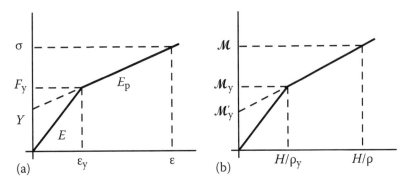

FIGURE 11.5 Stress–strain relation for a bilinear material (a) and moment–curvature for a TFS (b).

modulus E_p replacing its elastic counterpart E and with $\mathcal{M}'_y$ being somewhat smaller that the moment associated with the beginning of yield $\mathcal{M}_y$.

SECTION CAPACITY UNDER COMBINED LOAD COMPONENTS

The interaction diagram in Figure 11.4 merely visualizes the fact that if there is a certain axial load applied, the section is capable of carrying a lesser bending moment that $\mathcal{M}_y$. If a larger moment is applied, yielding will take place. The linear relationship is typical for all sections, which remain in the elastic range. For the stress distribution that follows a rigid-plastic (RP) model an example of a fully developed stress pattern is given in Figure 11.6a. This can be resolved into a symmetric component (P) and antisymmetric one ($\mathcal{M}$). The interaction equation for a rectangular section is

$$ r = \left(\frac{P}{P_y} \right)^2 + \left| \frac{\mathcal{M}}{\mathcal{M}_y} \right| = 1 \tag{11.22} $$

This is sometimes too awkward for practical applications, so it is simplified to the elastic-like relationship, by omitting the second power of the force ratio or, in effect, by using the set of dashed lines within the curves in Figure 11.6b. Another possibility, this time nonconservative, but very convenient, is to use a *rectangular yield criterion*, shown with dotted lines circumscribing the curves in Figure 11.6b. When given two values, P and $\mathcal{M}$, that act on a section, the section capacity check amounts to making certain that $\mathcal{M} < \mathcal{M}_y$ and $P < P_y$.* To make the procedure more rational, it is better to reduce the permissible values of the force and/or the moment, as appropriate for the circumstances.

For sections symmetric about the bending axis, other than a solid rectangle, the following relationship holds, according to Kaliszky [45]:

$$ r = \left(\frac{P}{P_y} \right)^\gamma + \left| \frac{\mathcal{M}}{\mathcal{M}_y} \right| = 1 \tag{11.23} $$

For some of the sections listed in Chapter 10, the exponent γ has the following values: $\gamma = 1$ for TFS, $\gamma = 2$ for solid circle and rectangle, $\gamma = 1.68$ for a thin-wall tube. The factor r is

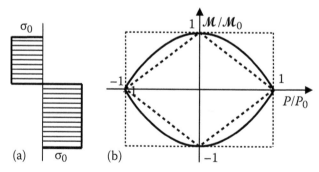

FIGURE 11.6   Stress distribution for RP material when P with $\mathcal{M}$ are applied (a) and interaction diagram for a rectangular section (b).

---

* The use of a rectangular yield criterion is equivalent to saying that the moment and the force are independently checked against their respective permissible values.

sometimes called a *usage ratio* and it is, in fact, the inverse of the safety factor against yielding. When bending and shear are combined, the following holds, for a rectangular section:

$$r = \left(\frac{Q}{Q_y}\right)^2 + \left(\frac{M}{M_y}\right)^2 = 1 \tag{11.24}$$

where Q_y is the yield force in shear.* On occasions, all three components have to be combined. A simplified, although not the exact relation may then be used:

$$r = \left(\frac{P}{P_y}\right)^2 + \left(\frac{Q}{Q_y}\right)^2 + \left(\frac{M}{M_y}\right)^2 = 1 \tag{11.25}$$

COMPRESSION OF IMPERFECT COLUMNS, YIELD STRENGTH, AND POST-YIELD RESPONSE

Every real column is imperfect in the sense that it is not straight, that it has some waviness. When it works as a beam, subjected to lateral loads only, those imperfections are of no consequence. It is the action of a compressive load, which makes those shape deviations important, as they often dictate the deflection response. The reason for mentioning *imperfect* columns is to highlight the importance of imperfections.

The initial, sinusoidal shape in Figure 11.7 has the maximum eccentricity δ_0 when the column is unloaded. The initial deflected shape is described by $\delta(x)$. The current, lateral deflection $y(x)$, measured from the straight axis, is assumed to follow the original shape:

$$\delta(x) = \delta_0 \sin\frac{\pi x}{L} \tag{11.26a}$$

and

$$y(x) = y_0 \sin\frac{\pi x}{L} \tag{11.26b}$$

in which the magnitude of y_0 is a function of P. When a convex line $y(x)$ is differentiated twice, the curvature, or $-1/\rho$ is obtained. Differentiating the above and subtracting sides gives

$$\frac{1}{\rho} - \frac{1}{\rho_0} = \frac{\pi^2}{L^2}(y_0 - \delta_0) \tag{11.27a}$$

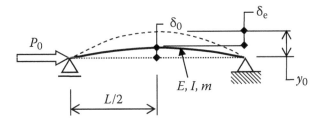

FIGURE 11.7 Initial (unloaded) shape with eccentricity δ_0 and deflected shape with y_0.

* Note that Equation 6.11, combining the direct and the shear stress may also be written in a similar way.

or

$$\frac{1}{\rho} - \frac{1}{\rho_0} = \frac{M}{EI} \tag{11.27b}$$

where the second equation was written on the basis of Equation 11.3 and it indicates that in the unloaded condition, with the initial curvature $1/\rho_0$, the moment is nil. Equating the right-hand sides of both equations and noting that $M = Py_0$, we can write the resulting expression as

$$y_0 = \frac{\delta_0}{1 - P/P_e} \tag{11.28}$$

which can be seen as a restatement of Equation 10.28 showing the effect of an axial force (compressive this time) on beam deflections in the elastic range. The final deflection y_0 caused by the application of P continues to grow until the inside fiber stress reaches the yield point, which is a transition point for the shape of a $P\text{--}y$ curve. The limiting form of Equation 11.28 is then*

$$y_1 = \frac{\delta_0}{1 - P_1/P_e} \tag{11.29}$$

Thus far the results presented here hold true for any beam section. From this point on, further development will be presented using the TFS profile only. For this section the transition point is reached when the inside flange force attains the yield point, $R_{iy} = P_y/2$. The load application (P and resulting M) in Figure 11.7 is opposite to that in Figure 11.4. Therefore, the flange forces are compressive, and in this section regarded as positive, as shown in Figure 11.8a. The limiting force P_1 is found from

$$R_{iy} = \frac{P_1}{2} + \frac{P_1 y_1}{H} \tag{11.30a}$$

or

$$P_1 = \frac{P_y}{1 + \dfrac{2y_1}{H}} \tag{11.30b}$$

Equations 11.29 and 11.30 have two unknowns, y_1 and P_1. They can be solved by assuming some $P_1 < P_e$ in the first expression, calculating y_1, then finding a new value of P_1 from the second equation. Repeating the procedure usually gives a quick convergence. (As mentioned before, $P \ll P_{cr}$ for relatively short columns (a common case), which makes the

---

* This relation can be obtained by solving the differential equation (Equation 11.3), where the applied moment is $M = P(\delta + u)$ and u is measured from the initially deflected position. See Ref. [110]. When a beam is in compression, the inside fiber, that is, the one closer to the original beam axis always experiences the higher stress. This is true regardless of whether the deflected shape is convex or concave.

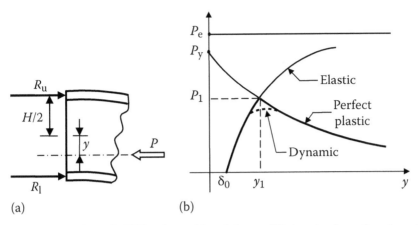

FIGURE 11.8 Mid-section of a TFS column (a) and the equilibrium load as a function of midpoint deflection $y = y_0$ (b). Under dynamic conditions the characteristic becomes smoother, as shown by a dotted line.

denominator of Equation 11.29 only slightly smaller than 1 and the second iteration is often unnecessary.) Once the transition load P_1 is exceeded, that is, $P > P_1$, Equation 11.29 is no longer valid. A more general view of the situation appears, when Equations 11.29 and 11.30b are rewritten, removing the index "1" for P_1 and y_1:

$$P = P_e\left(1 - \frac{\delta_0}{y}\right) \tag{11.31a}$$

and

$$P = \frac{P_y}{1 + 2y/H} \tag{11.31b}$$

The first equation relates to an elastic column with an eccentricity δ_0. It shows how P changes as a function of increasing y, as illustrated in Figure 11.8b. The second equation describes an initially straight column made of a perfectly plastic material. When y is controlled and when it is growing from the initially straight shape, $y = 0$, P decreases from its plastic strength P_y to whatever value is needed to attain equilibrium. The intersection of these two curves gives the exact, rather than an approximate solution. Equating both values of P from Equations 11.31 results in the expression for the deflection $y = y_1$ at the maximum force $P = P_1$. That expression is given in Case 11.24, while Case 11.25 provides analogous formulas for an unsymmetrical, two-flange section, TFUS.

One should also note that Equation 11.28 is valid in tension as well, except that P is then a negative number. Obviously, the case of tension is not of great consequence with regard to stiffness change for stout columns, but has some importance for slender members.

The quantification for an EPP material is simple in that the flange load $R_{iy} = P_y/2$ remains constant after yielding. When y increases past y_1, P decreases. For $y = H/2$, the compressive force $P = R_{iy}$ is in line with the inside flange and then the outer flange reaction is nil. For $y > H/2$ the outer flange force becomes tensile, but Equation 11.31b still holds. The load P applied to the column and needed for the deflected equilibrium is shown in

Figure 11.8b with a thick line. The dotted line shows that under dynamic conditions the peak may be bypassed.* A few comments will be made now to make the above derivations somewhat less restrictive.

1. The flange forces are statically determinate, which means their dependence on P is not material-related.
2. To associate deflections with loads, the materials of both chords need to be the same, in order to keep Equation 11.29 applicable. But the yield strength of each flange may be different, as R_{ly} and R_{uy}.
3. "Yielding" need not be taken literally. It simply means a load level at which structural capacity ceases to grow. This may be caused, for example, by local buckling of a thin wall of a flange.
4. The above is valid for small deflections of TFS sections. For other cross sections, one can get the first estimate by replacing it with a TFS of the same area A and, in case of a well-developed plasticity, the same first area moment Z.

The column in Figure 11.7 has only a shape imperfection of magnitude δ_0 associated with it. If that column is also loaded laterally in such a way that in absence of P an additional deflection δ_1 appears, and if that additional lateral loading is kept constant after P is applied, then the analysis can be carried out as if the column had an *initial* shape deviation of magnitude $\delta_0 + \delta_1$ and the lateral loading can be ignored.

DYNAMIC BUCKLING OF ELASTIC MEMBERS

The existence of higher mode buckling shapes, as indicated in Figure 11.1b, may be inferred by the following reasoning. Suppose the column is laterally restrained and the axial load attains some value P_0. That restraint is suddenly removed and the column is free to buckle. Assume that it buckled into n equal segments and a snapshot was taken at the instant when the extreme deflection was reached, giving a picture similar to what is seen in Figure 11.1b. The axial shortening of each segment l-long due to flexure is $P_0\lambda$ and the work of P_0 is $nP_0\lambda$. The energy of bending of each short segment is defined by Equation 11.9, with l replacing L while the coefficients λ and Γ are found in Table 11.1. Equating work and energy, one obtains

$$nP_0\lambda = n\Gamma EI \frac{\delta^2}{l^3} \quad \text{or} \quad P_0 \frac{\pi^2}{4} \frac{\delta^2}{l} = \frac{\pi^4}{4} EI \frac{\delta^2}{l^3} \tag{11.32a}$$

$$\text{which gives } P_0 = \frac{\pi^2 EI}{l^2} = n^2 \frac{\pi^2 EI}{L^2} \tag{11.32b}$$

As this is essentially the same as Equations 11.4, the conclusion is that when the column is subjected to a suddenly applied force $(n^2)P_e$, its length tends to become subdivided into n half-waves. The concept of the initial lateral restraint and its subsequent removal

---

* For other cross sections the dotted transition line may result from the moment being reduced when a transition from a perfectly plastic to an elastoplastic stress distribution is developed.

is needed to avoid the complications (stress waves) associated with a sudden application of an axial loading.*

Another way to look at dynamic buckling is to use some of the reasoning related to wave propagation in axial bars. In Chapter 5 it was shown that a bar moving with a velocity v_0 and impacting a rigid wall experienced the same stress as if a heavy and hard object moving with v_0 impacted the same, stationary bar. The event resulted in the stress of $\sigma_m = \rho c_0 v_0$ propagating along the bar from impact point. When this is set equal to the buckling stress defined by Equations 11.6 and 11.7, the buckling length l can be found as

$$l = \pi i \sqrt{\frac{E}{\rho c_0 v_0}} = \pi i \sqrt{\frac{c_0}{v_0}} \tag{11.33}$$

where $\mu = 1$, as the simply supported shape is anticipated. This equation shows that with the increasing impact speed, the length of a buckling half-wave is reduced, which is essentially the same result as presented by Equations 11.32. Johnson [40] quotes the test results confirming that the above is in line with experimental findings, save for scatter of results.

The loads necessary to induce the higher bending modes, as described here, are quite large, often exceeding the strength of material involved. Still, whatever happens initially, in the elastic range usually influences the final, plastically deformed shape (the residual shape.)

If one of the ends has a different constraint than a simple support, then the buckled shape will, in general, differ from a set of sine half-waves. However, if the load is large enough and the bar sufficiently slender to be capable of experiencing higher modes, the half-waves become closer to sinusoids as we move away from the end in question.

When an axial force is suddenly applied to the end of a column, it induces stress waves traveling back and forth along that column. While this complicates the analysis, it is of relatively little importance for short lasting loads, typical of impacts. The reason for it is that longitudinal stress waves travel much faster than the lateral displacement waves. The former are likely to be damped out before the latter develop into peak lateral displacements. (For a broader discussion of this subject refer to SRI [82].) Keeping lateral constraints in place until the axial force reaches its peak is a simple means to avoid dealing with a variable axial loading.

Higher-Order Elastic Buckling of Imperfect Columns

The equation of lateral motion of a beam–column differs from its static counterpart, Equation 11.1, mainly by the addition of the inertia term:

$$EI u^{IV} + P(u + u_0)'' = w - m\ddot{u}; \quad \text{where } \ddot{u} \equiv \frac{\partial^2 u}{\partial t^2} \tag{11.34}$$

The additional deflection component $u_0(x)$ is the shape of an initial imperfection, which has to be supplied. (This component does not appear in the first, curvature-related term.)

---

* One should keep in mind that the inertia of the beam also serves, in a way, as a physical restraint, as it delays the development of lateral deflections.

The problem is posed as before: The simply supported column is laterally restrained and the axial load attains some value P. The restraint is then suddenly removed and the resulting deflection $u(t)$ is to be found. No lateral load is present, that is, $w(x, t) = 0$.

Consider first a single-mode deviation, that is, $u_0 = \delta_0 \sin(\pi x/L)$. The solution is in the form

$$\delta_d = \delta(t) \sin(\pi x/L) \tag{11.35}$$

where δ_d is measured from the initially deflected position. After this is substituted in Equation 11.34, a differential equation for $\delta(t)$ is obtained. Depending on the magnitude of P, there are two distinct solutions, as detailed by Case 11.3:

$P < P_e$; harmonic vibrations

$P > P_e$; displacement δ_d becoming unbounded with time

Only the second one can be regarded as buckling. In case of a more complex initial shape, it may always be presented as a trigonometric series

$$\delta_d = \delta_{01} \sin(\pi x/L) + \delta_{02} \sin(2\pi x/L) + \cdots \tag{11.36}$$

with as many terms as necessary (although more than three would be unlikely). The solution of Equation 11.34 is then

$$\delta_d = \delta_1(t) \sin(\pi x/L) + \delta_2(t) \sin(2\pi x/L) + \cdots \tag{11.37}$$

Substituting the above equation in Equation 11.34, gives a set of differential equations, one for each function $\delta_i(t)$. Again the solution type depends on the magnitude of the applied load. Putting the nth order buckling force as

$$^nP_e = n^2 \frac{\pi^2 EI}{L^2} \tag{11.38}$$

one has harmonic vibration for $P < {}^nP_e$ and displacement growth with no bounds for $P > {}^nP_e$ for the mode of interest. Depending upon how large P is and how many modes are involved, some may be associated with vibrations and some with buckling. For a single mode n, where the second inequality holds, one has the following dynamic deflection as a function of time

$$\delta_n(t) = \beta \delta_{0n} \left[\cosh \Phi t - 1 \right] \tag{11.39}$$

where the amplitude multiplier β and the pseudofrequency Φ are, respectively

$$\beta = \frac{P/{}^nP_e}{P/{}^nP_e - 1} \tag{11.40a}$$

$$\Phi = \frac{\pi n}{L} \left(\frac{{}^nP_e}{m} \left(\frac{P}{{}^nP_e} - 1 \right) \right)^{1/2} \tag{11.40b}$$

and the maximum deflection of the nth mode is $\delta = \delta_{0n} + \delta_n(t)$. There is a different lateral response, depending on a mode number. If $P = P_0$ is suddenly applied to a column, then, with $^2P_e < P_0 < {}^3P_e$, that is, the force being somewhere between the second and the third level, harmonic vibrations are experienced for modes number 3, 4, 5, etc. A lateral movement, one-way only, as long as P is applied, takes place for modes 1 and 2. The extent of that movement depends, of course, on the size of coefficients δ_{0i}.

Another important question is the relation between the value of the compressive force P, and the mode number n for which the fastest magnification of deflection takes place. It turns out not to be $P = {}^nP_e$, but P found from

$$n_c \approx \sqrt{\frac{P}{2P_e}} \tag{11.41}$$

Let us suppose $P = 16P_e$, which is far in excess of static buckling. According to Equation 11.38, this corresponds to $n = 4$, or the length L being subdivided into four half-waves. But Equation 11.41 tells us that $n_c = \sqrt{8} \approx 3$ is the mode, whose amplitude grows the fastest under this load, not the mode with $n = 4$.

REINFORCED CONCRETE (RC) COLUMN

RC column analysis is a part of a broad subject of RC design. Our simplified coverage is limited to only two shapes: rectangular and circular. The sections, as presented in Cases 11.26 and 11.27, respectively, have steel uniformly placed around the sides. (Additionally, some simpler sections will be used to illustrate certain concepts.) Steel area A_s is only a small part of the entire section area A, but steel is stiffer then concrete according to the ratio of Young's moduli, $q = E_s/E_c$. Whenever an axial force is applied to a section, the strains in steel and concrete are the same, therefore the steel stress is q times larger. Concrete is weak in tension, as it can withstand only the stress level of

$$F_t = 0.6\sqrt{F_c'} \tag{11.42}$$

where F_c' is the nominal compressive strength, with both quantities in MPa.* As long as the resultant load produces stress below F_t, a section can be treated as monolithic. With steel area component of a cross section being small, as mentioned above, one can treat the whole RC section as consisting of concrete only, in the first approximation. For a more accurate approach one can employ the equivalent properties, as given in Cases 11.26 and 11.27.

The method of analysis here relates closely to ultimate strength design widely used in connections with RC structures. In line with the general approach in this book, the objective is to determine the most likely failure loads without making certain they are on the "safe" side. This includes, among other simplification, not using the capacity reduction factors φ, which are, in general, different for different elements.[†]

---

* F_t as given above is the flexural strength, a typical application. Pure tension on concrete members is quite rare.
[†] For tied columns, for example, the reduction factor often used is $\varphi = 0.7$.

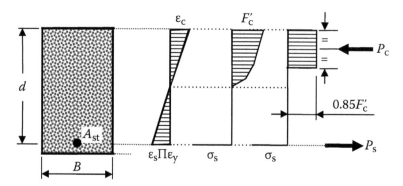

FIGURE 11.9   The cross section is shown with the strain diagram (ε), concrete stress (σ_c), and the equivalent concrete stress block.

A column is typically subjected to compression and bending due to eccentricity. When the compressive force is applied at the centroid, eccentricity is absent and the axial strength is then estimated by

$$P_0 = 0.85 F_c'(A - A_s) + A_s F_y \tag{11.43}$$

which means that the net concrete area, $A - A_s$, is capable of resisting 85% of the nominal material compressive strength.

Bending of a concrete beam in its limiting state is presented in Figure 11.9. There is only tensile reinforcement present, assumed to be capable of developing its full plastic strength F_y.* The "plane sections remain plane" assumption holds, which gives rise to a linear strain distribution along the depth. The concrete stress–strain relationship is nonlinear, but for computational purpose stress distribution is treated as rectangular block with the constant value of $0.85 F_c'$ and the resulting concrete compressive force P_c. This is pure bending; therefore the resultant steel tension P_s and concrete compression P_c are equal in magnitude. The bending moment capacity $\mathcal{M}_0$, resulting from this approach is[†]

$$\mathcal{M}_0 = A_{st} F_y d \left(1 - 0.6 \frac{A_{st}}{Bd} \frac{F_y}{F_c'} \right) \tag{11.44}$$

according to Darvall [21]. (The term in brackets simply shows that the effective moment arm is less than d.) A typical column is designed to carry an axial load P applied with a certain offset e from the centroid. This offset results in a moment Pe being applied at the same time. There is an important concept of a *balanced failure*, which means that yielding of steel and crushing of concrete takes place at the same load level. The approximate estimate of a balanced load capacity is made with the help of Figure 11.10, which shows a

---

[*] This is a configuration where steel yields before concrete fails. Such sections, which have only a moderate amount of steel, are called under-reinforced.

[†] The following equation applies for the $F_y \approx 400$ MPa reinforcing steel. The coefficient 0.6 is reduced for a stronger reinforcement.

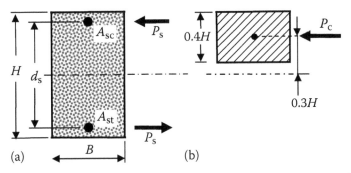

FIGURE 11.10 Simplified balanced condition for a beam with top and bottom reinforcement.

section having both tensile steel with section A_{st} and compression steel A_{sc}. These areas are equal, that is, $A_{st} = A_{sc}$, as usual for columns.

Assume that in a balanced condition, $0.4H$ of the entire section is concrete working to the full capacity of $0.85F_c'$. This gives a force of P_c acting on an offset of $0.3H$. Steel, on the other hand, gives only a moment, as the forces are $A_{st}F_y$ each. Thus we have the resultant force P_b and moment $\mathcal{M}_b$ as

$$P_b = 0.85F_c'(0.4BH) = 0.34(BH)F_c' \tag{11.45}$$

$$\mathcal{M}_b = A_{st}F_y d_s + (0.34BHF_c')(0.3H) = A_{st}F_y d_s + 0.102BH^2F_c' \tag{11.46}$$

Now we have all the components needed to sketch a rudimentary P–$\mathcal{M}$ diagram for a column, as done in Figure 11.11 with a thick line. The dashed line, on the other hand, makes it look a little closer to what might be seen in a concrete handbook.*

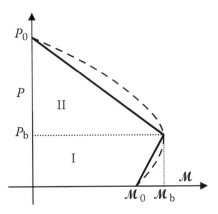

FIGURE 11.11 A simplified P–$\mathcal{M}$ diagram for a column (continuous line) and a more realistic version (dashed line).

---

* Numerous marginal caveats used in design make our dashed line an oversimplification, too.

If the applied axial load P is relatively small, that is, within region I in Figure 11.11, then the associated $\mathcal{M}$ at a limit state can be written as

$$\mathcal{M} = \mathcal{M}_0 + (\mathcal{M}_b - \mathcal{M}_0)\frac{P}{P_b} \tag{11.47}$$

For region II we have

$$\mathcal{M} = \frac{P_0 - P}{P_0 - P_b}\mathcal{M}_b \tag{11.48}$$

The above moment and force values are summarized in Case 11.26 and analogous expressions for round columns are given by Case 11.27. They have to be viewed as the first approximation to rather complex expressions found in specialty works on RC design. The accuracy can be improved by determining the characteristic points in Figure 11.11 according to the prevailing design rules rather than the simplified formulas proposed here.*

If a beam or a column section is a solid rectangle, the average shear stress in a section is $V/(BH)$ but the peak shear is 1.5× larger, according to Chapter 10. If there is a shear stress τ in a section under pure shear, that means there also is tension, $\sigma_t = \tau$, in another section through the same point as well as compression, $\sigma_c = \tau$ at 90° to the latter. (Refer to Appendix A.) If the maximum stress theory is invoked (Chapter 6), then we can expect the condition to avoid failure to read $\tau \le F_t$, where F_t is from Equation 11.42. This turns out to be conservative even for an un-reinforced beam, but to avoid complications inherent in the codes of practice, the criterion will be adopted here. For a rectangular section with or without shear reinforcement present the following shear strength results:

$$Q_0 \approx BHF_t \tag{11.49}$$

While this is quite conservative, it nevertheless provides an indicator of a proximity to failure.

All of the above considerations were concerned with the strength as dictated by a cross section and without any reference to the column length L. The magnifying effect of an axial load with respect to lateral response, such as deflection u or a bending moment $\mathcal{M}$ was presented in some detail in Chapter 2. The static effect of an eccentric compression was discussed in section "Section capacity under combined load components." A more general form of a magnification factor used in Equation 11.28 is

$$\eta = \frac{1}{1 - P/P_{cr}} \tag{11.50}$$

where
P is the sustained compressive load
P_{cr} is the buckling force of a column, calculated on elastic basis

The lateral effects can be established by first ignoring the presence of P and then multiplying the result by η. When P acts eccentrically and no lateral force is present, the lateral

---

* Still, the capacity reduction factor must be removed from any such improved calculation.

deflection u' is calculated as if the initial offset was constant and final deflection is again determined by multiplying u' by η.

Some codes of practice recommend that in calculating the buckling load the reduced modulus E_r be used:

$$E_r = E_c/3 \qquad (11.51)$$

where E_c is the nominal elastic modulus as defined in Chapter 4. The buckling force (or stress) is then found from elastic formulas after replacing E_c by E_r.* Most concrete columns are so stout that their axial strength P_0 is well below the buckling strength.

If the lateral action is strong enough to induce plastic hinges or shear slides in a column, then one can use the RP techniques presented in Chapter 10 in order to establish maximum displacement under such a loading. The column (or an RC beam for that matter) is regarded as failed if the discontinuity (kink) angle θ_m, resulting from joint rotation exceeds a certain value. (The angle is shown in Figure 10.11c for the SC end supports.) Henrych [32], for example, quotes θ_m from 0.05 to 0.13 as the value associated with failure. For supports that are built into other members, TM-5 [106] suggests that at least for some types of reinforcement, the support rotation $\theta = 5°$ is equivalent to failure. The limits used in the example problems will be of the same order of magnitude.

As for a permissible shear slide, there appears to be a lack of general criteria for failure. The following value will be tentatively employed:

$$u_s = 0.02H \qquad (11.52)$$

which means that calculated shear sliding should not exceed 2% of the section depth.

Consistent with our approach, no strength reduction factors (additional safety factors) are mentioned here. It is up to a designer to assign the SFs as appropriate for his practice.

RIGID STICK MODELS OF AXIALLY IMPACTED BEAMS

A good first estimate of response may often be obtained with the help of such models, typically for cases of a single-mode deformation. An example of a cantilever, as shown in Figure 11.12a, will be used to demonstrate the procedure. The first step is to determine the static buckling force. This can be found considering an initially undeflected beam and

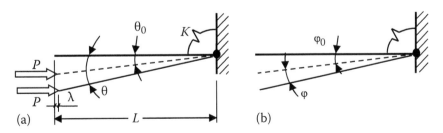

FIGURE 11.12 Rigid cantilever with an elastic angular spring. Notation for subcritical (a) and for supercritical compression (b).

---

* One may note that E_r is treated as an effective tangent modulus.

seeking a small displacement, at which the system can remain in equilibrium. Equating the moment of P with respect to the pivot point with that of the angular spring, one obtains:

$$P_{cr}\theta L = K\theta \quad \text{or} \quad P_{cr} = K/L \qquad (11.53)$$

The magnitude of the buckling force of a prototype elastic beam is known from Table 11.1 to be equal to $P_e/4$ and if this is to be preserved by the rigid stick model, the angular spring stiffness must be

$$K = \frac{P_e L}{4} = \frac{\pi^2 EI}{4L} \qquad (11.54)$$

To determine peak deflection due to an axial step load, the same approach as before will be employed. The first case considered is that of *subcritical compression*, that is, $P < P_{cr}$. The beam is assumed to have a small initial slope, $\theta_0 = u_0/L$ and lateral constraint is in place until the full axial load is reached. The constraint is then suddenly removed leaving the beam free to deflect laterally. Once the beam reaches the maximum deflection at some stationary point, there is no kinetic energy involved and the work of the external load, $P\lambda$, can be equated with the strain energy Π:

$$\Pi = \frac{K}{2}(\theta - \theta_0)^2 \qquad (11.55)$$

The apparent axial shortening λ depends on the slope θ:

$$\lambda = L(\cos\theta_0 - \cos\theta) \qquad (11.56a)$$

and

$$\lambda \approx \frac{L}{2}(\theta^2 - \theta_0^2) \qquad (11.56b)$$

where the last approximation holds for moderate deflections. Equating energy and the approximate work gives

$$\frac{K}{2}(\theta - \theta_0) = P\frac{L}{2}(\theta + \theta_0) \qquad (11.57)$$

from which the total maximum deflection can be recovered:

$$\theta = \frac{1 + P/P_{cr}}{1 - P/P_{cr}}\theta_0 \quad \text{for } P < P_{cr} \qquad (11.58)$$

For *supercritical compression*, a different approach must be taken, because the longer the loading lasts, the larger deflection will be induced and there is no upper bound one can anticipate. The equation of motion will be written for the deflected angle measured from the initial position of the beam:

$$J\ddot{\varphi} = PL\sin(\varphi + \varphi_0) - K\varphi \qquad (11.59)$$

where $J = mL^3/3$ is the mass moment of inertia about the end. To be able to solve this equation in a closed form, $\sin(\varphi + \varphi_0)$ must be replaced by $\varphi + \varphi_0$, thereby limiting the range of applicable deflections. The above can be rewritten as

$$\ddot{\varphi} - B\varphi = \frac{PL}{J}\varphi_0 \quad \text{with } B = \frac{K}{J}\left(\frac{P}{P_{cr}} - 1\right)$$ (11.60)

Omitting the term on the right side gives the homogeneous part of the equation, which is solved by $\cosh(\Phi t)$, where

$$\Phi = \sqrt{\frac{K}{J}\left(\frac{P}{P_{cr}} - 1\right)}$$ (11.61)

A particular solution is obtained by assuming it to be a constant multiple of φ_0. Substituting into Equation 11.59 and summing the terms yields

$$\varphi = \frac{P/P_{cr}}{P/P_{cr} - 1}(\cosh \Phi t - 1)\varphi_0 \quad \text{for } P > P_{cr}$$ (11.62)

For the subcritical case described above, it is not necessary to develop the equation of motion to obtain the peak response, a similar procedure gives

$$\varphi = \frac{P/P_{cr}}{1 - P/P_{cr}}(1 - \cos \Omega t)\varphi_0 \quad \text{for } P < P_{cr}$$ (11.63)

$$\Omega = \sqrt{\frac{K}{J}\left(1 - \frac{P}{P_{cr}}\right)}$$ (11.64)

At $\Omega t = \pi$, the maximum value of φ is obtained. When added to φ_0, the same total angle results, as predicted by Equation 11.58.

Extrapolation of this result to other end conditions can often be done without detailed calculations. For example, one can easily show that when a simply supported beam with length L is considered, then the spring constant K in the above must be replaced by $2K$ and the mass moment of inertia J refers to one-half of length (Case 11.5). Thus, for other end conditions analogous expressions result.

While rigid stick modeling offers clarity of quantification, some loss of accuracy must be expected. Consider, for example, axial shortening due to bending. When the cantilever is deflected by δ, its projected length is shorter by $\lambda = (\pi^2/16)\delta^2/L$, according to Table 11.1. From Case 11.4, one can deduce, that in absence of θ_0, the shrinking is

$$\lambda = \frac{L}{2}\theta^2 = \frac{L}{2}\left(\frac{\delta}{L}\right)^2 = \frac{1}{2}\frac{\delta^2}{L}$$ (11.65)

To get the beam result, one must therefore multiply this λ by the ratio of the two, namely $\pi^2/8$. The same ratio applies to other end supports, with which we are dealing here, except for SC, where the multiplier is 1.276.

The strain energy expression also needs a correction. Using Case 11.4, the energy in the stick model is

$$\Pi = \frac{1}{2}K\theta^2 = \frac{1}{2}\frac{L}{4}\frac{\pi^2 EI}{L^2}\left(\frac{\delta}{L}\right)^2 = \frac{\pi^2}{8}\frac{EI\delta^2}{L^3} \tag{11.66}$$

The corresponding coefficient in Table 11.1 is $\pi^4/64$, so the stick model value must be multiplied by $\pi^2/8$. Again, this holds for other end conditions as well except for SC, where the number is now 1.312.

When Equation 11.40b for time-dependent motion during supercritical compression is compared with Equation 11.61, one will notice that the latter has the function under the square root $12/\pi^2$ larger than the former. This, again is due to replacing a reasonable approximation of the deflected shape by a straight line. To improve the accuracy of Equations 11.61 and 11.58, both Ω and Φ should therefore be multiplied by $\pi/\sqrt{12} \approx 0.907$.

BEAMS WITH NO TRANSLATION OF END POINTS

In Chapter 9, the lateral resistance of ideal cables, having no bending stiffness, was discussed. The other way of resisting the lateral loads is, of course, by means of flexure, typical of a beam. However, any real, physical beam resists the load in both ways. When the end points are prevented from any translation, as in Figure 11.13, both stiffness components must be accounted for. This is the case where the influence of axial force and finite deflections are combined.* When the beam is initially straight and unstressed, a small external load is balanced by bending. Once the beam becomes curved and the axial load appears, the cable action develops. When the external load is big enough to cause the lateral deflection to be larger than the section depth, axial stretching usually begins to dominate the resistance.

To keep this presentation simple, it will be limited to the case in Figure 11.13, where the loading is the step function in time and the beam material is elastic. The solutions will be obtained by work–energy balance. The approximate expressions for deflected lines and strain energy associated with bending, were developed in section "Elastic strain energy and approximate deflected shapes." The strain energy of axial stretching as well as the work of external forces was quantified in section "Strain energy and work." The stretching or "cable" energy component is also shown in Case 9.1 as

$$\Pi_c = \frac{\pi^4}{32}\frac{EA\delta^4}{L^3} \tag{11.67}$$

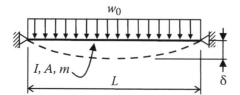

FIGURE 11.13   A simply supported beam with ends constrained against translation.

---

* The word "finite" rather than "large" is used here, because the maximum deflection is certainly not meant to exceed one-half of the beam length.

The energy associated with the flexural component comes from Equation 11.9 and Table 11.1:

$$\Pi_b = \frac{\pi^4}{4} EI \frac{\delta^2}{L^3}$$ (11.68)

The distributed load, $w = w_0 H(t)$, performs the work proportional to the area between the original axis and the maximum deflected line. Using the same, approximate deflection:

$$\mathcal{L} = \int_0^L uw\,dx = w\int_0^L \delta \sin\frac{\pi x}{L}\,dx = \frac{2}{\pi} w\delta L$$ (11.69)

For any elastic beam the general expressions for stretching energy, bending energy, and work are, respectively

$$\Pi_c = \Psi\frac{EA\delta^4}{L^3}$$ (11.70a)

$$\Pi_b = \Gamma\frac{EI}{L^3}\delta^2$$ (11.70b)

$$\mathcal{L} = \Theta Lw\delta$$ (11.70c)

Equating the total strain energy with the work of external load gives

$$\Psi EA\delta^3 + \Gamma EI\delta = \Theta L^4 w$$ (11.71)

The above equation, with $\Psi = \pi^4/32$, $\Gamma = \pi^4/4$, and $\Theta = 2/\pi$, is solved for the unknown maximum deflection δ. One should notice that when the δ is equal to

$$\delta_e = \sqrt{\frac{\Gamma I}{\Psi A}}$$ (11.72)

both energy terms are equal. If the indications are that the end result will be well in excess of this, there is no need to include the flexural energy term and the solution becomes simple. Also, the stretching energy can be ignored when $\delta \ll \delta_e$. Another way of simplifying the solution can be found in Example 11.9.

Once the peak deflection is found, the axial force P is calculated from Equation 9.2 or 9.6. The peak bending at the center of this beam is found from the approximate equation of the deflected line:

$$\mathcal{M} = EIu'' = \frac{\pi^2}{L^2} EI\delta$$ (11.73)

When a point force W is applied at the center instead of a distributed load discussed so far, the derivation of work–energy balance expression is similar and the results are given in Case 11.14.

When the ends are clamped, the procedure is essentially the same, except that the counterpart of Equation 11.73 gives an underestimate of the moment at a clamped end. The following approximation will be used based on static equations given in Roark [75] for a beam under lateral load and tension. The peak deflection is given as a function of the following parameter j and variable x:

$$j^2 = \frac{EI}{P} \tag{11.74a}$$

and

$$x = \frac{\pi}{2}\sqrt{\frac{P}{P_e}} \tag{11.74b}$$

The expression for a maximum deflection δ and bending moment $\mathcal{M}_e$ are given in terms of hyperbolic function, but for $x > 2$, which is relevant here, their simplified form may be written as follows:

$$\mathcal{M}_e = wj^2(x-1); \tag{11.75a}$$

$$\delta = \frac{wj^2}{8P}(3.952x^2 - 7.52x) \tag{11.75b}$$

After solving for δ and P, one seeks the equivalent distributed loading w and then the end moment $\mathcal{M}_e$. This can be done by substituting wj^2 from the second equation into the first. (The equivalent static w is strictly for the purpose of finding $\mathcal{M}_e$.) More details are shown in Cases 11.13 through 11.20.

When stress exceeds the yield point, analytical solutions (or rather approximations) can be found rather easily for a RP material with the yield point of σ_0. The axial load has then a constant value of $P_0 = A\sigma_0$. A kink mode of deflection is assumed, even though under uniform load there is a gradual curving of a beam near the center. (Figure 11.14.) The objective is to determine the peak deflection δ and provide an estimate of a peak strain ε_m. When a short impulse is applied, the initial kinetic energy E_k is equated with the maximum strain energy in tension plus the energy absorbed by plastic hinges (both components at the maximum deflection). This makes it possible to find θ or $\delta = \theta L/2$. The problem of the magnitude of strain at plastic hinges was discussed in section "Some cases of strain distribution along a beam" and demonstrated to depend on the hinge length involved. This quantity, however, cannot be

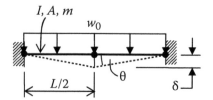

FIGURE 11.14 A clamped–clamped beam with ends constrained against translation. Kink mode of deformation.

readily estimated for a dynamic loading case. If the hinge length is taken as $H/2$, or one-half of the beam depth, the bending strain is equal to θ, or the angle of discontinuity, according to Equation 10.75:

$$\varepsilon_b = \frac{h_1}{\rho} = \frac{h_1\theta}{\ell_h} = \theta \qquad (11.76)$$

as long as a section is symmetric about the bending axis, that is, $2h_1 = H$. The component of strain due to extension of the element is $\varepsilon_t \approx 2\delta^2/L$, which gives the total as

$$\varepsilon_m = \varepsilon_b + \varepsilon_t = \theta\left(1 + \frac{\theta}{2}\right) \qquad (11.77)$$

When deflections of a beam-column grow to the level of a well-developed cable action, the bending strains tend to localize near supports. Equation 11.77 gives a marked underestimate past the elastic range, as can be found from some examples presented here. Cases 11.19 and 11.20 give better approximations in that range.

A section is not capable of resisting both of its limiting components, namely $\mathcal{M}_0$ and P_0 at the same time. The other way to say this is to state that $r \leq 1$ according to Equation 11.23. If it is desired, for ease of calculations, to treat these two quantities as independent, they should be replaced by smaller values, namely $\mathcal{M}_0'$ and P_0', making it certain that the above condition of the combined strength is fulfilled. Fortunately, things are easier in the type of problems discussed here. The solutions often fall into two categories: beam action or cable action dominating. In the first case, tension associated with axial load is small and so is the corresponding reduction of the allowable bending. When the cable action dominates, the situation reverses. Therefore, only an intermediate range of deflections requires a downward adjustment of $\mathcal{M}_0$ and P_0.

There is an important problem of strength of a laterally impacted column. Its solution is presented by Case 11.28 (constant impacting force) and Case 11.29 (mass impact). Two modes of failure are considered: Bending-with-shear (a combined mode) and a shear-dominated mode.

In both cases the column is assumed to be under a constant compressive load arising from the tributary mass being supported. There is another possibility of the upper end attachment namely, an axial constraint due to a large elastic resistance from the rest of the structure. In such a configuration the problem is of relevance for this section. If the axial restraint is very stiff, the net tension in the column may result in a higher overall strength in the combined deformation mode. There is no influence of this effect on the shear-dominated mode, however. The latter assumes that a short column segment is simply pushed out from the rest of the column by the external load W, as illustrated in Figure 11.15.

The calculations of such events, involving the lateral actions are illustrated by Examples 11.18 through 11.20. Some of the effects of a column impact on the upper parts of a building were described by Szuladziński [92].

CLOSING REMARKS

At this point one should reread the section "Closing remarks" of Chapter 10, where the disparity between ideal end conditions and those likely to be encountered in a physical experiment are discussed. What is meant to be rigid is always, to some extent, flexible with an exception provided by symmetry. Lateral supports, angular restraints, and an axial hold are unlikely to be (practically) rigid. The analytical difficulties usually grow in the sequence listed above, with the lateral movement easiest to correct for and the axial component being most troublesome.

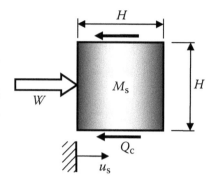

FIGURE 11.15   A fragment of the column, pushed out by an external action, resisted by shear forces.

This leads us to the problem of creating formulas predicting structural response in such a manner that the results agree with the experiment. If one uses ideal boundary conditions for that purpose, the effort is unlikely to be successful, unless coefficients are introduced, which compensate for imperfections of those conditions. With regard to the numerical methods, that is, finite element analysis (FEA), modeling must involve not only the structural components themselves, but at least the flexibility of constraining elements, if such elements are not explicitly modeled.

The term *ultimate limit state (method)* can often be seen in design rules. It is implied that if the ultimate load, calculated in accordance with the relevant design procedure is attained, a limit state, or a collapse follows. A collapse can indeed happen, but only under rare and unfortunate circumstances. First, the ultimate load is larger than the design load* by some specified factor. Second, the nominal member strength is reduced by a factor φ mentioned before. There is also another margin of safety, less visible. This is the difference between the nominal material strength and the most likely strength of a sample. (This aspect was discussed in Chapter 6.) These variables must be kept in mind by an analyst who wants to make a correct prediction, although he may choose a safety factor to apply in a practical situation.

---

* Design load means the largest load that can be reasonably expected to occur.

TABULATION OF CASES

COMMENTS

Cases 11.4 through 11.8: Lateral supports are active at the beginning, at $\theta=\theta_0$, until P reaches P_0. The supports (not visualized) are then suddenly removed. Shortening λ, as well as the strain energy Π, are measured from the unstressed position, at θ_0. The peak bending moment is calculated from the static formula. In supercritical compression (the applied force larger than the buckling force) there is unlimited growth of response with time. When deriving the displacement equations it is more convenient to use one angle designation for a subcritical range and another one for the supercritical range. This was not necessary in the summary results below.

Cases 11.10 through 11.12 relate to energy absorption when structural elements are subject to large deformations. The lack of inertia effects in those cases is of no consequence when the impinging objects are much heavier.

Cases 11.13 through 11.16 deal with beams in the elastic range, which are subjected to lateral loads and which have the ends prevented from translation. Cases 11.17 through 11.20, relate to the same beams in post-elastic range. The first group has results for step loading in time, while the second gives results for short impulses. In a postelastic range when a point force rather than a distributed load is applied, a better result is obtained then the kinetic energy is calculated using $0.15mL$ as an effective mass, placed at center, rather than the mass related to the kink mode assumption.

CASE 11.1   BASIC BEAM BUCKLING FORMULAS

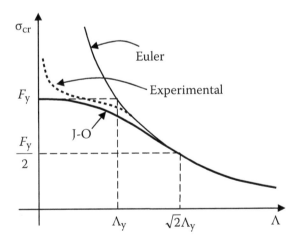

$i = \sqrt{\dfrac{I}{A}}$; radius of inertia

$\Lambda = \dfrac{L}{i\sqrt{\mu}}$; slenderness ratio

$$P_{cr} = \mu \frac{\pi^2 EI}{L^2}; \quad \sigma_{cr} = \frac{P_{cr}}{A} = \frac{\pi^2 E}{\Lambda^2}$$

elastic buckling force and stress

$$P_e = \frac{\pi^2 EI}{L^2}; \quad \sigma_e = \frac{P_e}{A} = \frac{\pi^2 E}{\Lambda^2}$$

elastic buckling for simply supported ends ($\mu = 1$)

$$\Lambda_y = \pi \sqrt{\frac{E}{F_y}}; \text{ limit value of } \Lambda, \text{ end of elastic range}$$

(The above relations are valid for $\sigma_{cr} < F_y$ or for $\Lambda > \Lambda_y$)

$$\sigma_{crp} = \frac{P_{crp}}{A} = \frac{\pi^2 E_p}{\Lambda^2} \text{ with } E_p = \frac{d\sigma}{d\varepsilon}; \text{ (*) buckling stress for } \sigma_{cr} > F_y \text{ based on tangent modulus,}$$
valid for $\Lambda < \Lambda_y$

$$\sigma_{cr} = F_u - (F_u - F_y)\left(\frac{\Lambda}{\Lambda_y}\right); \text{ (*) as above, when linear variation in } E_p \text{ is assumed}$$

$$\sigma_{crp} = \left(1 - \frac{F_y}{4\sigma_{cr}}\right) F_y; \text{ (**) Johnson–Ostenfeld (JO) buckling stress for } \sigma_{cr} > F_y; \text{ used for}$$

$$\Lambda < \sqrt{2}\Lambda_y$$

Note: Equation (**) is closer to the lower bound of test results, while Equations (*) is closer to their mean value. In theory, JO formula would apply above $F_y/2$, but this may be too conservative for most applications in the inelastic range. Coefficients μ for various support cases are given in the text.

CASE 11.2   MAXIMUM ECCENTRICITY ESTIMATE FOR A COMPRESSED COLUMN *L* LONG

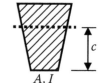

$$\delta_0 = \frac{I}{10Ac} + \frac{L}{k}; \text{ maximum imperfection or amplitude of eccentricity}$$
to be used as in Case 11.3

$250 < k < 1000$; typical range of k for various applications

Note: This estimate to be employed only when no specific data is available. Ref. [82].

CASE 11.3   THE EFFECT OF SUDDEN REMOVAL OF LATERAL RESTRAINTS ON A PRECOMPRESSED COLUMN

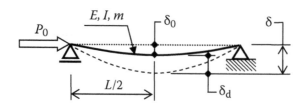

δ_0 = amplitude of the initial imperfection, in unloaded state, $u_0 = \delta_0 \sin(\pi x/L)$

P_0 = constant axial force, applied prior to constraint release, while δ kept at δ_0

$$P_e = \frac{\pi^2 EI}{L^2}$$

(a) Response to a sudden removal when $P_0 < P_{cr} = P_e$

$\delta_d = \alpha\delta_0\left[1 - \cos\Omega t\right]$; center dynamic deflection as function of time

$\alpha = \dfrac{P_0/P_e}{1 - P_0/P_e}$; amplitude multiplier

$\Omega = \dfrac{\pi}{L}\left(\dfrac{P_e}{m}\left(1 - \dfrac{P_0}{P_e}\right)\right)^{1/2}$; frequency of oscillation, period $\tau = 2\pi/\Omega$

$\delta_m = 2\alpha\delta_0$; maximum dynamic deflection at center; $\delta = \delta_0 + \delta_m$ is the total deflection

$\mathcal{M} = \dfrac{\pi^2 EI}{L^2}\delta$; maximum bending moment induced.

Note: Ω is the same as the natural frequency of a supported beam, per Chapter 2, additionally modified by the effect of an axial load P_0, according to Equation 2.39.

(b) Response to a sudden removal when $P_0 > P_{cr} = P_e$

$\delta_d = \beta\delta_0\left[\cosh\Phi t - 1\right]$; center dynamic deflection as function of time

$\beta = \dfrac{P_0/P_e}{P_0/P_e - 1}$; amplitude multiplier

$\Phi = \dfrac{\pi}{L}\left(\dfrac{P_e}{m}\left(\dfrac{P_0}{P_e} - 1\right)\right)^{1/2}$; multiplier of time, motion nonperiodic

$\delta = \delta_0 + \delta_d$ is the total deflection

$\mathcal{M} = \dfrac{\pi^2 EI}{L^2}\delta$; maximum bending moment induced. Ref. [43].

**CASE 11.4   A RIGID CANTILEVER BEAM (CF), WITH AN ELASTIC END JOINT, IMPACTED
BY AN AXIAL FORCE**

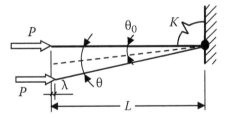

K is the elastic joint stiffness. Large deflections. Initial kink (with $P = 0$) is θ_0.

(a) Statics

$P = \dfrac{K\theta}{L\sin\theta}$; equilibrium load

$$K = \frac{P_e L}{4}; \quad P_{cr} = \frac{K}{L}$$

$\mathcal{M} = K(\theta - \theta_0)$; spring moment

$$\lambda = L(\cos\theta_0 - \cos\theta); \quad \Pi = K(\theta - \theta_0)^2/2; \quad u = L\sin\theta; \quad u_0 = L\sin\theta_0$$

Small deflections: $u = \dfrac{u_0}{1 - P/P_{cr}}$ $\lambda = L(\theta^2 - \theta_0^2)/2; u = L\theta$

(b) Subcritical step load $P_0 H(t)$ with $P_0 < P_{cr}$

$$\theta = \frac{1 + P/P_{cr}}{1 - P/P_{cr}}\theta_0$$

(b) Supercritical step load $P_0 H(t)$ with $P_0 > P_{cr}$

$J = mL^3/3$; mass moment of inertia about the pivot

$$\Phi = \sqrt{\frac{K'}{J}\left(\frac{P_0}{P_{cr}} - 1\right)}; \text{ time multiplier, } K' = 0.907\,K$$

$$\theta - \theta_0 = \frac{P_0/P_{cr}}{P_0/P_{cr} - 1}(\cosh\Phi t - 1)\theta_0$$

(c) Supercritical impulse load lasting t_1

$$\theta_1 - \theta_0 = \frac{P_0/P_{cr}}{P_0/P_{cr} - 1}(\cosh\Phi t_1 - 1)\theta_0; \text{ angle at } t = t_1$$

$\lambda_1 = L(\cos\theta_0 - \cos\theta_1)$; associated shortening

$$\theta_m = \sqrt{\frac{2P_0\lambda_1}{K}} + \theta_0; \text{ peak angle}$$

Note: Refer to General Comments. The static formula holds in dynamic cases as well.

CASE 11.5   A SIMPLY SUPPORTED RIGID BEAM (SS), WITH AN ELASTIC
                        CENTER JOINT, IMPACTED BY AN AXIAL FORCE

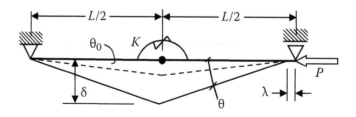

K is the elastic joint stiffness. Large deflections. Initial kink (with $P = 0$) is θ_0

(a) **Statics**

$$P = \frac{4K\theta}{L\sin\theta};\ \text{equilibrium load}$$

$$K = \frac{P_e L}{4};\quad P_{cr} = \frac{4K}{L}$$

$\mathcal{M} = 2K(\theta - \theta_0)$; spring moment

$$\lambda = L(\cos\theta_0 - \cos\theta);\quad \Pi = 2K(\theta - \theta_0)^2;\quad u = \frac{L}{2}\sin\theta;\quad u_0 = \frac{L}{2}\sin\theta_0$$

Small deflections; $u = \dfrac{u_0}{1 - P/P_{cr}}$; $\lambda = L(\theta^2 - \theta_0^2)/2$; $u = L\theta/2$

(b) **Subcritical step load $P_0 H(t)$ with $P_0 < P_{cr}$**

$$\theta = \frac{1 + P/P_{cr}}{1 - P/P_{cr}}\theta_0$$

(b) **Supercritical step load $P_0 H(t)$ with $P_0 > P_{cr}$**

$J = mL^3/24$; mass moment of inertia about an end, for half-length

$$\Phi = \sqrt{\frac{2K'}{J}\left(\frac{P_0}{P_{cr}} - 1\right)};\ \text{time multiplier, } K' = 0.907K$$

$$\theta - \theta_0 = \frac{P_0/P_{cr}}{P_0/P_{cr} - 1}(\cosh\Phi t - 1)\theta_0$$

(c) **Supercritical impulse load lasting t_1**

$$\theta_1 - \theta_0 = \frac{P_0/P_{cr}}{P_0/P_{cr} - 1}(\cosh\Phi t_1 - 1)\theta_0;\ \text{angle at } t = t_1$$

$\lambda_1 = L(\cos\theta_0 - \cos\theta_1)$; associated shortening

$$\theta_m = \sqrt{\frac{P_0\lambda_1}{2K}} + \theta_0;\ \text{peak angle}$$

Note: Refer to General Comments. The static formula holds in dynamic cases as well.

CASE 11.6   A CLAMPED-GUIDED (CG) RIGID BEAM, WITH ELASTIC END JOINTS, IMPACTED BY AN AXIAL FORCE

K is the elastic joint stiffness. Large deflections. Initial kink (with $P = 0$) is θ_0

(a) **Statics**

$$P = \frac{2K\theta}{L\sin\theta};\ \text{equilibrium load}$$

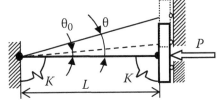

$$K = \frac{P_e L}{2}; \quad P_{cr} = \frac{2K}{L}$$

$\mathcal{M} = K(\theta - \theta_0)$; moment in each spring

$$\lambda = L(\cos\theta_0 - \cos\theta); \quad \Pi = K(\theta - \theta_0)^2; \quad u = L\sin\theta; \quad u_0 = L\sin\theta_0$$

Small deflections; $u = \dfrac{u_0}{1 - P/P_{cr}}$ $\lambda = L(\theta^2 - \theta_0^2)/2; u = L\theta$

(b) Subcritical step load $P_0 H(t)$ with $P_0 < P_{cr}$

$$\theta = \frac{1 + P_0/P_{cr}}{1 - P_0/P_{cr}} \theta_0$$

(b) Supercritical step load $P_0 H(t)$ with $P_0 > P_{cr}$

$J = mL^3/3$; mass moment of inertia about the pivot

$$\Phi = \sqrt{\frac{2K'}{J}\left(\frac{P_0}{P_{cr}} - 1\right)}; \text{ time multiplier, } K' = 0.907K$$

$$\theta - \theta_0 = \frac{P_0/P_{cr}}{P_0/P_{cr} - 1}(\cosh\Phi t - 1)\theta_0$$

(c) Supercritical impulse load lasting t_1

$$\theta_1 - \theta_0 = \frac{P_0/P_{cr}}{P_0/P_{cr} - 1}(\cosh\Phi t_1 - 1)\theta_0; \text{ angle at } t = t_1$$

$\lambda_1 = L(\cos\theta_0 - \cos\theta_1)$; associated shortening

$$\theta_m = \sqrt{\frac{P_0\lambda_1}{K}} + \theta_0; \text{ peak angle}$$

Note: Refer to General Comments. The static formula holds in dynamic cases as well.

CASE 11.7   A SUPPORTED-CLAMPED (SC) RIGID BEAM, WITH AN ELASTIC CENTER JOINT, IMPACTED BY AN AXIAL FORCE

K is the elastic joint stiffness. Large deflections. Initial kink (with $P = 0$) is θ_0

(a) Statics

$P = \dfrac{4K\theta}{L\sin\theta}$; equilibrium load

$K = 0.5118LP_e; \quad P_{cr} = \dfrac{4K}{L}$

$\mathcal{M} = 2K(\theta - \theta_0)$; spring moment

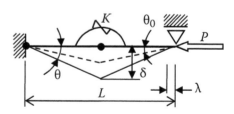

$$\lambda = L(\cos\theta_0 - \cos\theta); \quad \Pi = 2K(\theta - \theta_0)^2; \quad u = \frac{L}{2}\sin\theta; \quad u_0 = \frac{L}{2}\sin\theta_0$$

Small deflections: $u = \dfrac{u_0}{1 - P/P_{cr}}$; $\lambda = L(\theta^2 - \theta_0^2)/2$; $u = L\theta/2$

(b) Subcritical step load $P_0H(t)$ with $P_0 < P_{cr}$

$$\theta = \frac{1 + P_0/P_{cr}}{1 - P_0/P_{cr}}\theta_0$$

(b) Supercritical step load $P_0H(t)$ with $P_0 > P_{cr}$

$J = mL^3/24$; mass moment of inertia about an end, for half-length

$$\Phi = \sqrt{\frac{2K'}{J}\left(\frac{P_0}{P_{cr}} - 1\right)}; \text{ time multiplier, } K' = 0.907K$$

$$\theta - \theta_0 = \frac{P_0/P_{cr}}{P_0/P_{cr} - 1}(\cosh\Phi t - 1)\theta_0$$

(c) Supercritical impulse load lasting t_1

$$\theta_1 - \theta_0 = \frac{P_0/P_{cr}}{P_0/P_{cr} - 1}(\cosh\Phi t_1 - 1)\theta_0; \text{ angle at } t = t_1$$

$\lambda_1 = L(\cos\theta_0 - \cos\theta_1)$; associated shortening

$\theta_m = \sqrt{\dfrac{P_0\lambda_1}{2K}} + \theta_0$; peak angle

Note: Refer to General Comments. The static formula holds in dynamic cases as well.

CASE 11.8   A Rigid Beam with Clamped Ends (CC) and Elastic Joints, Impacted by an Axial Force

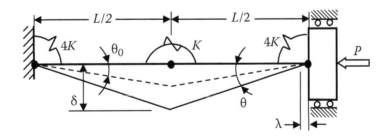

Large deflections. Initial kink (with $P = 0$) is θ_0

(a) Statics

$$P = \frac{12K\theta}{L\sin\theta}; \text{ equilibrium load}$$

$$K = \frac{P_e L}{3}; \quad P_{cr} = \frac{12K}{L}$$

$\mathcal{M}_c = 2K(\theta - \theta_0)$; $\mathcal{M}_b = 4K(\theta - \theta_0)$; center and base moment, respectively

$$\lambda = L(\cos\theta_0 - \cos\theta); \quad \Pi = 6K(\theta - \theta_0)^2; \quad u = \frac{L}{2}\sin\theta; \quad u_0 = \frac{L}{2}\sin\theta_0$$

Small deflections; $u = \dfrac{u_0}{1 - P/P_{cr}}$; $\lambda = L(\theta^2 - \theta_0^2)/2$; $u = L\theta/2$

(b) Subcritical step load $P_0 H(t)$ with $P < P_{cr}$

$$\theta = \frac{1 + P_0/P_{cr}}{1 - P_0/P_{cr}}\theta_0$$

(b) Supercritical step load $P_0 H(t)$ with $P_0 > P_{cr}$

$J = mL^3/24$; mass moment of inertia about an end, for half-length

$$\Phi = \sqrt{\frac{6K'}{J}\left(\frac{P_0}{P_{cr}} - 1\right)}; \text{ time multiplier, } K' = 0.907\,K$$

$$\theta - \theta_0 = \frac{P_0/P_{cr}}{P_0/P_{cr} - 1}(\cosh\Phi t - 1)\theta_0$$

(c) Supercritical impulse load lasting t_1

$$\theta_1 - \theta_0 = \frac{P_0/P_{cr}}{P_0/P_{cr} - 1}(\cosh\Phi t_1 - 1)\theta_0; \text{ angle at } t = t_1$$

$\lambda_1 = L(\cos\theta_0 - \cos\theta_1)$; associated shortening

$$\theta_m = \sqrt{\frac{P_0\lambda_1}{6K}} + \theta_0; \text{ peak angle}$$

Note: Refer to General Comments. The static formula holds in dynamic cases as well.

Case 11.9 A Sudden Eccentric Compression of a Column with a Supercritical Force

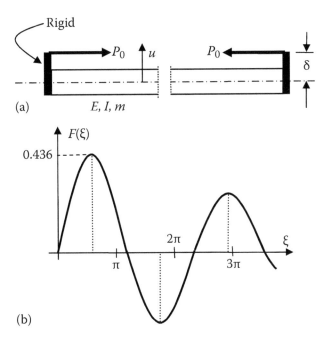

(a) Arrangement

(b) Deflected shape after time t, going from left end

P_0, a constant axial force of short duration

(a) Arrangement
(b) Deflected shape after time t, going from left end
    δ = initial eccentricity

$$P_e = \frac{\pi^2 EI}{L^2}; \quad \xi = x\sqrt{\frac{P_0}{EI}}$$

$$u(\xi,t) = 1.801\delta F(\xi) \cdot f(t)$$

$$F(\xi) = \frac{1}{\xi^2}\left(\sin\xi - \xi\cos\xi\right); \quad f(t) = \cosh\left(\frac{P_0 t}{2\sqrt{EIm}}\right) - 1$$

$u_m(t) = \dfrac{\pi}{4}\delta f(t)$; peak deflection as function of time, at $\xi \approx 2.1$

$\mathcal{M}_m(t) = 0.4285 P_0 \delta f(t)$; peak moment as function of time, at $\xi \approx 1.9$

$\sigma_b(t) = 2.571\dfrac{P_0}{BH}\dfrac{\delta}{H}f(t)$; peak bending stress, rectangular section $B \times H$

$\sigma(t) = \dfrac{P_0}{BH} + \sigma_b(t)$; compression plus bending

Note: The presentation is based on Ref. [82], except for a minor adjustment of coefficients. The experimental work supporting this theory was done using minimal eccentricities, much smaller than illustrated here. If the σ–ε curve of the material can be represented as RSH (as defined in Chapter 4) with the plastic modulus of E_p, then the above gives the first estimate of inelastic response after E is replaced by E_p.

CASE 11.10 A RIGID BEAM, WITH ONE PLASTIC JOINT AND PINNED ENDS, ACTED UPON BY A STATIC AXIAL FORCE WITH AN OFFSET ELEMENT e—LONG

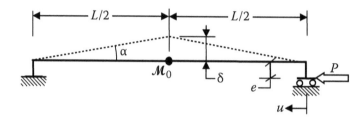

$\mathcal{M}_0 = Z\sigma_0$; bending capacity of the RP joint.

$P_0 = A\sigma_0 = BH\sigma_0$; axial load capacity of the section

$\dfrac{P}{P_0} = \dfrac{1}{2}\left\{\sqrt{B^2 + 4} - B\right\}$ with $B = \dfrac{P_0}{\mathcal{M}_0}(e + \delta)$; equilibrium load P for any deflection δ.

$P_m = P_0\left\{\sqrt{\dfrac{4e^2}{H^2} + 1} - \dfrac{2e}{H}\right\}$; peak P for $\delta = 0$, for δ growing, the load P decreases.

$\left(\dfrac{P}{P_0}\right)^2 + \dfrac{\mathcal{M}}{\mathcal{M}_0} = 1$; to calculate $\mathcal{M}$ after P is found. (Both $\mathcal{M}$ and P are positive)

$\Pi = 2\mathcal{M}_0\alpha$; strain energy stored, for $\delta > e$

$u = \delta \tan \alpha$; relation between lateral and axial displacements

$P \approx \dfrac{\mathcal{M}_0}{e + \delta}$; approximate expression for P, for $\delta \gg e$.

$u = \delta\alpha$; $\delta = L\alpha/2$; small-deflection approximations

Note: This may serve as a model of a real beam with simply supported ends, with the compressive load being applied via prescribed end offsets. The $\mathcal{M}$–P interaction equation is valid for a rectangular section of a beam, $B \times H$. Large deflections are allowed. No inertia effects included.

CASE 11.11 A RIGID BEAM, WITH THREE PLASTIC JOINTS, ACTED UPON BY A STATIC AXIAL FORCE WITH AN OFFSET ELEMENT e—LONG

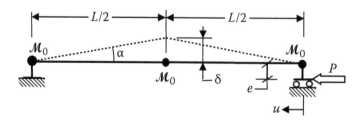

$$\frac{P}{P_0} = \frac{1}{2}\left\{\sqrt{B^2 + 4} - B\right\} \text{ with } B = \frac{P_0}{2\mathcal{M}_0}(e + \delta); \text{ equilibrium load } P \text{ for any deflection } \delta.$$

$$P_m = P_0\left\{\sqrt{\frac{e^2}{H^2} + 1} - \frac{e}{H}\right\}; \text{ peak } P \text{ for } \delta = 0; \text{ for } \delta \text{ growing, the load } P \text{ decreases.}$$

$$\left(\frac{P}{P_0}\right)^2 + \frac{\mathcal{M}}{\mathcal{M}_0} = 1; \text{ to calculate } \mathcal{M} \text{ after } P \text{ is found. (Both } \mathcal{M} \text{ and } P \text{ are positive)}$$

$u = \delta \tan \alpha$; relation between lateral and axial displacements

$\Pi = 4\mathcal{M}_0\alpha$; strain energy stored, for $\delta > e$

$$P \approx \frac{2\mathcal{M}_0}{e + \delta}; \text{ approximate expression for } P, \text{ for } \delta \gg e$$

$u = \delta\alpha$; $\delta = L\alpha/2$; small-deflection approximations

Note: This may serve as a model of a real beam with fixed ends, with the compressive load being applied via prescribed end offsets. The $\mathcal{M}$–P interaction equation is valid for a rectangular section of a beam, $B \times H$. Large deflections are allowed. No inertia effects included.

CASE 11.12 A RECTANGULAR, THIN-WALL SECTION SQUASHED BETWEEN TWO RIGID SURFACES

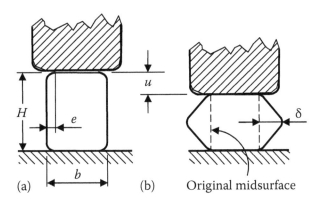

(a) (b) Original midsurface

Each vertical side is assumed acting as in Case 11.11. Quantities calculated are per unit length, normal to paper.

$\mathcal{M}_0 = Z\sigma_0$; bending capacity of the RP joint

$P_0 = A\sigma_0 = BH\sigma_0$; axial load capacity of the section

$$w_m = 2P_0\left\{\sqrt{\frac{e^2}{H^2} + 1} - \frac{e}{H}\right\}; \text{ peak resistance force reached at the initiation of squashing}$$

$u = \delta \tan \alpha$; relation between lateral and axial displacements

$\Pi = 8\mathcal{M}_0\alpha$; strain energy stored, for $\delta > e$

$w \approx \dfrac{4\mathcal{M}_0}{e+u}$; when squashing is in progress, for $\delta \gg e$.

$u = \delta\alpha$; $\delta = L\alpha/2$; small-deflection approximations

Note: This is an application of Case 11.11, with two vertical walls acting as columns. No inertia effects included.

CASE 11.13   SIMPLY SUPPORTED BEAM, NO TRANSLATION OF END POINTS

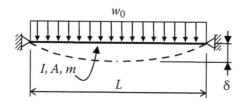

Distributed load.

Step loading: $w = w_0 H(t)$

$\dfrac{\pi^4}{32} EA\delta^3 + \dfrac{\pi^4}{4} EI\delta = \dfrac{2}{\pi} L^4 w_0$; equation to find peak deflection δ.

$\delta_e = \sqrt{\dfrac{8I}{A}}$; deflection, for which the bending and the stretching energy are equal.

$\lambda = \dfrac{\pi^2 \delta^2}{4L}$; axial elongation, $P = \dfrac{EA}{L}\lambda$; axial force.

$\mathcal{M} = \dfrac{\pi^2 EI}{L^2}\delta$; peak bending moment.

CASE 11.14   SIMPLY SUPPORTED BEAM, NO TRANSLATION OF END POINTS

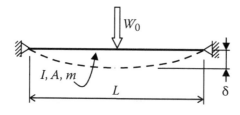

Point load

Step loading: $w = W_0 H(t)$

$2EA\delta^3 + 24EI\delta = L^3 W_0$; equation to find peak deflection δ.

$\delta_e = \sqrt{\dfrac{12I}{A}}$; deflection, for which the bending and the stretching energy are equal.

$\lambda = \dfrac{2\delta^2}{L}$; axial elongation, $P = \dfrac{EA}{L}\lambda$; axial force.

$\mathcal{M} = \dfrac{12EI}{L^2}\delta$; peak bending moment.

CASE 11.15   CLAMPED–CLAMPED BEAM, NO TRANSLATION OF END POINTS

Distributed load

Step loading: $w = w_0 H(t)$

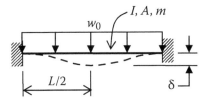

$\dfrac{\pi^4}{32}EA\delta^3 + \pi^4 EI\delta = \dfrac{1}{2}L^4 w_0$;   equation   to   find   peak deflection δ.

$\delta_e = \sqrt{\dfrac{32I}{A}}$; deflection, for which the bending and the stretching energy are equal.

$\lambda = \dfrac{\pi^2\delta^2}{4L}$; axial elongation; $P = \dfrac{EA}{L}\lambda$; axial force.

$\mathcal{M}_e = \dfrac{2\pi^2 EI}{L^2}\delta$; end bending (initial estimate).

The end bending moment:

$\mathcal{M} = \dfrac{16}{L^2}EI\delta\sqrt{\dfrac{P}{P_e}}$ with $P_e = \dfrac{\pi^2}{L^2}EI$; initial end-moment estimate.

$\mathcal{M}_e = \dfrac{8P(x-1)\delta}{3.952x^2 - 7.52x}$ with $x = \dfrac{\pi}{2}\sqrt{\dfrac{P}{P_e}}$; a better estimate.

CASE 11.16   CLAMPED–CLAMPED BEAM, NO TRANSLATION OF END POINTS

Point load

Step loading: $W = W_0 H(t)$

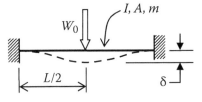

$2EA\delta^3 + 96EI\delta = L^3 W_0$; equation to find peak deflection δ.

$\delta_e = \sqrt{\dfrac{48I}{A}}$; deflection, for which the bending and the stretching energy are equal.

$\lambda = \dfrac{2\delta^2}{L}$; axial elongation, $P = \dfrac{EA}{L}\lambda$; axial force.

$$M_e = \frac{48EI}{L^2}\delta \; ; \text{ end bending moment, for } \delta < \delta_e.$$

Note: The above underestimates M for $\delta > \delta_e$.

CASE 11.17   INELASTIC BEAM, SIMPLY SUPPORTED, NO TRANSLATION OF END POINTS

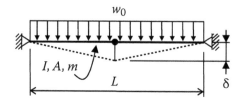

Distributed load. Short impulse loading: w_0 lasting t_0

$$\zeta = \frac{M_0}{P_0 L}$$

$E_k = \dfrac{3}{8}\dfrac{(w_0 t_0)^2 L}{m}$; kinetic energy

$\theta_m = 2\zeta\left[\left(1+\dfrac{E_k}{2M_0\zeta}\right)^{1/2} - 1\right]$; maximum slope attained

$\theta_e = \dfrac{4M_0}{P_0 L}$; rotation, for which bending and stretching have equal energies

$\delta = \dfrac{L}{2}\tan\theta \approx \dfrac{L\theta}{2}$; peak deflection

$\lambda = \left(\dfrac{1}{\cos\theta}-1\right)L \approx \dfrac{L\theta^2}{2}$; elongation

$\varepsilon_m = \theta\left(1+\dfrac{\theta}{2}\right)$; total strain at end

CASE 11.18   INELASTIC BEAM, SIMPLY SUPPORTED, NO TRANSLATION OF END POINTS

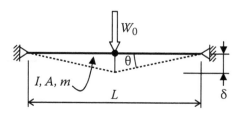

Point load

Short impulse loading: $S_0 = W_0 t_0$

$$\zeta = \frac{\mathcal{M}_0}{P_0 L} \; ; \; M_{\text{ef}} = 0.15mL; \text{ effective mass}$$

$$E_k = \frac{(W_0 t_0)^2}{2M_{\text{ef}}} \; ; \text{ kinetic energy}$$

$$\theta_m = 2\zeta \left[\left(1 + \frac{E_k}{2\mathcal{M}_0\zeta} \right)^{1/2} - 1 \right]; \text{ maximum slope attained}$$

$$\theta_e = \frac{4\mathcal{M}_0}{P_0 L} \; ; \text{ rotation, for which bending and stretching have equal energies}$$

$$\delta = \frac{L}{2}\tan\theta \approx \frac{L\theta}{2}; \text{ peak deflection}$$

$$\lambda = \left(\frac{1}{\cos\theta} - 1 \right) L \approx \frac{L\theta^2}{2}; \text{ elongation}$$

$$\varepsilon_m = \theta \left(1 + \frac{\theta}{2} \right); \text{ total strain at center}$$

CASE 11.19   INELASTIC BEAM, ENDS CLAMPED, NO TRANSLATION OF END POINTS

Distributed load

Short impulse loading: w_0 lasting t_0

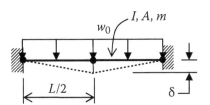

$$\zeta = \frac{\mathcal{M}_0}{P_0 L}$$

$$E_k = \frac{3}{8}\frac{(w_0 t_0)^2 L}{m}; \text{ kinetic energy}$$

$$\theta_m = 2\zeta \left[\left(4 + \frac{E_k}{2\mathcal{M}_0\zeta} \right)^{1/2} - 2 \right]; \text{ maximum slope attained}$$

$$\theta_e = \frac{8\mathcal{M}_0}{P_0 L}; \text{ rotation, for which bending and stretching have equal energies}$$

$$\delta = \frac{L}{2}\tan\theta \approx \frac{L\theta}{2}; \text{ peak deflection}$$

$$\lambda = \left(\frac{1}{\cos\theta} - 1 \right) L \approx \frac{L\theta^2}{2}; \text{ elongation}$$

$$\varepsilon_m = 7.623\frac{\delta}{L} - 0.1132; \text{ total strain at end (valid for } 0.05 \le 2\delta/L \le 0.2)$$

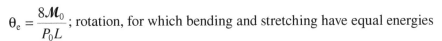

Note: For deflections larger than indicated the formula for ε_m gives an overestimate.

CASE 11.20   INELASTIC BEAM, ENDS CLAMPED, NO TRANSLATION OF END POINTS

Point load

Short impulse loading: $S_0 = W_0 t_0$

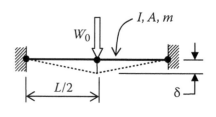

$\zeta = \dfrac{\mathcal{M}_0}{P_0 L}$; $M_{ef} = 0.15mL$; effective mass

$E_k = \dfrac{(W_0 t_0)^2}{2M_{ef}}$; kinetic energy

$\theta_m = 2\zeta \left[\left(4 + \dfrac{E_k}{2\mathcal{M}_0 \zeta} \right)^{1/2} - 2 \right]$; maximum slope attained

$\theta_e = \dfrac{8\mathcal{M}_0}{P_0 L}$; rotation, for which bending and stretching have equal energies

$\delta = \dfrac{L}{2} \tan\theta \approx \dfrac{L\theta}{2}$; peak deflection

$\lambda = \left(\dfrac{1}{\cos\theta} - 1 \right) L \approx \dfrac{L\theta^2}{2}$; elongation

$\varepsilon_m = 16.132 \left(\dfrac{\delta}{L} \right)^2 + 0.11$; maximum strain

Note: The equation for ε_m is valid for $0.01 \le 2\delta^2/L^2 \le 0.15$. For small-deflection strain at the clamped ends dominates, for moderate deflection the point of application is critical.

CASE 11.21   A RIGID CANTILEVER BEAM (CF), WITH A RP END JOINT, IMPACTED BY AN AXIAL LOAD

Initial kink (with $W_0 = 0$) is θ_0.

$\mathcal{M}_0$ is the bending capacity of the joint

$M = mL$ is beam mass

$J = ML^2/3$ is moment of inertia about pivot point

$W_1 = \dfrac{\mathcal{M}_0}{L \sin\theta_0}$; maximum static load, undeformed position

Short impulse $W_0 t_0$: $(W_0 \gg W_1)$

$S = W_0 t_0 \sin\theta_0$; effective impulse, $\lambda = \dfrac{SL}{J}$; initial angular velocity

$\theta_m = \dfrac{(SL)^2}{2J\mathcal{M}_0} + \theta_0$; $u_m = L\theta_m$; peak rotation and tip lateral displacement, respectively.

$\mathcal{L} = W_0 L(\theta^2 - \theta_0^2)/2$; work of the applied force (small deflection)

$\Pi = \mathcal{M}_0(\theta - \theta_0)$; accumulated strain energy

CASE 11.22   A SIMPLY SUPPORTED (SS) RIGID BEAM, IMPACTED BY AN AXIAL LOAD

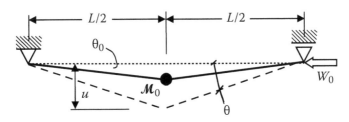

RP hinge at center. Initial kink (with $W_0 = 0$) is θ_0.

$\mathcal{M}_0$ is the bending capacity of the joint

$M = mL$ is beam mass

$J = ML^2/24$ is moment of inertia about pivot point for each half

$W_1 = \dfrac{2\mathcal{M}_0}{L \sin \theta_0}$; maximum static load, undeformed position

Short impulse $W_0 t_0$: $(W_0 \gg W_1)$

$S = W_0 t_0 \sin \theta_0$; effective impulse, each half, $\lambda = \dfrac{SL}{2J}$; initial angular velocity

$\theta_m = \dfrac{(SL)^2}{8J\mathcal{M}_0} + \theta_0$; $u_m = L\theta_m/2$; peak rotation and center lateral displacement, respectively.

$\mathcal{L} = W_0 L(\theta^2 - \theta_0^2)/2$; work of the applied force (small deflection)

$\Pi = 2\mathcal{M}_0(\theta - \theta_0)$; accumulated strain energy

CASE 11.23   A RIGID BEAM WITH CLAMPED ENDS AND THREE RP HINGES, IMPACTED BY AN AXIAL LOAD

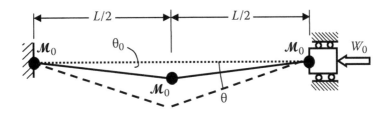

Initial kink (with $W_0 = 0$) is θ_0.

Use formulas for Case 11.22 except as noted.

$$W_1 = \frac{4\mathcal{M}_0}{L \sin \theta_0}\text{; maximum static load, undeformed position}$$

$$\theta_m = \frac{(SL)^2}{16 J\mathcal{M}_0} + \theta_0\text{; peak rotation}$$

$\Pi = 4\mathcal{M}_0(\theta - \theta_0)$; accumulated strain energy

CASE 11.24   TWO-FLANGE, SYMMETRIC BEAM (TFS)

P_e is Euler buckling load for the section as a whole

$P_y = F_y A$; yield load in absence of eccentricity

$R_1 = R_{1y}$ or $R_u = R_{uy}$; a condition for yielding to begin

$R_{1y} = P_y/2$

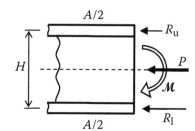

$$I = \left(\frac{H}{2}\right)^2 A;\ Z = \frac{AH}{2}$$

$$R_1 = \frac{P}{2} + \frac{M}{H};\ R_u = \frac{P}{2} - \frac{M}{H}\text{; flange forces}$$

$$\left(\frac{y}{H}\right)^2 + \left(\frac{1}{2} - \frac{\delta_0}{H} - \frac{R_{1y}}{P_e}\right)\left(\frac{y}{H}\right) - \frac{\delta_0}{2H} = 0\text{; to determine deflection } y = y_1 \text{ at the onset of yielding,}$$

when $R_1 = R_{1y}$ is attained.

The force P is then $P_1 = \dfrac{P_y}{1 + 2y_1/H}$

Special case: $P \ll P_e$: $P_1 = \dfrac{P_y}{1 + 2\delta_0/H}$; where δ_0 is the initial deviation

Note: Deflection y is measured from the initial line connecting ends, so $y > \delta_0$. See text for tension of such a beam with an initial offset. The flange under consideration has additive compression from P and $\mathcal{M}$. Also notice that "yielding" is not limited to material behavior only. The buckling load is applicable to simply supported ends.

CASE 11.25   TWO-FLANGE, UNSYMMETRIC BEAM (TFUS)

P_e is Euler buckling load for the section as a whole

$P_y = F_y A$; yield load in absence of eccentricity

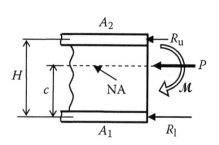

$$R_{1y} = \frac{A_1}{A_1 + A_2}P_y = \frac{1}{1+p}P_y\text{; lower flange force at yield}$$

Section properties:

$$A = A_1 + A_2 = A_1(1 + p)$$

$$p = A_2/A_1; \quad c = \frac{Hp}{1+p}; \quad I = \frac{A_2 H^2}{1+p}; \quad Z = \frac{2HA_2}{1+p}$$

$$R_1 = \frac{P}{1+p} + \frac{M}{H}; \quad R_u = \frac{pP}{1+p} - \frac{M}{H}; \text{ flange forces}$$

$$\left(\frac{y}{H}\right)^2 + \left(\frac{1}{1+p} - \frac{\delta_0}{H} - \frac{R_{1y}}{P_e}\right)\left(\frac{y}{H}\right) - \frac{\delta_0}{(1+p)H} = 0; \text{ to determine deflection } y_1 \text{ at the onset of}$$

yielding, when $R_1 = R_{1y}$ is attained.

The force P is then $P_1 = \dfrac{P_y}{1+(1+p)\dfrac{y_1}{H}}$; where δ_0 is the initial deviation

Special case: $P \ll P_e$: $P_1 = \dfrac{P_y}{1+(1+p)\dfrac{\delta_0}{H}}$

Note: See notes for Case 11.24. The sketch shows the case when the lower flange has smaller area.

CASE 11.26   RECTANGULAR COLUMN SECTION, RC

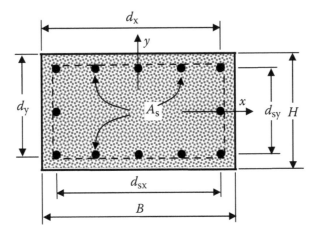

A_s = steel area; $q = E_s/E_c$

(a) Pre-cracking properties:

$A_{ef} = BH + (q-1)A_s$; stiffness – effective section area

$I_x = \dfrac{BH^3}{12} + (q-1)A_s \dfrac{d_{sy}^2}{6}$; effective moment of inertia in bending about x

$I_y = \dfrac{HB^3}{12} + (q-1)A_s \dfrac{d_{sx}^2}{6}$; effective moment of inertia in bending about y

(b) P–$\mathcal{M}$ for ultimate conditions, bending about x:

$$P_0 = 0.85F_c'(BH - A_s) + A_sF_y; \text{ pure compression}$$

$$\mathcal{M}_0 = \frac{7}{16}A_sF_yd_{sy} \text{ ; pure bending}$$

$$P_b = 0.34(BH)F_c'; \text{ axial force at balance condition}$$

$$\mathcal{M}_b = A_{sb}F_yd_{sy} + 0.102BH^2F_c' \text{ ; bending moment at balance condition}$$

Note: Most of the above are rough approximations only. A_{sb} is steel area along the bottom line only.

CASE 11.27   CIRCULAR COLUMN SECTION, RC

A_{st} = steel area; $q = E_s/E_c$.

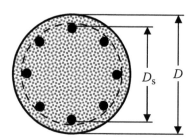

(a) Pre-cracking properties:

$$A_{ef} = \frac{\pi D^2}{4} + (q-1)A_{st}; \text{ stiffness – effective section}$$

area

$$I_x = I_y = \frac{\pi D^4}{64} + (q-1)A_{st}\frac{D_s^2}{8}; \text{ effective moment of}$$

inertia in bending

(b) P–$\mathcal{M}$ for ultimate conditions:

$$P_0 = 0.85F_c'(A - A_s) + A_sF_y; \text{ pure compression}$$

$$\mathcal{M}_0 = 0.4A_sF_yD_s; \text{ pure bending}$$

$$P_b = 0.34AF_c'; \text{ axial force at balance condition}$$

$$\mathcal{M}_b = 0.318A_sF_yD_s + 0.268P_bD; \text{ bending moment at balance condition}$$

Note: Most of the expressions in (b) are rough approximations only.

CASE 11.28   A RIGID COLUMN, WITH RP JOINTS, ACTED ON BY AN IMPULSE $S = W_0t_0$ IN PRESENCE OF A SUSTAINED AXIAL LOAD P_0. $(a < b)$

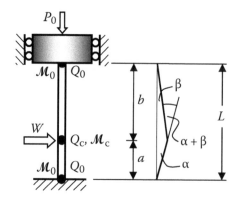

$Q_0 = A_{ef}\tau_0$; shear capacity (RP response)

$Q_c \approx 0.8Q_0$; shear strength near impact point

$W_s = 2Q_c = 1.6Q_0$; static collapse load in shear

$\mathcal{M}_c \approx 0.8\mathcal{M}_0$; bending strength near impact point

$$W_b = (\mathcal{M}_0 + \mathcal{M}_c)\left(\frac{1}{a} + \frac{1}{b}\right) = (\mathcal{M}_0 + \mathcal{M}_c)\frac{L}{ab}$$; static collapse load in bending

$$\frac{1}{W_f^2} = \frac{1}{W_s^2} + \frac{1}{W_b^2}$$; equation to find the static breaking force W_f for the combined mode

Short impulse applied: $S = W_0 t_0$. (W_0 exceeding a static collapse load)

$$v_0 = \frac{S}{M_{ef}}$$; initial velocity at impact point

$$E_k = \frac{1}{2}\frac{S^2}{M_{ef}}$$; kinetic energy supplied

(a) Bending-with-shear (combined) mode. $\Pi_b = W_f a\alpha$; strain energy

$$M_{ef} = \frac{mL}{3}$$; effective beam mass at impact point, combined mode

$$\mathcal{L}_P = \frac{La}{2b}P_0\alpha^2; \quad \mathcal{L}_W = Wa\alpha$$; work done by the static load P_0 and applied the load W, respectively.

$$P_0\frac{L}{b}\alpha^2 - 2W_f\alpha + \frac{S^2}{aM_{ef}} = 0$$; equation to find α

$$\alpha \approx \frac{S^2}{2W_f a M_{ef}}$$; special case when contribution of P_0 is small

(b) Shear-dominated failure. $\Pi_s = 2Q_c u_s$; strain energy

$M_s = Hm = HA\rho$; mass of a beam segment with length = section depth

$$u_s = \frac{S^2}{4Q_c M_s}$$; shear slide needed to arrest motion

Note: Moment capacity at the impact point is different from that at an end. The bending and shear strength near the impact point, as assumed here, are to be used only if more accurate data relating to the nature of impact is not available. See text for further comments.

CASE 11.29   A RIGID COLUMN, WITH RP JOINTS, IMPACTED BY A MOVING OBJECT IN PRESENCE OF A SUSTAINED AXIAL LOAD P_0. ($a < b$)

Follow Case 11.28 to determine load capacities and symbols missing below.

(a) Bending-with-shear (combined) mode. $\Pi_b = W_f a \alpha$; strain energy

$$M_{ef} = \frac{mL}{3} + M; \text{ effective combined mass, impact point}$$

$$V_0 = \frac{v_0}{\dfrac{mL}{3M} + 1}; \text{ post-impact velocity}$$

$$E_k = \frac{1}{2} M_{ef} V_0^2 = \frac{1}{2} \frac{(Mv_0)^2}{M_{ef}}; \text{ kinetic energy supplied}$$

$$P_0 \frac{L}{b} \alpha^2 - 2W_f \alpha + \frac{(Mv_0)^2}{aM_{ef}} = 0; \text{ equation to find } \alpha$$

$$\alpha \approx \frac{(Mv_0)^2}{2W_f a M_{ef}}; \text{ special case when contribution of } P_0 \text{ is small}$$

(b) Shear-dominated failure. $\Pi_s = 2Q_c u_s$; strain energy

$$M_s = Hm = HA\rho; \text{ mass of a beam segment with length = section depth}$$

$$M_{efs} = M_s + M; \text{ effective combined mass, impact point}$$

$$V_0 = \frac{v_0}{\dfrac{M_s}{M} + 1}; \text{ post-impact velocity}$$

$$E_k = \frac{1}{2} M_{efs} V_0^2 = \frac{1}{2} \frac{(Mv_0)^2}{M_{efs}}; \text{ kinetic energy supplied}$$

$$u_s = \frac{E_k}{2Q_c}; \text{ shear slide needed to arrest motion}$$

Note: This is similar to Case 11.28, except that the energy input is due to the flying mass. W_f is the smallest static load needed to overcome column resistance. The impact is plastic.

EXAMPLES

EXAMPLE 11.1   STATIC STABILITY OF A SIMPLY SUPPORTED COLUMN

A two-flange beam, with section geometry illustrated, is $L = 500$ mm long and is made of a bilinear material: $E = 60{,}000$ MPa, $F_y = 60$ MPa, $F_u = 300$ MPa, $\varepsilon_u = 0.25$ (ultimate elongation). The ends are simply supported. Determine Euler buckling stress and compare this with an estimate for inelastic buckling using a variable plastic modulus E_p.

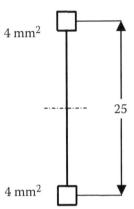

Section properties: $A = 8\,\text{mm}^2$, $I = AH^2/4 = 1250\,\text{mm}^4$

Slenderness from Equations 11.6: $\Lambda = L/(I/A)^{1/2} = 500/(1250/8)^{1/2} = 40$
(With $\mu = 1.0$)

Yield strength of the column: $P_y = AF_y = 480\,\text{N}$

Elastic buckling load: $P_{cr} = P_e = \pi^2 EI/L^2 = 2961\,\text{N}$

This corresponds to the average buckling stress of $\sigma_e = P_{cr}/A = 370.1$ MPa, far in excess of F_y. (This could have been also deduced from Equation 11.8, from which one can find that $\Lambda_y = 99.34$.)

From the data provided, find E_p: $\varepsilon_y = F_y/E = 0.001$

$$E_p \approx \frac{F_u - F_y}{\varepsilon_u - \varepsilon_y} = \frac{300 - 60}{0.25 - 0.001} = 963.9\,\text{MPa} \quad \text{(Secant modulus, refer to Chapter 4)}$$

If one attempts to use this secant value of E_p, the resulting buckling stress from Equation 11.12a is less than 1 MPa, obviously an incorrect result. Use Equation 11.13 instead:

$$\sigma_{cr} = F_u - (F_u - F_y)\left(\frac{\Lambda}{\Lambda_y}\right) = 300 - (300 - 60)\left(\frac{40}{99.34}\right) = 203.4\,\text{MPa}$$

Note: There is a large difference here between F_y, σ_{cr}, and F_u. This is typical of some high-temperature applications of steel.

EXAMPLE 11.2   ECCENTRIC COMPRESSION OF A MODERATELY SHORT COLUMN

Use the same column as in Example 11.1. The initial eccentricity is $\delta_0 = 0.01L = 5$ mm. Find the largest axial force P_1 that this column can resist without yielding. Compare axial shrinking due to compression and flexure when P_1 is applied. Determine the P–y curve, where y is the total lateral displacement and P is the compressive force inducing this displacement.

Section properties: $A = 8\,\text{mm}^2$, $I = 1250\,\text{mm}^4$

Yield compressive strength: $P_y = A\sigma_0 = 480\,\text{N}$

Yield bending strength: $\mathcal{M}_y = H(P_y/2) = 25(480/2) = 6000\,\text{N-mm}$

Elastic buckling load: $P_{cr} = P_e = 2961\,N$

The inside flange force at the onset of yield: $R_{iy} = P_y/2 = 240\,N$

The lateral deflection y at transition to yield, P_1, is found from Case 11.24:

$$\left(\frac{y}{H}\right)^2 + \left(\frac{1}{2} - \frac{\delta_0}{H} - \frac{P_y}{2P_e}\right)\left(\frac{y}{H}\right) - \frac{\delta_0}{2H} = 0$$

which gives $y = 5.63\,mm$. Then, from Equation 11.31a:

$$P_1 = P_e\left(1 - \frac{\delta_0}{y}\right) = 2961\left(1 - \frac{5}{5.63}\right) = 331\,N$$

The bending moment is $M_1 = P_1 y = 331 \times 5.63 = 1864$ N-mm

Shortening due to direct stress: $u_a = \dfrac{P_1 L}{EA} = \dfrac{331 \times 500}{60{,}000 \times 8} = 0.3448\,mm$

The effect of flexure from Table 11.1: $\lambda = \dfrac{\pi^2}{4}\dfrac{y_1^2}{L} = \dfrac{\pi^2}{4}\dfrac{5.63^2}{500} = 0.1564\,mm$

There is a larger deflection from direct compression than from bending. The shorter the column becomes, the larger the ratio of the first to the second component. The shortening due to flexure decreases in proportion to y^2. Also note that the assumed eccentricity in this problem was larger than one would expect in most engineered products. The latter is usually assumed between $L/1000$ and $L/250$.

EXAMPLE 11.3   DYNAMIC COMPRESSION OF A LONG COLUMN WITH A SHAPE IMPERFECTION

Use a bar with the same section and material as in Example 10.1, simply supported.

$L = 150\,mm$, $\phi 3.18$

$E = 69{,}000\,MPa$, $\rho = 0.0027\,g/mm^3$, $v = 0.3$, $G = 26{,}540\,MPa$

$A = 7.942\,mm^2$, $I = 5.02\,mm^4$, $A_s = 0.5A$, $m = \rho A = 0.02144\,g/mm$

Maximum shape deviation from the straight line is $\delta_0 = 0.4\,mm$. Assume that the column is pre-compressed to P_0 level and becomes activated by removal of lateral restraints. Find the peak bending moment when $P_0 = 0.8P_{cr}$ and the associated time point. Repeat the operation for $P_0 = 1.2P_{cr}$ using the same time point.

Use Case 11.3. The critical force: $P_{cr} = P_e = \dfrac{\pi^2 EI}{L^2} = 151.9\,N$

(a) Subcritical compression $P_0 = 0.8P_{cr}$

$$\alpha = \frac{P_0/P_e}{1 - P_0/P_e} = \frac{0.8}{1 - 0.8} = 4$$

$$\Omega = \frac{\pi}{L}\left(\frac{P_e}{m}\left(1-\frac{P_0}{P_e}\right)\right)^{1/2} = \frac{\pi}{150}\left(\frac{P_e}{0.02144}(1-0.8)\right)^{1/2} = 0.7884$$

Half-period $\tau = \pi/\Omega = 3.985\,\text{ms}$

Dynamic deflection $\delta_m = 2\alpha\delta_0 = 2 \times 4 \times 0.4 = 3.2\,\text{mm}$

$\delta = \delta_0 + \delta_m = 0.4 + 3.2 = 3.6\,\text{mm}$ is the total deflection

$M = P_e\delta = 151.9 \times 3.6 = 546.8\,\text{N-mm}$; maximum bending moment induced

(b) Supercritical compression $P_0 = 1.2P_{cr}$

$\beta = \dfrac{1.2}{1.2-1} = 6$; amplitude multiplier

$$\Phi = \frac{\pi}{L}\left(\frac{P_e}{m}\left(\frac{P_0}{P_e}-1\right)\right)^{1/2} = \frac{\pi}{150}\left(\frac{P_e}{0.02144}(1.2-1)\right)^{1/2} = 0.7884$$

Load duration $t = 3.985$ ms, as used above; $\Phi t = 3.1418$

$$\delta_d = \beta\delta_0\left[\cosh\Phi t - 1\right] = 6\times0.4\left[11.594-1\right] = 25.43\,\text{mm}$$

$\delta = \delta_0 + \delta_d = 0.4 + 25.43 = 25.83$ mm; The total deflection is now several times larger.

$M = P_e\delta = 151.9 \times 25.83 = 3924\,\text{N-mm}$

If the material has $F_y = 250\,\text{MPa}$, the onset of yield is at $M = IF_y/c = 631$ N-mm. Subcase (a) is still within the elastic range. In subcase (b) the response is so large, that the results are inaccurate and can be merely taken as an indication of an advanced plastic stage.

EXAMPLE 11.4   ECCENTRIC COMPRESSION OF A TFUS COLUMN

Use a non-symmetric version of the column in Example 11.1, namely $H = 25$ mm, $A_1 = 4\,\text{mm}^2$, and $A_u = 40\,\text{mm}^2$, but material properties remaining the same. The initial eccentricity is $\delta_0 = 0.01\,L = 5$ mm. Find the largest axial force P_1 that this column can resist without yielding and the associated lateral deflection.

Use Case 11.25:

$$A = A_1 + A_2 = 4 + 40 = 44\,\text{mm}^2; \quad p = A_2/A_1 = 40/4 = 10;$$

$$c = \frac{Hp}{1+p} = \frac{25\times10}{1+10} = 22.73\,\text{mm}; \quad I = \frac{A_2H^2}{1+p} = \frac{40\times25^2}{1+10} = 2273\,\text{mm}^4$$

Yield strength of the column: $P_y = AF_y = 44 \times 60 = 2640\,\text{N}$.

Yield strength of the flange: $R_{1y} = A_1F_y = 4 \times 60 = 240\,\text{N}$.

Elastic buckling load: $P_e = 2961\dfrac{2273}{1250} = 5384\,\text{N}$. (In proportion to I). Then

$$\left(\frac{y}{H}\right)^2 + \left(\frac{1}{1+p} - \frac{\delta_0}{H} - \frac{R_{1y}}{P_e}\right)\left(\frac{y}{H}\right) - \frac{\delta_0}{(1+p)H}$$

$$= \left(\frac{y}{H}\right)^2 + \left(\frac{1}{1+10} - \frac{5}{25} - \frac{240}{5384}\right)\left(\frac{y}{H}\right) - \frac{\delta_0}{11H} = \left(\frac{y}{H}\right)^2 - 0.1537\left(\frac{y}{H}\right) - 0.01818 = 0$$

which gives $y = 5.801$ mm.

Then, $P_1 = P_e\left(1 - \dfrac{\delta_0}{y}\right) = 5384\left(1 - \dfrac{5}{5.801}\right) = 743.6\,\text{N}$

EXAMPLE 11.5   ECCENTRIC COMPRESSION USING RIGID STICK MODEL

Solve Example 11.3 using a rigid stick model.

Use Case 11.5 and some of Example 11.3 results.
The buckling force: $P_{cr} = P_e = 151.9\,\text{N}$
$\theta_0 = 2u_0/L = 2 \times 0.4/150 = 1/187.5$
$K = P_{cr}L/4 = 5696.3\,\text{N-mm/rad}$
Half-period $\tau = \pi/\Omega = 3.985\,\text{ms}$

(a) Subcritical compression

$$\theta = \frac{1 + P/P_{cr}}{1 - P/P_{cr}}\theta_0 = \frac{1+0.8}{1-0.8}\frac{1}{187.5} = 0.048$$

The total deflection: $u = \theta L/2 = 3.6\,\text{mm}$
Peak moment: $M = 2K(\theta - \theta_0) = 2 \times 5696.3 \times (0.048 - 1/187.5) = 486.1\,\text{N-mm}$
Deflections are the same as in "original" beam in Example 11.3. The magnitude of bending moment is somewhat different due to the fact that this approach considers the peak moment to be related to the elastic rather than the total deflection.

(b) Supercritical compression

$$P/P_{cr} = 1.2\beta = \frac{P/P_{cr}}{P/P_{cr} - 1} = 6; \text{ amplitude multiplier}$$

$$J = mL^3/24 = 0.02144 \times 150^3/24 = 3015$$

$$\Phi = \left[\left(\frac{P_0}{P_{cr}} - 1\right)\left(\frac{2K}{J}\right)\frac{\pi^2}{12}\right]^{1/2} = \left(0.2\frac{2 \times 5696.3}{3015}0.8225\right)^{1/2} = 0.7884$$

Load duration $t = 3.985\,\text{ms}$, as used before; $\Phi t = 3.1418$
$\cosh \Phi t = 11.594$

$$\theta - \theta_0 = \beta\theta_0\left(\cosh(\Phi t) - 1\right) = (6/187.5)(11.594 - 1) = 0.339$$

$\theta = 0.339 + 1/187.5 = 0.3443$

$\delta = \theta L/2 = 0.3443 \times 150/2 = 25.83\,\text{mm}$

$M = 2K(\theta - \theta_0) = 2 \times 5696.3 \times 0.339 = 3862\,\text{N-mm}$

The total deflection is the same and the moment slightly below that in Example 11.3.

EXAMPLE 11.6   DYNAMIC COMPRESSION OF A SLENDER COLUMN WITH A SHAPE IMPERFECTION

Consider the same column as in Example 11.3. A short duration axial force $P = 1500\,\text{N}$ is applied to a column, which has unspecified deviations. Identify the modes of lateral motion, in which the beam will tend to vibrate and those, which will lead to buckling. If the flow strength of the material is $\sigma_0 = 300\,\text{MPa}$, find the maximum elastic force, which may be induced.

Critical forces, Equation 11.4:

$$P_{cr} = n^2 \frac{\pi^2 EI}{L^2} = 151.9 n^2$$

$$P_{cr1} \equiv P_e = 151.9\,\text{N}; \quad P_{cr2} = 607.8\,\text{N}; \quad P_{cr3} = 1367.5\,\text{N}; \quad P_{cr4} = 2430\,\text{N}$$

The applied force has a magnitude corresponding to between the third and the fourth mode. The modes 1, 2, and 3 will be associated with lateral motion, which the beam may or may not withstand due to a tendency to buckle. Mode 4 and higher modes will experience vibrations only.

The maximum elastic force is simply $P_m = A\sigma_0 = 7.942 \times 300 = 2383\,\text{N}$.

EXAMPLE 11.7   SUDDEN REMOVAL OF LATERAL SUPPORT FROM A COMPRESSED COLUMN

The column has a 450 mm × 450 mm section and is under 4455 kN compression, averaged along the height. The segment under consideration is 8.9 m high, but there is a floor slab at about its mid-height providing a lateral support. The maximum eccentricity of the segment is 18 mm. As a result of an accident that slab is broken and that lateral support suddenly vanishes. Treating the column as simply supported, estimate the resulting dynamic bending moment and stress. The material is 32 MPa RC, $E = 31{,}000\,\text{MPa}$, $\rho = 2400\,\text{kg/m}^3$, $F_c' = 38\,\text{MPa}$ (28-day compressive strength), and $F_t = 3.4\,\text{MPa}$.

Section properties: $A = 202{,}500\,\text{mm}^2$, $I = 3.417 \times 10^9\,\text{mm}^4$

Euler (buckling) force: $P_e = \dfrac{\pi^2\, 31{,}000 \times 3.417 \times 10^9}{8{,}900^2} = 13.2 \times 10^6\,\text{N}$

Axial load applied to the constrained column: $P_0 = 4.455 \times 10^6\,\text{N}$

The effects of a sudden removal of the intermediate restraint is quantified in Case 11.3

We have $P_0/P_e = 0.3375$ and the dynamic deflection δ_m is

$$2\alpha\delta_0 = \frac{P_0/P_e}{1 - P_0/P_e} 2\delta_0 = \frac{0.3375}{1 - 0.3375} 2 \times 18 = 18.34\,\text{mm}$$

Compressive stress: $\sigma_c = P_0/A = 22\,\text{MPa}$.

The resulting bending moment is calculated for the entire eccentricity, $\delta_0 + \delta_m = 18 + 18.34 = 36.34\,\text{mm}$

Bending moment: $\mathcal{M} = P_e(\delta_0 + \delta_m) = 13.2 \times 10^6 \times 36.4 = 479 \times 10^6\,\text{N-mm}$

This induces a bending stress of $\sigma_b = 6\mathcal{M}/h^3 = 31.58\,\text{MPa}$. This stress is much larger than F_t, even after reduction by σ_c. This means cracks will open on the tensile side and the section will rely on reinforcement to resist bending. On the other face of the column one has $31.58 + 22 = 53.58\,\text{MPa}$, even without including the magnifying effect of tensile cracking. Although the resisting moment depends largely on the amount of reinforcement, unspecified in this problem, it is unlikely that such a large moment can be withstood. A collapse is the expected outcome, unless a refined, computer-aided analysis, including NL effects, shows otherwise.

Example 11.8   Eccentric Dynamic Compression (End Impact) of a Long Column

Use a bar with the same section and material as in Example 11.3. The offset is $d/4$. Determine how long the compression $\sigma_c = 0.6F_y$ needs to be applied to bring the total stress to the F_y level, or to initiation of yielding. Follow the solution presented in Case 11.9.

$d = 3.18\,\text{mm}$, $E = 69,000\,\text{MPa}$, $\rho = 0.0027\,\text{g/mm}^3$

$A = 7.942\,\text{mm}^2$, $I = 5.02\,\text{mm}^4$, $m = \rho A = 0.02144\,\text{g/mm}$; $F_y = 300\,\text{MPa}$

The bending stress must reach $F_y - 0.6F_y = 0.4F_y = 120\,\text{MPa}$. This is a circular section, for which $c/I = (3.18/2)/5.02 = 0.3167$

The associated bending moment is found from $0.3167\mathcal{M}_m = 120$; then $\mathcal{M}_m = 378.9\,\text{N-mm}$

$P_0 = 0.6AF_y = 0.6 \times 7.942 \times 300 = 1429.6\,\text{N}$; the axial load applied. Then

$\mathcal{M}_m(t) = 0.4285P_0\delta f(t)$ or $378.9 = 0.4285 \times 1429.6(3.18/4)f(t)$; so $f(t) = 0.778$

Substituting

$$f(t) = \cosh\left(\frac{P_0 t}{2\sqrt{EIm}}\right) - 1 \quad \text{or} \quad 0.778 = \cosh\left(\frac{1429.6t}{2\sqrt{69,000 \times 5.02 \times 0.02144}}\right) - 1$$

which reduces to $\cosh(8.295t) = 1.778$. Solving for t gives $t = 0.142\,\text{ms}$.
For this instant, $f(t) = 0.778$. The associated peak lateral deflection:

$$u_m(t) = \frac{\pi}{4}\delta f(t) = \frac{\pi}{4}\frac{3.18}{4}0.778 = 0.486\,\text{mm}, \text{ quite a small deflection.}$$

EXAMPLE 11.9   SIMPLY SUPPORTED BEAM UNDER LATERAL, DISTRIBUTED STEP LOAD. NO END TRANSLATION

Consider the same beam as in Example 11.3, except that the ends are prevented from moving and the length is shorter, $L = 100\,$mm. The step load now has the magnitude of $w_0 = 15\,$N/mm. Find the peak deflection δ, the stretching force P_m, and the peak bending $\mathcal{M}_m$.

Use Case 11.13. The cable energy coefficient $\Psi = \pi^4/32 = 3.044$

The beam energy coefficient: $\Gamma = \dfrac{\pi^4}{4}$

The work: $\mathcal{L} = \dfrac{2}{\pi} w\delta L$, hence $\Theta = \dfrac{2}{\pi}$

Equation 11.71 had the initial form of $\Psi EA\delta^3 + \Gamma EI\delta = \Theta L^4 w$. Now we have
$\Psi EA = 1.668 \times 10^6$; $\Gamma EI = 8.435 \times 10^6$; $\Theta L^4 w = 954.9 \times 10^6$; $1.668u_m^3 + 8.435u_m = 954.9$
The solution gives $\delta = 8.1\,$mm. The above equation indicates that for

$$\delta_e = (8.435/1.096)^{1/2} = 2.77\,\text{mm}$$

the energy components are equal. The solution shows that the deflection is in the cable-dominated range. One could also deduce the approximate solution as follows: Omission of the second term on the left gives the cable solution as $u_c = 8.302\,$mm. Omitting of the other term gives the beam solution as $u_b = 113.2\,$mm. The inverse of a deflection, or stiffness, is additive, therefore

$$\frac{1}{\delta} \approx \frac{1}{u_c} + \frac{1}{u_b} = \frac{1}{8.302} + \frac{1}{113.2}; \text{ hence } \delta \approx 7.743\,\text{mm}$$

The "approximately equal to" sign is the reminder that we are in NL regime and therefore such a superposition is approximate. Still, the result is reasonably close. Using the more accurate value, $\delta = 8.1\,$mm:

$$\lambda = \frac{\pi^2}{4}\frac{\delta^2}{L} = 1.618\,\text{mm}$$

The cable force is then $P_m = \dfrac{EA}{L}\lambda = \dfrac{69{,}000 \times 7.942}{100}1.618 = 8{,}867\,\text{N}$
$\mathcal{M} = \dfrac{\pi^2}{L^2}EI\delta = 2769\,$N-mm; the peak bending moment
This problem was also solved using LS-Dyna and the results were

$$\delta = 8.171\,\text{mm}; \quad P_m = 8882\,\text{N}; \quad \mathcal{M} = 3320\,\text{N-mm}$$

EXAMPLE 11.10   CLAMPED BEAM UNDER POINT STEP LOAD. NO END TRANSLATION

Consider the same beam as in Example 11.9, except that the ends are clamped. The step load now is the force applied at center, $W_0 = 1500\,$N. Find the peak deflection δ, the stretching force P_m, and the peak bending $\mathcal{M}_m$.

Use Case 11.16. Repeating the calculation with new coefficients results in $1.096\delta^3 + 33.25\delta = 1500$, from which $\delta = 10.194$ mm.

Then $\lambda = 2.0784$ mm and $P_m = 11,389$ N

$$\delta_e = \sqrt{\frac{48I}{A}} = \sqrt{\frac{48 \times 5.02}{7.942}} = 5.51 \text{ mm; the cable action is dominant}$$

The estimate of the end moment

$$\mathcal{M}_e = \frac{48EI}{L^2}\delta = \frac{48 \times 69,000 \times 5.02}{100^2} 10.194 = 16,659 \text{ N-mm}$$

The same example, solved with LS-Dyna gave

$$\delta = 9.293 \text{ mm}; \quad P_m = 10,954 \text{ N}; \quad \mathcal{M} = 20,400 \text{ N-mm}$$

EXAMPLE 11.11   CLAMPED–CLAMPED BEAM UNDER DISTRIBUTED STEP LOAD. NO END TRANSLATION

The same beam as in Example 11.9, except that the ends are clamped, rather than supported. The step load now has the magnitude of $w_0 = 25$ N/mm. Find the peak deflection δ, the stretching force P_m, and the peak bending $\mathcal{M}_m$.

Use Case 11.15. The coefficients are

$$\frac{\pi^4}{32} EA = 1.668 \times 10^6$$

$$\pi^4 EI = \pi^4\, 69,000 \times 5.02 = 33.74 \times 10^6$$

$$\frac{1}{2}L^4 w = \frac{1}{2}100^4 \times 25 = 1250 \times 10^6; \text{ so that the equation for } \delta \text{ is}$$

$$17,125\delta^3 + 346,380\delta = 12.832 \times 10^6$$

The solution gives $\delta = 8.342$ mm. The above equation indicates that for $\delta_e = 4.5$ mm the energy components are equal. The solution also shows that the deflection is in the cable-dominated range. The elongation can now be calculated:

$$\lambda = \frac{\pi^2}{4}\frac{\delta^2}{L} = 1.717 \text{ mm}$$

The cable force is then $P = P_m = \dfrac{EA}{L}\lambda = \dfrac{69,000 \times 7.942}{100}1.717 = 9,409$ N

To estimate the end bending moment $P_e = \dfrac{\pi^2}{L^2}EI = 341.9$ and $\mathcal{M} = \dfrac{16}{L^2}EI\delta\sqrt{\dfrac{P}{P_e}} = \dfrac{16}{100^2}$

$$69{,}000 \times 5.02 \times 8.342 \sqrt{\frac{9{,}409}{341.9}} = 24{,}250 \,\text{N-mm; initial estimate}$$

$$x = \frac{\pi}{2}\sqrt{\frac{P}{P_e}} = 8.24; \quad M = \frac{8P(x-1)\delta}{3.952x^2 - 7.52x} = 22{,}030 \,\text{N-mm; better estimate}$$

This problem was also solved using LS-Dyna and the results were

$$\delta = 8.684 \,\text{mm}; \quad P_m = 10{,}111 \,\text{N}; \quad M = 19{,}700 \,\text{N-mm}$$

EXAMPLE 11.12   INELASTIC BEAM UNDER IMPULSIVE LOADS. NO END TRANSLATION

The same beam as in Example 11.9, $L = 100\,\text{mm}$, except that the ends are clamped, rather than supported. The beam is made of RP material, with $\sigma_0 = 200\,\text{MPa}$. Consider two load cases:

(a) Distributed load $w_0 = 50\,\text{N/m}$, applied for 0.05 ms.
(b) Point load $W_0 = 2000\,\text{N}$ at center, applied for 0.05 ms.

Find the peak deflection δ and peak strain ε_m, the latter based on half-height joint length.

The section capacity in tension is $P_0 = A\sigma_0 = 7.942 \times 200 = 1588\,\text{N}$. The limiting bending moment was found in Example 10.1 to be $M_0 = 1072\,\text{N-m}$.

$$\rho = 0.0027 \,\text{g/mm}^3, \quad A = 7.942 \,\text{mm}^2, \quad m = A\rho = 0.02144$$

(a) Use Case 11.19.

$$\zeta = \frac{M_0}{P_0 L} = \frac{1072}{1588 \times 100} = \frac{1}{148.1}$$

$$\theta_e = \frac{8M_0}{P_0 L} = \frac{8 \times 1072}{1588 \times 100} = 0.054; \text{ rotation when energy components are equal}$$

$$E_k = \frac{3}{8}\frac{(w_0 t_0)^2 L}{m} = \frac{3}{8}\frac{(50 \times 0.05)^2 100}{0.02144} = 10{,}932; \text{ kinetic energy}$$

$$\theta_m = 2\zeta\left[\left(4 + \frac{E_k}{2M_0\zeta}\right)^{1/2} - 2\right] = \frac{2}{148.1}\left[\left(4 + \frac{10{,}932 \times 148.1}{2 \times 1072}\right)^{1/2} - 2\right] = 0.345 = 19.77°$$

$$\delta = \frac{L}{2}\tan\theta = 50 \times 0.3594 = 17.97\,\text{mm; peak deflection}$$

$$\varepsilon_m = 7.623\frac{\delta}{L} - 0.1132 = 1.257; \text{ total strain at end}$$

This problem was also solved using LS-Dyna (100 elements, EPP material, $E = 69\,\text{GPa}$) and the results were

$$\delta = 18.97\,\text{mm}; \quad \varepsilon_m = 1.109$$

This is reasonably close for δ, but somewhat of an overestimate for peak strain. The reason is that the approximate formula is applied outside of the intended range, therefore the overestimate is anticipated. The resistance is cable-dominated.

(b) Use Case 11.20. ζ and θ_e are the same, but the kinetic energy is different:

$$M_{ef} = 0.15mL = 0.15 \times 0.02144 \times 100 = 0.3216 \, g; \text{ effective mass}$$

$$E_k = \frac{(W_0 t_0)^2}{2M_{ef}} = \frac{(2,000 \times 0.05)^2}{2 \times 0.3216} = 15,547; \text{ kinetic energy}$$

$$\theta_m = 0.4162; \quad \delta = 50 \times \tan 23.85° = 22.1 \, \text{mm}$$

$$\varepsilon_m = 16.132\left(\frac{\delta}{L}\right)^2 + 0.11 = 0.898$$

Using the same LS-Dyna model, it was found that

$$\delta = 22.82 \, \text{mm}; \quad \varepsilon_m = 0.952$$

The underestimate of peak strain results largely from the underestimate of δ.

EXAMPLE 11.13   FORCE–MOMENT OR P–M DIAGRAM OF AN RC COLUMN

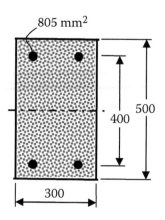

805 mm²

The column, shown in the sketch, is subjected to eccentric compression. Find the necessary section constants and create the basic P–M diagram. The materials are: $F_c' = 25 \, \text{MPa}$ concrete, $F_y = 410 \, \text{MPa}$ steel. Estimate the effect of slenderness, taking 3.0 as the end fixity factor.

Equations 11.43 through 11.46 are employed, as described in text.

$A = 300 \times 500 = 150,000 \, \text{mm}^2; \ A_{st} = 2 \times 805 = 1,610 \, \text{mm}^2 = A_{sc};$

$A_s = 3,220 \, \text{mm}^2$

$P_0 = 0.85 F_c'(A - A_s) + A_s F_y = 0.85 \times 25(150,000 - 3,220) + 3,220 \times 410 = 4.44 \times 10^6 \, \text{N}$

The distance from top to tensile steel: $d = 400 + (500 - 400)/2 = 450 \, \text{mm}$

$$M_0 = A_{st} F_y d\left(1 - 0.6\frac{A_{st}}{bd}\frac{F_y}{F_c'}\right) = 1610 \times 410 \times 450\left(1 - 0.6\frac{1610}{300 \times 450}\frac{410}{25}\right) = 262.2 \times 10^6 \, \text{N-mm}$$

For the balanced condition:

$$P_b = 0.34(BH)F_c' = 0.34 \times 150,000 \times 25 = 1.275 \times 10^6 \, \text{N}$$

$$M_b = A_{st} F_y d_s + 0.102 BH^2 F_c' = 1610 \times 410 \times 400 + 0.102 \times 300 \times 500^2 \times 25 = 455.3 \times 10^6 \, \text{N-mm}$$

One should note that an example with the same data was used by Darvall [21]. His more accurate results, obtained by more involved methods, gave approximately the same answers except for P_b, which was 1.42×10^6 N

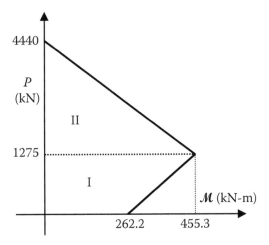

The interaction diagram, or the limiting state envelope, is shown in the sketch.

To find the buckling load first determine the reduced modulus:

$E_r = 27,000/3 = 9,000$ MPa, per Equation 11.51:

For this section, ignoring steel, one has $I = 3.125 \times 10^9$ mm⁴

$$P_{cr} = \mu \frac{\pi^2 E_r I}{L^2} = 3 \frac{\pi^2 \, 9000 \times 3.125 \times 10^9}{5000^2} = 33.3 \times 10^6 \text{ N}$$

The maximum axial force that can be developed is P_0, which is ~13% of the buckling force. Only a minor enhancement of response can be anticipated due to the column effect.

EXAMPLE 11.14   THE INFLUENCE OF SLENDERNESS ON LIMIT-STATE FORCES

Consider the same column section as in Example 11.13, subjected to compression with the eccentricity of $e = 0.5$ m and establish the limit-state components. If this belongs to a column that is 6 m tall and has end constraints such that $\mu = 1.8$, determine the necessary reduction in load capacity. Use $E_c = 27$ GPa.

The basic relation between the axial force and the moment is $M = Pe$, which can be drawn on the interaction diagram as a straight line, originating at the center. Its intersection with the envelope determines the limit components P and M. The slope of such a line in (P, M) coordinates is $1/e = 1/500$. For the balance point the slope of a line from the origin is $P_b/M_b = 1.275/453.3 = 1/356$. Our line will therefore pass below the balance point, in region I. The limit-state line is defined by Equation 11.47:

$$M = M_0 + (M_b - M_0)\frac{P}{P_b} \text{ ; in which, after } M \text{ is replaced by } Pe, \text{ one can write}$$

$$500P = 262.2 + (455.3 - 262.2)\frac{P}{1.275}; \text{ from which } P = 748{,}900\,\text{N}$$

The corresponding moment is $Pe = 748{,}900 \times 500 = 374.5 \times 10^6$ N-mm.

To find buckling load first determine the reduced modulus, per Equation 11.51:

$$E_r = 27{,}000/3 = 9{,}000\,\text{MPa}$$

For this section, ignoring steel, one has $I = 3.125 \times 10^9$ mm⁴

$$P_{cr} = \mu\frac{\pi^2 E_r I}{L^2} = 1.8\frac{\pi^2\, 9000 \times 3.125 \times 10^9}{6000^2} = 13.88 \times 10^6\,\text{N}$$

The applied load P is therefore associated with the following magnification of the moment:

$$\eta = \frac{1}{1 - P/P_{cr}} = \frac{1}{1 - 748{,}900/13.88 \times 10^6} = 1.057; \text{ per Equation 11.50.}$$

which is not very significant. Reducing the allowable P by the above ratio will correct the situation. One should note that if P acts on a smaller eccentricity, a larger P will be permissible by the same limit graph and the resulting factor η will be larger.

EXAMPLE 11.15   SECTION CONSTANTS OF A SQUARE RC COLUMN

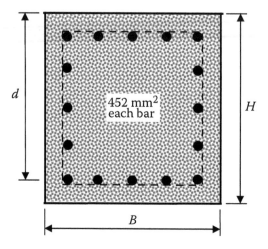

The column, shown in the sketch, is subjected to eccentric compression. Find the necessary section constants to create the basic P–M diagram. The materials are: $F_c' = 25$ MPa concrete and $F_y = 410$ MPa steel. Estimate the buckling strength using $E_c = 27$ GPa, taking the column to be 5 m long and having the end constraint factor $\mu = 3.0$. $B = H = 450$ mm, $d = 405$ mm.

Follow Case 11.26.

$$A = BH = 450 \times 450 = 202{,}500\,\text{mm}^2$$

$$A_s = 16 \times 452 = 7232 \, \text{mm}^2$$

$$d_s = 405 - (450 - 405) = 360 \, \text{mm}$$

$$P_0 = 0.85 F_c'(BH - A_s) + A_s F_y = 0.85 \times 25(202{,}500 - 7{,}232) + 7{,}232 \times 410 = 7.11 \times 10^6 \, \text{N}$$

$$\mathcal{M}_0 = \frac{7}{16} A_s F_y d_{sy} = \frac{7}{16} 7232 \times 410 \times 360 = 467 \times 10^6 \, \text{N-mm}$$

$$P_b = 0.34(BH)F_c' = 0.34(202{,}500)25 = 1.721 \times 10^6 \, \text{N}$$

$A_{sb} = 5 \times 452 = 2260 \, \text{mm}^2$, bottom line of bars only

$$\mathcal{M}_b = A_{sb} F_y d_{sy} + 0.102 BH^2 F_c' = 2260 \times 410 \times 360 + 0.102 \times 450^3 \times 25 = 565.9 \times 10^6 \, \text{N-mm}$$

When a graph from Concrete Handbook [15] was employed, the corresponding quantities were

$$P_0 = 6.8 \times 10^6, \quad \mathcal{M}_0 = 475.9 \times 10^6, \quad P_b = 1.331 \times 10^6, \quad \mathcal{M}_b = 577.1 \times 10^6$$

A marked overestimate of our simplification shows up only in P_b.

Example 11.16   Section Constants of a Round RC Column

The column, shown in the sketch, is subjected to eccentric compression. Find the necessary section constants to create the basic P–$\mathcal{M}$ diagram. The materials are: $F_c' = 25\,\text{MPa}$ concrete and $F_y = 400\,\text{MPa}$ steel. Light reinforcement, $A_s = 1964\,\text{mm}^2$.

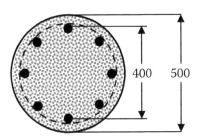

Follow Case 11.27:

$$D = 500 \, \text{mm}, \quad D_s = 400 \, \text{mm}, \quad \text{and} \quad A = 196{,}350 \, \text{mm}^2$$

$$P_0 = 0.85 F_c'(A - A_s) + A_s F_y = 0.85 \times 25(196{,}350 - 1{,}964) + 1{,}964 \times 400 = 4.916 \times 10^6 \, \text{N}$$

$$\mathcal{M}_0 = 0.4 A_s F_y D_s = 0.4 \times 1964 \times 400 \times 400 = 125.7 \times 10^6 \, \text{N-mm}$$

$$P_b = 0.34 A F_c' = 0.34 \times 196{,}350 \times 400 = 1.669 \times 10^6 \, \text{N}$$

$$\mathcal{M}_b = 0.318 A_s F_y D_s + 0.268 P_b D = 0.318 \times 1964 \times 400 \times 400 + 0.268 \times 1.669 \times 10^6 \times 500$$
$$= 308.8 \times 10^6 \, \text{N-mm}$$

When a graph from Ref. [15] was employed, the corresponding quantities were

$$\varphi P_0 = 3.358 \times 10^6, \quad \varphi \mathcal{M}_0 = 122.7 \times 10^6, \quad \varphi P_b = 1.374 \times 10^6, \quad \varphi \mathcal{M}_b = 196.4 \times 10^6$$

where φ is a capacity reduction factor used in design. As mentioned before, there is, in general, a different φ for each component.

EXAMPLE 11.17   LATERAL LOAD TO INITIATE CRACKING

Consider the same column as in Example 11.15. Suppose it is under a compressive load of $0.4P_0$ only, when a lateral loading is applied. What is the bending moment that will initiate cracking? For simplicity, omit the benefit of steel.

As long as the column is intact, it may be treated as elastic.

From Example 11.15: $P_0 = 7.11 \times 10^6$ N and the applied force is $0.4P_0 = 2.844 \times 10^6$ N

When this is divided by the area, a sustained stress of 14.04 MPa is obtained. In absence of any precompression, the flexural strength (on the tensile side) is given by Equation 11.42

$$F_t = 0.6\sqrt{F'_c} = 0.6\sqrt{25} = 3\,\text{MPa}$$

The applied tension (bending stress) must overcome the above and precompression stress, making the flexural strength to be $3 + 14.04 \approx 17.04$ MPa

(This might be called the apparent flexural strength.) From the basic bending formula applied to a square section, find the moment at the onset of cracking:

$$\sigma_b = \frac{6\mathcal{M}_{cr}}{H^3} \quad \text{or} \quad \mathcal{M}_{cr} = \frac{H^3\sigma_b}{6} = \frac{450^3 \times 17.04}{6} = 258.9 \times 10^6\,\text{N-mm}$$

This is quite a substantial moment, about 55% of the pure bending strength $\mathcal{M}_0$. The conclusion is simple: while the tensile strength of concrete is small, an axial preload makes it more difficult to induce tensile cracking.

EXAMPLE 11.18   LATERAL DYNAMIC PRESSURE APPLIED TO A COLUMN. ELASTIC APPROACH

Consider the same column as in Examples 11.15 and 11.17. Suppose it is under a compressive load of $0.4P_0$ only, prior to application of dynamic pressure. The height of column (between slabs) is 5 m and the ends should be treated as fixed. The pressure pattern is triangular decreasing, and results in a uniformly distributed load shown in the sketch as a function of time. Determine, on elastic basis, the peak response. $E = 27$ GPa for 25 MPa concrete, $\rho = 0.0024$ g/mm³.

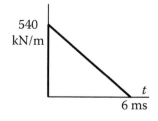

Properties: $L = 5000$ mm, 450×450 section

$$A = 202{,}500\,\text{mm}^2, \quad I = 3.417 \times 10^9\,\text{mm}^4, \quad m = \rho A = 486\,\text{g/mm}$$

The procedure of Chapter 10 is followed. The peak load is used for static calculations.

$$u_{st} = \frac{1}{384}\frac{540 \times 5{,}000^4}{27{,}000 \times 3.417 \times 10^9} = 9.527\,\text{mm; static deflection from Equation 10.12b}$$

The natural period from Equation 10.14:

$$\tau = \varepsilon L^2 \sqrt{\frac{m}{EI}} = \frac{5{,}000^2}{3.561} \sqrt{\frac{486}{27{,}000 \times 3.417 \times 10^9}} = 16.11 \text{ ms;} \quad \text{or} \quad \omega = 0.39 \text{ rad/ms}$$

$W_0 = 540 \times 5000 = 2700 \text{kN};$ the resultant force

$Q_{st} = W_0/2 = 1350 \text{kN};$ static shear at base

$\mathcal{M}_{st} = w_0 L^2/12 = 1.125 \times 10^9$ N-mm; static bending moments at the base

From Equation 10.22: $x = \omega t_0/4 = 0.39 \times 6/4 = 0.585$ and then

$$\frac{u_d}{u_{st}} = 2\frac{\sin^2 x}{x} = 2\frac{\sin^2 0.585}{0.585} = 1.043; \; u_d = 9.937 \text{ mm}$$

Estimate the peak shear and bending response by employing the above ratio:

$$Q_d = 1350(u_d/u_{st}) = 1408 \text{ kN}$$

$$\mathcal{M}_d = 1.125 \times 10^9 \; (u_d/u_{st}) = 1.173 \times 10^9 \text{ N-mm}$$

The reference strength constants of the column were calculated in Example 11.15, with the exception of the shear strength, which is, by Equation 11.49:

$$Q_0 = BHF_t = 202{,}500 \times 0.6 \times \sqrt{25} = 607{,}500 \text{ N}$$

Note that so far the effect of the axial loading was not even included. Because of a very large response based on elastic approach, another check, using an inelastic method is needed.

EXAMPLE 11.19 LATERAL DYNAMIC PRESSURE APPLIED TO A COLUMN. INELASTIC APPROACH

Consider the same column as in Example 11.18. Using column properties developed so far as well as the plastic joint concept, determine the likelihood and manner of failure under the applied lateral and sustained axial loading of $0.4P_0$. Also, refer to Examples 11.15 and 11.17 results. Treat the ends as clamped.

Properties: $L = 5000$ mm, 450×450 section

$$A = 202{,}500 \text{ mm}^2, \quad I = 3.417 \times 10^9 \text{ mm}^4, \quad m = \rho A = 486 \text{ g/mm}$$

The methods developed in Chapter 10 necessitate the use of plastic hinges in postelastic range.

According to Example 11.15, $P_0 = 7.11 \times 10^6$ N. Our sustained load is therefore $0.4P_0 = 2.844 \times 10^6$ N. This is more than $P_b = 1.433 \times 10^6$ N, therefore our loading state is in region II of Figure 11.11. The associated moment at the limit state is

$$M = \frac{P_0 - P}{P_0 - P_b} M_b = \frac{7.11 - 2.844}{7.11 - 1.433} 548.9 \times 10^6 = 412.5 \times 10^6 \text{ N-mm (Equation 11.48)}$$

The above will be used as the plastic hinge moment M_0. From Example 11.18: $Q_0 =$ 607,500 N. The limit-state distributed loads can now be found, using Equation 10.52 and Table 10.9:

$$w_b = 16 \, M_0/L^2 = 16 \times 412.5 \times 10^6/5000^2 = 264 \text{ N/mm}$$

$$w_s = 2Q_0/L = 2 \times 607,500/5,000 = 243 \text{ N/mm}$$

This indicates that plastic flow will take place when loading exceeds these limits. The column is more critical in shear, but only marginally so.

For a bending failure, use Case 10.21, denoting $L' = L/2 = 2500$ mm. (Half-beam length.) Also, adjust the expression for impulse S_0, so that it corresponds to a load varying in time, rather than constant.

$S_0 = w_0 L' t_0/2 = 540 \times 2500 \times 6/2 = 4.05 \times 10^6$ N-s; equivalent center impulse

$$M_{ef} = \frac{2mL'}{3} = \frac{2 \times 486 \times 2,500}{3} = 810,000 \, g$$

The impulse is regarded as short, therefore

$$E_k = \frac{S_0^2}{2M_{ef}} = \frac{(4.05 \times 10^6)^2}{2 \times 810,000} = 10.125 \times 10^6 \text{ N-mm; kinetic energy imparted to the beam}$$

$$\theta_m = \frac{E_k}{4M_0} = \frac{10.125 \times 10^6}{4 \times 412.5 \times 10^6} = 0.00614 = 0.352°; \text{ peak rotation}$$

This is the rotation angle at the support. The angle of discontinuity is twice that or 0.0123, which is much less than the allowable value of 0.10.

The shear slide is found from Equation 10.55:

$$u_s = \frac{S_0^2}{2dMQ_0} = \frac{(4.05 \times 10^6)^2}{2 \times 2(5,000 \times 486)607,500} = 2.78 \text{ mm}$$

The permissible shear slide of 2% of section depth is 9 mm. Failure will not take place in spite of permanent deformation resulting.

Finally, an estimate is made of the contribution of axial force to deformation energy to be absorbed by joints. According to Case 11.20, the axial end displacement is $u_x = L\theta^2/2$, so the energy input from P is

$$Pu_x = 2.844 \times 10^6 \times 5,000 \times 0.00614^2/2 = 268,040 \text{ N-mm}$$

This constitutes less than 3% of the energy input from the lateral load as calculated above. Ignoring that effect had little bearing on the results.

EXAMPLE 11.20   NEAR-FOOT IMPACT FAILURE OF AN RC COLUMN

Consider the column examined in Example 11.16, to which a horizontal force W is applied at $a = 0.8\,m$ above ground. First, determine the static limit value of W_f corresponding to breaking assuming that the local application of force has degraded the local strength by 20% in both shear and bending. The height is 4.4 m and the supported weight is $P_0 = 2000\,kN$. The dynamic force is applied as a triangular impulse lasting $t_0 = 0.9\,ms$. Find the peak value of that force needed to cause a collapse, with the latter defined as 0.1 rad rotation of the lower segment.

Follow Case 11.28:

$$a = 800\,mm, \quad b = 3,600\,mm, \quad L = 4,400\,mm; \quad A = 196,350\,mm^2$$

$\tau_0 \approx F_t = 0.6\sqrt{F_c'} = 0.6\sqrt{25} = 3\,MPa$; maximum shear stress

$Q_0 = A\tau_0 = 196,350 \times 3.0 = 589,050\,N$; $Q_c = 0.8Q_0 = 471,240\,N$, shear capacities

(Original strength and the strength at the impact point.)

$W_s = 2Q_c = 942,500\,N$; shear collapse load

$$\mathcal{M}_0 = \mathcal{M}_b = 308.8 \times 10^6\,N\text{-mm}; \quad \mathcal{M}_c = 0.8\,\mathcal{M}_0 = 247.1 \times 10^6\,N\text{-mm}$$

$$W_b = (\mathcal{M}_0 + \mathcal{M}_c)\left(\frac{1}{a} + \frac{1}{b}\right) = (308.8 + 247.1)10^6\left(\frac{1}{800} + \frac{1}{3,600}\right) = 849,300\,N; \text{ bending collapse}$$
load

Since $W_b < W_s$, the column is more critical in bending. However, the strength is comparable for both modes, therefore a combination must be taken into account.

$$\frac{1}{W_f^2} = \frac{1}{W_s^2} + \frac{1}{W_b^2} = \left(\frac{1}{0.9425^2} + \frac{1}{0.8493^2}\right)\frac{1}{10^{12}}$$

then $W_f = 630,900\,N$; the anticipated breaking load

Regardless of the exact shape of the triangular pulse, the magnitude of the impulse is $W_m t_0/2$.

We therefore have $S^2 = \left(\dfrac{W_m}{2}0.9\right)^2 = 0.2025 W_m^2$

$M_{ef} = \dfrac{mL}{3} = \dfrac{L}{3}A\rho = \dfrac{4400}{3}196,350 \times 0.0024 = 691,150\,g$; effective mass at impact point

The angle of rotation at failure from this event is prescribed as 0.1 rad = 5.73°. Substituting this into the equation for α in the combined mode, one has

$$P_0 \frac{L}{b} \alpha^2 - 2W_f \alpha + \frac{S^2}{aM_{ef}} = 2 \times 10^6 \frac{4,400}{3,600} 0.1^2 - 2 \times 630,900 \times 0.1 + \frac{0.2025 W_m^2}{800 \times 691,150}$$

Then $24,444\alpha^2 - 126,180\alpha + (W_0/52,254)^2 = 0$ or $W_m = 5.81 \times 10^6$ N

Check on shear-dominated failure.

$M_s = HA\rho = 500 \times 196,350 \times 0.0024 = 235,620$ g; mass for shear failure

$u_s = 0.02H = 0.02 \times 500 = 10$ mm; largest possible shear slide

$S^2 = 4Q_c M_s u_s = 4 \times 471, 240 \times 235, 620 \times 10$, the $S = 2.1074 \times 10^6$ g-m/s, the largest acceptable impulse. Impulse S can also be expressed by the unknown applied peak force W: $S = 0.45W$, as above. Then

$W = S/0.45 = 2.1074 \times 10^6/0.45 = 4.68 \times 10^6$ N; breaking force

This is less than the value associated with the combined mode, therefore the shear-dominated failure will govern. Note that this is a large multiple of the static collapse force W_f because of the short duration of the event. An example of a column impacted by a crushable object is given in Chapter 16.

12 Plates and Shells

THEORETICAL OUTLINE

Many formulas for plates are similar to those previously given for beams. The reason, of course, is that any rectangular plate section may be viewed as an assembly of beam sections of a unit width. The same designations of internal forces are used here as for beams, but dimensions are different in that M and Q stand for bending and shear, respectively, per unit width, rather than being referred to the total width B of a beam.

BASIC RELATIONSHIPS

Governing Equations

Consider Figure 10.1 showing bending of a beam and imagine this beam to have a large width in the direction normal to paper. The fact that the beam is wide constrains lateral deflection and makes such a beam stiffer by a factor of $1/(1 - v^2)$, without changing stress distribution. Since there are two directions along the surface of the plate, x and y, one can write two equations analogous to Equation 10.1:

$$\varepsilon_x = \frac{z}{\rho_x} \tag{12.1a}$$

$$\varepsilon_y = \frac{z}{\rho_y} \tag{12.1b}$$

A segment of a plate, with the midsurface curved due to applied bending moments, is shown in Figure 12.1a. A small cube cut out of that larger segment, shows bending moments and the associated stress components. The above equations were written with the assumption that the side walls of the small cube in Figure 12.1b remain plane and the midsurface undergoes no stretching as a result of bending. This represents an extension of similar assumptions used in the analysis of beams. A layer of the cube is shown as stressed in two dimensions. The generalized Hooke's law gives the relation between stress and strain:

$$\left\{ \begin{matrix} \sigma_x \\ \sigma_y \end{matrix} \right\} = \begin{bmatrix} \dfrac{E}{1-v^2} & \dfrac{vE}{1-v^2} \\ \dfrac{vE}{1-v^2} & \dfrac{E}{1-v^2} \end{bmatrix} \left\{ \begin{matrix} \varepsilon_x \\ \varepsilon_y \end{matrix} \right\} \tag{12.2}$$

When strains are substituted from Equation 12.1 and integration carried out along z, on a face of element one unit wide, the result is

$$M_x = \left(\frac{1}{\rho_x} + \frac{v}{\rho_y} \right) D \tag{12.3a}$$

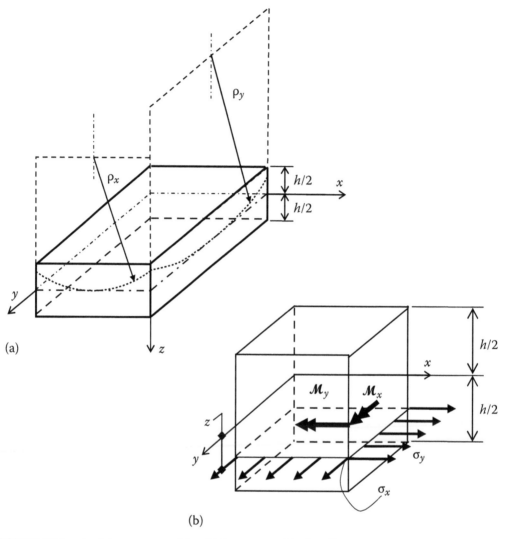

FIGURE 12.1 A plate segment with midplane curved by bending (a) and a smaller cube where bending moments and stress components at a layer at z away from the midplane are illustrated (b).

$$\mathcal{M}_y = \left(\frac{1}{\rho_y} + \frac{v}{\rho_x} \right) D \qquad (12.3b)$$

$$D = \frac{Eh^3}{12(1-v^2)} \qquad (12.3c)$$

Once these moments are known, stress can be calculated using beam formulas applied to a section one unit wide and h-deep. Some special cases deserve attention. When bending takes place only along x, with $\rho_y = 0$, then

$$\frac{1}{\rho_x} = \frac{\mathcal{M}_x}{D} \qquad (12.4a)$$

and

$$\mathcal{M}_y = \nu \mathcal{M}_x \tag{12.4b}$$

where the first equation is analogous to that for beams. When $\mathcal{M}_y = 0$, then $\rho_x = -\nu\rho_y$, which means, for example, that positive curvature along x is associated with a negative curvature along y. (Saddle-like surface results.) Another important case takes place when both moments are equal to $\mathcal{M}$. Then $\rho_y = \rho_x = \rho$

$$\frac{1}{\rho} = \frac{\mathcal{M}}{(1+\nu)D} \tag{12.5}$$

The plate assumes spherical shape. According to Timoshenko [109], the midsurface has always such a shape when a constant moment is applied to the edge, regardless of the outline of the plate itself.

Static Properties of Elastic Plates

There are four plates under detailed consideration, as shown in Figure 12.2: circular, square, rectangular (2:1 side ratio), and infinitely long. Each of them can have the edges supported or clamped. (Clamping means removing angular displacements, but the plate is left free to undergo in-plane displacement unless otherwise stated.) The plates are under distributed (pressure) load p (N/m²) or under center point load W (N).* The maximum static deflections can be written as a sum of flexural and shear components:

$$u = u_b + u_s \quad \text{or} \quad u = b\frac{pL^4}{D} + s\frac{pL^2}{Gh_s} \quad \text{or} \quad u = b\frac{WL^2}{D} + s\frac{W}{Gh_s} \tag{12.6}$$

where
  L is some characteristic dimension
  $h_s = 5h/6$ for uniform, elastic material

In a majority of applications, the plates are homogeneous across the thickness and not too thick, that is, $h/a \leq 10$. It is then reasonable to ignore the contribution of shear to overall deflections. For pressure load, one has the following displacements at the center:

$$u = b\frac{pa^4}{D} \tag{12.7a}$$

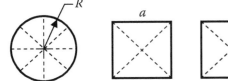
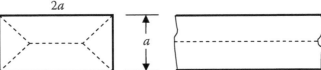

FIGURE 12.2 Shapes of plates under consideration. The dashed dividing lines are for inelastic analyses to follow.

---

* The limitations of the point load concept are discussed later.

or

$$u = b\frac{pR^4}{D}$$

(12.7b)

where
    a is a shorter side of a rectangular plate
    R is the radius of a circular plate

Analogously, for point loads:

$$u = b\frac{Wa^2}{D}$$

(12.8a)

or

$$u = b\frac{WR^2}{D}$$

(12.8b)

Coefficients b are given in Table 12.1 for pressure loading and in Table 12.2 for point loads. When pressure p is applied to the surface, the peak bending moments (N-mm/mm) can be expressed as

$$M = \beta pa^2$$

(12.9a)

or

$$M = \beta pR^2$$

(12.9b)

and, even simpler, when a point load is applied:

$$M = \beta W$$

(12.10)

With shear forces the situation is simpler in that the distribution is regarded as the same for supported as it is for clamped edges (just as for beams). For a circular plate, where shear is uniform along the edge, the equilibrium condition tells us that $Q = Rp/2$. For a square plate, according to Roark [75], $Q=0.42pa$, for the 2:1 side ratio $Q=0.503pa \approx 0.5pa$ and for the infinite side ratio,

---

TABLE 12.1

Deflection and Strength Coefficients, Pressure Loading p

	β	φ	b	ξ	ϑ
CIRC/SS	$\dfrac{3+v}{16}$	1/2	$\dfrac{5+v}{64(1+v)}$	6	1/4
CIRC/CL	1/8	1/2	1/64	12	1/8
Square/SS	0.0479	0.42	1/245.9	24	1/16
Square/CL	0.0513	0.42	1/791.3	48(42.84)	1/32
2:1/SS	0.1017	1/2	1/98.38	12(14)	1/10
2:1/CL	0.0829	1/2	1/394.2	24(24.14)	1/20
INFIN/SS	1/8	1/2	1/76.8	8	3/32
INFIN/CL	1/12	1/2	1/384	16	3/64

Note: Coefficients ξ and ϑ are for the plastic range, to be presented later.

TABLE 12.2
Deflection and Strength Coefficients, Point Load W

	β_1	β_2	b	ξ_m
CIRC/SS	$\frac{1}{4\pi}\left[(1+v)\ln\frac{R}{h}+1\right]$	0	$\frac{1}{16\pi}\frac{3+v}{1+v}$	2π
CIRC/CL	$\frac{1+v}{4\pi}\ln\left(\frac{R}{h}\right)$	$\frac{1}{4\pi}$	$\frac{1}{16\pi}$	4π
Square/SS	$\frac{1}{4\pi}\left[(1+v)\ln\frac{2a}{\pi h}+0.435\right]$	0	1/86.18	6
Square/CL	$\frac{1}{4\pi}\left[(1+v)\ln\frac{2a}{h}-0.238\right]$	0.1257	1/178.7	11.14
2:1/SS	$\frac{1}{4\pi}\left[(1+v)\ln\frac{2a}{\pi h}+0.958\right]$	0	1/60.5	4.21
2:1/CL	$\frac{1}{4\pi}\left[(1+v)\ln\frac{2a}{h}+0.067\right]$	0.168	1/138.6	8.66
INFIN/SS	$\frac{1}{4\pi}\left[(1+v)\ln\frac{2a}{\pi h}+1\right]$	0	1/59.0	4.11
INFIN/CL	$\frac{1}{4\pi}\left[(1+v)\ln\frac{2a}{h}+0.067\right]$	0.168	1/138.1	8.61

Notes: Coefficient ξ_m is to be used later, when the plastic range is discussed. Only the first two values of ξ_m were rigorously derived by Johnson [40]. The rest are to be viewed as approximations.

$Q=0.5pa$. (Note that if the distribution were uniform for a square plate, the coefficient would be 0.25 rather than 0.42.) The maximum shear load (N/m) can, therefore, be written for pressure load as

$$Q = \varphi pa \tag{12.11a}$$

or

$$Q = \varphi pR \tag{12.11b}$$

and, for point load as

$$Q = \frac{\varphi W}{a} \tag{12.12a}$$

or

$$Q = \frac{\varphi W}{R} \tag{12.12b}$$

where the coefficients φ for pressure loading are in Table 12.1.

The largest bending in the plates under consideration, subjected to pressure loading, occurs at the center, when the plate is supported. When the edge is clamped, the largest bending moment is at the edge. The coefficients β in Table 12.1 define those peak values for pressure loads. As for Table 12.2, relating to the point loads, coefficient β_1 refers to the center moment, while β_2 is for the edge moment calculation. The loading is applied over a small circle, with radius equal to the plate thickness h.

Shear Plate

The relation between shear and flexural deflections for plates is similar to that previously discussed for beams. The shear component of deflection may be visualized by introducing a concept of a *shear plate*, as shown in Figure 12.3.

The plate can be imagined as consisting of thin rings capable of shear distortion, but infinitely rigid in the radial direction. The center load W is applied through a rigid insert with a small, but finite-radius r_0. The shear angle γ is defined similarly as for a beam. The shear (stiffness) area is now referred to a circumferential ring section:

$$A_s = 2\pi r h_s \tag{12.13}$$

where h_s is a shear thickness. (For a solid plate with actual thickness h there is $h_s = h/1.2$, a rectangular section relationship.) The increment of center deflection originating from a ring with a radial width dr is

$$du_s = \gamma dr = \frac{W dr}{2\pi r G h_s} \tag{12.14}$$

integrating from r_0 to R gives

$$u_s = \frac{W}{2\pi G h_s} \ln \frac{R}{r_0} \tag{12.15}$$

The conditions under which this shear deflection is significant in comparison with the flexural component are similar to those previously spelled out for beams. For this shear

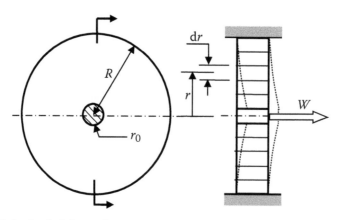

FIGURE 12.3 Point-loaded shear plate.

plate model, there is no difference between laterally supported and clamped edge. The rigid insert need not be small. Also note that loading and support may be reversed, that is, the rigid insert may be held while the outer edge is subjected to a ring load with a resultant W and the deflection is still expressed by Equation 12.15.

PRESSURE LOADING, ELASTIC RANGE

Response of Plates to Pressure Shock Loading in Elastic Range

This elementary approach is equivalent to treating a plate as an oscillator having mass M supported by a spring with stiffness k. This amounts to single-mode approximations, based on the same assumptions as were already spelled out for beams. Again, one can expect a decrease of accuracy of results with the decrease of load duration.

The peak dynamic displacement u_d resulting from a short duration rectangular impulse lasting t_0 has the amplitude determined by Equation 10.15:

$$u_d = 2u_{st}\sin(\pi t_0/\tau) = 2u_{st}\sin(\omega t_0/2) \tag{12.16}$$

Once the peak deflection u_d is calculated, the peak shear and moment are obtained by multiplying the static values by the ratio of the dynamic and static displacements, u_d/u_{st}:

$$Q_d = \frac{u_d}{u_{st}} Q_{st} \tag{12.17a}$$

$$\mathcal{M}_d = \frac{u_d}{u_{st}} \mathcal{M}_{st} \tag{12.17b}$$

The shears calculated in this manner are severely underestimated when load duration is only a small fraction of the natural period τ. A similar procedure, as used for beams, is employed to provide a higher estimate. The peak shear force is now

$$Q_d' = \left(c\frac{u_d}{u_{st}} + d\right)\frac{W_0}{a} \tag{12.18}$$

where
 W_0 is the resultant plate load
 a is replaced by R when a plate is circular

The two unknown coefficients, c and d, are calculated based on a known, accurate solution for a selected time point. (This does not imply, however, that for other, remote time points the solution will be nearly as accurate.) The coefficients have been determined for some cases of supports and summarized in Table 12.3. They are applicable for load duration not exceeding $\tau/2$.

Another solution is offered by Biggs [9] and presented in Table 12.4. His result for square plates is based on assuming a uniform shear along the edges, or uniform reaction per unit length of edge. For this reason the sum of c and d is 0.25 in Table 12.4, while in Table 12.3 it

TABLE 12.3

Shear Force Coefficients from an FEA Study, Pressure Load p

Shape	Edge	c	d
Circular	S	0.382	0.118
Square	S	0.238	0.182
Circular	C	0.320	0.180
Rectangular	C	0.538	−0.038

TABLE 12.4

Shear Force Coefficients from Ref. [9], Pressure Load p

Shape	Edge	Short Side, c	Short Side, d	Long Side, c	Long Side, d
Square	S	0.18	0.07	0.18	0.07
Square	C	0.15	0.10	0.15	0.10
Rectangular	S	0.09	0.04	0.28	0.09
Rectangular	C	0.08	0.05	0.25	0.12

Note: Similar to the previous table, the rectangular plates have 2:1 side ratio.

is 0.42, consistent with coefficient φ in Equation 12.11. There is a discrepancy between both solutions. The reader is encouraged to perform two calculations and use a larger answer.

The most serious underestimate of bending, calculated according to Equation 12.17b, seems to occur for rectangular plates. A higher estimate may be obtained by using the flexural wave concept from Chapter 10. For pressure loading, the plate edges are the source of such a wave spreading over the plate area with time. The distance traveled by the wave is l_1 for a supported edge and l_2 for the clamped edge, both quantities being defined by Equations 10.24a and b. When beam stiffness is replaced by plate stiffness, the equations become

$$l_1 = 1.496\left(\frac{D}{m}\right)^{1/4}\sqrt{t} \tag{12.19a}$$

and

$$l_2 = 1.887\left(\frac{D}{m}\right)^{1/4}\sqrt{t} \tag{12.19b}$$

Here, we set $t = t_0$ or pulse duration. Let us define an average reaction per edge of a rectangular plate as

$$q = \frac{pab}{2(a+b)} \tag{12.20}$$

The peak bending for a pulse lasting t_0 can then be expected to have the form of

$$\mathcal{M}_1 = \beta_1 q l_1 \tag{12.21a}$$

and

$$\mathcal{M}_2 = \beta_2 q l_2 \tag{12.21b}$$

for supported and clamped edges, respectively. With the help of FEA solutions, the coefficients are estimated by this author to be $\beta_1 = 0.81$ and $\beta_2 = 0.86$, respectively. This approach, like the previous one, is applicable for a load duration not exceeding $\tau/4$.

POINT LOADS, ELASTIC RANGE

Transient Plate Response to a Point Load

When a dynamic point load is applied laterally to a plate, the deflection gradually spreads over time until the entire plate surface is affected. This section presents an approach based on a flexural wave concept, similar to what was done for beams. At the same time, the approach illustrates the basic mechanism of spreading deflections.

A sudden, large point loading is, in many practical cases, associated with damage, or at least a local yielding and visible deformation of a plate. This is one of the reasons why the attention of researches has largely been focused on inelastic analyses using the RPP material model. The other obvious reason is that assuming a material model with a constant flow strength makes the investigation much simpler than the conventional methods of handling the elastic problem.

The concept of a prime flexural wave, introduced in Chapter 10, will now be extended to large radius plates subjected to flexure. In Figure 12.4 such flex waves have the lengths of l_1 (semi-infinite plate, edge or ring loading) and l_2 (infinite-radius plate, centre loading), respectively.

As before, we therefore conclude that if the flex waves so defined are indeed the predominant forms of early motion, the time needed by the wave to traverse the radial length of the plate should be close to one-quarter of the natural period of lateral vibrations. Such quarter-periods are given by Cases 2.40 and 2.33:

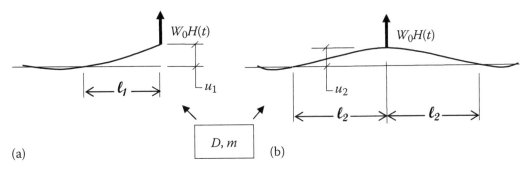

(a)   (b)

FIGURE 12.4 Flex waves in a semi-infinite plate, edge loading (a) and in an infinite-radius plate, center loading (b) resulting from application of a step load.

$$\frac{\tau}{4} = 0.1208 R^2 \sqrt{\frac{m}{D}} \qquad (12.22a)$$

and

$$\frac{\tau}{4} = 0.1537 R^2 \sqrt{\frac{m}{D}} \qquad (12.22b)$$

respectively, for plates (a) and (b) in Figure 12.4, where m is mass per unit area, D is plate stiffness, and h is plate thickness. The first one is an edge-loaded annular plate with $r_0/R = 0.5$ and the second is a center-loaded clamped plate. (One should note that the coefficient in Equation 12.22a is valid only for $r_0/R = 0.5$.)

Assuming that the wave uniformly sweeps the radius of the plate, one can generalize the above relations as follows: The quarter-period $\tau/4$ is replaced by the elapsed time t and the corresponding distance to be traversed ($R/2$ for the first plate and R for the other) by the flex wave length l, resulting in the following expressions:

$$l_1 = 1.4386 \left(\frac{D}{m} \right)^{1/4} \sqrt{t} \qquad (12.23a)$$

and

$$l_2 = 2.5507 \left(\frac{D}{m} \right)^{1/4} \sqrt{t} \qquad (12.23b)$$

Once the distance swept by the wave at a selected time point is known, the corresponding deflections may be calculated by static formulas:

$$u_1 = f(l_1) \frac{W_0 l_1^2}{D} \qquad (12.24a)$$

and

$$u_2 = \frac{W_0 l_2^2}{16 \pi D} \qquad (12.24b)$$

where Equation 12.24b comes from Table 12.2 and represents a plate of radius l_2, clamped at the edge.* For the configuration in Figure 12.5 described by Equation 12.24a, the relation is not so simple: As the wave progresses, the inner radius of a deflected area grows smaller, changing the proportions of the temporary configuration, which in turn changes the deflection

---

* For nondispersive waves, like one spreading along an axial bar, calculating displacements as if they were caused by a static load applied to a deformed portion of the structural member gives exact results. (It can be better visualized using a shear beam concept, rather than an axial bar.) For dispersive cases, like here, acting in the same manner is merely an approximation. (Refer to "Imposition of end disturbance" of Chapter 5 and "Transient beam response due to a suddenly applied point load" of Chapter 10.)

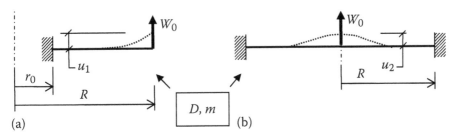

FIGURE 12.5 The same deflected patterns as in Figure 12.4 applied to finite-radius plates: annular plate with edge loading (a) and clamped plate loaded at center (b).

coefficient. The general deflection formula for the annular plate is quoted in Roark [75] in its full form. To avoid complexity, it is more practical to keep it in tabular form as in Young and Budynas [118]. Simplified equations, using $\alpha = r_0/R$, can also be developed:

$$u_1 = (0.0887\alpha^2 - 0.1404\alpha + 0.0562)\frac{W_0 R^2}{D} \quad \text{for } \alpha \le 0.6 \qquad (12.25a)$$

$$u_1 = \frac{1}{3\pi}\frac{(1-\alpha)^3}{1+\alpha}\frac{W_0 R^2}{D} \quad \text{for } \alpha > 0.6 \qquad (12.25b)$$

For a particular case of $r_0/R = 0.5$, one has $f(l_1) = 1/124.5$. The above deflection estimates are derived ignoring the limited plate size, as far as the spreading of the flex wave is concerned. While this wave may represent the bulk of the bending deflection, there are flexural components traveling faster than this primary wave. The reflection of those from a clamped boundary causes interference with the prime wave and loss of predictability in the boundary region. For this reason, one expects the accuracy of Equation 12.23b, for example, to degrade for l_2 approaching $R/2$.

The above theory is based on flexure only and it ignores the contribution of shear. This may lead to a serious underestimate of deflection in early stages, when the portion of a deflected radius is quite small. A simple, approximate way of adjusting the shear component is to assume that the shear-deformed radial length is the same as that due to bending and then calculate the correction on a static basis. For a shear plate with outer radius R, constrained against lateral deflections, rigid insert of radius r_0 at the center the shear deflection is given by Equation 12.15 with R replaced by l_2 from Figure 12.4.

Also, note that loading and support may be reversed, that is, the rigid insert may be held while the outer edge is subjected to a ring load with a resultant W and the deflection is still expressed by Equation 12.24a.

Global Response

As in the case of beams, the interest is limited to short, rectangular pulses of force W_0 acting over an interval of t_0. In the previous section, the displacement was determined for a period immediately after the load was applied. While that development has some limited application, the peak global response is always a major issue. It often takes place between $t = \tau/4$ and $t = \tau/2$, where τ is the fundamental period of vibrations.

The peak response is most readily found by using the SDOF approach, presented in "Response of plates to pressure shock loading in elastic range." The only difference is that now u_{st} is the static deflection attained due to the point load W_0 rather than the pressure p_0. The "point" loading was qualified before; it is actually a pressure over a small area with a radius comparable to the plate thickness. The SDOF approach implies the maximum dynamic factor, that is, $DF(u) = u_d/u_{st}$ does not exceed 2.0, which is usually sufficiently accurate for deflections.

The peak bending moment is often underestimated when the SDOF method is used. This is especially true for t_0 much shorter than the natural period and is more serious for rectangular rather than circular plates. The alternative estimate of bending can be obtained by first calculating the length of a flex wave emanating from the loaded area, by means of Equation 12.23b, setting $t = t_0$. Next, an average static force q per unit length of edge of the loaded area is found to be

$$q = W_0/(4a) \qquad (12.26)$$

if the load is applied to a square area with a side-length a. The moment estimate is

$$M_2 = \beta q l_2 \qquad (12.27)$$

where l_2 is defined by Equation 12.23b. The FEA studies by this author indicate that for a square, supported plate $\beta \approx 0.24$ while for a rectangular, clamped plate, $\beta \approx 0.20$. (For t_0 approaching a quarter-period, the estimate from Equation 12.27 becomes excessive.) The peak bending moment is expected to be the larger value of the two estimates given by Equations 12.17b and 12.27.

With regard to shear, quantification is simpler than for pressure loading. For a centrally loaded plate the shear magnitude diminishes, when going away from a zone of application. This is true both for statics and dynamics. The peak shear response appears not to exceed 1.4× the static value, as determined by Equation 12.15.

Mass–Plate Impact

The basic method for the mass-plate impact problem is a simple engineering approach where a plate is treated as a massless, flexible member. Its elementary form was already presented in Cases 3.27 and 3.28, where applying the energy balance equation gives the peak impact force. The next step in making the problem statement a little more realistic is to introduce a contact flexibility $1/k$, which is usually not negligible. (Even if the striker itself is very hard, there is a local deformation of the plate section, which is then represented by a spring with a stiffness k.) The procedure is exactly the same as that used for beams in the section "The lumped-parameter methods for mass-beam impact problem" of Chapter 10 and visualized in Figure 10.7. The peak contact (impact) force is

$$R_m = v_0\sqrt{k^* M} \qquad (12.28)$$

where
  v_0 is the impact velocity
  M is the mass of a striker
  k^* is the effective stiffness of two springs in series, one representing the plate and the other being the contact stiffness

The accuracy of this method is quite limited, especially with regard to the peak contact force.

A more accurate approach, developed by this author [102] will be outlined here. Typically, such an impact is characterized by a contact force, which rapidly decreases with time. At the first instant the shearing mode of deformation dominates, which suggests the use of a shear plate model presented earlier. The concept of a shear wave spreading from the impact point is implemented in a similar way as was done for beams, except that an average shear wave speed is used now:

$$c_s = \frac{2}{\pi} \sqrt{\frac{5Gh_s}{m}} \qquad (12.29)$$

where $h_s = h/1.2$ is the shear-effective thickness of a solid plate. The radius swept by the wave is

$$R(t) = r_0 + c_s t \qquad (12.30)$$

where r_0 is the radius of a rigid insert at the center of the plate. The speed of the lateral displacement at center, under a constant load W_0, is calculated with the help of Equation 12.15:

$$v_s = \frac{du_s}{dt} = \frac{Bc_s}{R(t)} \quad \text{with } B = \frac{W_0}{2\pi h_s G} \qquad (12.31)$$

The equivalent system for mass-plate impact is shown in Case 12.9, along with the formulas needed for quantification. The system consists of a mass M, a contact spring k, and a damper C representing the plate. The lateral velocity of the plate center described by Equation 12.31 decreases with time. Its average value is

$$v_{sav} = \frac{W}{4\pi h_s G} \frac{c_s}{r_0} \qquad (12.32)$$

The problem is now simplified to treating this velocity–force relationship as representative of the entire duration of impact. It can be put in a form of

$$v_{sav} = \frac{W_0}{C} \quad \text{with } C = 2\pi^2 r_0 \left(\frac{Gh_s m}{5} \right)^{1/2} \qquad (12.33)$$

Equation 12.33 simply tells us that the center of a plate, when pressed, moves like an end of a viscous damper with a constant C. This is the reason for placing a damper at the base of the contact spring. It also deserves mention that the same reasoning was used in finding the time-dependent plate deflection as was done before for beams: When the shear wave reaches the radius $R(t)$, its deflection can be found from statics of a shear plate of radius $R(t)$.

Once the damper constant C is available, the analysis becomes equivalent to searching for a response of a Maxwell element, already described in Chapter 3. The following procedure is employed for the sake of mathematical simplicity.

It is known from elementary dynamics that the peak contact force of a mass–spring system impacting a rigid wall is $P_{ms} = v_0 \sqrt{kM}$. This happens when the damper is rigid.

If the spring is assumed rigid, the force is $P_{md} = Cv_0$. The largest actual impact force P_m can now be estimated from

$$\frac{1}{P_m} \approx \frac{1}{P_{ms}} + \frac{1}{P_{md}} \tag{12.34}$$

which is a slightly modified equation for two elements in a series connection. The resulting P_m is always smaller than either of the two components. (The equation given in Case 12.9 marginally differs from the above, for the sake of improved accuracy.)

When P_m is attained, the spring is fully compressed and the associated speed is

$$v_e = P_m/C \tag{12.35}$$

The associated instant will be referred to as a *reversal point*, a concept, which is helpful in determining the rebound velocity. One should keep in mind that the analysis described makes sense only when the impact is relatively hard, so that the peak contact force may be attained before significant bending deflections develop. (See the section "A mass-beam impact approach involving transient beam deformation" of Chapter 10 for more details, as they essentially apply to plates as well.)

An alternative way of determining the impact parameters is to use the collision method already mentioned in Chapter 10. However, if only the plate moment response is of interest, the basic method described at the beginning of this section should be sufficient in most cases.

Limit Values of the Internal Forces and the Onset of Yielding

The bending moment capacity of a plate is related to the capacity of a rectangular beam of a unit width, such a fictitious beam being a part of a cross section of a plate. As the bending moment grows, a state is attained when the outer fiber stress reaches σ_0, the onset of yield. The associated bending moment is $\mathcal{M}_y$. With a further load growth a fully developed plastic stress distribution is reached and the moment attains its maximum capacity of $\mathcal{M}_0$. The Equations 10.5 and 10.6a through c hold here as well:

$$\mathcal{M}_y = I\sigma_0/c \tag{12.36a}$$

and

$$\mathcal{M}_0 = Z\sigma_0 \tag{12.36b}$$

For plates, where the sections are solid, homogeneous, of a unit width and thickness h, $Z = h^2/4I = h^3/12$, $c = h/2$ and the above simplifies to

$$\mathcal{M}_y = \frac{1}{6}h^2\sigma_0 \tag{12.37a}$$

and

$$\mathcal{M}_0 = \frac{1}{4}h^2\sigma_0 \tag{12.37b}$$

The yield shear stress, or flow strength in shear, $\tau_0 = \sigma_0/\sqrt{3}$, is consistent with the Huber–Mises theory. The corresponding onset of yield in shear and the shear capacity are, respectively

$$Q_y = h\tau_0/\alpha \tag{12.38a}$$

$$Q_0 = h_{ef}\tau_0 \tag{12.38b}$$

where h_{ef} is the section area of the unit width, effective for the limiting condition while h is the total area of this cross section. The ratio of peak elastic shear stress to the average shear stress, along thickness, designated by α can be found in Table 10.1. For a rectangle, $\alpha = 3/2$ and $h_{ef} \approx h$. The above plastic capacities are written for a rigid-plastic (RP) material.

The external load p_y corresponding to the static onset of yielding can be found by equating the value of the moment from Equation 12.37a with the general moment expressions, Equation 12.9:

$$p_y = \frac{\sigma_0}{6\beta}\left(\frac{h}{a}\right)^2 \tag{12.39a}$$

and

$$p_y = \frac{\sigma_0}{6\beta}\left(\frac{h}{R}\right)^2 \tag{12.39b}$$

$$W_y = \frac{h^2\sigma_0}{6\beta} \tag{12.40}$$

Equation 12.40 comes from conducting a similar operation for point loads. An analogous procedure applies for shear, where combining Equation 12.38a with Equations 12.11 and 12.12, while using $\alpha = 3/2$ one gets

$$p_{ys} = \frac{2\tau_0}{3\varphi}\left(\frac{h}{a}\right) \tag{12.41a}$$

and

$$p_{ys} = \frac{2\tau_0}{3\varphi}\left(\frac{h}{R}\right) \tag{12.41b}$$

$$W_{ys} = \frac{2\tau_0 h a}{3\varphi} \tag{12.42a}$$

and

$$W_{ys} = \frac{2\tau_0 h R}{3\varphi} \tag{12.42b}$$

for rectangular and circular plates, respectively. (φ comes from Table 12.1)

INELASTIC RANGE

Static Collapse Load under Pressure Loading

The reasoning applied to derive the collapse load is purely static, keeping in mind that there is no deformation until the load attains that limiting value. When a collapse begins, yielding is postulated to take place along dashed lines, as in Figure 12.2. At this instant, the external pressure applied to a plate fragment is balanced by the support reactions as well as by fully developed internal forces. (This is for simply supported plates. When the edges are clamped, the yield lines also appear along the edges.)

The simplest case is that of a circular plate, as in Figure 12.6a and b, where a radial segment, with bending moments along radial sections is depicted. For a small central angle α, the wedge is approximated by a triangle. The resultant force is $\alpha R^2 w_b/2$ and this force is located at $R/3$ from the edge. The moment of loading about the edge tangent is balanced by the internal moment, also projected onto that tangent:

$$\frac{1}{2}\alpha R^2 p_b\left(\frac{R}{3}\right) = \mathcal{M}_0 \alpha R \tag{12.43a}$$

or

$$p_b = \frac{6\mathcal{M}_0}{R^2} \tag{12.43b}$$

For a clamped edge, there is the same internal moment along the circumference, resisting pressure as well. This doubles the collapse load p_b. For the remaining shapes the procedure is similar.

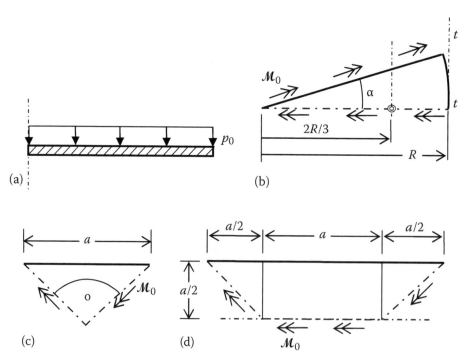

FIGURE 12.6 Circular plate section, elevation (a); plan view of one-half of a circular sector (b); a quarter of a square plate (c); and a fragment of a rectangular plate (d).

For the square shape, which fails along diagonals, one-quarter of a plate is shown in Figure 12.6c along with the internal bending moment. The results for all plates can be presented as

$$p_b = \xi \mathcal{M}_0 / a^2 \tag{12.44a}$$

$$p_b = \xi \mathcal{M}_0 / R^2 \tag{12.44b}$$

for rectangular and circular plates, respectively. The results obtained from such simple schemes should be related to the accurate results provided by Jones [43]. The coefficients ξ are listed in Table 12.1 and, in case of discrepancy, the results of the accurate method are given in brackets. With regard to an infinitely long plate, the procedure is to consider a unit-width beam spanning the supports. The limiting loads are the same as for the corresponding beams.

The determination of collapse loads in shear mode is made easier by assuming the same distribution of internal shears along the edge in both elastic and inelastic ranges. Equating the applied shear load Q from Equations 12.11a and b with the shear capacity from Equation 12.38b results in the following loads being associated with shear collapse by sliding:

$$p_s = \frac{\tau_0}{\varphi}\left(\frac{h}{a}\right) \tag{12.45a}$$

and

$$p_s = \frac{\tau_0}{\varphi}\left(\frac{h}{R}\right) \tag{12.45b}$$

for rectangular and circular plates, respectively.

The static collapse load can also be derived for the case of loading applied along a concentric ring, as shown in Cases 12.4 through 12.7. The same reasoning is applied to point loads, imposed at the plate center. The collapse load is

$$W_b = \xi_m \mathcal{M}_0 \tag{12.46}$$

where ξ_m is found in Table 12.2.

Estimates of Dynamic Deflection under Pressure

The dynamic load of interest has a form of a rectangular pulse; a constant magnitude p_m with a duration of t_0. The impulse applied to the plate is $S = Ap_m t_0$, where A is the surface area. If t_0 is very short, which for a prescribed impulse makes p_m large, this type of loading is called *impulsive*. (For a practical purpose of distinguishing impulsive loading one may say that p_m must be at least 10 times as large as the static capacity load.) In the following, impulsive loading is assumed to be in effect (Szuladziński [99]).

The first case is shear deformation or translational motion. The impulse applied to the plate with a mass Am can be equated with the momentum gained to find the initial velocity v_0:

$$Ap_m t_0 = Amv_0 \tag{12.47a}$$

or

$$v_0 = p_m t_0 / m \qquad (12.47b)$$

The maximum deflection is estimated by comparing the initial kinetic energy with the strain energy of the collapsing plate:

$$E_k = \Pi \qquad (12.48a)$$

or

$$(Am)v_0^2/2 = W_s u_s \qquad (12.48b)$$

where $W_s = A p_s$ is the resultant collapsing force. Substituting v_0 from Equation 12.47b into Equation 12.48b obtain the maximum sliding deflection as

$$u_s = \frac{Amv_0^2}{2W_s} \qquad (12.49)$$

After substitutions are made, one obtains

$$u_s = \frac{\varphi(Rp_m t_0)^2}{2m\tau_0 Rh} \qquad (12.50a)$$

$$u_s = \frac{\varphi(ap_m t_0)^2}{2m\tau_0 ah} \qquad (12.50b)$$

for a circular and any rectangular plate, respectively. (Note that the fact that motion starts only after the plate limit load of w_s is exceeded was neglected in the above derivation due to the assumption that $p_0 \gg p_s$.)

A similar approach is used for the case of joint rotation simulating bending motion. The kinetic energy of the triangular element in Figure 12.6b, which has the area $A = \alpha R^2/2$, is taken as*

$$E_k = \frac{1}{2}(Am)\left(\frac{p_0 t_0}{m}\right)^2 = \frac{1}{4}\alpha R^2 \frac{(p_0 t_0)^2}{m} \qquad (12.51)$$

Equating this initial kinetic energy to the strain energy of plastic deformation gives the maximum deflection:

$$\frac{1}{4}\alpha R^2 \frac{(p_0 t_0)^2}{m} = (M_0 R\alpha)\theta \qquad (12.52)$$

---

* Regardless of whether fixed or supported at the outer edge.

where, on the right is the strain energy absorbed in angular deformation equal to the work of the internal moment (projected on t–t axis) when the sector rotates by angle θ. This gives the maximum rotation θ and the associated tip translation $u_b = b\theta$ as

$$\theta = \frac{Rp_0^2 t_0^2}{4\mathcal{M}_0 m} \tag{12.53a}$$

or

$$u_b = \frac{(Rp_0 t_0)^2}{4\mathcal{M}_0 m} \tag{12.53b}$$

In general one can write

$$\theta_m = \frac{u_m}{R} \tag{12.54a}$$

and

$$u_m = \frac{\vartheta(Rp_0 t_0)^2}{\mathcal{M}_0 m} \tag{12.54b}$$

where R is replaced by a for noncircular plates. The remaining geometries are easier, as only finite-size parts of plates need to be considered. For the square plate the motion of one of the component triangles about the edge is handled in the same manner, as suggested by Figure 12.6c. For infinitely long plates, the procedure is to consider a unit-width plate spanning the supports. The resulting coefficients ϑ are displayed in Table 12.1. (Please note that this simplified method predicts deflections to be reduced by one-half, when edge fixity changes from supported to clamped.)

Also note that in what might be called a "traditional" method the approach is, in effect, to treat the plate as rigid from the outset. This results in the energy input of only 2/3 of what was found above. Consequently, if that method was used, all ϑ entries in Table 12.1 would be only 2/3 of our values.

The shear force (equal to the reaction per unit length of supporting edge) can be found from Table 12.5. This should be viewed as an underestimate due to reasons mentioned before, when discussing the analogous Table 12.4 for the elastic range.

TABLE 12.5

Shear Force Coefficients, Pressure Load p (Plastic Range)

Shape	Edge	Short Side, c	Short Side, d	Long Side, c	Long Side, d
Square	S	0.16	0.09	0.16	0.09
Square	C	0.15	0.10	0.15	0.10
Rectangular, 2:1	S	0.08	0.05	0.27	0.11
Rectangular, 2:1	C	0.08	0.04	0.27	0.11

Source: Szuladziński G., Transient response of circular, elastic plates to point loads, in press.

There is only a very limited treatment of plates of RP material acted upon by point loads included here. The reader should refer to Cases 12.11 and 12.12. The dynamic deflections can be found using methods given in previous chapters.

CYLINDRICAL AND SPHERICAL SHELLS

Long, Pressurized Cylindrical Shell with No End Constraints

The shell can be divided into *circumferential rings*, each defined by two planes perpendicular to the axis of symmetry and separated by a unit distance. It is enough to describe the deformation of such a single ring to have the description of the shell response to the applied pressure. Some results for such a shell were given by Case 3.29 before, except that now only a small thickness, $h \ll R$ is of interest.

The stress in each ring, also known as the *hoop stress*, is $\sigma = pR/h$. (This can easily be verified by cutting the ring in half and writing the equilibrium equation.) When pressure grows to the level p_y, associated with yield stress F_y, the onset of yield takes place:

$$p_y = F_y h / R \tag{12.55}$$

For an RP material, which cannot sustain a higher stress than the flow strength σ_0, this corresponds to collapse pressure p_c:

$$p_c = \sigma_0 h / R \tag{12.56}$$

A similar situation exists with respect to a pressurized, thin spherical shell. When it is cut in half and the equation of equilibrium is written, one obtains the membrane stress of $\sigma = pR/(2h)$. When this reaches σ_0 for an RP material, the related collapse pressure is

$$p_c = \frac{2h\sigma_0}{R} \tag{12.57}$$

Cylindrical Shell with Axisymmetric Load Varying along the Axis

Apart from circumferential rings described before, a shell may be viewed as a set of *generator beams*. One such beam is a longitudinal strip cut out of the shell by two planes passing through the axis of symmetry. Such an element, one unit wide along circumference, will serve to describe deflections of the shell as a function of a longitudinal coordinate, as illustrated in Figure 12.7.

A cylindrical shell of length L is shown in Figure 12.8 with two load types applied: a ring load q in a single plane and a varying pressure $p(x)$. If pressure is constant, then each ring is in the state of hoop tension, $\sigma = pR/h$. Consequently, the radial displacement of a point on the circumference is $u = \sigma R/E$, which means that the radial stiffness of the ring is

$$k_f = \frac{p}{u} = \frac{Eh}{R^2} \tag{12.58}$$

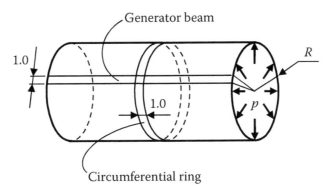

FIGURE 12.7   A cylindrical shell with idealized components.

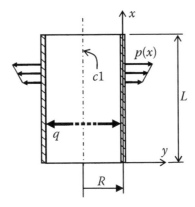

FIGURE 12.8   A thin, cylindrical shell with axisymmetric pressure load p (N/m²) and a ring load q (N/m).

(Case 2.45c agrees with the above for thin shells, except that the factor of $1-v^2$ is missing, because the load can be nonuniform along a generator.) The shell is now viewed as an assembly of generator beams attached to the row of the circumferential rings, the latter serving as the elastic foundation for the former. A bending stiffness of a beam of unit width is $Eh^3/(12(1-v^2))$, where the factor involving Poisson's ratio v reflects continuity of the beams in circumferential direction. The λ coefficient for beams on an elastic foundation, as used in Equations 10.84 and 10.85, becomes

$$\lambda^4 = \frac{k_f}{4EI} = \frac{3(1-v^2)}{R^2 h^2} \quad \text{or} \quad \lambda = \frac{1.285}{\sqrt{Rh}} \quad \text{for } v = 0.3 \qquad (12.59)$$

Consider now the ring load of intensity q (N/m) placed sufficiently far from any end so that the shell will act, for all practical purposes, as if it were infinitely long. The following refers to the generator beam of unit width, placed on the elastic foundation. The radial deflection u and the bending moment $\mathcal{M}$ in the plane of loading, per Equations 10.85a and c, are respectively:

$$u = \frac{q\lambda}{2k_f} = \frac{q\lambda}{2}\frac{R^2}{Eh} \qquad (12.60a)$$

and

$$M = \frac{q}{4\lambda} \qquad (12.60b)$$

where the substitution $q = 2W$ was made. The uniform axial stress in the ring is caused by the effective radial load p calculated from Equation 12.58:

$$p = \frac{Eh}{R^2} u = \frac{q\lambda}{2} \qquad (12.61)$$

For the unit-width ring $\sigma_h = pR/h$, therefore:

$$\sigma_h = \frac{qR\lambda}{2h} \qquad (12.62)$$

The peak bending stress in a generator beam is

$$\sigma_1 = \frac{6\mathcal{M}}{h^2} = \frac{3q}{2\lambda h^2} \qquad (12.63)$$

As long as the shell is relatively thin, the hoop stress is smaller than the beam bending or axial stress. If the load pushes outwards, the ring is loaded in tension. The beam, on the other hand, has tension on the outside and compression on the inside. According to Tresca's rule, the magnitude of the effective stress is largest on the inside, where the components act in opposite directions:

$$\sigma_e = \sigma_1 + \sigma_h = \frac{q}{2\lambda}\left(\frac{3}{h^2} + \frac{R\lambda^2}{h}\right) \qquad (12.64)$$

on the basis of Equations 12.62 and 12.63. Noting that from Equation 12.59

$$\frac{R\lambda^2}{h} = \frac{\sqrt{3(1-v^2)}}{h^2} = \frac{1.6523}{h^2} \quad \text{for } v = 0.3 \qquad (12.65)$$

and using Equation 12.64, one can find the load q_y, which makes $\sigma_e = F_y$ causing the onset of yield:

$$q_y = 0.43\lambda h^2 F_y \quad (\text{for } v = 0.3) \qquad (12.66)$$

Fixed-Ended Cylindrical Shell under Pressure Loading

Such as shell is shown in Figure 12.9. If the shell is sufficiently long, hoop stress dominates in the midplane, $x = L/2$. In a circumferential ring, the hoop stress is pR/h. The onset of yield is attained when the hoop stress reaches F_y, which means

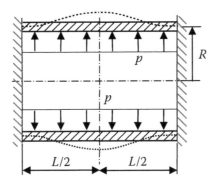

FIGURE 12.9 Cylindrical shell with ends fixed. Deflected shape marked by dotted line.

$$p_{y1} = \frac{hF_y}{R} \tag{12.67}$$

Another yield criterion is obtained when the end bending moment of a generator beam, $M_e = p/(2\lambda^2)$ according to Roark [75], is big enough for the stress in extreme fibers to reach F_y. This gives

$$\sigma_b = \frac{6M_e}{h^2} = \frac{6}{h^2}\frac{p}{2\lambda^2} = 1.817\frac{pR}{h} \quad \text{(for } v = 0.3) \tag{12.68}$$

$$\text{For } \sigma_b = F_y; \quad p_{y2} = 0.55\frac{hF_y}{R} \tag{12.69}$$

There also is another load component, namely the thrust load, the axial resultant of pressure, equal to $\pi R^2 p$ which results in axial stress of $pR/(2h)$ in beams.* When this is added with the bending stress from Equation 12.68, the resultant is

$$\sigma = 1.817\frac{pR}{h} + 0.5\frac{pR}{h} = 2.317\frac{pR}{h} \tag{12.70}$$

$$\text{For } \sigma = F_y; \quad p_{y3} = 0.432\frac{hF_y}{R} \tag{12.71}$$

The last criterion gives the earliest onset of yielding, as pressure grows. In absence of stress caused by thrust load, yielding will initiate at $p = p_{y2}$.

---

* Whether the shell experiences the thrust load or not, depends on the design arrangement. The latter is possible, for example, in Figure 12.9, if one of the ends is free to move axially.

Static Collapse Loads for Cylindrical Shells

This discussion will be limited to shells made of RP material with the flow stress of σ_0. The material response under simultaneous action of axial force P and bending moment $\mathcal{M}$, presented in Chapter 11, is of some relevance, although it is not directly applicable. While reasoning still revolves around a generator beam, the action of the hoop stress, rather than the axial load, is of a major concern. The hoop stress resultant acting on a unit length of the beam* is designated by P_θ. The limiting states of an element of a cylindrical shell are investigated in detail by Kaliszky [45]. The interaction diagram between the beam moment $\mathcal{M}_x$ and hoop force P_θ, consistent with Tresca's yield criterion, is shown in Figure 12.10. (This time only the first quadrant is shown, all the quadrants being mirror reflections.) There is a solid line between the vertical axis and point A, while the curved portion is described by

$$\frac{2P_\theta}{P_0}\left(\frac{P_\theta}{P_0}-1\right)+\frac{\mathcal{M}_x}{\mathcal{M}_0}=0 \tag{12.72}$$

where $\mathcal{M}_0$ and P_0 are the moment and the force capacities of a rectangular section $1 \times h$ with a stress distribution relevant for an RP material. ($P_0 = h\sigma_0$ relates to the collapse pressure p_c.) For practical applications, Equation 12.72 is found to be awkward. The curved part is often replaced by a straight dashed line on the inside of the curve. A bolder approximation is to use the vertical and a horizontal dashed line intersecting at point B. While this *rectangular yield criterion* circumscribing the true line is nonconservative, it is also very simple to use.

The RP collapse of a pressurized cylindrical shell is illustrated in Figure 12.11, using a single generator beam for this purpose. The figure implies that a *plastic hinge circle* forms at the center section and at each end of the shell. One might say that there is no essential difference between such a beam, as seen in the axial section here and what was discussed in Chapter 10, as the hinge moment is a source of resistance as well. The additional resistance here comes from the hoop action, namely p_c.

When force P_θ reaches P_0 in a circumferential ring, it is equivalent to the external resisting pressure p_c. Unlike in a beam on elastic foundation, where the reacting pressure is proportional

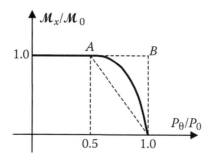

FIGURE 12.10 The solid line is the actual criterion and the dashed lines are some of the approximations.

---

* The axial thrust component is not usually included, perhaps because it may not always be present.

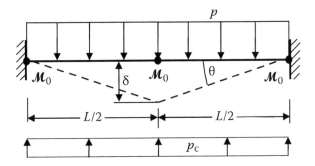

FIGURE 12.11 A single generator beam collapsing under pressure p.

to deflection by virtue of Equation 12.58, the reaction, from this point on, is constant and equal to p_c. The collapse pressure will be found by equating the work of p with the strain energy stored in the shell and the square yield criterion will be used.

Noting that the kink angle at the center is 2θ, the energy absorbed in the joints is $\Pi_b = 4\mathcal{M}_0\theta$. The strain energy associated with the overcoming of p_c (stretching of hoop rings with a constant stress σ_0) is $\Pi_a = P_0L\delta/(2R)$. The work of pressure p is $\mathcal{L} = pL\delta/2$.* The work–energy balance

$$\Pi_b + \Pi_a = \mathcal{L} \tag{12.73}$$

gives the following static collapse pressure, after replacing θ by $2\delta/L$ and writing $P_0 = h\sigma_0$ and $\mathcal{M}_0 = h^2\sigma_0/4$:

$$p_c = \frac{h\sigma_0}{R}\left(1 + \frac{4hR}{L^2}\right) \tag{12.74}$$

For a shell under a ring load q the procedure is the same, except that $\mathcal{L} = q\delta$. Equating work and energy gives

$$q_c = P_0\left(\frac{4\mathcal{M}_0}{P_0l} + \frac{l}{R}\right) \tag{12.75}$$

But this is a function of unknown l. A differentiation of q_c with respect to l shows that the former attains a minimum for

$$l = \sqrt{Rh} \tag{12.76}$$

which results in

$$q_c = 2P_0\sqrt{\frac{h}{R}} \tag{12.77}$$

---

* For the last two it is simply the area under the deflected line multiplied by p.

Note that at the center of the beam there is a full hoop force P_0 along with the full longitudinal moment $\mathcal{M}_0$, while only one-half of the former is permitted by Figure 12.10. But bending changes sign between center and the base, which means that the magnitude of the moment is smaller over much of that distance. Still, what is given by Equation 12.77 is an upper bound according to Jones [43], while the lower bound is 0.707 of the above.

In a beam with clamped ends, the ratio of the center bending moment to that at the end is 0.5. The same ratio is approached for a beam on an elastic foundation becoming very short, as can be readily confirmed by formulas given by Hetényi [33]. The same applies, of course, to a generator of a beam in Figure 12.11. When L/R grows, the ratio of moments decreases. For $\lambda L < \pi$ or $L^2 < 6Rh$ the moment ratio is more than 0.435, still reasonably close to 0.5. In order that the beam may plastically collapse in the indicated manner, the center bending moment magnitude should be much less, according to the criterion in Figure 12.10.* For this reason the condition $L^2 < 6Rh$ may be used to make certain that the static collapse in Figure 12.11 may take place. The dynamic collapse, however, is of a more complex nature. Even if the shell is longer than according to the above limit, the collapse will initiate at the ends, which then emit traveling plastic joints. Those meet at the center, result in the same final pattern as in Figure 12.11. This development was presented by Jones [43], according to whom the upper limit of length for the pattern to develop is $L^2 = 12Rh$.

Shells Stiffened by Equidistant Rings

This type of shells can be encountered in various industrial applications. This structural arrangement is illustrated in Figure 12.12 along with an approximate deflected line caused by pressure application. Due to symmetry, each ring enforces a zero slope condition on the adjacent shell. If the rings are infinitely stiff, then each segment is in a fixed-end condition, as in Figure 12.9. However, each ring is flexible and may have the section area A made of the same material as the shell. The end bending moment is then, according to Roark [75]:

$$\mathcal{M}_e = \frac{p}{2\lambda^2} \frac{A}{A + 2h/\lambda} \equiv \frac{p}{2\lambda^2} f(A) \qquad (12.78)$$

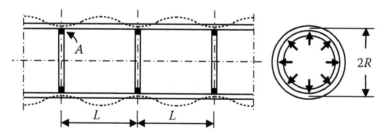

FIGURE 12.12   Cylindrical shell with stiffening rings.

---

* At the center there is a full hoop load P_0, while there is no hoop load at the fixed ends.

where $f(A)$ is the moment reduction factor due to ring flexibility. (The ring is assumed to be narrow.) The collapse pressure p_{cr} of this bending mode will be defined as the pressure needed to induce plastic moment $\mathcal{M}_0 = \sigma_0 h^2/4$ at the ring location. Equating this moment with $\mathcal{M}_e$ above and using the definition of λ from Equation 12.59 gives

$$p_{cr} = \frac{1}{2}\lambda^2 h^2 \left(1 + \frac{2h}{A\lambda}\right) = 0.826 \frac{\sigma_0 h}{R}\left(1 + \frac{2h}{A\lambda}\right) \tag{12.79}$$

where the term $p_c = \sigma_0 h/R$ is the collapse pressure for a plain, long shell. The condition $p_{cr} < p_c$, which means that yielding of beams is to precede the hoop yielding, gives

$$A > 9.47h/\lambda \tag{12.80}$$

The above derivation assumes that the rings are far enough so that there is no influence on their mutual deflections as well as on those at midpoints between rings. While those conditions are themselves only approximately fulfilled, it is expected that the overall approximation will be close. The rings themselves have to be checked for the effects of dynamic loads. The radial distributed load per ring, in the limit, is

$$w_r = 4\mathcal{M}_0 \lambda \text{ (N/m)} \tag{12.81}$$

after we set $\mathcal{M} = \mathcal{M}_0$ in Equation 12.60b. This is written on an elastostatic basis and will vary after the onset of yielding.

Dynamic Response of Shells

When the applied pressure exceeds the resistance, motion begins. For an elastic material the radial motion of a unit-width ring, which can be a unit-length segment of a shell does not differ from that of an oscillator with a mass $M = \rho h$ and stiffness $k = p/u_{st}$, with the details given in Case 12.13.* When the RP material model is involved, there is no motion until $p = p_c$ and for larger pressures the mass is driven by pressure p and resisted by p_c, Case 12.14. Finally, for an elastoplastic response, the motion is elastic until deflection is large enough to attain p_c and then the constant pressure p_c becomes the resisting force, Case 12.15.

The above refers to a cylindrical shell. The procedure is the same for a spherical shell, except simpler, as the length variable is not involved.

Closing Remarks

The effect of a sustained axial load on a shell or a plate is similar to what was presented in Chapter 11 for beam-columns. Tension makes the element stiffer, while compression works in the opposite direction. The knowledge of a critical compressive load is necessary to quantify the effect.

---

* Due to continuity between the adjacent rings, the radial stiffness is larger, by a factor of $1/(1 - \nu^2)$, than that of a single ring.

TABULATION OF CASES

CASE 12.1   CIRCULAR PLATE, CLAMPED, UNDER A SHORT RECTANGULAR PULSE LASTING t_0

(a) Distributed load p_0:

$$u_{st} = \frac{1}{64}\frac{p_0 L^4}{D}; \quad \mathcal{M}_{st} = p_0 R^2/8; \quad Q_{st} = p_0 R/2$$

(Edge moment and shear)

$$u_d = 2u_{st}\sin(\pi t_0/\tau); \quad \mathcal{M}_d = \frac{u_d}{u_{st}}\mathcal{M}_{st}; \quad Q_d = \frac{u_d}{u_{st}}Q_{st}$$

$$Q_d' = \left(0.12\frac{u_d}{u_{st}}+0.042\right)\frac{W_0}{R} \quad \text{(use for } t_0 < 0.25\tau \text{ if } Q_d' \text{ is larger than } Q_d)$$

(b) Point load W_0:

$$u_{st} = \frac{1}{16\pi}\frac{W_0 R^2}{D}; \quad \mathcal{M}_{st} = \frac{1+\nu}{4\pi}\ln\left(\frac{R}{h}\right)W_0; \quad Q_{st} = \frac{W}{2\pi r}$$

(Moment $\mathcal{M}$ at center. Q just outside the loaded point.)

$$u_d = 2u_{st}\sin(\pi t_0/\tau); \quad \mathcal{M}_d = \frac{u_d}{u_{st}}\mathcal{M}_{st}; \quad Q_d = 1.4Q_{st}$$

Note: The "point" load means that the loading is applied over a small circle with a radius equal to the plate thickness h. These peak dynamic response formulas are approximations only. For coefficients related to other shapes and support conditions refer to the text.

CASE 12.2   PEAK TRANSIENT PLATE RESPONSE DUE TO A POINT LOAD APPLIED FAR FROM EDGE

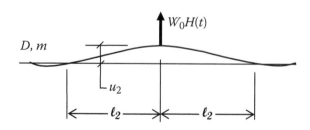

Local response:

$$l_2 = 2.5507 \left(\frac{D}{m}\right)^{1/4} \sqrt{t}; \quad u_2 = \frac{W_0 l_2^2}{16\pi D}$$

CASE 12.3 CIRCULAR PLATE OF RP MATERIAL, CLAMPED EDGES. SHORT PRESSURE PULSE LASTING t_0

Specific Impulse: $s = p_m t_0$

$\mathcal{M}_0 = h^2 \sigma_0 / 4$; moment capacity

$Q_0 = h\tau_0$; shear capacity

$p_b = 12\mathcal{M}_0 / R^2$; static collapse load in bending

$p_s = 2\tau_0 \left(\dfrac{h}{R}\right)$; static collapse load in shear

($p_m > p_b$ or $p_m > p_s$, whichever less)

$v_0 = \dfrac{p_m t_0}{m}$; initial velocity

$u_s = \dfrac{(R p_m t_0)^2}{4m\sigma_{os} R h}$; maximum shear deflection at shear failure

$\theta_m = \dfrac{R s^2}{12 m \mathcal{M}_0}$; maximum rotation angle at bending failure

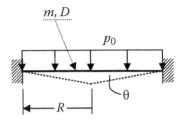

Note: The bending mode of failure is illustrated for $p_b < p_s$. For coefficients related to other support conditions refer to the text [99].

CASE 12.4 CIRCULAR PLATE WITH A HOLE. INNER EDGE SUPPORTED, OUTER FREE (SF)

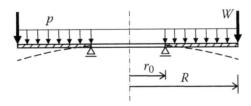

(a) Pressure load p:

$$u_{st} = 0.0826 \frac{p_0 R^4}{D}; \quad \mathcal{M}_{st} = 0.3414 p_0 R^2; \quad Q_{st} = \frac{(R^2 - r_0^2)p}{2r_0}$$

$$p_y = \frac{6\mathcal{M}_0}{(R - r_0)(2R + r_0)}; \quad p_y = \frac{3\mathcal{M}_0}{R^2} \text{ for } r_0 = 0; \text{ static collapse load (RP material)}$$

Short impulse $p_0 t_0$: $u_m = \dfrac{(p_0 t_0)^2}{\mathcal{M}_0 m} \dfrac{R^2 - r_0^2}{4}$

(b) Ring load with resultant W:

$$u_{st} = \frac{1}{16.25} \frac{W_0 R^2}{D}; \quad \mathcal{M}_{st} = 0.2469 W_0; \quad Q_{st} = \frac{W}{2\pi r_0}; \quad W_y = 2\pi \mathcal{M}_0$$

Short impulse $W_0 t_0$: $u_m = \dfrac{(W_0 t_0)^2}{2 M_e W_y}$

$$M_e = \frac{\pi m}{6} r_0 (R - r_0) \left(\frac{3R}{r_0} + 1 \right); \text{ effective edge mass}$$

Note: Coefficients of u_{st} and $\mathcal{M}_{st}$ are for the case $r_0/R = 1/2$ and for linear material only. $\mathcal{M}_{st}$ is for the circumferential moment, which dominates. For other ratios, refer to Young [118]. For a ring load, u_m is only the first approximation [40].

CASE 12.5   CIRCULAR PLATE WITH A HOLE. INNER EDGE FREE, OUTER SUPPORTED (FS)

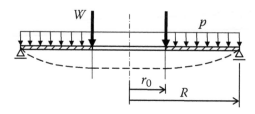

(a) Pressure load p:

$$u_{st} = 0.0624 \frac{p_0 R^4}{D}; \quad \mathcal{M}_{st} = 0.2404 p_0 R^2; \quad Q_{st} = \frac{(R^2 - r_0^2) p}{2R}$$

$$p_y = \frac{6\mathcal{M}_0}{(R - r_0)(R + 2r_0)}; \, p_y = \frac{6\mathcal{M}_0}{R^2} \text{ for } r_0 = 0; \text{ static collapse load (RP material)}$$

Short impulse $p_0 t_0$: $u_m = \dfrac{(p_0 t_0)^2}{\mathcal{M}_0 m} \dfrac{R^2 - r_0^2}{4}$

(b) Ring load with resultant W:

$$u_{st} = \frac{1}{16.25} \frac{W_0 R^2}{D}; \quad \mathcal{M}_{st} = 0.2469 W_0; \quad Q_{st} = \frac{W}{2\pi R}; \quad W_y = 2\pi \mathcal{M}_0$$

Short impulse $W_0 t_0$: $u_m = \dfrac{(W_0 t_0)^2}{2 M_e W_y}$

$$M_e = \frac{\pi m}{6} r_0 (R - r_0)\left(\frac{R}{r_0} + 3\right); \text{ effective edge mass}$$

Note: Coefficients of u_{st} and $\mathcal{M}_{st}$ are for the case $r_0/R = 1/2$ and for linear material only. $\mathcal{M}_{st}$ is for the circumferential moment, which dominates. For other ratios, refer to Young [118]. For a ring load, u_m is only the first approximation [40].

CASE 12.6   CIRCULAR PLATE WITH A HOLE. INNER EDGE CLAMPED, OUTER EDGE FREE (CF)

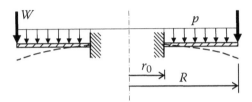

(a) Pressure load p:

$$u_{st} = 0.0086 \frac{p_0 R^4}{D}; \quad \mathcal{M}_{st} = 0.1736 p_0 R^2; \quad Q_{st} = \frac{(R^2 - r_0^2)p}{2r_0}$$

$$p_y = \frac{6 R \mathcal{M}_0}{(R - r_0)^2 (2R + r_0)}; \quad p_y = \frac{3 \mathcal{M}_0}{R^2} \text{ for } r_0 = 0; \text{ static collapse load (RP material)}$$

Short impulse $p_0 t_0$: $\quad u_m = \frac{(p_0 t_0)^2}{\mathcal{M}_0 m} \frac{R^2 - r_0^2}{4}\left(1 - \frac{r_0}{R}\right)$

(b) Ring load with resultant W:

$$u_{st} = \frac{1}{124.5} \frac{W_0 R^2}{D}; \quad \mathcal{M}_{st} = 0.1255 W_0; \quad Q_{st} = \frac{W}{2\pi r_0}$$

$$W_y = \frac{2\pi R \mathcal{M}_0}{R - r_0}; \quad W_y = 2\pi \mathcal{M}_0 \quad \text{for } r_0 = 0$$

Short impulse $W_0 t_0$: $\quad u_m = \frac{(W_0 t_0)^2}{2 M_e W_y}$

$$M_e = \frac{\pi m}{6} r_0 (R - r_0)\left(\frac{3R}{r_0} + 1\right); \text{ effective edge mass}$$

Note: Coefficients of u_{st} and $\mathcal{M}_{st}$ are for the case $r_0/R = 1/2$ and for a linear material only. $\mathcal{M}_{st}$ is for the radial moment, which dominates. For other ratios, refer to Young and Budynas [118]. For a ring load, u_m is only the first approximation [40].

CASE 12.7   **CIRCULAR PLATE WITH A HOLE. INNER EDGE FREE, OUTER EDGE CLAMPED (FC)**

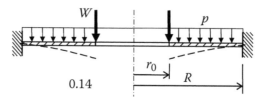

(a) Pressure load p:

$$u_{st} = 0.0053 \frac{p_0 R^4}{D}; \quad M_{st} = 0.08 p_0 R^2; \quad Q_{st} = \frac{(R^2 - r_0^2)p}{2R}$$

$$p_y = \frac{6(2R - r_0)M_0}{(R - r_0)^2(R + 2r_0)}; \, p_y = \frac{12M_0}{R^2} \text{ for } r_0 = 0; \text{ static collapse load (RP material)}$$

Short impulse $p_0 t_0$: $u_m = \dfrac{(p_0 t_0)^2}{M_0 m} \dfrac{R^2 - r_0^2}{4} \dfrac{R - r_0}{2R - r_0}$

(b) Ring load with resultant W:

$$u_{st} = \frac{1}{134.8} \frac{W_0 R^2}{D}; \quad M_{st} = \frac{W_0}{13.21}; \quad Q_{st} = \frac{W}{2\pi R}$$

$$W_y = \frac{2\pi(2R - r_0)M_0}{R - r_0}; \quad W_y = 4\pi M_0 \quad \text{for } r_0 = 0$$

Short impulse $W_0 t_0$: $u_m = \dfrac{(W_0 t_0)^2}{2M_e W_y}$

$$M_e = \frac{\pi m}{6} r_0 (R - r_0) \left(\frac{R}{r_0} + 3 \right); \text{ effective edge mass}$$

Note: Coefficients of u_{st} and M_{st} are for the case $r_0/R = 1/2$ and for linear material only. M_{st} is for the radial moment, which dominates. For other ratios, refer to Young and Budynas [118]. For a ring load, u_m is only the first approximation [40].

CASE 12.8   **CIRCULAR PLATE WITH A HOLE. INNER EDGE FREE, OUTER SUPPORTED. VARYING PRESSURE ALONG RADIUS**

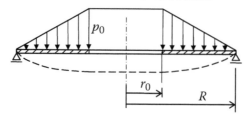

RP material.

$$p_y = \frac{12\mathcal{M}_0}{(R - r_0)(R + 3r_0)}$$

$p_y = \dfrac{12\mathcal{M}_0}{R^2}$ for $r_0 = 0$; static collapse load

Short impulse $p_0 t_0$ with $p_0 \gg p_y$: $u_m = \dfrac{(p_0 t_0)^2}{\mathcal{M}_0 m} \dfrac{R^2(1 + 3r_0/R)}{24}$ [43]

CASE 12.9 MASS-PLATE IMPACT

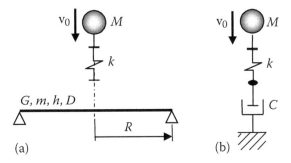

(a)   (b)

(From Szuladziński, G., *Acta Mech.*, 200(1–2), 111, 2008.)

  (a) Mass-impacted plate
  (b) Equivalent system

r_0 is the rigid insert radius at the center of the impacted plate. If there is no insert, use $r_0 = h/2$. (For a small contact area.)

$h_s = h/1.2$; shear thickness of a solid plate h thick

$C = 2\pi^2 r_0 \left(\dfrac{Gh_s m}{5}\right)^{1/2}$; equivalent damper constant

$P_{md} = Cv_0$; peak impact force when the spring is rigid

$P_{ms} = v_0 \sqrt{kM}$; peak impact force when the damper is rigid

$P_m \approx (P_{ms}^{-0.95} + P_{md}^{-0.95})^{-1.053}$; predicted peak impact force

When P_m is attained, the spring is fully compressed and velocity at that instant is

$v_e = P_m/C$; (reversal point)

$t_0 = \dfrac{\pi M v_0}{2 P_m}$; duration of loading phase when rebound is negligible

Estimate of contact duration $\tau/2$ when the damper is firm: $(v_e \ll v_0)$

$E_0 = \dfrac{1}{2} M v_0^2$; initial system energy

$E_e = \dfrac{P_m^2}{2}\left(\dfrac{M}{C^2} + \dfrac{1}{k}\right)$; system energy at the reversal point

$V = \left[\dfrac{2}{M}(2E_e - E_0)\right]^{1/2}$; rebound velocity

$k_e = k\left(\dfrac{P_m}{P_{ms}}\right)^2$; $\Omega = \sqrt{\dfrac{k_e}{M}}$; $\dfrac{\tau}{2} = \dfrac{\pi}{\Omega}$; equations to find the impact duration $\tau/2$

A more general approach to find rebound velocity:

$$\frac{1}{k_{ef}} = \frac{1}{k} + \frac{1}{k_d} \quad \text{with } k_d = \frac{C^2}{M}$$

$v_y = \dfrac{P_m}{\sqrt{k_{ef}M}}$; upward velocity jump, as illustrated

$V = v_y - v_e$; rebound velocity $\kappa = V/v_0$; coefficient of restitution

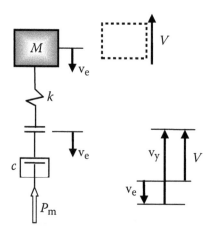

The equivalent system at the reversal point and a vector summation of rebound velocity V.

Note: Both methods of calculating the rebound have a definite limitation: they are meant for a single impact only. When the relative motion of the striker and the plate is such that a secondary impact takes place, the rebound velocity will be larger than predicted here. This does not, however, influence the peak contact force, related to the first impact. The coefficient κ, as calculated above, need not be positive [102].

CASE 12.10   REBOUND BASED ON PLATE FLEXURE

V is the velocity of rebound

$\kappa = V/v_0$; rebound coefficient

$$\lambda = 0.895 h^{0.3} \left(\frac{R_s}{h}\right)^2 \left(\frac{\rho_s}{\rho_p}\right)^{0.6} \left(\frac{\rho_p v_0^2}{D}\right)^{0.1} ; \text{ impact parameter}$$

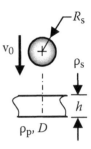

$\kappa \approx \exp(-1.731\lambda)$

Note: The approach relies only on flexural waves spreading from the impact point and does not include any energy dissipation via the shearing mechanism. The above equations are valid for a rigid striker only. It need not be a sphere; a spherical contact surface is sufficient. Our formula for κ is a major simplification of the relationships in Refs. [119] and [26].

CASE 12.11   CIRCULAR, POINT LOADED PLATE, OUTER EDGE SUPPORTED. LARGE DEFLECTIONS

$W_y = 2\pi \mathcal{M}_0$; static collapse load, small deflections

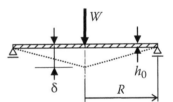

$\mathcal{M}_0 = h^2 \sigma_0/4$; moment capacity

$$\frac{W}{W_y} = 1 + \frac{1}{3}\left(\frac{\delta}{h_0}\right)^2 ; \text{ for } \delta \le h_0$$

$$\frac{W}{W_y} = \frac{\delta}{h_0} + \frac{1}{3}\left(\frac{h_0}{\delta}\right) ; \text{ for } \delta \ge h_0 \text{ [40]}$$

CASE 12.12   CIRCULAR, POINT LOADED PLATE, OUTER EDGE CLAMPED. LARGE DEFLECTIONS

$W_y = 4\pi \mathcal{M}_0$; static collapse load, small deflections

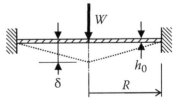

$\mathcal{M}_0 = h^2 \sigma_0/4$; moment capacity

$$\frac{W}{W_y} = 1 + \frac{5}{12}\left(\frac{\delta}{h_0}\right)^2 \text{ [40]}$$

CASE 12.13   LONG CYLINDRICAL SHELL, PRESSURIZED

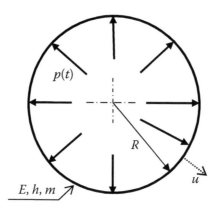

(a) **Static response to pressure p_0:** stress, radial deflection, and strain, respectively:

$$\sigma_{st} = \frac{p_0 R}{h}; \quad u_{st} = \frac{p_0 R^2}{E''h}; \quad \varepsilon_{st} = \frac{u_{st}}{R}; \quad E'' = \frac{E}{1 - v^2}$$

$$\omega = \frac{1}{R}\sqrt{\frac{E''}{\rho}}; \text{ natural frequency}$$

(b) **Impulse $s = p_0 t_0$**

$$u_m = 2u_{st} \sin\frac{\omega t_0}{2}; \text{ maximum when } t_0 = \tau/2$$

(c) **Short impulse, $s = p_0 t_0$:** $u_m = \dfrac{v_0}{\omega} = v_0 R\left(\dfrac{m}{E''h}\right)^{1/2}$ at $t = \tau/4$

(d) **Step load, $p = p_0 H(t)$:** $u_m = 2u_{st}$ at $t = \tau/2$

Note: Refer to Case 3.30 for a moderately thick shell.

CASE 12.14   LONG CYLINDRICAL SHELL, RP MATERIAL, UNIFORM PRESSURE

(Refer to the illustration in Case 12.13)

$$p_c = \frac{P_0}{R} = \frac{h\sigma_0}{R}; \text{ static collapse load}$$

$\Pi = p_c u_m$; strain energy per unit surface area

(a) Initial velocity v_0: $u_m = \dfrac{mv_0^2}{2p_c}$

(b) Rectangular pulse of p_m during t_0: $u_m = \dfrac{1}{2}\dfrac{p_m t_0^2}{m}\dfrac{p_m - p_c}{p_c}$; $t_m = \dfrac{mv_0}{p_c}$

(c) Impulsive loading, as above, but $p_m \gg w_0$: $u_m = \dfrac{1}{2}\dfrac{p_m^2 t_0^2}{mp_c}$; $t_m = \dfrac{mv_0}{p_c}$

(d) Exponential load $p = p_m(1 - t/t_0)\exp(-t/t_0)$

$$u_m = \frac{p_m t_0^2}{m}\left[1 - \frac{p_c}{p_m}\left(1 + \ln\frac{p_m}{p_c} + \frac{1}{2}\ln^2\frac{p_m}{p_c}\right)\right]; \quad t_m = t_0 \ln\frac{p_m}{p_c} \quad [45]$$

Note: Refer to Case 4.2.

CASE 12.15   LONG CYLINDRICAL SHELL, EPP MATERIAL, UNIFORM PRESSURE

$u_y = \dfrac{\sigma_0 R}{E'}$; deflection at onset of yield

(a) Initial velocity v_0: $u_m = \dfrac{1}{2}\left(\dfrac{\sigma_0 R}{E''} + \dfrac{\rho v_0^2 R}{\sigma_0}\right)$; $E'' = \dfrac{E}{1 - v^2}$; $t_m = \dfrac{mv_0}{p_c}$

(b) Step pressure $p = p_0 H(t)$: $(p_c/2 < p_0 < p_c)$

$$u_m = \frac{R^2 p_c^2}{2E'' h(p_c - p_0)} \qquad t_m \approx \sqrt{\frac{8\rho h u_m}{3p_c}}$$

Note: In case (a) when $u_m > 2u_y$, there are decaying oscillations about the permanent deflection equal to $u_m - u_y$. p_c defined as in Case 2.14.

CASE 12.16   RIGID-PLASTIC SHELL UNDER IMPULSIVE PRESSURE

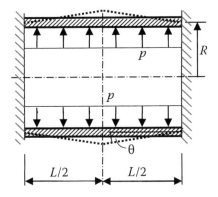

$p_y = 0.55\dfrac{h\sigma_0}{R}$; initiation of yielding

$$p_c = \frac{h\sigma_0}{R}\left(1 + \frac{4hR}{L^2}\right); \text{ static collapse pressure}$$

Short pulse, $p_0 t_0$ with $p_0 \gg p_c$:

$$E_k = \frac{(p_0 t_0)^2 L}{2m}; \text{ kinetic energy}$$

$$\theta_m = \frac{(p_0 t_0)^2 L}{8\mathcal{M}_0 m}\bigg/\left\{1 + \frac{L^2}{4hR}\right\}; \quad t_m = \frac{p_0}{p_c}t_0$$

Note: This gives somewhat larger response than derived by Jones [43].

CASE 12.17   RIGID-PLASTIC SHELL UNDER RING LOAD q (N/m)

$$\lambda^4 = \frac{3(1-\nu^2)}{R^2 h^2} \quad \text{or} \quad \lambda = \frac{1.285}{\sqrt{Rh}} \quad \text{for } \nu = 0.3 \text{ (shell parameter)}$$

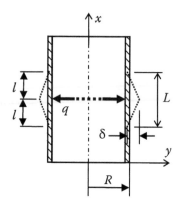

$q_y = 0.43\lambda h^2 \sigma_0$ (for $\nu = 0.3$); initiation of yield

$$q_c = \frac{4\mathcal{M}_0}{l}; \text{ static collapse load}$$

$l = \sqrt{Rh}$; distance from the plane of loading to the far plastic hinge

Short pulse, $q_0 t_0$ with $q_0 \gg q_c$:

$$E_k = \frac{3}{2}\frac{(q_0 t_0)^2}{mL}; \text{ kinetic energy}$$

$$\theta_m = \frac{3}{8}\frac{(q_0 t_0)^2}{\mathcal{M}_0 mL}\bigg/\left\{1 + \frac{q_c L^2}{16\mathcal{M}_0}\right\}$$

Note: The load is assumed to be applied far enough from the ends, so that end conditions do not matter.

CASE 12.18   THIN SPHERICAL SHELL, PRESSURIZED

(Refer to the illustration in Case 12.13)

(a) Static response to pressure p_0: stress, radial deflection, and strain, respectively:

$$\sigma_{st} = \frac{p_0 R}{2h}; \quad u_{st} = \frac{p_0 R^2}{2E'h}; \quad \varepsilon_{st} = \frac{u_{st}}{R}; \quad E' = \frac{E}{1-\nu}$$

$$\omega = \frac{1}{R}\sqrt{\frac{2E'}{\rho}}; \text{ natural frequency}$$

(b) Impulse $s = p_0 t_0$

$$u_m = \frac{2p_0}{\omega^2 m}\sin\frac{\omega t_0}{2}; \text{ maximum when } t_0 = \tau/2$$

(c) Short impulse, $s = p_0 t_0$: $u_m = \dfrac{v_0}{\omega} = v_0 R\left(\dfrac{m}{2E'h}\right)^{1/2}$ at $t = \tau/4$

(d) Step pressure, $p = p_0 H(t)$: $u_m = 2u_{st}$ at $t = \tau/2$

Note: Refer to Case 3.31 for a moderately thick shell.

CASE 12.19 THIN SPHERICAL SHELL OF RP MATERIAL, PRESSURIZED

$$p_c = \frac{2P_0}{R} = \frac{2h\sigma_0}{R}; \text{ static collapse load}$$

$\Pi = p_c u_m$; strain energy per unit surface area

All formulas are the same as in Case 12.14, except for a change in p_c.

CASE 12.20 THIN, SPHERICAL SHELL, EPP MATERIAL, UNIFORM PRESSURE

$$u_y = \frac{\sigma_0 R}{E'}; \text{ deflection at onset of yield, } E' = \frac{E}{1-\nu}$$

(a) Initial velocity v_0: $u_m = \dfrac{1}{2}\left(\dfrac{\sigma_0 R}{E'} + \dfrac{\rho v_0^2 R}{2\sigma_0}\right)$

(b) Step pressure $p = p_0 H(t)$: $(p_c/2 < p_0 < p_c)$

$$u_m = \frac{R^2 p_c^2}{4E'h(p_c - p_0)}; \quad t_m \approx \sqrt{\frac{8\rho h u_m}{3p_c}}$$

Note: In case (a), after u_m is reached, when $u_m > 2u_y$, there is a decaying oscillation about the permanent deflection position.

CASE 12.21   THIN CYLINDER OF RP MATERIAL, AXIALLY CRUSHED

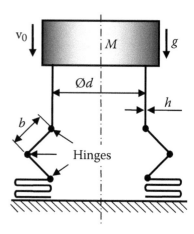

$$P_{cr} = \left(5.98\sqrt{\frac{d}{h}} \pm 1.814\right)h^2\sigma_0;$$

crushing force, "+" for outside forming folds, "−" for inside

$$b = 0.952\sqrt{hd};\text{ fold length}$$

(a) Rigid block of mass M subjected to gravity.

$$u_m = \frac{Mv_0^2}{2(P_{cr} - Mg)};\text{ peak displacement or shortening}$$

(b) A pipe flying against a rigid wall

Use M = pipe mass in the above [40].
For small deflections and inelastic material the following quasistatic relations govern for a thin shell in compression:

$$\sigma_{cr} = \frac{2h\sqrt{EE_t}}{d\sqrt{3(1-v^2)}};\text{ critical stress, }\sigma_{cr} = \frac{P_{cr}}{\pi dh}$$

$$2b = 1.72\sqrt{\frac{dh}{2}}\sqrt{\frac{E_t}{E}};\text{ half-wave length, for }v = 0.3\text{ [110]}$$

Note: E_t is tangent modulus corresponding to σ_{cr}. In the elastic range, $E = E_t$.

CASE 12.22   THIN-WALL RECTANGULAR TUBE

Section sides a and b, thickness h, subjected to axial compression. Rigid-plastic material with σ_0.

$$P_{cr} = 12.16\left(\frac{c}{h}\right)^{0.37}h^2\sigma_0;\text{ crushing force, rectangular section}$$

$$P_{cr} = 20.23\left(\frac{c}{h}\right)^{0.4}h^2\sigma_0;\text{ crushing force, hexagonal section [115]}$$

$c = (a + b)/2$

Note: This case is similar to Case 12.21, except the folds are more complex.

CASE 12.23   THIN CYLINDRICAL TUBE OF RP MATERIAL, LATERALLY DEFORMED WITH KNIFE-EDGE LOAD

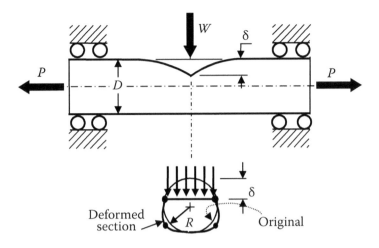

h is the wall thickness

$M_0 = hD^2\sigma_0$; bending capacity

$P_y = \pi Dh\sigma_0$; axial capacity

P is positive when tension

W–δ characteristic:

$$\frac{W}{M_0} = \frac{4}{D}\sqrt{\frac{\pi h \delta}{D^2}\left[1 - \frac{1}{2}\left(\frac{P}{P_y} - 1\right)^2\right]}$$

$$\frac{M_p}{M_0} = 1 - \frac{\delta}{D}$$

M_p is the reduced bending capacity of a deformed center section.

Note: Knife-edge load means that the crease formed in the application zone is a straight line. Use $P/P_y = 0$ for a free-end tube and $P/P_y = 1$ when ends are fixed [70].

CASE 12.24   THIN CYLINDER OF RP MATERIAL, COMPRESSED BETWEEN RIGID DIES

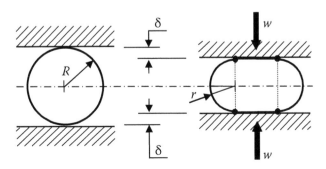

w–δ characteristics. w is the resultant load per unit length on cylinder.

$$\mathcal{M}_0 = (h^2/4)\,\sigma_0$$

h is the wall thickness

$$w = \frac{2\pi\mathcal{M}_0}{R\left(1 - \dfrac{\delta}{R}\right)} \text{ (N/m)}$$

This is a dynamic solution for the hinges moving outward. There is also a static solution, with two hinges on the horizontal axis and two on the vertical axis, not illustrated:

$$w = \frac{4\mathcal{M}_0}{R}\left[1 - \left(\frac{\delta}{2R}\right)^2\right]^{1/2}$$

Note: This is per unit length of cylinder, length normal to paper. Four joints appear [115].

CASE 12.25 BUCKLING OF A THIN CYLINDRICAL SHELL SUBJECTED TO AN IMPULSIVE, RADIAL COMPRESSIVE PRESSURE $p(t)$

The shell is L long with simply supported ends

$m = \rho h$ is mass density

$i = mv_0$ is the applied specific impulse

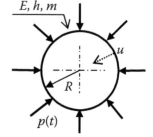

$\tilde{\varepsilon}_y$ is the yield strain when σ–ε curve is idealized to a bilinear shape

p_m, i_m are peak values of pressure and impulse, respectively.

K is the coefficient tabulated as follows:

$$\varepsilon_y = F_y/E$$

$$r = \left(\frac{p_m}{p_e} - 1\right)\left(\frac{i_m}{i_e} - 1\right) \geq 1.0; \text{ condition for buckling to take place in the elastic range}$$

$$p_e = 0.92E\left(\frac{R}{L}\right)\left(\frac{h}{R}\right)^{5/2}; \text{ elastic range, pressure limit}$$

$$i_e = 3\rho c_0 R\left(\frac{h}{R}\right)^2; \text{ elastic range, impulse limit; } c_0 \text{ is the sonic velocity}$$

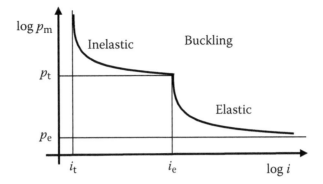

$$r = \left(\frac{p_m}{p_t} - 1\right)\left(\frac{i_m}{i_t} - 1\right) \geq 1.0; \text{ condition for buckling to take place, plastic range}$$

$p_t = 0.75\sigma_y h/R$; plastic range, pressure limit

$$i_t = 1.15\rho c_0 R\left(\frac{h}{R}\right)^2 \quad \text{for } \frac{R}{h} \geq 0.405\frac{K^{1/2}}{\tilde{\varepsilon}_y}$$

$$i_t = 1.807\rho c_0 R\left(\frac{h}{R}\right)^{3/2}\frac{\sqrt{\tilde{\varepsilon}_y}}{K^{1/4}} \quad \text{for}$$

$$\frac{R}{h} \leq 0.405\frac{K^{1/2}}{\tilde{\varepsilon}_y}; \text{ plastic range, impulse limit}$$

Static buckling pressure of a long shell (for reference only):

$$p_{crs} = \frac{3D}{R^3}; \quad \text{where } D = \frac{Eh^3}{12(1-v^2)}$$

Properties of Selected Alloys (tested up to $\varepsilon = 0.05$)

Alloy	ρ (kg/m³)	E (GPa)	K	ε_y	$\tilde{\varepsilon}_y$
Al 6061-T6	2730	69	30	0.0039	0.0043
Al 6061-T6 (in compression)	2730	69	15	0.0038	0.0045
Magnesium AZ31B	1820	41	14	0.0048	0.0059
Magnesium ZK60A	1820	41	30	0.0065	0.0069
Titanium 6Al-4V	4470	106	30	0.0070	0.0075

Note: If $r < 1$, the point (p, i) lies below the critical level and no buckling takes place. The above relationships have a limited accuracy [82].

EXAMPLES

EXAMPLE 12.1 ANNULAR PLATE UNDER RING LOAD

The plate is of annular type, $r_0 = 100$, $R = 200$ mm, $h = 3$ mm. The inside edge is clamped and the outer edge is free. Find its peak deflection as a function of time when a ring load with the resultant of $W = 1000H(t)$ N is applied to the outer edge. Conduct the calculation for up to $\tau/4$ and check the results using an FEA code. The material properties are

$$E = 69{,}000\,\text{MPa}, \quad \rho = 0.0027\,\text{g/mm}^3, \quad \nu = 0.3, \quad G = 26{,}540\,\text{MPa}$$

Mass per unit area: $m = \rho h = 0.0081$ g/mm^2

$$D = \frac{Eh^3}{12(1-\nu^2)} = \frac{69{,}000 \times 3^3}{12(1-0.3^2)} = 170{,}600 \text{ N-mm, plate stiffness}$$

From Equation 12.22a: $\dfrac{\tau}{4} = 0.1208 \times 200^2 \sqrt{\dfrac{0.0081}{170{,}600}} = 1.053$ ms

Equation 12.19a becomes $l_1 = 97.46t^{1/2}$.
When such a ring load is applied to the outer edge, the static deflection is, according to Case 12.6:

$$u_{\text{st}} = \frac{1}{124.5}\frac{1{,}000 \times 200^2}{170{,}600} = 1.883 \text{ mm}$$

The peak dynamic deflection is calculated by the same formula, except it has a varying coefficient, dependent on time, or on l_1, per Equation 12.24a:

$$u_{\text{d}} = f(l_1)\frac{1{,}000 \times 200^2}{170{,}600} = 234.46 f(l_1)$$

However, the coefficient $f(l_1)$ depends now on the ratio l_1/R, l_1 being an instantaneous inner radius. The history of displacement is shown in the table, along with FEA results, which were obtained as follows. A model of the plate was developed using ANSYS v.10 code [2]. There were 100 elements (SHELL51) placed along the radius of this axisymmetric model. This code was set to work in a small-deflection mode. The results of the simulation are given in the table and the graph to follow.

Annular Plate Deflections u (from FEA) and u_d (Calculated)					
t (ms)	u (mm)	$1/f\ (l_1)$	l_1 (mm)	u_d	u_d/u
0.0026	0.00018	1.21E + 06	4.97	0.00019	1.068
0.0105	0.00163	147401	9.99	0.00159	0.978
0.0421	0.01351	17920	20.00	0.01308	0.968
0.168	0.1133	2139.1	40.00	0.10961	0.967
0.379	0.3743	609	60.00	0.38499	1.029
1.053	1.917	124.5	100.00	1.88321	0.982

Deflection u is obtained from the simulation, while u_d is hand calculated. The correlation between the two is quite good most of the time, even when approaching the boundary, $R = 200$ mm, although a much worse result could be anticipated. A correction is certainly needed at the first time point, where l is only about $5\times$ (element length). However, when a static run was made using a sensible number of elements corresponding to this segment, the answer was $u = 0.000197$ mm, only a little larger. A different problem here is that the shell elements involved usually underestimate shear deflection. A static run made with 2D solids and $l = 5$ gave $u = 0.000263$, a substantial improvement.

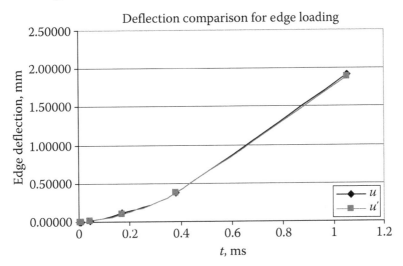

It is left to the reader to check if there is a good agreement for the remaining case of the plate loaded at the center. (The multiplier of l_2^2 is constant, which simplifies the problem.) The fact is that modeling with shell elements leads to an underestimate of deflection for very small values of t. This becomes quite visible in this example.

EXAMPLE 12.2   CIRCULAR PLATE UNDER SHORT PRESSURE IMPULSE

The plate radius is $R = 50$ mm, $h = 3$ mm, and has clamped edges. The material is aluminum alloy, with the same properties as in Example 12.1, also including $\tau_0 = 180/\sqrt{3} = 103.9$ MPa.

Determine the onset of yielding, the static collapse pressure, and peak deflection due to a pressure impulse of 3 MPa magnitude and 0.08 ms duration treating the material as RP.

Case 12.3 governs, but some additional equations are needed. Moments and shear forces related to onset of yield and to ultimate capacity (Equations 12.37a and b), respectively:

$$M_y = \frac{1}{6}h^2\sigma_0 = \frac{1}{6}3^2 \times 180 = 270 \text{ N} \quad \text{and} \quad M_0 = \frac{1}{4}h^2\sigma_0 = 405 \text{ N}$$

The corresponding values for shear from Equations 12.38a and b:

$$Q_y = h\tau_0/\alpha = 3 \times 103.9(2/3) = 207.8 \text{ N/mm} \quad \text{and} \quad Q_0 = h_{ef}\tau_0 = 3 \times 103.9 = 311.7 \text{ N/mm}$$

Pressures associated with the onset of yield, Equations 12.39b and 12.41b:

$$p_y = \frac{\sigma_0}{6\beta}\left(\frac{h}{R}\right)^2 = \frac{180}{6(1/8)}\left(\frac{3}{50}\right)^2 = 0.864 \text{ MPa}$$

$$p_{ys} = \frac{2\tau_0}{3\varphi}\left(\frac{h}{R}\right) = \frac{2 \times 103.9}{3(1/2)}\left(\frac{3}{50}\right) = 8.312 \text{ MPa}; \ (8.312 > 0.864), \text{ bending governs}$$

$p_b = \xi M_0/R^2 = 12 \times 405/50^2 = 1.944 \text{ MPa}$; static collapse

Circular frequency, per Case 2.33: $\omega = \dfrac{10.22}{R^2}\left(\dfrac{Eh^3}{12m(1-v^2)}\right)^{1/2} = 18.76 \text{ rad/ms}$

Natural period: $\tau = 2\pi/\omega = 0.335$ ms (to get an idea of the time scale)

Rectangular, uniform loading pulse, $p_m = 3$ MPa, $t_0 = 0.08$ ms, Equation 12.54b:

$$u_m = \theta_m R = \frac{\vartheta(Rp_0t_0)^2}{M_0m} = \frac{1}{8}\frac{(50 \times 3 \times 0.08)^2}{405 \times 0.0081} = 5.487 \text{ mm}; \text{ the peak displacement}$$

EXAMPLE 12.3   CIRCULAR PLATE UNDER A SHORT-DURATION POINT LOAD

Consider the same plate as in Example 12.2, but with the edges supported. Determine the elastic response due to an applied center load of $W = 10,000$ N lasting 0.069 ms. Set up an FEA model to verify the answer.

Circular frequency, per Case 2.33: ($D = 170,600$ per Example 12.1)

$$\omega = \frac{4.977}{R^2}\left(\frac{D}{m}\right)^{1/2} = \frac{4.977}{50^2}\left(\frac{170,600}{0.0081}\right)^{1/2} = 9.136 \text{ kHz} \quad \text{or} \quad \tau = 0.688 \text{ ms}$$

Following Case 12.1 but with coefficients from Table 12.2: (Set $r_0 = 3\,\text{mm}$)

$$u_{\text{st}} = \frac{1}{16\pi}\frac{3+\nu}{1+\nu}\frac{W_0 R^2}{D} = \frac{1}{16\pi}\frac{3.3}{1.3}\frac{10{,}000\times 50^2}{170{,}600} = 7.4\text{ mm (at center)}$$

$$M_{\text{st}} = \frac{1+\nu}{4\pi}\ln\left(\frac{R}{h}\right)W_0 = \frac{1+0.3}{4\pi}\ln\left(\frac{50}{3}\right)10{,}000 = 3{,}706\text{ N-mm (at center)}$$

$$Q_{\text{st}} = \frac{W_0}{2\pi r} = \frac{10{,}000}{2\pi 3} = 530.5\text{ N \ (just outside } r_0)$$

$$u_{\text{d}} = 2u_{\text{st}}\sin(\pi t_o/\tau) = 2\times 7.4\ \sin(\pi 0.069/0.688) = 4.587\text{ mm}$$

$$M_{\text{d}} = \frac{u_{\text{d}}}{u_{\text{st}}}M_{\text{st}} = \frac{4.587}{7.4}3{,}706 = 2{,}297\text{ N-mm}$$

$$Q_{\text{d}} = 1.4Q_{\text{st}} = 1.4\times 530.5 = 742.7\text{ N}$$

An FEA model of the plate was developed using ANSYS v.10 code [2]. There were 100 elements (SHELL51) placed along the radius of this axisymmetric model with the nodal spacing of 0.5 mm. This code was set to work in a small deflection mode. Instead of a point load, an equivalent pressure was applied on a circle of radius r_0:

$$p = P/(\pi r_0^2) = 10{,}000/(\pi 3^2) = 353.7\text{ MPa}$$

The results of simulation were as follows:

Statics: $u = 7.345\,\text{mm}$, $M = 3727\,\text{N-mm}$, and $Q = 530.5\,\text{N}$

(The moment value was read at the center of plate. It was close to our hand-calculated value at 3 mm radius.)

The corresponding dynamic values were 4.827, 1779, and 701.9, which gives a good or a reasonable agreement with u_{d}, M_{d}, and Q_{d} as determined above.

EXAMPLE 12.4   STEEL SPHERE IMPACTING LUCITE PLATE

Reference [81] presents a set of test results obtained by dropping steel spheres of several diameters on a lucite plate, $h=12.7\,\text{mm}$ thick. The drop height was 635 mm resulting in impact velocity of 3.53 m/s. The diameters of $d=6.35$, 12.7, and 17.5 mm can be used as representative of the tested range. Assume the contact to be of plastic type with the effective yield stress of $F_y=70\,\text{MPa}$. Treating the spheres as undeformable determine the rebound velocities. The mechanical properties of lucite are

$$E = 2500\text{ MPa}, \quad \rho = 0.00118\text{ g/mm}^3, \quad \nu = 0.3, \quad G = 961.5\text{ MPa}$$

Follow Case 12.9. For the plate, $m = \rho h = 0.015\,\text{g/mm}^2$.

For a small contact area, the equivalent insert radius is assumed as $h/2 = 6.35\,\text{mm}$.

The shear thickness is $h_{\text{s}} = h/1.2 = 10.58\,\text{mm}$.

The detailed calculations will be shown for 12.7 mm diameter only.

$$M = \pi d^3 \rho / 6 = 8.419 \text{g}$$

$$C = 2\pi^2 r_0 \left(\frac{Gh_s m}{5} \right)^{1/2} = 2\pi^2 \times 6.35 \left(\frac{961.5 \times 10.58 \times 0.015}{5} \right)^{1/2} = 692.4 \text{ N-s/mm}$$

$P_{md} = Cv_0 = 692.4 \times 3.53 = 2444 \text{ N}$; damper force limit

$k = 2\pi R_s F_y = 2\pi 6.35 \times 70 = 2793 \text{ N/mm}$; contact stiffness per Case 7.26

$P_{ms} = v_0 \sqrt{kM} = 3.53\sqrt{2793 \times 8.419} = 541.3 \text{ N}$; spring force limit:

$P_m \approx (2444^{-0.95} + 541.3^{-0.95})^{-1.053} = 433 \text{ N}$; peak contact force

$v_e = 433/692.4 = 0.6254 \text{ m/s}$; velocity at the reversal point

$E_0 = \frac{1}{2} M v_0^2 = \frac{1}{2} \times 8.419 \times 3.53^2 = 52.45 \text{ N-mm}$; initial system energy

$E_e = \frac{P_m^2}{2} \left(\frac{M}{C^2} + \frac{1}{k} \right) = \frac{433^2}{2} \left(\frac{8.419}{692.4^2} + \frac{1}{2793} \right) = 35.21 \text{ N-mm}$; energy at the reversal point

$$V = \left[\frac{2(2E_e - E_0)}{M} \right]^{1/2} = \left[\frac{2(2 \times 35.21 - 52.45)}{8.419} \right]^{1/2} = 2.066 \text{ mm/ms}$$

The rebound calculations for all three diameters are tabulated as follow:

D (mm)	P_{ms}	P_m	v_e	E_e	E_0	V	κ	V (S)
6.35	135.32	126.99	0.183	5.791	6.556	3.091	0.83	2.930
12.7	541.30	432.97	0.625	35.206	52.451	2.066	0.58	2.047
17.5	1027.79	702.22	1.014	75.395	137.233	1.109	0.36	1.271

The last two columns refer to the test results, giving the restitution coefficient and the associated velocity V(S). The coefficient was read from a small-scale plot, so the accuracy of the data is limited. Yet, overall agreement is quite good except, possibly, for the largest mass, where this method underestimates the speed of rebound.

EXAMPLE 12.5   REBOUND CALCULATION BASED ON PLATE FLEXURE

Using the data from Example 12.4 determine the rebound velocity of a steel striker with $\emptyset d = 12.7$ mm when impacting a lucite plate with $v_0 = 3.53$ m/s. This time apply a flexure-based approach of Case 12.10.

$R_s = d/2 = 12.7/2 = 6.35$ mm.

$$D = \frac{2500 \times 12.7^3}{12(1 - 0.3^2)} = 468{,}950; \text{ plate stiffness}$$

$$\lambda = 0.895 h^{0.3} \left(\frac{R_s}{h} \right)^2 \left(\frac{\rho_s}{\rho_p} \right)^{0.6} \left(\frac{\rho_p v_0^2}{D} \right)^{0.1}$$

$$= 0.895 \times 12.7^{0.3} \left(\frac{6.35}{12.7} \right)^2 \left(\frac{0.00785}{0.00118} \right)^{0.6} \left(\frac{0.00118 \times 3.53^2}{468{,}950} \right)^{0.1}$$

$$= 0.2657; \text{ impact parameter}$$

$\kappa \approx \exp(-1.731\lambda) = \exp(-1.731 \times 0.2657) = 0.6314; \text{ coefficient of restitution}$

$V = 0.6314 \times 3.53 = 2.229$ m/s; rebound velocity

This comes out somewhat less accurate than our $V = 2.066$ m/s in Example 12.4, against measured 2.047 m/s. According to Ref. [81] it is typical for this method to overestimate the rebound speed.

EXAMPLE 12.6   ANNULAR PLATE UNDER A SHORT PRESSURE IMPULSE

The plate properties are as in Example 12.2 except the outer radius is $R = 100$ mm (free) and the inside radius is $r_0 = 50$ mm (clamped). Determine static collapse load and peak deflection due to a pressure impulse of 2 MPa magnitude and 0.08 ms duration treating the material as RP.

Repeat the calculation for a ring load with the resultant of one-quarter of the above.

Use Case 12.6. Mass per unit area: $m = \rho h = 0.0081$ g/mm^2

Moments and shear forces related to yield capacity:

$\mathcal{M}_0 = 405$ N-mm/mm; $Q_0 = 311.7$ N/mm (Example 12.2)

Static collapse takes place at the following pressures:

$$p_y = \frac{6R\mathcal{M}_0}{(R - r_0)^2 (2R + r_0)} = \frac{6 \times 100 \times 405}{(50)^2 (2 \times 100 + 50)} = 0.3888 \text{ MPa; bending}$$

$$p_{ys} = \frac{2r_0 Q_0}{R^2 - r_0^2} = \frac{2 \times 50 \times 311.7}{100^2 - 50^2} = 4.156 \text{ MPa; shear, bending governs}$$

Rectangular, uniform loading pulse, $p_m = 2\,\text{MPa}$, $t_0 = 0.08\,\text{ms}$.

$$u_m = \frac{(p_0 t_0)^2}{\mathcal{M}_0 m} \frac{R^2 - r_0^2}{4}\left(1 - \frac{r_0}{R}\right) = \frac{(2 \times 0.08)^2}{405 \times 0.0081} \frac{100^2 - 50^2}{4}\left(1 - \frac{50}{100}\right) = 7.316\,\text{mm};$$

peak displacement

EXAMPLE 12.7 RING-STIFFENED CYLINDRICAL SHELL; MEMBRANE RESPONSE

A long shell, with rings at even intervals, has the following properties:

$$R = 100\,\text{mm}, \quad h = 3\,\text{mm}, \quad \sigma_0 = 250\,\text{MPa}, \quad \rho = 0.0027\,\text{g/mm}^3; \quad v = 0.3$$

The rings, each with area $A = 150\,\text{mm}$, are uniformly spaced at $L = 100\,\text{mm}$. Check if the rings are capable of enforcing the bending mode of failure. Find the displacement response due to a rectangular pressure pulse of $p_0 = 20\,\text{MPa}$ lasting for 0.05 mms. Treat the material according to the RP model.

Shell parameter $\lambda = \dfrac{1.285}{\sqrt{Rh}} = \dfrac{1}{13.48}$; $m = \rho h = 0.0081\,\text{g/mm}^2$

Plain shell collapse pressure: $p_c = \sigma_0 h/R = 250 \times 3/100 = 7.5\,\text{MPa}$ (Equation 12.56)

$$p_{cr} = 0.826 p_c\left(1 + \frac{2h}{A\lambda}\right) = 0.826 \times 7.5\left(1 + \frac{2 \times 3}{150}13.48\right) = 9.535\,\text{MPa}\quad\text{(Equation 12.79)}$$

Membrane-type collapse will happen first. The ring section area is insufficient. According to Equation 12.80, it must be $A > 9.47h/\lambda = 9.47 \times 3 \times 13.48 = 383\,\text{mm}^2$ to force a bending mode of collapse. Employ Case 12.14 to find the parameters of interest:

Peak deflection: $u_m = \dfrac{1}{2}\dfrac{p_m t_0^2}{m}\dfrac{p_m - p_c}{p_c} = \dfrac{1}{2}\dfrac{20 \times 0.05^2}{0.0081}\dfrac{20 - 7.5}{7.5} = 5.144\,\text{mm}$

Duration: $t_m = \dfrac{mv_0}{p_c} = \dfrac{p_m t_0}{p_c} = \dfrac{20 \times 0.05}{7.5} = 0.133\,\text{ms}$

For design purposes the presence of such "soft" rings can be ignored. However, the effect of ends in the form of a traveling plastic hinge will cause some increase in the maximum deflection attained.

EXAMPLE 12.8 BUCKLING OF A THIN CYLINDRICAL SHELL SUBJECTED
TO AN IMPULSIVE, RADIAL PRESSURE

The shell is 500 mm long with simply supported ends. $h = 1\,\text{mm}$, $R = 50\,\text{mm}$. The material is Al 6061-T6 with $F_y = 262\,\text{MPa}$. The exponentially decaying pressure is prescribed as

$p = p_0 \exp(-t/t_x)$, where $p_0 = 2\,\text{MPa}$ and $t_x = 0.54\,\text{ms}$. Determine whether this pressure and impulse, applied outside the shell, are sufficient to induce buckling.

Follow Case 12.25 by finding the parameters first:

$$p_e = 0.92E\left(\frac{R}{L}\right)\left(\frac{h}{R}\right)^{5/2} = 0.92 \times 69,000\left(\frac{50}{500}\right)\left(\frac{1}{50}\right)^{5/2} = 0.359\,\text{MPa}$$

$$c_0 = (69,000/0.00273)^{1/2} = 5027\,\text{m/s}$$

$$i_e = 3\rho c_0 R\left(\frac{h}{R}\right)^2 = 3 \times 0.00273 \times 5027 \times 50\left(\frac{1}{50}\right)^2 = 0.8234\,\text{MPa-ms}$$

$p_t = 0.75F_y h/R = 0.75 \times 262 \times 1/50 = 3.93\,\text{MPa}$

$$0.405\frac{K^{1/2}}{\tilde{\varepsilon}_y} = 0.405\frac{15^{1/2}}{0.0045} = 348.6$$

In our case $R/h = 50/1 < 348.6$; so the longer equation for i_t holds:

$$i_t = 1.807\rho c_0 R\left(\frac{h}{R}\right)^{3/2}\frac{\sqrt{\tilde{\varepsilon}_y}}{K^{1/4}}$$

$$= 1.807 \times 0.00273 \times 5027 \times 50\left(\frac{1}{50}\right)^{3/2}\frac{\sqrt{0.0045}}{15^{1/4}}$$

$$= 0.1195\,\text{MPa-ms}$$

According to Case 3.5, the impulse of the prescribed pressure curve is $i_m = p_m t_x = 2 \times 0.54 = 1.08\,\text{MPa-ms}$. The peak applied pressure p_m is between p_e and p_t, which tells us that buckling, if any, will take place in the elastic range. Then

$$r = \left(\frac{p_m}{p_e} - 1\right)\left(\frac{i_m}{i_e} - 1\right) = \left(\frac{2}{0.359} - 1\right)\left(\frac{1.08}{0.8234} - 1\right) = 1.424$$

As $r > 1$, buckling will indeed eventuate. As a matter of interest, the static buckling load for this shell, without the benefit of end constraints would be

$$p_{crs} = \frac{3D}{R^3} = \frac{3}{R^3}\frac{Eh^3}{12(1-v^2)} = \frac{3}{50^3}\frac{69,000 \times 1^3}{12(1-0.3^2)} = 0.1516\,\text{MPa only}$$

13 Dynamic Effects of Explosion

THEORETICAL OUTLINE

The main interest in this chapter is the external effect of a chemical explosive, which, upon detonation, is almost immediately converted into a hot and highly pressurized gas. The gas expands violently pushing the air in front of it. As the speed of expansion (outflow speed) far exceeds the sonic velocity of air, a shock wave (a blast wave) forms, traveling with the expanding surface of the gas volume. The pressure inside the gas decreases during the expansion, but even when it drops below the atmospheric pressure, the expansion still continues due to inertia, later stopping and possibly reversing the direction of movement. During an early phase of explosion, the blast wave separates from the explosive gas, and continues independently, becoming gradually slower and weaker.

Numerous references are made in this chapter to results of physical experiments. Such results are characterized by a large coefficient of variation, as it is not unusual to read about a 30% difference from two seemingly identical setups. It is, therefore, pointless to spend plenty of time striving for an analytically exact answer.

After characterizing an explosion in terms of its pressure and impulse output in air, the interaction of the blast wave with physical objects and structures is studied. A new concept of that interaction as a collision of a blast wave and an object in its path is presented as a complement to the well-entrenched procedure of using a prescribed pressure–time relationship.

PHYSICAL EFFECTS OF AN EXPLOSION

The effects will be briefly discussed with reference to Figure 13.1, illustrating pressure history following a detonation of an explosive charge suspended in air. The explosion itself takes place at $t = 0$. The emanating shock wave reaches an observer at a distance from the explosion source after the *arrival time* t_a. He can see pressure rising, almost instantly, to the level of p_s, usually called *incident* or *side-on pressure*. Then, over the interval t_s the overpressure drops to zero, or pressure drops to the atmospheric level p_a. This ends the *positive phase* of the pressure history. The decrease of pressure may continue in time so that the *negative phase* is entered with the peak suction Δp_n reaching a fraction of the atmospheric pressure level.

In estimating the structural effects, there are two reasons for the attention to be usually focused on the positive phase. First, at a close distance from a source the negative phase does not develop. Second, the Δp_n is usually much smaller than p_s, although this argument is offset, to a certain degree, by a long duration of the negative phase. The *incident* or *side-on impulse* i_s is the hatched area in Figure 13.1a.

The pressure history in Figure 13.1a is a free-field diagram, which means there is no significant interference between the moving blast wave and the objects in its way. For structural loading, this side-on pressure is only one of several parameters that needs to be considered. When a plate, for example, is impacted by the wave, the reflected pressure and its

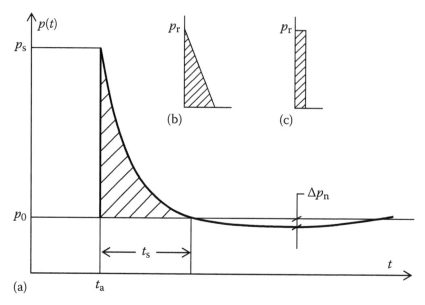

FIGURE 13.1 Free-field pressure history (a), simplified reflected pressure history (b), and the same shown as a rectangular pulse (c).

duration is what really matters. This is another way of saying that the *reflected impulse* i_r, or the area under reflected pressure, when plotted against time, is of essential importance.

One should also mention that the peak reflected pressure p_r is typically several times larger than the peak incident pressure p_s and the same relation applies to impulses, as $i_r > i_s$. The wave parameters are summarized in Case 13.3. The wave speed U decreases with the decrease of the incident pressure and in the limit it equals the sound velocity c_0.

As Figure 13.1 indicates, the side-on pressure is really an *overpressure*, or the excess of the total over the ambient pressure p_0. This is a minor ambiguity, because sometimes a distinction between the two quantities is needed in the course of a calculation. If this is the case, the symbol $\Delta p_s = p_s - p_0$ will be used for overpressure, with p_s standing for the absolute value.

The blast wave information, such as pressures, durations, and impulses are available in a number of documents, one of them being TM5 [106]. Those parameters serve to create simplified pressure diagrams, typically in the form of a triangular decreasing pulse, such as one in Figure 13.1b. (A more involved diagram may be needed, if the natural period of the impacted structure is long enough to be comparable to t_s.) The rectangular pulse (with the same shaded area) in Figure 13.1c used instead of a true pressure history causes only a small decrease in accuracy provided the pulse duration is much shorter than the natural period.

The above comments were made with reference to a *normal impact* of the wave, that is, the direction of propagation being normal to the impacted surface. Also, said surface is assumed to be infinite, which is another way of saying that within the observation time the extent of the surface plays no role. The case of inclined impact as well as the influence of the finite size of an impacted object will be discussed later.

When a beam or a plate is subjected to loads of large magnitude, but short duration, making the calculated stress vastly exceed the yield point of material, this by itself does not imply failure. In such cases the main response quantity, to which anticipated failure is often related, is the peak deflection or, sometimes, the maximum slope of a surface.

INSTANTANEOUS EXPLOSION CONCEPT; γ-LAW

The explosive is treated as a polytropic gas, confined to the initial charge volume. To fully characterize that gas, three constants are needed, that is, initial density ρ_0, energy content per unit mass q, and the polytropic exponent n. Usually, the basic explosive data provided by a manufacturer are ρ_0, q, and D, the latter being the velocity of detonation, also designated by VOD. The following equation, proposed by Henrych [32], provides the initial pressure of explosion:

$$p_0 = \frac{\rho_0 D^2}{2(n+1)} \tag{13.1}$$

When Equation 4.29 is written as

$$p_0 = q_0 \rho_0 (n-1) \tag{13.2}$$

the equivalent polytropic exponent n can be determined by equating p_0 from Equations 13.1 and 13.2:

$$n^2 = \frac{D^2}{2q_0} + 1 \tag{13.3}$$

This exponent n is analogous to γ in the exact theory. Meyers [61], for example, uses in his Table 10.3, an expression for γ, similar to Equation 13.3.

Here, one should also note that the peak pressure of explosion, designated by p_{CJ} is $2p_0$, as one can find, for example, in Zukas et al. [122]. The approach is analogous to what is known as γ-law method, with possibly two differences: (1) the initial pressure,* which here is p_0, is elsewhere often set as p_{CJ} and (2) $\gamma = 3$ is often employed, instead of the physically consistent value defined above. With regard to the latter, Figure 9.9 in Meyers [61] clearly indicates that $\gamma = 3$ is somewhat excessive.

The instant of explosion takes place when the confinement of the compressed gas is suddenly removed. There are three geometrically distinct cases to be considered: uniaxial, cylindrical, and spherical. The first describes an explosion in a rigid-wall tube. The pressure–volume relationship for this geometry was discussed and tabulated in Chapter 4, where properties of a polytropic process were also presented. The remaining two cases are tabulated here, as Cases 13.1 and 13.2.

The most popular method in recent years of including the explosive action in computer simulations is sometimes called a progressive burn. A conversion from a solid to explosive gas is conducted gradually, from a designated initiation point. (The term "gradually" is somewhat misleading in the sense that the same detonation velocity is involved and this is measured in km/s.) The γ-law could be seen as a complementary process in the sense that it inputs

---

* The initial pressure p_0, as defined here, has a specific physical meaning. It is estimated by Persson [66], for example, to be close to the maximum borehole pressure in rock, provided the explosive completely fills the hole radius.

the same amount of energy, but instead of doing it very fast, it does it suddenly. In practice the two methods give different result, but typically the differences are small [91]. The advantage of the progressive burn is that the role ignition point can be properly accounted for. However, this does not hold true in many practical situations. (See "Closing remarks.")

Basic Properties of Explosives

Those properties, consistent with the above formulation, are summarized in the Table 13.1. All of them are military explosives except for ammonium nitride and fuel oil (ANFO). The properties are based on Ref. [52] as well as other sources. As those properties vary somewhat based on method of manufacture and other factors, the aim was to quote typical values.

The constants ρ, q, and D are the basic data, available from manufacturers. The conversion from a solid explosive material to a high-pressure gas takes place as the detonation wave travels across the volume of explosive. Because of the high speed (D) associated with it, the conversion from the solid to the gaseous state may be treated as instantaneous, at least as most external mechanical effects are concerned.

There is a qualitative concept of *brisance* of an explosive, a term used quite frequently but often vaguely defined. According to some, the rapidity with which an explosive reaches its peak pressure is a measure of its brisance. The term is often related to the extent of damage in the immediate surrounds of a charge. If one imagines a charge acting like a sort of 3D spring, according to the γ-law concept, the more *brisant* an explosive, the stiffer the spring (larger n) and that spring is preloaded to a higher pressure p_0.

Spherical Charge Suspended in Air

This is a reference configuration, which had been a subject of extensive physical testing. The most frequently cited source of experimental results appears to be TM-855-1. The

TABLE 13.1
Basic and Derived Explosive Properties

Explosive	ρ	$q/10^6$	D	n	p_0	$q_G/10^6$	q_G/q
TNT	1.60	4.61	6800	2.4526	10714.3	3.58	0.7766
RDX	1.77	5.37	7320	2.4473	13756.0	4.28	0.7970
HMX	1.89	5.46	9110	2.9326	19943.0	4.44	0.8132
Tetryl	1.62	4.51	7550	2.7055	12460.5	3.14	0.6962
PETN	1.76	6.09	8260	2.5694	16821.0	4.31	0.7077
CompB	1.72	5.21	7990	2.6696	14961.5	3.64	0.6987
ANFO(1)	0.80	3.72	4500	1.9292	2765.3	2.30	0.6183
ANFO(2)	0.80	3.72	2500	1.3565	1060.9	2.30	0.6183

Note: ρ, density in g/cm^3; q, energy content (4.61×10^6 N-mm/g or 4.61 MJ/kg for TNT); D, detonation velocity, m/s; q_G, Gurney energy (to be defined later); p_0, initial pressure of inst. explosion, MPa (from Equation 13.1); n, polytropic exponent (from Equation 13.3).

main quantities of interest with regard to structural effects are incident pressure p_s, reflected pressure p_r, and reflected impulse i_r. The accurate values of experimental averages can be extracted from the reference, but it is convenient to have simple, approximate formulas for those as well. The *range*, or the *scaled distance Z* is defined as

$$Z = \frac{r}{M^{1/3}} \qquad (13.4)$$

where
    r is the distance from the center of charge to the target (meters)
    M is the charge mass (kg)*

The pressures for a TNT explosion and the associated impulses are specified in Case 13.4. The results are also available, in the cited references, for *cased* explosives, like bombs, artillery shells, etc. If an explosive different then TNT is used, then the mass *M* used in Equation 13.4 is calculated from

$$M = M_e \frac{q_e}{q_{TNT}} \qquad (13.5)$$

where
    M_e is the actual mass of the explosive used
    q_e is its specific energy content
    M calculated in this way is called a *TNT-equivalent mass*

Positive phase duration t_s, shown in Figure 13.1, also called the *overpressure duration*, is an important quantity in calculating the structural effects of explosions. Henrych [32] gives several formulas for this parameter, with the simplest being

$$t_s = 1.5 M^{1/6} r^{1/2} \ (ms) \qquad (13.6)$$

when $M \sim kg$ and $r \sim m$. While not very accurate, this gives a reasonable first estimate. The negative phase peak suction Δp_n and the specific impulse i_n of that phase, after Smith [80], are

$$\Delta p_n = \frac{0.035}{Z} \sim MPa, \quad Z > 1.6 \qquad (13.7)$$

$$i_n \approx \left(1 - \frac{1}{2Z}\right) i_s \qquad (13.8)$$

where i_s is a side-on impulse.
    A few words of the mechanics of explosion may be in order with reference to the basic case of a spherical charge initiated at center. The detonation wave, which is a shock wave,

---

* One can note that the radius of a spherical charge is proportional to $M^{1/3}$, therefore the range is a multiple of a ratio of distance to the charge radius. That multiple is unit-dependent.

spreads radially, leaving behind a hot, highly compressed gas. When this wave reaches the surface of the charge, a shock wave forms and propagates in the air. The highest pressure is at the surface of charge; about 62 MPa according to Henrych [32]. For $Z \leq 1.6$, the explosive gas is a part of the shock wave. For $Z > 1.6$, the wave contains only air.

HEMISPHERICAL CHARGE ON A SOLID SURFACE

The charge is shown in Figure 13.2b, along with a fictitious complementary hemisphere drawn in a dashed line, below the surface. The previous type of explosive charge discussed, namely a charge suspended in air, is shown in (a). Both charges have the same radius. One can draw a horizontal plane across the center of the spherical shape in (a) and make the following observation: As long as the plane is rigid and frictionless, it does not interfere with the sudden expansion of the sphere. For the space above the charge there is no difference between (a) and (b) as long as the ground surface is rigid and frictionless.

Therefore, if the ground is idealized as described, the effect of the explosion of a hemispherical charge on its surface is the same as the free-air explosion of the charge of the same radius. This is equivalent to saying that a ground charge with mass M, under conditions described, has the same effect as the air charge with the mass of $2M$. The actual ground surface may sometimes approach such an "ideal reflector" but in most real situations a surface explosion leaves a crater, which is associated with energy dissipation. The net effect is such that if the mass of the surface charge is M, then, instead of being equivalent to $2M$ charge in the air, it is equivalent, most frequently, to about $1.7-1.8M$. That multiplier of is only the first approximation, a guide for a typical condition. Even for a perfectly hemispherical shape of explosive, the multiplier visibly depends on the type of soil involved. The surface explosion is often called a *ground burst* and has been even more extensively tested than the free-air burst. The problem with the latter is that the height of suspension has its practical limits and if the height is not sufficient, then reflections of the blast wave take place.

The incident pressure p_s from the ground burst may therefore be estimated from the air burst data (presented in Case 13.4) by inserting $1.75M$ in place of M. The importance of p_r and i_r requires, however, that the formulas with the actual charge mass M involved in the ground burst be separately specified, as was done in Case 13.5.

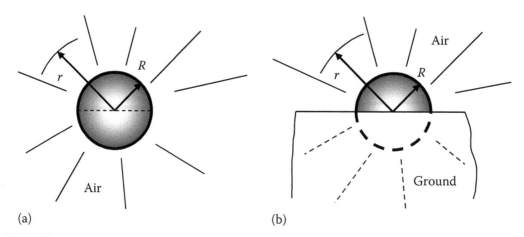

(a)   (b)

FIGURE 13.2   Free-air explosion (a) and surface burst (b).

PROPELLING OF SOLID CONTAINER PIECES BY EXPLOSIVE ACTION

Detonating a high-explosive charge filling a thick-wall metal container had been done under laboratory conditions many times in the past. The resulting metal fragments fly away often with speeds similar to those of a rifle bullet. The same event takes place when a slab of explosive is lined on both sides with a metal plate and detonated. When resulting fly-off velocity is measured, it gives one the ability to estimate the mechanical energy of the explosive.

The best-known such experiments were conducted during the Second World War and reported by Gurney in 1943 [28]. A summary of what was done and the reasoning behind it is presented, in a number of texts. The effective energy content is q_G, known as Gurney energy, is given for selected explosives in Table 13.1.

The ratio of this to the total energy content, namely q_G/q, may be termed the mechanical efficiency of the explosive. For the above examples, we find this efficiency in the range of 60%–80%. If the industrial explosive ANFO is not included, the efficiency range becomes 70%–80%. (The net mechanical energy of ANFO is based on data from mining industry, where it is called the useful energy.)

Five specific metal explosives configurations are presented by Cases 13.15 through 13.19. They are a long tube, a sphere, and three sandwich-type combinations of different degrees of symmetry. In each case, the fly-off velocity v_f is found from

$$v_f = \sqrt{\frac{2q_G}{M_m/M_e + \beta}} \tag{13.9}$$

where
M_m is the mass of metal
M_e is the mass of explosive

The coefficient β is 1/2 for the tube, 3/5 for the sphere and 1/3 for a symmetric sandwich, while the nonsymmetric sandwiches lead to more complex formulations. The expression for v_f comes from equating the effective energy of explosive ($M_e q_G$) with the peak kinetic energy of the flying metal plus the kinetic energy of the expanding explosive. (The velocity of the latter is assumed to vary linearly between the center and the outside.)

Consider the initial configuration of an explosive filling a cylindrical shell, as illustrated in Case 13.15. A unit element of such a shell (unit depth normal to paper and unit length measured along R) has a mass of ρh. If an internal explosive pressure p is applied, the driving force is $p r_i/R$, somewhat less than p because pressure is applied to r_i rather than R. If the material strength is ignored, the overall acceleration of the element is

$$a = \frac{p r_i}{\rho h R} \tag{13.10}$$

The mechanical resistance of the metal is, in many cases, sufficiently small compared to a driving pressure p, so that the above expression gives a good approximation for estimating the kinetic aspects of the event. This relation is, of course, relevant for small deflections only, as it ignores the fact that the internal radius grows and that thickness h diminishes during a large deformation process.

CONTACT EXPLOSION ON A HARD SURFACE

Consider a layer of explosive, h_e thick, overlaying a hard surface and examine what happens within a small column, cut out of the charge away from the edges, having a unit area base and a height h_e. The presence of the rest of the material around it, which behaves in the same manner, makes it possible to treat this column as a channel with rigid walls. According to the simple approach used by Gurney, the particle velocity within the column is linearly distributed, reaching its maximum at the top. It is easy to find out that for such a column the kinetic energy is

$$E_{ke} = M_e v_p^2/6 \tag{13.11}$$

where
   v_p is the peak outflow velocity
   M_e is the mass of the column

To calculate v_p of the charge material, one assumes that the entire effective energy content $Q_e = M_e q_G$ of the explosive charge converts into kinetic energy, which gives

$$v_p = \sqrt{\frac{6Q_e}{M_e}} = \sqrt{6q_G} \tag{13.12}$$

To find the maximum impulse applied to the surface of the metal, apply the equation for an open sandwich, Case 13.19, to a configuration where the metal mass M_m is much larger than the explosive mass M_e. In such a case the fly-off velocity of metal is

$$v_f \approx \sqrt{2q_G}\sqrt{\frac{3}{4}\frac{M_e}{M_m}} \tag{13.13}$$

When this is multiplied by M_m, the result is the impulse $S_s = M_m v_f$ applied to the metal. Applying this to a column, cut out of an explosive layer:

$$S_s = Mv_f = 0.866\sqrt{2q_G}\,\rho_e h_e \tag{13.14}$$

where
   h_e is the thickness of explosive layer
   S_s is the impulse is per unit surface area

The above relates to a continuous layer of explosive with $\rho_e h_e$ being the mass per unit area. If a different charge, of a compact shape, is involved, one can write, in a more general fashion:

$$S = \lambda M_e\sqrt{2q_G} \tag{13.15}$$

where the coefficient, $\lambda < 1$, depends on the geometry of the charge. If, for an arbitrary mass the fly-off velocity v_f is experimentally found, then the impulse $S = M_e v_f$ becomes known and the coefficient λ can be found from Equation 13.15.

As for the peak outflow velocity of the gas itself, it may be obtained from the impulse–momentum relation as well. The metal velocity in the open-sandwich case, for which $M_m > M_e$, is quite small, therefore the average speed of the explosive gas may be taken as a half of its maximum, $v_{av} \approx v_p/2$. Equating the momentum $M_e v_{av}$ with the impulse from Equation 13.14 results in Equation 13.12 again.

The other part of the evidence is that of a cylindrical charge used to propel a metal plate. There are several references to such experiments, among others in Cooper [19]. The effective part of a cylindrical charge standing on a flat plate was found to be the cone with 60° apex angle. The ratio of the mass within the cone to the total mass of explosive would then be our λ per Equation 13.15. This estimate, however, is of a historical interest only. In his recent work, Cooper [20] examined a large amount of experimental data for plates driven by cylindrical contact charges. As it turned out, the above approximation overestimates the fly-off velocity, especially for relatively large explosive/metal mass ratio. A good fit to experimental data is obtained by

$$\lambda = 0.22\sqrt{\frac{d}{h}} \tag{13.16}$$

with d=cylinder diameter. For a case with $d=h$, for example, one finds λ=0.22 from Equation 13.16 while λ=0.289 would be found on geometrical grounds, a substantial overestimate.

The four basic shapes (flat charge, hemisphere, cube, and cylinder) placed on a rigid surface were investigated using FEA simulations by this author [100] and the results are given as Cases 13.20 through 13.23. It was determined, among other things that the output, namely the impulse and the peak reaction force vary, sometimes substantially, depending upon how far the ignition point is from the rigid surface. Typically, the results were slanted towards a midpoint of ignition, as the most probable in frequent cases where the ignition point is unknown. There was quite a discrepancy between the findings for a cylinder of $d/h = 1$, which resulted in $\lambda = 0.27$ against 0.22 from Equation 13.16. The most probable reason for this may have been that while the last figure was relevant for flyer plates; those plates were not sufficiently thick so as to approach the reaction of a rigid surface.

CLOSE PROXIMITY IMPULSE

Equation 13.15 tells us, that in an idealized case of an explosive acting on a rigid surface, with the entire energy content being effective, we may set $\lambda = 1$ and replace q_G with q, obtaining $M_e\sqrt{2q}$. Using a similar form, Baker [5] proposed that a specific reflected impulse i_r on a rigid surface surrounding a spherical charge be approximated by

$$i_r = \frac{[2(M_e + M_a)M_e q]^{1/2}}{4\pi r^2} = \frac{M_e}{4\pi r^2}\left[\left(1 + \frac{M_a}{M_e}\right)2q\right]^{1/2} \tag{13.17}$$

where
   r is the radius of the surrounding surface
   M_a is the mass of air enclosed within r

As long as $r > r_s$, where r_s is the charge radius, then, with the density of air ρ_a one has $M_a \approx (4/3)\pi r^3 \rho_a$. Equation 13.17 was experimentally demonstrated to work well up to about 20 charge radii. A similar expression may be written for a long cylindrical charge.

ABOVEGROUND EXPLOSION

If a charge explodes over a surface, as in Figure 13.3, then the largest reflected impulse i_{r0} is experienced at the point C directly below the charge. This is the point, for which the formulas given before prescribe the reflected pressure p_{r0} and impulse i_{r0} and where the height of burst h simply replaces the distance r. We speak of a plane wave or a *planar wave* of explosion in the vicinity of that point, implying constant reflection parameters. For other points, where $\alpha > 0$, those parameters result from a rather complex interaction of the incident and the reflected spherical wave. The ratio of $\Delta p_r / \Delta p_s$, or reflected and incident overpressure, may vary as α grows and sometimes exceed the value for $\alpha = 0$, as a graph in Henrych [32] shows. The effect of angle α on the reflected values can be seen in the families of curves provided in TM-5 [106]. The simplest way of summarizing the effect is to compare the quantities obtained for a plane wave reflection at r (p_r and i_r)* and the corresponding quantities for the actual inclined wall ($p_{r\alpha}$ and $i_{r\alpha}$). The following approximate equations

$$\ln(p_{r\alpha}) \approx \ln(p_r) \cdot \cos(0.8\alpha); \quad (0^\circ < \alpha < 80^\circ) \tag{13.18a}$$

$$\ln(i_{r\alpha}) \approx \ln(i_r) \cdot \cos(0.7\alpha); \quad (0^\circ < \alpha < 80^\circ) \tag{13.18b}$$

give a rough, but a reasonable picture of the relationships. The important aspect of those observations is a slow change in the magnitude of the reflected impulse in the vicinity of point C.

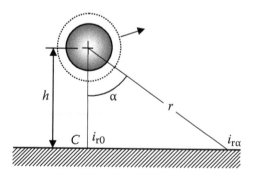

FIGURE 13.3   A spherical charge exploding over a hard surface.

---

* The quantities p_r and i_r are the reflected values for a wall normal to radius r placed at r away.

One has to keep in mind that the proximity of a hard surface always magnifies the effect of an explosion. If, say, a 100 kg charge of TNT is suspended high in the air, it applies a certain impulse to surrounding objects. If the same charge is placed on an ideally hard ground, it becomes equivalent to a 200 kg charge with respect to objects at the same distance as before. For intermediate positions, that is, at some heights above ground, the magnification effect depends on not only on the relative distances, but also on angles involved.

LOADS ON FIXED, SOLID OBJECTS FROM A DISTANT BLAST

The most relevant object of attention is, of course, a building, assumed to be solid, without such complications as door and window openings. The word "distant" means that the source of the blast is far enough away so that the nearest wall experiences an almost uniform, initial pressure effect.

Let us focus our attention, for a moment, on the front wall of the building illustrated in Figure 13.4 and let us treat that wall as a kind of rigid plate, fixed in space. At the instant the wave front impacts this wall, the pressure jumps to p_r. Simultaneously, a complex process of interaction along the edges, called *diffraction*, begins. The edge becomes a source of an unloading or relief wave, moving into the surface with the velocity of c_U, the sound speed in the reflected wave front, as illustrated in Figure 13.4b. Those waves cause the pressure to drop from the reflected value of p_r to the side-on level of overpressure Δp_s plus the drag pressure q_d. It is generally assumed that it takes

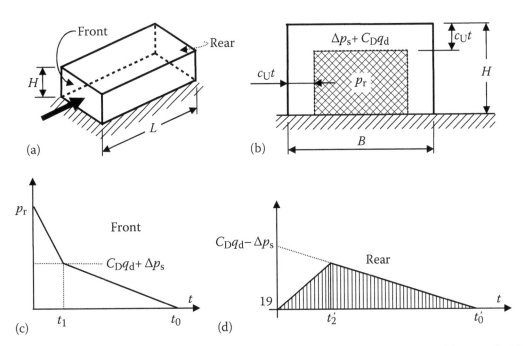

FIGURE 13.4 (a) Blast wave impacting a box-like building, (b) partially relieved front wall, (c) overpressure history on the front wall, and (d) net suction on rear wall, time being counted from the wave arrival at that wall.

$$t_1 = \frac{3H}{c_U} \quad \text{or} \quad t_1 = \frac{3B}{2c_U} \tag{13.19}$$

whichever less for this pressure drop to eventuate. After this time there is no trace of the wave impact, although the air still keeps moving against the plate giving rise to drag pressure. Time t_1 is frequently referred to as the *clearing time*. Afterward, the action of the moving air behind the wave front is almost the same as that of a wind with gradually decreasing velocity.* More details can be found in Case 13.6.

The above explains why it is important to distinguish between two different load mechanisms involved. One is the application of a shock wave, corresponding to an interval of time between t_a in Figure 13.1 and $t_a + t_1$, where t_1 designates the duration of wave action. ($t_a = 0$ was assumed in Figure 13.4c.) After the wave clears, the second mechanism, namely the drag loading, or, in plain words, wind loading, defined in Appendix C governs. (The above applies to the front wall, that is, the wall first experiencing the shock wave.) When the overpressure duration (t_s) is long, as typical of relatively large distances from the explosion source, the second mechanism may become more important than the first.

Now let us return to what is happening to the rest of the building. While reflection and subsequent relief is taking place on the front wall the rest of the wave continues to "slide" along the side walls and the roof. During this time the side walls and the roof are subjected to overpressure Δp_s. It takes the wave $\Delta t = L/U$, moving with the speed U, to reach the rear wall and initiate the pressure growth process. Then it takes

$$t_2' = \frac{5}{3} t_1' \tag{13.20}$$

to build up the pressure to the level of $C_D q + \Delta p_s$. (Time t_1' is calculated as t_1 before, except that the current, rather than reflected pressure is involved.) One should be mindful of the fact that the wind or drag pressure gives suction along all faces of the building except the front wall thereby causing the reduction of the resultant pressure. (See Appendix C.) For the rear wall dynamic pressure $C_D = -0.3$ is sometimes used, although the actual value depends on the building geometry. More details on this subject can be found in Baker [4] and Henrych [32].

In practical situation the loading of an object is often resolved into two simplified impulses, as illustrated by the second sketch of Case 13.6. It is often helpful to refer to Table 13.2.

BLAST WAVE INTERACTION WITH MOVABLE SOLID BODIES

An air blast wave is usually treated as a pressure–time event, which affects solid bodies in its path. There is another way of looking at such an encounter, however, as presented by this author [98], namely, as a collision of two distinct bodies, one being the wave (M_1) and

---

*The above description and illustrations are simplified in several ways. The most obvious one is to show a pressure drop as taking place along a straight line, in time.

TABLE 13.2

Blast Wave Parameters

Side Overpressure Δp_s (kPa)	Wave Front Speed U (m/s)	Particle Velocity V (m/s)	Density ρ (kg/m³)	Sound Speed c_U (m/s)	Reflected Overpressure Δp_r (kPa)	Dynamic Pressure q_d (kPa)
20	367.97	44.38	1.393	349.15	43.29	1.37
40	393.70	82.96	1.552	357.01	92.82	5.34
60	417.85	117.25	1.703	364.16	148.08	11.70
80	440.67	148.24	1.846	370.81	208.66	20.28
100	462.38	176.60	1.982	377.08	274.16	30.90
140	502.98	227.28	2.235	388.79	418.50	57.71
180	540.54	271.92	2.465	399.70	578.65	91.10
220	575.65	312.07	2.675	410.04	752.56	130.23
260	608.74	348.76	2.868	419.94	938.53	174.39
300	640.13	382.68	3.046	429.48	1135.13	222.97
350	677.32	421.95	3.249	440.97	1393.99	289.16
400	712.57	458.37	3.434	452.08	1665.57	360.65
450	746.15	492.46	3.603	462.84	1948.23	436.76
500	778.29	524.58	3.758	473.30	2240.59	516.91
600	838.88	584.03	4.032	493.45	2849.99	687.50
700	895.38	638.37	4.268	512.70	3486.44	869.35
800	948.52	688.70	4.472	531.19	4144.56	1060.23
900	998.84	735.75	4.651	549.01	4820.32	1258.47
1000	1046.74	780.09	4.809	566.24	5510.62	1462.76
1100	1092.55	822.13	4.949	582.94	6213.05	1672.10
1200	1136.51	862.18	5.075	599.15	6925.69	1885.71
1300	1178.83	900.49	5.188	614.92	7647.04	2102.93
1400	1219.68	937.28	5.291	630.29	8375.84	2323.27
1600	1297.53	1006.90	5.469	659.93	9851.94	2771.64
1800	1370.97	1072.09	5.619	688.27	11347.80	3228.25
2000	1440.67	1133.58	5.747	715.48	12859.03	3691.26
2400	1570.82	1247.59	5.953	766.97	15915.76	4631.57
2800	1690.99	1352.09	6.112	815.19	19005.15	5585.48
3200	1803.16	1449.11	6.239	860.70	22117.17	6548.82

Note: The above quantities are calculated according to Case 13.3. They define the parameters just behind the progressing wave front. They correspond to a distant source, in the manner defined before or, to a small patch on a surface surrounding a spherical source.

the other the impacted object (M_2). The collision problem was quantified in Chapter 8 and presented, for this particular case, in Figure 13.5.

A particular case is considered here, when M_2 is stationary (or $v_2 = 0$) at the beginning and the rebound is elastic. For this case, one has

$$v_0 = \frac{M_1 v_1}{M_1 + M_2} \qquad (13.21)$$

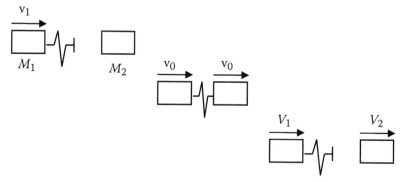

FIGURE 13.5   Three states of motion during collision of the air wave M_1 with an (initially) stationary solid object M_2.

The final velocities are

$$V_1 = 2v_0 - v_1$$

$$V_2 = 2v_0 \tag{13.22}$$

The impulse transferred during collision can simply be written as

$$S = M_2 V_2 = 2M^* v_1 \tag{13.23}$$

$$\frac{1}{M^*} = \frac{1}{M_1} + \frac{1}{M_2} \tag{13.24}$$

When M_2 is so large that M_1 is, in fact, impacting against a rigid wall, then $M^* = M_1$ and

$$S_m = 2M_1 v_1 \tag{13.25}$$

In case M_1 and v_1 are not separately known, but S_m and S are, one can write

$$\frac{S_m}{S} = \frac{M_1}{M_2} + 1 \tag{13.26}$$

which makes it possible to extract M_1 and then to determine v_1 from Equation 13.25:

$$M_1 = M_2(S_m/S - 1) \tag{13.27}$$

$$v_1 = \frac{S_m}{2M_1} \tag{13.28}$$

In the referenced work, the impulses involved, namely S_m and S are determined by a semiexperimental way, using FE simulation. The impacting mass is then found from Equation 13.27. One of the necessary conditions for verification of this hypothesis is that when M_2

is varied, the calculated mass M_1 of the impacting air wave remains constant, or nearly so. When the study was carried out, with the ratio of the heaviest to the lightest M_2 as 25, the largest calculated mass M_1 was 1.15 times its mean value, a relatively modest change. This indicates that the "two-body collision" hypothesis is a plausible one.

One aspect of virtual experiments mentioned above is reported here as Case 13.13. Body M_2 in Figure 13.5 is a piston or a plate sliding in a shock tube. The piston is impacted by a blast wave generated by an explosive. A set of material densities assigned to the piston gives a corresponding set of impulses S. The peak impulse S_m is found when the plate is fixed in place. The end result gives the reduction of impulse as a function of the decrease of the mass of the impacted body. This is a purely inertial effect, as the flow around the impacted body is not allowed.

A similar investigation is reported as Case 13.12. Eight basic profiles are presented there with material properties assigned varying from heavy to light solids. They are successively subjected to a plane blast wave while unconstrained in the direction of flow. The medium (air) is free to flow around the profiles. The profiles may be thought of as cross sections of long beams, placed at right angles to the flow. The problem is therefore two-dimensional. The impulse coefficient is defined as

$$C_i = S/S_m \tag{13.29}$$

where
S is the impulse experienced by the unconstrained section
S_m is the nominal reflected impulse corresponding to the height of the profile

The impulse S applied to the body is equal to the momentum gained:

$$S = Mv_m \tag{13.30}$$

where M is the profile mass. There are a few interesting observations, that can be made from inspection of results in Case 13.12. Thin sections, like 1, 5, and 6 have their C_i considerably reduced when their density is decreased. On the other hand, large-area profiles, like 2, 3, and 4 show only a slight decrease in their resistance. Also, a more "aerodynamic" or "nose forward" shape, the smaller C_i displays. This is most evident between "sister" sections like 5 and 6 as well as 7 and 8. Evidently, there are two components of resistance, as measured by C_i, namely the shape and the inertia.

The approach culminating in Equations 13.29 and 13.30 is a "united" treatment in the sense that no distinction is made between the wave action and drag action. The distinct action of the two phases can be simply treated only in case of fixed objects. An indirect confirmation of the united approach is presented in Example 13.10. One should keep in mind that the method was quantified on the basis of a plane wave impact.

One of the relevant investigations of how the shock wave interacts with three-dimensional bodies was conducted by Shi et al. [77] by means of virtual experiments. When comparing the reflected wave from a circular- and rectangular-section columns (with the diameter the same as the side length across the flow), it was determined that the ratio of peak reflected pressures was 1.28 and for the reflected impulses it was 1.24.* The material was concrete,

---

* The numbers for the square section were larger.

which was not specifically included in Case 13.12. Interpolating between Material I and II therein gives the ratio of impulses of 1.28 (when sections 3 and 2 are compared), reasonably close to that obtained by Shi et al. This relates mainly to the shape of the section while the ability to deflect, which makes a cross section a "movable" body, is of less relevance here.

The increase of inertia of a body impacted by a blast wave increases C_i, or the pushing forces. This indicates again that the interaction between the two is certainly not the case of a prescribed pressure vs. time, applied to the impacted body, but it is more like a collision of two bodies.

At the time of this writing, the awareness of this attribute of blast wave action does not appear to be widespread. In the early 2000s some apparent paradoxes were reported as encountered in experiments related to blast wave mechanics. In one paper the author notices, with a surprise, that a layer of energy-absorbing material attached to a tested body leads to an increase in the impulse applied to that body. In another report the attachment of a protective layer of material to a structure had a more dramatic effect by causing a larger failure then in a structure with no protection. It seems that in both quoted cases the increase of the mass of the target resulted in the increase of the applied impulse.

FORCE–IMPULSE DIAGRAMS

A concise way describing the action of a force is to state its peak magnitude along with the impulse that it applies to a body. From a structural effects viewpoint, it is best to have a graph of some response parameter as a function of all possible combinations of a force magnitude and its impulse magnitude. It will then become evident, which conditions lead to exceeding the permissible response thereby leading to an onset of damage.

Another way of expressing the above is to say that one needs to see the effect of a force, when its duration varies from a very short to a very long time. The first example of it was done in Case 3.5, where a characteristic time t_x, representative of the impulse magnitude, was an independent variable. For a very short t_x the effect was that of an impulsive load and for a large t_x the response approached that of a step load.

Let us first consider two extreme ways of applying the load. Under an impulsive load, equivalent to applying an initial velocity v_0 to a body, the peak displacement is obtained by equating peak kinetic energy with peak strain energy:

$$\frac{1}{2} M v_0^2 = \frac{1}{2} k u_m^2 \tag{13.31a}$$

or

$$u_m = \frac{S}{\sqrt{kM}} \tag{13.31b}$$

where $S = M v_0$ is the abrupt, applied impulse. When, on the other hand, a step load W_0 is involved, its work is equated with strain energy:

$$W_0 u_m = \frac{1}{2} k u_m^2 \tag{13.32a}$$

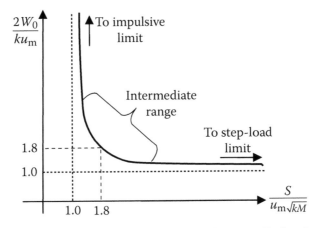

FIGURE 13.6 Nondimensional force vs. nondimensional impulse. Each point of the curve shows combinations of force and impulse that induce the same deflection.

or

$$u_m = \frac{2W_0}{k} \qquad (13.32b)$$

To be more specific, let us create a plot of W_0–S, a force–impulse diagram, associated with the same peak displacement u_m, that represents the maximum permissible deflection, related to an onset of damage. This will be created for a basic oscillator, although the procedure is similar for any other system. One can create it in a "raw" form, placing W_0 on the vertical axis and S on the horizontal, as is done in Example 13.5. However, this type of graph is seen more often presented in a nondimensional form as shown in Figure 13.6.*

The horizontal asymptote, $2W_0/(ku_m) = 1$, corresponds to a step-load limit. For any load, that does not instantly grow to W_0, a larger S will have to be applied than that prescribed by that asymptote, so the curve must have larger values, when going towards the origin. A similar reasoning holds for the other axis; if the load is less than instantaneous, a larger magnitude of W_0 is needed to achieve the same u_m. (For an elaboration on that refer to section "Response of oscillator to impulsive load" of Chapter 3.)

So far the reasoning has been quite general. In order to determine a specific intermediate point on the curve, choose the condition $x = y$ in Figure 13.6, that is, let it be a point on a 45° median line. Let the magnitude of an impulse associated with that point be $S = W_0 t_0$. Then, equating u_m from Equations 13.31b and 13.32b, one gets

$$\frac{2}{k} = \frac{t_0}{\sqrt{kM}} \quad \text{or} \quad \omega t_0 = 2$$

---

* The purpose of the normalization is to deal with nondimensional force and impulse ratios, which is better for bringing up similarities between a variety of structural phenomena.

Let us now return to Case 3.5 mentioned earlier and deal with an exponentially decaying pulse, often used as an approximation of a blast loading. The total impulse applied by that load was also $W_0 t_x$. For the load duration corresponding to* $\omega t_x = 2$ we had $u_m = 1.111 u_{st}$. Noting that the ordinate of Figure 13.6 contains $W_0/k = u_{st}$, we can therefore write the value of the ordinate y as

$$y = \frac{2u_{st}}{u_m} = \frac{2u_{st}}{1.111 u_{st}} = 1.8$$

which gives a position of an intermediate point and permits us, along with the previous information, to sketch the normalized W_0–S curve. This figure contains essentially the same information as one in Case 3.5: there is a part approaching the impulsive limit, a part close to the step-load limit and an intermediate range. Finding the intermediate point in Figure 13.6 must be done in reference to a specific load case, but the application of this methodology is quite general.

A graph of the type in Figure 13.6 is sometimes referred to as an *isodamage curve* of a structural member the curve is representing. The use of force–impulse diagrams extends to much larger systems than the single dynamic degree of freedom (SDOF) used here. When more than one mode of failure is involved, the shape of the corresponding force–impulse diagram becomes more complicated.

BLAST SCALING BASICS

When blast parameters are specified using range Z defined as above, this approach is called Hopkinson's scaling. It implies that the quantities with the dimensions of pressure and velocity are the same as long as Z is unchanged. This scaling principle is illustrated in Figure 13.7, where two charges are shown, one with radius R and the other with a larger radius ζR.

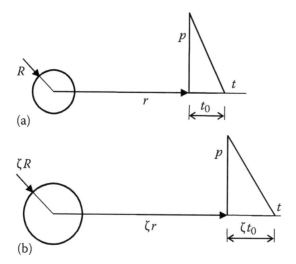

FIGURE 13.7   The original explosive effect (a) and another one, scaled up by ζ (b).

---

* t_x rather than t_0 was used in the reference case.

The location of target points, where the explosion pressure is measured, is also scaled by the same factor ζ, but the range Z is the same at both points. Consequently, the measured pressure is the same. However, the duration of the positive phase is ζ times larger for the second point and so is the impulse. This helps to explain the structure of formulas for impulse, as given in Case 13.4.

As noted by Baker [5], if the model is also scaled dimensionally by ζ, the stress and strain in the model will remain the same, while deflections will change by ζ. As demonstrated in Examples 13.6 and 13.7, this holds true for both elastic and elastic-perfectly plastic (EPP) materials.

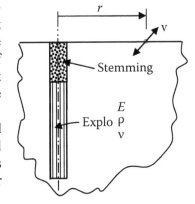

FIGURE 13.8 A cylindrical charge placed in a borehole in rock.

THE EFFECTS OF EXPLOSIONS IN GROUND

The effects can vary substantially depending on the physical properties of the medium. There are essentially two ground types: rock and soil. Their principal difference is the material strength, or cohesion. Soil is, typically, a weak material, which can be broken up by a human hand, while rock needs tools to be disintegrated.* Rock can often be treated as an elastic, homogeneous body, while soil may act like a liquid or as an inelastic solid. The water content and the amount of entrapped air in soil substantially affect its response to an explosion.

Regardless of the medium properties, however, certain aspects of an internal explosion are similar. When a blast takes place deep underground and the charge is of a compact shape, a near-spherical *cavity* results. When, on the other hand, the charge is placed sufficiently close to a free surface a *crater* is formed by means of ejecting some of the material between the charge and the surface. At some intermediate charge positions a *camouflet* may be created; a cavity that is vertically elongated.[†] The attention in this chapter is focused on the response to an in-ground explosion, or *groundshock*, while the cratering problem is discussed in Chapter 15.

One of the typical rock mining applications is to drill a hole normal to the rock surface, fill it with an explosive, put some inert material (stemming) in the upper part of the hole and detonate, Figure 13.8. The purpose of stemming is to retain the explosive gas and a hole is usually a part of an array of holes, being fired one after the other. Strong shock waves emanating from such a borehole break up the rock in the immediate vicinity, but at some distance away regular P-waves appear, which can travel large distances. The disturbance to the medium has a complex character, arising mainly from the interaction with the free surface and appearance of shear waves and Rayleigh waves. At a distance an order of magnitude larger than the length of hole the peak particle velocity (PPV) is

$$v = \frac{KM^\alpha}{r^\beta} \text{ (m/s)} \tag{13.33}$$

---

* The natural materials are not that easy to classify, because boundaries can be quite fluent. A shale, for example, can resemble a grainy soil or a weak rock.
[†] Provided the free surface is horizontal.

when the explosive charge mass $M \sim$kg and distance $r \sim$m. According to Persson et al. [66], the typical values for hard rock are: $K = 0.7$, $\alpha = 0.7$, and $\beta = 1.5$.* In the vicinity of the cylindrical charge, a different relationship holds; this is presented by Case 13.30.

A number of equations are given in cases listing with regard to the output of spherical and cylindrical charges in rock and soil. Whenever overpressure is quoted, it can be converted to PPV by dividing it by impedance, according to Chapter 5:

$$v_m = \frac{p_m}{\rho c_p} \tag{13.34}$$

where c_p is the pressure wave speed.

CLOSING REMARKS

In many publications relating to the explosion effects on slender targets, like beams or columns, the following procedure is employed. The reflected impulse i_r, corresponding to the distance between the charge and the target, is calculated. The load applied to the target, or the impulse per unit length of the target, is simply Bi_r, where B is the target width. One must bear in mind that the procedure is conservative. First, due to the finite target width, the full value of the reflected pressure cannot develop. Second, the reflected pressure will be smaller near the edge than near the center line of the affected surface. Third, the impacted element can deflect, which further decreases the reflected impulse.[†] These influences come through quite clearly in the work of Shi et al. [77], quoted before.

Sometimes the location of the ignition point of the explosive charge can have an influence on the output. This may happen in close proximity explosions. Two observations are in order: Firstly, the ignition point location is frequently not quoted, often in otherwise thorough descriptions of experiments. Secondly, when malevolent explosions take place, the ignition point is practically never known. This reduces the possibility of benefiting from that information and make the γ-law a tool just as good as the progressive burn in such situations.

---

[*] The reality is always more complex than simple schemes often used. The boreholes may be drilled at an inclined rather than a straight angle with respect to the surface. And the "surface" may be a floor of an underground cavern, rather than literally the top of the rock mass.

[†] In case of heavy elements, like a solid concrete section, the effect of mobility will be small. Not so when the blast wave is acting on thin-wall metallic sections.

TABULATION OF CASES

COMMENTS

Cases 13.4 and 13.5 offer simple, approximate formulas for the basic blast wave parameters. Cases 13.12 and 13.13 as well as Cases 13.20 through 13.23 are to be used with caution, as they are based on a limited amount of FE simulations.

Case 13.14 describes necessary conditions to overturn a vehicle subjected to a blast wave. It is derived from Case 7.15, which presents a more general setting.

Cases relating to explosion in rock and soil show a lot of variability of results, the latter related to both the explosive and even to a larger extent to the medium. A more reliable prediction is expected for a good quality rock than for a soil.

CASE 13.1 LONG CYLINDRICAL CHARGE, γ-LAW EXPANSION

Volume per unit length $V = \pi r^2$, expansion from r_h to r. (From p_0 to p)

$p_0 r_h^{2n} = p r^{2n}$ polytropic law

$p = p_0 \left(\dfrac{r_h}{r_h + u} \right)^{2n}$ pressure vs. radial displacement $u = r - r_h$

$u = \left[\left(\dfrac{p_0}{p} \right)^{1/2n} - 1 \right] r_h$ displacement when expanding from p_0 to p

$Q = \dfrac{\pi p_0 r_h^2}{n-1}$ or $Q = \pi r_h^2 \rho q$; energy content per unit-thick layer

$q = \dfrac{Q}{\pi r_h^2 \rho}$; energy content per unit mass; $e = \rho q$ energy content per unit volume

$\mathcal{L} = \dfrac{\pi p_0 r_h^2}{n-1} \left[1 - \left(\dfrac{r_h}{r} \right)^{2(n-1)} \right]$; work performed by gas in going from state 0 to the current state

$p_0 = (n-1)q\rho$ initial pressure

Note: All of the above is per unit axial length.

CASE 13.2 SPHERICAL CHARGE, γ-LAW EXPANSION

Charge volume $V = 4\pi r_s^2/3$, expansion from r_s to r. (From p_0 to p)

$p_0 r_s^{3n} = p r^{3n}$ polytropic law

$p = p_0 \left(\dfrac{r_s}{r_s + u} \right)^{3n}$ pressure vs. radial displacement $u = r - r_s$

$Q = \dfrac{4}{3} \dfrac{\pi p_0 r_s^3}{n-1} = \dfrac{4}{3} \pi r_s^3 \rho q$ energy content $\quad u = \left[\left(\dfrac{p_0}{p} \right)^{1/3n} - 1 \right] r_s$ displacement when expanding from p_0 to p

$$q = \frac{Q}{(4/3)\pi r_s^3 \rho}; \text{ energy content per unit mass; } e = \rho q \text{ energy content per unit volume}$$

$$\mathcal{L} = Q\left[1 - \left(\frac{r_s}{r}\right)^{3(n-1)}\right]; \text{ work performed by gas in going from state 0 to the current state}$$

$p_0 = (n-1)q\rho$ initial pressure

CASE 13.3   AIR SHOCK WAVE PARAMETERS

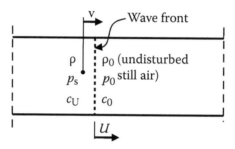

$\gamma = 1.4$; polytropic exponent for air

$p_0 = 0.1013\,\text{MPa}$; ambient pressure at 15°C

$\rho_0 = 1.225\,\text{kg/m}^3 = 1.225 \times 10^{-6}\,\text{g/mm}^3$; ambient density at 15°C

$$c_0 = \sqrt{\frac{\gamma p_0}{\rho_0}} = 340.3\,\text{m/s}; \text{ sonic velocity}$$

All wave parameters below are at (just behind) the wave front.

Expressions are valid up to $p_s = 5\,\text{MPa}$

$$\left(\frac{U}{c_0}\right)^2 = 1 + \frac{6}{7}\frac{\Delta p_s}{p_0} = \frac{1}{7}\left(1 + 6\frac{p_s}{p_0}\right); U \text{ is wave front velocity, second equation valid for } \Delta p_s > p_0$$

$$\frac{v}{U} = \frac{\Delta p_s}{1.4 p_0 + 1.2\Delta p_s} = \frac{(p_s/p_0 - 1)}{0.2 + 1.2(p_s/p_0)}; v \text{ is particle velocity at wave front, second equation}$$
valid for $\Delta p_s > p_0$

$$\frac{\rho}{\rho_0} = \frac{7 + 6(\Delta p_s/p_0)}{7 + (\Delta p_s/p_0)} = \frac{1 + 6(p_s/p_0)}{6 + (p_s/p_0)}; \rho \text{ is density at wave front, second equation valid for}$$
$\Delta p_s > p_0$

$$c_U = \sqrt{\frac{\gamma p_s}{\rho}}; \text{ sonic velocity in shock-compressed air}$$

$$\Delta p_r = 2(\Delta p_s) + \frac{6(\Delta p_s)^2}{\Delta p_s + 7p_0}; \text{ overpressure, reflected from rigid obstacle}$$

$\Delta p_r = p_r - p_0$; p_r is reflected pressure, $\Delta p_s = p_s - p_0$; side-on overpressure

$$q_d = \frac{1}{2}\rho v^2 = \frac{2.5(\Delta p_s)^2}{\Delta p_s + 7p_0}; \text{ dynamic pressure in the flow behind the wave front}$$

$$\frac{T}{T_0} = \frac{6 + (p_s/p_0)}{6(p_s/p_0) + 1}\left(\frac{p_s}{p_0}\right)T; \text{ absolute temperature [5,32]}$$

Note: p_s is referred to as side-on pressure, as it is measured at right angles to the flow. The side overpressure is $\Delta p_s = p_s - p_0$. For stronger shocks, when $p_s \gg p_0$, then $\Delta p_s \approx p_s$. See Table 13.2.

CASE 13.4   SPHERICAL CHARGE SUSPENDED IN AIR

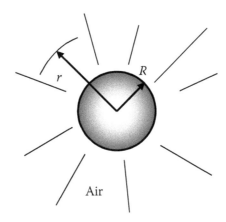

(Bare charge)

$Z = \dfrac{r}{M^{1/3}}$; range, or scaled distance

r is distance from the center of charge to the target (m)

M is charge mass (kg)

$p_s = 1610Z^{-1.4064}$ for $0.2 < Z \leq 0.4$; side-on pressure (kPa)

$p_s = 909.6Z^{-2.17}$ for $0.4 < Z \leq 2$

$p_s = 668Z^{-1.88}$ for $2 < Z \leq 7$

$p_s = 248.3Z^{-1.3471}$ for $7 < Z \leq 39$

$p_r = 11{,}290Z^{-1.65}$ for $0.2 < Z \leq 0.4$; reflected pressure (kPa)

$p_r = 4880Z^{-2.67}$ for $0.4 < Z \leq 1.5$

$p_r = 3300Z^{-2.4}$ for $1.5 < Z \leq 5$

$p_r = 600Z^{-1.3955}$ for $5 < Z \leq 39$

$i_r = 523\dfrac{M^{1/3}}{Z^{1.533}}$ for $0.2 < Z \leq 0.7$; reflected impulse (kPa-ms)

$i_r = 575\dfrac{M^{1/3}}{Z^{1.2}}$ for $0.7 < Z \leq 5$

$i_r = 452.8\dfrac{M^{1/3}}{Z^{1.0562}}$ for $5 < Z \leq 39$

$i_s \approx (p_s/p_r)i_r; \; i_r \approx (p_r/p_s)i_s{}^*)$

first approximation when the data is incomplete

Note: The equations for pressure and impulse are accurate to within a few percent and usually err on the side of caution. The formulas are for TNT. For other explosives output, see text. Equation*) gives an underestimate of i_s for $Z > 6$ and an overestimate for $Z < 6$.

CASE 13.5   HEMISPHERICAL CHARGE PLACED ON GROUND

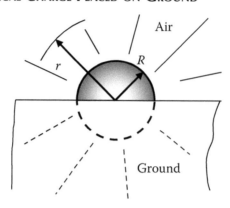

(Bare charge)

$$Z = \frac{r}{M^{1/3}}; \text{ range, or scaled distance}$$

r is the distance from the center of charge to the target (m)

M is charge mass (kg)

$p_r = 13{,}450 Z^{-1.63}$ for $0.2 < Z \le 0.4$; reflected pressure (kPa)

$p_r = 9050 Z^{-2.1}$ for $0.4 < Z \le 1.0$

$p_r = 7625 Z^{-2.7}$ for $1.0 < Z \le 5$

$p_r = 880 Z^{-1.3955}$ for $5 < Z \le 38$

$$i_r = 826 \frac{M^{1/3}}{Z^{1.581}} \text{ for } 0.2 < Z \le 0.7; \text{ reflected impulse (kPa-ms)}$$

$$i_r = 910 \frac{M^{1/3}}{Z^{1.232}} \text{ for } 0.7 < Z \le 5$$

$$i_r = 681 \frac{M^{1/3}}{Z^{1.0537}} \text{ for } 5 < Z \le 39$$

Note: The equations for pressure and impulse are accurate to within a few percent and usually err on the side of caution. The formulas are for TNT. For other explosives output, see text.

CASE 13.6   STRUCTURAL LOADING PARAMETERS, FRONT WALL OF A RIGID OBJECT

$\gamma = 1.4$; polytropic exponent for air

$p_0 = 0.1013 \text{ MPa}$; ambient pressure at 15°C

$\rho_0 = 1.225 \text{ kg/m}^3 = 1.225 \times 10^{-6} \text{ g/mm}^3$; ambient density at 15°C

$$c_0 = \sqrt{\frac{\gamma p_0}{\rho_0}} = 340.3 \text{ m/s}; \text{ sonic velocity}$$

$\Delta p_s = p_s - p_0$; overpressure, difference between pressure behind blast front wave and ambient

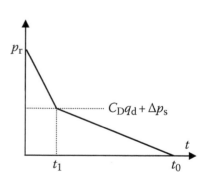

$$t_1 = \frac{3H}{c_U} \text{ or } t_1 = \frac{3B}{2c_U} \text{ whichever less; clearing time for wave-impacted face}$$

$$q_d = \frac{1}{2}\rho v^2 = \frac{2.5(\Delta p_s)^2}{7p_0 + \Delta p_s}; \text{ dynamic pressure}$$

$$i_w = \frac{1}{2}(p_r + C_D q_d + \Delta p_s)t_1; \text{ specific, wave-applied impulse}$$

$$i_d = \frac{1}{2}(C_D q_d + \Delta p_s)(t_s - t_1); \text{ specific, drag-applied impulse}$$

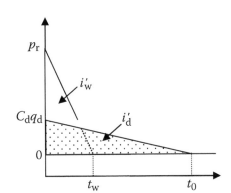

$S_w = Ai_w$; total wave impulse on a flat wall of area A, at right angle to flow

$S_d = Ai_d$; drag impulse

$i'_w \approx p_r t_w/2$; approximate wave impulse

$i'_d \approx C_D q_d t_0/2$; approximate drag impulse

Note: p_s is referred to as side-on pressure, as it is measured at right angles to the flow. The second sketch illustrates a separate treatment of a wave pulse and a drag pulse; a frequent approximation [32,4]. Typically, $t_0 \gg t_w$, which makes the approximation very close.

CASE 13.7   SPHERICAL CHARGE EXPLODING OVER A HARD SURFACE

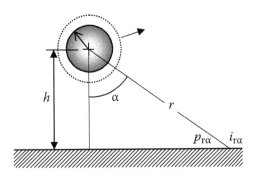

p_r ~MPa and $i_{r\alpha}$ ~MPa-ms are the reflected pressure and impulse, respectively, at distance r from the charge, on the reflecting surface:

$\ln(p_{r\alpha}) \approx \ln(p_r) \cdot \cos(0.8\alpha)$; $(0° < \alpha < 80°)$ natural log of reflected pressure

$\ln(i_{r\alpha}) \approx \ln(i_r) \cdot \cos(0.7\alpha)$; $(0° < \alpha < 80°)$ natural log of reflected impulse

p_r and i_r refer to pressure and impulse, respectively, for a normal reflection at r

Note: The above is the first approximation only. The reflected impulse will, in general, decrease with distance r, but not as fast as the reflected impulse for a plane wave (normal reflection) would.

CASE 13.8   FLAT CHARGE EXPLODING NEAR A HARD SURFACE

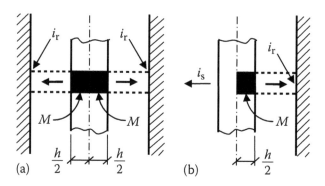

(a)   (b)

(a) Symmetrical arrangement

   $M = h\rho_e/2$; explosive mass per unit surface area

   $i_r = M\sqrt{2q}$; impulse reflected from a hard surface (Pa-s)

(b) Asymmetric arrangement

   The same i_r on the hard surface side, but only free-field impulse on the other.

Note: The point where the reflected pressure is measured should be sufficiently far from the end of explosive plate.

CASE 13.9   SPHERICAL CHARGE EXPLODING OVER A CIRCULAR PLATE

$M_e = \left(4\pi r_s^3/3\right)\rho_e$; explosive mass

$S_r = \dfrac{M_e}{4}\sqrt{2q}\,\dfrac{1}{1+(h/R)^2}$; impulse reflected from a heavy plate

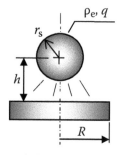

Note: This is in line with the formulation in Ref. [32], except that the velocity term is significantly reduced. See Example 13.4 for an alternative procedure.

CASE 13.10   IMPULSE APPLIED TO UNCONSTRAINED OBJECTS CLOSE TO A SPHERICAL SOURCE OF RADIUS r_s

i_r is specific impulse reflected from target

$S = Ai_r$; impulse applied to the target by explosion

A is the projected target area

R_t is the target radius

$$i_r = 1.078 \times 10^6 \beta r_s \left(\frac{R_t}{r_s} \right)^{0.158} \left(\frac{r}{r_s} \right)^{-1.4} \left(\frac{\text{N-s}}{\text{m}^2} \right)$$

when $r_s \sim \text{m}$

Valid for $0.13 \leq r/r_s \leq 5.07$ and for Composition B characterized by density ρ and specific energy q.

For another explosive, with q_e instead of q and ρ_e instead of ρ the impulse is i_e instead of i_r:

$$i_e = \sqrt{\frac{q}{q_e} \left(\frac{\rho}{\rho_e} \right)} \, i_r$$

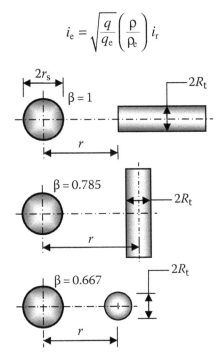

Note: This is a modification of an experimentally derived formula quoted in Ref. [4]. Values of q and ρ can be found in Table 13.1.

CASE 13.11   EXPLOSION IN A CLOSED CHAMBER WITH DIMENSION RATIOS CLOSE TO 1.0

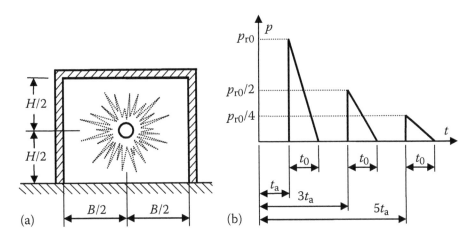

(a) Layout, (b) pressure response

t_a is the arrival time of the first blast wave to the nearest wall (or ceiling)

$i_{r0} = p_{r0}t_0/2$; the first reflected (specific) impulse

t_0 is pulse duration (equivalent)

Subsequent two impulses are as shown in (b).

Note: This is an approximate treatment of the wave action according to Ref. [4]. The first reflected impulse is calculated in the usual way, Case 13.4. When the natural frequency of an impacted wall is such that duration of the three impulses is less than $\tau/4$, where τ is the natural period of the wall, the action of those three impulses may be further simplified by applying a single impulse of magnitude $1.75i_{r0}$ instead.

A lightly vented chamber is defined by $0 \le A_v/V^{2/3} \le 0.022$, where $A_v \sim m^2$ is vented area and $V \sim m^3$ is the air mass volume in the chamber.

Let $x = M/V$, where $M \sim$ kg is the charge mass, then

$$p_g = 2.8153x^{0.8132} \quad \text{for } 0.016 < x < 0.64$$

$$p_g = 1.9416x^{0.7967} \quad \text{for } 0.64 < x < 64$$

where $p_g \sim$ MPa is the gas pressure [4,106].

Note: Shock wave action is followed by gas pressure loading p_g. This pressure usually lasts longer than the structural response; therefore a constant p_g can be assumed acting. The air mass volume V is the net value, after subtracting solid contents. The gas pressure results are valid for any shape of chamber.

CASE 13.12   IMPULSE APPLIED TO UNCONSTRAINED PROFILES
             PLACED IN THE PATH OF A PLANE WAVE

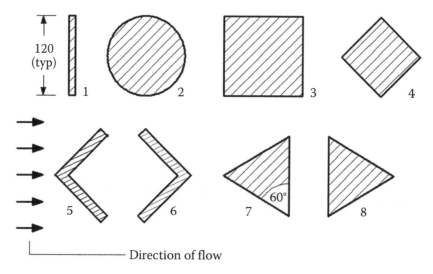

S_m is the reflected impulse experienced when the entire flow is blocked.

$S_m = Hi_r$

where $H = 120\,$mm is the profile height and i_r is the specific reflected impulse.

S is the peak impulse experienced by a profile of specified density, free to translate

M is the mass of profile, per unit length normal to paper

$v = S/M$; peak velocity attained; $C_i = S/S_m$; impulse coefficient

---

CASE TABLE 13.1
Material Properties Used

	ρ (g/cm³)	E (MPa)
Material I	7.85	200,000
Material II	1.57	200,000
Material III	0.80	8,000

CASE TABLE 13.2
Impulse Coefficient C_i for Three Materials Investigated

Profile	1	2	3	4	5	6	7	8
Material I	0.615	0.630	0.812	0.596	0.527	0.745	0.476	0.835
Material II	0.338	0.605	0.773	0.557	0.336	0.427	0.447	0.761
Material III	0.220	0.590	0.798	0.526	0.241	0.283	0.425	0.702

Note: The results are obtained from FE simulations described in Ref. [98]. Each profile was independently set in the flow. The boundaries of the stream were far enough to allow a free flow around the profile. The plane wave used had $p_s = 1\,$MPa, but another one, with $p_s = 0.1\,$MPa, gave quite close figures and the tabulated values are the averages of the two. Material I is steel. The density of Material II is five times less, which corresponds to a typical, hollow thick-wall steel section, whose density was averaged over the entire section area. Material III is wood.

CASE 13.13   IMPULSE REDUCTION COEFFICIENT FOR A PLATE SUBJECTED
                       TO A PLANE WAVE

S_m is an impulse for an infinitely heavy plate

S is the peak impulse for a plate of mass M_2

M_a is the mass of air that fits in the plate volume

$v = S/M_2$; peak velocity gained by plate

$$\frac{S}{S_m} = 0.173\ln\left(\frac{M_2}{M_a}\right) - 0.901 = 0.173\ln\left(\frac{\rho_2}{1.225}\right) - 0.901$$

where ρ_2 is plate material density ~kg/m³

Note: This formula is a generalization of results described in Case 13.12. The plate is long in the direction normal to paper. The side ratio is 4/120. The formula should be used with caution for other proportions. Also, the results may differ for longer or shorter duration waves compared to what was used in [91]. For solid steel, $M_2/M_a = 6408$ and $S/S_m = 0.663$.

CASE 13.14   VEHICLE OVERTURNING CAUSED BY A BLAST WAVE

The vehicle is standing on a no-slip surface.

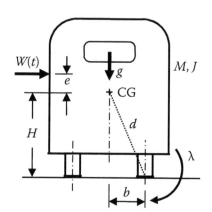

$W_0 > \dfrac{Mgb}{e+H}$; applied force needed for overturning

$S > \dfrac{1}{e+H}[2MgJ_B(d-H)]^{1/2}$ applied impulse needed to overturn

($S = W_0t_0$ for a short rectangular pulse)

$\lambda = S(e+H)/J_B$; angular velocity gained

$J_B = J + Md^2$

Note: J is the moment of inertia about the centroidal axis, normal to the paper. For the overturning to take place, both inequalities must be satisfied.

CASE 13.15   METAL TUBE FILLED WITH EXPLOSIVE

Mass per unit length

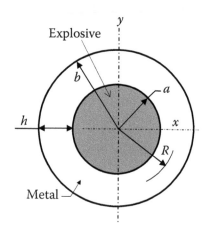

$M_m = \pi(b^2 - a^2)\rho_m$ (metal)

$M_e = \pi a^2 \rho_e$ (explosive)

$v_f = \sqrt{\dfrac{2q_G}{(M_m/M_e)+\dfrac{1}{2}}}$; fly-off speed

$R = (a + b)/2$

CASE 13.16   METAL SPHERE FILLED WITH EXPLOSIVE

(Refer to the illustration in Case 13.15)

$$\text{Mass: } M_m = \frac{4}{3}\pi(b^3 - a^3)\rho_m$$

$$M_e = \frac{4}{3}\pi a^3 \rho_e$$

$$v_f = \sqrt{\frac{2q_G}{(M_m/M_e)+\dfrac{3}{5}}}\ ;\ \text{fly-off speed}$$

CASE 13.17   SYMMETRICAL METAL-EXPLOSIVE SANDWICH

Mass of metal: $M_m = 2M_1$

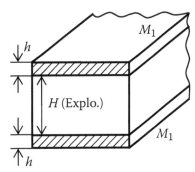

$$v_f = \sqrt{\frac{2q_G}{(M_m/M_e) + \frac{1}{3}}}\ ; \text{fly-off speed}$$

CASE 13.18   UNSYMMETRICAL METAL-EXPLOSIVE SANDWICH

(Tamper, i.e., inert material placed on one side of explosive.)

M_m mass of metal

M_e mass of explosive

M_t mass of tamper

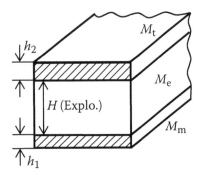

$$v_f = \sqrt{2q_G}\left[\frac{1 + A^3}{3(1 + A)} + \frac{M_t}{M_e}A^2 + \alpha\right]^{-1/2}$$

with $\alpha = M_m/M_e$ and $A = \dfrac{1 + 2\alpha}{1 + 2M_t/M_e}$

CASE 13.19   OPEN-FACE, METAL-EXPLOSIVE SANDWICH

$$v_f = \sqrt{2q_G}\left[\frac{(1 + 2\alpha)^3 + 1}{6(1 + \alpha)} + \alpha\right]^{-1/2}$$

with $\alpha = M_m/M_e$

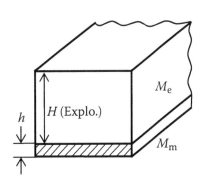

CASE 13.20   FLAT EXPLOSIVE CHARGE ON RIGID SURFACE

$h \ll b, h \ll L$

$M_e = bLh\rho_e$; explosive mass

$A = bL$; footprint area

$S = 0.87 M_e \sqrt{2q_G}$; impulse

$R = 1.49 A p_0$; peak reaction force [100]

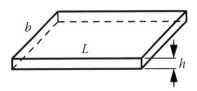

Note: See comments in text.

CASE 13.21   HEMISPHERICAL CHARGE ON RIGID SURFACE

$M_e = 2\pi r_s^3/3\rho_e$; explosive mass

$A = \pi r_s^2$; footprint area

$S = 0.44 M_e \sqrt{2q_G}$; impulse

$R = 1.00 A p_0$; peak reaction force

Note: See comments in text [100].

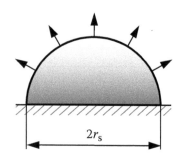

$2r_s$

CASE 13.22   CUBIC CHARGE ON RIGID SURFACE

$M_e = a^3 \rho_e$; explosive mass

$A = a^2$; footprint area

$S = 0.25 M_e \sqrt{2q_G}$; impulse

$R = 1.14 A p_0$; peak reaction force

Note: See comments in text [100].

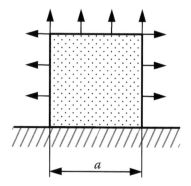

a

CASE 13.23   CYLINDRICAL CHARGE ON RIGID SURFACE

$M_e = \pi r^2 h \rho_e$; explosive mass

$A = \pi r^2$; footprint area

$S = 0.27 M_e \sqrt{2q_G}$; impulse

$R = 1.51 A p_0$; peak reaction force

This is for $2r = h$

Note: See comments in text [100].

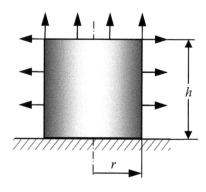

h

r

CASE 13.24   AN SDOF SYSTEM RESPONSE TO AN EXPONENTIALLY DECAYING PULSE

$W_0 = p_0 A$; load from peak pressure p_0 applied to area A

$W = W_0 \exp\left(-\dfrac{t}{t_x}\right)$; current load, $u_{st} = W_0/k$

$S(t) = W_0 t_x \left\{1 - \exp\left(-\dfrac{t}{t_x}\right)\right\}$; current impulse

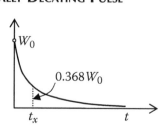

W_0

$0.368 W_0$

t_x t

$S_m = W_0 t_x$; total impulse

(a) $\omega t_x < 0.4$; $u_m \approx (\omega t_x) u_{st}$; maximum response close to that of initial velocity $v_0 = \omega^2 u_{st} t_x$

(b) $\omega t_x > 40$; $u_m \approx 2u_{st}$; maximum response approaching that of step loading with W_0

(c) For intermediate values, $0.4 < \omega t_x < 40$, use

$$u_m \approx u_{st} \frac{2.222}{1 + 2/(\omega t_x)}$$

Note: For more details see Case 3.5. Note that a decaying impulse of a finite duration S can be approximated by the above when the appropriate value of parameter t_x is used $(t_x = S/W_0)$.

CASE 13.25 SPHERICAL PETN CHARGE EXPLODING UNDERGROUND IN HARD ROCK

The target point at r away is also deep underground.

$$p_m = 98,100 \frac{B_1}{\breve{r}^3} + 981 \frac{B_2}{\breve{r}^2} + 9.81 \frac{B_3}{\breve{r}} \text{ (MPa); peak pressure; } \breve{r} = \frac{r}{r_s}$$

	B_1	B_2	B_3
Limestone	−1.51	21.33	−3.90
Marble	1.67	4.71	46.70
Granite	1.27	20.18	38.59

$$v_m = \frac{8761}{\breve{r}^3} + \frac{1392}{\breve{r}^2} + \frac{26.62}{\breve{r}} \text{ (mm/ms or m/s); PPV for granite, based on } \rho c_p = 14.22$$

$$\frac{i_m}{r_s} = 8005 \times 10^6 \frac{1}{\breve{r}^3} - 201.6 \times 10^6 \frac{1}{\breve{r}^2} + 2.693 \times 10^6 \frac{1}{\breve{r}} \left[\frac{N-s}{m^3} \right]; \text{ specific impulse } i_m$$

where r_s is in meters. This is valid for $15 < \breve{r} < 120$ [32].

Note: Calculation of PPV for the remaining two rock types may use the same ρc_p if specific data are not available. The charges were buried sufficiently deep to exclude the effect of the free surface. If the target point is on a free surface rather than within the rock mass, the calculated PPV must be doubled.

CASE 13.26 SPHERICAL TNT CHARGE EXPLODING UNDERGROUND IN SANDY SOIL

$$\Delta p_m = \frac{A_p}{9.81} \left(\frac{1}{\breve{r}} \right)^a \text{ (MPa); overpressure}$$

$\breve{r} = r/r_s$ relative distance

Tabulated values of A_p and a

	A_p	a	
WS, $\beta = 0$	600	1.05	WS means water-saturated sand
WS, $\beta = 0.0005$	450	1.50	β is relative air content, by volume
WS, $\beta = 0.01$	250	2.00	
WS, $\beta = 0.04$	45	2.50	
WU, $\rho = 1.6–1.7$	15	2.80	WU means water-unsaturated sand
WU, $\rho = 1.52–1.6$	7.5	3.00	ρ is density (g/cm³)
WU, $\rho = 1.45–1.5$	2.5	3.50	

$$i_m = 0.0981 A_i M_e^{1/3} \left(\frac{1}{\tilde{r}}\right)^b \left(\frac{\text{N-s}}{\text{mm}^2}\right); \text{ peak impulse}$$

Tabulated values of A_i and b

	A_i	b
WS, $\beta = 0$	1/127.4	1.05
WS, $\beta = 0.0005$	1/135.9	1.10
WS, $\beta = 0.01$	1/226.5	1.25
WS, $\beta = 0.04$	1/291.2	1.40
WU, $\rho = 1.52–1.6$	1/339.8	1.50

Note: The above are quoted in Ref. [32] based on experiments by G.M. Lyakhov. The charges were buried sufficiently deep to exclude the effect of the free surface.

CASE 13.27   SPHERICAL OR COMPACT CYLINDRICAL TNT CHARGE EXPLODING IN SANDY LOAM

Two types of explosion quantified: contained (deep underground) and contact (placed at surface)

Two sets of experiments, (1) and (2), each in different locations.

$$t_m = 0.0435 M^{1/3} Z^{1.64} \text{ (s); time from explosion to soil stress reaching peak value (Set 1)}$$

$$U = 145.3 Z^{-0.64} \text{ (m/s); shock wave speed (Set 1)}$$

$$\Delta p_m = \frac{1.089}{Z^{2.7}}; \text{ radial overpressure (Contained, Set 1)}$$

$$\Delta \sigma_\theta = 0.427(\Delta p_m); \text{ hoop overpressure (Contained, Set 1)}$$

$$\Delta p_m = 1.089 \left[\frac{r}{(0.28M)^{1/3}}\right]^{-2.7}; \text{ radial overpressure (Contact, Set 1)}$$

$$\Delta \sigma_\theta = 0.427(\Delta p_m); \text{ hoop overpressure (Contact, Set 1)}$$

$$\Delta p_m = \frac{0.7848}{Z^3}; \text{ radial overpressure (Contained, Set 2)}$$

$\Delta\sigma_\theta = 0.45(\Delta p_m)$; hoop overpressure (Contained, Set 2)

$$\Delta p_m = 0.7848\left[\frac{r}{(0.3M)^{1/3}}\right]^{-3}; \text{ radial overpressure (Contact, Set 2)}$$

$\Delta\sigma_\theta = 0.45(\Delta p_m)$; hoop overpressure (Contact, Set 2)

(Pressure ~ MPa; Z ~m/kg$^{1/3}$)

$t_0 = 0.001M^{1/3}(0.13 + 7.8Z)$ (s); pressure duration (Contained, Set 1)

$t_0 = 0.001M^{1/3}(5.5 + 5.64Z)$ (s); pressure duration (Contact, Set 1)

$t_0 = 0.001M^{1/3}(10.7 + 9.81Z)$ (s); pressure duration (Contained, Set 2)

$t_0 = 0.001M^{1/3}(7 + 9.5Z)$ (s); pressure duration (Contact, Set 2)

$v_m = 4.72Z^{-2.06}$ (m/s); PPV, contained, Set 1

$v_m = 1.08Z^{-1.65}$ (m/s); PPV, contact, Set 1 [32]

Note: Set 1 involved a soil of larger density. For contact explosion the distance is measured along a vertical, underneath the explosive.

CASE 13.28   BURIED WEAPON EXPLOSION

U (m/s) is shock wave speed; c_p is the P-wave speed
$U \approx c_p$ for fully saturated clays

$$U = 0.6c_p + \left(\frac{n+1}{n-2}\right)v \text{ for saturated clays}$$

$$U = c_p + \left(\frac{n+1}{n-2}\right)v \text{ for sand}$$

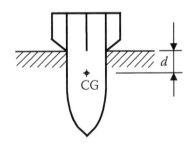

n is the decay coefficient, M is charge mass ~kg

Soil Type	Decay Coefficient, n	c_p
Heavy saturated clay and clay shale	1.5	>1500
Saturated sandy clay and sand with less than 1% air voids	2.25–2.5	1500
Wet sandy clay with air voids greater than 4%	2.5	550
Dense sand, dry or wet	2.5	500
Sandy loam, dry sand, and backfill	2.75	300
Loose, dry, poorly graded sand	3.0–3.5	200

$\kappa = -1.884\delta^2 + 1.991\delta + 0.428$; coupling factor

$$\delta = \frac{d}{M^{1/3}} \sim \frac{m}{\text{kg}^{1/3}}; \text{ scaled depth of burst } (\delta \leq 0.5)$$

Free-field response:

$v_m = 48.8\kappa(2.52Z)^{-n} \text{(m/s)}$; PPV

$$u_m = \frac{60\kappa}{c_p} M^{1/3}(2.52Z)^{1-n} \text{(m)}; \text{ peak displacement}$$

$p_m = \rho U v \text{ (N/m}^2)$ if $\rho \sim \text{kg/m}^3$ and U, $v \sim \text{m/s}$; peak pressure [80,105].

Note: The projectile can enter at any angle; only the depth of center of gravity (CG) matters. The above is limited to the near-field, $Z \leq 5 \text{ m/kg}^{1/3}$.

CASE 13.29   RESPONSE OF AN UNDERGROUND STRUCTURE TO AN UNDERGROUND EXPLOSION

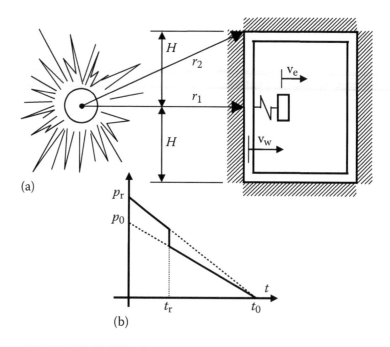

(a) Layout and **(b)** pressure vs. time at the wall center.

U (m/s) is shock wave speed; c_p is the P-wave speed

p_s is the free-field pressure

$p_r \approx 1.5p_s$; reflected pressure, where peak value of p_s is p_m per Case 13.28.

$t_0 = 2i_s/p_s$; duration of applied pressure; i_s is the free-field impulse

$$t_r = \frac{1}{c_p}(r_2 + H - r_1); \text{ duration of reflected pressure near the center of the wall}$$

The following are the peak displacement, velocity, and acceleration of a point on the wall:

$$u_w = 2u_m; \quad v_w = 2v_m; \quad a_w = 2p_m/m$$

where v_m, u_m, and p_m are the corresponding free-field responses from Case 13.28, while m is the mass of wall per unit surface.

The floor and ceiling horizontal response is to be assumed the same as the free-field.

The following are the first estimates of the equipment mounted on the wall:

$$u_e = 1.2u_w; \quad v_e = 1.5v_w; \quad a_e = 2a_w \ [80,105]$$

Note: To calculate t_r use the vertical dimension as shown, or a corresponding horizontal dimension, if smaller. For an intermediate point reflected pressure use linear interpolation between the edge, where only p_s is applied and center, where the above p_r is given.

CASE 13.30 CYLINDRICAL CHARGE IN ROCK, UNDER SURFACE. PPV PREDICTION IN THE VICINITY OF CHARGE

m is the distributed explosive mass (kg/m)

$$v = K\left(\frac{m}{r}\right)^\alpha \left\{ \arctan\left(\frac{H + z_t - z_0}{r}\right) + \arctan\left(\frac{z_0 - z_t}{r}\right) \right\}^\alpha$$

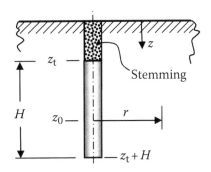

The value of coefficients can often be

$$K = 0.7, \quad \alpha = 0.7, \quad \text{and} \quad \beta = 1.5 \ [66]$$

This is the peak response *after* the wave, described in Case 13.31, passes away.

CASE 13.31 CYLINDRICAL CHARGE IN ROCK, UNDER SURFACE. DETONATION WAVE IN EARLY STAGES AFTER IGNITION

D is the detonation speed of explosive and ρ_e is its density.

$$p_D = \frac{\rho_e D^2}{\gamma + 1}; \text{ pressure in explosive wave front}$$

$$\gamma = 3.0 \quad \text{for } \rho_e > 1.2\,\text{g/cm}^3$$

$$\gamma = 2.1 \quad \text{for } \rho_e < 1.2\,\text{g/cm}^3$$

For waves in rocks

$$p_m = p_D\left(\frac{R_c}{r}\right)^a = \sigma_{rm} = -\sigma_{\theta m}$$

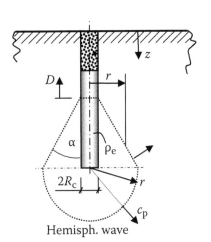

Hemisph. wave

peak radial pressure or radial stress reached, equal to hoop stress, in absolute value.

$1 < a < 2$; often $a \approx 1.5$

$\tan \alpha = \dfrac{c_p}{D}$; slope of the wave front

For charges exploded in sandy loams: (radial, vertical, and hoop stress)

$$p_m = \frac{864.1}{\breve{r}^{1.44}}; \quad \sigma_{zm} = \frac{145.1}{\breve{r}^{1.2}}; \quad \sigma_{\theta m} = \frac{497.4}{\breve{r}^{1.6}} (\text{MPa}); \quad \breve{r} = r/R_c$$

$$U = \frac{1.207}{(\breve{r} - 1)^{0.57}} \ (\text{m/s}); \text{ wave speed}$$

Pressure duration:

$t_0 = (0.448\,\breve{r} - 6.06)\sqrt{m}$ (s); if $m \sim$ kg/m where m is charge mass per unit length [32].

EXAMPLES

EXAMPLE 13.1   BLAST WAVE PARAMETERS

Calculate parameters of the wave characterized by a peak overpressure of $\Delta p_s = 70\,\text{kPa}$, namely U, v, ρ, c_U, p_r, and q_d. Assume standard conditions, 15°C ambient temperature.

The side-on pressure in the wave: $p_s = p_0 + \Delta p_s = 0.1013 + 0.070 = 0.1713\,\text{MPa}$; $p_s/p_0 = 1.691$

Our constants, from Case 13.3:

$$\gamma = 1.4$$
$$p_0 = 0.1013\,\text{MPa}$$
$$\rho_0 = 1.225 \times 10^{-6}\,\text{g/mm}^3$$
$$c_0 = 340.3\,\text{m/s}$$

Then

$$\left(\frac{U}{c_0}\right)^2 = \frac{1}{7}\left(1 + 6\frac{p_s}{p_0}\right) = \frac{1}{7}(1 + 6 \times 1.691); \quad U = 340.3 \times 1.262 = 429.4\,\text{m/s}$$

$$\frac{v}{U} = \frac{p_s/p_0 - 1}{0.2 + 1.2(p_s/p_0)} = \frac{1.691 - 1}{0.2 + 1.2(1.691)}; \quad v = 0.31U = 133.1\,\text{m/s}$$

$$\frac{\rho}{\rho_0} = \frac{1 + 6(p_s/p_0)}{6 + (p_s/p_0)} = \frac{1 + 6 \times 1.691}{6 + 1.691}; \quad \rho = 1.449\rho_0 = 1.775 \times 10^{-6}\,\text{g/mm}^3$$

$$c_U = \sqrt{\frac{\gamma p_s}{\rho}} = \sqrt{\frac{1.4 \times 0.1713}{1.775 \times 10^{-6}}} = 367.6\,\text{m/s}$$

$$\Delta p_r = 2(\Delta p_s) + \frac{6(\Delta p_s)^2}{\Delta p_s + 7p_0} = 2(0.07) + \frac{6(0.07)^2}{0.07 + 7 \times 0.1013} = 0.1777\,\text{MPa}$$

$$p_r = 0.1013 + 0.1777 = 0.279\,\text{MPa}$$

$$q_d = \frac{1}{2}\rho v^2 = \frac{1}{2} \times 1.775 \times 10^{-6} \times 133.1^2 = 15.72 \times 10^{-3}\,\text{MPa}$$

EXAMPLE 13.2   SPHERICAL CHARGE AIRBLAST

A 230 kg suspended charge of emulsion is exploded in air. Calculate the side-on pressure at 30 m from the charge as well as the reflected impulse at that distance. The emulsion has the same energy content as ANFO in Table 13.1, but its density is 1.0 g/cm³. Also, determine the masses and responses of energy-equivalent charges of TNT and PETN.

The equivalent masses of the other two explosives are found using a form of Equation 13.5:

$$M = M_e \frac{q_e}{q}$$

where, this time, index e relates to the emulsion. Taking ρ and q from Table 13.1, one can calculate the equivalent masses and then, from geometry, determine the sphere radius r_s, which is given here only for the sake of completion. The calculation is set out in the table below.

	ρ (g/mm³)	q (10^6 N-mm)	M (kg)	r_s (mm)
PETN	0.00176	6.09	140.5	267.1
TNT	0.00160	4.61	185.6	302.5
Emulsion	0.00100	3.72	230.0	380.1

All of the above charges are TNT-equivalent, therefore it is sufficient to calculate the response using TNT-based formulas, as quoted in Case 13.4:

$$Z = \frac{30}{(185.6)^{1/3}} = 5.26; \text{ range}$$

$$p_s = 668 Z^{-1.88} = 29.47 \text{ kPa; side-on pressure}$$

$$i_r = 452.8 \frac{M^{1/3}}{Z^{1.0562}} = 452.8 \frac{(185.6)^{1/3}}{5.26^{1.0562}} = 447.3 \text{ kPa-ms; reflected impulse}$$

A more accurate computation, using Conwep, gives

$$p_s = 28.74 \text{ kPa}; \quad i_r = 449.8 \text{ kPa-ms}$$

EXAMPLE 13.3 EXPLOSIVE CHARGE EFFECTS IN A NEAR-FIELD

An RDX charge of $M = 100$ kg is placed on the ground and detonated. The shape of the charge is unknown, therefore a hemisphere may be assumed. Estimate the side-on pressure, the reflected pressure and the specific impulse at a distance of 15 m from the center of the charge.

From Table 13.1 the RDX energy content is $q = 5.37$ against 4.61 for TNT

Since the explosive is not TNT, the mass must be adjusted per Equation 17.1:

$$M = M_e \frac{q_e}{q_{TNT}} = 100 \frac{5.37}{4.61} = 116.5 \text{ kg}$$

Another adjustment, needed to use the suspended charge formulas is to allow for the surface placement, by multiplying the mass by the magnification factor of 1.7:

$$M = 1.7 \times 116.5 \text{ kg} = 198 \text{ kg}$$

From Case 13.4

$$Z = \frac{r}{M^{1/3}} = \frac{15}{198^{1/3}} = 2.574; \quad p_s = 668 \times 2.574^{-1.88} = 112.9 \text{ kPa}$$

$$p_r = 3300 \times 2.574^{-2.4} = 341.2 \text{ kPa}; \quad i_r = 575 \frac{198^{1/3}}{2.574^{1.2}} = 1077 \text{ kPa-ms}$$

An alternative approach would be to avoid including the 1.7 factor and use Case 13.5, intended for a hemispherical surface charge:

$$Z = \frac{r}{M^{1/3}} = \frac{15}{116.5^{1/3}} = 3.072$$

$$p_r = 7625 \times 3.072^{-2.7} = 368.3 \text{ kPa}; \quad i_r = 910 \frac{116.5^{1/3}}{3.072^{1.232}} = 1115 \text{ kPa-ms}$$

A more accurate estimate may be obtained from Conwep:

$$p_s = 106.8 \text{ kPa}; \quad p_r = 299.1 \text{ kPa}; \quad i_r = 1033$$

When using our single-term approximations, an overestimate may usually be expected.

EXAMPLE 13.4   KINETIC EFFECT OF A PROXIMITY CHARGE

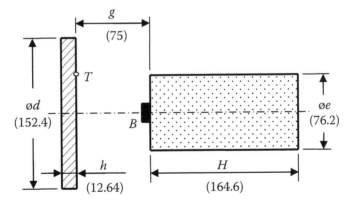

A set of experiments reported by Held [31] was carried out by using an Octol charge, containing 70% of HMX and 30% of TNT. The typical setup shows a cylindrical charge detonated close to a steel disk ($\rho = 0.00785$ g/mm³). The initiation was done at point B. Estimate the density and energy content of Octol using Table 13.1. With the explosive mass of $M_e = 1390$ g and disk mass of $M_d = 1810$ g determine the (reflected) impulse experienced by the disk and the velocity attained by the disk.

Based on the Table 13.1 as well as the percentage of components:

$0.7 \times 1.89 + 0.3 \times 1.60 = 1.803$ so $\rho_e = 0.001803$ g/mm³; mass density

$0.7 \times 5.46 + 0.3 \times 4.61 = 5.205$; $q = 5.205 \times 10^6$ N-mm; energy content

Following Case 13.9 and pretending we deal with a spherical charge, one has

$$R = 152.4/2 = 76.2 \, \text{mm}; \quad h = 75 + 164.6/2 = 157.3 \, \text{mm}.$$

$$S_r = \frac{M_e}{4}\sqrt{2q}\,\frac{1}{1+(h/R)^2} = \frac{1390}{4}\left[2\times5.205\times10^6\right]^{1/2}\frac{1}{1+(157.3/76.2)^2} = 213{,}100 \text{ N-ms}$$

The velocity acquired by the disk: v = S_r/M_d = 213,100/1810 = 117.7 m/s.

There is an alternative procedure, however, where Equation 13.17 is used and the air mass is ignored.

A "representative" point at $R/2$ is selected for calculating the effective distance:

$$r = \sqrt{((152.4/4)^2 + 157.3^2)} = 161.8 \text{ mm}$$

The simplified Equation 13.17 gives

$$i_r = \frac{M_e}{4\pi r^2}[2q]^{1/2} = \frac{1390}{4\pi 161.8^2}\left[2\times5.205\times10^6\right]^{1/2} = 13.63 \text{ MPa-ms}$$

The mass of the disk, per unit surface is $m = \rho h$. The velocity gained should, therefore, be

$$v = i_r/m = 13.63/(0.00785\times12.64) = 137.4 \text{ m/s}.$$

The experimentally found speed was $v_e = 133$ m/s, therefore the second result is more accurate. This was one of a number of similar tests; it is shown as a single line of the table below. This procedure was used to obtain the values of v. The agreement is quite good except the last line, which was a major deviation.

M_e	M_d	g	Øe	H	r	i_r	m	v	v_e	v/v_e
1390	3630	100	76.2	164.59	186.23	10.292	0.199	51.72	58	0.89
790	3630	100	63.5	137.16	172.83	6.792	0.199	34.13	33	1.03
1390	1810	75	76.2	164.59	161.84	13.627	0.099	137.34	133	1.03
790	3630	75	63.5	137.16	148.55	9.194	0.199	46.20	40	1.15
1724	1819	50	82.6	178.42	144.33	21.254	0.100	213.13	201	1.06
790	3630	50	63.5	137.16	124.55	13.078	0.199	65.72	47	1.40

v_e, experimentally measured velocity.

EXAMPLE 13.5 FORCE–IMPULSE DIAGRAM FOR A RIGID-PLASTIC SYSTEM

A rigid-plastic cantilever with a tip mass, acted upon by a rectangular pulse, was previously presented in Example 10.15. Here $L=3$ m, tip mass $M=368$ kg, distributed mass $m=461$ kg/m, and plastic moment capacity $\mathcal{M}_0=1.246\times10^6$ N-m. The prescribed, damage-related deflection is $\theta_m=5°=0.0873$ rad. The diagram is to depict a set of force–impulse pairs giving the same θ_m.

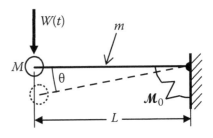

The force, needed to start rotation, on static basis, is $W_p = \mathcal{M}_0/L = 415{,}330$ N. Any larger load can be sustained for a limited time only. The effective mass for a rigid-body rotation about the base joint is $M_{ef} = M + mL/3 = 368 + 461\times3/3 = 829$ kg.

For the elastic case, the impulsive asymptote was obtained from Equation 13.31a. Now, with the rigid-plastic system, the expression for strain energy is $\mathcal{M}_0\theta$ and when the angle is set at the maximum value of θ_m, the following form of Equation 13.31b is obtained:

$$S_1 = \sqrt{2M_{ef}\mathcal{M}_0\theta_m} = (2\times829{,}000\times1.246\times10^9 \times0.0873)^{1/2} = 13.43\times10^6 \text{ N-ms}$$

The step-load amplitude was previously obtained from Equation 13.32a. This expression for strain energy gives, instead

$W_0u_m = \mathcal{M}_0\theta_m$, or, with $u_m = L\theta_m$ one gets $W_0 = \mathcal{M}_0/L = 415{,}300$ N $= W_p$

The fact that the step-load amplitude comes as load identical with the plastic limit should be no surprise as this load must be supported infinitely long. To calculate coordinates of some intermediate point, choose $W_x = 10W_0 = 4.153\times10^6$ N and find the corresponding impulse. In the equation for θ_m from Case 10.24b symbol W_0 is replaced by W_x. After a minor reformulation, we find the rotation to be

$$\theta_m = \frac{1}{2}\frac{W_x}{LM_{ef}}t_0^2 \ \frac{W_x - W_p}{W_p} = \frac{1}{2}\frac{10W_p}{LM_{ef}}t_0^2 \ \frac{10W_p - W_p}{W_p} = \frac{90W_p}{2LM_{ef}}t_0^2$$

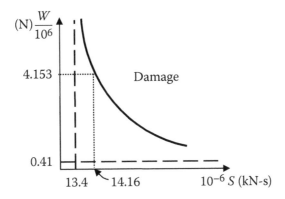

After substituting, the corresponding value of t_0 is found from $0.0873 = t_0^2/133.1$, then $t_0 = 3.409$ ms. This gives $S_x = W_xt_0 = 14.16\times10^6$ N, as shown on the graph.

EXAMPLE 13.6   SCALING OF AN ELASTIC STRUCTURE SUBJECTED TO BLAST LOADING

A large cantilever beam that may be exposed to blast loading is field-tested by means of a model scaled down by the factor $\zeta = 1/5$. Applying Hopkinson's scaling to blast effects and using the same material in the model as in the prototype determine how the maximum stress and deflection are going to differ between the two.

Our concern is the ratio of the corresponding quantities, for example, deflections u/u', where the first variable relates to the prototype and the second to the tested value. Instead of addressing the variables using their full expressions, it is, therefore, enough to state to what they are proportional ($\sim$). Case 2.24 shows that the natural period of a cantilever is

$$\tau = 1.787 L^2 \sqrt{\frac{m}{EI}};\ \text{which means}\ \tau \sim L^2 \sqrt{\frac{m}{EI}} \sim L^2 \sqrt{\frac{A\rho}{EI}} \sim L^2 \sqrt{\frac{A}{I}} \sim L$$

where material properties were suppressed on account of being the same and $A \sim L^2$ while $I \sim L^4$.

The general expression for dynamic deflection is given by Case 10.4:

$$u_d = 2u_{st}\sin(\pi t_o/\tau)$$

If the scale factor is ζ, duration t_0 becomes $t_0' = \zeta t_0$. The same thing happens to the period, since according to the above, $\tau \sim L$, which gives $t_0' = \zeta \tau$. In the trigonometric term above ζ cancels out and $u_d \sim u_{st}$. As for the latter

$$u_{st} = \frac{1}{8}\frac{w_0 L^4}{EI} \sim Bp_0$$

where

B is the width of the beam
p_0 is the peak applied pressure

Since the latter remains the same, the final result is that the peak dynamic deflection is proportional to the width B, which, in turn, $\sim\zeta$. The tested deflection will therefore be 1/5th of that of the prototype.

A much simpler procedure is needed to show the stress relation. Using the bending stress expression and the first line of Table 10.2, one can write bending stress as

$$\sigma = \frac{Mc}{I} = \left(\frac{w_0 L^2}{2}\right)\frac{c}{I} \sim \frac{w_0}{L} \sim p_0$$

because $c \sim L$ and $w_0 = Bp_0$. This means the prototype and the test piece model will experience the same stress level.

EXAMPLE 13.7   SCALING OF A RIGID-PLASTIC STRUCTURE SUBJECTED TO BLAST LOADING

The formulation of the problem is the same as in Example 13.6, except that the beam is rigid and it has a rigid-plastic base joint, as in Case 10.15. Determine how the maximum angle of rotation θ and the tip deflection will differ between original beam (prototype) and its model.

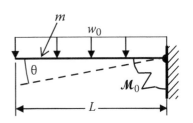

The reference case gives us a formula for the angle of rotation: $\theta_m = \dfrac{3(w_0 t_0)^2 L}{8\mathcal{M}_0 m}$

The new distributed loading is $w_0' = B'p_0' = \zeta B p_0$ as pressure remains unchanged, but the width is scaled. Time is also scaled, $t_0' = \zeta t_0$ and so is length, $L' = \zeta L$.

According to Chapter 10, the full plasticity moment is expressed as $\mathcal{M}_0 = Z\sigma_0$ where Z is proportional to L^3, which means $\mathcal{M}_0' = \zeta^3 \mathcal{M}_0$. The distributed mass $m = A\rho$, which means $m' = \zeta^2 m$. After substituting these variables in the new equation for θ, one obtains $\theta' = \theta$.

As the deflection is $\sim L$, the tested deflection will therefore be 1/5th of that of the prototype. So far, the scaling is the same for the elastic beam. The bending moment here is proportional to $w_0 L^2$, so it will change by ζ^3. This is consistent with the elastic case.

EXAMPLE 13.8   EXPLOSIVE ENERGY AND THE WORK OF SHELL EXPANSION

In Example 4.8 a long, thick-wall steel tube with $a = 19\,\mathrm{mm}$ and $b = 38\,\mathrm{mm}$ was analyzed to find the amount of work needed to expand it plastically to the extent of doubling its outer diameter. What is the Gurney energy in such a tube filled with TNT as compared with the expansion work? Calculate the corresponding work and energy for a thinner tube, $R = 45.8$, $h = 13.2\,\mathrm{mm}$, and the same $\sigma_0 = 500\,\mathrm{MPa}$. Also, find the corresponding numbers for spherical shells of the same dimensions and properties.

Thick-wall tube

With the properties of TNT from Table 13.1, the explosive energy per unit length is

$$Q_e = M_e q_G = (\pi 19^2 \times 0.0016)3.58 \times 10^6 = 6.496 \times 10^6 \text{ N-mm}$$

The previously calculated expansion work was $\mathcal{L} = 1.782 \times 10^6$ N-mm

Thick-wall sphere

$$Q_e = \left(\frac{4}{3}\pi 19^3 \times 0.0016\right)3.58 \times 10^6 = 164.57 \times 10^6 \text{ N-mm}$$

against $\mathcal{L} \approx 180 \times 10^6$ N-mm.

Thinner tube

$$R = 45.8, \quad h = 13.2\,\mathrm{mm}, \quad \text{then } a = 39.2\,\mathrm{mm} \quad \text{and} \quad b = 52.4\,\mathrm{mm} \text{ or}$$

$$Q_e = \pi 39.2^2 \times 0.0016 \times 3.58 \times 10^6 = 27.65 \times 10^6 \text{ N-mm}$$

The expansion work may be calculated according to Case 4.19. The first step is to find the mean radius r and thickness h after the outer radius doubles to 104.8 mm, using section area preservation:

$104.8^2 - a^2 = 52.4^2 - 39.2^2$ gives $a = 98.86$ mm in the deformed condition.

Then $h = 104.8 - 98.86 = 5.94$ mm; $r = 104.8 - 5.94/2 = 101.83$ mm

$$h_{av} = (13.2 + 5.94)/2 = 9.57, \quad r_{av} = (45.8 + 101.83)/2 = 73.82 \text{ mm}$$

$$\mathcal{L}_{pl} = 2\pi R(h_0\sigma_0)\left(1 + \frac{h_{av}}{2r_{av}}\right)\ln(r/R) = 2\pi 45.8(13.2 \times 500)\left(1 + \frac{9.57}{2 \times 73.82}\right)\ln(101.83/45.8)$$

$$= 1.616 \times 10^6 \text{ N-mm}$$

Thinner sphere

Repeating the calculation for a spherical shell, per case of the same dimensions, one obtains

$$M_e = 96.38 \text{ g}; \quad Q_e = 345 \times 10^6 \text{ N-mm}$$

$$h = 2.61, \quad r = 103.5 \text{ mm}, \quad \mathcal{L}_{pl} = 283.68 \times 10^6 \text{ N-mm}$$

For the thick cylindrical shell the explosive energy is several times larger than the expansion work or the strain energy absorbed by metal. For the spherical shell with the same dimensions both quantities are about equal. For the thinner version the first variable is larger in both cases. To keep the results in perspective one should remember that the expansion so large as specified here is an extreme case and shells usually fail at a fraction of this expansion ratio. The other way to put it is to state that assuming the shell to double its outer radius just prior to failure overestimates the shell strength in most cases.

EXAMPLE 13.9   TIME ESTIMATE FOR A SHELL BREAKUP

A long tube with $R = 45.8$, $h = 13.2$ mm, is filled with TNT. Estimate time t_f needed for the fragments to achieve the fly-off velocity V_f and the distance traveled by a fragment prior to attaining that velocity. In doing so assume a constant driving force in the acceleration phase and ignore the material strength.

Table 13.1 tells us that the relevant properties of TNT are $\rho = 0.0016$ g/mm^3 and $q_G = 3.58 \times 10^6$ N-mm. The other tube radii are $r_i = 39.2$ and $r_0 = 52.4$ mm. The explosive mass and the metal mass, per unit length, are

$$M_e = \pi 39.2^2 \times 0.0016 = 7.724; \quad M_m = \pi(52.4^2 - 39.2^2)0.00785 = 29.819 \text{ g}$$

$$V_f = \sqrt{\frac{2q_G}{(M_m/M_e) + \beta}} = \sqrt{\frac{2 \times 3.58 \times 10^6}{(29.819/7.724) + 0.5}} = 1281.4 \text{ m/s; fly-off velocity}$$

The pressure–time plot is likely to be close to a decreasing triangular pulse, with the effective initial pressure being $p_0 = 10{,}714$ MPa, per Table 13.1. The average pressure will, therefore, be one-half of that, namely 5357 MPa. When a fragment of a unit size is subjected to pressure p, its acceleration, according to Equation 13.10 is

$$a = \frac{pr_i}{\rho hR} = \frac{5,357 \times 39.2}{0.00785 \times 13.2 \times 45.8} = 44,249 \text{ mm/ms}^2$$

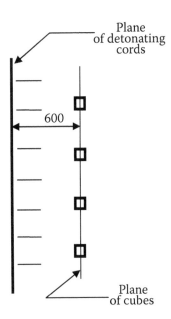

Plane of detonating cords

600

Plane of cubes

From simple kinematics: $V_f = at_f$ or $1281.4 = 44,249t_f$, which gives $t_f = 0.029$ ms.

Again, from kinematics, the distance traveled is $u = at_f^2/2 = 18.55$ mm. This tells us that at the instant of full velocity and most of disintegration achieved the mean radius of the shell has grown by over 40%. As a footnote, the dimensions given here are close to those of a 4″ artillery shell at the maximum cross section.

EXAMPLE 13.10   BLAST WAVE PUSHING A CUBE

A row of detonating cords was suspended as shown, in the plane normal to paper. The explosive content of each cord was 40 g/m of TNT and the spacing of cords was 200 mm. The purpose of this arrangement was to create a plane blast wave pushing a set of cubes located in a parallel plane, 600 mm away. Each cube had the edge length of 40 mm and was made of aluminum. Find the impulse applied to a cube by the detonation.

At some distance from the plane of cords their exact layout becomes irrelevant and they can be treated as a uniform layer with the following mass density:

$$m = 40/0.2 = 200 \text{ g/m}^2 \text{ of TNT}$$

The reflected impulse is found from Case 13.8, where M stands for a mass corresponding to a half-thickness and a unit surface:

$$M = 0.5 \times 200 \text{ g/m}^2 = 0.5 \times 200/1000^2 = 0.0001 \text{ g/mm}^2$$

$i_r = M\sqrt{2q} = 0.0001\sqrt{2 \times 4.61 \times 10^6} = 0.3036$ g-m/s; with q from Table 13.1

If the face of a cube acted as a part of an infinite, rigid plane, the impulse applied would be

$$S_r = Ai_r = 40 \times 40 \times 0.3036 = 485.8 \text{ g-m/s}$$

A reduction factor applied in accordance with Case 13.12 (3) would be about 0.8 for this material of density 2.7 g/cm³:

$$S_r' = 0.8 \times 485.8 = 388.6 \text{ g-m/s}$$

The description of the experiment comes from Ref. [13], which gives a range of answers for the actual momentum gained, from 290 to 634 g-ms with the average of 462 g-m/s. It is interesting to see that the average is closer to the calculated impulse of ~486 g-m/s obtained with no shape reduction.

EXAMPLE 13.11   TRUCK OVERTURNING BY A BLAST WAVE

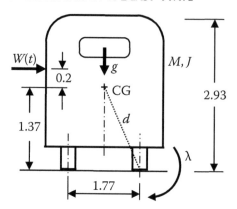

A truck with properties shown in the sketch is subjected to a blast wave with $\Delta p_s = 310\,\text{kPa}$ and $i_s = 1210\,\text{Pa-s}$. The side area, on which the wave acts is $A = 14.8\,\text{m}^2$. The mass is $M = 5430\,\text{kg}$. Determine whether a wave so described is capable of overturning the truck.

Refer to Case 13.14. An estimate of the moment of inertia about the CG axis normal to paper gives

$$J \approx \frac{M}{12}(2.93^2 + 1.77^2) = 5302\ \text{kg-m}^2;\quad \text{then } d = \left[((1.77/2)^2 + (2.93/2)^2)\right]^{1/2} = 1.712\ \text{m}$$

$$J_B = J + Md^2 = 5{,}302 + 5{,}430 \times 1.712^2 = 21{,}210\,\text{kg-m}^2$$

The remaining wave parameters are found with Case 13.3:

$$\Delta p_r = 2(\Delta p_s) + \frac{6(\Delta p_s)^2}{\Delta p_s + 7p_0} = 2(310) + \frac{6(310)^2}{310 + 7 \times 101.3} = 1186\ \text{kPa}$$

$$\frac{\rho}{\rho_0} = \frac{7 + 6(\Delta p_s/p_0)}{7 + (\Delta p_s/p_0)} = \frac{7 + 6(310/101.3)}{7 + (310/101.3)} = 2.521;\ \rho \text{ is density at wave front}$$

$$c_U = \sqrt{\frac{\gamma p_s}{\rho}} = \sqrt{\frac{1.4(0.1013 + 0.310)}{2.521 \times 1.225 \times 10^{-6}}} = 431.8\ \text{m/s; sonic velocity in shock-compressed air}$$

The minimum force needed for overturning is

$$W_{\min} = \frac{Mgb}{e + H} = \frac{5430 \times 9.81(1.77/2)}{0.2 + 1.37} = 29{,}320\ \text{N, while}$$

$A(\Delta p_r) = 14.8\ \text{m}^2 \times 1.186 \times 10^6\ \text{Pa} = 17.55 \times 10^6\ \text{N}$ is the peak applied force

The above shows that the force requirement is comfortably met. The following minimum impulse is needed to achieve overturning:

$$S_{\min} = \frac{1}{e + H}[2MgJ_B(d - H)]^{1/2} = \frac{1}{1.57}[2 \times 5430 \times 9.81 \times 21{,}210(1.712 - 1.37)]^{1/2} = 17{,}707\ \text{N-s}$$

The reflected impulse was not specified. The first estimate of this may be obtained from Case 13.4:

$i_r \approx (p_r/p_s)i_s = (1186/310)1210 = 4629$ Pa-s, then

$Ai_r = 14.8 \times 4629 = 68{,}510$ N-s is the impulse applied to the vehicle by the wave action

This would assure an overturning, *provided* the wave action may last long enough for the above impulse to be realized. To ascertain that, the clearing time must be evaluated:

$$t_1 = \frac{3h_v}{c_U} = \frac{3 \times 2930}{431.8} = 20.36 \text{ ms}$$

This is a rather long time with regard to wave action. To demonstrate the point, consider a specific triangular impulse s with the same reflected pressure and duration of t_1:

$$s = p_r t_1/2 = 1.186 \times 10^6 \times 20.36/2 = 12{,}071 \text{ Pa-s}$$

This vastly exceeds i_r, which indirectly confirms that the clearing time is long enough for the wave action to be fully developed. The data for this problem was taken from Baker [4], whose solution employs a different methodology, although the results are similar.

EXAMPLE 13.12   ABOVEGROUND EXPLOSION—INFLUENCE OF ANGLE

A spherical RDX charge of $M = 100$ kg is placed 4 m above the ground and detonated. Estimate the reflected pressure and impulse at a ground point located at a 70° angle.

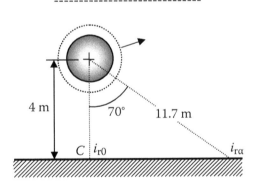

Estimate the parameters at 11.7 m from the center of charge using Conwep:

$$p_s = 126.9 \text{ kPa}; \quad p_r = 373.8 \text{ kPa}; \quad i_r = 901.6 \text{ kPa-ms}$$

Follow Case 13.7 in determining the reflected parameters on the inclined surface.

$\ln(p_{r\alpha}) \approx \ln(p_r) \cdot \cos(0.8\alpha) = 5.924 \cos (0.8 \times 70) = 3.313$; then $p_{r\alpha} = 27.6$ kPa

$\ln(i_{r\alpha}) \approx \ln(i_r) \cdot \cos(0.7\alpha) = 6.804 \cos(0.7 \times 70) = 4.464$; then $i_{r\alpha} = 86.82$ kPa-ms

A more accurate estimate can be obtained from Tables 2.9 and 2.10 in TM-5 [106]:

$\ln(p_r)/\ln(p_{r\alpha}) = 7.313/4.174 = 1.752$ vs. 1.788 above

$\ln(i_r)/\ln(i_{r\alpha}) = 4.5/3.0 = 1.5$ vs. 1.524 above

(Note that units are not involved in this approach.) In many cases the accuracy of this method will be worse than suggested by the above example. One of the reasons for it is the irregularity of slope of some curves representing scaled charge weights.

EXAMPLE 13.13   PULSE TRANSMISSION FROM EXPLOSIVE CHARGES WITHIN GROUND

Calculate the peak pressure at 30 m from the center of 230 kg charge of emulsion (as defined in Example 13.2) for (a) granite, (b) water-saturated sand with no entrapped air, and (c) sandy loam.

(a) The solution for PETN is given in Case 13.25. The equivalent masses of PETN for this charge was found in Example 13.2 as $M = 140.5$ kg. This corresponds to the PETN charge radius of $r_s = 267.1$ mm, or to $\check{r} = 30,000/267.1 = 112.34$. A substitution gives

$$v_m = \frac{8761}{112.34^3} + \frac{1392}{112.34^2} + \frac{26.62}{112.34} = 0.354 \text{ m/s}; \quad p_m = \rho c_p v_m = 14.22 \times 0.354 = 5.03 \text{ MPa}$$

(b) Case 13.26 provides the answer. The equivalent TNT, as calculated in Example 13.2, has

$M = 185.6$ kg and $r_s = 302.5$ mm, which gives $\check{r} = 30,000/302.5 = 99.17$ and

$$\Delta p_m = \frac{A_p}{9.81}\left(\frac{1}{\check{r}}\right)^a = \frac{600}{9.81}\left(\frac{1}{99.17}\right)^{1.05} = 0.49 \text{ MPa}$$

(c) In Case 13.27 there are two answers, originating from two different sets of experiments.

For $Z = 30/185.6^{1/3} = 5.26$, the first set of equations gives

$$\Delta p_m = \frac{1.089}{Z^{2.7}} = \frac{1.089}{5.26^{2.7}} = 0.0123 \text{ MPa}; \text{ while from the second set}$$

$$\Delta p_m = \frac{0.7848}{5.26^3} = 0.0054 \text{ MPa}$$

A large difference between the last two pressures illustrates how sensitive the results are to the soil type and water saturation.

EXAMPLE 13.14   EXPLOSION IN ROCK AND LONG DISTANCE PULSE TRANSMISSION

A 510 kg, spherical charge of PETN is exploded in granite, deep underground. Calculate the PPV at several locations in accordance with an empirical formula. For granite use

$F_{cu} = 100$ MPa (uniaxial compressive strength), $E = 80$ MPa, $\rho = 0.0026$, and $v = 0.2$.

Construct a simple FEA model, consisting of a string of brick elements, to simulate the explosion. Using an EPP material model select the yield strength F_y in such a way that there is a reasonable match between the empirical formula and simulation results. Model to be made in a reduced scale, 1:10.

Taking density of PETN as 0.00176 g/mm³ from Table 13.1, one finds the charge radius to be r_s = 410.2 mm. The PPV resulting from explosion of PETN in granite is given in Case 13.25:

$$V_m = \frac{8761}{\check{r}^3} + \frac{1392}{\check{r}^2} + \frac{26.62}{\check{r}} \quad \text{(mm/ms or m/s)}$$

This is valid for $15 < \check{r} < 120$, or for 6.15 m $< r <$ 49.2 m. The FE model intended to approach this result is made in 1/10 scale, which means the mass of the charge is reduced by the factor of 1000. The sketch shows a fragment of the model near the origin. The model is to represent a small-angle cone originating from the center, but the use of eight-node brick elements made it an elongated pyramid. Instead of placing the explosive material at the center, it was made into a layer at some distance from it, with the four internal nodes being fixed. The mass of explosive was preserved. The purpose of this reshaping was to avoid a thin apex, which sometimes brings about a distortion of results.

The simulation was carried out using LS-Dyna, treating the explosive according to Case 13.2. The effective energy was taken as $q_G = 4.31 \times 10^6$ N-mm, per Table 13.1. LS-Dyna input requires energy per unit volume, $e = q_G \rho = 7586$. The brick elements were capable of yielding and the yield strength F_y was varied to see how it influences the results.

The table column "Formula" shows the value of PPV according to the above expression. The remaining columns give the same variable when the stated F_y is input. Only the first three lines illustrate the values within the "permissible" range, while the other three are associated with distances more than an order of magnitude larger, as such distances were also of interest.

r (mm)	$\check{r}$	Formula	F_y = 300	F_y = 400	F_y = 500
1,170	28.52	3021.9	3060	3246	3850
2,570	62.65	815.1	1401	1657	1960
4,720	115.07	342.2	694.4	505.3	452.8
41,170	1003.66	27.9	41	40.3	37.3
81,170	1978.79	13.8	14.3	11.2	10.4
124,920	3045.34	8.9	8.5	7.7	7.1

All values of PPV are in m/s. This is a scaled model, which means that r = 1,170 mm corresponds in reality to 11,700 mm. Yet, such scaling should not, in principle, have an effect on PPV.

As the table shows, varying the material properties does not bring about a smooth variation of results. This is typical of a situation, when one wants to create a model, reproducing experimental results over a wide range of distances. Especially poor results show for r = 2570 mm. F_y = 300 MPa is relatively best for the first two locations, but F_y = 500 MPa excels for $r \geq$ 4720 mm.

There are several ways in which such a simulation could be improved. The most obvious one would be to decrease the explosive strength. This would be very sensible from a physical viewpoint, as strong explosives pulverize the rock, which is associated with energy loss. Using q_G intended for metallic application means using more energy than is likely to be left in PETN outside the crush zone. (Refer to Chapter 15.) The other way is to use a more sophisticated material model, if the first measure proves inadequate.

If FEA results agree well with experiment in the tested range, they should also give a good indication outside the range. This was the rationale behind going to such far distances, as in the last three lines of the table.

EXAMPLE 13.15   BLAST PARAMETERS AND STRUCTURAL RESPONSE FEATURES

A 453.6 kg hemispherical charge of TNT placed on ground is exploded. At a distance of 207.3 m from its center a test cantilever (shown) is placed so that its face is normally impacted. The cantilever is made of Al alloy. After establishing the blast wave parameters, determine the relative importance of the wave action and the drag action on the response of this beam.

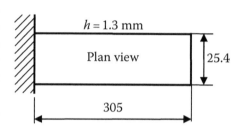

$h = 1.3$ mm

Plan view

25.4

305

The following is obtained from Conwep [113] for this charge and distance:

$\Delta p_s = 4.119$ kPa; $\Delta p_r = 8.383$ kPa; overpressures

$i_s = 89.91$ kPa-ms; $i_r = 161.1$ kPa-ms; impulses

$t_0 = 49.4$ ms; positive phase duration

$U = 346.1$ m/s; blast wave speed

These figures indicate that the wave action is not very pronounced. First, the wave speed is only slightly larger than the sound speed of 340.3 m/s. Also, the ratio of $\Delta p_r/\Delta p_s$ is only marginally larger than 2.0, the value that holds for a sound wave. Follow Case 13.3:

$$\frac{\rho}{\rho_0} = \frac{7 + 6(\Delta p_s/p_0)}{7 + (\Delta p_s/p_0)} = \frac{7 + 6(4.119/101.3)}{7 + (4.119/101.3)} = 1.029; \ \rho \text{ is density at wave front}$$

$$c_U = \sqrt{\frac{\gamma p_s}{\rho}} = \sqrt{\frac{1.4(0.1013 + 0.0041)}{1.029 \times 1.225 \times 10^{-6}}} = 342.1 \text{ m/s; sonic velocity in shock-compressed air}$$

From Case 13.6 one can find the time for shock wave action to vanish:

$$t_1 = \frac{3B}{2c_U} = \frac{3 \times 25.4}{2 \times 342.1} = 0.1138 \text{ ms}$$

$$q_d = \frac{2.5(\Delta p_s)^2}{\Delta p_s + 7p_0} = \frac{2.5(4.119)^2}{4.119 + 7 \times 101.3} = 0.0595 \text{ kPa}$$

To have an idea of the cantilever reaction, one needs to find the natural period from Case 2.24. The material constants from Table 5.2: $E = 69,000$ MPa, $\rho = 0.0027$ g/mm^3.

$$A = 25.4 \times 1.3 = 33.02 \text{ mm}^2; \quad I = 4.65 \text{ mm}^4$$

$$\tau = 1.787L^2 \sqrt{\frac{A\rho}{EI}} = 1.787 \times 305^2 \sqrt{\frac{33.02 \times 0.0027}{69,000 \times 4.65}} = 87.62 \text{ ms}$$

The deflection will reach its peak between $\tau/4$ and $\tau/2$, which is a substantial portion of the positive phase duration t_0. The pressure associated with this is $\Delta p_s = 4.119\,\text{kPa}$ plus $C_D q_d$ with q_d rather small. A larger pressure is involved the wave action, $\Delta p_r = 8.383\,\text{kPa}$ but it lasts only 0.1138 ms. It appears that the drag action will prevail. A detailed calculation is carried out in Example 13.16.

EXAMPLE 13.16   STRAIN RESPONSE OF A CANTILEVER IN A BLAST WAVE

The blast and the cantilever were described in Example 13.15. Using those results and treating the beam as an SDOF system, determine

(a) Peak bending moment at the base due to wave action
(b) The same due to drag action, with $C_D = 1.4$ when pressure applied as described.
(c) Combined moment
(d) Peak strain at 25.4 mm from the base
(e) The same, but using a reduction factor on impulse, as presented in Case 13.13

An application of static pressure of $1\,\text{kPa} = 0.001\,\text{MPa}$ to the surface of the cantilever gives a distributed loading of $w_0 = 25.4 \times 0.001 = 0.0254\,\text{N/mm}$, with the assumption of pressure being uniformly distributed across the width. The base (unit) moment:

$$M_a = w_0 L^2/2 = 0.0254 \times 305^2/2 = 1181 \text{ N-mm}$$

(a) The triangular impulse applied by the reflected pressure lasts only $t_1 = 0.1138\,\text{ms}$. The pressure $\Delta p_r = 8.383\,\text{kPa}$ gives the static moment of

$$M_{st} = 1181 \frac{8.383}{1} = 9900 \text{ N-mm}$$

In order to use Case 10.4, the equivalent rectangular impulse must be applied, instead of the original triangular impulse. With Δp_r selected as the reference, the duration must be halved to $t_i = 0.1138/2 = 0.0569\,\text{ms}$. The peak moment is

$$M_i = 2M_{st}\sin(\pi t_i/\tau) = 2 \times 9900 \sin(\pi 0.0569/87.62) = 40.4\,\text{N-mm}$$

(b) The total initial pressure of the drag phase is $p_d = C_D q_d + \Delta p_s = 1.4 \times 0.0595 + 4.119 = 4.202\,\text{kPa}$. The incident impulse, from Δp_s alone, was given before as 89.91 kPa-ms. The inclusion of dynamic pressure increases it somewhat

$$s_m = 89.91(4.202/4.119) = 91.72\,\text{kPa-ms}$$

Owing to the SDOF assumption, the moment changes in time in the same way displacement does. Let us first estimate the dynamic factor for displacements for the drag phase using Case 13.24. (The notation will change slightly, because pressures rather than loads are involved).

From $s_m = p_d t_x$ or $91.72 = 4.202 t_x$ find the characteristic time $t_x = 21.83\,\text{ms}$

The parameter $\omega t_x = \dfrac{2\pi}{\tau} t_x = \dfrac{2\pi}{87.62} 21.83 = 1.5654$

Since $0.4 < \omega t_x < 40$, use $u_m \approx u_{st}\dfrac{2.222}{1+2/\omega t_x} = u_{st}\dfrac{2.222}{1+2/1.5654} = 0.9756 u_{st}$

The static bending moment resulting from 4.202 kPa is obtained from the unit value calculated before: $M_{st} = (4.202/1.00)1181 = 4963\,\text{N-mm}$. The peak moment will therefore be

$$M_d = 0.9756; \quad M_{st} = 0.9756 \times 4963 = 4842\,\text{N-mm}$$

(c) The combination of the initial impulse and the drag loading is handled using Case 10.12. The latter is written for displacements, but our SDOF approach makes displacements combined in the same way as moments are, so we have

$$M_m \approx (M_d^2 + M_i^2)^{1/2} = (4842^2 + 44^2)^{1/2} = 4842.2\,\text{N-mm}$$

The influence of the wave action is therefore negligible.
(d) In a uniformly loaded cantilever the bending moment is proportional to the square of the distance from the tip. The reduction factor to be applied to the peak moment calculated above is, therefore, $[(305 - 25.4)/305]^2 = 0.84$. The strain is found using the basic formula for a rectangular section (Equation 10.4b):

$$\varepsilon = \frac{M}{EI}\frac{H}{2} = \frac{0.84 \times 4842}{69,000 \times 4.65}\frac{1.3}{2} = 8.24 \times 10^{-3}$$

(e) Making use of Case 13.13 gives the following impulse reduction factor for the Al plate:

$$\frac{S}{S_m} = 0.1928\ln\left(\frac{\rho}{1.225}\right) - 1.0568 = 0.1928\ln\left(\frac{2700}{1.225}\right) - 1.0568 = 0.4274$$

When the reduction is applied to the moment and the resulting strain, the new value is

$$\varepsilon = 0.4274 \times 8.24 \times 10^{-3} = 3.52 \times 10^{-3}.$$

The description of the problem and the experimental results come from Baker et al. [3]. The authors have also carried out calculations, in a totally different manner, obtaining $\varepsilon = 0.155 \times 10^{-3}$, while measurements gave $\varepsilon = 0.535 \times 10^{-3}$. The discrepancies are obvious. While the response reduction formula brings us closer to the true answer, there is a big gap between prediction and the test result.

EXAMPLE 13.17   EXPLOSIVE CHARGE TO BREAK A ONE-WAY SLAB

A long slab has a width $2L = 8.5$ m, thickness $h = 300$ mm, and the long edges simply supported.

The reinforcement is such that the bending strength is $M_0 = 180,000$ N-mm/mm of length. The specific mass is $\rho = 2500$ kg/m^3. A suspended explosive charge is detonated at an average distance of 5 m from the surface. Determine the mass of a TNT charge necessary to cause the failure of the slab in a symmetric bending mode. As a criterion of failure assume the angle of rotation of $5°$ at the support.

Since the slab is long, one can look at it as at a beam of unit width, which spans the long edges. The solution for such a beam, uniformly loaded, is given in Case 10.17 and quoted here replacing the distributed load w_0 by pressure p_0 and using $L = 4250$ mm. The angle of rotation θ is found from

$$\theta = \frac{S_0^2}{4 M_0 M_{ef}}; \text{ where } S_0 = p_0 t_0 L \text{ and } M_{ef} = \frac{2mL}{3} \text{ is the effective mass}$$

($p_0 t_0$ is a short specific impulse applied to the surface.) With $m = A\rho$ and for a unit-wide section

$$M_{ef} = \frac{2}{3}(300 \times 0.0025)4250 = 2125 \text{ g}$$

When the end-rotation angle is $\theta = 5° = 0.08727$, the necessary impulse is found from

$$0.08727 = \frac{S_0^2}{4 \times 180,000 \times 2,125} \text{ or } S_0 = 11,555 \text{ N-s/mm of slab length}$$

Then, our specific surface impulse to cause failure is

$$i_r = p_0 t_0 = S_0/L = 11,555/4250 = 2.7188 \text{ MPa-ms} = 2,719 \text{ kPa-ms}$$

Assume the explosive mass to be $M = 100$ kg and use Case 13.4 to check on this guess:
The scaled distance: $Z = \dfrac{r}{M^{1/3}} = \dfrac{5}{100^{1/3}} = 1.077$

$$i_r = 575 \frac{M^{1/3}}{Z^{1.2}} = 575 \frac{100^{1/3}}{1.077^{1.2}} = 2441 \text{ kPa-ms} \quad \text{for } 0.7 < Z \le 5$$

This is somewhat less than the required value. Taking the effect to be mass-proportional would give a charge of $100(2719/2441) = 111.4$ kg. When 110 kg charge is used and the calculation repeated, $i_r = 2618$ kPa-ms results. Taking the charge to be 115 kg will be sufficiently accurate.

The "average distance" in the problem statement requires a comment. In practical terms, this would be the distance to the midpoint of the nearest and the farthest boundary of the slab. This estimate relates mainly to locations measured in the short direction, that is, along $2L$. If there is a symmetry deviation in the lengthwise direction, one should limit his reasoning to a length of $2L$ in the long direction as well.

Example 13.18   The Influence of Gravity on Explosive Breaking of a Slab

The slab in Example 13.17 was investigated without reference to gravity-induced bending. The negligible effect of gravity is typical of walls, but not for floor slabs, where gravity provides most of the working load. Consider the same slab again, but now with the constant gravity loading inducing bending in the middle, reaching one-half of the bending capacity, that is, $\mathcal{M}_0/2$. Determine the new mass of explosive charge capable of breaking the slab.

A slab, which is being gradually bent into plastic range, can be modeled as rigid-plastic, since during most of its deformation path the resistance grows only marginally. Instead of using Case 10.17 as before, one can apply Case 10.19, with θ_m given as

$$\theta_m = \frac{S_0^2}{4 M_{ef} \mathcal{M}_0 (1 - \mathcal{M}_{st}/\mathcal{M}_0)}$$

From this one can determine the magnitude of S_0 needed to attain θ_m. If $\mathcal{M}_{st}/\mathcal{M}_0 = 0.5$ according to the problem statement, then, instead of 4 in the denominator, as in the previous problem, one has 2 and the associated value of S_0^2 is halved. From the previous result

$$S_0 = 11,555/\sqrt{2} = 8,171\,\text{N-s/mm} \quad \text{or} \quad i_r = S_0/L = 8,171/4,250 = 1.923\,\text{MPa-ms}$$
$$= 1,923\,\text{kPa-ms}$$

is now needed to break the slab. As was found before, 100 kg of explosives gives 2441 kPa-ms. Using a proportion as the first approximation, the mass of 80 kg could do the job. A more detailed approach gives

$$Z = \frac{r}{M^{1/3}} = \frac{5}{80^{1/3}} = 1.16 \quad \text{and} \quad i_r = 575\frac{80^{1/3}}{1.16^{1.2}} = 2073\,\text{kPa-ms}$$

which confirms that the selected mass is more than sufficient. Note that the selected level of gravity loading may not be accurate for the slab in question, but is probably somewhere near the gravity effect in many typical situations.

14 Penetration and Perforation

THEORETICAL OUTLINE

A fast-moving object, capable of impacting structures on its way, is sometimes referred to as a *kinetic energy threat*. Such an object can be a round of ammunition, purposely shot to inflict damage, but it can also result from an accidental event; e.g., a roofing sheet hurled by strong wind. At least some of the effects of those objects on structural elements belong to *terminal ballistics*, a branch of military engineering. While the terms like *penetration* and *perforation* are qualitatively understood by the readers, the exact definition of the latter depends on the area of application.

Only some of the material developed in this book can be directly linked to the formulas used here. One could say that the mechanical penetration and perforation are almost subjects of their own.

IMPACT AGAINST A SEMI-INFINITE MEDIUM

One way to quantify this type of event is to relate the speed of impact to plastic deformation, which the medium is undergoing. A nondimensional quantity, $\rho v_0^2/F_y$, is called the *damage number*, as it is a relative measure of the degree of plastic deformation resulting from impact. The numerator is proportional to the kinetic energy per impacted surface area (v_0^2) while the denominator is proportional to the average flow strength of the target. The projectile density ρ is used to make the number nondimensional. Johnson [40] presents a table (Table 14.1) with the damage number for mild steel, which, upon some formal adjustments, is given below.

Those speed limits were written in reference to steel, but they are believed to be relevant for other materials, too. A vinyl–ester resin, for example, where $\rho = 0.0016$ and $F_y = 5$, when impacted with a medium–high speed bullet of the same material, at $750\,\text{m/s}$, will have $\rho v^2/F_y = 180$, which puts it somewhere between advanced plastic flow and hypervelocity in the scale of this table. As Johnson remarks, in a low-velocity range, an alternative damage number, $\rho v_0 c_0/F_y$, may be more useful for comparative studies.

Another way to briefly describe how the impact velocity influences material property is this: when impact speed exceeds $1\,\text{km/s}$, fluid-like behavior is experienced by some metals. For speed greater than $10\,\text{km/s}$, the impact may be associated with vaporization and explosion of the colliding objects.

When two similar objects impact a medium surface, their shape differences do matter. An object that delivers more kinetic energy per unit area of the impacted surface will cause more damage. In case of ductile media, the imprint of that object will be deeper, i.e., a larger penetration.

With a sufficiently high impact speed, or with a sufficiently weak medium, we have not only a surface damage from impact but also a penetration of a striker into a medium.

TABLE 14.1

Damage Number for Various Materials

v_0 (m/s)	$\rho v_0^2/F_y$	
0.75	1.6×10^{-5}	Quasistatic
7.5	1.6×10^{-3}	Noticeable plasticity begins
75	0.16	Well-developed plasticity
750	1.6	Plastic deformation predominates
7500	1600	Hypervelocity phenomena

$\rho = 0.00785\,\text{g/mm}^3$; $F_y = 275\,\text{MPa}$.

Consider a case of an undeformable projectile of mass M moving within a solid medium and write its equation of its motion as

$$M\frac{dv}{dt} = -F_2 v^2 - F_1 v - F_0 \tag{14.1}$$

where the first term on the right is a hydrodynamic resistance, the second is a viscous resistance and the third term is velocity independent and related to the medium properties. Mathematically simplest is a motion with a constant resistance F_0:

$$M\frac{d^2u}{dt^2} = -F_0 \tag{14.2a}$$

or

$$M\frac{dv}{dt} = -F_0 \tag{14.2b}$$

In this case, velocity changes linearly from the initial v_0 to nil when the projectile attains the full penetration depth X. The acceleration is constant, $a = F_0/M$. The simplest way to find X is to equate the initial kinetic energy to the work of the resisting force, namely, $F_0 X$. This gives

$$X = \frac{Mv_i^2}{2F_0} \tag{14.3}$$

When the resistance includes both hydrodynamic component F_2 as well as the constant F_0, the maximum penetration is

$$X = \frac{M}{2F_2}\ln\left(\frac{F_2\,v_0^2}{F_0} + 1\right) \tag{14.4}$$

A more detailed expression for the hydrodynamic component is

$$F_2 v^2 = \frac{1}{2}C_d A\rho v^2 \tag{14.5}$$

with its terms defined in Appendix C.

IMPACT AGAINST A FINITE-THICKNESS TARGET

With a sufficiently high impact speed, or with a sufficiently weak target material, the projectile can traverse the entire thickness and emerge at the other side, thereby resulting in *perforation*. Figure 14.1 illustrates what happens to the stricken object in terms of physical damage.

One should notice that all these forms of damage are associated mainly with ductile materials, except Figure 14.1c, which is the result of brittleness. As the sketches imply, the projectile is treated as rigid. The speed of the projectile, after parting with the target, is called the *exit velocity* or *residual velocity*.

Ballistic velocity limit, designated by v_B is the smallest velocity at which the projectile can still perforate the target.* The status before perforation and after is shown in Figure 14.2, assuming that the projectile and the plug fly out as one. The initial velocity is v_0 and the exit speed is V_r. Consistent with this, the corresponding kinetic energies are E_{k0} and E_{kr}, respectively. When the projectile perforates, but stops just afterward, this means $v_0 = v_B$ and the corresponding initial energy can be designated by E_{kB}. In a general case, one can write $E_{k0} = E_{kB} + E_{kr}$, or

$$\frac{1}{2}Mv_0^2 = \frac{1}{2}Mv_B^2 + \frac{1}{2}(M + M_p)V_r^2 \tag{14.6}$$

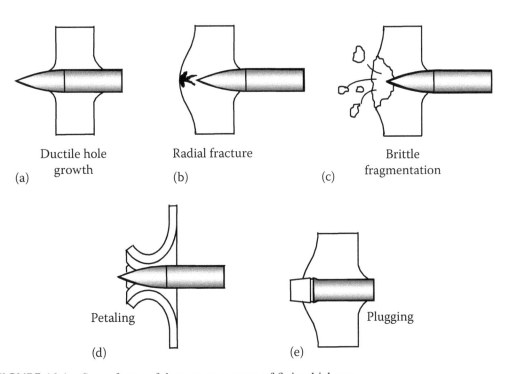

Ductile hole growth

(a)

Radial fracture

(b)

Brittle fragmentation

(c)

Petaling

(d)

Plugging

(e)

FIGURE 14.1   Some forms of damage to a target of finite thickness.

---

* This is often encountered in literature as a *ballistic velocity* or a *ballistic limit*.

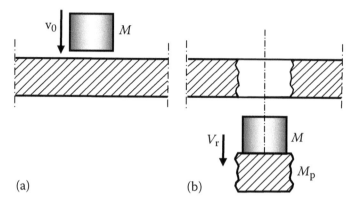

(a)   (b)

FIGURE 14.2 A high-velocity striker impacting a plate (a) and flying out along with a plug torn out (b).

If the plug mass M_p is known, then v_B is the only unknown in this equation. But often M_p is not known; therefore writing the expression for V_r can be more useful:

$$V_r^2 = \frac{M}{M + M_p}(v_0^2 - v_B^2) = \frac{1}{1 + \mu}(v_0^2 - v_B^2) \tag{14.7}$$

where $\mu = M_p/M$. Writing the above expression (similar to Recht–Ipson equation) for the results of two separate tests allows us to find v_B and M_p.

An accurate determination of v_B is difficult from the experimental viewpoint and is certainly not made easier by a large dispersion of results encountered in physical testing. For this reason a statistical counterpart of this quantity was devised and designated by v_{50}. This stands for the ballistic velocity limit corresponding to penetration in 50% of tests.

SOIL PENETRATION

Soil penetration problem is somewhat simpler than that of other media, because soil has a small or sometimes negligible cohesion. If a hard projectile, as depicted in Figure 14.3, is moving relatively slowly, its equation of motion is often expressed by Equation 14.2, while a hydrodynamic term shows up at larger velocities.

A simple expression for a "standard projectile," as shown in Figure 14.3b, is offered by TM5 [106]. The penetration of this steel object into a large volume of sand is

$$X = 3.53d \ln\left(1 + \frac{v_0^2}{43}\right) \tag{14.8}$$

where $X \sim$m, $d \sim$m, (diameter), and $v_0 \sim$m/s. If the sand layer thickness H is less then X calculated above, it means that the projectile slows down, rather than stops. The residual velocity is found from

$$\frac{V_r}{v_0} = \left(1 - \frac{H}{X}\right)^{0.555} \tag{14.9}$$

An expression for penetration into various soil types is provided as Case 14.7.

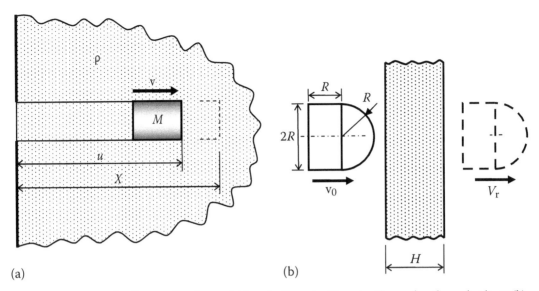

FIGURE 14.3   Projectile traveling in sand (a) and a "standard" projectile passing through a layer (b).

CONCRETE PENETRATION

Concrete penetration formulas have to be more involved than those for typical soils, because the strength of the medium must be brought into the picture. The equations associated with the names of ACE, NDRC, and Sandia are quoted as Cases 14.3, 14.4, and 14.6, respectively. They can all be presented as velocity-dependent expressions for $X - X_0$, where X_0 is a constant.* In the first one the net penetration $(X - X_0)$ is proportional to $v^{1.5}$, in the second to $v^{1.8}$, and in the third to v. A common feature of these three equations is that penetration is inversely proportional to $\sqrt{F'_c}$. It was stated in Chapter 4 that the concrete tensile strength is also proportional to $\sqrt{F'_c}$. In fact, when the numbers given for the natural rock in Chapter 15 are examined, a similar trend is found. This implies that the net penetration is also inversely proportional to the tensile strength F_t of brittle materials.

Hansson [29] conducted a broad study of experimental data and came to the following conclusions (in additions to those given in Case 14.4):

NDRC equation, with a modified constant (Case 14.4), provides a reasonable prediction (within ±20%) of concrete penetration for a broad range of impact velocities and for projectiles with length/diameter ratio between 6 and 10. The average test figures usually exceed those obtained from the equation.

Modern geopenetrators can reach up to twice the depth predicted by the NDRC equation.

Often the formulas used for a conventional concrete give inaccurate results for a high-strength concrete. According to Ref. [29] the accuracy improves when F'_c in the equations is limited to 65 MPa. Alternatively one can, instead of limiting the unconfined strength, develop new formulas, as was demonstrated by Chen and Li [16].

Bulson [14] notes that penetration into plain concrete is about 10% greater than into reinforced concrete. He also quotes a simple formula for a minimum concrete thickness X_p necessary to prevent perforation:

---

* To make this possible with Sandia equations, the impact velocity must be larger than 61 m/s.

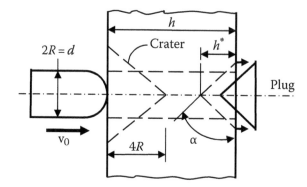

FIGURE 14.4   Typical damage zones in a brittle material following a projectile impact.

$$X_p = \frac{4}{3}X \tag{14.10}$$

where X is the maximum penetration of the same projectile into a very thick slab. This comes from the fact that when there is no resisting material behind the impacted layer, the rear face spalling may occur. That effect is related to the action of stress waves due to impact. Another factor is that when a projectile comes close to a rear wall, it may simply push out the remaining, already precracked material.

The outline of a perforation process of a brittle plate is presented in Figure 14.4. Near the impact point the surface damage has a form of a crater, about $4R$ deep. The motion past the crater is known as *tunneling*. Finally, at a distance of h^* from the back surface a cone of material is pushed out,* forming a rear crater. The initial cratering and/or the tunneling stage may not clearly show up if the plate thickness is too small. According to Seah et al. [76] the angle α and the plug resistance F_p are

$$\alpha = 45° + \theta/2 \tag{14.11}$$

$$F_p = \pi\sigma_t h^* \tan\alpha \tag{14.12}$$

where
   θ is the friction angle
   σ_t is the tensile strength according to the Mohr–Coulomb criterion

To find h^* one needs the experimental data, which were quoted in Ref. [76] for a certain type of granite.

The observation by Bulson mentioned above, regarding the decrease of penetration when a conventional reinforcement is placed should be understood in statistical sense. Other studies, like those by Nyström and Leppänen [64] suggest that there is little difference between the plain and the RC unless a projectile impacts a reinforcing bar. Positioning of steel elements is also important. If the pattern is such that it increases the confinement of

---

* That "cone" is rather of irregular shape.

the volume subjected to penetration, the reduction of the latter shows up; in fact it reached up to 15% in the above study.

Suppose now that for a particular concrete plate the thickness to prevent penetration X_p is known. In case the actual thickness of the concrete slab H is smaller than X_p, the residual velocity of the striker may be calculated from equations provided by TM-1300:

$$\frac{v_r}{v_i} = \left[1 - \left(\frac{H}{X_p} \right)^2 \right]^{0.555} \quad \text{for } X < 2d \tag{14.13}$$

$$\frac{v_r}{v_i} = \left(1 - \frac{H}{X_p} \right)^{0.555} \quad \text{for } X > 2d \tag{14.14}$$

Lok et al. [57] conducted experiments with penetration into concrete panels that not only had a conventional steel mesh, but were also reinforced with steel fibers. The penetration was, typically, less than one-half of what was obtained from the ACE formula. In those experiments the length of the projectile was comparable to the panel thickness. A marked decrease in penetration seems to be caused by a large increase of F_t, due to presence of steel fibers.

METAL PENETRATION

The constant (average) resistance of the impacted object is often used, at least to get the first estimate of the penetration depth X. The resisting force F_0 (Equation 14.2) can be presented as

$$F_0 = \pi R^2 \beta F_y \tag{14.15}$$

where
   R is the projectile radius
   β is an empirical coefficient

According to Wijk et al. [117], $\beta = 5$ often gives a good prediction of X for metallic objects, especially for a hard steel.

This number has some relation to the deformability of an elastoplastic cavity described by Cases 4.27 and 4.28. To bring this concept into the picture, one should imagine a as in Figure 14.3a, but with a hemispherical nose. Each time the projectile advances by $a = R$, a new, hemispherical cavity of the same radius is created in the medium, by means of crushing and expanding the penetrated material. A spherical part of that cavity around the axis of the striker can be regarded as a part of a sphere that had been expanded from zero to the radius R.

For simplicity, this forward part of the spherical nose is treated as a representative of a full, spherical cavity. By setting $u_r = a = R$ in Case 4.28 one can find the relation between R and the radius c of the plastic zone around the cavity. That radius is several times larger than R, so the expression can be simplified. Once this ratio is established, the internal pressure to expand the cavity to radius R can be found.

If the expansion to radius R of a cylindrical cavity, described by Case 4.27 is postulated, one can find that a relatively smaller pressure is needed to accomplish the task, but the plastic zone radius becomes larger. The elastoplastic cavity descriptions, specialized to the expansion problem, are presented in this chapter as Cases 14.1 and 14.2. The cylindrical expansion governs at the transition of the shape from a hemisphere to a cylinder. In the nose tip area the spherical expansion is dominant, as described above. The plastic zone size may be expected to vary smoothly between the two. One sensible approach is to use the average plastic zone radius c as representative of the whole hemispherical area. This makes it possible to find the equivalent internal pressure of the hemispherical cavity and thereby establish the coefficient β. The procedure is illustrated by Example 14.1.

Goldsmith [26] quotes a compact formula for small metal projectiles shot into a volume of metal. This is given later as Case 14.13.

KINEMATICS OF A HARD PROJECTILE MOVEMENT

Typically, projectiles or strikers have a tapered nose part, so the resistance of the penetrated body increases as the nose progresses. In order to remove this variable from the picture, one should use instead a striker with the equivalent cylindrical shape as in Figure 14.5. A simple way to do so is to replace the nose part with a cylinder of the same volume and diameter d of the basic striker body. This leads to shortening of the striker by ΔL as well as an apparent shortening of the penetration, by a comparable length, depth from X to u_m. In effect, we have

$$u_\mathrm{m} = X - \Delta L \qquad (14.16)$$

or whatever the geometry of the nose dictates. Experimental formulas usually give us X as a function of impact velocity v_0. But it is often of interest to determine the event duration as well as the peak acceleration, to which the projectile is subjected.

Once the change from the actual to a modified displacement, as described by Equation 14.16, is made, then the displacement u, velocity v, and acceleration a are treated as functions

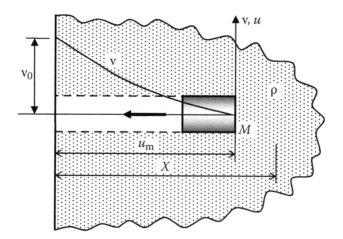

FIGURE 14.5 The motion of a striker inside the medium is treated in a reverse way, i.e., as accelerating towards the entry point.

of time. The derivation becomes simpler, if motion is reversed, i.e., the projectile is accelerated from rest to velocity v_0. This is illustrated in Figure 14.5.

An experimental formula of the form $v_0(u_m)$ can be written more generally as $v = f(u)$. The latter has an implicit time dependence in it, which can be seen if one writes

$$v = \frac{du}{dt} = f(u) \tag{14.17a}$$

or

$$\frac{du}{f(u)} = dt \tag{14.17b}$$

provided that an equation in the style of 14.17b can be written without difficulty. Having the variables separated one can integrate both sides:

$$\int_0^{u_m} \frac{du}{f(u)} = t \tag{14.18}$$

From this $u(t)$ can be found and differentiated to obtain $v(t)$ and then acceleration $a(t)$.

A relatively simple solution for $u(t)$ is obtained when $u = Gv^n$, where G is a constant. In particular, when

$$u = G_1 v^{1.8} \tag{14.19a}$$

and

$$u = G_2 v^{1.5} \tag{14.19b}$$

the integration in Equation 14.18 and then differentiating with respect to time yields, for the first function:

$$u = G_1 \left(\frac{4t}{9G_1} \right)^{9/4} \tag{14.20a}$$

$$v = \left(\frac{4t}{9G_1} \right)^{5/4} \tag{14.20b}$$

$$a = \frac{5}{4} \left(\frac{4}{9G_1} \right)^{5/4} t^{1/4} \tag{14.20c}$$

where $v \sim m/s$ is velocity and $a \sim mm/(ms)^2$ is acceleration. Similarly, for the second one:

$$u = G_2 \left(\frac{1}{3G_2} \right)^3 t^3 \tag{14.21a}$$

$$v = \left(\frac{1}{3G_2}\right)^2 t^2 \qquad\qquad (14.21\text{b})$$

$$a = 2\left(\frac{1}{3G_2}\right)^2 t \qquad\qquad (14.21\text{c})$$

One would normally first calculate X from a penetration equation, then u_m and, if duration of motion t_m is of interest, find t_m corresponding to u_m from the appropriate formula above.

The t_m so calculated has to be increased by the duration of the nose penetration, approximately $\Delta L/v_0$. The impacted object is subjected to a force of magnitude $Ma(t)$.

CLOSING REMARKS

All models presented in this chapter use a rigid striker assumption. This is valid as long as the dynamic strength of the penetrator exceeds the dynamic resistance of the medium. When the impact speed is sufficiently large so that both quantities are equal, the penetrator enters a semihydrodynamic regime.

TABULATION OF CASES

COMMENTS

All formulas given here are based on a rigid projectile concept. This leaves the mass and shape as the only variables of the projectile, which are involved.

CASE 14.1   CYLINDRICAL CAVITY EXPANSION IN A CONTINUOUS, ELASTOPLASTIC MEDIUM

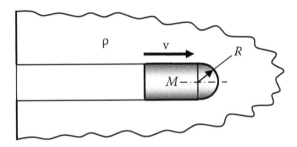

The cavity is expanded from zero to R by the incoming, undeformable projectile.

(a) EPP medium material with $F_y \equiv \sigma_0$

$$c = R\sqrt{\frac{2\sqrt{3}}{\sigma_0}\frac{E}{5-4v}}\;;\;\text{radius of the plastic zone around cavity.}$$

$$p = \frac{\sigma_0}{\sqrt{3}}\left(1+2\ln\frac{c}{R}\right);\;\text{internal pressure to keep the cavity open}$$

The equations are derived from Case 4.27 by assuming $c \gg a$ and setting $u_r = a = R$. Also, $p \equiv -\sigma_r$. Ref. [45].

An alternative expression for cavity radius from Ref. [34]:

$$c = R\sqrt{\frac{\sqrt{3}}{\sigma_0}\frac{E}{5-4v}}$$

(b) Bilinear material specified by E, F_y, F_u, E_p, and $v = 0.5$ is assumed. Ref. [34]. Cavity expanding from zero radius to R.

$$c = R\sqrt{\frac{2}{F_y}\frac{E}{5-4v}}\;;\;\text{radius of plastic zone}$$

$$p = \frac{F_y}{\sqrt{3}}\left(1+\ln\frac{E}{\sqrt{3}F_y}\right) + \frac{\pi^2 E_p}{18}\;;\;\text{pressure inside cavity}$$

Note: Although a projectile is shown as a reference, the equations are written for an infinitely long cylindrical cavity on its own.

CASE 14.2   SPHERICAL CAVITY EXPANSION IN A CONTINUOUS, ELASTOPLASTIC MEDIUM

(Refer to the illustration in Case 14.1)

The cavity is expanded from zero to R by the incoming, undeformable projectile.

(a) EPP medium material with $F_y \equiv \sigma_0$

$$c = R\left\{\frac{E}{(1-v)\sigma_0}\right\}^{1/3} ; \text{ radius of the plastic zone around cavity.}$$

$$p = \frac{2\sigma_0}{3}\left(1 + 3\ln\frac{c}{R}\right); \text{ internal pressure to keep the cavity open}$$

The equations are derived from Case 4.28 by assuming $c \ll a$ and setting $u_r = a = R$. Ref. [45].

Also, $p \equiv -\sigma_r$.

An alternative expression for cavity radius from Ref. [34]:

$$c = R\left\{\frac{E}{3(1-v)\sigma_0}\right\}^{1/3}$$

(b) Bilinear material specified by E, F_y, F_u, E_p, and $v = 0.5$ is assumed. Ref. [34]:

$$c = R\left\{\frac{2}{3}\frac{E}{F_y}\right\}^{1/3} ; \text{ radius of plastic zone}$$

$$p = \frac{2F_y}{3}\left(1 + 3\ln\frac{c}{R}\right) + \frac{2\pi^2 E_p}{27}; \text{ pressure inside cavity}$$

Note: Although a projectile is shown as a reference, the equations are written for a perfect spherical cavity on its own.

CASE 14.3   PENETRATION OF A PROJECTILE INTO CONCRETE (ACE EQUATIONS)

$$X = 282\left(\frac{W}{d^{1.785}\sqrt{F_c'}}\right)\left(\frac{v_0}{1000}\right)^{1.5} + 0.5d; \text{ for } X \sim \text{in., } v_0 \sim \text{ft/s, } F_c' \sim \text{psi}$$

W and d are weight ~lb and diameter ~in. of the striker, respectively

$h_p = 1.32d + 1.24X$; perforation thickness of the target

$h_s = 2.12d + 1.36X$; spall thickness of the target

Metric units:

$$X = \frac{1}{12.61}\frac{M}{\sqrt{F_c'}}\frac{v_0^{1.5}}{d^{1.785}} + 0.5d; \text{ for } X \sim \text{mm, } d \sim \text{mm, } M \sim \text{g, } v_i \sim \text{m/s, } F_c' \sim \text{MPa}$$

Note: The English units formula is quoted in essentially the same manner as in Ref. [6]. The expressions give X for plain, conventional concrete. The second term is an additional penetration due to nose of the projectile. This term vanishes when the impacting end is flat. No matter how strong the concrete is, F_c' in the equation should not be taken larger than 65 MPa.

A hard aggregate, granite, or equivalent, is needed for the prediction to be reasonable. For other aggregate materials, like limestone, penetration increases by up to 20%.

CASE 14.4 PENETRATION OF A PROJECTILE INTO CONCRETE (NDRC EQUATIONS)

$$X = \frac{180}{\sqrt{F_c'}} NW \left[\frac{v_0}{1000d} \right]^{1.8} + d; \text{ for } X > 2d \text{ with } X \sim\text{in., } W \sim\text{lb, } v_0 \sim\text{ft/s, } F_c' \sim\text{psi}$$

$N = 0.72$ for a flat nose, 0.84 for a blunt nose, 1.00 for a hemispherical nose, 1.14 for a sharp nose

$h_p = 1.32d + 1.24X$; perforation thickness of the target, $h_p > 3d$

$h_s = 2.12d + 1.36X$; spall thickness of the target, $h_s > 3d$

$$\frac{h_p}{d} = 3.19 \left(\frac{X}{d} \right) - 0.718 \left(\frac{X}{d} \right)^2 ; \text{ the same quantities for thinner targets}$$

$$\frac{h_s}{d} = 7.91 \left(\frac{X}{d} \right) - 5.06 \left(\frac{X}{d} \right)^2$$

Metric units:

$$X = \frac{1}{\sqrt{F_c'}} \frac{NM}{104} \left[\frac{v_0}{d} \right]^{1.8} + d; \text{ for } X > 2d \text{ with } X \sim\text{mm, } v_0 \sim\text{m/s, } M \sim\text{g, } d \sim\text{mm}$$

Note: Notes for Case 14.3 apply. Hansson [29], who made a more recent study, quotes the above equation with a constant of 85 rather than 104.

CASE 14.5 PENETRATION OF A PROJECTILE INTO ROCK (SANDIA EQUATIONS)

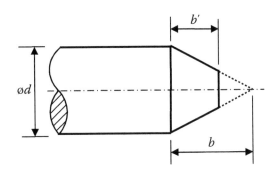

$$X = 0.3KSN \left(\frac{W}{A} \right)^{0.7} \ln \left[1 + \frac{v_0^2}{50,000} \right]; \text{ for } v_0 < 200 \text{ft/s}; \ v_0 \sim \text{ft/s}, \ A \sim \text{in.}^2$$

$$X = 0.00178KSN \left(\frac{W}{A} \right)^{0.7} (v_0 - 100); \text{ for } v_0 > 200 \text{ft/s}$$

$K = 1$, but when $W < 400 \text{lb}$, $K = 0.4W^{0.15}$

$$S = \frac{12}{(QF_c')^{0.3}}; \text{ when } F_c' \sim \text{psi, index of penetrability}$$

W is the weight of the striker $\sim$lb

Metric units:

M is the mass of the striker $\sim$kg

$$X = \frac{KSN}{12,910} \left(\frac{M}{A} \right)^{0.7} \ln \left[1 + \frac{v_0^2}{4,651} \right]; \text{ for } v_0 < 61 \text{ m/s}; \ v_0 \sim \text{m/s}, \ A \sim \text{m}^2$$

$$X = K \frac{SN}{663,230} \left(\frac{M}{A} \right)^{0.7} (v_0 - 30.5); \text{ for } v_0 > 61 \text{ m/s}$$

$K = 1$, but when $M < 181$ kg, $K = 0.45M^{0.15}$

$$S = \frac{2.7}{(QF_c')^{0.3}}; \text{ when } F_c' \sim \text{MPa}$$

Rock Quality	Q
Very good	0.9
Good	0.7
Fair	0.5
Poor	0.3
Very poor	0.1
Slightly weathered	0.7
Moderately weathered	0.4
Highly weathered	0.2

Nose Shape	N
Conic, full	$0.25b/d + 0.56$
Ogive, full	$0.18b/d + 0.56$
Conic, blunt	$0.125(b + b')/d + 0.56$
Ogive, blunt	$0.09 (b + b')/d + 0.56$

Note: The accuracy depends substantially on how well the factor S describes the condition of the medium. A typical deviation from true result of using the above formulas is likely to be within 20%. Equations valid for $v_0 < 1200 \text{m/s}$, for penetrators not lighter than 2 kg and for $X \geq 3d$.

CASE 14.6 PENETRATION OF A PROJECTILE INTO CONCRETE (SANDIA EQUATIONS)

The equations for X and N are the same as for rock, but the expression for S differs.

$S = 0.085K_e(11 - P) (t_cT_c)^{-0.06} (5000/F_c')^{0.3}; F_c' \sim \text{psi}$

$K_e = \left(\dfrac{F}{W_1}\right)^{0.3}$; where $F = 30$ for plain concrete and $F = 20$ for RC. For thin targets, less

than $2d$, use $F = 15$ and $F = 10$, respectively (d is projectile diameter)

W_1 is target width, as a multiple of d. When $W_1 > F$, then $K_e = 1$

P = (volume of all reinforcing steel)/(volume of RC)

t_c is cure time, expressed as a fraction of 1 year. For older concrete use $t_c = 1$

T_c is the target thickness expressed as a multiple of d (projectile diameter)

F_c' is the unconfined concrete strength at test time

If the data is insufficient to use the formula, assume $S = 0.9$.

$S = 0.085K_e (11 - P) (t_cT_c)^{-0.06} (34.5/F_c')^{0.3}; F_c' \sim \text{MPa}$

Note: The notes for Case 14.5 apply.

CASE 14.7 PENETRATION OF A PROJECTILE INTO SOIL (SANDIA EQUATIONS)

The equations for X and N are the same as for rock, but the expression for K differs.

$K = 1$, but when $W < 60\,\text{lb}$, $K = 0.2W^{0.4}$

$K = 1$, but when $M < 27\,\text{kg}$, $K = 0.27M^{0.4}$

The values of S also differ:

Soil Type	S
Frozen soil. Dense, dry cemented sand. Massive gypsite and selenite.	2–4
Hard, dry clay. Sand, dense, w/out cementation. Gravel deposits.	4–6
Topsoil, loose to very loose. Silt and clay, moist to wet.	10–20
Soft and saturated clay of low shear strength.	20–30
Clay marine sediments.	30–60

CASE 14.8 PENETRATION OF A PROJECTILE INTO CONCRETE FOR LONG, OGIVE-NOSED PROJECTILES

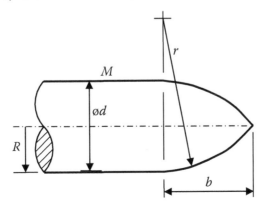

$\sigma_r = Q\sigma_0 + B\rho v^2$; radial stress during spherical cavity expansion as a function of the expansion velocity v.

$$V_1^2 = \frac{Mv_0^2 - 4\pi R^3 Q\sigma_e}{M + 4\pi R^3 B\rho N}; \text{ initial tunneling velocity (squared)}$$

$$\sigma_e = \left(1 - \frac{\lambda}{3}\right)\sigma_0; \text{ effective medium stress}$$

$$N = \frac{1}{3\psi}\left(1 - \frac{1}{8\psi}\right); \psi = r/d$$

$$X = \frac{M}{2\pi R^2 \rho BN}\ln\left(1 + \frac{\rho BNV_1^2}{Q\sigma_e}\right) + 4R; \text{ penetration}$$

The above is based on Ref. [24]. For a concrete with $F_c' = 16.2$ MPa, the following constants were used: $E = 12.15$ GPa, $\rho = 0.00231$ g/mm^3, $\lambda = 0.25$, $v = 0.23$, $\sigma_0 = 164$ MPa, $\sigma_e = 150$ MPa, $Q = 2.8$, and $B = 1.37$.

CASE 14.9   PENETRATION OF A PROJECTILE INTO CONCRETE (BASED ON DYNAMIC CAVITY EXPANSION THEORY)

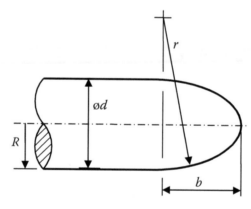

$F_r = BX$; axial resisting force for $X < 2Rk$ (during cratering)

$F_r = \pi R^2(SF_c' + N^*\rho v^2)$; axial resisting force for $X > 2Rk$ (during tunneling)

$$B = \frac{\pi R}{2k}\frac{(SF_c' + \rho N^* v_0^2)}{\left(1 + (2\pi kR^3/M)\rho N^*\right)}; \text{ apparent medium resistance coefficient}$$

$$S = \frac{72}{\sqrt{F_c'}} \text{ or } S = \frac{82.8}{(F_c')^{0.544}}; F_c' \sim \text{MPa}$$

$k = 0.707 + \dfrac{b}{2R}$; when $X/R < 10$, $k = 2$ when $X/R > 10$; for flat, hemispherical, and conical noses

$k = 2.367$ for the ogive shape with $r = 3d$ and $k = 2.77$ for ogive shape with $r = 4.5d$

$N^* = \dfrac{1}{1 + 4\psi^2}$; for a conical nose, $\psi = b/d$

$N^* = 1 - \dfrac{1}{8\psi^2}$; for a nose curved with a radius R, $\psi = R/d$

$N^* = 1$; for a flat nose and $N^* = 1/2$ for a hemispherical nose

$N^* = \dfrac{1}{3\psi} - \dfrac{1}{24\psi^2}$; for an ogive nose curved with a radius r, $\psi = r/d$

$I = \dfrac{M v_0^2}{8 S R^3 F_c''}$; nondimensional kinetic energy

$N = \dfrac{1}{N^*}\left(\dfrac{M}{8\rho R^3}\right)$; nondimensional mass ratio

$X = 1.628 d \left[\dfrac{4kI}{\pi(1 + I/N)}\right]^{1.395}$; for $X < R$, shallow penetration

$X = d \left[\dfrac{4kI}{\pi(1 + I/N)}\right]^{1/2}$; for $R < X < 2Rk$

$X = \dfrac{2Nd}{\pi} \ln\left[\dfrac{(1 + I/N)}{1 + \pi k/4N}\right] + 2kR$; for $X > 2Rk$, Refs. [16,55].

CASE 14.10   PERFORATION OF A METAL PLATE BY A PROJECTILE FLYING WITH VELOCITY v_0

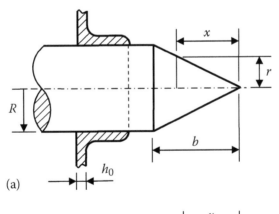

(a)

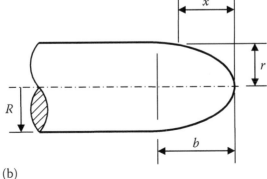

(b)

(a) Cone end, $r = Rx/L$

$$\mathcal{L} = \pi h_0 R^2 \left[\frac{\sigma_0}{2} + \rho \left(\frac{Rv_0}{b} \right)^2 \right];$$

Work to deform plate

(b) Ogive end, $r = R \sin \dfrac{\pi x}{2b}$ (approx.)

$$\mathcal{L} = \pi h_0 R^2 \left[\frac{\sigma_0}{2} + 1.86 \rho \left(\frac{Rv_0}{b} \right)^2 \right]; \text{ Ref. [26]}$$

$$v_B \approx \sqrt{\frac{2\mathcal{L}}{M}} \text{ (both cases)}$$

Note: Material model is EPP. The plastic part of the total work (first term) is due to hoop stretching of material. No radial cracking is assumed.

CASE 14.11   PERFORATION OF A METAL PLATE BY A COMPACT PROJECTILE OF MASS M

ρ_p and R; density and equivalent radius of the projectile, respectively

h, ρ_t and F_y; thickness, density, and yield strength of the target plate, respectively

$$R = \left(\frac{3M}{4\pi\rho_p} \right)^{1/3}; \text{ equivalent projectile radius}$$

$$v_{50} = 3.52 \frac{h}{R} \frac{\sqrt{F_y \rho_t}}{\rho_p}; \text{ ballistic velocity to perforate the plate (with 50\% probability)}$$

Note: Material model is EPP. The experimental v_{50} values are in the $\pm 16\%$ band, relative to the formula results. Ref. [4].

CASE 14.12   SOLID WOODEN CYLINDERS PERFORATING THIN, MILD STEEL PLATES

ρ_p, L, d; density, length, and diameter of projectile, respectively. h and F_y; thickness and yield strength of the target plate, respectively.

$$v_{50}^2 = \frac{F_y}{\rho_p} \frac{h}{L} \left[1.751 + 144.2 \frac{h}{d} \right]; \text{ ballistic velocity to perforate the plate (with 50\% probability)}$$

Note: Material model is EPP. The striker impacts normally. Limits of validity: $5 \leq L/d \leq 31$; $0.05 \leq h/d \leq 0.10$; $0.01 \leq \dfrac{\rho_p v_{50}^2}{F_y} \leq 0.05$, Ref. [4].

CASE 14.13   SMALL METAL STRIKERS PENETRATING METAL

$$X = Bd \left(\frac{v_0}{c_0} \right)^{1.4} \quad \text{for } 0.1 < v_0/c_0 < 1.0$$

$B = 2.5$ for most metals

c_0 is sonic velocity Ref. [26]

d is diameter

Note: Two sets of experiments are reported: one for a block of aluminum (with sonic velocity $c_0 = 5102\,\text{m/s}$) impacted with aluminum strikers and one for steel ($c_0 = 4962\,\text{m/s}$) impacted with steel, brass, and lead strikers. Those projectiles were 0.125 in. spheres and axially shot cylinders with $L/d = 1$, either 0.22 or 0.50 in. in diameter. There was a significant dispersion of results, especially for the steel block.

CASE 14.14   METALLIC PLATE PERFORATION DUE TO IMPACT BY A STRIKER WITH MASS M AND VELOCITY v_0 (SIMPLE THEORY)

Refer to Figure 14.4 for thickness definition. $2R = d$ is projectile diameter

$h^* = \dfrac{d}{2\pi}(2\beta - \gamma)$; distance from the rear face at which perforation starts. (A plug is only one possibility.)

β is the resistance coefficient in $F = \pi R^2 \beta \sigma_0$ ($\beta \approx 5$ for hard steel)

γ is the nose bluntness coefficient. $\gamma \approx 1$ for conical noses with apex angle of 55° and $\gamma \approx 2$ for spherical ends

$$\mathcal{L}_p = \frac{\pi d h^*}{8}(\pi h^* + \gamma d)\sigma_0 + \frac{\pi d^2}{4}(h - h^*)\beta\sigma_0; \text{ perforation energy when } h > h^*$$

$$\mathcal{L}_p = \frac{\pi d h}{8}(\pi h + \gamma d)\sigma_0; \text{ perforation energy when } h < h^*$$

$M_t = \pi R^2 h \rho_t$; shattered target mass, ρ_t is target density

$$V_r^2 = \frac{Mv_0^2 - 2\mathcal{L}_p}{M + M_t}; \text{ residual velocity (squared)}$$

Note: The case is based on Ref. [117]. See Example 14.1 for determination of coefficient β. The last equation can be used for estimation of the ballistic limit.

CASE 14.15   PLATE PERFORATION BY SHEAR PLUGGING DUE TO IMPACT BY A STRIKER WITH MASS M AND VELOCITY v_0

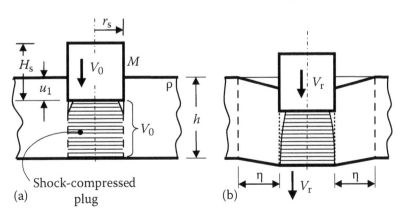

(a) Shock-compressed plug

(b)

$c_{pp} = \sqrt{\dfrac{\sigma_0}{\rho}}$; notional plastic wave velocity

(a) Status at the end of Stage 1. The shock wave with speed U has reached the opposite face of the plate.

$M_t = \rho\pi(r_s + \eta)^2 h$; effective target mass

$\eta = h/2$; additional tributary plate radius

$V_0 = \dfrac{Mv_0}{M + M_t} = \dfrac{v_0}{1+\mu}$; common velocity of M and plug, $\mu = M_t/M$

$u_1 = \dfrac{\ln(1+\mu)}{\mu}\dfrac{v_0}{U}h$; depth of indentation

$t_1 = h/U$; duration of this phase

(b) Status at the end of Stage 2. The crack has penetrated full depth. The plug is beginning to separate.

$u_{cr} = K^{-1/\lambda}\dfrac{hV_0}{c_{pp}}$; displacement of striker head during Stage 2.

$\left(\dfrac{V_r}{c_{pp}}\right)^2 = \left(\dfrac{V_0}{c_{pp}}\right)^2 - \dfrac{8}{\sqrt{3}}\dfrac{\mu}{1+\mu}\dfrac{\lambda}{1+\lambda}\dfrac{u_{cr}}{d}$; residual velocity of striker, attained when the plug

becomes separated. v_b is the ballistic velocity, i.e., the impact velocity when V_r from the above equation becomes zero. K and λ are material constants.

$t_2 \approx \dfrac{2u_{cr}}{V_0 + V_r}$; duration of this phase

(c) Stage 3 is from the end of Stage 2 until the passage of the striker through the beam is complete.

$t_3 = \dfrac{1}{V_r}(h + H_s - u_1 - u_{cr})$

Then $t_1 + t_2 + t_3$ is the total transit time of the striker.

$c_r \approx \dfrac{h}{u_{cr}}\dfrac{V_0 + V_r}{2}$; average crack speed (Stage 2)

Available material constants to be used with the following power law:

$\sigma = A + B\varepsilon^n$; stress–strain for quasistatic conditions, both σ and ε are the effective values.

Weldox 460E steel:

$K = 0.5$, $\lambda = 4$, $A = 490\,\text{MPa}$, $B = 383\,\text{MPa}$, $n = 0.45$, $c_{pp} = 321$ m/s, $\nu = 0.33$, $\sigma_0 = 809\,\text{MPa}$

2024-T351 Aluminum:

$K = 10$, $\lambda = 4$, $A = 352\,\text{MPa}$, $B = 440\,\text{MPa}$, $n = 0.42$, $c_{pp} = 458$ m/s, $\nu = 0.3$, $\sigma_0 = 566\,\text{MPa}$

c_{pp} and σ_0 values are representatives for an impact with v_0 of several hundred m/s.

Note: The case is based on Teng's results [107]. The major change is to include some of plate material outside the plug in M_t, which gives a larger v_b. The striker is treated as rigid.

Case 14.16   The Effect of Obliquity with Respect to the Target

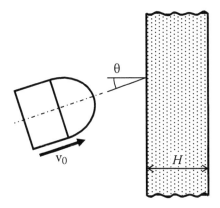

H is the minimum thickness of target to stop the projectile while impacting normally, at $\theta = 0$. H_θ is the corresponding minimum thickness when impacting at θ (measured in the same way as H).

$$H_\theta = H\left(1 - \frac{\theta}{90}\right); \theta \text{ measured in degrees}$$

(Valid up to $\theta \approx 75°$) Ref. [80]

Note: The above is a slightly unconservative.

Case 14.17   Penetration of Composite Armor

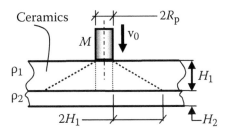

H_1, H_2; Thickness of ceramic layer and backing plate, respectively (~m)

$$V_{50}^2 = \frac{H_2 F_{tu} \varepsilon_u}{0.91 \cdot g(R)}; \text{ ballistic velocity limit (squared)}, M \sim \text{kg Ref. [80]}$$

$$g(R) = \frac{\left(\dfrac{M}{\pi R^2}\right)^2}{\left[\left(\dfrac{M}{\pi R^2}\right) + H_1\rho_1 + H_2\rho_2\right]} \; ; \text{ mass coefficient, consistent units}$$

$R = R_p + 2H_1 \; (\sim\text{m})$

F_{tu} and ε_u; tensile strength ($\sim$Pa) and ultimate elongation of the backing plate, respectively

Note: Owing to the ceramic layer, the effective contact area between both plates is the base of the cone, outlined with the dotted lines (radius R).

EXAMPLES

EXAMPLE 14.1   THE RESISTING FORCE ESTIMATE IN METAL PENETRATION

A projectile with a 9 mm diameter impacts a thick, hard-steel slab, which has $E = 208\,\text{GPa}$, $v = 0.3$, and $\sigma_0 = 1000\,\text{MPa}$. Treating the target as EPP material, estimate the resisting force during a well-developed penetration process. (The target material data is the same as in Ref. [117].)

To find the plastic zone radius c the spherical cavity and the pressure, use Case 14.1:

$$c \approx R\left\{\frac{E}{(1-v)\sigma_0}\right\}^{1/3} = R\left\{\frac{208,000}{(1-0.3)1,000}\right\}^{1/3} = 6.673R;\ \text{plastic zone radius}$$

$$p = \frac{2\sigma_0}{3}\left(1+3\ln\frac{c}{R}\right) = \frac{2\times\sigma_0}{3}(1+3\ln 6.673) = 4.463\sigma_0;\ \text{internal pressure}$$

Repeating the operations for a cylindrical cavity:

$$c \approx R\sqrt{\frac{2\sqrt{3}}{\sigma_0}\frac{E}{5-4v}} = R\sqrt{\left(\frac{2\sqrt{3}}{1,000}\right)\frac{208,000}{5-4\times0.3}} = 13.77R$$

$$p = \frac{\sigma_0}{\sqrt{3}}\left(1+2\ln\frac{c}{r}\right) = \frac{\sigma_0}{\sqrt{3}}(1+2\ln 13.77) = 3.606\sigma_0$$

Apply the average of the two radii and calculate pressure based on a spherical cavity:

$$c_{av} = 0.5(6.673 + 13.77)R = 10.222\,R,\quad \text{then}$$

$$p = \frac{2\sigma_0}{3}(1+3\ln 10.222) = 5.316\sigma_0$$

This identifies our pressure coefficient, $\beta = 5.316$. (Note that out of the three values of p calculated above the first two are for reference only.)
$F = \pi R^2\beta\sigma_0 = \pi(9.0/2)^2 \times 5.316 \times 1,000 = 338,170\,\text{N};\ \text{resisting force.}$

EXAMPLE 14.2   STEEL BULLET PERFORATING ALUMINUM PLATE

A steel bullet with a hemispherical nose, $R = 5\,\text{mm}$ and $M = 8.22\,\text{g}$ is perforating a 3 mm thick aluminum plate. Determine the velocity change of the bullet (treated as rigid) when the initial speed is $v_0 = 100\,\text{m/s}$ and then $v_0 = 500\,\text{m/s}$. Use Case 14.10b with the kinetic coefficient of 2.0. Plate material to be treated as EPP, with a velocity-independent flow stress σ_0.
$E = 69,000\,\text{MPa};\ \rho = 0.0027\,\text{g/mm}^3;\ \sigma_0 = 200\,\text{MPa}$ (Aluminum)

The perforation work:

$$\mathcal{L} = \pi h_0 R^2\left[\frac{\sigma_0}{2} + 2\rho\left(\frac{Rv_0}{L}\right)^2\right] = \pi 3\times 5^2\left[\frac{200}{2} + 2x0.0027\left(\frac{5\times100}{5}\right)^2\right]$$

$$= 23,562 + 12,723 = 36,285\ \text{N-mm}$$

At this speed, the strength component is larger than the inertial one. The kinetic energy difference between the impact and the exit velocities is

$$0.5M(v_0^2 - v_1^2) = \mathcal{L} \quad \text{or} \quad 0.5 \times 8.22(100^2 - v_0^2) = 36,285; \quad \text{then } v_1 = 34.2 \text{ m/s}$$

The other impact velocity, where the second component is 25× bigger, is left as an exercise.

EXAMPLE 14.3   PROJECTILE PENETRATING A LAMINATED PLATE

When a rifle bullet was shot with a velocity of 1005 m/s against an experimental plate, it penetrated the plate and flew out with a residual speed of 327 m/s. When the impact velocity was reduced to 810 m/s, the exit speed was 112 m/s. Estimate the ballistic velocity v_B.

From Equation 14.7: $V_r^2 = \dfrac{1}{1+\mu}(v_0^2 - v_B^2)$

Writing this equation twice gives us

$$327^2 = \frac{1}{1+\mu}(1005^2 - v_B^2)$$

$$112^2 = \frac{1}{1+\mu}(810^2 - v_B^2)$$

Dividing sides, obtain $\left(\dfrac{327}{112}\right)^2 = \dfrac{1005^2 - v_B^2}{810^2 - v_B^2}$; or

$5.59 \times 10^6 - 8.524v_B^2 = 1.01 \times 10^6 - v_B^2$; or $4.58 \times 10^6 = 7.524v_B^2$ or $v_B \approx 780$ m/s

EXAMPLE 14.4   BALLISTIC LIMIT FOR A HARD-STEEL PLATE

A projectile with a $d = 4$ mm diameter and mass $M = 2$g impacts a 9 mm thick, hard steel (armor) plate, which has $E = 208$ GPa, $v = 0.3$, and $\sigma_0 = 1670$ MPa. Treating the target as EPP material, estimate the ballistic velocity limit, v_B. (The data is the same as in Ref. [117].) Assume $\beta = 5$ and $\gamma = 1$.

Follow Case 14.14.

$h^* = \dfrac{d}{2\pi}(2\beta - \gamma) = \dfrac{4}{2\pi}(2 \times 5 - 1) = 5.73$ mm; distance from the rear face at which perforation starts. We have $h > h^*$.

$$\mathcal{L}_p = \left[\frac{h^*}{8}(\pi h^* + \gamma d) + \frac{d}{4}(h - h^*)\beta\right]\pi d\sigma_0 = \left[\frac{5.73}{8}(\pi 5.73 + 1 \times 4) + \frac{4}{4}(9 - 5.73)5\right]\pi 1670 x4$$

$$= 32.108 \times \pi 1670 \times 4 = 673,820 \text{ N-mm}$$

The ballistic velocity v_B is attained when the above is equal to the kinetic energy $Mv_B^2/2$:

$$v_B^2 = \frac{2\mathcal{L}_p}{M} = \frac{2 \times 673,820}{2} \quad \text{or} \quad v_B = 821 \text{ m/s}$$

The experimental result, per Ref. [117], is $v_B \approx 800$ m/s, which is close enough.

EXAMPLE 14.5 STEEL PROJECTILE PENETRATING A LARGE CONCRETE BLOCK, 1 m THICK

Consider a "standard" steel bullet, as shown in Figure 14.3, with $R = 7.25$ mm and $M = 15.66$ g. Determine the depth of penetration when the initial speed is $v_0 = 500$ m/s and then $v_0 = 1000$ m/s. Use ACE, NDRC, and Sandia formulas. The concrete block has $F_c' = 32$ MPa.

Solve in detail for $v_0 = 500$ m/s. From Case 14.3, ACE: $(d = 2 \times 7.25 = 14.5$ mm)

$$X = \frac{M}{12.61} \frac{1}{\sqrt{F_c'}} \left[\frac{v_0^{1.5}}{d^{1.785}} \right] + 0.5d = \frac{15.66}{12.61} \frac{1}{\sqrt{32}} \left[\frac{500^{1.5}}{14.5^{1.785}} \right] + 0.5 \times 14.5 = 28.00 \text{ mm}$$

Case 14.4, NDRC equation with $N = 1.0$:

$$X = \frac{1}{\sqrt{F_c'}} \frac{NM}{104} \left[\frac{v_0}{d} \right]^{1.8} + d = \frac{1}{\sqrt{32}} \frac{1.0 \times 15.66}{104} \left[\frac{500}{14.5} \right]^{1.8} + 14.5 = 30.09 \text{ mm}$$

To use Sandia equations, employ Case 14.6: v_0 ~m/s, M ~kg, A ~m^2
$A = \pi R^2 = \pi 7.25^2 = 165$ mm^2 = 0.000165 m^2
$K_e = 1$, as the target is large
$P = 0$, as no reinforcing steel is mentioned
With $T_c = 1.0$ m and $t_c = 1$
$S = 0.085 K_e (11 - P) (t_c T_c)^{-0.06} (34.5/F_c')^{0.3} = 0.085 \times 1 \times (11 - 0) (1.0)^{-0.06} (34.5/32)^{0.3} = 0.9563$
If the nose is treated as an ogive, with $L_n = 7.25$ mm, the nose coefficient becomes
$N = 0.18 L_n/d + 0.56 = 0.18 \times 7.25/14.5 + 0.56 = 0.65$, then

$$X = K_e \frac{SN}{663,230} \left(\frac{M}{A} \right)^{0.7} (v_0 - 30.5) = 1.0 \times \frac{0.9563 \times 0.65}{663,230} \left(\frac{0.01566}{0.000165} \right)^{0.7} (500 - 30.5)$$

$$= 0.01066 \text{ m}$$

$$= 10.66 \text{ mm (Equation not applicable, } X < 3d = 43.5 \text{ mm)}$$

Calculated penetrations

v_0 (m/s)	ACE (mm)	NDRC (mm)	Sandia (mm)
500	28.00	30.09	N/A
1000	65.94	68.79	N/A

The higher speed gave only $X = 22.01$ mm per Sandia equation, rendering it inapplicable again. This projectile is of a very compact shape. A more elongated outline would give a larger energy per unit of impacted surface and result in a larger penetration.

EXAMPLE 14.6   STEEL PROJECTILE PENETRATING A LARGE CONCRETE BLOCK, CAVITY EXPANSION

Use the same projectile and concrete block as in Example 14.5. Determine the penetration for $v_0 = 1000$ m/s using the cavity expansion theory. Compare with the previous results.

From Example 14.5:

$R = 7.25$ mm, $d = 14.5$ mm, $M = 15.66$ g, $F_c' = 32$ MPa, and concrete density $\rho = 0.0024$ g/mm³

Then, using Case 14.9:

$$S = \frac{72}{\sqrt{F_c'}} = \frac{72}{\sqrt{32}} = 12.73; \quad k = 0.707 + \frac{b}{2R} = 0.707 + \frac{7.25}{2 \times 7.25} = 1.207; \quad N^* = 1/2$$

$$N = \frac{1}{N^*}\left(\frac{M}{8\rho R^3}\right) = \frac{1}{0.5}\left(\frac{15.66}{8 \times 0.0024 \times 7.25^3}\right) = 4.281; \text{ non-dimensional mass ratio}$$

$$I = \frac{Mv_0^2}{8SR^3F_c'} = \frac{15.66 \times 1000^2}{8 \times 12.73 \times 7.25^3 \times 32} = 12.61; \text{ non-dimensional kinetic energy}$$

Try the formula for $X < 2Rk = 2 \times 7.25 \times 1.207 = 17.5$ mm

$$X = d\left[\frac{4kI}{\pi(1 + I/N)}\right]^{1/2} = 14.5\left[\frac{4 \times 1.207 \times 12.61}{\pi(1 + 12.61/4.281)}\right]^{1/2} = 32.14 \text{ mm, larger than expected,}$$

use the other equation:

$$X = \frac{2Nd}{\pi}\ln\left[\frac{(1 + I/N)}{1 + \pi k/4N}\right] + 2kR$$

$$= \frac{2 \times 4.281 \times 14.5}{\pi}\ln\left[\frac{(1 + 12.61/4.281)}{1 + 1.207\pi/4 \times 4.281}\right] + 2 \times 1.207 \times 7.25$$

$$= 39.52 \times \ln(3.23) + 17.5 = 63.84 \text{ mm}$$

This is a smaller penetration than predicted in Example 14.5.

Example 14.7   Duration of Motion During Penetration

Use the same projectile and concrete block as in Example 14.5. Assume the NDRC result, $X = 68.8$ mm for $v_0 = 1000$ m/s to be the true value. Find the time needed to stop the striker.

The NDRC equation can be found in Case 14.4. Modify it as follows:

$$X - d = \frac{1}{\sqrt{F_c'}} \frac{NM}{104} \left[\frac{v_0}{d} \right]^{1.8} = \frac{1}{\sqrt{32}} \frac{1.0 \times 15.66}{104} \left[\frac{v_0}{14.5} \right]^{1.8} = \frac{v_0^{1.8}}{4627}$$

Eliminating time from the first two Equation 14.20 yields: $u = G_1 v^{9/5}$. This tells us that $G_1 = 1/4627$.

The termination time can be found from a rewritten Equation 14.20b:

$$t_m = \frac{9G_1}{4} v_0^{4/5} = \frac{9}{4 \times 4627} 1000^{4/5} = 0.1222 \, \text{ms}$$

To account for the fact that the above time relates to the travel by $d = 14.5$ mm shorter than X, add a correction $\Delta t \approx 14.5/1000 = 0.0145$ mm, with the total travel time being

$$t_m' = 0.1222 + 0.0145 = 0.1367 \, \text{ms}$$

A simpler way of estimating the duration of motion is to employ Equation 7.24 with the appropriately modified notation:

$$t_1 = \frac{\pi X}{2v_0} = \frac{\pi 68.8}{2 \times 1000} = 0.1081 \, \text{ms}$$

While this second number is less accurate, it can be quickly estimated.

15 Damage, Failure, and Fragmentation

THEORETICAL OUTLINE

The terms used in this title are familiar; therefore attempting a formal definition might be somewhat artificial. *Damage* is usually associated with a mechanical degradation. When damage is extensive, a loss of integrity or *failure* takes place. This means the structural member no longer participates in resisting the external loading. The term *fragmentation* usually suggests a breakup of an element into more than two pieces. The criteria of failure used here are based on stress and strain and are consistent with those described in Chapter 6. Many approaches developed elsewhere in this book come together in this chapter.

PRESSURIZED RINGS AND SHELLS

Static vs. Dynamic Breakup

In typical configuration of those members, hoop stress predominates. The pressure grows slowly until the shell begins to yield and then breaks. If the pressure drops suddenly, because of the loss of fluid, for example, a single break is the only effect. On the other hand, pressure may persist sufficiently long for a fracture to take place at the diametrically opposite side, as shown in Figure 15.1. This would be, most likely, the effect of combined hoop tension and bending at the second location.

When the pressure growth is rapid, on the other hand, the container may break into several pieces. The quicker the energy release inside the container, the faster is the pressure growth and more broken pieces result, as numerous experiments tell us. As in many applications of structural dynamics, the terms "fast" and "slow" should be taken in relation to the natural period, as explained and defined in Chapter 2.

The short study to follow is limited to an investigation of a disintegrating ring, which may be thought of as forming a segment of a cylindrical container. It is assumed that the strength of the ring material varies randomly along its circumference between an upper and the lower limit. When such a ring is intended to represent the whole finite-length shell and when the problem is set up using finite elements, a number of interesting questions arise with regard to the interpretation of the experimental formulas. Further reading may be found in this author's publications, Refs. [90,95,96].

Mechanism of Multiple Fracturing

This phenomenon can be explained on the basis of nonuniform material strength along the circumference. If such strength could be measured and therefore plotted, it would show a series of weak spots, each exhibiting a higher strength than the previous one, except they would not necessary be in the proximity of one another.

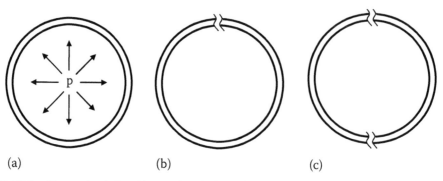

FIGURE 15.1 Pressurized ring (a), quasistatic breaking at a single point (b), and two points (c).

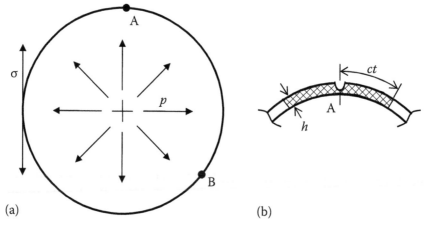

FIGURE 15.2 A ring with only two weak spots (a) and a ring fragment showing the unloading wave spreading from a fracture point (b).

The ring in Figure 15.2 is subjected to a dynamically applied pressure p, which stretches it, thereby inducing a hoop stress σ_h. For the sake of this explanation, it is sufficient to assume that there are only two points of the ring, where a disintegration can take place: the weakest point A with the tensile strength F_1 and an additional point B with F_2, while $F_1 < F_2$. When the hoop stress level becomes as large as F_1, the fracture takes place at A and afterwards, when σ_h reaches F_2, point B breaks.

After a fracture at A, as illustrated in the top part of Figure 15.2, an unloading wave emanates from that spot, relieving stress at the adjoining segments. The wave moves with speed c, therefore the length of ring equal to ct on either side is relieved and cannot crack. Point B is at a distance x along the ring from A. It takes time $t_w = x/c$ for the unloading wave to reach point B after a break at A. On the other hand, it takes $t_S = (\sigma_2 - \sigma_1)/\dot{\sigma}$ for the stress level at B to increase sufficiently for the break to eventuate. ($\dot{\sigma}$ is a uniform rate of stress growth.) Time is calculated from the onset of the first cracking.) The condition $t_S < t_w$ is necessary for the fracture to occur at B. The faster the stress growth, the more likely it will happen and the ring will fail in two, rather than in one spot. Because the failure at B is a possibility, rather than a certainty, B will be referred to as a *virtual break point*. The ring material is, at least initially, regarded as linearly elastic, with a brittle failure at the end of the stretching process.

The above reasoning can be extended to a large number of points along the ring. Every real material has some nonuniformity in strength, which is an underlying cause for a multiple fracture to occur. For a normal engineering design only a minimum guaranteed strength is used. For the evaluation of a dynamic multiple failure, it is helpful to also know the mean strength and the standard deviation of this quantity.

Case 15.2 gives a simple method of estimating the anticipated number of fragments in a thin ring, provided the duration of breakup is known. The simplification is in assuming that all resulting pieces are of equal length. If the end product is eight fragments, this implies that cracking starts at four points and another four cracks appear at the end of the process, at where the unloaded segments reach. The breakup duration t_b can be found by measurement or by indirect means.

Maximum Number of Fragments from Experiment

This overview is based on the results compiled and published by N.F. Mott as reported by TM5 [106], but adjusted to suit metric units. Physical testing was done on bombs and shells. The results are meaningful for moderately thick as well as thin cylinders subjected to action of high explosives. The fragment size distribution formula developed by Mott is

$$N(M) = \frac{M_0}{2M_k^2} \exp\left(\frac{-M^{1/2}}{M_k}\right) \tag{15.1}$$

where
   $N(M)$ is the number of fragments, each having a mass larger than M
   $M_0 = \pi h d_m L \rho$ is the mass of the shell
   M_k is a distribution factor

$$M_k = B h^{5/6} d_i^{1/3} (1 + h/d_i) \tag{15.2}$$

where
   h is thickness of shell
   L is length
   ρ is density
   h and d_i stand for thickness and inner diameter, respectively

The associated units are presented in Case 15.1. The experimental coefficient B is relevant for a given combination of metal and explosive. As the original author was casual with units, the dimension of B came out as $lb^{1/2}/in.^{7/6}$. The largest number of fragments is obtained when the fragment mass tends to zero

$$N_M = \frac{M_0}{2M_k^2} \tag{15.3}$$

The above results are valid for mild steel shells. The B coefficients are provided for a number of explosives in Ref. [106]. For other explosive materials B may be assumed to vary in proportion to p_{CJ} pressure or to p_0, as defined in Chapter 13.

Our discussion so far has been limited to thin rings. Such a ring may be thought of as a very short cylinder placed between two rigid and frictionless planes. From this perspective, cracks that appear are longitudinal. Now, let us go to another extreme and think of a long, thin wall cylinder. The virtual crack points are distributed differently in each cross section. When breakup process sufficiently progresses, the cracks originating at different spots in adjacent sections coalesce to form finite-size wall fragments.

A ring or a cylinder floating between two walls has no longitudinal stress as the walls serve to retain pressure. While such a configuration may be created for test purposes, every real container must have its ends closed. This fixed closure gives rise to a longitudinal stress in a pressurized cylinder. (In a thick cylinder the situation is more complex, but essentially the same rule holds.) The circumferential cracks near a closed end, which may appear, are a secondary effect.

From observation of detonated shells, as evidenced by Lamborn [51], for example, it appears that often the resulting pieces are compact in the sense that both main dimensions are comparable, i.e., a typical ratio of length to width is about 1.0, which may be approximated by a square, especially when deformation prior to rupture is small. If such proportions are assumed, one can relate the total number of fragments N_M to the number n_M that would be noticed in a single cross section of a shell

$$N_M = \frac{\pi d_m}{s} \frac{L}{s} \tag{15.4}$$

where s is the average fragment size. On the other hand, $s = \pi d_m / n_M$, therefore

$$n_M = \left(\frac{\pi N_M d_m}{L} \right)^{1/2} \tag{15.5}$$

If we are interested in a number of fragments seen in a cross section of a long cylinder, an important question is how to select the reference length L. An obvious choice is a length associated with a repetition of the breakup pattern. If we assume square-like pieces then the longitudinal and circumferential directions should be treated equally and for this reason $L = \pi d_m$ should be selected. Substituting in Equation 15.5 we have

$$n_M = (N_M)^{1/2} \tag{15.6}$$

Radial Motion in Cylindrical Shell under Explosive Pressure

When dealing with a ring or a segment of a cylinder pushed by pressure radially outward, one has to account for two main sources of resistance: inertia and strength of material. It will be easy to check if the strength component has little significance, at least in early stages, when pressures associated with high explosives are directly applied. The radial acceleration, to which the ring is subjected, is then

$$a_0 = \frac{W}{M} = \frac{p_0 r_i}{\rho h_0 R} \tag{15.7}$$

as was discussed in section "Closing remarks." This acceleration is expected at the mean radius R. With the passage of time pressure will decrease, but the internal radius r_i will grow, so at least for the initial stage of motion the approximation of $p_0 r_i$ being constant is reasonable. As the mean radius grows, the product $h_0R = hr$ should also remain constant, so that its mass is preserved. Using the above acceleration, the radial velocity and displacement are, respectively:

$$v_0 = a_0t \tag{15.8a}$$

$$u_0 = \frac{a_0t^2}{2} \tag{15.8b}$$

and the associated strain is $\varepsilon_0 = u_0/R$. The above was written with the assumption that the force exerted on radian of a ring is constant. This is a permissible approximation only for moderate movements.

Fragmentation Caused by Wave Reflection

Spalling Effects in an Axial Bar

In Example 5.5 the action of a triangular decreasing pulse* with the peak compression of σ_m traveling along a bar was examined. It was determined that the pulse retains its shape after it reflects from a free end, except that it changes from compression to tension. It was also found that the following tensile peak stress is experienced by the bar, as a function of the distance x from the end:

$$\sigma(x) = \frac{2x}{b}\sigma_m \quad \text{for } x \le b/2 \tag{15.9}$$

where b denotes the pulse length. In other words, the maximum attainable stress grows as a function of x until the value of $x = b/2$ is reached. At that point the peak stress attains σ_m and remains constant for $x > b/2$. This picture is valid as long as the material is strong enough to withstand the resulting tension. If not, a tensile failure, or *spalling* takes place and a piece of the bar flies off. This happens as soon as the stress reaches F_{tt} or the dynamic tensile strength. The latter quantity, experimentally found by Rinehart and Pearson [74] for some metals, is shown in Table 15.1. Next to F_{tt} the static value of F_t is quoted[†] as well as the ratio of the two. This ratio varies widely depending on material involved.

TABLE 15.1
Metal Strength for Spalling Calculations

	F_{tt} (MPa)	F_t (MPa)	F_{tt}/F_t
Copper	2828	260	10.9
Brass	2138	500	4.28
4130 Steel	3035	520	5.84
1020 Steel	1103	425	2.60

---

* The pulse was decreasing in time, when applied at the end of a bar. When viewed moving from left to right (viewed in space rather than time) it appeared increasing.

[†] The condition of the metals involved in testing is unknown, therefore the average of the range is given as F_t.

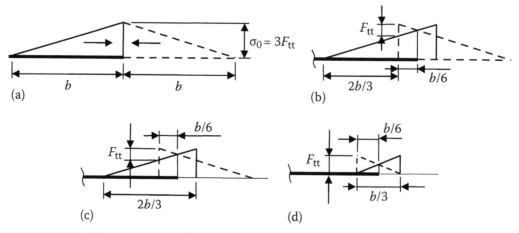

FIGURE 15.3 Triangular decreasing pulse (with respect to application at the left end) and its reflection at the right end of the bar. As soon as the resultant tension F_{tt} is reached at the right end, a piece $b/6$ long breaks off. This repeats another two times.

If the triangular pulse is increasing (when referred to the end, where it is applied) then the distance from the end, needed to develop the full σ_m in tension is b rather than $b/2$. (This is shown in detail by Johnson [40].)

When the compressive strength F_{cc} is a multiple of tensile strength F_{tt}, then it is possible to send along a bar a strong compressive pulse causing multiple spalls at the free end. This phenomenon is illustrated in Figure 15.3, for a pulse of amplitude $\sigma_0 = 3F_{tt}$. The material is elastic and exhibits brittle failure, which is assumed instantaneous. The history of stress near the free end is shown essentially in the same manner as in Chapter 5, namely, as a superposition of two waves moving in opposite directions. A slight difference in presentation is to plot both tension and compression on the same side of the bar and finding their resultant by subtracting the ordinates.

The state illustrated in Figure 15.3b corresponds to the instant when the pulse had moved by $b/6$ past the end of the bar and the tensile stress had grown to F_{tt} at $x = b/6$ from the end. The fracture takes place at this instant, which means that a piece $b/6$ long flies off to the right. The remaining portion of the bar is still swept by a compressive wave, but the length of the new pulse is only 2/3 of the original length. The superposition process is applied to the new free end, taking place in Figure 15.3c and resulting in another segment, $b/6$ long, flying off. A new free end is formed with the pulse being only 1/3 of original length and the third and the final spall is ready to take place, as shown in Figure 15.3d.

Breaking of Solids with Stress Waves

A simple example in the form of a thick rod with a square section is shown in Figure 15.4. A hole is drilled along the rod, filled with an explosive and detonated. An outgoing compressive, cylindrical wave is reflected from the outer surface as a tensile wave. At 45° corner planes of symmetry the tensile waves from the adjacent sides meet and reflect doubling their magnitude. This gives rise to tensile cracks along those planes. The result is counterintuitive to someone not familiar with wave mechanics; cracking takes place along the strongest, rather than the weakest longitudinal sections.

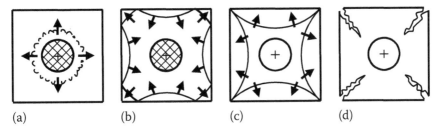

FIGURE 15.4   Section of a four-sided rod with an explosive charge at center. (a) initial compressive wave after detonation, (b) the original wave reflected as tensile, (c) after tensile waves were reflected from 45° planes, and (d) resulting corner cracks.

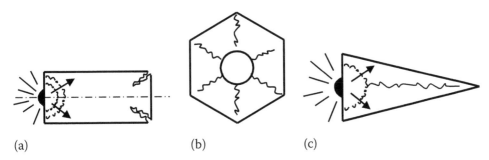

FIGURE 15.5   (a) Conical cracking at the opposite end from a surface impulse, (b) hexagonal rod cracked along stronger sections, and (c) internal crack in a wedge.

Further examples are shown in Figure 15.5. In Figure 15.5a a charge is detonated at one end of a cylinder and the resulting waves reflecting from the opposite end and the side surface superpose and cause a conical cracking. Figure 15.5b is similar to that in Figure 15.4; cracking appears along stronger sections.

Finally, in Figure 15.5c there is a wedge with an explosive discharge on the short face. The waves reflecting from the long faces cause a fracture along the plane of symmetry.

In Chapter 13 the quantification of surface explosions was presented. The cratering effects associated with such events are discussed later in this chapter. The examples illustrated in this section suggest metallic object, for which cratering is poorly defined in geometrical sense. When there is a need to quantify the effects of a contact explosion on the volume of metal, one can use an idea of a notional crater, a hemisphere of a radius equal to the contact radius of the explosive charge.

This crater will emit a pulse as high as (the instantaneous explosion) pressure p_0 shown in Table 13.1, but a lower pressure may result, consistent with the strength of the medium.

FAILURE OF BEAMS

The Four Basic Modes of Failure

Those modes, for a beam with clamped ends are visualized in Figure 15.6. The numbering of the first three is well established in literature and the fourth one is just another fact of

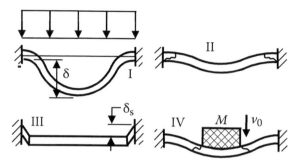

FIGURE 15.6   Failure modes of a beam with clamped ends. Mode I: Excessive deformation. Mode II: Tensile tearing at supports. Mode III: Shearing off at supports. Mode IV: Shearing off near projectile.

life. Mode I is not associated with a breakup, as the remaining ones, but is related to the loss of functionality by a badly distorted structural member. Mode II failure comes from a superposition of string action and bending at the base points. Mode III is associated with relatively large, short-lasting loading and Mode IV with relatively high velocity impacts. It is also possible to induce Mode III with a projectile impact, but the likelihood of this becomes smaller as the impinging width of the projectile decreases. The same or analogous failure modes can take place for beams of other end conditions, for example, a simply supported beam can fail in Mode II, but without the assistance of bending. When Mode IV results in more vertical cracks, than what is shown in the sketch, the failure is referred to as *shear plugging.*

There is a problem in interpretation of test results, especially acute for beams with clamped and axially constrained ends. This was discussed in section "Closing remarks" of Chapters 10 and 11, where the imperfection of real constraints was mentioned as a source of additional deflections to be encountered in experiments. There is also reduction of stress at the clamped end arising from the fact that a zero slope condition cannot be fully enforced. A comparison between a theoretical solution, which assumes a perfect constraint and test results becomes less meaningful, if no allowance is made for an increase of deflection and decrease in bending near ends.*

A more practical method was used here, as shown by Cases 15.5, 15.6, and 15.8. A formula is written, with certain terms present in it, as well as with a coefficient, which is a function chosen to fit the test results. This approach allows, to some extent, a generalization of experimental findings without artificial modification of material properties, for example.

As a good example of the difficulties mentioned above, one should reread Example 2.4 where a cantilever is built into a continuous medium and even provided with generous fillet radii at the base. The fact of imperfect fixity shows itself as a reduction in natural frequency, but one should keep in mind that the relative increase of deflections is much larger than the decrease of frequency. Also, the example did not involve an axial restraint, which seems to cause the most acute problem. For this reason, most of experimental setups can be referred

---

* This point seems to have been lost on most writers so far.

to as *nominally* clamped or fixed, meaning that the constraint is less than perfect. This is also significant with respect to Exercise 15.10, where a beam with substantially thickened ends was used in a test with obvious intent to make the end conditions more decisive. Yet, those "built-in ends" were even less favorable, as far as fixity is concerned, than those of Exercise 2.4.

In Chapters 10 and 11 a number of inelastic load cases were given, often without any reference to failure. Such a failure may usually be inferred by using either a strain or a stress criterion.

Dynamic Crack Propagation

To say that a member disintegrated is usually equivalent to saying that a number of cracks appeared breaking the member up into smaller fragments. When studying a structural breakup, it is usual to ignore the duration of a cracking process itself. Such a simplification may, or may not be warranted.

A cracking phenomenon may exhibit one of the two basic forms. The first would be a quasistatic process, whereby there is an incremental increase of the crack length arising from one of a series of periodical, relatively small load pulses. The other form, referred to as dynamic, involves a complete tearing of a section when a strong, single pulse is experienced. This is the only form, in which we are interested.

The crack speed depends on many factors, the magnitude of an external load being an important one. Generally, the larger load, the larger crack speed. But there is a limit, namely $c_r = c_R$, where c_R is the surface wave speed, as defined in Chapter 5. This upper limit is rarely approached in practice. Meyers [61] quotes a set of experiments, where samples having pronounced shape discontinuities were impactively loaded. Under such severe conditions the highest ratio c_r/c_R observed was 0.905 (Figure 15.7). He also quotes an extensive testing program, involving metal, glass, and plastic, in which the range of speed ratios was found to be between 0.3 and 0.66. From this and other examples in literature, it seems that the ratio of $c_r/c_R = 0.4$ will be close enough for most situations. If this is used a guide, then one has

$$c_r = 0.4c_R = 0.369c_s = 0.229c_0 = 0.197c_p \qquad (15.10)$$

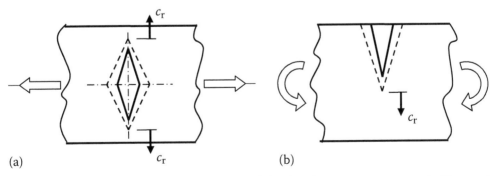

(a)   (b)

FIGURE 15.7 (a) Crack in a tensile member, originating from center, symmetrically growing, (b) edge crack due to bending. In either case c_r stands for the crack tip speed.

according to the relations presented in Chapter 5. In deriving the above, $v = 0.3$ was used, as representative for metals, for which most results are available. Symbols c_s, c_0, and c_p stand for the speeds of a shear wave, uniaxial (sonic) wave, and a pressure wave in a continuous medium, respectively.

In steel, for example, where c_0 is 5048 m/s, according to Table 5.1, it takes less than 0.005 ms to crack a shell, 5 mm thick according to Equations 15.10. For a member 600 mm deep, as often seen in building construction, the time is 0.52 ms. Depending on the situation, those intervals of time may or may not be negligible. In very rapid processes inclusion of cracking time is likely to have a noticeable effect on a total duration.

COLLAPSE

Dynamic Destruction of a Structural Element

Such a destruction caused by an accidental or a malevolent act can, in the most unfortunate situation, begin a chain of events leading to a total collapse of a structure. To assess the degree of danger one has to carry out an investigation comprising several steps. Let us take a loss of a column, at the base of a building, as a relevant example. The general procedure is the same, regardless of whether a hand check or a computer simulation is involved.

Step 1. Static analysis, when the member is missing. In Figure 15.8a a schematic picture of an intact structure is given and in Figure 15.8b the same structure is presented, less a member that may be critically damaged. This first simple test answers the following question: Will the structure remain standing without that column? If the answer is "no," then the collapse that follows, which means the end of the analysis. Only simple statics is needed to find out if there is a sufficient strength margin left to accommodate the increased loading on the remaining members.

Step 2. Dynamic analysis estimating the effect of a sudden disappearance of the member in question. (This, of course, makes sense only if the answer in Step 1 is "yes.") If a computer model is ready and if the program has the capability to "kill," i.e., to suddenly remove the elements, then little time is needed to set the analysis in motion. If any of these two conditions is not fulfilled, then the approach is schematically shown in Figure 15.8c and d. Instead of the column itself, the interface force W_{st} is displayed. This force is the axial column load, in the initial, static equilibrium condition. The effect of a sudden removal is given as $W_{st}H(t)$, i.e., the same interface force, but dynamically applied downward. One can therefore expect this to be equivalent to $2W_{st}$ applied statically downward, while the upward W_{st} is maintained. It is obvious that the sudden column removal not only takes away any support provided by the column, but also adds a short-lasting column load W_{st} directed downward!.... While this approach is somewhat idealized, the dynamic removal is definitely more menacing than the gentle removal equivalent to Step 1. If the analysis shows that this event is not serious enough to cause failure at any location, then this means that the structure is ready for the final check that follows.

Step 3. Full dynamic simulation of a forcible member removal, as illustrated in Figure 15.8e and f. The column is tied to the structure, therefore a sideways push $W(t)$, as illustrated, will induce a set of time-dependent forces at the attachment point or the upper end. Those forces are related not only to the rapidity of $W(t)$ application, but also on the properties of column itself. A large tensile strength of the column will exert an equally large downward pull on the

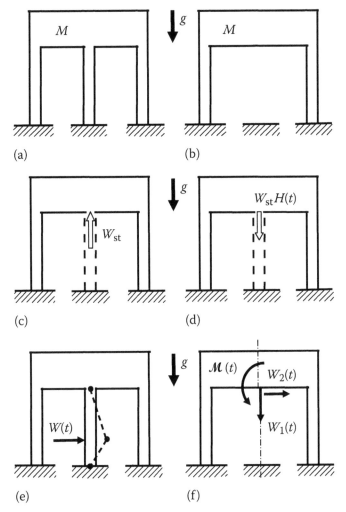

FIGURE 15.8 (a) Original structure, (b) the same, but without the element under consideration, (c) original structure, but with the member replaced by the interface force, (d) element removal as imposition of downward force, (e) destructive action of $W(t)$, and (f) that action replaced by the forces imposed on the remainder by the element being torn out.

structure. One could figuratively say that the element being destroyed attempts to demolish the rest of the structure.

A successful passing of the test in Step 1 fulfills the *necessary* condition for the structure to survive. The *sufficient* condition, on the other hand, is to be able to withstand the complete set of loads imposed by the removal, in accordance with Step 3. One could argue that the test in Step 2 is too severe, because every removal requires time and is not, strictly speaking, instantaneous. On the other hand, Step 2 does not introduce any external actions, which means that there is gentler treatment of the structure from this viewpoint. It is unlikely that Step 2 will be exactly equivalent to Step 3 in any particular project, but the degree of simplicity associated with the action involved in the former is encouraging. In some configurations and element properties used it may be clear that replacing Step 3 with Step 2 does not make much difference. Furthermore, in the opinion of this author, a structure not capable of surviving Step 2 is unlikely to withstand Step 3.

Progressive Collapse

If one builds a simple structure from playing cards (the proverbial *house of cards*) and then removes a single card from the bottom level of that structure, everything will fall down. This is, by far, the simplest practical example of a *progressive collapse*, where elements fall (or fail) one by one, with motion spreading upward, until an assembly standing upright eventually changes into a heap of its basic elements.

On a structural level, the act of demolition of a building gives an example of such a process. When the base columns are cut by the explosive charges placed there for the purpose, the building descends, crushing the base columns first. The first ceiling impacts the ground followed by the remaining floors crushing the columns below and piling onto the ground.

Cases 8.16 and 8.17 bear some resemblance to the analytical problem encountered here. They show two masses aligned vertically, with supports between them, resisting the action of gravity. The supports can be either ductile or frangible (easily breakable). Those masses can represent floors of a building and the supports can symbolize columns. The main difference here is that multistory buildings will be considered, rather than two-story only. Figure 15.9 illustrates a subproblem of determining a post-impact V_c velocity when an assembly of 3 slabs collides with a single stationary slab in a plastic manner. As a result, a moving assembly of 4 slabs is created.

If we designate by p the number of slabs involved ($p = 4$ in this case) then the equations provided in Case 8.16 make it possible to determine V_c and a loss of the current kinetic energy ΔE_k as follows:

$$V_c = \left(\frac{p-1}{p} \right) v_0 \tag{15.11}$$

$$\Delta v = v_0 - V_c = \frac{v_0}{p} \tag{15.12}$$

$$\frac{\Delta E_k}{E_k} = \frac{1}{p} \tag{15.13}$$

This aspect of slab collision is the same for both types of columns between floors. The action of gravity is also the same: If a group of p floor slabs descends by h, then it absorbs $\Delta U_g = pMgh$ of the potential energy. This increment is converted into a change of kinetic energy, provided

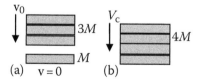

FIGURE 15.9   (a) A group of floor slabs moving with v_0 impacts a slab below and (b) after a plastic collision, the common velocity of the assembly is V_c.

column ductility is not involved. For ductile columns the strain energy gain, $\Pi = Rh$, where R is the average resistance of the yielding member, reduces the current kinetic energy. The following energy relation for the motion between floors can be written as

$$\Delta E_k = \Delta U_g - \Pi \tag{15.14}$$

or

$$\frac{1}{2} M_a (v_0^2 - V_c^2) = M_a g h - R h \tag{15.15}$$

This describes the change of condition during motion between floors, ending with a terminal speed V_c, which becomes the initial velocity for the next segment of travel. For the downward motion to continue, the absorbed strain energy Π must not be too large. If it is, then the speed will decrease or motion will stop altogether. The frangible and the ductile configuration is described in detail as Cases 15.27 and 15.28, respectively. One must keep in mind that h is the *squash* height, or the clear height between two floors after the thickness of crushed material between the two is subtracted. (This *unobstructed* height is sometimes estimated as 80% of the clear height.)

Our formulas are set up as if all floors had the same column strength. This is not quite the case, as the columns become stronger on the way down, but not in a continuous manner. At the transitions, where columns become stronger, one should check if there is enough energy left to continue the fall.

If the top story weakens, so that it can no longer support the roof, the roof begins to descend. If the right side of Equation 15.15 is positive, the roof will impact the floor below. In spite of some loss of kinetic energy, the downward motion will continue, into the story below. However, the set of columns being squashed is stronger now, which may cause the energy condition for complete squashing not to be fulfilled. In this case the motion will stop. If the first undamaged floor cannot stop the motion, it will then continue until at least a sufficiently strong floor is encountered.

So far the discussion has been limited to the top-initiated failure. The Cases 8.16 and 8.17 also describe, on a small scale, a failure originated at the bottom, where the ground story collapses first and then the successive slabs "pancake," one on top of the previous one. This is described in detail as Cases 15.29 and 15.30, which deal with a frangible and a ductile configuration, respectively. As in the top-initiated failure, the process may stop if the kinetic plus the potential energy is insufficient to exceed strain energy associated with failure of the columns of a story in question.

Even such a simple scheme as this has quite a variety to it. An intermediate story can fail, causing the structure above it to travel downward, as detailed by Cases 15.31 and 15.32. If a continuous motion all the way down is assured, then after the damaged floor crashes against the ground, the intact part of the structure may in turn begin to collapse.

The progressive collapse may also be presented as a continuous process, by "smearing" the mass and resistance along the length. Unfortunately, the differential equation of motion must be solved by numerical means, as explained by Bažant and Verdure [7]. Those equations hold only over a segment with a constant column strength.

One has to remember that the "weakening of a story" as mentioned above may mean "one column in a group was damaged and the rest of columns in that group were incapable of supporting the structure above it." To further expand the applications of the simple schemes described, it is advisable to think of all vertical, load-bearing members (for example, walls) instead of strictly columns, as such.

Finally, it seems that a progressive collapse of a vertical segment of a building (like in Ronan Point collapse and the Oklahoma bombing effects) is more likely than a total collapse as described by our schemes. If a portion of a building falls down, the floor slabs break off at the boundary. This means that the strain energy involved in such breaks will enter the energy balance, in addition to what is absorbed by columns. In case of brittle connections between vertical and horizontal elements the chance of a progressive collapse is greatly increased.

ROCK BREAKING

Two types of damage around a cavity in a brittle medium are considered. One is associated with a high-pressure event and is characterized by local crushing around a cavity along with radial cracking at some distance away. This will be referred to as a *CC damage mode*. The other type takes place under lower initial pressures and involves radial cracking only. This will be called *cracking mode*. (The terms "high-pressure" and "low-pressure" are relative to rock strength.) The crushed zone is treated here as an incompressible liquid while the cracked zone is assumed to consist of separate radial elastic wedges. Both cylindrical and spherical cavities are considered. In CC mode pressure is confined to the expanding cavity. In cracking mode pressure fully penetrates the opening cracks. Although the phenomena are of dynamic nature, they are treated here as quasistatic.

Deformability of Cracked Material around Cavity

Consider a cylindrical cavity in Figure 15.10 and radially cracked rock surrounding it. The first case of interest is when pressure is not allowed to enter cracks. The cracking radius R, which is the inner boundary of the surrounding intact medium, can be viewed as the inner surface of the cavity within the medium. When radial stress p_c is applied at R, the circumferential stress has the same magnitude.* Cracking continues when the latter stress component becomes equal to the tensile strength F_t. As there is no hoop stress over the radial distance between a and R, the relationship between pressures is

$$pa = p_c R \qquad (15.16)$$

The largest value p_0 can attain is the confined, compressive, dynamic strength of the rock material, F_{cc}. For many rocks and concrete as well, the (static) basic compressive strength F_c is often about 10 times larger than the basic tensile strength F_t. The dynamic ratio is anticipated to be of the same order of magnitude, but little experimental data is available on the benefits of the elevated strain rate on F_t.

---

* This is for a cylindrical cavity of radius R, inside elastic medium. When p_c is applied inside, the peak hoop stress is also p_c. For a spherical cavity, when p_c is inside, the hoop stress is $p_c/2$. See relationships given in Cases 3.32 and 3.33.

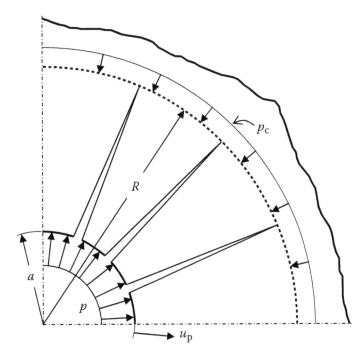

FIGURE 15.10   Cylindrical cavity surrounded by cracked material. (From Szuladziński, G. and Saleh, A., *Acta Mech.*, 115, 79, 1996.) The radial wedges, into which the medium separates, are seen along the axis. The powdered material in the hole, exerting pressure, is not shown.

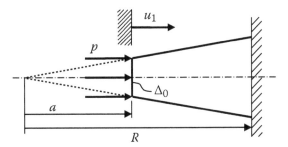

FIGURE 15.11   A wedge of material seen looking along the axis of a cylindrical cavity with pressure p_0 applied at cavity radius. (From Szuladziński, G. and Saleh, A., *Acta Mech.*, 115, 79, 1996.)

In Figure 15.11 there is a wedge, of a unit angular size, loaded with a radial force, the pressure resultant. The radial deflection of the inside edge, with respect to its fixed outside edge is found from Hooke's law:

$$u_1 = \frac{pa}{E} \ln\left(\frac{R}{a}\right) \qquad (15.17)$$

The "fixed end" in the above picture is at the cracked radius R, at which, as at the boundary of the intact medium, the deflection is, according to Case 3.32:

$$u_c = \frac{1+v}{E} p_c R = \frac{1+v}{E} pa \tag{15.18}$$

This gives the following edge displacement total:

$$u_p = \frac{pa}{E}\left[\ln\left(\frac{R}{a}\right) + (1+v) \right] \tag{15.19}$$

So far, the action of pressure was confined to the hole itself. If full pressure is applied to the entire surface of wedge (in cracking mode as previously defined), one can use Hooke's law again, to obtain the new wedge component of displacement:

$$u_1 = \frac{pa}{E}\left(\frac{R}{a} - 1 \right) \tag{15.20}$$

Pressure level p is the same now in the cavity as well as at radius R, so the medium contribution to displacement is

$$u_c = \frac{1+v}{E} pR \tag{15.21}$$

The wall of cavity therefore moves by $u_p = u_1 + u_c$:

$$u_p = \frac{p_0 a}{E}\left[(2+v)\left(\frac{R}{a}\right) - 1 \right] \tag{15.22}$$

The forces per radian for the above cases are as follows: When pressure is confined inside the hole, then $W = p_0 a$. For full pressure: $W = p_0 R$. (As will be explained later, not necessarily the same magnitude of p_0 should be used in both cases.) The distinction is made because in a practical situation the flow of expanding gas into opening cracks may be impeded or prevented.

The other aspect of interest is the volumetric growth of the void space induced by pressure. This increase in volume is due to compression of the wedges and the intact medium deflection. The latter is given by Equation 15.21. As $u_c \ll R$, the increase in volume of the cavity due to displacement on radius R can be written as

$$\Delta V_1 \approx 2\pi R u_c \tag{15.23}$$

The squashing of the wedges may be seen as a uniform compression with p_0 applied to the volume within radius R, which is $V \approx \pi R^2$, as $R \gg a$. With the use of bulk modulus K, one has

$$\Delta V_2 = \frac{F_t}{K} \pi R^2 \qquad (15.24)$$

For the sake of simplicity we set $v = 0.25$, which gives $K = 2E/3$. The sum of both volumetric components is

$$\Delta V = 2\pi R \frac{1.25}{E} F_t R + \frac{3F_t}{2E} \pi R^2 = 4\pi R^2 = \frac{F_t}{E} \qquad (15.25)$$

The development of equations for a spherical cavity is analogous to that for a cylinder. The wedges now have the form of radial elements. The cracked surface of a spherical cavity should resemble a soccer ball, when viewed from the cavity center. The quantitative difference is that now the wedge area grows in proportion to the square of the radial distance, rather than to the first power. The relationship between pressures at the beginning and at the end of wedge is

$$pa^2 = p_c R^2 \qquad (15.26)$$

Deformability of Crushed and Cracked Medium in a Cylindrical Cavity

When the applied pressure equals or exceeds the compressive strength F_{cc}, the medium becomes crushed or pulverized. Consider the expansion of cylindrical hole under the $p = F_{cc}$, as illustrated in Figure 15.12a. Ignoring the inertia effect the crushed zone is assumed to be pressurized up to F_{cc}. On the inside of the cracked zone pressure is still F_{cc}, but on the outside it is equal to F_t. As the deflection u at r_0 may be quite sizeable as compared to

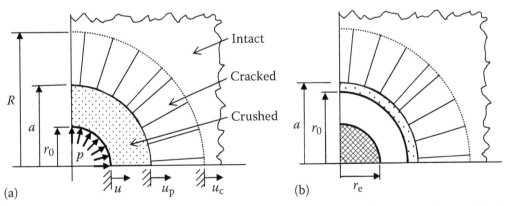

FIGURE 15.12   Damage around cavity in a CC mode. In (a) the explosive (not shown) fills the entire cavity r_0. In (b) there is decoupling; explosive radius is only r_e.

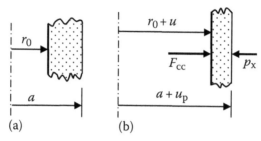

FIGURE 15.13 Crushed zone, effective dimensions reduced to original position (a) and at deformed, balanced position (b).

r_0 itself, the relation between u and u_p is NL, in general. Ignoring the compressibility of the crushed material* one has

$$(a + u_p)^2 = (r_0 + u)^2 + a^2 - r_0^2 \tag{15.27}$$

The cavity wall deflection u can be found as corresponding to the expansion of gas from the initial pressure to the prescribed value of F_{cc}. When the crushed zone radius a is found by independent means, the above expression serves to find u_p.

For a spherical cavity with crushed and cracked medium, the same Figure 15.13 is applicable. Treating the crushed material as incompressible,[†] one has

$$(a + u_p)^3 = (r_0 + u)^3 + a^3 - r_0^3 \tag{15.28}$$

Explosive Energy Content

Explosive energy content was previously expressed by means of the initial pressure p_0, as given in Cases 13.1 and 13.2:

$$Q = \frac{\pi p_0 r_0^2}{n-1} \quad \text{(cylindrical)} \tag{15.29a}$$

$$Q = \frac{4}{3} \frac{\pi p_0 r_0^3}{n-1} \quad \text{(spherical)} \tag{15.29b}$$

In the cylindrical case, the energy is given for a unit thick layer, normal to the axis. Sometimes, energy losses are approximately known and therefore it is necessary to input a smaller energy into a predictive model. This is done by reducing the initial pressure from p_0 to p_h, leaving the exponent n unchanged.

---

* Shrinkage due to compressive stress is one aspect of crushing. The other is swelling due to appearance of hollows.

† The balance of forces acting on a radial sector indicates $p_x < F_{cc}$. On the other hand, treating the crushed zone as an incompressible liquid sets $p_x = F_{cc}$. This type of inconsistency is difficult to avoid in quasistatic approximation of a continuum.

An explosive can have a form of a prepackaged cartridge, so that its net radius r_e is smaller than the borehole radius r_0 as in Figure 15.12b. This is referred to as a *decoupled* configuration. There is some loss of energy associated with decoupling, because an expansion takes place prior to pressure reaching the wall. In the experience of this author, the net energy Q_h after that partial expansion can be expressed as

$$Q_h = \frac{Q}{2}\left(1 + \frac{r_e^2}{r_0^2}\right) \quad \text{(cylindrical)} \tag{15.30a}$$

$$Q_h = \frac{Q}{2}\left(1 + \frac{r_e^3}{r_0^3}\right) \quad \text{(spherical)} \tag{15.30b}$$

for a cylindrical and a spherical cavity, respectively. The associated cavity pressure p_h will initially act against the rock mass. According to γ-law expansion, p_h is proportional to Q_h, as per Case 13.1.

Compressive Strength of Rock and an Estimate of Crushed Zone Radius

The largest resisting pressure the rock can offer is F_{cc}, the dynamic, compressive, confined strength of the rock material. Let us designate the strength ratio as $k = F_{cc}/F_c$, with factor k to account both for the strain-rate effect and confinement. In the experience of this author, k should be taken as about 3.0 when mining-type explosives are used and $k = 6$ for stronger, military explosives. According to Esen et al. [23], for example, a set of tests carried out on 12 rock types using a Split Hopkinson Bar revealed that dynamic-to-static strength ratio was in the range of 2.5–4.6 *without* including the confinement effect. The tests were conducted using strain rates typical for mining applications.

An exception is made for an apparent compressive strength F_{cc}' in the immediate vicinity of the charge, where the crushed zone is formed. Due to high strain rates involved, $F_{cc}' = 8F_c$ should be assumed, unless more precise data are available.

An estimate of the outer radius a of this zone, per Figure 15.12a, can be obtained by assuming that one-half of the net explosive energy Q_h to be converted into crushing. (Refer to Case 13.1) For a cylindrical shape, this leads to

$$a^2 = \frac{Q_h}{2\pi F_{cc}'} + r_0^2 \tag{15.31}$$

When only the energy content q of the explosive is known and there is a full coupling in the borehole, a simpler relation can be written, based on somewhat different reasoning:

$$a = \sqrt{\frac{2Q_{ef}}{\pi F_{cc}'}} = r_0 \sqrt{\frac{\rho q}{6F_c}} \tag{15.32}$$

where Q_{ef} is taken as $2Q/3$. The latter form was used by this author in Ref. [88] and later favorably compared with experimental results by Esen et al. [23].* Those experiments were carried out using mining-type explosives in a hard rock (basalt) as well as in a soft variety of limestone. For the conditions associated with the use of Equation 15.32 setting $F'_{cc} = 8F_c$ worked quite well. In Example 15.19 where a high explosive is used in a weak mortar, $k = 17.5$ is calculated, but this is not a usual result. As far as the maximum size of the crushed zone is concerned, a number of researchers have indicated that the a/r_0 ratio rarely exceeds 5.0.

Cylindrical Cavity Expanded by Explosive Pressure, CC Mode

Consider again a quasistatic expansion of a cylindrical hole under the $p = F_{cc}$, as illustrated in Figure 15.12a. The crushed zone, shown with its undeformed volume, has pressure F_{cc} on the inside, the resisting pressure during the deformation process. When the initial gas pressure relaxes from p_0 to a final value of F_{cc}, associated with the static balance, we can rewrite the appropriate equation from Case 13.1:

$$u = \left[\left(\frac{p_0}{F_{cc}} \right)^{\frac{1}{2n}} - 1 \right] r_0 \tag{15.33}$$

which determines the radial movement u between the initial and the balanced position. The above relationship is not limited to small deflections. From Equation 15.27, one can then find u_p. As the deflection u may be quite sizeable as compared to r_0 the relation between u and u_p becomes NL. The radial balance of the crushed zone gives

$$F_{cc}(r_0 + u) = p_x(a + u_p) \tag{15.34}$$

Noting that p_x is balanced by F_t on radius R, one gets

$$R = \frac{F_{cc}}{F_t}(r_0 + u) \tag{15.35}$$

This is analogous to Equation 15.16, except for taking into account the deformed position of the crushed boundary.

The same procedure is employed for a spherical cavity, except that equilibrium is written with reference to the surface, rather than the circumference. Also, the outward pressure at R, inducing the limiting hoop stress, is $2F_t$, according to Case 3.33. As a result, the equation to determine R becomes

$$F_{cc}(r_0 + u)^2 = 2F_t R^2 \tag{15.36}$$

The expressions for a cylindrical and a spherical cavity failing in this mode are summarized as Cases 15.16 and 15.17, respectively.

---

* This approach gave better estimates than other theories, which, typically, predicted quite excessive values. The authors of Ref. [23] stated that their new theory gives even better predictions, but there were not enough details in the document to enable a verification of that statement.

Cylindrical Cavity Expanded by Explosive Pressure, Cracking Mode

Consider a quasistatic expansion of a cylindrical hole as illustrated in Figure 15.10, except that now gas is allowed to penetrate cracks and the crushed medium is absent. This gives a constant pressure at balance p_b anywhere, including the cracked radius R. As long as $p_b = F_t$, or tensile strength, the hoop stress at R is also equal to F_t and cracking progresses. When the cavity expansion is sufficiently large for the gas pressure to drop below F_t, the balance position is attained. For the gas expanding from p_0 to F_t, Case 13.1 gives

$$\frac{\pi a^2 + \Delta V}{\pi a^2} = \left(\frac{p_0}{F_t} \right)^{\frac{1}{n}} \tag{15.37}$$

in which ΔV designates the increase of the original volume πa^2, where a is the cavity radius prior to cracking. The increase in volume of the cracked part was calculated as a sum of wedge quashing and medium expansion. With these components being proportional to R^2, the cracked radius R is found from

$$R^2 = \frac{E a^2}{4 F_t} \left\{ \left(\frac{p_{ef}}{F_t} \right)^{\frac{1}{n}} - 1 \right\} \tag{15.38}$$

with $v = 0.25$. The initial pressure p_0 is replaced in the last formula by p_{ef} in recognition of the fact that there are large energy losses due to crack penetration. Lacking better information, one can apply $p_{ef} = 0.5 p_0$, which amounts to reducing the input energy by one-half. If a sizable decoupling is involved, p_h consistent with Q_h in Equation 15.30a replaces p_{ef} in the above. The development for a spherical cavity is quite similar, except the third rather than the second power of a is involved. The equations for a cylindrical and a spherical cavity in the cracking mode are summarized as Cases 15.14 and 15.15, respectively.

Boulder Breaking

The term *boulder* is used in a loose way. It can be a large chunk of rock or a concrete block; a solid body of brittle material. The formulation is done for two idealized cases: a cylindrical and a spherical shape, along the lines developed by this author, Ref. [89], and presented here as Cases 15.18 through 15.21. Initially, the cylindrical shape is discussed here.

Two modes of failure are distinguished, as for the infinite medium: A CC mode and a cracking mode. The conditions, under which one or the other takes place are the same as for cavities in the infinite medium. Regardless of which mode eventuates, the boulder strength is the same, equal to

$$W_m = 0.6 b F_t \tag{15.39}$$

which is the force resulting from projecting the breaking pressure onto the related diameter. As the radial load from pressure grows during the expansion process, so does the length of the radial cracks. The above strength, or the largest resistance is reached when $R = 0.486b$. The modes of failure differ in the manner of pressure application. In CC mode, the wedges, created by cracking are loaded on the inside surface of the hole. In a cracking mode the pressure is uniformly applied within the center hole and the cracks.

The change of volume is calculated in relation to the radial growth, on radius R, of an imaginary tube with an internal/external radius R/b, using the thick ring formula from Appendix D. This change of volume is induced by some pressure p_{br}, applied on radius R. For CC, this pressure is a/R times smaller that the associated hole pressure, but for the cracking mode it is the same. However, the effective pressure of the latter is reduced by half, in the same manner as for a cavity, due to large expected pressure losses when penetrating cracks. Once the necessary radial load needed to cause failure is known, the initial hole pressure can be calculated from the polytropic gas expansion formula.

Having the boulder strength determined by Equation 15.39, one has to compare it with the largest explosive force that can be applied. Consistent with the assumption that the maximum internal pressure that the boulder material can withstand is F_{cc}, the largest force trying to split the boulder in two, in CC mode, per unit length of axis, is

$$W'_e = 2aF_{cc} \qquad (15.40)$$

owing to the action of the crushed zone on radius a. For the cracking mode the estimate is based on the initial geometry:

$$W''_e = 2r_0 p_h \qquad (15.41)$$

The cracking mode takes place when the initial gas pressure is relatively low compared with the rock compressive strength, as mentioned before. This is a very effective mode for boulder breaking, as pressure is applied to a large cross-section area (Figure 15.10). The estimated radius of the crushed zone a is an indicator of which breaking mode is likely to take place. The ratio of $a/r_0 > 2$ will certainly impede the flow of gas into cracks and make the crushing mode the predominant one. A thin layer of crushed material, on the other hand, indicates the likelihood of the cracking mode. (See Cases 15.18 and 15.20 for a cylindrical boulder.) The important thing to remember is that with our quasistatic approach pressure must decrease as cracking progresses in order that equilibrium may be preserved.

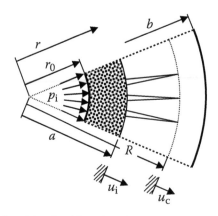

FIGURE 15.14   Sector of a breaking boulder, showing crushed, cracked and intact zones.

A spherical boulder analysis is conducted along the same lines with the explosive splitting forces applied to circular areas:

$$W_e' = \pi a^2 F_{cc} \quad \text{(CC mode)} \tag{15.42}$$

$$W_e'' = \pi R^2 p_h \quad \text{(cracking mode)} \tag{15.43}$$

$$W_m = 0.397\pi b^2 F_t \quad \text{(strength, either mode)} \tag{15.44}$$

Cases 15.19 and 15.21 provide all the relevant details for a spherical boulder.

Ground Cratering due to Subsurface Explosion

Contained explosion inside the ground is probably the best starting point to begin a discussion of this subject. Such an explosion takes place when a charge is detonated deep under the surface, so that ground movement is not readily visible at the surface.

The local effect is a creation of an underground cavity. A spherical charge in a homogeneous ground will create a cavity of a spherical shape. If the ground is solid rock, then one can expect the cavity radius to be as described in Case 15.17. A layer of crushed material would be found around a void previously occupied by the charge and then, outside of it, a radially cracked zone would be located. If the explosive is weak, or there is an initial void between the medium and the charge (uncoupling), then crushing may not eventuate and only the cracked zone may exist, as quantified in Case 15.15.

The other type of medium response takes place when the ground is noncohesive, like sand. When a deep, underground explosion occurs, the only resistance to the explosive gas expansion, other than the inertia of sand, comes from what might be called a locked-in pressure, consisting of the ambient pressure of air, p_a plus the hydrostatic pressure $p_{hyd} = \rho g h$, where h is the depth of the charge center.* Again, a cavity is created and the radius of that cavity may be estimated as

$$R = r_0 + u = r_0 \left(\frac{p_h}{p_a + p_{hyd}} \right)^{\frac{1}{3n}} \tag{15.45}$$

which is merely a rewriting of the corresponding formula of Case 15.17 with the medium resistance of F_{cc} replaced by $p_a + p_{hyd}$. Similarly, for a long cylindrical charge deep underground:

$$R = r_0 \left(\frac{p_h}{p_a + p_{hyd}} \right)^{\frac{1}{2n}} \tag{15.46}$$

---

* There also may be a geological prestress in the ground. This is unknown, unless special measurements are made.

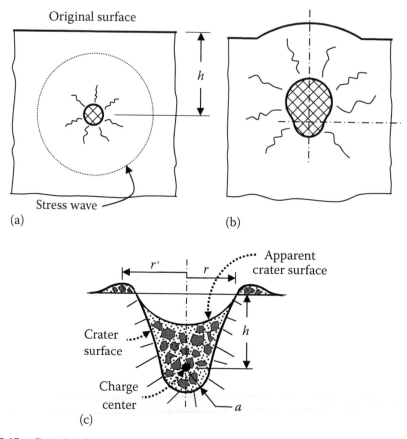

Original surface

h

Stress wave

(a)

(b)

r' r Apparent
crater surface

Crater
surface

h

Charge
center a

(c)

FIGURE 15.15 Cratering from a spherical charge. (a) Shock wave after detonation with cracking to follow, (b) the cavity deforms after the stress wave reflection and return, and (c) final configuration.

A *cratering* explosion takes place when the charge is close enough to the free surface, so that the gas can breakup the material separating the charge from the surface. This is illustrated in Figure 15.15, where the process is shown in stages. In Figure 15.15a a shock wave, which follows the explosion, is sent into the medium and causes local damage around the charge. Once the compressive wave reaches the free surface, it reflects as a tensile wave, travels back to the explosion source stretching the cavity upward. At the same time the gas pressure begins to lift the material radially, toward the free surface, as the returning tensile wave is effectively reducing the upward resistance. The process continues until the material is broken up and lifted away from the hole. After some time the gravity makes the rubble/dust material fall back bringing the crater to the final position in Figure 15.15c. The upper surface, visible to a ground observer, is the *apparent* crater, while the *true* crater is marked by a solid line.

The quantification of a cratering explosion of a spherical is given by Case 15.24, as a relation between the depth of burial h and the mass of the charge M_c for various grounds. The use of it is illustrated by Example 15.21.

A cratering shot can be also conducted with respect to a vertical surface. In this event the force of gravity is not significantly involved and there is no apparent crater obscuring the real one. Also, inverted crater blasting is used, when gravity helps in material removal.

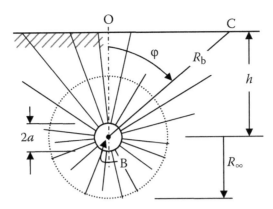

FIGURE 15.16 Cracking stage of subsurface explosion of a cylindrical charge. R_∞ is the crack length corresponding to the infinite medium. R_b is the longest crack between the medium and the surface.

If the charge is placed at an intermediate depth, so that forming a crater is not possible, but the ground surface is permanently moved upward, a cavity forms, similar to what is shown in Figure 15.15b. This cavity is called a *camouflet* and is characterized by asymmetry with regard to a horizontal plane passing through it.

When a long cylindrical charge is placed in a rock medium, parallel to a free surface, with the purpose of cratering the medium, we speak of *bench blasting*, as schematically shown in Figure 15.16. A suitable estimate of the removed *burden h*, can be obtained by treating the problem in a similar manner as that of disintegration of a cylindrical boulder, presented before. As this author showed by a simple geometric reasoning, Ref. [89], a bench with a burden h is likely to be equivalent to a boulder with the radius $b = 1.2h$. The estimate of pressure that must be applied to a borehole wall in order to attain breaking, comes from Case 15.20:

$$p_h = 0.3 \frac{b}{r_0} F_t = 0.36 \frac{h}{r_0} F_t \tag{15.47}$$

The largest borehole pressure that may be effectively applied will not exceed F_{cc} of rock material. Substituting F_{cc} for p_h one obtains

$$h = \frac{r_0}{0.36} \frac{F_{cc}}{F_t} \tag{15.48}$$

As stated before, we can anticipate the ratio of F_c/F_t to be about 10 and, when using mining explosives, F_{cc}/F_c may be assumed ≈ 3. This gives a ratio $F_{cc}/F_t \approx 30$. Broadening this somewhat to a range between 20 and 40, the following range of burdens can be expected:

$$55.6r_0 < h < 111.1r_0 \tag{15.49}$$

There is a well-known predictive formula in mining, derived by Konya [47] by using a totally different methodology. After converting to consistent units, it reads

$$h = 75.6 r_0 \left(\frac{\rho_e}{\rho_r} \right)^{1/3}$$ (15.50)

where indices "e" and "r" refer to explosive and rock, respectively. For a frequently used combination of granite-like rock and ANFO—like explosive, one has the density ratio of $0.8/2.6$, which results in $h = 51 r_0$. Equation 15.50 is therefore a little more conservative than Equation 15.48 in that it predicts a smaller cracked radius. Although Equation 15.50 is supposed to predict a minimum charge needed to break a prescribed burden, it is really "the minimum charge that will for sure break it," accounting for the variations as usually seen when rock and explosives are involved. In this context, our Equation 15.48 can be deemed quite satisfactory.

In bench blasting, the borehole with the center at point B in Figure 15.16 is usually vertical or nearly so. As a result of a blast a wedge of rock, corresponding to the angle of 2φ, separates from the medium. This happens along the radial crack BC that can still reach the surface of the medium. The length of the radius BC can be estimated by noting that in the infinite medium, according to Case 15.16 the cracking radius is

$$R = R_\infty \approx \frac{F_{cc}}{F_t} a$$ (15.51)

when the enlargement of a is ignored. For the boulder, on the other hand, the radial stress on the cracking radius is $0.6179 F_t$, per Case 15.20. The radial equilibrium of a wedge gives $F_{cc} a = (0.6179 F_t) R$, which results in

$$R = 1.62 \frac{F_{cc}}{F_t} a$$ (15.52)

Treating the bench-blasting configuration as the average of the two, the estimate of the longest crack length reaching free surface is

$$R_b = 1.3 \frac{F_{cc}}{F_t} a$$ (15.53)

The reasoning has been carried out using the concept of a crushed zone filling the equivalent hole of radius a. It is more desirable to have the final result in terms of the initial configuration, namely, the borehole radius r_0. Let us consider a range of values of a/r_0, from 3 to 6. Using the average ratio of $F_{cc}/F_t = 30$, the following bounds are obtained: For $a/r_0 = 3$, $R_b = 117 r_0$ and for $a/r_0 = 6$, $R_b = 234 r_0$. Equation 15.48 gives $h = 83.33 r_0$. Noting that $\cos \varphi = h/R_b$, the following bounds of the breakout angle result:

For $a/r_0 = 3$, $\varphi = 44.6°$ and for $a/r_0 = 6$, $\varphi = 69.1°$

Not many field test data are available, as the breakout angle is not often systematically measured. One series of experimental results is quoted by Bilgin [10] in Table 2 in his paper. His measured range of breakout angles is $47.5° < \varphi < 74°$, which is in line with the above estimates, or more precisely, indicates that our equations are slightly pessimistic for the rock involved.

An important aspect of Equations 15.48 and 15.53 along with Figure 15.16 is that they explain one of the basic features of the resulting crater. The maximum radius R_b according to the second equation is constant for a rock type being used, once the explosive is known. Suppose that one selects a burden based on Equation 15.48, a number typically smaller than h. This corresponds to some angle φ. If a decision is then made to reduce the burden, the angle will become larger due to the fact that R_b remains the same. Conversely, placing the charge farther away from the free surface, results in a deeper crater. Both formulas are based on the assumption that the CC mode is involved, which implies a well-developed crushed zone. This was certainly true in the quoted experiments, involving ANFO in fully charged boreholes drilled in a rather soft limestone. If the condition of a marked crushed zone is not fulfilled, the quantification would look differently, as it had to be based on the cracking mode.

Suppose that the maximum cracking radius R_b is known, either from Equation 15.53 or from a different formula. The question is the optimum DOB, or h_{opt}, so that maximum amount of rock material can be removed. To answer this, one can write the equation for the volume V of the removed cone as a function of h/R_b or cos φ. By differentiation, it is easy to find that the maximum volume is reached when

$$h_{opt} = R_b \cos \varphi_0 = \frac{R_b}{\sqrt{3}} \approx 0.577 R_b \qquad (15.54)$$

which corresponds to $\varphi_0 = 54.74°$. Interestingly, Persson [66] arrives at a coefficient of 0.58 by means of experiment. This is, practically, an identical result.

Sometimes a concept of a *critical DOB* is used to designate the value of $h = h_{opt}$, at which the effect of an underground explosion is barely noticeable on the surface. This coincides with our R_b, because a crater can be created only for $h < R_b$. When applied to soil, rather than rock, one has to replace the concept of a *cracking radius* by a *breaking radius*. For practical purposes, finding h_{opt} is even more important. In Case 15.24 various soil constants are quoted, permitting one to associate the depth h with the charge mass necessary for cratering. For a soil of interest, h_{opt} should be the function of the charge size only. If cos φ = √3, at optimum depth, then tan φ = √2. This last variable, tan φ, is designated by α in Case 15.24. In the first approximation one needs to retain only one coefficient, k_3, in the expression for M_c, which gives

$$M_c = k_3 h_{opt}^3 \left(\frac{\alpha^2 + 1}{2} \right)^2 = k_3 h_{opt}^3 \left(\frac{2+1}{2} \right)^2 = 2.25 k_3 h_{opt}^3 \qquad (15.55)$$

For the soils shown in the table k_3 varies from 0.9 to 2.0. This sets the limits for h_{opt}:

$$0.606 M_c^{1/3} < h_{opt} < 0.79 M_c^{1/3} \quad (h_{opt} \sim m \text{ if } M_c \sim kg) \qquad (15.56)$$

Interestingly, when we look up the assumptions for cratering calculations in TM-855-1 [105], we find the optimum DOB to be set as $1.5 W^{1/3}$ ($h_{opt} \sim$ft if $W \sim$lb), which translates to $0.595 M_c^{1/3}$ in metric units.* This is quite close to our lower bound of h_{opt}.

---

* The above applies for an explosive somewhat weaker than TNT. Adjusting for TNT would result in marginally lower coefficients.

Another cratering problem exists when the effect (the crater itself) can be measured, but the contact charge that created the crater is unknown. This is addressed by Case 15.25, which may be used for *post mortem* investigations.

CLOSING REMARKS

Failure has been treated in most of this chapter as synonymous with the material attaining its maximum capacity, be it strain or strength. This is not the full story, of course, because the internal forces must persist long enough for the cracking process to be completed. The difference between such an implied failure and the actual one is rather fine considering all other uncertainties encountered in the evaluation process.

TABULATION OF CASES

Case 15.1   Number of Fragments in Disintegrating, Metallic Cylindrical Shell

$$N(M) = N_m \exp\left(\frac{-M^{1/2}}{M_k}\right)$$; number of fragments larger than M

$M_0 = \pi h d_m L \rho$; shell mass

$$M_k = B h^{5/6} d_i^{1/3} (1 + h/d_i)$$

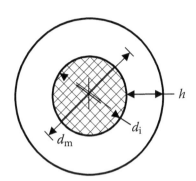

Metric: dimensions in mm, mass in g; B in table

English units: dimensions in inches, mass in lb; use B' instead of B and use N_a instead of N_m.

$N_m = \dfrac{M_0}{2M_k^2}$; total number of fragments (length ~mm, mass ~g; B in table)

$N_a = \dfrac{8M_0}{M_k^2}$; total number of fragments (length ~in., mass ~lb; use B' in table)

	B'	B
RDX	0.212	0.02593
Tetryl	0.272	0.03327
TNT	0.312	0.03816

Note: $B = 0.1223 B'$. See text for additional information.

$M_b = \left[M_k \ln\left(\dfrac{M_0}{2M_k^2}\right) \right]^2$; the largest fragment size mass (metric units)

$n_M = (N_M)^{1/2}$; number of fragments seen in a cross section for square fragments and $L = d_m$

Case 15.2   Estimating Number of Fragments in a Ring When Breakup Duration Is Known

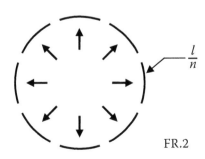

FR.2

$$n = \frac{\beta l}{c t_b}$$

n is the resulting number of fragments

l is the length of circumference

t_b is breakup duration

c is the unloading wave speed (sonic speed)

$1 < \beta < 2$, often $\beta \approx 1.5$

Note: In the illustrated case, $n = 8$. t_b counts from the initiation of the first crack to the completion of the last one. See text for comments. Ref. [96].

CASE 15.3   FRAGMENT SPEEDS IN DISINTEGRATING CYLINDRICAL AND SPHERICAL VESSELS

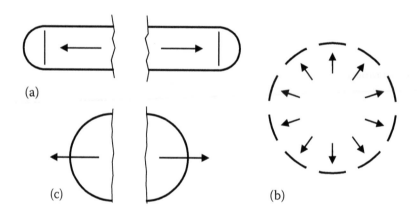

γ is the polytropic constant

R_m is the gas constant $\left(\dfrac{m^2}{s^2 \cdot {}^0K} \right)$

θ_0 is the absolute temperature at burst, K

V_0 is the internal volume (m³)

M_c is the mass of vessel (kg)

$x = \dfrac{(p - p_0)V_0}{M_c \gamma R_m \theta_0}$; nondimensional overpressure in the vessel

$y = \dfrac{v}{\sqrt{\gamma R_m \theta_0}}$; nondimensional fly-off speed v

(a) $y = 1.438x^{0.825}$; cylindrical vessel breaking up into two (2) parts
(b) $y = 1.677x^{0.594}$; cylindrical vessel breaking up into ten (10) parts
(c) $y = 0.916x^{0.75}$; spherical vessel breaking up into two (2) parts

(Equations are valid for $0.003 < x < 0.3$)

Note: A cylindrical vessel has length/diameter ratio of 10, including hemispherical ends. On breaking into two pieces, it separates along the axis, but for 10 fragments it is regarded as breaking up into longitudinal strips. The fly-off speed for a spherical vessel, when number of fragments exceeds two is only marginally larger than for two fragments. Ref. [4] gives more details.

Gas	γ	R_m
Hydrogen	1.4	4124.0
Air	1.4	287.0
Argon	1.67	208.1
Helium	1.67	2078.0
Carbon dioxide	1.225	188.9

CASE 15.4 SPALLING IN REINFORCED CONCRETE PANELS DUE TO PIPE MISSILE IMPACT

M is pipe mass ~g.

$K_e = Mv_0^2/2$ is the kinetic energy ~N-mm needed for spalling

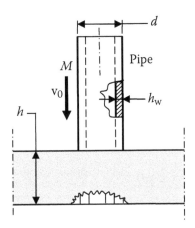

Note: These are experimental results as reported in Ref. [4]. The above values are given as a function of a pipe geometry. Another ratio, h/d is a parameter, on which there is some weak dependence of results. The above table is for $h/d = 2$, which represents the average for the tested range of h/d.

$\dfrac{2h_w}{d}$	$\dfrac{K_e}{h_w^3}$ (MPa)
0.06	4.90
0.08	4.15
0.125	3.00

CASE 15.5 END-TEARING FAILURE OF A BEAM WITH BUILT-IN ENDS DUE TO A PROJECTILE IMPACT (MODE II)

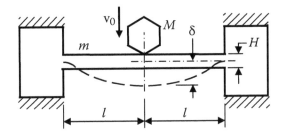

Rigid plastic (RP) material.

δ is deflection at the instant of beginning of tearing

$\mathcal{M}_0 = BH^2\sigma_0/4$; moment capacity, rectangular section

$\theta \approx \delta/l$

$$\Pi_{\mathrm{f}} = \frac{116.5\varepsilon_u}{H^{0.6523}}\,\mathcal{M}_0\theta; \; (\Pi_{\mathrm{f}} \sim \text{N-mm if } H \sim \text{mm and } \mathcal{M}_0 \sim \text{N-mm}); \text{ strain energy absorbed prior}$$

to failure.

$$E_{\mathrm{k}} = \frac{1}{2}M v_0^2 \geq \Pi_{\mathrm{f}}; \text{ condition for tearing failure to occur}$$

Note: The term "built-in ends" has to be used with some reservations. The result shown here is an interpretation of experiments reported in Ref. [56] on rectangular-section beams impacted with speeds up to 10 mm/s. (Examples 15.10 and 15.12.) The flow strength σ_0 must correspond to energy involved in breaking of the material. This is Mode II failure.

CASE 15.6   END-TEARING FAILURE (MODE II) OF A BEAM WITH CLAMPED ENDS DUE TO A DISTRIBUTED SHOCK LOAD

Rigid plastic (RP) material. Rectangular, $B \times H$ section.

$p_0 t_0$ is a short, specific impulse on top surface

Bp_0 is a distributed load per unit length

$i_{\mathrm{r}} = 0.8827 H^{0.7935}\sqrt{\sigma_0\rho}$; specific reflected impulse to break the beam

($i_{\mathrm{r}} \sim$ MPa-ms when $\sigma_0 \sim$ MPa, $H \sim$ mm, and $\rho \sim$ g/mm³)

Bi_{r} is the impulse per unit length

$$v_0 = \frac{i_{\mathrm{r}}}{\rho H} \; ; \text{ corresponding initial velocity}$$

Note: The beam is *nominally* clamped in the sense explained in the text. The above is an interpretation of tests, as reported in Ref. [42], conducted on aluminum beams of 6061-T6 alloy with $\sigma_0 = 266$ MPa and $\varepsilon_u = 0.09$.

CASE 15.7   END-SHEARING FAILURE (MODE III) OF A BEAM WITH SUPPORTED ENDS DUE TO A DISTRIBUTED SHOCK LOAD

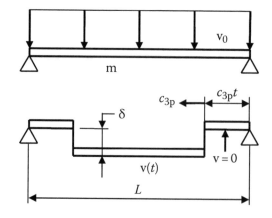

Rigid, strain-hardening (RHS) material.

Initial velocity v_0 is prescribed, as resulting from the lateral impulse $w_0 t_0$,

$$\text{Wave speed}: c_{3p} = \left(\frac{G_p A_s}{m}\right)^{1/2} = \left(\frac{G_p A_s}{\rho A}\right)^{1/2}$$

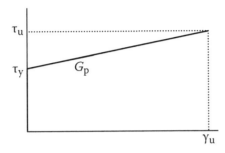

$Q_m = Q_y + m(c_{3p} v_0)$; maximum shearing force, at $t = 0$

$Q_y = A_s \tau_y$

$\gamma_m = v_0 / c_{3p}$; maximum shear strain, at $t = 0$

$$t_s = \frac{L}{2c_{3p}}\left[1 - \exp\left(-\frac{mc_{3p} v_0}{Q_y}\right)\right]; \text{ when motion ceases}$$

$$M_m = \frac{Q_y L}{4}\left[1 + \frac{2c_{3p} t_s}{L}\right]; \text{ maximum moment at center, } t = t_s$$

Failure takes place when $\gamma_m = \gamma_u$ or $Q_m = Q_u = A_s \tau_u$

Velocity to break off the end: $v_e = \dfrac{Q_u - Q_y}{mc_{3p}}$

Note: This approach is based on traveling shear hinge concept as presented in Ref. [114].

CASE 15.8   END-SHEARING FAILURE (MODE III) OF A BEAM WITH CLAMPED ENDS DUE TO A DISTRIBUTED SHOCK LOAD

Rigid plastic (RP) material. Rectangular, $B \times H$ section.

$p_0 t_0$ is a short, specific impulse on top surface

Bp_0 is a distributed load per unit length

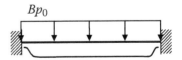

$i_r = \left(0.4385\dfrac{H}{B} + 1.238\right)\sqrt{2\sigma_0 \rho}$; specific reflected impulse to break the beam

(i_r ~MPa-ms when σ_0 ~MPa, H ~mm and ρ ~g/mm³)

Bi_r is the impulse per unit length

$v_0 = \dfrac{i_r}{\rho H}$; corresponding initial velocity

Note: The beam is *nominally* clamped in the sense explained in the text. The above is an interpretation of tests, as reported in Ref. [42].

CASE 15.9   BEAM FAILURE BY SHEAR PLUGGING (MODE IV) DUE TO IMPACT BY A STRIKER WITH MASS M AND VELOCITY v_0

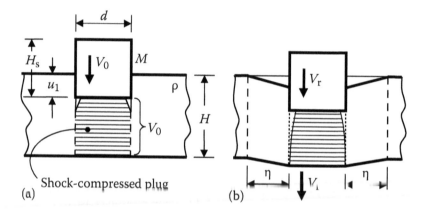

(a)   Shock-compressed plug   (b)

B is beam width normal to page

$c_{pp} = \sqrt{\dfrac{\sigma_0}{\rho}}$; notional plastic wave velocity

(a) Status at the end of Stage 1. The shock wave with speed U has reached the opposite face of the beam.

$M_t = BH\rho(d + 2\eta)$; effective target mass

$2\eta = H\sqrt{3} \approx 1.73H$; additional tributary beam length

$V_0 = \dfrac{Mv_0}{M + M_t} = \dfrac{v_0}{1 + \mu}$; common velocity of M and plug, $\mu = M_t/M$

$u_1 = \dfrac{\ln(1 + \mu)}{\mu} \dfrac{v_0}{U} H$; depth of indentation

$t_1 = H/U$; duration of this phase

(b) Status at the end of Stage 2. The crack has penetrated full depth. The plug is beginning to separate.

$u_{cr} = K^{-1/\lambda} \dfrac{HV_0}{c_{pp}}$; displacement of striker head during Stage 2.

$$\left(\frac{V_r}{c_{pp}}\right)^2 = \left(\frac{V_0}{c_{pp}}\right)^2 - \frac{4}{\sqrt{3}}\frac{\mu}{1+\mu}\frac{\lambda}{1+\lambda}\frac{u_{cr}}{d} \; ; \text{ residual velocity of striker, attained when the}$$

plug becomes separated

v_b is the ballistic velocity, i.e., the impact velocity when V_r from the above equation becomes zero. K and λ are material constants.

$$t_2 \approx \frac{2u_{cr}}{V_0 + V_r}; \text{ duration of this phase}$$

(c) Stage 3 is from the end of Stage 2 until the passage of the striker through the beam is complete.

$$t_3 = \frac{1}{V_r}(H + H_s - u_1 - u_{cr})$$

Then $t_1 + t_2 + t_3$ is the total transit time of the striker.

$$c_r \approx \frac{H}{u_{cr}}\frac{V_0+V_r}{2} \; ; \text{ average crack speed (Stage 2)}$$

Available material constants to be used with the following power law:

$\sigma = A + B\varepsilon^n$; stress–strain for quasistatic conditions, both σ and ε are the effective values.

Weldox 460E steel:

$K = 0.5$, $\lambda = 4$, $A = 490\,\text{MPa}$, $B = 383\,\text{MPa}$, $n = 0.45$, $c_{pp} = 321$ m/s, $\nu = 0.33$, $\sigma_0 = 809\,\text{MPa}$.

2024-T351 Aluminum:

$K = 10$, $\lambda = 4$, $A = 352\,\text{MPa}$, $B = 440\,\text{MPa}$, $n = 0.42$, $c_{pp} = 458$ m/s, $\nu = 0.3$, $\sigma_0 = 566\,\text{MPa}$

c_{pp} and σ_0 values are representative for an impact with v_0 of several hundred m/s.

Note: The case is based on Teng's results [107]. The major change is to include some of beam material outside the plug in M_t, which gives a larger v_b. The striker is treated as rigid.

CASE 15.10 CANTILEVER FAILURE DUE TO BLAST LOAD

Ductile or brittle material.

V_f is the fly-off velocity of the broken piece (m/s)

Ψ is material toughness, as defined in Chapter 4 (MPa)

B is the beam width, normal to page (m)

s is the adjusted impulse per unit length (N-s/m)

A is the section area (m²)

ρ is density (kg/m³)

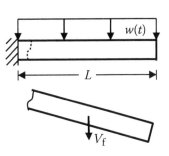

$$V_f = \sqrt{\frac{\Psi}{\rho}}\left\{0.436\left(\frac{s}{A\sqrt{\rho\Psi}}\right)\left(\frac{2L}{B}\right)^{0.3} - 0.2625\right\}$$

$$\text{for } \left(\frac{s}{A\sqrt{\rho\Psi}}\right)\left(\frac{2L}{B}\right)^{0.3} \geq 0.602$$

otherwise, no failure, or $V_f = 0$. Ref. [4].

Note: The above holds for shear and for bending failures, any cross section, any material. If the width B is large, so that the beam resembles a plate, then $s = Bi_r$, where i_r is the specific reflected impulse. Otherwise, s may be reduced, as per Case 13.12. If the effective loading per unit length $w(t)$ is a rectangular pulse with magnitude w_0 and duration of t_0, then $s = w_0 t_0$.

CASE 15.11 BREAKING OFF OF A CANTILEVER TIP DUE TO IMPACT BY A SQUARE BLOCK OF MASS M

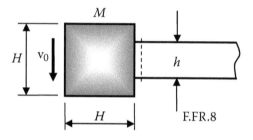

Rigid-plastic beam material with moment capacity $\mathcal{M}_0$.

$e_f = \dfrac{E_{k0}}{\mathcal{M}_0} = \dfrac{Mv_0^2}{2\mathcal{M}_0}$; nondimensional kinetic energy of the flying block needed to achieve failure

$\alpha = H/h$

(a) Pure shear failure

$$e_f = 2 + \frac{8}{3}\alpha^2; \quad \text{for } \alpha \geq 1$$

$$e_f = \alpha\left(2 + \frac{8}{3}\alpha^3\right); \quad \text{for } \alpha < 1$$

(b) Shear–bending interaction at failure

$$e_f = 5 + \frac{3}{\alpha} + \frac{4}{3}\alpha^2; \quad \text{for } \alpha \geq 2.65$$

$$e_f = 5 + \frac{3\mu}{\alpha} + \frac{8\alpha^2}{3(1+\mu)}; \quad \text{for } \alpha < 2.65,$$

$$\mu \approx 0.808\alpha - 1.141; \quad \text{for } 1 < \alpha < 2.65$$

Note: The block, flying with velocity v_0 becomes attached to the cantilever tip and then breaks it off, provided v_0 is large enough. Shear failure dominates for $\alpha < 2$, shear with bending above this value [84].

CASE 15.12   CANTILEVER TIP BREAK OFF DUE TO A HIGH-SPEED PROJECTILE IMPACT. SHEAR FAILURE

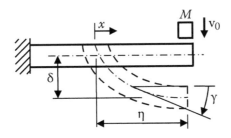

Rigid plastic (RP) material.

τ_0 is the limit shear stress, γ_u is the ultimate shear angle

$Q_0 = A_s\tau_0$; limit shear force

Ψ_s is the material toughness or the area under $\tau-\gamma$ curve (Defined in Chapter 4)

$$\eta = \frac{Mv_0^2}{A_s\tau_0\gamma_u} = \frac{Mv_0^2}{A_s\Psi_s}; \text{ deformed length associated with onset of failure}$$

$$\eta \approx \frac{Mv_0^2}{A_s\Psi_s}; \text{ deformed length, any material characteristic}$$

$$\delta = \frac{1}{2}\gamma_u\eta; \text{ tip deflection at failure}$$

$$\Pi_{br} \approx \frac{1}{2}A_s\Psi_s\eta = \frac{1}{\gamma_u}A_s\Psi_s\delta; \text{ strain energy associated with failure}$$

Note: For determination of $\tau-\gamma$ curve from $\sigma-\varepsilon$ data refer to Case 4.8. For a rectangular section in the limit state $A \approx A_s$. There is no experimental verification of the above results.

CASE 15.13   METAL CUTTING WITH EXPLOSIVES

Index e designates a minimum charge needed for breaking. A is cross-sectional area, H is plate thickness

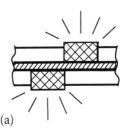

(a) Steel beam shearing (TNT)

$$M_e = 0.375\ A\ (M \sim\text{lb when } A \sim\text{in.}^2) \qquad \text{(a)}$$

$$M_e = A/3793\ (M \sim\text{kg when } A \sim\text{mm}^2)$$

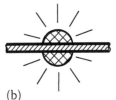

(b) Cutting of steel cable, bar, or chain (TNT)

$$M_e = A \text{ in air} \quad \text{and} \quad M_e = 2A \text{ in water} \qquad \text{(b)}$$
$$(M \sim\text{lb when } A \sim\text{in.}^2)$$

$$M_e = A/1422 \text{ in air} \quad \text{and} \quad M_e = A/711 \text{ in water}$$
$$(M \sim kg \text{ when } A \sim mm^2)$$

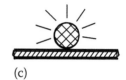

(c)

(c) Cutting steel sheet with explosive chord (blasting gelatin)

$$m = 0.0054H^2 - 0.0328H + 2.378 \ (m \sim kg/m \text{ when } H \sim mm)$$

Note: 1 kg of blasting gelatin is energy-equivalent to 0.9 kg of TNT. The charges must be attached to the objects to be broken. Ref. [73].

CASE 15.14   CYLINDRICAL CAVITY, INFINITE MEDIUM. CRACKING MODE OF FAILURE

(Refer to the illustration in Case 15.18 with b = ∞)

Quasistatic gas expansion from initial pressure p_{ef} to $p = F_t$. Pressure penetrating cracks.

$p_{ef} = \dfrac{p_h}{2} = \dfrac{1}{2}q\rho_e(n-1)$; effective initial pressure, coupled configuration

$p_c = F_t$; radial stress at radius R at initiation of cracking

$R^2 = \dfrac{Ea^2}{4F_t}\left[\left(\dfrac{p_{ef}}{F_t}\right)^{\frac{1}{n}} - 1\right]$; to calculate cracked radius R

$u_c = \dfrac{(1+v)}{E}F_t R$; displacement at R

$u_p = u_c + \dfrac{F_t a}{E}\left(\dfrac{R}{a} - 1\right)$; displacement at cavity wall a

Note: Cracking is assumed to be caused by hoop stress exceeding F_t in the intact cavity. The initial pressure of explosion p_h, used in CC mode, is reduced to $p_{ef} = 0.5p_0$. R should be limited to $100a$ for materials with small F_t.

CASE 15.15   SPHERICAL CAVITY, INFINITE MEDIUM. CRACKING MODE OF FAILURE

(Refer to the illustration in Case 15.18 with b = ∞)

Quasistatic gas expansion from initial pressure p_{ef} to $p = 2F_t$. Pressure penetrating cracks.

$p_{ef} = \dfrac{p_h}{2} = \dfrac{1}{2}q\rho_e(n-1)$; effective initial pressure, coupled configuration

$p_c = F_t$; radial stress at R at initiation of cracking

$R^3 = \dfrac{4Ea^3}{27F_t}\left[\left(\dfrac{p_{ef}}{2F_t}\right)^{\frac{1}{n}} - 1\right]$; to calculate cracked radius R

$$u_c = \frac{(1+v)}{E} F_t R; \text{ displacement at } R$$

$$u_p = u_c + \frac{2F_t a}{E}\left(\frac{R}{a} - 1\right); \text{ displacement at cavity wall}$$

Note: Cracking is assumed to be caused by hoop stress exceeding F_t in the intact cavity. The initial pressure of explosion p_h, used in CC mode, reduced to $p_{ef} = 0.5p_0$. R should be limited to $30a$ for materials with small F_t.

CASE 15.16   CYLINDRICAL CAVITY IN INFINITE MEDIUM. CRACKING AND CRUSHING (CC) FAILURE MODE

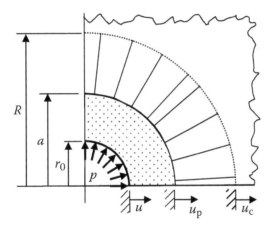

p_h is effective initial gas pressure
Q_h is effective charge energy

$$a^2 = \frac{Q_h}{2\pi F'_{cc}} + r_0^2; \text{ to calculate crushed zone radius}$$

$$u = \left[\left(\frac{p_h}{F_{cc}}\right)^{\frac{1}{2n}} - 1\right] r_0; \text{ wall displacement}$$

$$u_p = \sqrt{(r_0 + u)^2 + a^2 - r_0^2} - a; \text{ displacement at } a$$

$$R = \frac{F_{cc}}{F_t}(a + u_p); \text{ cracked zone radius}$$

$$u_c = \frac{1+v}{E} F_t R; \text{ displacement at } R$$

CASE 15.17   SPHERICAL CAVITY IN INFINITE MEDIUM. CRACKING AND CRUSHING (CC) FAILURE MODE

(Refer to the illustration in Case 15.16)

p_h is effective initial gas pressure, Q_h is effective charge energy

$$a^3 = \frac{3Q_h}{8\pi F_{cc}'} + r_0^3; \text{ to calculate crushed zone radius } a$$

$$u = \left[\left(\frac{p_h}{F_{cc}} \right)^{\frac{1}{3n}} - 1 \right] r_0; \text{ wall displacement}$$

$$u_p = \left[(r_0 + u)^3 + a^3 - r_0^3 \right]^{1/3} - a; \text{ displacement at } a$$

$$R = \sqrt{\frac{F_{cc}}{2F_t}} (a + u_p); \text{ cracked zone radius; } u_c = \frac{1+\nu}{E} F_t R; \text{ displacement at } R$$

CASE 15.18   CYLINDRICAL BLOCK WITH INTERNAL PRESSURE. BRITTLE MATERIAL. CRACKING FAILURE MODE. ($a \equiv r_0$)

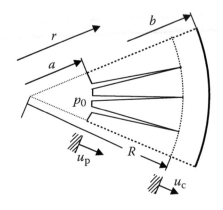

p_h is effective initial gas pressure

$$p_h = \frac{1 - (a/b)^2}{1 + (a/b)^2} F_t; \text{ hole pressure to initiate cracking}$$

$$p_{br} = \frac{1 - \alpha^2}{1 + \alpha^2} F_t; \text{ radial stress at } R \text{ to cause breakup}$$

$$\alpha = \frac{R}{b}; \text{ boulder strength } W_m \text{ reached at } \alpha \approx 0.486, \text{ or } R = 0.486b$$

$$W_m = 2Rp_{br} = 0.6bF_t; \text{ boulder strength, N/mm}$$

$p_{br} = 0.3 \dfrac{b}{R} F_t = 0.6173 F_t$; pressure breaking boulder, uniform inside, cracked condition

$u_c = 0.5606 \dfrac{bF_t}{E}$; displacement on R at W_m

$u_p = u_c + \dfrac{F_t a}{E}\left(\dfrac{R}{a} - 1\right)$; displacement at cavity wall a

$p_{hu} = 2 p_{br}\left(1 + 0.545 \dfrac{b^2 F_t}{a^2 E}\right)^n$; initial hole pressure needed for breakage

Note: Cracking is assumed to be caused by hoop pressure σ_h exceeding F_t in the intact part of the block. During quasistatic expansion, after the initial cracking, the equilibrium pressure is decreasing. R is best called a *virtual* radius, because the boulder may break before attaining the associated load. Ref. [88].

Case 15.19   Spherical Block with Internal Pressure. Brittle Material. Cracking Failure Mode. ($a \equiv r_0$)

(Refer to the illustration in Case 15.18)

p_h is effective initial gas pressure

$p_h = \dfrac{1 - (a/b)^3}{1/2 + (a/b)^3} F_t$; hole pressure to initiate of cracking

$p_c = \dfrac{1 - \alpha^3}{1/2 + \alpha^3} F_t$; radial stress at R to cause breakup

$\alpha = \dfrac{R}{b}$; boulder strength W_m reached at $\alpha \approx 0.63$, or $R_m = 0.63b$

$p_{br} = 0.397 \dfrac{b^2}{R^2} F_t = F_t$; pressure breaking boulder, uniform inside, cracked condition

$W_m = \pi R^2 F_t = 0.397 \pi b^2 F_t$; boulder strength (N)

$u_c = 0.63 \dfrac{bF_t}{E}$; displacement on R at W_m

$p_{hu} = 2 p_{br}\left(1 + 0.75 \dfrac{b^3 F_t}{a^3 E}\right)^n$; initial hole pressure needed for breakage

Note: The notes for Case 15.18 are applicable.

CASE 15.20   CYLINDRICAL BLOCK WITH INTERNAL PRESSURE. BRITTLE MATERIAL.
              CC FAILURE MODE

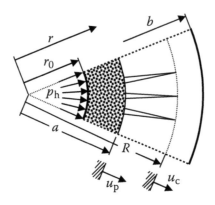

p_h is effective initial gas pressure

$p_c = (r_0/R)p_h$; radial stress at $r = R$

Find radius a as in Case 15.16.

$$p_h = \frac{1-(r_0/b)^2}{1+(r_0/b)^2} F_t; \text{ hole pressure to initiate of cracking}$$

$$p_c = \frac{1-\alpha^2}{1+\alpha^2} F_t; \text{ radial stress at } R \text{ to continue breakup}$$

$$\alpha = \frac{R}{b}; \text{ boulder strength } W_m \text{ reached at } \alpha \approx 0.486, \text{ or } R = 0.486b$$

Then $p_c = p_{br} = 0.6179F_t$ and the hole pressure to break boulder is $p_h = 0.3\dfrac{b}{r_0}F_t$

$W_m = 2r_0 p_{br} = 0.6bF_t$; boulder strength, N/mm

$u_c = 0.692\dfrac{bF_t}{E}$; displacement on R at W_m

$u_p = u_c + 0.618\dfrac{RF_t}{E}\ln\left(\dfrac{R}{a}\right)$; displacement on a at W_m

$u = \sqrt{(a+u_p)^2 - a^2 + r_0^2} - r_0$; displacement on radius r_0 when p_{br} is reached

$W'_e = 2aF_{cc}$; applied explosive force N/mm

Note: The same comments as for Case 15.18. As an initial approximation $u_p = u_c$ can be assumed. Ref. [88].

CASE 15.21 SPHERICAL BLOCK WITH INTERNAL PRESSURE. BRITTLE MATERIAL. CC FAILURE MODE

(Refer to the illustration in Case 15.20)

p_h is effective initial gas pressure

Find radius a as in Case 15.17.

$$p_c = \left(\frac{r_0}{R}\right)^2 p_h; \text{ radial stress at } r = R$$

$$p_h = \frac{1-(r_0/b)^3}{1/2+(r_0/b)^3} F_t; \text{ hole pressure to initiate cracking}$$

$$p_{br} = \frac{1-\alpha^3}{1/2+\alpha^3} F_t; \text{ radial stress at } R \text{ to continue breakup}$$

$\alpha = \dfrac{R}{b}$; boulder strength W_m reached at $\alpha \approx 0.63$, or $R = 0.63b$

Then: $p_c = p_{br} = F_t$ and the hole pressure to break boulder: $p_h = \left(\dfrac{R}{r_0}\right)^2 F_t$

$W_m = \pi r_0^2\, p_{br} = 0.397\pi b^2 F_t$; boulder strength, N

$u_c = 0.63\dfrac{bF_t}{E}$; displacement at p_{br}

$u_p = u_c + \dfrac{R^2 F_t}{Ea}\left(1-\dfrac{a}{R}\right)$; displacement on a at W_m,

$u = \left[(a+u_p)^3 - a^3 + r_0^3\right]^{1/3} - r_0$; hole wall displacement when p_{br} is reached

$W'_e = \pi a^2 F_{cc}$; applied explosive force, N

Note: The same comments as for Cases 15.18 and 15.19. As an initial approximation $u_p = u_c$ can be assumed. Ref. [88].

CASE 15.22 DAMAGE TO ROCK MASS FROM EXPLOSIONS

F_{cc} is the dynamic compressive strength

F_{tt} is the dynamic tensile strength (both to be compared with stress induced by rock blasting)

v_r is the limiting PPV of ground movement, mm/s, related to pressure or compressive stress

Rock strength parameters

	F_{cc} (MPa)	F_{tt} (MPa)	v_r (m/s)
Limestone I	44.1	6.9	6.2
Limestone II	158.8	—	11.5
Granite	147.1	17.6	12.4

(continued)

	F_{cc} (MPa)	F_{tt} (MPa)	v_r (m/s)
Shale I	45.1	4.9	9.2
Shale II	172.5	3.9	12.2
Quartzite	145.1	—	9.3
Marble, white	73.5	14.7	7.3
Marble, red	117.6	24.5	8.8
Marble, black	73.5	20.6	5.9
Gneiss	115.2	—	7.6
Dolomite	184.8	33.3	12.6
Diabase	154.9	10.8	9.6
Porphyrite	156.9	19.6	11.2
Gabbro-diabase	259.8	13.7	16.3

Rock crushing takes place when $p > F_{cc}$ or $v > v_r$.

Rock cracking when $\sigma_t > F_{tt}$. (p is the applied pressure and σ_t the applied tension, while v is PPV).

Source: Henrych, J., *The Dynamics of Explosion and Its Use*, Elsevier, Amsterdam, 1979.

Note: Tensile stress arises at some distance from the explosive source.

CASE 15.23   DAMAGE TO RESIDENTIAL BUILDINGS FROM GROUND VIBRATIONS

v_0 is the limiting peak particle velocity of ground vibrations, mm/s

v_0 values for buildings on different grounds, mm/s

	No Visible Cracking	Fine Cracks, Plaster Falling	Visible Cracking	Major Cracking
Sand, gravel, clay	18	30	40	60
Moraine, slate, soft limestone	35	55	80	115
Granite, hard limestone, quartzite sandstone	70	110	160	230

If the seismic velocity of the ground c_p (m/s) is known, then $v_0 = c_p/65$ replaces the first column values and the remaining columns are modified in the same proportion [32].

Note: The above values are for typical residential building according to Swedish experience. Those are brick buildings, plastered or cement-rendered inside and placed on concrete foundations. The information comes from Ref. [66], where modifying factors relating to condition of buildings, their age, and duration of imposed vibration are also quoted. Specific local recommendations often give much lower v_0 than what is shown above.

CASE 15.24   CRATERING FROM A SPHERICAL CHARGE PLACED UNDER SURFACE

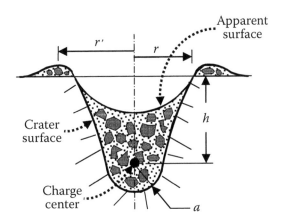

(From Henrych, J., *The Dynamics of Explosion and Its Use*, Elsevier, Amsterdam, 1979.)

M_c is charge mass (kg)

r is the crater radius (m); h is the depth where the charge center is located (DOB) ~m

$\alpha = r/h$; nondimensional crater radius

$$M_c = \left(k_2 h^2 + k_3 h^3 + k_4 h^4\right)\left(\frac{\alpha^2 + 1}{2}\right)^2 ; \text{ explosive charge (kg) as a function of DOB}$$

$k_2 = 0$ for moderately hard soils and $k_2 = 0.35\,\text{kg/m}^2$ for hard rocks

$k_4 = 0.026\,\text{kg/m}^4$ for moderately hard soils and $k_4 = 0.0022\,\text{kg/m}^4$ for hard rocks

Coefficients k_3 empirically found for standard crater shapes, $\alpha = 1$.

Ground Type	k_3 (kg/m³)
Dry sand	1.8–2.0
Compact or moist sand	1.4–1.5
Heavy sandy loam	1.2–1.35
Compact loams	1.2–1.5
Loess	1.1–1.5
Chalk	0.9–1.1
Gypsum	1.2–1.5
Shell limestone	1.8–2.1
Sandstone, clay shale, limestone	1.35–1.65
Dolomite, limestone, sandstone	1.5–1.95
Limestone, sandstone	1.5–2.4
Granite	1.8–2.55
Basalt	2.1–2.7
Quartzite	1.8–2.1
Porphyrite	2.4–2.55

The above figures are based on the use of ammonite with $q = 4.19\,\text{kJ/kg}$ (4.19×10^6 N-mm).

If another charge material is used, adjust the calculated M_c in proportion to its q. For materials not listed, use $k_3 \approx \rho/1300$, where $\rho \sim kg/m^3$ is the material density.

h' is the depth of an apparent crater (m)

$$h' = \frac{2\alpha - 1}{3}h; \text{ (soil, medium brisance explosives, say } n < 2)$$

$$h' = \frac{r}{2}; \text{ (soil, high-brisance explosives, } n > 2)$$

$$h' = (0.2\alpha + 0.6)h; \text{ (rock) (valid for } h > 1 \text{ m)}$$

$$V' = 1.2h'r^2; \text{ apparent crater volume (m}^3)$$

When h is as large as tens of meters (deep crater), the apparent crater depth is somewhat smaller:

$$k_h = \frac{1}{1 + ah^{2/3}}; \text{ reduction factor for deep craters; } a = 0.013 \text{ for soils of medium strength}$$

$H^* = k_h h'$; apparent crater depth for deep craters

$V^* = k_h V'$; apparent crater volume for deep craters

Note: The data comes from Ref. [32].

CASE 15.25   SOIL CRATERING FROM A HEMISPHERICAL CHARGE PLACED ON SURFACE

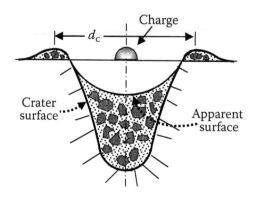

M_c is the charge mass $\sim kg$

d_c is the apparent crater diameter $\sim m$

$M_c = p_2 d_c^2 + p_1 d_c + p_0$; explosive charge of TNT as a function of the resulting crater radius

Coefficients for charge mass M_c calculation

Ground Type	p_2	p_1	p_0
Dry sand	34.84	−230.4	466.5
Dry sandy clay	28.65	−208.9	466.3
Wet sand	22.21	−183.7	465.4
Dry clay	18.01	−165.3	464.6
Wet sandy clay	15.05	−150.8	462.8
Wet clay	8.02	−110.0	462.3

The above figures on TNT-equivalent charge. If another charge material is used, adjust the calculated M_c in proportion to its q.

Note: The data comes from the information in Ref. [105] and is based on action of bare charges. The purpose is to determine the explosive charge from the crater size it leaves. If the charge is detonated above ground, using the formula will underestimate M_c.

CASE 15.26   CRATERING OF CONCRETE SURFACE BY A CONTACT CHARGE

M_c is charge mass (kg)

R is the crater radius (m), close to the charge

r_c is the radius of an equivalent hemispherical charge (m)

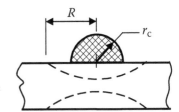

(a) Common concrete, standard reinforcement, $F_c = 46\,\text{MPa}$

$$R = 3.65r_c$$

This comes from tests on a 300 mm thick slab, with standard top and bottom reinforcement and bar spacing of 160 mm. This spacing was larger than the largest diameter of an equivalent hemispherical charge; 137 mm. Ref. [48].

(b) High-strength concrete, $F_c = 200\,\text{MPa}$ $(F_t = 18\,\text{MPa})$

$$R = 5.4r_c$$

This is unreinforced concrete. Ref. [120].

Note: The above should be viewed as quotations of selected test results rather than of a systematic study. The results are applicable for TNT.

CASE 15.27   PROGRESSIVE COLLAPSE OF A BUILDING WITH FRANGIBLE COLUMNS. TOP-INITIATED FAILURE

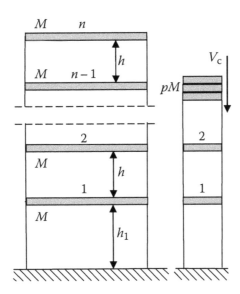

There are n stories, all identical except the ground story being taller.

Top story columns fail and the slabs sequentially "pancake." The sketch shows

(a) Original status
(b) p slabs from top "pancaked" and moving down after top story failure. ($p = 3$ shown above.)

The following is for $(p-1)$th collision:

$$U_g = \frac{p(p-1)}{2} Mgh; \text{ gravitational energy or kinetic energy absorbed}$$

$$M^* = \frac{p-1}{p} M; \text{ effective mass in collision}$$

$$V_c = \left(\frac{p-1}{p}\right) v_0; \text{ relation between pre-impact speed } v_0 \text{ and post-impact speed } V_c \text{ after a}$$

collision in which p slabs form an assembly

$$\frac{\Delta E_k}{E_k} = \frac{1}{p}; \text{ portion of the incoming kinetic energy loss at the collision}$$

$$E_k' = \frac{p-1}{p} E_k; \text{ portion of the incoming kinetic energy retained after the collision}$$

$$v_{0i}^2 = V_{c,i-1}^2 + 2gh; \text{ relation between post-impact speed } V_{c,i-1} \text{ of the previous collision and}$$

the pre-impact velocity v_{0i} of the ith collision.

First collision, $p = 2$: $E_k = Mgh$; $E_k' = \frac{1}{2} E_k = \frac{1}{2} Mgh$

Second, $p = 3$: $E_k = \frac{5}{2} Mgh$; $E_k' = \frac{2}{3} E_k = \frac{5}{3} Mgh$

Third, $p = 4$: $E_k = \frac{14}{3} Mgh$; $E_k' = \frac{3}{4} E_k = \frac{7}{2} Mgh$

Fourth, $p = 5$: $E_k = \frac{15}{2} Mgh$; $E_k' = \frac{4}{5} E_k = 6Mgh$

$$E_k \approx \frac{1}{2}(p-1)(p-2)Mgh = \left(1 - \frac{2}{p}\right)U_g; \text{ approximate kinetic energy for } p > 5$$

$$E_{kn} \approx \frac{1}{2}(n-1)(n-2)Mgh; \text{ kinetic energy before descent from height } h_1$$

$$E_{kf} = E_{kn} + nMgh_1; \text{ final kinetic energy before ground impact}$$

$$v_n = \sqrt{\frac{2E_{kn}}{nM}}; v_f = \sqrt{\frac{2E_{kf}}{nM}}; \text{ velocity before and after the descent from height } h_1$$

$$S = \sqrt{(2E_{kf})nM}; \text{ impulse applied to the ground}$$

$$t = \frac{2(n-1)h}{v_n} + \frac{2h_1}{v_n + v_f}; \text{ event duration}$$

Note: The failure at the top floor causes a gravity fall of the roof slab, which initiates the event. This is a limiting case when no energy absorption by failing columns is assumed.

CASE 15.28 PROGRESSIVE COLLAPSE OF A BUILDING WITH DUCTILE COLUMNS. TOP-INITIATED FAILURE

(Refer to the illustration in Case 15.27)

Top story columns fail and the slabs sequentially "pancake." Collision-related definitions of U_g, M^*, V_c, ΔE_k, and E'_k remain the same as in Case 15.27.

Π is the energy lost by squashing by h of columns between two stories

$\eta = \dfrac{\Pi}{Mgh}$ is a nondimensional energy loss between stories.

After a collision in which p slabs form an assembly with a total mass of M_a

$v_{0i}^2 = V_{c,i-1}^2 + 2gh - 2\Pi_i/M_a$; for movement between collisions: $V_{c,i-1}$ is post-impact speed of the previous collision, v_{0i} is the pre-impact velocity of the ith collision and Π_i is strain energy.

$v_{0i} > 0$ or $V_{c,i-1}^2 + 2gh > 2\Pi_i/M_a$; energy condition for the motion to continue

The state described below is after p slabs come in contact, at $(p-1)$th collision. ($p=3$ illustrated)

First collision, $p = 2$: $E_{k1} = Mgh - \Pi = (1-\eta)Mgh$; $E'_{k1} = \dfrac{1}{2} E_{k1}$

Second, $p = 3$: $E_{k2} = E'_{k1} + (2-\eta)Mgh$; $E'_{k2} = \dfrac{2}{3} E_{k2}$

Third, $p = 4$: $E_{k3} = E'_{k2} + (3-\eta)Mgh$; $E'_{k3} = \dfrac{3}{4} E_{k3}$

Fourth, $p = 5$; $E_{k4} = E'_{k3} + (4-\eta)Mgh$; $E'_k = \dfrac{4}{5} E_k$

$E'_k = \dfrac{p-1}{p} E_k$; kinetic energy, incoming (E_k) and retained after a collision (E'_k)

$E_{kf} = E_{k,n} + nMgh_1 - \Pi_1$; final kinetic energy before ground impact ($n = p - 1$)

$v_n = \sqrt{\dfrac{2E_{kn}}{nM}}$; $v_f = \sqrt{\dfrac{2E_{kf}}{nM}}$; velocity before and after the descent from height h_1, respectively

$S = \sqrt{(2E_{kf})nM} = nMv_f$; impulse applied to the ground

$t = \dfrac{2(n-1)h}{v_n} + \dfrac{2h_1}{v_n + v_f}$; event duration

Note: For most tall buildings at least a several stories will have the same strength. When story columns change at some level, the energy Π suddenly becomes bigger. The stories below remain intact as long as their strength is not exceeded by impact forces.

CASE 15.29   PROGRESSIVE COLLAPSE OF A BUILDING WITH FRANGIBLE COLUMNS. BOTTOM-INITIATED FAILURE

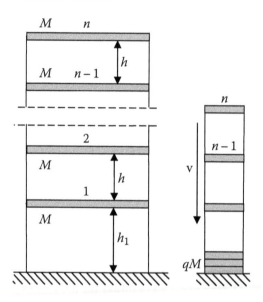

There are *n* stories, all identical except the ground story being taller.

The ground story columns fail and the slabs sequentially "pancake." The sketch shows

(a) Original status
(b) *q* bottom slabs "pancaked" and the remainder of a building continuing to move down. (*q* = 3 shown on sketch.)

$E_{k1} = nMgh_1$; kinetic energy of the moving building after descending by h_1

$v_1 = \sqrt{2gh_1}$ is the first impact speed

$S_1 = Mv_1$ is the first impulse applied to the ground

$H = h_1 + (n-1)h$ is the effective drop height

$v_n = \sqrt{2gH}$ is the top floor impact speed

$t = \sqrt{\dfrac{2H}{g}}$ is the duration of event

$S_n = Mv_n$ is the final impulse applied to the heap on the ground

Note: The failure of the ground level columns causes a simple gravity fall of the entire building. This is a limiting case when no energy absorption by failing columns is assumed.

CASE 15.30   PROGRESSIVE COLLAPSE OF A BUILDING WITH DUCTILE COLUMNS. BOTTOM-INITIATED FAILURE

(Refer to the illustration in Case 15.29)

The formulation in similar to Case 15.29 except for column-absorbed energy Π.

$$c = \sqrt{\frac{EA}{m}} = \sqrt{\frac{EAh}{M}}$$; notional sound speed, where A is the total section area of the second

story columns while M and h are typical for the vertical segment of building above the ground floor.

$E_{k1} = nMgh_1 - \Pi_1$; kinetic energy of the moving building after descending by h_1.

$$v_1 = \sqrt{\frac{2E_{k1}}{nM}}$$; velocity at that instant (impact speed)

$S_1 = nMv_1$ is the first impulse applied to the ground

$P_1 = A\rho cv_1$ is the total peak impact force, ρ being the average density of column material

$$t \approx \sqrt{\frac{2h_1}{g}} + \frac{2(n-1)h}{v_1}$$ is the duration of event, until the roof is on the ground.

Note: For most tall buildings at least a several stories will have the same strength. When story columns change at some level, the energy Π suddenly becomes bigger, which may cause the motion to be arrested. The stories above remain intact as long as their strength is not exceeded by impact forces.

CASE 15.31   PROGRESSIVE COLLAPSE OF A BUILDING WITH FRANGIBLE COLUMNS. MIDDLE-INITIATED FAILURE

r intact stories moving down. ($r = 2$ shown)

There are n stories, all identical except the ground story being taller. An intermediate story columns fail and the slabs sequentially "pancake" below the intact part. The sketch shows $p = 3$ slabs pancaked and $p + r$ slabs in post-impact assembly.

The following is for $(p-1)$th collision:

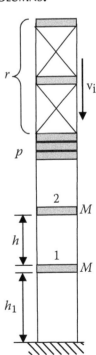

$$U_g = \frac{2r + p}{2}(p-1)Mgh$$; gravitational energy absorbed

$M_a = (r + p - 1)M$; striking assembly mass before $(p-1)$th collision:

$$M* = \frac{r + p - 1}{r + p}M$$; effective mass in collision

$$E_k' = \frac{r + p - 1}{r + p}E_k$$; portion of the incoming kinetic energy retained after

the collision

Before first collision, $p = 2$, $M_a = (r + 1)M$, $E_{k1} = (r + 1)Mgh$;

After first collision, $E_{k1} = \dfrac{r+1}{r+2} E_{k1}$, $M_a = (r+2)M$

Before second collision, $p = 3$, $M_a = (r+2)M$, $E_{k2} = E'_{k1} + (r+2)Mgh$;

After second collision, $E'_{k2} = \dfrac{r+2}{r+3} E_{k2}$; $M_a = (r+3)M$

When the collapsing mass impacts the slab at h_1:

$n - r - 1$ impacts have taken place

$U_g = \dfrac{r+n}{2}(n-r-1)Mgh$; gravitational energy absorbed

$\Delta E_k \approx \dfrac{2U_g}{n-r+2}$; energy lost by impacts, for $n - r > 4$

$E_{kn} = U_g - \Delta E_k$; net energy before final descent

$E_{kf} = E_{kn} + nMgh_1$; final kinetic energy before ground impact

$v_n = \sqrt{\dfrac{2E_{kn}}{nM}}$; $v_f = \sqrt{\dfrac{2E_{kf}}{nM}}$; velocity before and after the descent from height h_1

$S = \sqrt{(2E_{kf})nM}$; impulse applied to the ground

$t = \dfrac{2(n-r-1)h}{v_n} + \dfrac{2h_1}{v_n + v_f}$; event duration

Note: The upper part of r stories remains intact as long as their strength is not exceeded by the impact forces. This is a limiting case when no energy absorption by failing columns is assumed.

CASE 15.32 PROGRESSIVE COLLAPSE OF A BUILDING WITH DUCTILE COLUMNS. MIDDLE-INITIATED FAILURE, r INTACT STORIES MOVING DOWN. (r = 2 SHOWN)

(Refer to the illustration in Case 15.31)

An intermediate story columns fail and the slabs sequentially "pancake" below the intact part. Collision-related definitions of U_g, M_a, M^*, and E'_k/E_k remain the same as in Case 15.31. Π is the energy lost by squashing of columns by h between two stories.

Before 1st collision, $p = 2$, $M_a = (r+1)M$, $E_{k1} = (r+1)Mgh - \Pi$;

After 1st collision, $E'_{k1} = \dfrac{r+1}{r+2} E_{k1}$, $M_a = (r+2)M$

Before 2nd collision, $p = 3$, $M_a = (r+2)M$, $E_{k2} = E'_{k1} + (r+2)Mgh - \Pi$

After 2nd collision, $E'_{k2} = \dfrac{r+2}{r+3} E_{k2}$; $M_a = (r+3)M$, etc.

When the collapsing mass impacts the slab at h_1:

$n - r - 1$ impacts have taken place

$U_g = \dfrac{r+n}{2}(n-r-1)Mgh$; gravitational energy absorbed

$\Delta E_k \approx \dfrac{2U_g}{n-r+2} + (n-r-1)\Pi$; energy lost, for $n - r > 4$

$E_{kn} = U_g - \Delta E_k$; net energy before final descent to ground

$E_{kf} = E_{kn} + nMgh_1$; final kinetic energy before ground impact

$v_n = \sqrt{\dfrac{2E_{kn}}{nM}}$; $v_f = \sqrt{\dfrac{2E_{kf}}{nM}}$; velocity before and after the descent from height h_1

$S = \sqrt{(2E_{kf})nM}$; impulse applied to the ground

$t = \dfrac{2(n-r-1)h}{v_n} + \dfrac{2h_1}{v_n + v_f}$; event duration

Note: For most tall buildings at least a several stories will have the same strength. When story columns change at some level, the energy Π suddenly becomes bigger which may cause the motion to be arrested. The upper r stories remain intact as long as their strength is not exceeded by impact forces.

CASE 15.33   SPECIFIC IMPULSE *s* TO BREAK A MASONRY WALL SUPPORTING A HEAVY WEIGHT

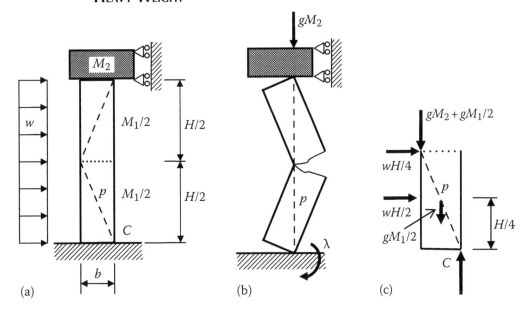

(a)   (b)   (c)

(a) Original position with the impulsive load applied
(b) Uppermost position, just prior to collapse
(c) Equilibrium of the lower half, beginning of movement

$s = wt_0$ with t_0 assumed short

$$p = \left(b^2 + \frac{H^2}{4} \right)^{1/2}; \text{ diagonal length}$$

$J_C = J + M_1 p^2/2$; moment of inertia of the lower half about the pivot point C

$$w \geq \frac{4bg}{H^2} \left(M_2 + \frac{3}{4} M_1 \right); \text{ minimum distributed load needed } \sim \text{N/m}$$

$$\Pi = (2p - H)g \left(M_2 + \frac{M_1}{2} \right); \text{ potential energy at configuration (b)}$$

$$\lambda = \frac{wH^2}{8J_C} t_0; \text{ angular velocity gained, each half}$$

$E_k \approx (J_C + 2M_2 b^2)\lambda^2$; kinetic energy due to the impulse ($E_k > \Pi$ when overturning)

$$wt_0 = \frac{8J_C}{H^2} \left[\frac{(2p - H)g \left(M_2 + \dfrac{M_1}{2} \right)}{J_C + 2M_2 b^2} \right]^{1/2}; \text{ the smallest impulse needed to overturn}$$

Note: M_2 can represent a concrete slab and a part of a building resting on it. J is the moment of inertia about the centroidal axis (normal to page) of a half of the wall. No credit for the mortar joint strength was taken. Both the load and the impulse criterion must be fulfilled to achieve failure. For a nonrectangular pulse take w as time-average. If the masonry material is weak (e.g., concrete block, about 12 MPa, not filled with concrete) then crushing of corners may take place, which will significantly reduce the wall strength.

EXAMPLES

EXAMPLE 15.1 EXPLOSIVE STRETCHING OF A CYLINDRICAL SHELL

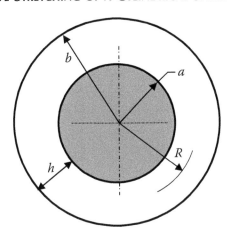

The results of high-speed tensile testing conducted on a number of steels indicate that particle velocities in stretching motion attain a range of 600–1200 m/s before breakup begins. Using those as a guide, estimate radial elongation prior to failure for a 4″ artillery shell with the following dimensions of its main section: $a = 39.2$ and $b = 52.4$ mm (or $R = 45.8$; $h_0 = 13.2$) while density $\rho = 0.00785$. The explosive is TNT, for which the initial pressure is $p_0 = 10,714$ MPa.

From Equation 15.7, which describe motion with a constant acceleration, the velocity expression can be written as

$$v = a_0 t = \frac{p_0 a_i}{\rho_m h_0 R} t = \frac{10,714 \times 39.2}{0.00785 \times 13.2 \times 45.8} t = 88,497t \ \text{(m/s when } t \sim \text{ms)}$$

This is the radial component of velocity, but it is easy to determine that the circumferential component, which is of main interest here, is the same. The lower limit of possible breaking, namely, 600 m/s is associated with time point.

$$t_1 = 600 / 88,497 = 1/147.5 \, \text{ms}$$

The distance traversed in that time is $u_1 = a_0 t_1^2 / 2 = 2.034$ mm. This corresponds to the strain of $\varepsilon_1 = u_1 / R = 0.0444$, quite a small value. Achieving the upper end of the range, namely, 1200 m/s, involves doubling of stretching time to $2/147.5$ ms. This, in turn, gives 4× larger strain at failure: $\varepsilon_2 = u_2 / R = 0.1776$, which is definitely in the ductile range.

EXAMPLE 15.2 NUMBER OF FRAGMENTS IN AN EXPLODING SHELL

A mild steel shell described as in Example 15.1, $L = 25.4$ mm long, is filled and detonated using TNT. Determine the total number of fragments and compare the calculation with one carried out in English units.

Use Case 15.1. The relevant diameters are: $d_i = 78.4$, $d_m = 91.6$ mm

First, calculate $M_0 = \pi h d_m$, $L\rho = 757.4$ g (shell mass), then

$$M_k = Bh^{5/6}d_i^{1/3}(1+h/d_i) = 0.03816 \times 13.2^{5/6}78.4^{1/3}(1+13.2/78.4) = 1.6384$$

$$N_m = \frac{M_0}{2M_k^2} = \frac{757.4}{2 \times 1.6384^2} = 141.1 \approx 141 \text{ total number of fragment}$$

Now, in English units

$$d_i = 3.09', \quad h = 0.5197', \quad \therefore d_m = 3.61', \quad L = 1.0', \text{ (length) with } \rho_c = 0.2834 \text{ lb/in.}^3,$$
it gives $M_0 = 1.67$ lb (shell mass)

$$M_k = 0.312 \times 0.5197^{5/6}3.09^{1/3}(1+0.5197/3.09) = 0.3077 \text{ lb}$$

$$N_a = \frac{8M_0}{M_k^2} = \frac{8 \times 1.67}{0.3077^2} = 141.1, \text{ the same result}$$

The length was arbitrarily selected and no assumption was made with regard to the shape of fragments. A more detailed approach to such a problem is illustrated in Example 15.5.

EXAMPLE 15.3   ENERGY ABSORBED PRIOR TO FRAGMENTATION OF A STEEL SHELL

Consider the same 4″ shell as in Example 15.1, expanding and fracturing when its inner radius grows by 30%. Assume elastic behavior up to fracture and calculate the mechanical work needed. Then assume an EPP material with a flow stress of $\sigma_0 = 1000$ MPa and again find the necessary work.

Compare those answers with the energy stored in the explosive material (TNT) filling the shell.

Take a unit wide section, therefore $A = h$. The initial r_i of the shell is 39.2 mm, so the expanded radius at fracture is $1.3 \times 39.2 = 50.96$ mm. The outer expanded radius b can be found from material preservation and by noting that the original outer radius was 52.4 mm:

$$b^2 - 50.96^2 = 52.4^2 - 39.2^2; \quad \text{hence } b = 61.69 \text{ mm.}$$

The expanded thickness: $h' = 61.69 - 50.96 = 10.73$ mm. The inner radius expands by $50.96 - 39.2 = 11.76$ mm. The new mean radius: $(61.69 + 50.96)/2 = 56.33$ mm, up from 45.8 mm.

1. The first, rough approximation is to use a small-deflection equation, Case 3.29, rewritten to express work in terms of displacement u, rather than surface loading:

$$\mathcal{L} = \frac{\pi}{R}EAu^2 = \frac{\pi}{45.8}200,000 \times 13.2 \times 11.76^2 = 25.044 \times 10^6 \text{ N-mm}$$

2. A better answer for the linear material is to use a large-deflection approach. One needs the mean radius displacement $u = 56.33 - 45.8 = 10.53$ mm and $e = u/R = 10.53/45.8 = 0.2299$. Then, following the developments in section "Large deflections of simple elements" one can write

$$\mathcal{L} = 2\pi EA_0R[e - \ln(1 + e)] = 2\pi \times 200{,}000 \times 13.2 \times 45.8[0.2299 - \ln(1.2299)]$$
$$= 17.45 \times 10^6 \text{ N-mm}$$

3. For an EPP material use Case 4.15. The expanded A is the same as expanded thickness $h' = 10.73$. The average area during the expansion process $A_{av} = (13.2 + 10.73)/2 = 10.97$ mm^2. The average mean radius $r_{av} = (45.8 + 56.33) = 51.07$ mm. The main, plastic component of work:

$$\mathcal{L}_{pl} = 2\pi R(A_0\sigma_0)\left(1 + \frac{A_{av}}{2Br_{av}}\right)\ln(r/R)$$

$$= 2\pi \times 45.8(13.2 \times 1{,}000)\left(1 + \frac{10.97}{2 \times 1 \times 51.07}\right)\ln(56.33/45.8) = 878{,}510 \text{ N-mm}$$

To check if the elastoplastic energy component is substantial, calculate u_u, the inner surface displacement when full yielding is attained by the shell (Case 4.18):

$$u_u = \frac{\sqrt{3}\sigma_0}{2E}a\left(\frac{b}{a}\right)^2 = \frac{1{,}000\sqrt{3}}{2 \times 200{,}000}39.2\left(\frac{52.4}{39.2}\right)^2 = 0.303 \text{ mm}$$

This is miniscule compared to the inside radius expansion of 11.76 mm and therefore the elastoplastic energy component can be ignored.

Taking data from Table 13.1 and referring to a cylinder of unit width, find the entire energy content Q of explosive and its useful part, Q_G:

$$Q = \pi r^2\rho q = \pi \times 39.2^2 \times 0.0016 \times 4.61 \times 10^6 = 35.61 \times 10^6 \text{ N-mm}, \quad \text{and}$$

$$Q_G = 0.7766Q = 27.65 \times 10^6 \text{ N-mm}$$

When the shell material acts like a EPP model, the expansion work is negligible compared with the available explosive energy, $\mathcal{L} \ll Q_G$, as calculated in Item 3 above. If the elastic response could last through the end of the expansion process, most of energy would be lost in overcoming metal resistance, as was found in Item 2. This would visibly reduce the kinetic energy of fragments, which is not a desirable result when designing weapons.

EXAMPLE 15.4   DURATION OF EXPLOSIVE STRETCHING OF A CYLINDRICAL SHELL

In Example 15.1 the time to the initiation of breakup t_1 was deduced from general limits of stretching speeds observed in tensile testing. A more direct approach would be to calculate a fly-off velocity v_f, as prescribed in Chapter 13, and assume that t_1 is when about 90% of

v_f is attained. Using this information as well as the gas properties, calculate t_1 for the shell of Example 15.1, when filled with (a) TNT and (b) RDX. Also find displacements, strains, and strain rates just prior to the initiation of fracture.

Use Case 13.15, and geometry of Example 15.1. Take the explosive data from Table 13.1:

$$M_m = \pi(b^2 - a^2)\rho_m = \pi(52.4^2 - 39.2^2)0.00785 = 29.82\,g\,(\text{metal})$$

$$M_e = \pi 39.2^2 \times 0.0016 = 2.459\,g\,(\text{TNT}) \quad M_e = 2.72\,g\,(\text{RDX})$$

(a) TNT: $v_f = \sqrt{\dfrac{2q_G}{\dfrac{M_m}{M_e} + \dfrac{1}{2}}} = \sqrt{\dfrac{2\times 3.58\times 10^6}{\dfrac{29.82}{2.459} + \dfrac{1}{2}}} = 753\,m/s$; fly-off speed

From Example 15.1, for this explosive: $v = 88,497\,t$. With $0.9\,v_f$ reached at t_1 or breakup, we have

$$0.9 \times 753 = 88,497\,t_1 \quad \text{or} \quad t_1 = 1/130.58\,ms$$

By Equation 15.8b, the distance traversed in that time is $u_1 = 88,497\,t_1^2/2 = 2.595\,mm$. This corresponds to the strain of $\varepsilon_1 = u_1/R = 0.0567$.
    The strain rate at the breakup initiation is $\dot\varepsilon = 0.9\,v_f/R = 0.9 \times 753/45.8 = 14.8/ms = 14,800\,1/s$

(b) RDX: $q_G = 4.28 \times 10^6$ N-mm, $v_f = 864.1$ m/s

The velocity equation will differ somewhat:

$$v = \frac{p_0 r_i}{\rho_m h_0 R}t = \frac{13,756 \times 39.2}{0.00785 \times 13.2 \times 45.8}t = 113,624\,t;$$

Then $0.9 \times 864.1 = 113,624\,t_1$ or $t_1 = 1/146.1$ ms
    Consequently, $u_1 = 113,624\,t_1^2/2 = 2.662$ mm, $\varepsilon_1 = 0.0581$.

$$\dot\varepsilon = 0.9\,v_f/R = 0.9 \times 864.1/45.8 = 16.98/ms = 16,980\,1/s$$

With this approach the predicted strains at the beginning of fracture are between 5% and 6%. The predicted duration of breakup is similar to what was found in Example 15.1.

EXAMPLE 15.5   NUMBER OF FRAGMENTS IN AN EXPLODING SHELL, A CLOSER ANALYSIS

A mild steel shell described as in Example 15.1, is filled and detonated using TNT. Assuming the shape of fragments to be approximately square determine the number of fragments to be seen in a cross section. What will be the number of sizeable pieces, in excess of 1 g in mass? Repeat the calculation for RDX. Also, find the number of fragments with $M < 1$ g using Case 15.2 with $\beta = 1.5$

Use Case 15.1. Consistent with the assumed fragment shape take $L = 2\pi R = 287.8$ mm

As in Example 15.2: $d_i = 78.4$, $h = 13.2$, $d_m = 91.6$, $d_0 = 104.8$ mm (d_0 is the outer diameter) The shell mass is now $M_0 = 8582$ g

TNT: $M_k = 1.6384$ (as in Example 15.2)

$$N_m = \frac{M_0}{2M_k^2} = \frac{8582}{2 \times 1.6384^2} = 1598.5 \approx 1599; \text{ total number of fragments}$$

In a cross section, Equation 15.6, $n_M = (N_M)^{1/2} = 40$ fragments. To get the number of pieces with $M > 1$ g use Case 15.1:

$$N(M) = N_m \exp\left(\frac{-M^{1/2}}{M_k}\right) = N_m \exp\left(\frac{-1}{M_k}\right) = 1599 \exp(-0.6104) = 868.5$$

Therefore $n_M = (868.5)^{1/2} = 29.5 \approx 30$ fragments

RDX: $M_k = 1.1133$ ($B = 0.02593$ instead of 0.03816)

$$N_m = \frac{8582}{2 \times 1.1133^2} = 3462; \quad n_M = 58.8 \approx 59 \text{ fragements}$$

$$N(1) = 3462 \exp(-0.8982) = 1410; \quad n_M \approx 38$$

To make the estimate by means of our Case 15.2, one needs the breakup time t_b. Unfortunately, only the time t_1, based on a guess, is available from Example 15.4. A more realistic assumption would be that $t_b = t_1/2$, which gives the following number of fragments:

$$\text{TNT}: t_b = t_1/2 = 3.83 \times 10^{-3} \text{ ms}; \quad n = \frac{\beta l}{c t_b} = \frac{1.5 \times 287.8}{5048 \times 3.83 \times 10^{-3}} = 22.33 \approx 22$$

where the wave speed $c = c_0 = 5048$ (Table 5.2).

$$\text{RDX}: t_b = 3.422 \times 10^{-3} \text{ ms}; \quad n = 24.9 \approx 25$$

Ignoring very small pieces by setting $M > 1$ g, the preceding numbers were 30 and 38, respectively. Our formula, with $\beta = 1.5$, gives an underestimate in a case described here, where small pieces can be generated.

EXAMPLE 15.6   CIRCUMFERENTIAL BREAKUP OF A CYLINDRICAL VESSEL

A cylindrical vessel (10 m total length including spherical ends) of 1 m in diameter, could possibly crack around the circumference and as a result one part could fly-off like a rocket. The vessel contains air with an internal pressure of 2 MPa and temperature 100°C. The wall thickness is 8 mm. Assuming the breakup to be into two equal parts, calculate the fly-off speed. Simplify the problem by using mean, rather than internal and external dimensions.

Put $L = 10 - 2 \times 0.5 = 9\,\text{m}$, length of cylindrical part

Vessel mass: $M_c = (\pi DL + 4\pi R^2)\rho h = (\pi \times 1 \times 9 + 4\pi \times 0.5^2)7850 \times 0.008 = 1973\,\text{kg}$

Volume of air: $V_0 = \pi(R^2 L + 4R^3/3) = \pi\,(0.5^2 \times 9 + 4 \times 0.5^3/3) = 7.592\,\text{m}^3$

From Case 15.3 one has for air: $\gamma = 1.4$, $R_m = 287$, while $\theta_0 = 273 + 100 = 373\text{K}$, then

$$x = \frac{(p - p_0)V_0}{M_c \gamma R_m \theta_0} = \frac{(2.0 - 0.1013)10^6 \times 7.592}{1973 \times 1.4 \times 287 \times 373} = 0.04875$$

Then $y = 1.438 x^{0.825} = 1.438 \times 0.04875^{0.825} = 0.1189$

But, $y = \dfrac{v}{\sqrt{\gamma R_m \theta_0}}$, which means $y\sqrt{\gamma R_m \theta_0} = v$, or $v = 0.1189\sqrt{1.4 \times 287 \times 373} = 46.05\,\text{m/s}$

EXAMPLE 15.7   END SHEAR FAILURE (MODE III) OF A CLAMPED–CLAMPED BEAM UNDER UNIFORM LOADING

The beam with a rectangular section, shown in the sketch, is made of mild steel of $\sigma_0 = 250\,\text{MPa}$ and $\varepsilon_u = 50\%$. The lateral pressure, as shown, is very large: $p_0 = 10.2\sigma_0$, but it lasts only $t_0 = 1\,\text{ms}$. Determine if the lateral velocity caused by this pressure impulse is large enough to cause shear failure according to methodology developed in Chapter 5. Assume the material has a small plastic modulus, $E_p = 0.01E = 2000\,\text{MPa}$. (The data comes from Horoshun [36], who performed finite element analysis (FEA) simulation with $t_0 = 4\,\text{ms}$ and obtained an almost instant shear failure.) Estimate the duration of fracture.

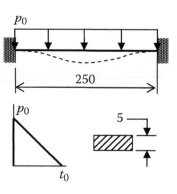

For a rectangular section the problem can be solved using the relevant quantities per unit width.

$$A = 5\,\text{mm}, \quad m = \rho A = 0.00785 \times 5 = 0.03925\,\text{g/mm}.$$

The triangular pressure diagram gives the following impulse per unit length:

$$s = 0.5 p_0 t_0 = 0.5(10.2)1.0 = 5.1\,\text{N-ms/mm (per unit width)}$$

This, results in the initial velocity imparted to the beam: $v_0 = s/m = 5.1/0.03925 = 130\,\text{m/s}$.

This can also be treated as the end velocity suddenly applied to a stationary beam, as initially explained for a cable in Chapter 9.

Material properties come from Table 5.2: $G = 76.92\,\text{Gpa}$, $c_s = 3130\,\text{m/s}$. For simplicity assume $A = A_s$, so that no correction is needed to get beam shear speed; $c_3 = c_s$.

Use Case 5.10 to determine the (initial) breaking velocity for a very ductile material, $v_u = v_e$:

$$v_u \approx c_{3p}\gamma_u$$

where $c_{3p} = c_3 \left(\dfrac{G_p}{G} \right)^{1/2} = c_3 \left(\dfrac{E_p}{E} \right)^{1/2} = 3130(0.01)^{1/2} = 313\,\text{m/s},$

From Case 4.8: $\gamma_u = \varepsilon_u \sqrt{3} = 0.5\sqrt{3} = 0.866$, then $v_u = 313 \times 0.886 = 271\,\text{m/s}$.

For a so ductile material, a larger applied speed is needed than the calculated $v_0 = 130\,\text{m/s}$. The condition is $v_0 > v_u$, for the beam to fail in shear.

To estimate duration of tearing itself, read the sonic velocity $c_0 = 5048\,\text{mm/ms}$ from Table 5.2.

The cracking velocity is, approximately, $0.23c_0 = 0.23 \times 5048 = 1161\,\text{m/s}$. The time of cracking through the 5 mm thickness is therefore $5/1161 = 0.0043\,\text{ms}$. This is an additional time needed to disintegrate during which a substantial load must persist.

EXAMPLE 15.8   BASE FAILURE OF A CANTILEVER UNDER UNIFORM LOADING

Consider a cantilever made of one-half of the beam in Example 15.7, $L = 125\,\text{mm}$, where the impulse applied was $s = 5.1\,\text{N-ms/mm}$. Using Case 15.10, which quantifies a cantilever failure in any mode, determine if there is a failure and if so, what is the fly-off velocity. Let the material be bilinear with $F_u \approx F_y + E_p\varepsilon_u = 250 + 2000 \times 0.5 = 1250\,\text{MPa}$

The first task is to calculate the material toughness, or the area under the stress–strain curve. For this bilinear material

$$\Psi \approx 0.5(F_y + F_u)\,\varepsilon_u = 0.5(250 + 1250)0.5 = 375\,\text{MPa}$$

(Refer to Case 4.3 for a more accurate expression.) Now, one can substitute into Case 15.10:

$$\left(\frac{s}{A\sqrt{\rho\Psi}} \right)\left(\frac{2L}{B} \right)^{0.3} = \left(\frac{5.1}{5\sqrt{0.00785 \times 375}} \right)\left(\frac{2 \times 125}{1.0} \right)^{0.3} = 3.116 > 0.602; \quad \text{failure}$$

$$V_f = \sqrt{\frac{375}{0.00785}}\,\{0.436 \times 3.116 - 0.2625\} = 239.5\,\text{m/s}$$

This predicts the same order of the breaking speed as v_u in Example 15.7, but the failure mode is unspecified.

EXAMPLE 15.9   MODE III FAILURE OF CLAMPED–CLAMPED BEAM

The beam is made of Al alloy 6061-T6 with $F_y = 241$, $F_u = 290\,\text{MPa}$, $\varepsilon_u = 0.09$. The load is a uniformly distributed pressure p giving a short-duration impulse. Using a frequently employed criterion, which says that shear failure is achieved when shear displacement with respect to a fixed end reaches the section depth H, find the necessary impulse to induce such failure. The

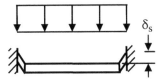

beam has a rectangular section with $H = 6.35\,mm$. Find the magnitude of a specific pressure impulse i_r to cause failure for this beam when $L = 203.2\,mm$ and for one with $L = 101.6\,mm$. Compare the results to what is obtained from Case 15.8.

As the beam is rectangular, the solution can be made "per unit width." Equation 10.55 gives shear deflection when a beam is under a uniform impulse $S = Lw_0t_0$:

$$u_s = \frac{S^2}{4MQ_0} = \frac{(Lwt_0)^2}{4(Lm)Q_0} = \frac{L(wt_0)^2}{4mQ_0}$$

when $d = 2$. Here $wt_0 = pt_0 = i_r$ designates the impulse per unit length of the beam. Noting that $m = \rho H$, $u_s = H$, as postulated and $Q_0 = \tau_u A = \tau_u H$ in the limiting condition, while $\tau_u = F_u/\sqrt{3}$, one has

$$i_r^2 = \frac{4\rho h^3 F_u}{\sqrt{3}L} = \frac{4\times0.0027\times6.35^3\times290}{203.2\sqrt{3}} \quad \text{or } i_r = 1.509\,\text{MPa-ms} = 1509\,\text{Pa-s}$$

For a shorter beam, $L = 101.6\,mm$, the result is $i_r = 2135\,\text{Pa-s}$. Unfortunately, the experimental result is $4800\,\text{Pa-s}$ *for both* beams, which shows the criterion to be quite conservative.
Compare this with the result given by Case 15.8 with $\sigma_0 = (290 + 241)/2 \approx 266\,\text{MPa}$:

$$i_r = \left(0.4385\frac{H}{B}+1.238\right)\sqrt{2\sigma_0\rho} = \left(0.4385\frac{6.35}{1.0}+1.238\right)\sqrt{2\times266\times0.0027}$$

$$= 4.821\,\text{MPa-ms}$$

which is the same as $4821\,\text{Pa-s}$. The initial velocity, that must be developed:

$$v_0 = \frac{i_r}{\rho H} = \frac{4.821}{0.0027\times6.35} = 281.2\,\text{m/s}$$

One can conclude that at least for the set of circumstances, on which Case 15.8 is based, the previous criterion is somewhat pessimistic. To balance the issue one must remember that a perfectly clamped end condition can rarely be obtained, therefore the test conditions and the criterion were not quite compatible.

EXAMPLE 15.10   BUILT-IN END BREAKING IN A BEAM UNDER A PROJECTILE IMPACT (MODE II)

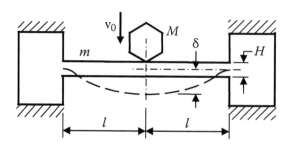

The following is a listing of data and results of some experiments described in Ref. [56].

A rectangular-section beam with enlarged ends, as shown in the sketch, is impacted at center with a projectile of $M = 5\,kg$, flying with $v_0 = 5.73\,m/s$. Beam width, normal to paper, is 10.16 mm. The beam tears off at the supports when its center deflection reaches $\delta = 14.8\,mm$. The material is Al alloy with the average flow strength $\sigma_0 = 415\,MPa$.

Use this result to find a formula for energy-absorbing capability of a beam of this thickness and material. *Hint:* Note that in cases describing a rigid cantilever with a rigid-plastic resistance $\mathcal{M}_0$ at the base, Chapter 10, the strain energy absorbed is $\mathcal{M}_0\theta$, where θ is the angle of rotation.

Also, consider another beam of the same material and ad dimensions, except 7.62 mm thick. This one breaks when impacted by the striker flying with $v_0 = 6.66\,m/s$ when the center deflection is 15.77 mm. Repeat the calculation and find how the strain energy absorbed prior to failure relates to section depth H.

Beam data: $\mathcal{M}_0 = BH^2\sigma_0/4 = 10.16 \times 6.35^2 \times 415/4 = 42{,}503\,N\text{-}mm$

$E_k = Mv_0^2/2 = 5{,}000 \times 5.73^2/2 = 82{,}082\,N\text{-}mm$; kinetic energy of striker

Using the result for a rigid-plastic cantilever as a pattern, anticipate the strain energy in this case as $F\mathcal{M}_0\theta$, where F is an unknown coefficient. As a measure of rotation θ for moderate deflection, one can employ $\theta \approx \delta/l$:

$$\Pi_f = F\mathcal{M}_0\theta = 42{,}503(14.8/50.8)F = 12{,}383F$$

This can be equated with E_k to obtain the unknown coefficient:

$$12{,}383F = 82{,}082 \quad \text{or} \quad F = 6.629$$

For the second beam: $\mathcal{M}_0 = 10.16 \times 7.62^2 \times 415/4 = 61{,}205\,N\text{-}mm$

$E_k = 5{,}000 \times 6.66^2/2 = 110{,}890\,N\text{-}mm$; kinetic energy of striker

$$\Pi_f = F\mathcal{M}_0\theta = 61{,}205\,\frac{15.77}{50.8}\,F = 19{,}000F$$

$$19{,}000F = 110{,}890 \quad \text{or} \quad F = 5.836$$

If we now postulate that Π_f is proportional to $(1/H)^\alpha$, where α is the exponent to be determined from the above numbers, we find that

$$\Pi_f = \frac{22.136}{H^{0.6523}}\mathcal{M}_0\theta$$

gives the correct value of the coefficient F for both cases. One should keep in mind that the above is limited to the beams made of the material under consideration, which ruptures at $\varepsilon_u = 0.19$.

EXAMPLE 15.11   MODE IV FAILURE OF AN ALUMINUM BEAM (SHEAR PLUGGING)

The beam is made of the Al alloy described in Case 15.9, with the shock wave speed of $U = 7000\,m/s$. It is impacted by a striker with $d = 6.35\,mm$, $H_p = 20\,mm$ in plane, and $B = 10\,mm$, normal to the page and the same width as the beam. Find the ballistic speed v_b and an approximate cracking speed of the beam, which has rectangular section with $H = 6.35\,\text{mm}$.

$M = 6.35 \times 20 \times 10 \times 0.00785 = 9.97$ g, striker mass

$M_t = BH\rho(d + 1.73H) = 10 \times 6.35 \times 0.0027(6.35 + 1.73 \times 6.35) = 2.97\,\text{g}$; effective target mass

$\mu = M_t/M = 2.97/9.97 = 0.2981$

$V_0 = \dfrac{v_0}{1+\mu} = \dfrac{v_0}{1+0.2981} = 0.7703v_0$; common velocity at the end of Stage 1.

The important material data from Case 15.9:

$K = 10$, $\lambda = 4$, $c_{pp} = 458\,\text{m/s}$, then

$$u_{cr} = K^{-1/\lambda}\frac{HV_0}{c_p} = 10^{-1/4}\frac{6.35 \times 0.7703v_0}{458} = \frac{v_0}{93.63}$$

The ballistic limit is attained when $V_r = 0$ or $v_0 = v_b$:

$$\left(\frac{V_0}{c_{pp}}\right)^2 = \frac{4}{\sqrt{3}}\frac{\mu}{1+\mu}\frac{\lambda}{1+\lambda}\frac{u_{cr}}{d} = \frac{4}{\sqrt{3}}\frac{0.2981}{1+0.2981}\left(\frac{4}{1+4}\right)\frac{v_b}{128.3 \times 6.35} = \frac{v_b}{1920}$$

This can be further expressed as

$$\left(\frac{0.7703v_b}{458}\right)^2 = \frac{v_b}{1920} \quad \text{or} \quad \left(\frac{0.7703}{458}\right)^2 v_b = \frac{1}{1920} \quad \text{or} \quad v_b = 184.1\,\text{m/s}$$

and the common velocity is $V_0 = 0.7703 \times 184.1 = 141.8\,\text{m/s}$

To determine the cracking speed, u_{cr} is needed:

$$u_{cr} = \frac{v_0}{93.63} = \frac{184.1}{93.63} = 1.966\,\text{mm};\ \text{depth gained during cracking}$$

When $V_r = 0$, the average cracking speed is

$$c_r \approx \frac{H}{u_{cr}}\frac{V_0}{2} = \frac{6.35}{1.966}\frac{141.8}{2} = 229\,\text{m/s}$$

EXAMPLE 15.12   BUILT-IN END BREAKING IN A STEEL BEAM UNDER PROJECTILE IMPACT (MODE II)

The formula for the energy absorbed at failure Π_f, derived in Example 15.10 is valid for an Al alloy with the ultimate strain of $\varepsilon_u = 0.19$. Generalize this to be representative of any ductile material.

Check, if the following event, reported in Ref. [56] is consistent with the expression derived:

A beam with enlarged ends, as shown and described in Example 15.10, is impacted at center with a projectile of $M = 5\,\text{kg}$, flying with $v_0 = 8.88\,\text{m/s}$. Beam properties are as before, except the material is now steel, with $\varepsilon_u = 0.31$ and $\sigma_0 = 448\,\text{MPa}$. There is no failure and the permanent center deflection reaches $\delta = 16.19\,\text{mm}$.

$M_0 = BH^2\sigma_0/4 = 10.16 \times 6.35^2 \times 448/4 = 45{,}884\,\text{N-mm}$; moment capacity
One can assume that the energy absorption capacity is directly proportional to the ultimate strain. If so, one can simply multiply the previous expression by $\varepsilon_u/0.19$, obtaining

$$\Pi_f = \frac{116.5\varepsilon_u}{H^{0.6523}} M_0 \theta$$

$E_k = Mv_0^2/2 = 5{,}000 \times 8.88^2/2 = 197{,}136\,\text{N-mm}$; kinetic energy of striker
Suppose that the entire kinetic energy is absorbed and failure results. The kink angle θ_f that can be reached is then found from

$$197{,}136 = \frac{116.5 \times 0.31}{6.35^{0.6523}} 45{,}884\theta_f \quad \text{or} \quad \theta_f = 0.3972\,\text{rad}$$

The angle that was actually reached was $\theta \approx \delta/l = 16.19/50.8 = 0.3188$. This means that failure was not attained, which is consistent with the experimental finding.

EXAMPLE 15.13   CANTILEVER TIP BREAK OFF DUE TO STRIKER IMPACT. STRONGE'S METHOD

A cantilever beam $L = 305\,\text{mm}$ and a $6.35\,\text{mm} \times 6.35\,\text{mm}$ cross section has its tip impacted by a striker of mass $M = 10\,\text{g}$. The striker is a block with $H = 9\,\text{mm}$. Find the impact velocity v_0, which causes tip failure, according to Case 15.11. The material is steel, treated as rigid-plastic with $\sigma_0 = 400\,\text{MPa}$. (The data comes from the experiments by Parkes [65], except σ_0, which is assumed.)

For the square section, $m = \rho A = 0.00785 \times 6.35^2 = 0.3165\,\text{g/mm}$

$$M_0 = Z\sigma_0 = (h^3/4)\sigma_0 = (6.35^3/4)\,400 = 25{,}605\,\text{N-mm}$$

$$\alpha = H/h = 9/6.35 = 1.417$$

Find the kinetic energy ratio ε_f to achieve failure:

(a) Pure shear failure

$$e_f = 2 + \frac{8}{3}\alpha^2 = 7.357$$

(b) Shear–bending interaction at failure

$$\mu \approx 0.808\alpha - 1.141 = 0.004$$

$$e_f = 5 + \frac{3\mu}{\alpha} + \frac{8\alpha^2}{3(1+\mu)} = 5 + \frac{3\times0.004}{1.417} + \frac{8\times1.417^2}{3(1+0.004)} = 10.342$$

Clearly, mode (a) dominates by giving smaller e_f. Since $e_f = \dfrac{Mv_0^2}{2\mathcal{M}_0}$, then

$$v_0^2 = \frac{2\mathcal{M}_0 e_f}{M} = \frac{2\times25,605\times7.357}{10}; \quad \text{or} \quad v_0 = 194.1\,\text{m/s}$$

EXAMPLE 15.14 CANTILEVER TIP BREAK OFF DUE TO STRIKER IMPACT. EXPERIMENT AND THEORY

In experiments reported by Stronge [84], the following strain energy was found to be associated with failure:

$$D_a = 1.54\left(\frac{1}{2}F_y AH\right); \text{Al alloy}, \quad D_a = 1.04\left(\frac{1}{2}F_y AH\right); \text{Mild steel}$$

where H and A was the section depth and area, respectively, of a square bar. The basic material data are

$$F_y = 115, \quad F_u = 320\,\text{MPa}, \quad \varepsilon_u = 0.49, \text{Al alloy}$$

$$F_y = 245, \quad F_u = 485\,\text{MPa}, \quad \varepsilon_u = 0.90, \text{steel}$$

Estimate the deflection at failure taking into account only the shear mode.

From Case 15.12, the deformation energy to bring the beam to an onset of cracking is

$$\Pi_{br} \approx \frac{1}{2}A\Psi_s\eta = \frac{1}{\gamma_u}A\Psi_s\delta$$

When this is equated to D_a, one gets, with the help of Case 4.8, treating the material as RHS:

$$\frac{\delta}{H} = \alpha\frac{\gamma_u F_y}{2}\frac{1}{\Psi_s} = \alpha\frac{\sqrt{3}F_y}{F_y + F_u}$$

where $\alpha = 1.54$ for Al and $\alpha = 1.04$ for steel. For both materials, exhibiting large strains, an RSH material approximation is quite satisfactory. Using material properties listed here and $F_y = \sqrt{3}F_s$:

$$\frac{\sqrt{3}F_y}{F_y + F_u} = \frac{\sqrt{3}\times115}{115+320} = 0.4579 \text{ for Al alloy and 0.5813 for steel.}$$

$$\frac{\delta}{H} = \alpha \frac{\sqrt{3}F_y}{F_y + F_u} = 1.54 \times 0.4579 = 0.7052 \text{ for Al alloy and } 0.6046 \text{ for steel.}$$

In hand analyses it is often assumed that the deflection associated with failure is close to the section depth H. The above confirms, that it is indeed the order of magnitude, although the deformation is spread over a segment η, rather than being strictly localized. Another observation made after testing was the value of δ/H, just prior to cracking, in the thickness range $5\,\text{mm} < H < 40\,\text{mm}$:

$$\frac{\delta}{H} = 0.16 + \frac{1.19}{H} \text{ for Al alloy}$$

$$\frac{\delta}{H} = 0.27 + \frac{0.94}{H} \text{ for mild steel } (H \sim \text{mm})$$

The above gives the following results, for selected $H = 10\,\text{mm}$ and $H = 25\,\text{mm}$, with the "theoretical" values being those previously derived.

Experimental vs. theoretical δ/H, for selected thicknesses

	$H = 10\,\text{mm}$	$H = 25$	Theory
Al alloy	0.279	0.208	0.705
Mild steel	0.364	0.308	0.605

The conclusion is simple: The deflection at failure, as predicted by shear deformation alone, is far too large. The bending mode, which is activated soon after the contact is made, acts simultaneously to break the beam and makes the resultant deflection at failure much smaller, than predicted by the shear mode alone.

EXAMPLE 15.15   ANFO-FILLED CYLINDRICAL CAVITY IN GRANITE, CC MODE OF BREAKING

Using ANFO(1) from Table 13.1 quantify the breakup around a cylindrical bore-hole, ø102 mm. Granite properties: $E = 61{,}580\,\text{MPa}$, $v = 0.22$, $F'_{cc} = 8 \times 180\,\text{MPa}$, $F_{cc} = 3 \times 180\,\text{MPa}$, $F_t = 18\,\text{MPa}$. Repeat calculations for a spherical charge with the same radius.

Follow Case 15.16. The explosive properties are $p_0 = 2765\,\text{MPa}$, $q = 3.72 \times 10^6\,\text{N-mm}$, and $n = 1.929$. With radius $r_0 = 51\,\text{mm}$, the energy, per unit depth of hole (Case 13.1) is:

$$Q = Q_h = \pi \rho r_0^2 q = \pi \times 0.0008 \times 51^2 \times 3.72 \times 10^6 = 24.32 \times 10^6 \text{ N-mm/mm}$$

$$u = \left[\left(\frac{p_h}{F_{cc}} \right)^{\frac{1}{2n}} - 1 \right] r_0 = \left[\left(\frac{2765}{540} \right)^{\frac{1}{2n}} - 1 \right] 51 = 26.88\,\text{mm; wall displacement}$$

$$a^2 = \frac{Q_h}{2\pi F'_{cc}} + r_0^2 = \frac{24.32\times10^6}{2\pi\times1440} + 51^2; \quad a = 72.73\,\text{mm}; \text{crushed zone radius}$$

$$u_p = \sqrt{(r_0+u)^2 + a^2 - r_0^2} - a = \sqrt{(51+26.88)^2 + 72.73^2 - 51^2} - 72.93 = 20.83\,\text{mm}$$

$$R = \frac{F_{cc}}{F_t}(a+u_p) = \frac{540}{18}(72.73+20.83) = 2807\,\text{mm}; \text{cracked zone radius}$$

$$u_c = \frac{1+v}{E}F_t R = \frac{1.22}{61,580}18\times2807 = 1.0\,\text{mm}; \text{displacement at } R$$

When Case 15.17 for a spherical cavity is employed, the following is obtained:

$$u = 16.63\,\text{mm}, \quad a = 51.24\,\text{mm}, \quad u_p = 16.5\,\text{mm}, \quad R = 261.9\,\text{mm}, \quad \text{and} \quad u_c = 0.093\,\text{mm}$$

The thickness of the crushed zone, $a - r_0 = 0.24\,\text{mm}$, is so small that cracking mode is more likely for the spherical charge.

EXAMPLE 15.16   ANFO-FILLED CYLINDRICAL CAVITY IN GRANITE, CRACKING MODE

Using the same data as in Example 15.15 perform the calculation for the cracking mode and compare results.

Follow Case 15.14. The explosive properties are: $\rho = 0.0008$, $q = 3.72\times10^6$ N-mm, and $n = 1.929$. The effective pressure: $p_{ef} = p_0/2 = 1383\,\text{MPa}$

$$R^2 = \frac{Ea^2}{4F_t}\left[\left(\frac{p_{ef}}{F_t}\right)^{\frac{1}{n}} - 1\right] = \frac{61,580\times51^2}{4\times18}\left\{\left(\frac{1,383}{18}\right)^{\frac{1}{1.929}} - 1\right\}; \quad R = 4,346\,\text{mm}$$

$$u_c = \frac{(1+v)}{E}F_t R = \frac{(1+0.22)}{61,580}18\times4,346 = 1.55\,\text{mm}; \text{displacement at } R$$

$$u = u_c + \frac{F_t a}{E}\left(\frac{R}{a} - 1\right) = 1.55 + \frac{18\times51}{61,580}\left(\frac{4,346}{51} - 1\right) = 2.81\,\text{mm}; \text{wall displacement}$$

When the spherical cavity Case 15.15 is employed, the following is obtained:

$$R = 223.7\,\text{mm}; \quad u_c = 0.08\,\text{mm}; \quad \text{and} \quad u = 0.181\,\text{mm}$$

For a cylindrical cavity, the cracked radius R is considerably larger in the cracking mode.

EXAMPLE 15.17 CRUSHED ZONE RESULTING FROM BLASTING IN COAL

A strong mining charge (emulsion) in a ø160 mm borehole resulted in the crushed radius of between 760 and 830 mm, according to Esen [23] (two experiments). Coal had static unconfined strength F_c = 20 MPa. The explosive properties reported were $\rho = 1.2$ g/cm³, VOD $\equiv D$ = 5364 m/s and the borehole pressure of p_0 = 6878 MPa. Determine the effective compressive strength F'_{cc} in accordance with Equation 15.31 and the applicability of our theory to such test results.

As the hole is completely filled, $r_e = r_0 = 80$ mm and $a = 760$ mm in the first test. From Equation 13.1:

$$n+1 = \frac{\rho D^2}{2p_0} = \frac{0.0012 \times 5364^2}{2 \times 6878}; \quad \text{or} \quad n = 1.51$$

Then, from Equation 20.1a: $Q = \dfrac{\pi p_0 r_0^2}{n-1} = \dfrac{\pi \times 6878 \times 80^2}{0.51} = 271.1 \times 10^6$ N-mm

No specific energy content q was stated in the data provided. To make certain that the information is consistent, check on q using Equation 13.2:

$$q = \frac{p_0}{\rho(n-1)} = \frac{6878}{0.0012 \times 0.51} = 11.24 \times 10^6 \text{ N/mm}^3$$

A look in Table 13.1 shows that there are no explosives with such a large q. It seems from the Reference that $q = 4.2 \times 10^6$ N/mm² is the most likely the energy output of this explosive type. Evidently, the reference meant $p_{CJ} = 2p_0$ and not p_0, or borehole pressure. The calculation is now repeated using $q = 4.2 \times 10^6$ N/mm² and the stated value of ρ.

$$Q = \pi \rho r_0^2 q = \pi \times 0.0012 \times 80^2 \times 4.2 \times 10^6 = 101.3 \times 10^6 \text{ N-mm}$$

From Equation 15.31, with $Q_h = Q$, the ratio $k = F_{cc}/F_c$ is

$$k = \frac{Q}{2\pi F_c(a^2 - r_0^2)} = \frac{101.3 \times 10^6}{2\pi 20(760^2 - 80^2)} = 1.411$$

Using $a = 830$ mm, one gets $k = 1.181$. The average value is therefore $k \approx 1.3$ and $F_{cc} = 1.3 \times 20 = 26$ MPa. This is a very low k value, which could be explained by poor confinement and by gas pockets in the immediate vicinity of boreholes. Our theory does not apply under such circumstances.

The amount of data in papers reporting test results is usually insufficient to allow one to make a computer simulation. But this was not a problem here; it was a terminology issue. A cross-check of data helps to guard against this source of errors.

**EXAMPLE 15.18   BLOCKS OF BRITTLE MATERIAL WITH INTERNAL PRESSURE,
                   CRACKING MODE ASSUMED**

A cylindrical block with $E = 60,000\,\text{MPa}$, the tensile strength of $F_t = 10\,\text{MPa}$ has $50\,\text{mm}$ internal and $150\,\text{mm}$ external radius. The length is $H = 200\,\text{mm}$. Find the initial cracking pressure, peak static load it can resist and how far the cracking radius can advance before failure. For polytropic gas expansion assume $n = 1.5$. Find corresponding numbers for a spherical boulder with the same radial dimensions.

Use Case 15.18.

$$p_h = \frac{1-(a/b)^2}{1+(a/b)^2}F_t = \frac{1-(50/150)^2}{1+(50/150)^2}10.0 = 8\,\text{MPa}; \text{ pressure at onset of cracking}$$

$$R = 0.486b = 0.486 \times 150 = 72.9\,\text{mm}; \text{ cracked radius at peak load}$$

$$p_{br} = 0.6173F_t = 6.17\,\text{MPa}; \text{ breaking pressure, after expansion}$$

$$W_m = 0.6bHF_t = 0.6 \times 150 \times 200 \times 10 = 180,000\,\text{N}; \text{ corresponding global strength}$$

$$p_{hu} = 2p_{br}\left(1+0.545\frac{b^2 F_t}{a^2 E}\right)^n = 2 \times 6.17\left(1+0.545\frac{150^2 \times 10}{50^2 \times 60,000}\right)^{1.5} = 12.37\,\text{MPa}; \text{ initial hole}$$

pressure needed for breakage. The second term is not much larger than unity, because the ratio of radii is relatively small. Cracking will therefore start at higher pressure than the failure, but the resultant load is, of course, much smaller at the initiation of cracking.

It is easy to check that a spherical block with the same radial dimensions will start cracking at $p_c = 17.93\,\text{MPa}$ $(140,830\,\text{N})$ and can withstand $p_{br} = 10\,\text{MPa}$. $(280,620\,\text{N})$, but needs $\sim 20\,\text{MPa}$ of the initial pressure for breakage.

EXAMPLE 15.19   BREAKING A MORTAR BLOCK BY AN EXPLOSIVE CHARGE

The test described below was conducted using mortar blocks resembling cubes, with the shortest side of $280\,\text{mm}$, compressive strength of $F_c = 48.95\,\text{MPa}$ and tensile strength of $F_t = 2.62\,\text{MPa}$. A $\varnothing 9\,\text{mm}$ hole is drilled in the top face and $\varnothing 6.2\,\text{mm}$ charge of tetryl inserted in that hole. The hole is stemmed and the charge is detonated. The crushed zone is measured to be $\varnothing 14.2\,\text{mm}$. Estimate the dynamic, confined crushing strength F_{cc} of the mortar. Also, treating the block as a cylinder with $b = 140\,\text{mm}$ radius, calculate the breaking strength W_m in CC mode of failure.

As radii are easier to handle, we have $r_e = 3.1$, $r_0 = 4.5$, and $a = 7.1\,\text{mm}$. For tetryl, Table 13.1 gives q and ρ, which then yields the energy content per unit length, Case 13.1:

$$Q = \pi\rho r_0^2 q = \pi\,0.00162 \times 3.1^2 \times 4.51 \times 10^6 = 220,580\,\text{N-mm}$$

The effective energy for the decoupled configuration is, per Equation 15.30a:

$$Q_h = \frac{Q}{2}\left(1 + \frac{r_e^2}{r_0^2}\right) = 0.7323Q = 162,630\,\text{N-mm}$$

Equation 15.31 is used, suitably rewritten, to show the ratio $k = F_{cc}'/F_c$:

$$k = \frac{Q_h}{2\pi F_c(a^2 - r_0^2)} = \frac{162,630}{2\pi 48.95(7.1^2 - 4.5^2)} = 17.53$$

The strength of the cylindrical boulder is found from Case 15.20:

$$W_m = 0.6bF_t = 0.6 \times 140 \times 2.62 = 220.1\,\text{N/mm}$$

The explosive is of military type, so $F_{cc} = 6F_c = 293.7\,\text{MPa}$ should be used as the rock resistance. The first estimate of the explosive force in CC mode:

$$W_e' = 2aF_{cc} = 2 \times 7.1 \times 293.7 = 4171\,\text{N/mm};\ (\text{an underestimate})$$

$W_e'/W_m \approx 19$; which is more than sufficient to break the boulder.

At this point one may notice that the above would be meaningful only if the charged cavity extended over most of the depth of the boulder, 300 mm. This is not the case, however, because the charge is only 25 mm deep. Consequently, the calculation merely presents a methodology. A more sensible approximation is that of a spherical boulder with a spherical charge inside. This is done in Example 15.20.

EXAMPLE 15.20   EXPLOSIVE BREAKUP OF A CONCRETE BLOCK

As already mentioned in Example 15.19, a spherical cavity in a spherical boulder is a realistic approach to the problem posed there. The spherical charge must be enclosed in a somewhat larger spherical cavity and the block will be treated as a sphere of ø280 mm. Estimate of the initial explosive pressure to break the block. Consider both failure modes.

According to Table 13.1, the explosive data for tetryl are: $p_0 = 12,461\,\text{MPa}$ and $n = 2.706$.

The energy per unit length from Example 15.19: $Q = 220,580\,\text{N-mm}$.

The total energy content of a charge 25 mm long: $Q = 220,580 \times 25 = 5.515 \times 10^6\,\text{N-mm}$.

The volume of the charge is $V_e = \pi r_e^2 h = \pi 3.1^2 \times 25 = 754.8\,\text{mm}^3$. A sphere of the same volume has the radius of

$$r_e' = \left(\frac{3V_e}{4\pi}\right)^{1/3} = 5.648\,\text{mm}$$

The cylindrical cavity with $r_0 = 4.5\,\text{mm}$ and the same length gives $V_0 = 1590.4\,\text{mm}^3$ and results in $r_0' = 7.241\,\text{mm}$, the equivalent spherical cavity wall radius. The experimental setup fits somewhere between a spherical and a cylindrical charge. The thickness of the crushed zone measured was, in effect, $7.1 - 4.5 = 2.6\,mm$. When this is applied to our spherical model, the effective cavity radius becomes $a = 7.241 + 2.6 = 9.841$ mm.

Find the Young's modulus from Equation 4.7:

$$E = 20 + 0.25F_c = 20 + 0.25 \times 48.95 = 32.24\,GPa$$

The initial hole pressure is reduced by decoupling according to Equation 15.30b:

$$Q_h = \frac{Q}{2}\left(1 + \frac{V_e}{V_0}\right) = \frac{5.515 \times 10^6}{2}\left(1 + \frac{754.8}{1590.4}\right) = 4.066 \times 10^6$$

The initial pressure is therefore reduced to $p_h = p_0(Q_h/Q) = 12{,}461(4.066/5.515) = 9{,}188\,MPa$
For the *cracking mode*, use Case 15.19.

$R_m = 0.63b = 0.63 \times 140 = 88.2$ mm; the largest cracking radius prior to failure

$W_m = 0.397\pi b^2\ F_t = 0.397\pi \times 140^2 \times 2.62 = 64{,}050\,N$; tensile failure strength

$p_{br} = F_t$; pressure breaking boulder, uniform inside, in cracked condition

$$p_{hu} = 2p_{br}\left(1 + 0.75\frac{b^3 F_t}{a^3 E}\right)^n = 2 \times 2.62\left(1 + 0.75\frac{140^3 \times 2.62}{4.5^3 \times 32{,}240}\right)^{2.706} = 87.92\,MPa;\ \text{initial hole}$$

pressure needed for breakage. The ratio of the applied pressure to breaking pressure: $9188/87.92 = 104.5$; two orders of magnitude larger.

Consider now the *CC mode*, according to Case 15.21. The boulder strength remains the same, but the applied explosive force is different.

$$p_h = \left(\frac{R}{r_0}\right)^2 F_t = \left(\frac{88.2}{4.5}\right)^2 2.62 = 1006.5\,MPa;\ \text{hole pressure to break boulder}$$

This time the ratio of the applied pressure to breaking pressure is $9188/1006.5 = 9.13$, an order of magnitude larger than necessary minimum.

The above clearly demonstrates that the initial hole pressure needed to break a boulder is much smaller in cracking than in CC mode. Stronger rocks are more likely to be in cracking mode, as less powder is created in an explosion. This explains a mining paradox, according to which it takes a smaller charge to break a boulder of stronger material.

EXAMPLE 15.21   SUBSURFACE CRATERING EXPLOSION

A 6.8 kg charge of TNT is placed 1.5 m below the surface of wet sand. Calculate crater parameters per Case 15.24 and compare the results with those offered by Conwep.

TNT has a larger specific energy content than the explosive on which the table is based, namely, 4.61 kJ/kg, according to Table 13.1. The equivalent mass of ammonite is therefore

$$M_c = 6.8 \times 4.61/4.19 = 7.48 \, \text{kg}$$

For sand $k_2 = 0$, so the expression for the charge mass is

$$M_c = (k_3 h^3 + k_4 h^4) \left(\frac{\alpha^2 + 1}{2} \right)^2 = (1.45 h^3 + 0.026 h^4) \left(\frac{\alpha^2 + 1}{2} \right)^2 ; \text{ or, with } h = 1.5 \, \text{m}:$$

$$7.48 = 5.025 \left(\frac{\alpha^2 + 1}{2} \right)^2 ; \text{ Then } \alpha = 1.2$$

$r = \alpha h = 1.2 \times 1.5 = 1.8 \, \text{m}$; crater radius

$h' = \dfrac{r}{2} = 0.9 \, \text{m}$, apparent crater depth in soil, for a high-brisance explosive

$V' = 1.2 h' r^2 = 1.2 \times 0.9 \times 1.8^2 = 3.5 \, \text{m}^3$; apparent crater volume

The Conwep results are as follows:
  Total depth: 1.801 m (includes expanded zone around explosion center), $r = 1.89 \, \text{m}$,

$$h' = 1.231 \, \text{m}, \quad r' = 1.66 \, \text{m}, \quad V' = 5.305 \, \text{m}^3$$

A substantial difference in the apparent crater results may be, at least in part, related to a loose definition of a soil involved.

EXAMPLE 15.22   ANFO-FILLED CYLINDRICAL CAVITY IN WET SAND

The problem is the same as in Example 15.15, except the medium is now wet sand, 1800 kg/m³ and the cylindrical charge is placed at the average depth of 15 m. Find the typical cavity radius after detonation of ø102 mm cylindrical ANFO charge.

The ambient pressure is $p_0 = 0.1013 \, \text{MPa}$ and the hydrostatic pressure at $h = 15 \, \text{m}$ is $p_{hyd} = \rho g h = 0.0018 \times 0.00981 \times 15,000 = 0.2649 \, \text{MPa}$. For this explosive: $p_h = 2,765 \, \text{MPa}$ and $n = 1.929$.

Now, follow Equation 15.46: $R = r_0 \left(\dfrac{p_h}{p_0 + p_{hyd}} \right)^{\frac{1}{2n}} = 51 \left(\dfrac{2765}{0.1013 + 0.2649} \right)^{\frac{1}{2 \times 1.929}} = 516 \, \text{mm}.$

This compares with the size of crushed zone of ~73 mm in granite. If the shot were executed closer to the surface, so that a crater could arise, one would expect the cavity radius on the lower side to be equal to R. However, this would be obscured by an apparent crater.

EXAMPLE 15.23   PROGRESSIVE COLLAPSE OF A CONCRETE BUILDING

Consider again the building previously analyzed in Example 2.6. As a result of a strong, vertical ground movement the roof becomes detached and falls on the slab below creating a cascading effect. Assume the columns to be frangible, i.e., disintegrating on impact without absorption of any significant energy. Find the speed of impact of the assembly against the ground, the associated impulse and duration of the event. The squashing height of each floor is $h = 2.7$ m, except the ground floor, for which we have $h_1 = 4.4$ m. The mass of each story is $M = 103,450$ kg.

This is, effectively, the elevation rotated by 90°.

Refer to Case 15.27.

After the lowest slab is impacted by the mass falling from above, one has $p=6$ and

$$U_g = \frac{p(p-1)}{2} Mgh = \frac{6(6-1)}{2} Mgh = 15Mgh$$ would be the kinetic energy in absence of internal impact losses. After the losses, the actual kinetic energy is then the initial

$$E_{k6} \approx \left(1 - \frac{2}{p}\right)U_g = \left(1 - \frac{2}{6}\right)15Mgh = 10Mgh$$ velocity of the slab group is

$$v_6 = \sqrt{\frac{2E_{k6}}{nM}} = \sqrt{\frac{2 \times 10gh}{6}} = \sqrt{\frac{2 \times 10 \times 9.81 \times 2.7}{6}} = 9.4 \text{ m/s}$$

After adding to the above the potential energy of the lowest story:

$$E_{kf} = E_k + nMgh_1 = (10h + 6h_1)Mg = (10 \times 2.7 + 6 \times 4.4)Mg = 53.4Mg$$

The velocity of impact against ground:

$$v_f = \sqrt{\frac{2E_{kf}}{nM}} = \sqrt{\frac{2 \times 53.4Mg}{nM}} = \sqrt{\frac{2 \times 53.4 \times 9.81}{6}} = 13.21 \text{ m/s}; \text{ then } S = (6M)\, v_f = (6 \times 103,450)$$

$13.21 = 8.2 \times 10^6$ kg-m/s is the impulse applied to ground.

$$t = \frac{2(n-1)h}{v_n} + \frac{2h_1}{v_n + v_f} = \frac{2(5)2.7}{9.4} + \frac{2 \times 4.4}{9.4 + 13.21} = 3.26\,\text{s}; \text{ event duration}$$

For the sake of comparison note that a free fall from $H' = 23$ m would result in the impact velocity of 21.2 m/s and would last 2.165 s. This, however, is somewhat misleading, because it is the unobstructed height $H = 5 \times 2.7 + 4.4 = 17.9$ m that matters. For that height $v_f = 18.74$ m/s and $t = 1.91$ s. It is evident that the series of impacts on the way has a definite slowing effect.

Example 15.24   Progressive Collapse with Column Ductility Involved

The building is the same as in Example 15.23. Consider ground-floor initiation of collapse. The squashing height of the ground floor is $h_1 = 4.4\,$m. Compare the first impact velocity of two configurations: One with frangible columns and one where column-squashing energy is 40% of the potential energy.

Refer to Case 15.29 dealing with frangible columns.

$v_1 = \sqrt{2gh_1} = \sqrt{2\times 9.81 \times 4.4} = 9.29\,$m/s   is the first impact speed.

According to Case 15.30, the kinetic energy of the falling building is

$$E_{k1} = nMgh_1 - \Pi_1 = \left(1 - \frac{\Pi_1}{nMgh_1}\right)nMgh_1 = (1-0.4)6Mgh_1 = 3.6Mgh_1$$

which gives the impact speed of

$$v_1 = \sqrt{\frac{2E_{k1}}{nM}} = \sqrt{\frac{2\times 3.6Mgh_1}{6M}} = \sqrt{1.2\times 9.81 \times 4.4} = 7.20\,\text{m/s}$$

This ratio of velocities could be anticipated, as the impact speed is a square root of kinetic energy. Still, the impact effects, in terms of column forces above the squashed floor, are velocity-proportional. The frangible column concept may therefore be used as the first, conservative estimate of impact speeds involved.

16 Selected Examples

THEORETICAL OUTLINE

The examples presented here, while useful, would not fit well in other chapters, because they are either too long, or not of great interest to some readers.

EXAMPLE 16.1 ELASTIC BEAM UNDER SUDDENLY APPLIED, DISTRIBUTED LOAD

Find the dynamic response to a load $w_0 = 0.15$ N/mm, applied as a rectangular impulse over $t_0 = 0.35$ ms. The beam is a cantilever (CF), a slender aluminum rod with a round solid section and with properties specified below. (The units are, as usually, g, mm, and ms.)

$$L = 100 \, \text{mm}, \, \phi 3.18$$

$$E = 69,000 \, \text{MPa}, \quad \rho = 0.0027 \, \text{g/mm}^3, \quad \nu = 0.3, \quad G = 26,540 \, \text{MPa}$$

$$A = 7.942 \, \text{mm}^2, \quad I = 5.02 \, \text{mm}^4, \quad A_s = 0.5A, \quad m = \rho A = 0.02144 \, \text{g/mm} \, (EI/L^3 = 0.34638)$$

The beam is slender enough to be able to ignore shear deformation.

Follow Case 10.4. The tip static deflection is

$$u_{st} = \frac{1}{8} \frac{w_0 L^4}{EI} = \frac{1}{8} \frac{0.15 \times 100^4}{69,000 \times 5.02} = 5.413 \, \text{mm}$$

The circular frequency (Example 10.2): $\omega = 1.4132$ rad/ms.
The static shear force and the bending moments at the base are

$$Q = w_0 L = 15 \, \text{N} \quad \mathcal{M} = 0.5 \, w_0 L^2 = 750 \, \text{N-mm}$$

The resultant force: $W_0 = 0.15 \times 100 = 15$ N, then

$$u_d = 2u_{st} \sin(\omega t_0/2) = 2 \times 5.413 \times \sin(1.4132 \times 0.35/2)$$
$$= 2.65 \, \text{mm; peak dynamic displacement}$$

$$u_d/u_{st} = 2.65/5.413 = 0.4896; \text{dynamic factor, DF}(u)$$

$$Q_d = Q(u_d/u_{st}) = 15 \times 0.4896 = 7.344 \, \text{N; basic shear estimate}$$

$$Q'_d = \left(0.68\frac{u_d}{u_{st}}+0.32\right)w_0L = (0.680\times0.4896+0.32)15.0 = 9.794\,\text{N}$$

This second value is our final estimate. Bending moment:

$$M_d = (u_d/u_{st})M_{st} = 0.4896\times750 = 367.2\,\text{N-mm; basic bending estimate.}$$

To get an upper bound, find the flex wave length at $t = t_0$:

$$l_2 = 1.887\left(\frac{EI}{m}\right)^{1/4}\sqrt{t_0} = 1.887\left(\frac{69{,}000\times5.02}{0.02144}\right)^{1/4}\sqrt{t_0} \quad\text{or}\quad l_2 = 119.6\sqrt{0.35} = 70.76\,\text{mm}$$

Then, with the help of Table 10.5: $M_d = 0.47W_0l_2 = 0.47\times15\times70.76 = 499\,\text{N-mm}$ (This larger value should be used).

The calculation using an FEA program gave, the following extreme values, respectively:

$$u_d = 2.813; \quad Q_d = 11.367; \quad M_d = 443.9$$

The approximate results are, therefore, not very accurate, but quite sufficient as first estimates.

To systematically check on results, an FEA model of the beam, with the properties as described above, made up of 100 elements, was constructed and run a number of times. (The code was LS-Dyna, v960. The elements were specified as Hughes–Liu beams with cross-section integration.)

The flow of calculations and the results are shown in the Table 16.1. The responses (u_d, Q_d, and M_d) are found first, according to the method used above. Next, the ratios of those results to the accurate values are presented, the latter based on the finite element analysis (FEA) calculations. (These ratios are given by symbols $\%u_d$, $\%Q_d$, and $\%M_d$.) In retrospect, the above example illustrates how a single line in the table is determined. The upper-bound moment estimation is not shown in the table. Still, there is a need for it, since the moments obtained from the elementary method are definitely on the low side.

With regard to shear, the second result is much better result (0.866 of true value, expressed as $\%Q'_d = 0.866$ in table) than by our basic method, which gave only 0.646. In fact, this alternative method gives better results in most cases, especially for shorter duration pulses. However, for triangular pulses it is not necessarily true.

Rectangular impulse response, distributed loading

CF beam, $u_{st} = 5.4125$, $Q_{st} = 15$, $M_{st} = 750$, $\tau = 4.446$

t_0	$\pi t_0/T$	sin()	u_d	Q_d	M_d	$\%u_d$	$\%Q_d$	$\%M_d$	t_0/τ	Q'_d	$\%Q'_d$
0.1000	0.0707	0.0706	0.7643	2.118	105.90	0.925	0.360	0.647	0.02249	6.2403	1.0627
0.2000	0.1413	0.1409	1.5247	4.226	211.28	0.932	0.499	0.747	0.04498	7.6734	0.9113
0.3500	0.2473	0.2448	2.6500	7.344	367.20	0.942	0.646	0.827	0.07872	9.7939	0.8660
0.7000	0.4946	0.4747	5.1387	14.241	712.06	0.985	1.274	1.075	0.15744	14.4840	1.3002

SS beam; $u_{st} = 0.564$, $Q_{st} = 7.5$, $M_{st} = 187.5$, $\tau = 1.5839$

0.1000	0.1983	0.1970	0.2223	2.956	73.89	0.977	0.765	0.797	0.06314	3.9555	1.0242
0.2000	0.3967	0.3864	0.4358	5.796	144.89	0.996	1.115	0.949	0.12627	6.1705	1.1866
0.3000	0.5950	0.5605	0.6323	8.408	210.20	0.993	1.047	0.919	0.18941	8.2083	1.0222
0.6000	1.1901	0.9284	1.0472	13.926	348.15	0.994	1.111	0.944	0.37881	12.5122	0.9986

SC beam; $u_{st} = 0.234$, $Q_{st} = 9.375$, $M_{st} = 187.5$, $\tau = 1.014$

0.0500	0.1549	0.1543	0.0722	2.893	57.86	0.955	0.662	0.809	0.04931	4.8404	1.1071
0.1000	0.3098	0.3049	0.1427	5.717	114.33	0.971	0.853	0.967	0.09862	6.7831	1.0122
0.2000	0.6196	0.5807	0.2718	10.889	217.78	0.982	1.133	1.043	0.19724	10.3416	1.0759
0.3000	0.9295	0.8013	0.3750	15.024	300.49	0.987	1.305	1.120	0.29586	13.1868	1.3719

CC beam; $u_{st} = 0.1128$, $Q_{st} = 7.5$, $M_{st} = 125$, $\tau = 0.6987$

0.0250	0.1124	0.1122	0.0253	1.683	28.04	0.914	0.542	0.748	0.03578	3.3115	1.0661
0.0500	0.2248	0.2229	0.0503	3.344	55.73	0.933	0.740	0.847	0.07156	4.5076	0.9977
0.1000	0.4496	0.4346	0.0981	6.520	108.66	0.965	1.009	1.001	0.14312	6.7941	1.0516
0.2000	0.8993	0.7829	0.1766	11.743	195.72	0.970	1.234	1.037	0.28625	10.5550	1.1094

Note: u_{st}, Q_{st}, and M_{st} are the static responses due to uniformly distributed load of 0.15 N/mm.

The alternative formulas for shear and especially for bending are intended to be used for rectangular pulse duration not larger than $\tau/4$.

EXAMPLE 16.2   A BEAM UNDER SUDDENLY APPLIED POINT LOAD

A tip load of $W_0 = 10\,\text{N}$ is applied to a cantilever (CF) as a pulse lasting $t_0 = 0.045\,\text{ms}$.
   The beam is in a shape of a slender aluminum rod with a round solid section and with properties specified in Example 16.1.

Follow Case 10.4. The static shear force and the bending moments at the base are

$$Q_{st} = W_0 = 10\,\text{N} \quad M_{st} = W_0 L = 10 \times 100 = 1000\,\text{N-mm}$$

For this pulse duration, the affected length is

$$l_1 = 1.496 \left(\frac{EI}{m}\right)^{1/4} \sqrt{t_0} = 1.496 \left(\frac{69{,}000 \times 5.02}{0.02144}\right)^{1/4} \sqrt{t_0} \quad \text{or} \quad l_1 = 94.84\sqrt{t_0} = 20.12\,\text{mm}$$

The tip deflection at t_0 and the tip static deflection, respectively, are (Equations 10.25a and 10.12):

$$u_1 = \frac{W_0 l_1^3}{3EI} = \frac{10 \times 20.12^3}{3EI} = 0.07838 \, \text{mm}; \quad u_{st} = \frac{W_0 L^3}{3EI} = \frac{10 \times 100^3}{3EI} = 9.623 \, \text{mm}$$

where the first value is estimated on the basis of the flex wave motion.

The natural period, from Example 10.2, is $\tau = 4.446 \, \text{ms}$.

$$u_d = 2 u_{st} \sin(\pi t_0 / \tau) = 2 \times 9.623 \sin(\pi \times 0.045/4.446)$$
$$= 0.6119 \, \text{mm peak dynamic displacement}$$

$$u_d / u_{st} = 0.06359; \text{ dynamic factor, DF}(u)$$

$$\mathcal{M}_d = (u_d / u_{st}) \mathcal{M}_{st} = 0.06359 \times 1000 = 63.59 \, \text{N-mm; peak bending moment}$$

$$\mathcal{M}_d = W_0 l_1 = 10 \times 20.12 = 201.2 \, \text{N-mm; an alternative estimate}$$

Similarly, two estimates are performed for the shear force:

$$Q_d = (u_d / u_{st}) Q_{st} = 0.06359 \times 10 = 0.636 \, \text{N}$$

$$Q_d = (6.013 \times 0.0636 + 1.415)10 = 17.97 \, \text{N}$$

This time the first or "basic" estimate is negligibly small.

To indirectly check the results, an FEA model of the beam, with the properties as described above, made up of 100 elements, was constructed and run. (The code was Ansys, v10 [2]. The elements were specified as BEAM3 elastic beams.) The simulation gave the following extreme values, respectively:

$$u = 0.689; \quad \mathcal{M} = 220.4; \quad \text{and} \quad Q = 17.97.$$

The approximate results are, therefore, not very accurate, with the exception for Q, where the calculation method was set just for this load duration, i.e., $t_0 / \tau = 0.1$.

The flow of calculations and the FEA results for all cases considered is shown in the table. The responses (u_d, Q_d, and $\mathcal{M}_d$) are found first, according to the above equations. Next, the ratios of those responses to the accurate values (from FEA simulations) are presented, namely, ra(u), ra($\mathcal{M}$), and ra(Q). In summary, the above example illustrates how a single line in the table is obtained, but this applies only to the first four lines of each case. The last line is for the load application lasting for one-half of the natural period. This is certainly not a "short duration"; therefore bending and shear equations do not apply. For this reason, the last line is not typical. The values with (*) in column ra($\mathcal{M}$) indicate the DF($\mathcal{M}$), i.e., the ratio of peak dynamic response from FEA to the static value. Those values are quite close to 2.5 for all cases except the SS case. The values with (*) in column Q_d show the global DF(Q) for the support condition. Those factors are based on the peak values from FEA, the latter in column Q_a. (Interestingly, an earlier load removal can cause somewhat larger peak shear response.)

Rectangular impulse response. Point load $W_0 = 10$ N applied.

Cantilever, CF: $1/b = 3$, $u_{st} = 9.623$, $M_{st} = 1000$, $Q_{st} = 10$, $\tau = 4.446$.

t_0	l_1	M_d	ra(M)	u_d	ra(u)	Q_d	Q_a	ra(Q)
0.011	9.947	99.47	1.539	0.1496	0.886	15.08	11.68	1.292
0.045	20.12	201.20	0.913	0.6119	0.888	17.97	17.97	1.000
0.18	40.24	402.40	0.613	2.4413	0.910	29.40	29.4	1.000
1.1115	100	1414.21	0.709	13.6090	1.048	98.51	48.83	2.017
2.223			*2.43	19.2460	1.023	*4.88	41.33	

Guided, CG: $1/b = 12$, $u_{st} = 2.406$, $M_{st} = 500$, $Q_{st} = 10$, $\tau = 2.795$.

0.007	10.01	50.05	1.190	0.0379	0.872	10.46	5.33	1.962
0.028	20.02	100.10	1.189	0.1514	0.870	11.68	11.68	1.000
0.112	40.04	200.20	0.978	0.6042	0.894	16.54	16.54	1.000
0.699	100	707.31	1.267	3.4036	1.045	46.59	33.62	1.386
1.398			*2.56	4.8120	1.008	*4.21	42.06	

Supported, SS: $1/b = 48$, $u_{st} = 0.6015$, $M_{st} = 250$, $Q_{st} = 5$, $\tau = 1.584$.

0.007	10.01	50.05	2.400	0.0167	0.898	5.07	2.357	2.151
0.028	20.02	100.10	1.850	0.0668	0.904	5.22	5.623	0.999
0.175	50.05	250.25	1.652	0.4092	1.010	6.22	6.22	1.000
0.396	75.28	376.40	1.174	0.8507	1.002	7.51	11.26	0.667
0.792			*1.88	1.2030	1.001	*3.52	17.59	

Supp-clamped, SC: $1/b = 107.3$, $u_{st} = 0.2691$, $M_{st} = 187.5$, $Q_{st} = 6.875$, $\tau = 1.014$.

0.007	10.01	50.05	2.165	0.0117	0.949	6.97	4.354	1.600
0.028	20.02	100.10	1.302	0.0466	0.958	7.53	7.53	1.000
0.175	50.05	250.25	1.111	0.2777	1.059	11.23	11.19	1.000
0.254	60.24	301.20	0.708	0.3812	1.043	12.89	20.32	0.634
0.507			*2.43	0.5382	1.022	*3.49	24.02	

Clamped, CC: $1/b = 192$, $u_{st} = 0.1504$, $M_{st} = 125$, $Q_{st} = 5$, $\tau = 0.699$.

0.007	10.01	50.05	1.825	0.0095	0.884	5.87	3.662	1.602
0.028	20.02	100.10	1.322	0.0378	0.901	7.23	7.228	1.001
0.086	35.09	175.45	1.159	0.1134	0.986	10.89	10.88	1.001
0.175	50.05	250.25	1.034	0.2130	0.983	15.69	15.09	1.040
0.35			*2.47	0.3008	1.009	*3.73	18.64	

Note: 1 = flex wave length; the shear force from FEA is designated by Q_a.

The coefficients in the second equation for the shear force are based on FEA results for lines 2 and 3 for each support case. This method gave an overprediction for very short impulses (line 1). In line 4 (quarter period) there was either an under- or overprediction resulting. One should also note that the peak shear response for the cantilever (CF) gave a large DF of 4.88.

In reviewing the result, one should keep in mind that some of the impulses applied here have a very short duration, on relative basis. (For example, the first case for the CG beam the load is applied for only 1/400 of the fundamental period.) For those short impulses (and for much of the rest) the calculated deflection is under-predicted and shear forces are overpredicted.

For cases designated as SS, SC, and CC the load is applied at mid-length and, therefore, $l = 50\,\text{mm}$ is the larger distance for the flex wave to travel, without a reflection, in a beam with $L = 100\,\text{mm}$. On occasions where l was exceeding that, the results have to be viewed with caution.

A different estimate of the peak shear is found using Equation 10.19 along with the coefficients from Table 10.4:

$$Q_d' = \left(c\frac{u_d}{u_{st}} + d \right)W_0 = (1.35 \times 0.06359 - 0.35)10 = -0.264\,\text{N}$$

This result must be rejected, as the calculated value should be positive. The formula does not work for point loads applied for such a short time.

EXAMPLE 16.3 CONCRETE BEAM IMPACTED BY A HEAVY STRIKER

A cylindrical-face striker with $M = 193\,\text{kg}$ and the contact-face diameter of $2R = 40\,\text{mm}$ impacts the top surface with a velocity of $v_0 = 4\,\text{m/s}$. The beam has $B = 150$ and $H = 180$ and is simply supported. The width of the striker extends over the full beam width B. Concrete strength is $F_c' = 27.6\,\text{MPa}$ and density of $2350\,\text{kg/m}^3$. Estimate the force of impact.

Refer to sections "Collision method for a mass–beam impact problem," and "The lumped-parameter methods for mass–beam impact problem," as well as Case 7.26, and Case 8.18b to have a complete picture of methodology. The maximum pressure that can be applied at the contact surface is $F_{cc} = 3F_c' = 82.8\,\text{MPa}$. ($f_c \equiv F_{cc}$ is the confined, dynamic compressive strength.) The maximum possible bearing load W_0 is experienced when the striker is submerged to the depth of $u = R$ in concrete:

$$W_0 = (2R)bF_{cc} = (2 \times 20)150 \times 82.8 = 496,800\,\text{N}$$

where $2Rb$ is the largest imprint area. The next step is to check if the initial velocity v_0 can induce such a strong impact. The maximum surface penetration, for the beam fixed in space is

$$u_m = R\left(\frac{3Mv_0^2}{4\sqrt{2}RW_0} \right)^{2/3} = 20\left(\frac{3 \times 193,000 \times 4^2}{4\sqrt{2} \times 20 \times 496,800} \right)^{2/3} = 6.012\,\text{mm}$$

and the corresponding impact force

$$W_{\mathrm{m}} = 2bRf_{\mathrm{c}}(2x_{\mathrm{m}})^{1/2} = 2 \times 150 \times 20(3 \times 27.6)\ (2 \times 6.012/20)^{1/2} = 385{,}200\,\mathrm{N}$$

However, the beam is not fixed (held), except at the ends, which means the above number is an upper bound. To get an answer appropriate for the center impact of a simply supported beam, one has to estimate the effective mass of the beam during the collision, using the above derived W_{m} for the first estimate of the contact area:

$$A_0 = W_{\mathrm{m}}/F_{\mathrm{cc}} = 385{,}200/82.8 = 4{,}652\,\mathrm{mm}^2.$$

This gives $s = \sqrt{A_0} = 68.2\,\mathrm{mm}$, therefore the effective beam mass is

$$M_{\mathrm{e}} = \rho bh(s + H) = 0.00235 \times 150 \times 180(68.2 + 180) = 15{,}748\,\mathrm{g}$$

The effective mass M^* due to impact with a relative velocity v_{r} and the loss of kinetic energy:

$$\frac{1}{M^*} = \frac{1}{193} + \frac{1}{15.75}; \quad \text{then } M^* = 14.56\,\mathrm{kg}$$

$$\Delta E_{\mathrm{k}} = \frac{1}{2}M^*\,v_{\mathrm{r}}^2 = \frac{1}{2} \times 14{,}560 \times 4^2 = 116{,}480\,\mathrm{N\text{-}mm}$$

But this loss is equal to strain energy, or crushing energy of concrete. Using the equality

$$\Pi = \frac{2}{3}\sqrt{2}Rx_{\mathrm{m}}^{3/2}W_0; \text{ from Case 7.26, one gets}$$

$$116{,}480 = \frac{2}{3}\sqrt{2} \times 20x_{\mathrm{m}}^{3/2}\,496{,}800 \quad \text{or} \quad x_{\mathrm{m}} = u_{\mathrm{m}}/R = 0.05367\,\mathrm{mm}, \quad \text{and}$$

$$W_{\mathrm{m}} = 2Rbf_{\mathrm{c}}(2x_{\mathrm{m}})^{1/2} = 2 \times 20 \times 150 \times 82.8(2 \times 0.05367)^{1/2} = 162{,}770\,\mathrm{N}$$

This is much less than the test result of 200 kN, according to the publication by this author [94]. The disparity is larger than in other similar tests. It may be related to the radius R, whose size was somewhat uncertain. The calculation repeated for a larger radius, namely, 40 mm, gives

$$W_{\mathrm{m}} = 208.5\,\mathrm{kN}$$

EXAMPLE 16.4   LOCAL BENDING IN BEAM DUE TO MASS IMPACT

Goldsmith [26] reports a set of experiments where spherical, hard metal strikers were impacted against steel beams. At each impact the peak stress on the far face of the beam, just opposite the impact point, was measured. The four configurations selected for this example are listed below.

	M (lb)	R (in.)	v_0 (ft/s)	$B \times H$ (in.)
1	1.32	2	8.33	1×1
2	4.01	2	5.83	1×1
3	5.29	3.94	15.6	1.97×1.97
4	0.18	0.25	150	0.75×0.75

Note: M is striker mass, R is striker radius, v_0 is impact speed, and $B \times H$ is the section size.

All beams were impacted in the middle. Try to predict the peak stress treating the impact as a local phenomenon.

One possible approach would be to use Case 10.8, expecting that for a low-velocity impact the peak bending could be predicted from v_0 alone. Taking line 2 from the table for a detailed solution, the section properties are

$$B = H = 25.4\,\text{mm}; \quad A = 25.4^2 = 645.2\,\text{mm}^2; \quad I = 25.4^4/12 = 34{,}690\,\text{mm}^4$$

$$m = A\rho = 645.2 \times 0.00785 = 5.065\,\text{g/mm}$$

Bending moment from Case 10.8:

$$\mathcal{M} = v_0\sqrt{EIm} = 5.83 \times 0.3048\sqrt{200{,}000 \times 34{,}690 \times 5.065} = 333{,}110\,\text{N-mm}$$

$$\sigma = 6\mathcal{M}/H^3 = 6 \times 333{,}110/25.4^3 = 122\,\text{MPa}; \text{ bending stress.}$$

The measured stress was 144.8 MPa. The prediction is not very accurate, but reasonable. If this is extended to the remaining three examples, the following numbers result:

M (g)	R (mm)	H (mm)	I (mm⁴)	m (g/mm)	$(EIm)^{1/2}$	v_0 (m/s)	$\mathcal{M}$	σ (MPa)	σ, exp
598.8	50.8	25.4	34686.0	5.065	1,87,439	2.54	4,75,905	174.25	110.3
1818.9	50.8	25.4	34686.0	5.065	1,87,439	1.78	3,33,076	121.95	144.8
2399.5	100.0	50.0	520833.3	19.625	14,29,780	4.75	67,98,432	326.32	158.6
81.6	6.35	19.05	10974.9	2.849	79,076	45.72	36,15,350	3137.74	171

The second line, of course, corresponds to the preceding calculation. The last column gives the experimentally measured stress. The discrepancy between this and the calculated figures (σ, MPa) increases with the impact velocity. This initially proposed mechanism must be unrelated to what happens at a relatively fast impact, at ~46 m/s.

Let us examine if a more elaborate wave approach gives a chance to come a little closer to what was measured, while keeping the simplifications used so far. This involves a local nature of impact and the assumption, implied by Case 10.8, that the impact is hard, i.e., the contact flexibility is negligible. (Not an unreasonable assumption in view of the fact that it is a solid sphere contacting a solid beam.)

The larger the impact speed, the faster the decrease of the contact force. Accordingly, the last line in the table will be examined. The section properties are now

$$A = 19.05^2 = 362.9\,\text{mm}^2; \quad A_s = (5/6)A = 302.4\,\text{mm}^2; \quad I = 10{,}975\,\text{mm}^4; \quad m = 2.849\,\text{g/mm}$$

The estimate of the initial peak is provided by Case 10.13. The damper constant

$$C = 2\sqrt{GA_s m} = 2\sqrt{76,920 \times 302.4 \times 2.849} = 16,280$$

In absence of a contact spring, the peak impact force is

$$R_{md} \equiv W_m = Cv_0/2 = 16,280 \times 45.72/2 = 372,170 \, \text{N}$$

Let us arbitrarily assume that the peak moment appears when the contact force decreases to one-half of its initial value. When treating the impacted beam as a damper, one can refer to Case 3.7. The time needed to reduce the contact force by one-half is designated there as t_{05}:

$$t_{05} = 0.6931 M/C = 0.6931 \times 81.6/16,280 = 3.474 \times 10^{-3} \, \text{ms}$$

From the initiation of contact, there also is a flex wave spreading from the contact point. After time t_{05} the half-length of that wave, according to Equation 10.24b, is

$$l_2 = 1.887 \left(\frac{EI}{m} \right)^{1/4} \sqrt{t_{05}} = 1.887 \left(\frac{200,000 \times 10,975}{2.849} \right)^{1/4} \sqrt{3.474 \times 10^{-3}} = 18.53 \, \text{mm}$$

The moment estimate according to Equation 10.26b is

$$\mathcal{M}_d = \frac{W_m l_2}{2} = \frac{372,170 \times 18.53}{2} = 3.448 \times 10^6 \, \text{N-mm}, \quad \text{then}$$

$$\sigma_b = 6\mathcal{M}_d/H^3 = 6 \times 3.448 \times 10^6/19.05^3 = 2993 \, \text{MPa}$$

This is quite close to what was obtained by the initial, simple method and is very far from the correct result. A better estimate of the moment can be obtained using the basic method of mass–beam impact described in Chapter 10 and demonstrated in Example 10.6. Our beam is simply supported and has $L = 762$ mm. There is no contact spring involved. Recalculating Example 10.6 is left to the reader.

EXAMPLE 16.5   SQUASHING COLLAPSE OF A STEEL COLUMN, INITIAL PHASE

In figures (a) and (b) a probable collapse mode of hollow steel columns in the WTC towers is shown, as proposed by Bažant [7]. This problem was also discussed by Szuladziński [97,104]. Figure (c) shows force–deflection characteristics, as proposed by this author: P_{cr} is the basic buckling force while P_0 is the resisting force based on plastic compression with a large offset. The cross section is shown in figure (d). Calculate the strain energy absorbed by the column as it deflects to its near-terminal point u_1 defined by $\theta = 75°$, as well as the squashing load at that point. Steel properties, when treated as a elastic-perfectly plastic (EPP) material are $E = 200,000$ MPa and $\sigma_0 = 500$ MPa. The clamped-ends column is $L = 3.63$ m long. (The segment of the characteristic past u_1 is analyzed in Example 16.6.)

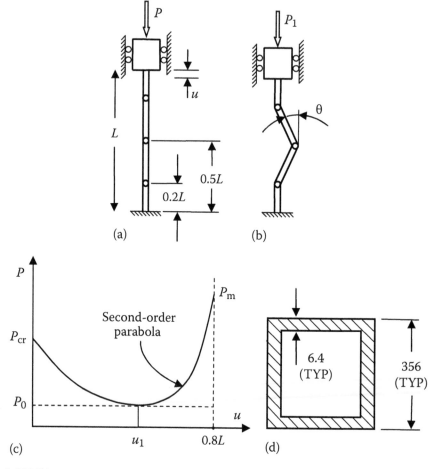

(See Ref. [104].)

Section properties, per Table 10.1:

$$A = 8950 \text{ mm}^2, I = 182.37 \times 10^6 \text{ mm}^4; Z = 1.173 \times 10^6 \text{ mm}^3$$

Column strength, when edges are clamped, according to Case 11.1, ($\mu = 4$):

$$i = \sqrt{\frac{182.37 \times 10^6}{8950}} = 142.75 \quad \text{and} \quad \Lambda = \frac{3630}{\sqrt{4} \times 142.75} = 12.71$$

$$\sigma_{cr} = \frac{\pi^2 E}{\Lambda^2} = \frac{\pi^2 200,000}{12.71^2} = 12,220 \text{ MPa; elastic buckling stress}$$

$$\sigma_{crp} = \left(1 - \frac{\sigma_0}{4\sigma_{cr}}\right)\sigma_0 = \left(1 - \frac{500}{4 \times 12,220}\right)500 = 494.9 \text{ MPa or } P_{cr} = 4.429 \times 10^6 \text{ N}$$

The initial squashing phase is from $\theta = 0°$ to $\theta = 75°$. This corresponds to vertical displacement of

$$u_1 = 0.6L - 0.6L \cos 75° = 0.4447L = 1614 \text{ mm}$$

To evaluate column resistance during squashing and energy absorbed, use Case 11.23. The moment capacity is

$M_0 = Z\sigma_0 = 1.173 \times 10^6 \times 500 = 586.5 \times 10^6$ N-mm. The rotation angle is $\theta_2 = 75° \approx 1.309$:

$$\prod \approx 4 M_0 \theta_2 = 4 \times 586.5 \times 10^6 \times 1.309 = 3071 \times 10^6 \text{ N-mm; energy absorbed up to } u_1.$$

The resistance at the terminal angle:

$$P_0 \approx \frac{2M_0}{(0.3L)\sin\theta_2} = \frac{2 \times 586.5 \times 10^6}{0.3 \times 3630 \times 0.9659} = 1.115 \times 10^6 \text{ N}$$

Needless to say, this axial resistance is only a fraction of axial capacity, namely $P_0/P_{cr} = 0.2517$. (Note that using Equation 11.23 would result in a slightly smaller P_0.) The eccentricity of the column was not stated, which means it was negligible. This was the reason why the P–u characteristic for $u < u_1$ was similar to the descending curve (originating at P_y) in Figure 11.8b.

EXAMPLE 16.6   SQUASHING COLLAPSE OF A STEEL COLUMN, PROCESS CONTINUED

Consider the same hollow steel column as in Exercise 16.5. The second, stiffening segment of the P–u characteristic is now of interest. Let the peak squashing force be attained when the original height is reduced to $0.2L$. Let this peak be limited to $2P_{cr}$, with P_{cr} being the buckling load previously calculated. The P–u curve on this segment is assumed to be a second-order parabola, tangent to the horizontal line at $u = u_1$. Calculate the total strain energy absorbed by the column as it deflects from the initial configuration to its terminal point, $u_2 = 0.8L$.

The data in this example as well as in 16.5 relates to the 95th story of the WTC North Tower. The mass above level 95 was that of 15 floors plus an additional 2 kton for the roof. This results in $M = 78.26$ ton supported by one outer column. ($P_b = 767.7$ kN/column). Determine if this mass was sufficient to initiate the collapse the story in question keeping in mind that the strength of this story must be degraded to the level of the supported gravity load. If the story is squashed, assume that the 94th story below was degraded to one-half of its original strength and check on the possibility of the continued collapse.

The final compacted length is taken as $0.2L$, which means that squashing will continue from $u_1 = 1614$ mm to $u_2 = 0.8L = 0.8 \times 3630 = 2904$ mm, giving the top end travel in this phase as:

$$u_2 - u_1 = 2904 - 1614 = 1290 \text{ mm}.$$

The values of the squashing force or the column resistance as determined in Exercise 16.5, were

$$P_0 = 1.115 \times 10^6 \text{ N at } u_1 \text{ and } P_m = 2P_{cr} = 8.858 \times 10^6 \text{ N at } u_2$$

The simplest expression for a second-order parabola is $y = ax^2$. This line is tangent to the x-axis at the origin and its value for $x_2 = c$ is $y_2 = ac^2$. It is easy to check by integration that the area under y is $cy_2/3$. (See Appendix B.) In this case $c = u_2 - u_1$ and $y_2 = P_m - P_0$. Adding to that the rectangular area cP_0, we find the area total area under the P–u curve, to the right of u_1, to be

$$\Pi_2 = \frac{u_2 - u_1}{3}(2P_0 + P_m) = \frac{1290}{3}(2\times1115 + 8858)10^3 = 4.768\times10^9 \text{ N-mm}$$

Total absorbed energy, after including Π_1 from Exercise 16.5:

$$\Pi = \Pi_1 + \Pi_2 = (4.768 + 3.071)10^9 = 7.839 \times 10^9 \text{ N-mm}$$

The quoted values of P_0 and Π refer to undamaged story near that level of the building.

Various influences, like a direct loss of some columns, impact effects and subsequent fire had to reduce the column strength to the level of applied load, i.e., to the weight of 767.7 kN/column. (This was necessary for the downward motion to begin.) This means that the original strength was degraded by the factor of

$$4.429/0.7677 = 5.77$$

when the failure process began. Not only the axial strength, but also the strain energy have to be reduced accordingly:

$$P_{cr1} = 767,700 \text{ N; new axial strength}$$

$$\Pi = 7.839 \times 10^9 \text{ N-mm}/5.77 = 1.359 \times 10^9 \text{ N-mm; new energy absorption capacity}$$

The potential energy of the part of the building above level 95 is

$$U = Mg(0.8L) = 78.26\times10^6 \times0.00981\times0.8\times3630 = 2.229\times10^9 \text{ N-mm}$$

This is more than the compressed column can absorb. The critical story will be squashed and the kinetic energy at the end of the process will be

$$E_k = U - \Pi = (2.229 - 1.359)10^9 = 0.87\times10^9 \text{ N-mm}$$

The story below is assumed to retain 50% of its original strength, which means

$$P_{cr2} = P_{cr}/2 = 4.429\times10^6/2 = 2.215\times10^6 \text{ N}$$

$$\Pi = 7.839 \times 10^9 \text{ N-mm}/2 = 3.92 \times 10^9 \text{ N-mm}$$

If the largest force that can be developed by the 95th story is assumed not to exceed $2P_{crl}$ or twice its reduced buckling strength, then the force available to squash the next story is $2P_{crl} = 2 \times 767,700 = 1.5354 \times 10^6$ N and we conclude that $2P_{crl} < P_{cr2}$ which means that a sufficient compressive force cannot be developed and the column will remain standing. To check on the energy criterion (ignoring the accreted mass increase) use the same energy U as before. We find the total energy available for continued crushing to be $U + E_k = (2.229 + 0.87)10^9 = 3.10 \times 10^9$ N-mm, but then $U + E_k < \Pi = 3.92 \times 10^9$ N-mm

The strain energy Π (as a measure of resistance to be overcome), which is associated with a collapse of the column, is larger than the potential plus the kinetic energy available. The conclusion is that the motion will be arrested after the damaged story collapse and the building will stand. This, is, of course, based on the assumption that the story above level 94 has its strength degraded by a half. If there was a larger degradation, say by ¾, this story would be squashed too and the next story below would have to be considered.

EXAMPLE 16.7   COLUMN–BUILDING INTERACTION DUE TO NEARBY EXPLOSION

One of steel columns supporting a building experiences an explosive pressure of 6 MPa and a specific impulse of 4.8 MPa-ms. Both the pressure and the impulse quoted are the average values along the length, along the projected width of the column as well as in time. The column is a tube with the outer diameter of 406.4 mm and wall thickness of 9.5 mm. The yield strength is 350 MPa and the ultimate strength is 430 MPa. The architectural finish of the column increases its diameter to 460 mm and the unit mass to 181 kg/m. The net length of the column is 2.6 m and it may be regarded as clamped at both ends. The weight of the building induces the axial compression of 1750 kN.

Treating the ends as axially fixed, i.e., the building as immovable, find the axial load induced in the column by the described lateral action and the associated lateral deflection. For the sake of simplicity treat the gravity preload as independent of the axial effect sought.

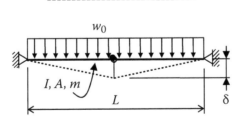

Follow Case 11.17. To act in accordance with the rigid-plastic material model the value of σ_0 must be nominated. For simplicity set $\sigma_0 = (350 + 460)/2 = 405$ MPa. Determine the plastic bending parameters using Table 10.1:

$$R_m = (406.4 - 9.5)/2 \approx 198.5 \text{ mm}; \quad h = 9.5 \text{ mm}$$

$$P_0 = A\sigma_0 = 2\pi R_m h \sigma_0 = 2\pi \, 198.5 \times 9.5 \times 405 = 4.8 \times 10^6 \text{ N}; \text{ axial load capacity}$$

$$\mathcal{M}_0 = Z\sigma_0 = 4R_m^2 h\sigma_0 = 4 \times 198.5^2 \times 9.5 \times 405 = 606.1 \times 10^6 \text{ N-mm; bending capacity}$$

$$w_0 = p_0 B = 6 \times 460 = 2760 \text{ N/mm; load per unit length}$$

$$t_0 = i/p = 4.8/6 = 0.8 \text{ ms; effective impulse duration}$$

$$\zeta = \frac{\mathcal{M}_0}{P_0 L} = \frac{606.1}{4.8 \times 2600} = \frac{1}{20.59}$$

$$E_k = \frac{3}{8}\frac{(w_0 t_0)^2 L}{m} = \frac{3}{8}\frac{(2760 \times 0.8)^2 2600}{181} = 26.26 \times 10^6 \text{ N-mm; kinetic energy}$$

$$\theta_m = 2\zeta\left[\left(1+\frac{E_k}{2\mathcal{M}_0\zeta}\right)^{1/2} - 1\right] = \frac{2}{20.59}\left[\left(1+\frac{26.26 \times 10^6 \times 20.59}{2 \times 606.1 \times 10^6}\right)^{1/2} - 1\right]$$

$$= 0.0197; \text{ maximum slope attained}$$

$$\delta \approx \frac{L\theta}{2} = \frac{2600 \times 0.0197}{2} = 25.61 \text{ mm; peak deflection}$$

No special effort is needed to find the peak axial load induced. This is the same as P_0 = 4800 kN, as long as the interaction between the axial and the bending component is ignored. The result shows that in spite of a very small peak deflection, the maximum axial capacity is developed. This is larger than the gravity preload of 1750 kN, which means that the column will be in the resultant tension, significantly exceeding the preload value. This means that the building will experience a downward pull of a large magnitude, but of short duration. In a real setting the vertical flexibility will tend to reduce that pull while in the elastic range, but the upper limit of the applied tension is dictated by the axial capacity of the beam.

EXAMPLE 16.8   A CONCRETE FRAGMENT IMPACTING A RC COLUMN

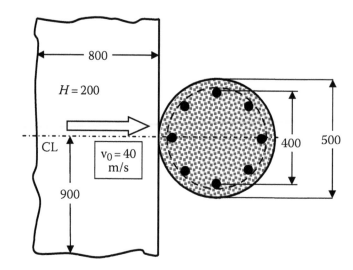

The same column, as described in Example 11.16 is being impacted by a fragment of a concrete slab, 200 mm thick, ejected by a nearby explosion and moving horizontally, in the plane of the sketch. The slab has $F'_c = 20\,\text{MPa}$, while for the column has $F'_c = 25\,\text{MPa}$. The column to be impacted has the same height of 0.8 m as in Example 11.20 and its end conditions are to remain the same. Evaluate the peak impact force as follows:

a. Assume the column to be rigid and the edge of the slab crushing against it.
b. Assume the slab to be breaking in half along the CL. Its reinforcement is such that the initial break will take place when the elastic bending stress of the entire section reaches the peak of 5 MPa. A total failure will eventuate when the angle of opening between the halves reaches 10°.
c. Assume a fragment of the column, 500 mm high, to be shearing off the rest. Treat this as a collision between the impacting slab and the column fragment. The edge of the slab undergoes crushing in the process of collision. Once the three checks are made, evaluate whether the column will fail.

(a) Calculate the slab mass first: $M_1 = 0.8(2 \times 0.9)0.2 \times 2400 = 691.2\,\text{kg}$
The peak contact load W_m between the two objects can be estimated using Case 7.27, except that the column will act like an impactor, because it is made of a stronger material.

$$f_c = 3F'_c = 3 \times 20 = 60\,\text{MPa, dynamic bearing strength of the slab}$$

$$R = 500/2 = 250\,\text{mm; radius of the cylindrical impactor}$$

$$W_m = v_0(2M\pi R f_c)^{1/2} = 40(2 \times 0.6912 \times 10^6 \times \pi \times 250 \times 60)^{1/2}$$
$$= 10.21 \times 10^6\,\text{N; peak contact load}$$

This assumes no column deformation or deflection, which is quite conservative.

(b) Bending of a 200×800 section:

$$\sigma_b = 6M/(bh^2) = 6M/(200 \times 800^2) = M/21.33 \times 10^6$$

Setting $\sigma_b = 5\,\text{MPa}$, obtain $M = M_0 = 106.7 \times 10^6$ N-mm as the breaking moment. From Case 10.32:

$$W_0 = \frac{2M_0}{L} = \frac{2 \times 106.7 \times 10^6}{900} = 237{,}040\,\text{N}$$

This means that $W_m = 2W_0 = 474{,}080\,\text{N}$, a minimum quasistatic contact force to break the slab in the prescribed manner. This is also a sustained resistance of the slab after cracking, if we treat the opening of the crack as having a rigid-plastic character. Is this resistance sufficient to arrest the motion?

The kinetic energy of the projectile is $E_k = M_1 v_0^2/2 = 691{,}200 \times 40^2/2 = 553 \times 10^6$ N-mm . When the resisting W_m is maintained past the initial breaking, then the displacement, accompanying this force is found from

$$W_m u = E_k \quad \text{or} \quad u = E_k/W_m = 553 \times 10^6/474{,}080 = 1166 \, \text{mm}$$

Such a large displacement needed to absorb E_k hints to a decisive failure of the system.

(c) The breaking load of the column, according to Example 11.20 is $W_f = 630{,}900 \, \text{N}$. This is the most likely quasistatic resistance of the column. The slab breaking with a $W_m = 474{,}080 \, \text{N}$ calculated above is more critical mode of failure of the system. Suppose that while impacting the column the slab opens up by $\alpha = 10° = 0.1745 \, \text{rad}$ prior to failure. The energy absorbed in this deformation mode is therefore

$$\mathcal{M}\alpha = 106.7 \times 10^6 \times 0.1745 = 18.62 \times 10^6 \, \text{N-mm} \ll E_k$$

The slab will, therefore, fail while retaining most of its kinetic energy. If there is no impacted edge reinforcement, the slab will disintegrate leaving the column intact. A different result might be obtained if the edge has a reinforcing bar. The following will lead to the failure estimate.

The loading phase of collision between M_1, the slab mass and M_2, the tributary column mass imposes the loss of kinetic energy according to Case 8.8. We have

$$M_2 = Ad\rho = 196{,}350 \times 500 \times 0.0024 = 235{,}620 \, \text{g}$$

$$\frac{1}{M^*} = \frac{1}{M_1} + \frac{1}{M_2} = \frac{1}{691.2} + \frac{1}{235.6}; \quad \text{or} \quad M^* = 175.7 \, \text{kg}$$

$$\Delta E_k = \frac{1}{2}M^* v_r^2 = \frac{1}{2} \times 1.757 \times 10^6 \times 40^2 = 140.56 \times 10^6;$$ kinetic energy loss in the loading phase of collision. The remaining kinetic energy to be absorbed is

$$E_{kr} = 553 \times 10^6 - 18.62 \times 10^6 - 140.56 \times 10^6 = 393.82 \times 10^6 \, \text{N-mm}$$

From Example 11.20 it is clear that the column impact at this level is shear-critical, with the shear resistance of $W_s = 942{,}500 \, \text{N}$. The following shear slide u_s is needed for the absorption of the impact energy:

$$u_s = E_{kr}/W_s = 393.82/0.9425 = 417.8 \, \text{mm}$$

The shear slide associated with failure is $0.02d = 0.02 \times 500 = 10 \, \text{mm}$. The failure of the column will be decisive.

EXAMPLE 16.9   SQUARE, SUPPORTED PLATE UNDER PRESSURE IMPULSE

Dimensions: $a = 100 \, \text{mm}$, $h = 3 \, \text{mm}$.

The material is aluminum alloy with the following properties:

$$E = 69{,}000 \, \text{MPa}, \quad \rho = 0.0027 \, \text{g/mm}^3, \quad \nu = 0.3, \quad G = 26{,}540 \, \text{MPa}$$

Mass per unit area: $m = \rho h = 0.0081$ g/mm².

A rectangular, uniform loading pulse, $p_0 = 3$ MPa, $t_0 = 0.172$ ms is applied to the plate. Determine the dynamic response.

Bending stiffness: $D = \dfrac{Eh^3}{12(1-v^2)} = 170{,}604.$

Static deflection from Equation 12.6: $u_{st} = b\dfrac{p_0 a^4}{D} = \dfrac{1}{245.9}\dfrac{3\times 100^4}{170{,}604} = 7.151$ mm.

Static shear from Equation 12.11a: $Q_{st} = \varphi p a = 0.42 \times 3.0 \times 100 = 126$ N/mm.

Static moment from Equation 12.9a: $\mathcal{M}_{st} = \beta p_0 a^2 = 0.0479 \times 3.0 \times 100^2 = 1437$ N-mm/mm.

Circular frequency, per Case 2.34: $\omega = \dfrac{19.74}{a^2}\left(\dfrac{D}{m}\right)^{1/2} = 9.0594$ rad/ms.

Natural period: $\tau = 2\pi/\omega = 0.6936$ ms.

The center dynamic peak displacement from Equation 12.16:

$$u_d = 2u_{st}\sin(\pi t_0/\tau) = 2\times 7.151 \times \sin(\pi \times 0.172/0.6936) = 10.049 \text{ mm.}$$

The displacement ratio is $u_d/u_{st} = 1.4053$.

The initial dynamic response estimates, Equations 12.17:

$$Q_d = 126(u_d/u_{st}) = 177.1 \text{ N/mm}; \quad \mathcal{M}_d = 1437(u_d/u_{st}) = 2019 \text{ N-mm.}$$

An alternative estimate of Q from Equation 12.18 and Table 12.3:

$$Q'_d = (0.238 \times 1.4053 + 0.182)\dfrac{3\times 100^2}{100} = 129.2 \text{ N/mm.}$$

To obtain an alternative moment estimate note that $q = pa/4 = 75$ N/mm per Equation 12.20. The flex wave length from Equation 12.19a:

$$l_1 = 1.496\left(\dfrac{D}{m}\right)^{1/4}\sqrt{t} = 1.496\left(\dfrac{170{,}604}{0.0081}\right)^{1/4}\sqrt{0.172} = 42.03$$

Then, from Equation 12.21a: $\mathcal{M}_d = 0.81 \times 75 \times 42.03 = 2553$ N-mm/mm.

To indirectly check the results, a plate FEA model was constructed, with the properties as described and with 1 mm × 1 mm mesh. The FEA code was Ansys, v10 [2]. The elements were specified SHELL63 (square). The simulation gave the following extreme values, respectively: $u = 10.58$ mm; $\mathcal{M} = 2545$ N; $Q = 160.3$ N/mm.

In conclusion, the simplified calculations give quite reasonable results, but the alternative methods performed somewhat better. This, however, need not always be true.

EXAMPLE TABLE 16.1

Rectangular Impulse Response of the Plate. Uniform Pressure $p_0 = 3$ MPa

| | Square, Supported; $u_{st} = 7.151$, $Q_{st} = 126$, $M_{st} = 1437$, $\tau = 0.6936$ | | | | | | | | $W = 30000$ | |
t_0	u_d	Q_d	M_d	$u(A)$	$Q(A)$	$M(A)$	$r(u)$	$r(Q)$	$r(M)$	Q_d'
0.0347	2.2386	39.444	449.85	2.264	50.48	809.40	0.989	0.781	0.556	76.95
0.0690	4.3974	77.481	883.66	4.304	88.42	1306.00	1.022	0.876	0.677	98.51
0.1720	10.0487	177.057	2019.30	10.580	160.30	2545.00	0.950	1.105	0.793	154.9
0.3440	14.3008	251.980	2873.77	14.290	248.30	3247.00	1.001	1.015	0.885	197.4

Note: $r(u) = u_d/u(A)$; $r(Q) = Q_d/Q(A)$; $r(M) = M_d/M(A)$.

EXAMPLE TABLE 16.2

Upper Bounds of Bending for a Square, Supported Plate

t_0	ℓ	q	M_d'
0.0347	18.870	75	1146
0.0690	26.609	75	1617
0.1720	42.012	75	2552

The flow of calculations and the FEA results for all the duration times considered are shown in Example Table 16.1. The responses (u_d, Q_d, and M_d) are found first, according to the SDOF approach. Next, the corresponding values (u_d, $Q(A)$, and $M(A)$) from FEA simulations are quoted. Then the ratios of the first to the second of those quantities are presented, namely, ra(u), ra(M), and ra(Q). In summary, the above example illustrates how a single line in the Example Table 16.1 is obtained.

The successive values of t_0 correspond to $\tau/20$, $\tau/10$, $\tau/4$, and $\tau/2$. When $t_0 = \tau/2$, one could expect the dynamic magnification factor, DF = 2.0. This holds reasonably well for most cases and response components. For bending up to DF = 2.26 eventuates.

The SDOF method gives a reasonable approximation of displacement response, but is deficient in predicting shear values for low t_0/τ ratios. The last column gives shear values Q_d' per Equation 12.18. Example Table 16.2 shows the moment calculations according to Equations 12.20 and 12.21. The shorter t_0, the bigger overestimate of actual bending takes place.

EXAMPLE 16.10   MASS–PLATE IMPACT

A circular plate has $r_0 = 2$ mm insert, $R = 100$ mm, $h = 3$ mm, and the edge is supported. The material is aluminum alloy with the following properties. (The units are g, mm, and ms.) The impacting mass is $M = 127$ g and its velocity is $v_0 = 100$ m/s. The contact stiffness is 338 N/mm. $E = 69,000$ MPa, $\rho = 0.0027$ g/mm³, $\nu = 0.3$, $G = 26,540$ MPa. Find the peak contact force, duration of contact, and the rebound velocity.

The mass per unit surface and the plate stiffness, respectively:

$$m = \rho h = 0.0081 \text{ g/mm}^2; \quad D = \frac{Eh^3}{12(1-v^2)} = 170{,}604.$$

Natural period, Case 2.23:

$$\tau = 1.2624 R^2 \sqrt{\frac{m}{D}} = 1.2624 \times 100^2 \sqrt{\frac{0.0081}{170{,}604}} = 2.7508 \text{ ms.}$$

The remaining parameters are from Case 12.9:

$$C = 2\pi^2 r_0 \left(\frac{Gh_s m}{5} \right)^{1/2} = 2\pi^2 \times 2 \left(\frac{26{,}540 \times 3 \times 0.0081}{5} \right)^{1/2}$$

$$= 448.4; \text{ equivalent damper constant}$$

(Note that $h_s = 3$ instead of $h_s = 3/1.2$ was used. This will be explained below.)

$P_{md} = Cv_0 = 448.4 \times 100 = 44{,}840 \text{ N}$; damper force limit

$P_{ms} = v_0 \sqrt{kM} = 100\sqrt{338 \times 127} = 20{,}719 \text{ N}$; spring force limit

$P_m \approx (20{,}719^{-0.95} + 44{,}840^{-0.95})^{-1.053} = 13{,}757 \text{ N}$; anticipated peak contact force

$v_e = 13{,}757/448.4 = 30.68 \text{ mm/ms}$; velocity at the reversal point

$$t_0 = \frac{\pi M v_0}{2 P_m} = \frac{\pi \times 127 \times 100}{2 \times 13{,}757} = 1.45 \text{ ms}$$; duration of the loading phase

$$E_e = \frac{13{,}757^2}{2} \left(\frac{127}{448.4^2} + \frac{1}{338} \right) = \frac{13{,}757^2}{2} \frac{1}{278.53} = 339{,}734 \text{ N-mm;} \quad \text{system energy, reversal}$$

point

$$E_0 = \frac{1}{2} \times 127 \times 100^2 = 635{,}000 \text{ N-mm}$$; initial system energy

$$V = \left[\frac{2(2E_e - E_0)}{M} \right]^{1/2} = \left[\frac{2(2 \times 339{,}734 - 635{,}000)}{127} \right]^{1/2} = 26.46 \text{ mm/ms}$$; rebound speed

The second method can also be tried. With $k_{ef} = 278.53 \text{ N/mm}$ (in the expression for E_e)

$$v_y = \frac{P_m}{\sqrt{k_{ef} M}} = \frac{13{,}757}{\sqrt{278.53 \times 127}} = 73.15 \quad \text{and} \quad V = 73.15 - 30.68 = 42.47 \text{ mm/ms.}$$

The latter answer is expected more likely, as the ratio of $v_e/v_0 = 0.307$ is quite high.

To indirectly check the results, an axisymmetric FEA model of the plate made up of 100 elements, with the properties as described above, was constructed and run. The code was Ansys, v10, Ref. [2]. The elements were specified as PLANE42 axisymmetric solids, which means the thickness of the plate was explicit, rather than implied. The simulation gave the

EXAMPLE TABLE 16.3

Ansys-Computed Contact Force (N) for Different Values of M and k

v_0 (m/s)	M = 5 k = 338	M = 5 k = 3380	M = 25 k = 338	M = 25 k = 3380	M = 127 k = 338	M = 127 k = 3380	M = 255 k = 338	M = 255 k = 3380
10	371.8	979.8	750.3	1,641	1,252	2,348	1,402	2,939
100	3,718	9,798	7,503	16,410	12,520	23,480	14,020	29,390

peak force of 12,520 N with a secondary impact of about the same magnitude as the first. This duality of impact invalidated our rebound predictions. The peak force, however, is a reasonable approximation.

Several other combinations of contact stiffness k, impacting mass M, and impact velocity v_0 were run, as shown in Example Table 16.3. The overall agreement with what was found by our approximate method (Example Table 16.4) is quite good, considering the nature of the contact force.

The detailed example was showing, in fact, the steps in calculating a single entry in Example Table 16.4. With the plate mass of $M_p = 254.5$ g, the impactor masses have the following fractions of the former: ~1/50, 1/10, 1/2, and 1. Example Table 16.5 shows that the accuracy is very good for smaller fractions. When $M/M_p = 1/2$, the overprediction comes close to 10% and for the mass fraction of 1.0 it becomes 23%. Obviously, the mass ratio should not be too big for the method to be applicable.

The FEA model had only a single layer of elements, which implied a constant shearing stress along the depth. (The spacing was variable and near the center it was only ~0.5 mm.) The hand-calculated example results were to be compared to the FEA output. For the sake of a reasonable comparison of two sets of results the factor of 1.2 was suppressed when calculating h_s.

The plate stiffness in this approach is based on shear only. If so, there is no distinction between fixed and supported edge and no difference in results. The reason for ignoring the flexural component of deflection in evaluating how the plate reacts to the impacting mass was the presumed brevity of the event. Yet, as the example shows, when the impactor mass is substantial, like one-half of the plate mass, the calculated duration of contact, up to

EXAMPLE TABLE 16.4

Hand-Calculated Contact Force (N) for the Same Parameters as in Example Table 16.3

10	371.4	982.0	746.1	1,706.8	1,374.6	2,576.1	1,717.0	2,932.6
100	3,717.5	9,827.5	7,467.3	17,081.8	13,756.6	25,781.4	17,183.7	29,349.4

EXAMPLE TABLE 16.5

Ratio of Values in Example Table 16.4 to Those in Example Table 16.3

10	0.999	1.002	0.994	1.040	1.098	1.097	1.225	0.998
100	1.000	1.003	0.995	1.041	1.099	1.098	1.226	0.999

peak load, is about one-half of the natural period, which is not short. The analysis of shear deflections has shown that the stiffness increases as the motion progresses. In reality, however, that is more than offset by the growth of the flexural component. For this reason the averaged shear stiffness, as used here, has a good chance of giving a reasonable approximation of the total resistance during downward movement.

The Ansys output of displacement, velocity, and contact force history (for the hand-calculated configuration) is shown in the enclosed figures. As it turned out, the selected example is not typical in the sense that there are two peaks of contact force. Example Figure 16.1c shows that the decay of the first lobe of the curve is definitely terminated at about 1.3 ms, and that a new growth begins at that time point. This is roughly one-half of the natural period, which means the impact is sufficiently long for the flexural wave originated

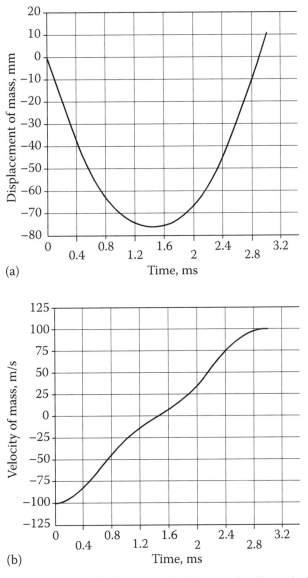

(a)

(b)

FIGURE 16.1   (a) History of striker displacement. (b) History of striker velocity.

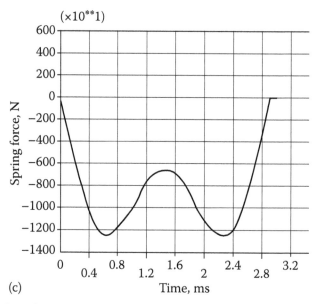

FIGURE 16.1 (continued) (c) History of contact force. Multiply ordinate by 10.

at the center to reflect from the edge and return to the origin. (Refer to Szuladziński [103] for further explanations of this aspect.) The rebound is almost perfect, with nearly a full reversal of the impact velocity. It is revealed in Figure 16.1c that this is due to a secondary impact. Clearly, our rebound calculation method fails in this case.

The impacted mass did not rebound sufficiently to lose contact with the spring, which is a typical response for a larger mass and a moderately stiff contact spring. The duration of contact was over 2.8 ms, about the same as the period of flexural vibration τ. There was sufficient time for the mass to be pushed back upward by the flexural plate motion.

EXAMPLE 16.11   EXPLOSION IN ROCK AND PULSE TRANSMISSION

A 510 kg, spherical charge of PETN is exploded in granite, deep underground. Calculate the PPV at several locations in accordance with an empirical formula. For granite use

$$F_{cu} = 100\,\text{MPa (uniaxial compressive strength)}, \quad E = 80\,\text{GPa}, \quad \rho = 0.0026, \quad \text{and} \quad \nu = 0.2.$$

Construct a simple FEA model, consisting of a string of brick elements, to simulate the explosion. Using an EPP material model select the yield strength F_y in such a way that there is a reasonable match between the empirical formula and simulation results. Model to be made in a reduced scale, 1:10.

Taking density of PETN as 0.00176 g/mm³ from Table 13.1, one finds the charge radius to be $r_s = 410.2$ mm. The PPV resulting from explosion of PETN in granite is given in Case 13.25:

$$v_m = \frac{8761}{\tilde{r}^3} + \frac{1392}{\tilde{r}^2} + \frac{26.62}{\tilde{r}} \quad \text{(mm/ms or m/s)}$$

This is valid for $15 < \check{r} < 120$, or for $6.15\,\mathrm{m} < r < 49.2\,\mathrm{m}$. The FE model intended to approach this result is made in 1/10 scale, which means the mass of the charge is reduced by the factor of 1000. The sketch shows a fragment of the model near the origin, where the radial distance between the nodes is about 5 mm. The model is to represent a small-angle cone originating from the center, but the use of 8-node brick elements made it an elongated pyramid. Instead of placing the explosive material at the center, it was made into a layer at some distance from it, with the 4 internal nodes being fixed. The mass of explosive was preserved. The purpose of this reshaping was to avoid a thin apex, which sometimes brings about a distortion of results.

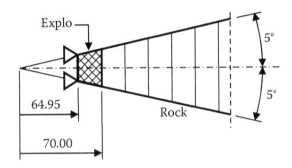

The simulation was carried out using LS-Dyna, treating the explosive according to Case 13.2. The effective energy was taken as $q_G = 4.31 \times 10^6$ N-mm, per Table 16.4. LS-Dyna input requires energy per unit volume, $e = q_G \rho = 7586$. The brick elements were capable of yielding and the yield strength F_y was varied to see how it influences the results.

The table column "Formula" shows the value of PPV according to the above expression. The remaining columns give the same variable when the stated F_y is input. Only the first 3 lines illustrate the values within the "permissible" range, while the other three are associated with distances more than an order of magnitude larger, as such distances were also of interest.

r (mm)	$\check{r}$	Formula	$F_y = 300$	$F_y = 400$	$F_y = 500$
1170	28.52	3021.9	3060	3246	3850
2570	62.65	815.1	1401	1657	1960
4720	115.07	342.2	694.4	505.3	452.8
41170	1003.66	27.9	41	40.3	37.3
81170	1978.79	13.8	14.3	11.2	10.4
124920	3045.34	8.9	8.5	7.7	7.1

All values of PPV are in m/s. This is a scaled model, which means that $r = 1170\,\mathrm{mm}$ corresponds in reality to $11,700\,\mathrm{mm}$. Yet, such scaling should not, in principle, have an effect on PPV.

As the table shows, varying the material properties does not bring about a smooth variation of results. This is typical of a situation, when one wants to create a model, reproducing experimental results over a wide range of distances. Especially poor results show for $r = 2570\,\mathrm{mm}$. $F_y = 300\,\mathrm{MPa}$ is relatively best for the first two locations, but $F_y = 500\,\mathrm{MPa}$ excels for $r \geq 4720\,\mathrm{mm}$.

There are several ways in which such a simulation could be improved. The most obvious one would be to decrease the explosive strength. This would be very sensible from physical viewpoint, as strong explosives pulverize the rock, which is associated with energy loss. Using q_G intended for metallic application means using more energy than left in PETN outside the crush zone. (Refer to Chapter 15.) One can also use a more sophisticated material model, if the first measure proves inadequate.

If FEA results agree well with experiment in the tested range, they should also give a good indication outside the range. This was the rationale behind going to such far distances, as in the last three lines.

The radial node spacing requirement was presented in Chapter 5 in reference to a rectangular pulse. The character of the impulse applied to the rock by the explosive gas may be regarded as asymptotically decreasing, at least in the initial phase. In Case 3.5 a relationship between the initial magnitude of such a pulse W_0 and the impulse S associated with it was shown as $S = W_0 t_x$, where t_x would be found from the other two variables. One can choose t_x so defined as a relevant time of application of the force. In our situation this translates to

$$i_r = p_0 t_x$$

where
    i_r is the specific impulse reflected from the medium surface
    p_0 is the initial pressure value
    t_x is the relevant duration

To find the close-proximity impulse use Equation 13.17, omitting the air component:

$$i_r = \frac{M_e[2q_G]^{1/2}}{4\pi r_s^2} = \frac{510{,}000\left[2\times4.31\times10^6\right]^{1/2}}{4\pi 410.2^2} = 708.1\,\text{MPa-ms}$$

The initial pressure is proportional to the specific energy content q, or q_G used in this case. Taking the figures for PETN from Table 13.1:

$$p_0 = 16{,}821\times0.7077 = 11{,}904\,\text{MPa}.$$

Now, t_x can be found from $708.1 = 11{,}904 t_x$; which gives $t_x = 0.0595$ ms. To translate this into a distance, over which the relevant part of the wave spreads, the wave speed is needed. Using, for simplification, c_0 instead of c_p, one has the length of the pulse as

$$b = c_0 t_x = (E/\rho)^{1/2} t_x = (80{,}000/0.0026)^{1/2}\, 0.0595 = 330\,\text{mm}.$$

There is close to $330/5 = 66$ elements over that distance, which is much more than the minimum of 12 required, according to Chapter 5.

Appendix A   Mohr Circles

PLANE STRESS

The state of stress at a point, in a two-dimensional problem, is defined by three components. These components are applied to the edges of a small square drawn around the point of interest.

The following question is often posed: if that square is rotated, i.e., if a new square is drawn, what will the new components of stress be? The answer is provided by a Mohr circle, which was initially intended to be a graphical means of finding answers, but which nevertheless provides a good visual and mnemonic tool.

Positive axial stress corresponds to tension and positive shear is in accordance with Figure A.1d: when the shear vector tries to rotate the element (or the surface normal n) clockwise. Point A on the circle is located by the positive stress σ_x and negative τ, according to the convention. (Point A corresponds to the edge marked by A in Figure A.1a.) The angular distance α between two sections in Figure A.1a corresponds to points, which are 2α away in Figure A.1c. For this reason point B is $180°$ away from point A on the Mohr circle. At point B a positive σ_y is accompanied by a positive τ.

The circle can now be constructed by joining points A and B by a straight line, which intersects the σ axis at a point corresponding to the average stress. The latter point is the center of the circle. Thus the circle is defined by the position of the center at σ_{av} and by radius equal to maximum shear τ_m:

$$\sigma_{av} = \frac{1}{2}(\sigma_x + \sigma_y) \tag{A.1a}$$

and

$$\tau_m = \sqrt{\left(\frac{\sigma_x - \sigma_y}{2}\right)^2 + \tau^2} \tag{A.1b}$$

Suppose now that the principal direction needs to be found, i.e., direction which is associated with absence of shear. This corresponds to two diametrically opposed points on the σ axis: σ_{max} and σ_{min}. In order to go from point A to point σ_{max} on the Mohr circle, the counterclockwise rotation needs to be made by the angle 2α:

$$|\sin 2\alpha| = \frac{\tau}{\tau_m} \tag{A.2}$$

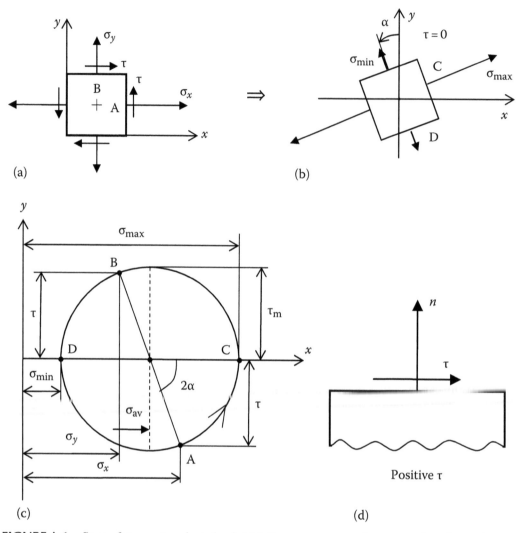

FIGURE A.1 State of stress at a selected point (a), the same stress at the same point when reference coordinates are rotated by angle α to coincide with principal directions (b), corresponding Mohr circle (c), and positive sign convention for shear stress (d).

This means that in the physical plane the rotation is counterclockwise by the angle α. If the normal to edge A is rotated accordingly, its new position determines the direction of the larger principal stress, as shown in Figure A.1b.

Another important direction, namely that associated with the extreme shear τ_m is located on an apex of the circle. To go from A to the lower apex, the clockwise rotation of $90° - 2\alpha$ is needed. This means rotating the x-axis by $45° - \alpha$ (also clockwise) to arrive at the maximum shear orientation in the physical plane.

It is important to remember that the above relationships are based on equilibrium only, which means they are material-independent. The maximum and minimum stress components, as shown in Figure A.1, are usually called principal stresses and designated by σ_1 and σ_2.

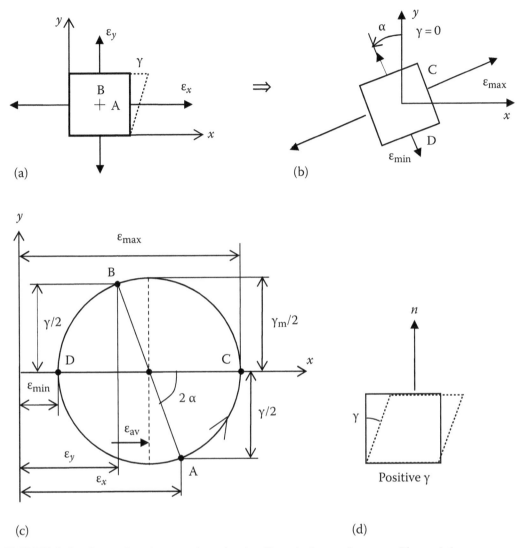

FIGURE A.2 State of strain at a selected point. Description analogous to Figure A.1.

PLANE STRAIN

There is a complete analogy between strain and stress. Three components of strain are defined for a plane problem: extension ε_x along x, then ε_y along y, and shear distortion γ. When a coordinate change takes place, the change of components can be established using a Mohr circle, almost identical to the previous one, as Figure A.2 shows. The only difference is that instead of shear stress τ, a one-half shear angle, namely $\gamma/2$ is transferred.

THREE-DIMENSIONAL STATE OF STRESS

In the most general case, the state of stress is defined by six components. Three of them represent direct stress: σ_x, σ_y, and σ_z, while the remaining three are shears: τ_{xy}, τ_{yz}, and τ_{zx}. It can be proven, that when a cube of material, visualizing the stressed state is suitably oriented, it will coincide with three principal directions, characterized by direct stress only, as in Figure A.3a.

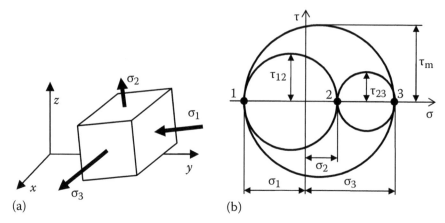

FIGURE A.3   (a) A cube cutout of stressed material, oriented along principal directions and (b) three Mohr circles built on three principal stresses.

If a pair of directions is selected, for example, 1 and 2, then a Mohr circle may be constructed based on the two principal stresses in the same manner as it was done for a plane stress case. This circle describes a complete state of stress in 1–2 plane. For all three direction pairs three Mohr circles can be constructed, as in Figure A.3b. If the state of stress is truly two-dimensional, it simply means that the third principal stress is nil and it corresponds to 0,0 point in the σ–τ. The three circles can still be constructed and the radius of the largest one determines the maximum shearing stress.

Appendix B   Shortcuts and Approximations

FILLING FACTORS

An area under a curve described by $y(x)$ can be defined as

$$A = \int_0^a y(x)\mathrm{d}x = \chi a y_\mathrm{m} \tag{B.1}$$

where
  a is the length of segment of integration
  y_m is the peak value of the function

The term χ, which is called here a filling factor, shows what portion of the circumscribing rectangle $a y_\mathrm{m}$ is occupied by the area under the curve $y(x)$, as illustrated in Figure B.1.

The above will now be illustrated for a few functions, frequently encountered in engineering applications, with some of those illustrated in Figure B.2. The first group is power-type relations, given here along with their derivatives and integrals and resulting χ's:

$$y = bx^n; \quad y' = nbx^{n-1}; \quad A = \int_0^a y\,\mathrm{d}x = \frac{a^{n+1}}{n+1}b = \frac{y_\mathrm{m}a}{n+1}; \quad \text{for } n \geq 1 \tag{B.2}$$

$$\chi = \frac{1}{n+1} \tag{B.3}$$

$$y = bx^{1/n}; \quad y' = \frac{b}{n}x^{1/n-1}; \quad A = \frac{b}{1/n+1}a^{1/n+1} = \frac{y_\mathrm{m}a}{1/n+1}; \quad \text{for } n \geq 1 \tag{B.4}$$

$$\chi = \frac{n}{n+1} \tag{B.5}$$

Trigonometric functions are also frequently encountered. As it can be readily checked

$$y = b\sin x \quad \text{or} \quad y = b\cos x; \quad A = \int_0^{\pi/2} y\,\mathrm{d}x = b \tag{B.6}$$

which gives $\chi = 2/\pi$ \tag{B.7}

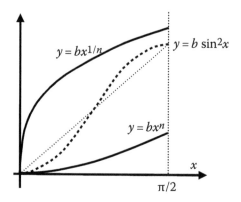

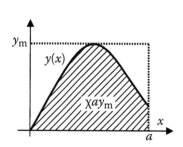

FIGURE B.1   Filling factor χ.

FIGURE B.2   Some frequently encountered functions.

$$y = b\sin^2 x \quad \text{or} \quad y = b\cos^2 x; \quad A = \int_0^{\pi/2} y\,\mathrm{d}x = (\pi/4)b \tag{B.8}$$

which gives

$$\chi = \frac{1}{2} \tag{B.9}$$

for every quarter-period. The χ coefficients derived from these will hold over any whole
multiple of a quarter-period. The arguments of these functions have a variety of forms, as
for example in $\sin(\pi x/L)$. In this case a quarter-period is $x = L/2$ long.

Simpson's integration is a very helpful tool in cases where a graph of a function over a
segment of interest does not differ much from a second-order parabola. The simplest case is
when the range of integration consists of two segments only, as illustrated in Figure B.3. The
integral can then be approximated by

$$\int_a^b y(x)\,\mathrm{d}x \approx \frac{b-a}{6}(y_0 + 4y_1 + y_2) \tag{B.10}$$

The importance of this approximation is visible when a
complex expression has a smooth graph over the range of
interest.

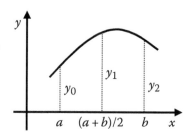

FIGURE B.3   Simpson integration over the segment ab.

APPROXIMATE FUNCTIONS FOR SMALL ARGUMENT VALUES

The original function $f(x)$ can be replaced with its approximation $\varphi(x)$ provided $x \ll 1$. The last column in the table gives an error $e_{0.1}$ defined as

$$e_{0.1} = \frac{\varphi(x) - f(x)}{f(x)} \quad \text{for } x = 0.1 \tag{B.11}$$

$f(x)$	$\varphi(x)$	$e_{0.1}$
$\sin x$	x	0.2%
$\cos x$	$1 - \dfrac{x^2}{2}$	$\sim 0\%$
$\tan x$	x	-0.3%
$\sinh x$	x	-0.2%
$\cosh x$	$1 + \dfrac{x^2}{2}$	$\sim 0\%$
$\dfrac{1}{1+x}$	$1 - x$	-1%
$\sqrt{1+x}$	$1 + \dfrac{x}{2}$	0.1%
$\dfrac{1}{\sqrt{1+x}}$	$1 - \dfrac{x}{2}$	-0.4%
$(1+x)^n$	$1 + nx$	-4.4% for $n = 4$
e^x	$1 + x$	-0.5%
e^{-x}	$1 - x$	-0.5%
$\ln(a + x)$	$\ln a + \dfrac{x}{a}$	4.9% for $a = 1$
$\ln \dfrac{a + x}{a - x}$	$\dfrac{2x}{a}$	-0.3% for $a = 1$

In the algebraic functions above x can have either sign.

Appendix C   Aerodynamic Drag Coefficients

When a body is immersed in a steady fluid flow with a velocity v, the resultant aerohydro-dynamic force, applied in the direction of the undisturbed flow is

$$W_D = \frac{1}{2}\rho v^2 C_D A \qquad (C.1)$$

where
   ρ is the fluid density
   C_D is the drag coefficient
   A is the reference area, usually the frontal area or section area normal to the flow

The drag coefficients for common shapes are tabulated below.

Drag Coefficient C_D for Common Shapes	
Object	**C_D**
Sphere	0.47
Cube, face-on	1.10
Cube, edge-on	0.80
Circular cylinder, side-on	0.75
Square cylinder, side-on	2.00
Square cylinder, long edge-on	1.60
Triangular, 60° cylinder, face-on	2.20
Triangular, 60° cylinder, edge-on	1.39
Long circular cylinder, side-on, laminar	1.20
Long circular cylinder, side-on, turbulent	0.30
Long circular cylinder, end-on	1.00
Long, thin plate, face-on	1.40
Long, thin plate, edge-on	1.50
Open hemisphere, concave against wind	1.40
Open hemisphere, convex against wind	0.40
Disk, face-on	1.10

Sources: Douglas, J.F. et al., *Fluid Mechanics*, 2nd Edn. Longman Scientific & Technical, 1985; Streeter, V.L. et al., *Fluid Mechanics*, 9th Edn. WCB, McGraw-Hill, New York, 1998.

Note: There is a dependence of C_D on Reynolds number *Re*, most visible, perhaps, in case of a long circular cylinder, side-on, with a low value of 0.3 and a high of 1.2. The average is given above. For other shapes, typically much less sensitive, the values of C_D were given for $Re = 10,000$. For a long, thin plate the value is quoted for the side ratio of 15.

The drag coefficients for buildings, which, in the simplest case, are rectangular blocks attached to ground, is a rich subject in itself. Bangash [6], for example, suggests the range of $0.8 \leq C_D \leq 1.6$ for a front (windward) face of a building, while the rear face is estimated to have $0.25 \leq C_D \leq 0.5$. (The rear face is associated with suction, therefore a negative sign of C_D is used sometimes.)

The above is only the pressure component of the drag force. There is also a friction component, which is typically smaller or much smaller, when smooth surfaces are involved. It is, however, quite difficult to evaluate for most structural applications, when surface parallel to the flow (e.g., roof, or even walls) may be quite rough and therefore associated with a substantial resistance. The term

$$q = \frac{\rho v^2}{2} \tag{C.2}$$

is often referred to as *dynamic pressure*, but it has more to do with convention rather than being a universal truth. One should recall from textbooks on hydrodynamics that if a jet of fluid acts on a rigid surface, at right angle to it, and then spreads sideways, it applies the pressure of ρv^2, which is the real dynamic pressure in this case. Also, a large plate with a flow normal to its surface, has $C_D = 2$. This means, that according to Equation C.1 the force applied to it is $\rho v^2 A$, where A is the surface area. The reason why C_D is less than 2.0 for finite-sized objects is that pressure cannot build up to such a high level near the edge as it does near the center.

The pressures and drag coefficients described here are focused on the front face of the object immersed in the flow. This does not give a complete picture of pressure distribution around such box-like objects like buildings. It is well known that walls parallel to the flow can experience suction of a magnitude approaching q while suction in parts of a roof can exceed q. The wind-pressure effects are additive with the ambient pressure p_0 in the medium, the total being

$$p = p_0 + C_D q \tag{C.3}$$

When dynamic pressure is applied to a surface during time t, the body experiences a drag impulse of magnitude:

$$S = W_D t = C_D A(qt) \tag{C.4}$$

where the product qt is analogous to a specific impulse in blast wave mechanics. The equation is strictly applicable to an object fixed in space. However, if the pressure duration is short enough, so that an unconstrained, free to move object does not gain an appreciable velocity during t, Equation C.4 can also be used.

The steady, incompressible flow, which the above relationships describe and which is often named *drag loading*, has only a limited use in deriving structural loads related to blast waves. The extent of its use is described in Chapter 13.

Appendix D Lamé Equations

THICK, PRESSURIZED RING

A ring of unit width, with radii a and b, as in Figure D.1, is pressurized with p. The displacements at the inner surface, $r = a$:

$$u = \frac{pa}{E}\left(\frac{\beta^2 + 1}{1 - \beta^2} + \nu\right) \tag{D.1}$$

The radial and hoop stress at a distance r from the center are designated by σ_r and σ_h, respectively:

$$\sigma_r = \frac{p}{(1 - \beta^2)}\left(\beta^2 - \frac{a^2}{r^2}\right) \tag{D.2}$$

$$\sigma_h = \frac{p}{(1 - \beta^2)}\left(\beta^2 + \frac{a^2}{r^2}\right) \tag{D.3}$$

where $\beta = a/b$. When b grows to infinity or β tends to nil, the case of a pressurized hole in an unbounded medium is obtained:

$$u = \frac{pa}{E}(1 + \nu) \tag{D.4}$$

$$\sigma_r = -p\left(\frac{a}{r}\right)^2 \quad \sigma_h = p\left(\frac{a}{r}\right)^2 \tag{D.5}$$

At this limit the deflection at r is given by

$$u_r = \frac{1 + \nu}{E}\frac{pa^2}{r} \tag{D.6}$$

The above results hold for a unit-thick layer of a medium, whose thickness, measured normal to paper, is infinite. In practice the results describe deformation of a medium around a sufficiently long, cylindrical hole.

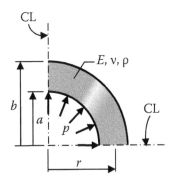

FIGURE D.1 One quarter of a pressurized ring or a quarter section of a spherical shell.

THICK, PRESSURIZED SPHERICAL SHELL

Consider a hollow sphere with radii as shown in Figure D.1. The displacement at the inner radius is

$$u = \frac{pa}{E(1-\beta^3)}\left[(1-2\nu)\beta^3 + \frac{1+\nu}{2}\right] \tag{D.7}$$

The radial stress and the hoop stress are, respectively

$$\sigma_r = \frac{p}{1-\beta^3}\left(\beta^3 - \frac{a^3}{r^3}\right) \tag{D.8a}$$

$$\sigma_h = \frac{p}{1-\beta^3}\left(\beta^3 + \frac{a^3}{2r^3}\right) \tag{D.8b}$$

The case of a cavity in an infinite medium is obtained when b grows to infinity or β tends to nil:

$$u = \frac{pa}{E}\frac{1+\nu}{2} \tag{D.9}$$

$$\sigma_r = -p\left(\frac{a}{r}\right)^3 \tag{D.10a}$$

$$\sigma_h = \frac{p}{2}\left(\frac{a}{r}\right)^3 \tag{D.10b}$$

The deflection at r at this limit is given by

$$u_r = \frac{1+\nu}{E}\frac{pa^3}{2r^2} \tag{D.11}$$

References

1. Alves, M. and Jones, N. Impact failure of beams using damage mechanics. *Int. J. Impact Eng.*, 27, 837–861, 2002.
2. *ANSYS General-Purpose Structural Analysis Code, Version 10*. Ansys Inc, Pittsburgh, PA, 2005.
3. Baker, W.E. et al. The elastic and plastic response of cantilevers to air blast loading. *Proceedings of the Fourth U.S. National Congress of Applied Mechanics* (pp. 853–866). ASME, New York, 1962.
4. Baker, W.E. *Explosion Hazards and Evaluation*. Elsevier Scientific, Amsterdam, 1983.
5. Baker, W.E. *Explosions in Air*. Wilfred Baker Engineering, San Antonio, TX, 1983.
6. Bangash, M.Y.H. *Impact and Explosion*. CRC Press, Boca Raton, FL, 1993.
7. Bažant, Z.P. and Verdure, M. Mechanics of progressive collapse: Learning from World Trade Center and building demolitions. *J. Eng. Mech.*, 133(3), 308–319, 2007.
8. Belayev, V.I. et al. *High-Speed Deformation*. Academy of Sciences, USSR. Response of some metallic materials to dynamic extension (pp. 54–56). Science, Moscow, 1971 (in Russian).
9. Biggs, J.M. *Introduction to Structural Dynamics*. McGraw-Hill, New York, 1964.
10. Bilgin, H.A. et al. Optimum burden determination and fragmentation evaluation by full scale slab blasting. *Proceedings of the Fourth International Symposium on Rock Fragmentation by Blasting. (Fragblast-4)*. Vienna, July 1993.
11. Blevins, R.D. *Formulas for Natural Frequency and Mode Shape*. Van Nostrand Reinhold, New York, 1979.
12. Brach, R.M. *Mechanical Impact Dynamics*. Wiley, New York, 1991.
13. Britner, A. and Saal, H. Performance of blast with planar shock wave front. *Proceedings of the Sixth Asia-Pacific Conference on Shock and Impact Loads on Structures*. Perth, Australia, December 2005.
14. Bulson, P.S. *Explosive Loading of Engineering Structures*. E&FN Spon, London, U.K., 1997.
15. Cement and Concrete Association of Australia. *Concrete Design Handbook*, 2nd Edn. Cement and Concrete Association of Australia, July 1991.
16. Chen, X.W. and Li, Q.M. Penetration into high-strength concrete target by a non-deformable projectile. *Proceedings of the Fourth Asia-Pacific Conference on Shock and Impact Loads on Structures* (pp. 157–163). Singapore, November 2001.
17. Clough, R.W. and Penzien, J. *Dynamics of Structures*. McGraw-Hill, New York, 1975.
18. Cristescu, N. *Dynamic Plasticity*. North-Holland Publishing Co., Amsterdam, 1967.
19. Cooper, P.W. *Explosives Engineering* (pp. 394–399). Wiley-VCH, New York, 1996.
20. Cooper, P. Explosive weight correction for untamped flyer-plates. *Proceedings of the 20th International Symposium on Ballistics*. Orlando, FL, September 2002.
21. Darvall, P.L. and Brown, H.P. *Fundamentals of Reinforced Concrete Analysis and Design*. Macmillan, Australia, 1976.
22. Douglas, J.F. et al. *Fluid Mechanics*, 2nd Edn. Longman Scientific & Technical, New York, 1985.
23. Esen, S., Onederra, I., and Bilgin, H.A. Modeling the size of the crushed zone around a blast-hole. *Int. J. Rock Mech. Mining Sci.*, 40, 485–495, 2003.
24. Forrestal, M.J. et al. Penetration of concrete targets. Sandia Report 92–2513C. Sandia Laboratories, Albuquerque, NM, August 1993.
25. Fuchs, H.O. and Stephens, R.I. *Metal Fatigue in Engineering*. Wiley-Interscience, New York, 1980.
26. Goldsmith, W. *Impact*. Edward Arnold Ltd., London, 1960.
27. Grybos, R. *Theory of Impact in Discrete Mechanical Systems*. PWN, Warsaw, 1969 (in Polish).

28. Gurney, R.W. The initial velocity of fragments of bombs, shells, and grenades. Report No. 405. U.S. Army Ballistic Research Laboratory, Aberdeen Proving Ground, MD, 1943.

29. Hansson, H. A note on empirical formulas for the prediction of concrete penetration. Technical Report, Swedish Defence Research Agency (FOI), November 2003.

30. Harris, C.M. and Crede, C.E. (Editors). *Shock and Vibration Handbook*, 2nd Edn. McGraw-Hill, New York, 1976.

31. Held, M. Blast moments of detonating HE charges in the near field. *Proceedings of the Fourth Asia-Pacific Conference on Shock and Impact Loads on Structures*. Singapore, November 2001.

32. Henrych, J. *The Dynamics of Explosion and Its Use*. Elsevier, Amsterdam, 1979.

33. Hetényi, M. *Beams on Elastic Foundation*. The University of Michigan Press, Ann Arbor, MI, 1946.

34. Hill, R. *The Mathematical Theory of Plasticity*. Oxford University Press, London, U.K., 1950.

35. Honma, H. The necessary amount of explosive and rock strength in blasting. *Proceedings of the Fourth International Symposium on Rock Fragmentation by Blasting*. Vienna, Austria, 1993.

36. Horoshun, G. Deflection and rupture of beams subjected to impulsive loads. *Proceedings of the Fourth Asia-Pacific Conference on Shock and Impact Loads on Structures*. Singapore, November 2001.

37. Huber, M.T. *Energy of Distortion as a Measure of Material Straining*. Czasopismo Techniczne, Lwów, 1904 (in Polish).

38. Irvine, M. *Cable Structures*. Dover Publications, New York, 1992.

39. Jayatilake, I.N. et al. Influence of setbacks on performance of high-rise buildings under blast loading. *Proceedings of the 2nd International Conference on Design and Analysis of Protective Structures*. Singapore, November 2006.

40. Johnson, W. *Impact Strength of Materials*. Edward Arnold, London, U.K., 1972.

41. Johnson, K.L. *Contact Mechanics*. Cambridge University Press, Cambridge, U.K., 1985.

42. Jones, N. On the dynamic inelastic failure of beams. Chapter 5. In *Structural Failure*. Wierzbicki, T. and Jones, N. (Editors). Wiley, New York, 1989.

43. Jones, N. *Structural Impact*. Cambridge University Press, Cambridge, U.K., 1997.

44. Kaliski, S. (Editor) *Vibrations and Waves*. Polish Academy of Science, Warsaw, 1966 (in Polish).

45. Kaliszky, S. *Plasticity. Theory and Engineering Applications*. Elsevier, Amsterdam, 1989.

46. Kolsky, H. *Stress Waves in Solids*. Dover Publications, New York, 1963.

47. Konya, C.J. and Walter, E.J. *Surface Blast Design*. Prentice Hall, NJ, 1990.

48. Kraus, D. et al., The interaction of high explosive detonations with concrete structures. Computational modeling of concrete structures. EURO-C 1994, Insbruck, Austria, 1994.

49. Kroon, R.P. Turbine-blade vibrations due to partial admission. *ASME J. Appl. Mech.*, June, A-161–A-166, 1940.

50. Kutter, H.K. and Fairhurst, C. On the fracture process in blasting. Doctoral thesis, University of Queensland, Australia, September 1970.

51. Lamborn I.R. The natural fracture and fragmentation of internally detonated steel cylinders; The influence of microstructure and mechanical properties. Department of Supply, Australian Defence Scientific Service. March 1997.

52. Lawrence Livermore National Laboratory. *LLNL Explosives Handbook*. Livermore, CA, January 1985.

53. Li, Q.M. and Jones, N. Blast loading of fully clamped beams with transverse shear effects. *Mech. Struct. Mach.*, 23(1), 59–86, 1995.

54. Li, Q.M. and Meng, H. About the dynamic strength enhancement of concrete-like materials in a split Hopkinson pressure bar test. *Int. J. Solids Struct.*, 40, 343–360, 2003.

55. Li, Q.M. Penetration of a hard missile. *Proceedings of the Sixth Asia-Pacific Conference on Shock and Impact Loads on Structures* (pp. 49–62). Perth, Australia, December 2005.

56. Liu, J. and Jones, N. Experimental investigation of clamped beams struck transversely by a mass. *Int. J. Impact Eng.*, 6(4), 303–335, 1987.

57. Lok, T.S. et al. Steel fibre reinforced concrete panels subjected to multiple ballistic impacts. *Proceedings of the Third Asia-Pacific Conference on Shock and Impact Loads on Structures* (pp. 241–256). Singapore, November 1999.

58. Low, K.H. and Zhang, X. On the coefficients of restitution during impact of slender bars with external surfaces. *Proceedings of the Fourth Asia-Pacific Conference on Shock and Impact Loads on Structures*. Singapore, November 2001. pp. 443–450.

59. *LS-Dyna Keyword Users Manual*. Version 960. Livermore Software Corporation, Livermore, CA, 2001.

60. Masuya, H. et al. Experimental study of some parameters for simulation of rock fall on slope. *Proceedings of the Fourth Asia-Pacific Conference on Shock and Impact Loads on Structures*. Singapore, November 2001.

61. Meyers, M.A. *Dynamic Behavior of Materials*. Wiley, New York, 1994.

62. Mises, R.v. Mechanik der festen Körper im plastisch deformablen Zustand. Göttingen Nachrichten. *Math.-Phys. Klasse*, 582–592, 1913.

63. Nonaka, T. Shear and bending response of a rigid-plastic beam to blast-type loading. *Ingenieur-Archiv*, 46, 35–52, Springer-Verlag, 1977.

64. Nyström, U. and Leppänen, J. Numerical studies of projectile impacts on reinforced concrete. *Proceedings of the First International Conference on Design and Analysis of Protective Structures Against Impact and Shock Loads* (pp. 310–319). Tokyo, December 2003.

65. Parkes, E.W. The permanent deformation of a cantilever struck transversely at the tip. *Proc. R. Soc., Ser. A*, 228, 462, 1955.

66. Persson, P.A., Holmberg, R., and Lee, J. *Rock Blasting and Explosives Engineering*. CRC Press, Boca Raton, FL, 1994.

67. Peterson, R.E. *Stress Concentration Factors*. John Wiley & Sons, New York, 1974.

68. Pilkey, W.D. *Formulas for Stress, Strain and Structural Matrices*, 2nd Edn. Wiley, New York, 2005.

69. Popov, E.P. *Engineering Mechanics of Solids*, 2nd Edn. Prentice Hall, NJ, 1998.

70. Reid, S.R. and Goudie, K. Denting and bending of tubular beams under local loads. Chapter 10. In *Structural Failure*, Wierzbicki and Jones (Editors). Wiley, New York, 1989.

71. Reinhardt, H.W. Structural behavior of high performance concrete. *Otto-Graf J.*, 11, 9–17, 2000.

72. Richart, F.E. et al. *Vibrations of Soils and Foundations*. Prentice Hall, Englewood Cliffs, NJ, 1970.

73. Rinehart, J.S. and Pearson, J. *Explosive Working of Metals*. Pergamon Press, New York, 1963.

74. Rinehart, J.S. and Pearson, J. *Behavior of Metals Under Impulsive Loads*. Dover Publications, New York, 1965.

75. Roark, R.J. *Formulas for Stress and Strain*, 4th Edn. McGraw-Hill, New York, 1965.

76. Seah, C.C. et al. Projectile penetration into granite targets of finite thickness. *Proceedings of the 2nd International Conference on Design and Analysis of Protective Structures*. Singapore, November 2006.

77. Shi, Y. et al. Numerical study of interactions between shock wave and RC columns. *Proceedings of the 2006 RNSA Security Technology Conference*. Melbourne, Australia, 2006.

78. Silva, de, C.W. (Editor). *Vibration and Shock Handbook*. CRC Press, Boca Raton, FL, 2005.

79. Smith, C.S. Behavior of composite and metallic superstructures under blast loading. Chapter 13. In *Structural Failure*. Wierzbicki and Jones (Editors). Wiley, New York, 1989.

80. Smith, P.D. and Hetherington, J.G. *Blast and Ballistic Loading of Structures*. Butterworth & Heinemann, London, 1994.

81. Sondergaard, R. et al. Measurements of solid spheres bouncing off flat plates. *J. Appl. Mech.*, 57, 694, September 1990.

82. SRI International. *Dynamic Pulse Buckling—Theory and Experiment*. SRI International, Menlo Park, CA, 1983.

83. Streeter, V.L. et al. *Fluid Mechanics*, 9th Edn. WCB, McGraw-Hill, New York, 1998.

84. Stronge, W.J. and Yu, T.X. *Dynamic Models for Structural Plasticity*. Springer-Verlag, London, 1993.

85. Szuladziński, G. Moment-curvature for elastoplastic beams. Technical Note. *J. Struct. Div., Proc. ASCE*, 101, 1641–1644, July 1975.

86. Szuladziński, G. Bending of beams with nonlinear material characteristics. *J. Mech. Des. Trans. ASME*, 102, 776–780, October 1980.

87. Szuladziński, G. *Dynamics of Structures and Machinery. Problems and Solutions*. Wiley-Interscience, New York, 1982.

88. Szuladziński, G. Response of rock medium to explosive borehole pressure. *Proceedings of the Fourth International Symposium on Rock Fragmentation by Blasting. (Fragblast-4)*. Vienna, July 1993.

89. Szuladziński, G. and Saleh, A. Analysis of cylindrical boulder breaking with application to bench blasting. In *Acta Mechanica*, Vol. 115 (p. 79). Springer-Verlag, Vienna, 1996.

90. Szuladziński, G. Multiple fragmentation of a ring under explosive pressure. *Proceedings of the Parari'99—Fourth Australian Explosive Ordnance Symposium*. Canberra, Australia, November 1999.

91. Szuladziński, G. Computer simulation of propelling of solid pieces by explosive action. *Proceedings of the Third Asia-Pacific Conference on Shock and Impact Loads on Structures*. Singapore, November 1999.

92. Szuladziński, G. Dynamic response of a multi-story building to a shock damage of a base column. *Proceedings of the Third Asia-Pacific Conference on Shock and Impact Loads on Structures*. Singapore, November 1999.

93. Szuladziński, G. Stress and strain magnification effect in rapidly loaded structural joints. *Proceedings of the Seventh International Symposium on Structural Failure and Plasticity (Implast 2000)*. Melbourne, Australia, October 2000.

94. Szuladziński, G. Parameters of impact of a heavy mass falling onto a concrete surface. *Proceedings of the Fourth Asia-Pacific Conference on Shock and Impact Loads on Structures*. Singapore, November 2001.

95. Szuladziński, G. Disintegration of cylindrical shells under explosive pressure. *Proceedings of the Parari 2001—Fifth Australian Explosive Ordnance Symposium*. Canberra, Australia, November 2001.

96. Szuladziński, G. Plasticity of steel under explosive loading and the effect on cylindrical shell fragmentation. *Proceedings, APEA2002 (The Sixth Asia-Pacific Symposium on Engineering Plasticity and Its Applications)*. Sydney, December 2002.

97. Szuladziński, G. Mechanism and sequence of core collapse of the North Tower of WTC. *Proceedings of the First International Conference on Design and Analysis of Protective Structures*. Tokyo, December 2003.

98. Szuladziński, G. Collision method for airblast waves interaction with movable solid bodies. *Proceedings of the Sixth Asia-Pacific Conference on Shock and Impact Loads on Structures*. Perth, Australia, December 2005.

99. Szuladziński, G. Response of inelastic plates to blast loading. *Proceedings of the Second International Conference on Design and Analysis of Protective Structures*. Singapore, November 2006.

100. Szuladziński, G. Mechanical output of contact explosive charges. *Proceedings of the 2007 RNSA Security Technology Conference*. Melbourne, Australia, 2007.

101. Szuladziński, G. Response of beams to shock loading: Inelastic range. *J. Eng. Mech., ASCE*, 133(3), 320–325, March 2007.

102. Szuladziński, G. Mass-plate impact parameters for the elastic range. *Acta Mech.*, 200(1–2), 111–125, 2008.

103. Szuladziński, G. Transient response of circular, elastic plates to point loads. pp. 171–179.

104. Szuladziński, G. Discussion of "Mechanics of Progressive Colllapse: Learning from World Trade Center and Building Demolitions" by Z.P. Bazant and M. Verdure. In press.

105. Technical Manual TM5-855-1. Fundamentals of Protective Design for Conventional Weapons. Headquarters, Department of the Army, Department of Defense, 1986.

106. Technical Manual TM5-1300. Structures to resist the effects of accidental explosions. Explosives Safety Board, Department of Defense, 1992.

107. Teng, X. High Velocity Impact Fracture. Doctoral Dissertation. Massachusetts Institute of Technology, MA, February 2005.

108. Timoshenko, S. *Theory of Elasticity*. McGraw-Hill, New York, 1951.

109. Timoshenko, S. *Strength of Materials*, Part II. Van Nostrand, Princeton, NJ, 1956.

110. Timoshenko, S. *Theory of Elastic Stability*, 2nd Edn., McGraw-Hill, New York, 1961.

111. Timoshenko, S. *Vibration Problems in Engineering*. 4th Edn. Wiley, New York, 1974.

112. Tullis, J.P. *Hydraulics of Pipelines*. Wiley, New York, 1989.

113. U.S. Army Engineers Waterways Experiment Station. *Conwep* – Conventional weapons effects (computer program), 1991.

114. Vang, L. and Jones, N. An analysis of the shear failure of rigid—linear hardening beams under impulsive loading. *Acta Mech. Sin.*, 12(4), 338–348, November 1996.

115. Wierzbicki, T. Abramowicz, W., and Jones, N. The Mechanics of Deep Plastic Collapse of Thin-Walled Structures. Chapter 9. In *Structural Failure*. Wierzbicki and Jones (Editors). Wiley, New York, 1989.

116. Wierzbicki, T. and Hoo Fatt, M.S. Impact response of a string-on-plastic foundation. *Int. J. Impact Eng.*, 12(1), 21–36, 1992.

117. Wijk, G. et al. A model for rigid projectile penetration and perforation of hard steel and metallic targets. FOI (Swedish Defence research Agency) Report FOI-R-1617-SE. April 2005.

118. Young, W.C. and Budynas, R.G. *Roark's Formulas for Stress and Strain*, 7th Edn. McGraw-Hill, New York, 2002.

119. Zener, C. The intrinsic inelasticity of large plates. *Phys. Rev.*, 59, 669, 1941.

120. Zhou, X.Q. et al. Modeling of dynamic damage of concrete slab under blast loading. *Proceedings of the Sixth Asia-Pacific Conference on Shock and Impact Loads on Structures*. Perth, Australia, December 2005.

121. Zukas, J.A. et al. *Impact Dynamics*. Krieger Publishing Company, Malabar, FL, 1992.

122. Zukas, J.A. and Walters, P.W. (Editors). *Explosive Effects and Applications*. Springer, New York, 1998.

Index